Peter T. Bobrowsky

Hans Rickman

Comet/Asteroid Impacts and Human Society

An Interdisciplinary Approach

Peter T. Bobrowsky
Hans Rickman
(Editors)

Comet/Asteroid Impacts and Human Society

An Interdisciplinary Approach

With 85 Figures, 46 in Color

 Springer

Editors

Dr. Peter T. Bobrowsky

Geological Survey of Canada
Landslides and Geotechnics
ESS/GSC-CNCB/GSC-NC/EDS
Natural Resources Canada
601 Booth Street
K1A 0E8 Ottawa, ON
Canada
E-mail: Peter.Bobrowsky@nrcan-rncan.gc.ca

Dr. Hans Rickman

Uppsala Astronomical Observatory
Box 515
SE-751 20 Uppsala
Sweden
E-mail: Hans.Rickman@astro.uu.se

Library of Congress Control Number: 2006934201

ISBN-10   3-540-32709-6   **Springer Berlin Heidelberg New York**
ISBN-13   978-3-540-32709-7   **Springer Berlin Heidelberg New York**

This work is subject to copyright. All rights are reserved, whether the whole or part of the material is concerned, specifically the rights of translation, reprinting, reuse of illustrations, recitations, broadcasting, reproduction on microfilm or in any other way, and storage in data banks. Duplication of this publication or parts thereof is permitted only under the provisions of the German Copyright Law of September 9, 1965, in its current version, and permission for use must always be obtained from Springer. Violations are liable to prosecution under the German Copyright Law.

Springer is a part of Springer Science+Business Media
springeronline.com
© Springer-Verlag Berlin Heidelberg 2007

The use of general descriptive names, registered names, trademarks, etc. in this publication does not imply, even in the absence of a specific statement, that such names are exempt from the relevant protective laws and regulations and therefore free for general use.

Cover design: Erich Kirchner, Heidelberg
Typesetting: Klaus Häringer, Stasch · Bayreuth (stasch@stasch.com)
Production: Agata Oelschläger

Printed on acid-free paper   30/2132/AO – 5 4 3 2 1 0

Preface

The International Council for Science (ICSU) recently recognized that the societal implications (social, cultural, political and economic) of a comet/asteroid impact on Earth warrants an immediate consideration by all countries in the world. Given the paucity of information on this important issue, ICSU thus contacted the International Astronomical Union (IAU) and the International Union for Geological Sciences (IUGS) to address the topic on behalf of the global science community.

This volume provides a summary of opinions regarding the controversy of fact vs. fiction in dealing with comet and asteroid impacts. Each contribution provides a timely state-of-the-art and state-of-the-science synthesis regarding the likelihood and implications of past, present and future comet/asteroid impacts and their effect on human society. Individual chapters represent a wide range of disciplines, specialties and topics which are either directly or indirectly related to impact events. In this way, this book differs considerably from previous comet/asteroid impact books as well as most other natural hazard volumes that commonly focus on a single discipline of study. Our goal in compiling this volume was to ensure that representatives from ancillary disciplines (anthropology, archaeology, economics, geography, atmospheric sciences, political science, psychology and so on) had the opportunity to contribute to the discussion by astronomers and geologists and therefore broaden the restrictive vision normally accorded to topical discussions of natural hazards. Our aim is to widen the appeal of the subject of natural hazards to include specialists that deal with the subject but lack an appreciation of the related implications surfacing from other disciplines. Moreover, the papers were written with the non-scientist in mind, with the expectation to better inform and educate decision makers, politicians and the general public at large about the diverse nature of the physical and social consequences which have in the past, and will in the future, arise from an impact of a comet or asteroid with our planet Earth.

This volume is clustered into three parts comprising 33 chapters. The focus of this book provides those individuals interested in multi-hazard interdisciplinary research a concise appraisal of what is currently known regarding the threat of comet/asteroid impacts, the likelihood and magnitude of such events in the future, an historic review of past impacts based on geological, archaeological and anthropological evidence, an elaboration on the likely physical effects of a significant impact, the ecological and atmospheric effects following an impact, the psycho-sociological implications associated with risk, hazards and disasters as well as the financial, economic and insurance consequences of a catastrophic impact on our planet.

Part one covers the ancient (geology), prehistoric (archaeology) and historic (anthropology) record of comet and asteroid events. This includes papers on popular culture and the use of tree ring studies in modern research as well as a review of the analogies of mega catastrophes resulting from volcanic eruptions. Part two contains contributions focused on the status of near-earth object (NEO) surveys, current knowledge of NEO populations in space, physical properties of NEOs, the quantitative risk of impacts and risk reduction scenarios, the physical terrestrial effects of impacts, the atmospheric and oceanic (tsunami) effects of impacts, case studies including the Kaali meteorite and Tunguska events and cryometeors. Part three examines the social science of near-earth objects, perceptions of risk, dynamic risk assessment, social perspectives on hazards, social vulnerability, the potential collapse of society, disaster planning, insurance coverage, economic consequences, communicating impact risk to the public, impact risk communication management, international policies on NEOs and the future of NEO research.

In April 2004 Hans Rickman of the International Astronomical Union (IAU) and Peter Bobrowsky of the International Union for Geological Sciences (IUGS) met with a few key representatives of the comet/asteroid professional community in Paris under the auspices of the International Council for Science (ICSU). At that time, the group was encouraged by ICSU to consider collaboration in an interdisciplinary effort on the subject of comet/asteroid impacts and human society. ICSU was very interested in supporting a research proposal relevant to the topic that explicitly included individuals in broadly allied fields of study that were not normally included in discussions on this subject. The intent of the proposal was to provide an open platform of discussion and interaction between astronomers, geologists, anthropologists, archaeologists, economists, sociologists, geographers, psychologists, journalists and many others interested in natural hazards, disaster management, risk assessment and ancillary fields of study, but focussed specifically on the potential psycho-social and physical consequences of a catastrophic comet or asteroid impact on Earth. Following the initial meeting in April of 2004, IAU and IUGS coordinated a formal proposal submission to ICSU for a Class II grant. Representatives from allied unions including IUGG (International Union of Geodesy and Geophysics), IGU (International Geographic Union) and IUPsyS (International Union of Psychological Science) agreed to contribute to the working efforts of the project. Similarly, specialists in other disciplines including anthropology, archaeology, medicine, and so on, but not official representatives of their respective ICSU unions also agreed to contribute to such a project. Shortly thereafter, ICSU approved the grant proposal. An Advisory group consisting of the following individuals was struck: Harry Atkinson (UK NEO Task Force), Clark Chapman (Member at Large), Viacheslav Gusiakov (IUGG), Wing-Huen Ip (COSPAR), Michael MacCracken (SCOR) and Stefan Michalowski (OECD). Invitations were then sent to noted specialists in varied disciplines to participate in a week long retreat which included technical presentations, breakout group discussions, interactive debates and a local field trip. The retreat was held in early December 2004 in La Laguna, Tenerife, Spain with the local support of Mark Kidger and the Instituto de Astrofisica de Canarias. The Editors are most grateful to Dr. Kidger and the staff and management of the institute for their kind support in facilitating this important meeting.

As an outcome of the workshop, a summation of the current state of the art and science on the subject and a discussion of related key political questions on the hazard lead to the development of a "white paper". This compilation, aimed as a background document for politicians, is to appear as a separate published document. At the same time, all invited participants were asked to submit a technical manuscript summarizing their specialty, in a format that addressed the multi-disciplinary nature of the meeting. This volume represents the end product of this effort and thus addresses the outputs identified in the original proposal to ICSU.

This volume represents the collective efforts of a great number of individuals. Most importantly, the Editors recognize the hard work of the contributing authors to clearly capture the key issues of their field of expertise and structure this information in a broadly informative nature readable by others outside their field of interest. The Editors also appreciate the support and work of the editorial staff at Springer Verlag who helped them deal with the difficult process of managing modern techniques in copy-editing. Finally the Editors wish to thank all those individuals who kindly provided their time and effort as critical reviewers for the submitted papers; in some cases reviewing several different papers. The critical reviews were important to us and the book, as they add a level of technical acceptability even when some of the opinions of some of the authors were contentious. Each manuscript was initially reviewed by Peter Bobrowsky and/or Hans Rickman and at least two other impartial persons. As a consequence of this referee process, several papers originally submitted to this volume were rejected and are not included in the published volume. The list of reviewers in alphabetical order were: Johannes Andersen, Joe Arvai, Mark Bailey, Elizabeth Barber, Tony Berger, John Birks, Bill Bottke, Edward Bryant, Andrea Carusi, David Carusi, Alberto Cellino, Clark Chapman, Rejean Couture, Curt Covey, John Davis, Robert Dimand, Eric Elst, David Etkin, Marten Geertsema, John Grattan, Richard Grieve, Peter Horn, David Huntley, Monica Jaramillo, Ruthann Knudson, David Kring, Howard Kunreuther, Jose Lozano, Brian Marsden, Bruce Masse, Jay Melosh, Patrick Michel, Millan Millan, Urve Miller, David Morrison, Jon Nott, Andrei Ol'khovatov, Effim Pelinovsky, Benny Peiser, Juri Plado, Alex Rabinovich, Barrie Raftery, Marko Robnik, Paul Slovic, Richard Spalding, Doug Stead, Duncan Steel, John Twigg, Juha Uitto, Giovanni Valsecchi, Don Yeomans, Fumi Yoshida, Ben Wisner, and Colin Wood.

We acknowledge the support of our respective institutes (Geological Survey of Canada and Uppsala Astronomical Observatory), Unions (International Union of Geological Sciences and International Astronomical Union) and families for providing us the valuable time needed to pursue this important activity.

Peter Bobrowsky
Hans Rickman

November 2006

Contents

Part I · Anthropology, Archaeology, Geology 1

1 **The Geologic Record of Destructive Impact Events on Earth** 3
1.1 Introduction 3
1.2 General Character of the Record 3
    1.2.1 Spatial Distribution 4
    1.2.2 Age Distribution 4
    1.2.3 Size Distribution 4
    1.2.4 Terrestrial Cratering Rate 5
    1.2.5 Periodic Impacts 5
1.3 Recognition of Terrestrial Impact Structures 6
    1.3.1 Morphology 6
    1.3.2 Geology of Impact Structures 7
    1.3.3 Geophysics of Impact Structures 11
1.4 Impacts in the Stratigraphic Record 12
1.5 Impacts and the Biosphere 13
    1.5.1 Early Life 13
    1.5.2 Coupling through the Atmosphere and Hydrosphere 13
    1.5.3 Local and Mass Extinctions 17
    1.5.4 Threat to Humanity 18
1.6 Concluding Remarks 18
    Acknowledgments 20
    References 20

2 **The Archaeology and Anthropology of Quaternary Period Cosmic Impact** 25
2.1 Introduction 25
2.2 The Quaternary Period Cosmic Impact Record 27
    2.2.1 Documented Impact Structures 27
    2.2.2 Validated Holocene Crater-Forming Impact Events 29
    2.2.3 Airbursts, Tektites, and Impact Glass Melts 32
    2.2.4 A Sample of Current Studies of Potential Late Quaternary–Holocene Period Terrestrial Impact Sites 34
    2.2.5 Oceanic Impacts 38

2.3		Oral Tradition, Myth, and Cosmic Impact	39
	2.3.1	The Nature and Principles of Myth and Oral Tradition	40
	2.3.2	Using Myth to Identify and Model South American Cosmic Impacts	42
	2.3.3	Modeling the Flood Comet Event – a Hypothesized Globally Catastrophic Mid-Holocene Abyssal Oceanic Comet Impact	46
2.4		Epilog and Conclusions	61
	2.4.1	Candidate Abyssal Impact Structure	61
	2.4.2	Post-Workshop Final Thoughts	63
		Acknowledgments	64
		References	65
3		**The Sky on the Ground: Celestial Objects and Events in Archaeology and Popular Culture**	**71**
3.1		Introduction	71
3.2		The Archaeological Record	72
	3.2.1	Architecture	72
	3.2.2	Artifacts and Rock Art	72
	3.2.3	Oral Tradition	73
3.3		Celestial Objects in Popular Culture	74
	3.3.1	Astrology in Popular Culture	74
	3.3.2	Art and Literature	76
	3.3.3	Other Examples	82
3.4		Garnering Public Support	83
	3.4.1	Public Awareness and Support through Cinematic Film	83
	3.4.2	Public Education	84
3.5		Conclusions	85
		Acknowledgments	85
		References	86
4		**Umm Al Binni Structure, Southern Iraq, As a Postulated Late Holocene Meteorite Impact Crater**	**89**
4.1		Introduction	89
4.2		Geological Setting	90
4.3		Origin of the Umm Al Binni Structure	94
4.4		New Satellite Imagery	95
		References	101
5		**Tree-Rings Indicate Global Environmental Downturns That Could Have Been Caused by Comet Debris**	**105**
5.1		Introduction	105
5.2		The Historical Record	108
5.3		Mythology	110
5.4		What Actually Happened – the Global Consequences?	112
5.5		The Dust and Corrupted Air	112

5.6	The Scientific Prior Hypothesis	114
5.7	The AD540 Symptoms	115
5.8	Linkages to Other Events	118
5.9	Conclusion	119
	Acknowledgments	120
	References	120
6	**The GGE Threat: Facing and Coping with Global Geophysical Events**	**123**
6.1	Introduction	123
6.2	Volcanic Super-Eruptions	125
6.3	The Toba Super-Eruption	127
6.4	Reassessment of the Super-Eruption Threat	128
6.5	Collapsing Ocean-Island Volcanoes and Mega-Tsunami Formation	129
6.6	Volcano Instability and Structural Failure	129
6.7	Environmental Triggers of Ocean-Island Volcano Collapse	130
6.8	Tsunami Generation from Ocean-Island Volcano Collapses	131
6.9	Contemporary North Atlantic Mega-Tsunami Risk	133
6.10	High-Frequency GGEs	134
6.11	Addressing the GGE Threat	136
	References	138

Part II · Astronomy and Physical Implications ... 143

7	**The Asteroid Impact Hazard and Interdisciplinary Issues**	**145**
7.1	Introduction	145
7.2	Near-Earth Asteroids (NEAs)	147
7.3	Consequences of NEA Impact	150
7.4	Mitigation: Deflection and/or Disaster Management and Response	154
7.5	Perceptions of the Impact Hazard	156
7.6	Societal Impacts	157
	7.6.1 The News Media	158
	7.6.2 Religion	159
	7.6.3 The Military	159
	7.6.4 Science	160
7.7	Hazards Research/Disaster Management	161
	References	162
8	**The Impact Hazard: Advanced NEO Surveys and Societal Responses**	**163**
8.1	Background	163
8.2	The Spaceguard Survey	164
8.3	Sub-Kilometer Impacts	166
8.4	Communication and Miscommunication	168
8.5	Public Policy Issues	169
	Acknowledgments	171
	References	172

9	**Understanding the Near-Earth Object Population: the 2004 Perspective**	175
9.1	Introduction	175
9.2	Dynamical Origin of NEOs	176
	9.2.1 Near-Earth Asteroids	176
	9.2.2 Near-Earth Comets	178
	9.2.3 Evolution in NEO Space	179
9.3	Quantitative Modeling of the NEO Population	180
9.4	The Debiased NEO Population	181
9.5	Nearly Isotropic Comets	183
9.6	NEA Size-Frequency Distribution	184
9.7	Conclusion	185
	References	185
10	**Physical Properties of NEOs and Risks of an Impact: Current Knowledge and Future Challenges**	189
10.1	Introduction	189
	10.1.1 Key Questions before Impact	189
	10.1.2 The True Nature of NEOs	189
10.2	Densities: from Feather to Lead?	190
	10.2.1 Determining Mass and Density	190
	10.2.2 Typical Results on Densities	190
	10.2.3 Open Questions	191
10.3	Structure: from Monoliths to Rubble Piles?	191
	10.3.1 Determining the Structure	191
	10.3.2 Outer Shape and Structure	192
	10.3.3 Porosity and Structure	192
	10.3.4 Comets Disruption and Fragmentation	193
	10.3.5 Open Questions	194
10.4	Surface Properties: from Sand Dunes to Concrete?	195
	10.4.1 Estimating the Surface Properties	196
	10.4.2 Typical Results on Surface Properties	196
	10.4.3 Open Questions	197
10.5	Knowledge Expected from Future Science	198
	10.5.1 Remote Observations and Simulations under Development	198
	10.5.2 Future Space Missions	198
10.6	Conclusion	199
	References	199
11	**Evaluating the Risk of Impacts and the Efficiency of Risk Reduction**	203
11.1	Introduction	203
11.2	Near-Earth Objects Surveys	204
	11.2.1 The Problem of Orbit Determination	205
11.3	Checking for Impact Possibilities	206
11.4	Eliminating Virtual Impactors	207
	11.4.1 Decrease of the Risk Estimate	208

11.5	Deflection	208
	11.5.1 Kinetic Energy Deflection	208
11.6	Conclusions	209
	References	210

12 Physical Effects of Comet and Asteroid Impacts: beyond the Crater Rim ... 211

12.1	Introduction: the Impact Hazard	211
12.2	Local and Regional Devastation by Impacts	212
	12.2.1 Thermal Radiation	213
	12.2.2 Seismic Shaking	214
	12.2.3 Ejecta Deposition	215
	12.2.4 Airblast	216
	12.2.5 Tsunamis from Oceanic Impacts	217
12.3	Global Devastation?	218
	12.3.1 The Thermal Pulse from Ejecta Rain Back	218
	12.3.2 Dust Loading of the Atmosphere	219
	12.3.3 Injection of Climatically Active Gases	220
	12.3.4 Indirect Effects of Biological Extinctions	221
12.4	Conclusion	221
	References	222

13 Frequent Ozone Depletion Resulting from Impacts of Asteroids and Comets ... 225

13.1	Introduction	225
13.2	Physical Interactions with the Atmosphere	225
13.3	Chemical Perturbations of the Upper Atmosphere	227
	13.3.1 Nitric Oxide Production	227
	13.3.2 Lofting of Water	229
	13.3.3 Fate of Salt Particles	230
	13.3.4 Activation of Halogens from Sea Salt Particles	231
	13.3.5 Catalytic Cycles for Ozone Depletion	232
	13.3.6 Estimates of Asteroid Impact and Ozone Depletion Frequency	233
	13.3.7 Model of Coupled Chemistry and Dynamics of the Upper Atmosphere	235
	13.3.8 Model Results for Injections of Nitric Oxide and Water Vapor	236
	13.3.9 Possible Test of the Impact-Induced Ozone Depletion Hypothesis	243
	Acknowledgments	244
	References	244

14 Tsunami As a Destructive Aftermath of Oceanic Impacts ... 247

14.1	Introduction	247
14.2	Geographical and Temporal Distribution of Tsunamis	249
14.3	Basic Types of Tsunami Sources	251
14.4	Tsunamigenic Potential of Oceanic Impacts	254

14.5	Operational Tsunami Warning	257
14.6	Detection of Impact Tsunamis by Tide Gauge Network	258
14.7	Geological Traces of Tsunamis	259
14.8	Conclusions	260
	Acknowledgments	261
	References	261

15 The Physical and Social Effects of the Kaali Meteorite Impact – a Review 265

15.1	Introduction	265
15.2	The Meteorite	266
15.3	Age of the Impact	268
15.4	Effects of the Meteorite Impact	271
	Acknowledgments	273
	References	273

16 The Climatic Effects of Asteroid and Comet Impacts: Consequences for an Increasingly Interconnected Society 277

16.1	Introduction	277
16.2	The Global Climatic Effects of Large Asteroid or Comet Impacts	280
	16.2.1 Injection of Asteroidal and Cometary Material	281
	16.2.2 Injection of Dust	281
	16.2.3 Injections from Fires	282
	16.2.4 Injection of Water	283
	16.2.5 Injection of Sulfur Dioxide	283
	16.2.6 Injection of Nitrogen Oxides	284
16.3	Potential Weather and Climate-Related Impacts of Small to Modest-Sized Asteroids and Comets	285
	16.3.1 Asteroid and Comet Impacts That Do Not Involve a Surface Impact	285
	16.3.2 Modest-Sized Asteroid and Comet Impacts That Do Involve a Surface Impact	286
16.4	Discussion	287
	References	288

17 Nature of the Tunguska Impactor Based on Peat Material from the Explosion Area 291

17.1	Introduction	291
17.2	Search for the TCB Remnants in the Epicenter Area	291
17.3	Platinum Group Elements (PGE) Investigation	292
17.4	Isotopic Investigations of Light Elements in the Peat	295
17.5	Discussion	297
17.6	Conclusions	298
	Acknowledgments	299
	References	299

18	The Tunguska Event	303
18.1	Introduction	303
18.2	The Hypotheses	303
	18.2.1 Comet or Asteroid?	303
	18.2.2 "Non-Traditional" Hypotheses	305
	18.2.3 Alternative Approaches	305
18.3	Known Data	309
	18.3.1 Objective Data	309
	18.3.2 Eyewitnesses Testimonies	314
18.4	Parameters Deduced	316
	18.4.1 Explosion Time	316
	18.4.2 Coordinates of the Epicenter	317
	18.4.3 Trajectory Parameters, Height of the Explosion and Energy Emitted	317
18.5	Tunguska-Like Impacts	320
	18.5.1 Recent Models and Impact Frequency	320
	18.5.2 Global and Local Damages	324
18.6	Concluding Remark	324
	Acknowledgments	325
	References	325

19	Tunguska (1908) and Its Relevance for Comet/Asteroid Impact Statistics	331
19.1	What Happened North of the Stony Tunguska River in the Early Morning of 30 June 1908?	331
19.2	The Tectonic Interpretation of the Tunguska Catastrophe	333
19.3	(Other) Recorded Impact Events	335
19.4	(Likely) Tectonic Outbursts	336
19.5	How to Discriminate between Impacts and Outbursts?	336
19.6	Conclusions	338
	Acknowledgments	338
	References	338

20	Atmospheric Megacryometeor Events Versus Small Meteorite Impacts: Scientific and Human Perspective of a Potential Natural Hazard	341
20.1	Introduction	341
20.2	Megacryometeors	343
	20.2.1 Textural, Hydrochemical and Isotopic Characteristics	344
	20.2.2 Theoretical Modeling	345
20.3	Megacryometeors Versus Small Meteorite Impacts	346
	20.3.1 Comparison of the Rate of Falls during Human Times (Historical Record)	347
20.4	Final Remarks	349
	Acknowledgments	350
	References	350

Part III · Socio-Economic and Policy Implications ... 353

21 Social Science and Near-Earth Objects: an Inventory of Issues ... 355
21.1 Introduction ... 355
21.2 Globally Relevant Disasters ... 355
21.3 Preparation and Response: General Issues ... 357
21.4 Preparation and Recovery: Planning ... 359
21.5 Preparation and Response: the Problem of Trust ... 362
21.6 Preparation and Response: the Problem of Panic ... 363
21.7 Conclusions ... 365
Acknowledgments ... 366
References ... 366

22 Perception of Risk from Asteroid Impact ... 369
22.1 Early Work: Decision Processes, Rationality, and Adjustment to Natural Hazards ... 369
22.2 Stage 2: Psychometric Studies of Risk Perception ... 371
22.3 Perceptions Have Impacts: the Social Amplification of Risk ... 373
22.4 Stage 3: Risk As Feelings ... 374
22.5 Public Perceptions of the Impact Hazard ... 377
    22.5.1 Will the Public Be Concerned about the Impact Hazard? ... 377
    22.5.2 Exploratory Research on Public Attitudes and Perceptions ... 379
22.6 Where Next? ... 380
Acknowledgment ... 381
References ... 381

23 Hazard Risk Assessment of a Near Earth Object ... 383
23.1 Background ... 383
23.2 Defining Risk ... 384
23.3 Ontology of NEO Hazards ... 386
    23.3.1 Level 1 NEOs ... 386
    23.3.2 Level 2 NEOs ... 388
    23.3.3 Level 3 NEOs ... 389
    23.3.4 Level 4 NEOs ... 390
23.4 Dynamic Hazard Risk Assessment and Possible Mitigation and Preparedness Strategies ... 391
23.5 Potential Mitigation, Data Needs, Response, and Prognosis ... 395
References ... 397

24 Social Perspectives on Comet/Asteroid Impact (CAI) Hazards: Technocratic Authority and the Geography of Social Vulnerability ... 399
24.1 Introduction ... 399
24.2 The Perspective of Social Vulnerability ... 400
24.3 Regional and Comparative Aspects of CAI Hazards ... 402
    24.3.1 Regional CAI Risks and the Role of Secondary Hazards ... 403
    24.3.2 Comparative Threat Evaluations ... 405
    24.3.3 Uncertain Uncertainties ... 408

24.4	Conceptual Issues	408
	24.4.1 Limitations of the Agent-Specific Approach	410
	24.4.2 Organizational Risk	411
24.5	Concluding Remarks	414
	References	415

25	**May Land Impacts Induce a Catastrophic Collapse of Civil Societies?**	**419**
25.1	Introduction	419
25.2	Medium–Small Scale Impacts on a European Country: a Case Study	420
	25.2.1 Probability of Impact and Objects Properties	421
	25.2.2 Level of Damage	421
25.3	The Civil Society As a Complex System	424
	25.3.1 Recent Developments in the Science of Complexity	424
	25.3.2 What Is a Complex System?	426
	25.3.3 The Phase of Catastrophe	427
	25.3.4 Main Structures of the Country Social System	427
25.4	Results	429
	25.4.1 Consequences of the 13 MT Impact on the Three Points	429
	25.4.2 Point 1: Consequences of the 1 000 MT Impact	429
	25.4.3 Point 2: Consequences of the 1 000 MT Impact	431
	25.4.4 Point 3: Consequences of the 1 000 MT Impact	431
25.5	Discussion	433
	Acknowledgments	435
	References	435

26	**The Societal Implications of a Comet/Asteroid Impact on Earth: a Perspective from International Development Studies**	**437**
26.1	A Mighty Heuristic: Scale, Space and Time	437
	26.1.1 Assumptions	438
26.2	Do CAI-Scale Events Have Any Precedents?	438
	26.2.1 Adaptation and Resilience	439
26.3	The Perspective of International Development Studies	440
	26.3.1 Would "Sustainable Development" Be Enough?	441
	26.3.2 A Remaining Big Worry	443
26.4	Some Tentative Conclusions	443
	Notes	445
	References	446

27	**Disaster Planning for Cosmic Impacts: Progress and Weaknesses**	**449**
27.1	Introduction	449
27.2	Probabilities	452
27.3	Goal Setting	454
27.4	Risk Mapping	454
27.5	Safety by Improved Design	456
27.6	Disaster Simulation and Prediction	456

27.7	Warning Systems	459
27.8	Disaster Planning	461
27.9	Reconstruction	463
27.10	Summary and Conclusions	465
	References	466

28 Insurance Coverage of Meteorite, Asteroid and Comet Impacts – Issues and Options ... 469

28.1	Introduction	469
28.2	A Brief History of Insurance	469
28.3	Insurance and Natural Hazards	470
28.4	Do Asteroid Impacts Fit within the Principles of Insurance?	470
	28.4.1 Scenario 1: Asteroid Impact	471
	28.4.2 Scenario 2: Meteoroid Impact (Meteorite)	472
28.5	Insurance Coverage of Asteroid and Meteorite Damage	472
28.6	Assessing the Potential for Damage	474
28.7	Insurers Need to Prepare	475
28.8	The Cost of an Impact	475
28.9	Insurers' Capacity to Pay	476
28.10	Conclusions	477
	References	477

29 The Economic Consequences of Disasters Due to Asteroid and Comet Impacts, Small and Large ... 479

29.1	Introduction	479
29.2	Necessary Conditions	481
29.3	Scenario Construction	482
	29.3.1 Scenario 1	482
	29.3.2 Scenario 2	483
	29.3.3 Scenario 3	485
	29.3.4 Scenario 4	485
	29.3.5 Scenario 5	487
	29.3.6 Scenario 6	490
29.4	Summary and Conclusions	492
	Acknowledgments	493
	References	493

30 Communicating Impact Risk to the Public ... 495

30.1	Introduction	495
30.2	Our Present World: Brief Considerations	495
30.3	Principal Characteristics of NEO Impact Risks	497
30.4	Previous Experiences in Disaster Prevention	499
30.5	To Communicate or to Educate?	499
30.6	A Scheme for Transmission of Information	500
30.7	Conclusions	502
	References	503

31	**Impact Risk Communication Management (1998–2004): Has It Improved?**	505
31.1	Introduction	505
31.2	1997 XF11	505
31.3	1999 AN10	508
31.4	2000 SG344	509
31.5	2002 MN	510
31.6	2002 NT7	510
31.7	2004 AS1	511
31.8	2004 MN4	515
31.9	2003 QQ47	516
31.10	Purgatorio Ratio	517
	References	519
32	**Towards Rational International Policies on the NEO Hazard**	521
32.1	Introduction	521
32.2	"The 1997 XF11 Affair"	521
32.3	Putting the Astronomers' House in Order	522
	32.3.1 The Minor Planet Center	523
32.4	From Pure Science into the Real World	524
32.5	Epilog: the True Mess	526
	References	526
33	**A Road Map for Creating a NEO Research Program in Developing Countries**	527
33.1	Introduction	527
33.2	The Crisis	528
33.3	The Opportunity	530
33.4	Conclusion	531
	Acknowledgments	532
	References	532
	Index	533

Contributors

Johannes Andersen

Nordic Optical Telescope Scientific Association and Astronomical Observatory
University of Copenhagen, Juliane Maries Vej 30
2100 Copenhagen, Denmark
E-mail: ja@astro.ku.dk

M. G. L. Baillie

School of Archaeology and Palaeoecology, The Queen's University of Belfast
Belfast, BT7 1NN, Northern Ireland, UK
E-mail: m.baillie@qub.ac.uk

John W. Birks

Department of Chemistry and Biochemistry and Cooperative Institute for Research in Environmental Sciences (CIRES), University of Colorado UCB 215
Boulder, CO 80309-0215, USA

William F. Bottke, Jr.

Southwest Research Institute, Suite 400, 1050 Walnut Street
Boulder, CO 80302, USA
E-mail: bottke@boulder.swri.edu

Alessandro Carusi

Castelvecchi Publishing House, Rome, Italy
E-mail: carusi@libero.it

Andrea Carusi

Istituto di Astrofisica Spaziale e Fisica Cosmica, INAF, Area Ricerac Tor Vergata, Via Fosso del Cavaliere 100
00133 Rome, Italy
E-mail: andrea.carusi@iasf-roma.inaf.it

Clark R. Chapman

Southwest Research Institute, Suite 400, 1050 Walnut Street
Boulder, CO 80302, USA
E-mail: cchapman@boulder.swri.edu

Lee Clarke

Department of Sociology, Rutgers University
New Brunswick, NJ 08903, USA
E-mail: lee@leeclarke.com

Paul J. Crutzen

Max-Planck-Institute for Chemistry, Joh.-Joachim-Becher-Weg 27
55128 Mainz, Germany
and Scripps Institution of Oceanography, UC San Diego, 9500 Gilman Drive
La Jolla, CA 92093-0221, USA

Mohammed H. I. Dore

Climate Change Laboratory, Department of Economics, Brock University
St. Catharines, ON L2S 3A1, Canada
E-mail: dore@brocku.ca

Harold D. Foster

Department of Geography, University of Victoria, PO Box 3050 STN CSC
Victoria, BC V8W 3P5, Canada
E-mail: hfoster@office.geog.uvic.ca

Richard A. F. Grieve

Earth Sciences Sector, Natural Resources Canada
Ottawa, Ontario, K1A 0E8, Canada
E-mail: RGrieve@NCRan.gc.ca

V. K. Gusiakov

Tsunami Laboratory, Institute of Computational Mathematics and Mathematical Geophysics
Siberian Division, Russian Academy of Sciences, prospect Akademika Lavrentjeva, 6
Novosibirsk 630090, Russia
E-mail: gvk@sscc.ru

Andrew Hallak

Institute for Catastrophic Loss Reduction, University of Western Ontario, 20 Richmond Street
East Toronto M5C 2R9, Canada
E-mail: ahallak@pacicc.ca

William T. Hartwell

Desert Research Institute, Division of Earth and Ecosystem Sciences, 755 E. Flamingo Rd.
Las Vegas, Nevada 89119, USA
E-mail: Ted.Hartwell@dri.edu

Atko Heinsalu

Institute of Geology, Tallinn University of Technology, Ehitajate tee 5
19086 Tallinn, Estonia

Michel Hermelin

Universidad EAFIT, Carrera 49 # 7 Sur - 50
Medellin, Colombia
E-mail: hermelin@eafit.edu.co

Kenneth Hewitt

Cold Regions Research Center and Department of Geography and Environmental Studies
Wilfrid Laurier University
Waterloo, Ontario N2L 3C5, Canada
E-mail: khewitt@wlu.ca

Quanlin Hou

Centre of Earth System Science, Graduate School of Chinese Academy of Sciences
100039 Beijing, China

Wing-Huen Ip

Institutes of Astronomy and Space Science, National Central University
32054 Chung-Li, Taiwan
E-mail: wingip@astro.ncu.edu.tw

Evgeniy M. Kolesnikov

Department of Geochemistry, Geological Faculty, Lomonosov Moscow State University
119899 Moscow, Russia
E-mail: k_e_m@mail.ru, evgenkol@geol.msu.ru

Natal'ya V. Kolesnikova

Department of Geochemistry, Geological Faculty, Lomonosov Moscow State University
119899 Moscow, Russia
E-mail: k_e_m@mail.ru, evgenkol@geol.msu.ru

Paul Kovacs

Institute for Catastrophic Loss Reduction, University of Western Ontario
20 Richmond Street
East Toronto M5C 2R9, Canada

David A. Kring

Lunar and Planetary Laboratory, Department of Planetary Sciences
University of Arizona
Tucson, Arizona, 85721, USA

Wolfgang Kundt

Argelander-Institut for Astronomy, Department of Astrophysics, Bonn University
Auf dem Hügel 71
53121 Bonn, Germany
E-mail: wkundt@astro.uni-bonn.de

A. Chantal Levasseur-Regourd

Université P. & M. Curie (Paris VI), Aéronomie CNRS-IPSL, B.P. 3
91371, Verrières, France
E-mail: Chantal.Levasseur@aerov.jussieu.fr

Giuseppe Longo

Dipartimento di Fisica, Università di Bologna, Via Irnerio 46
40126 Bologna, Italy
E-mail: longo@bo.infn.it

Michael C. MacCracken

Climate Institute, 1785 Massachusetts Avenue, N.W.
Washington DC 20036, USA
E-mail: mmaccrac@comcast.net

Brian G. Marsden

Harvard-Smithsonian Center for Astrophysics
Cambridge, MA 02138, USA
E-mail: bmarsden@cfa.harvard.edu

Jesús Martínez-Frías

Planetary Geology Laboratory, Centro de Astrobiologia (CSIC/INTA)
associated to the NASA Astrobiology Institute, Ctra. de Ajalvir km. 4
28850 Torrejón de Ardoz, Madrid, Spain
E-mail: martinezfrias@mncn.csic.es

W. Bruce Masse

Cultural Resources Team
ENV-EAQ Ecology and Air Quality Group, Mail Stop J978
Los Alamos National Laboratory
Los Alamos, New Mexico 87545, USA
E-mail: wbmasse@lanl.gov

Sharad Master

Impact Cratering Research Group, Economic Geology Research Institute, School of Geosciences
University of the Witwatersrand
Johannesburg, South Africa
E-mail: masters@geosciences.wits.ac.za

W. J. McGuire

Benfield UCL Hazard Research Centre
Department of Earth Sciences, University College London
Gower Street
London WC1E 6BT, UK
E-mail: w.mcguire@ucl.ac.uk

H. Jay Melosh

Lunar and Planetary Lab, 429E Space Sciences Building, University of Arizona
Tucson AZ 85721-0092, USA
E-mail: jmelosh@lpl.arizona.edu

Andrea Milani Comparetti

Department of Mathematics, University of Pisa
via Buonarroti 2
56127 Pisa, Italy

David Morrison

14660 Fieldstone
Saratoga CA 95070, USA
E-mail: dmorrison@arc.nasa.gov

Anneli Poska

Institute of Geology, Tallinn University of Technology, Ehitajate tee 5
19086 Tallinn, Estonia

Luca Pozio

University of Rome III, Department of Economics
Rome, Italy
E-mail: lucapozio@libero.it

Kaare L. Rasmussen

Department of Chemistry, University of Southern Denmark
Campusvej 55
5230 Odense, Denmark
E-mail: klr@chem.sdu.dk

Raymond G. Roble

High Altitude Observatory, National Center for Atmospheric Research
PO Box 3000
Boulder, CO 80307-3000, USA
E-mail: jwbirks@hotmail.com

José Antonio Rodríguez-Losada

Departamento de Edafología y Geología, Universidad de La Laguna
38206 La Laguna, Tenerife (Islas Canarias), Spain
E-mail: jrlosada@ull.es

Leili Saarse

Institute of Geology, Tallinn University of Technology, Ehitajate tee 5
19086 Tallinn, Estonia

Roy C. Sidle

Slope Conservation Section, Geohazards Division, Disaster Prevention Research Institute
Kyoto University
Gokasho, Uji Kyoto 611-0011, Japan
E-mail: sidle@slope.dpri.kyoto-u.ac.jp

Paul Slovic

Decision Research, 1201 Oak Street, Suite 200
Eugene, Oregon 97401, USA
E-mail: pslovic@darkwing.uoregon.edu

Giovanni B. Valsecchi

INAF-IASF, via Fosso del Cavaliere 100
00133 Roma, Italy
E-mail: giovanni@rm.iasf.cnr.it

Jüri Vassiljev

Institute of Geology, Tallinn University of Technology, Ehitajate tee 5
19086 Tallinn, Estonia

Siim Veski

Institute of Geology, Tallinn University of Technology, Ehitajate tee 5
19086 Tallinn, Estonia
E-mail: Veski@gi.ee

Ben Wisner

Oberlin College, 173 West Lorain Street
Oberlin, OH 44074, USA
and Crisis States Programme, Development Studies Institute, London School of Economics
Benfield Greig Hazard Research Centre, Gower Street, University College
London WC1E 6BT, UK
E-mail: bwisner@igc.org

T. Woldai

International Institute for Geoinformation Sciences & Earth Observation (ITC)
Hengelosestraat 99
P.O. Box 6
7500 AA Enschede, The Netherlands
E-mail: Woldai@itc.nl

Liewen Xie

Institute of Geology, Chinese Academy of Sciences
100029 Beijing, China

Part I Anthropology, Archaeology, Geology

Chapter 1 The Geologic Record of Destructive Impact Events on Earth

Chapter 2 The Archaeology and Anthropology of Quaternary Period Cosmic Impact

Chapter 3 The Sky on the Ground: Celestial Objects and Events in Archaeology and Popular Culture

Chapter 4 Umm Al Binni Structure, Southern Iraq, As a Postulated Late Holocene Meteorite Impact Crater

Chapter 5 Tree-Rings Indicate Global Environmental Downturns That Could Have Been Caused by Comet Debris

Chapter 6 The GGE Threat: Facing and Coping with Global Geophysical Events

Chapter 1

The Geologic Record of Destructive Impact Events on Earth

Richard A. F. Grieve · David A. Kring

1.1
Introduction

The Earth is the most geologically active of the terrestrial planets and it has retained the poorest sample of the record of hypervelocity impact by interplanetary bodies throughout geologic time. Although the surviving sample of impact structures is small, the terrestrial impact record has played a major role in understanding and constraining cratering processes, as well as providing important ground-truth information on the three dimensional lithological and structural character of impact structures (Grieve and Therriault 2004). Recently, there has been a growing awareness in the earth-science community that impact is also potentially important as a stochastic driving force for changes to the terrestrial environment. This has stemmed largely from: the discovery of chemical and physical evidence for the involvement of impact at the Cretaceous-Tertiary (K/T) boundary and the associated mass extinction event (e.g. Alvarez et al. 1980; Smit and Hertogen 1980; Bohor et al. 1984), and their relation to the Chicxulub impact structure in the Yucatan Peninsula, Mexico (Hildebrand et al. 1991), the recognition of the resource potential of impact structures, some of which are related to world-class ore deposits, both spatially and genetically (Grieve and Masaitis 1994; Grieve 2005), and the recognition of the potentially disastrous consequences of impacts for human civilization (Gehrels 1994).

1.2
General Character of the Record

The known record of hypervelocity impact on the Earth consists of approximately 170 individual impact structures or crater fields, in the case of small impacting bodies, which broke up in the atmosphere. In addition, there are over 20 impact events registered as depositional events in the stratigraphic record, some of which are related to known impact structures (Grieve 1997; Koeberl 2001). A listing of currently known terrestrial impact structures and some of their salient characteristics can be found at *http://www.unb.ca/passc/ImpactDatabase/index.html*. The terrestrial impact record contains a number of biases, reflecting modification and obliteration of terrestrial impact structures by post-impact, terrestrial geologic processes.

1.2.1
Spatial Distribution

The spatial distribution of known terrestrial impact structures is biased towards the stable cratonic areas of the crust, as they are the best available surfaces for the preservation of impact structures in the terrestrial environment. Approximately 30% of known terrestrial impact structures are eventually buried by post-impact sediments. Most buried impact structures were detected initially as geophysical anomalies and later drilled, for scientific or economic purposes, thus confirming their impact origin. A small number of impact structures are completely submerged beneath the sea. They occur, however, on continental shelves and no impact structures are known from the true oceanic crust. This reflects the relatively young age (< 200 Ma) and the generally poor resolution of geological knowledge of the ocean floors. Meteoritic debris, however, is known over a distance of at least 500 km in the south-east Pacific, where a 1–4 km diameter, stony-iron asteroid impacted in the Late Pliocene but apparently failed to impact the 2.5–5.0 km deep ocean floor (Kyte et al. 1988; Gersonde et al. 1997).

1.2.2
Age Distribution

Approximately 40% of the known terrestrial impact structures have been dated isotopically; generally from the analysis of impact melt rocks. Most of the materials (90%) involved in an impact event, however, are subjected to insufficient shock pressures and postshock temperatures to significantly disturb isotopic dating systems (Deutsch and Schärer 1994). The remainder of known terrestrial impact structures have biostratigraphic or stratigraphic dates, which, in some cases, provide only upper limits, based on the age of the target rocks. There is a general bias in the ages of known terrestrial impact structures, since more than 60% are < 200 Ma old (Grieve and Shoemaker 1994), which reflects the problems of preservation and, to a lesser extent, recognition in the highly active geological environment of the Earth.

1.2.3
Size Distribution

In most cases, original rim diameters (D) for terrestrial impact structures are reconstructed estimates. Individual diameter estimates can be different and controversial so that quantitative interpretations based on data compilations of rim diameters of terrestrial impact structures should be regarded with some caution. Problems can also occur with buried structures, where rim diameter estimates are based on the interpretation of geophysical data, e.g. the initial estimate for the rim diameter of Chicxulub, Mexico was 180 km (Hildebrand et al. 1991) but estimates have ranged from 130 km (Morgan et al. 1997) to 300 km (Sharpton et al. 1993). The most recent interpretation of reflection seismic data suggests that the ~180 km estimate is the most accurate (Morgan and Warner 1999; Snyder et al. 1999; Morgan et al. 2002). A recent analysis of the disparities in estimates of rim diameters and the implications for energy scaling and resulting potential environmental degradation of specific terrestrial impact events is given in Turtle et al. (2005). It can also be shown that there is a bias in the sizes of surviving terrestrial impact structures, since at larger diameters, the cumulative size-frequency distribution can be approximated by a power law; whereas, at diameters below 20 km the cumulative size-frequency falls off the power law, with an increasing deficit of structures at smaller diameters (Grieve and Shoemaker 1994). This drop-off is an inherent property of the terrestrial record, as it has remained through the addition of new structures to the known record. The deficit of small craters is due to a combination of atmospheric crushing of weaker impacting bodies and the greater difficulty in recognizing smaller eroded and/or buried structures.

1.2.4
Terrestrial Cratering Rate

With these biases, it is clear that care must be exercised when estimating an average cratering rate from the terrestrial impact record. To reduce the effects of the loss of older and smaller structures, the sample of structures used to calculate a rate can be restricted to only relatively young and large structures. The net result is that the estimated average cratering rate for the last approximately 100 Ma is $5.6 \pm 2.8 \times 10^{-15}$ km^{-2} a^{-1} for $D \geq 20$ km (Grieve and Shoemaker 1994). The relatively high ($\pm 50\%$) uncertainty attached to this estimate reflects concerns of small number statistics and the completeness of search for existing impact structures. Although the bulk of the larger impact structures have likely been recognized on the better-searched areas of the Earth, e.g. the North American craton, this estimated average cratering rate illustrates just how poorly the record is known in other areas. For example, although none are known from Africa, the average cratering rate suggests that approximately 17 ± 8 structures with $D \geq 20$ km should have been formed in an area the size of Africa (approximately 30×10^6 km^2) in the last 100 Ma.

1.2.5
Periodic Impacts

When Raup and Sepkoski (1984) reported evidence for a periodicity in the marine extinction record, a number of others claimed a similar periodicity in the terrestrial cratering record (e.g. Alvarez and Muller 1984; Davis et al. 1984; Rampino and Stothers 1984), as a result of periodic cometary showers. Grieve et al. (1988) argued against these conclusions, noting that, if the uncertainties in crater age estimates are taken into account, periodicities in the cratering record are questionable. Heisler and Tremaine (1989) reached a similar conclusion based on different statistical arguments and Baski (1990) detected no periodicity, if the selected impact structures were restricted to those with age estimates of sufficient accuracy and precision. Weissman (1990) also found no evidence for periodic cometary showers and challenged the proposed mechanisms for producing periodic cometary showers. Despite such arguments, periodic cometary showers, as defined by time-series analysis of the terrestrial cratering record, are still featured (e.g. Yabushita 1992; 2004), and suggested as a causative agent for various geologic phenomena on Earth (e.g. Stothers and Rampino 1998; Rampino and Haggerty, 1996). Recently, Jetsu and Pelt (2000) reanalyzed both the terrestrial impact and mass extinction record and detected no periodicity, apart from a spurious "human-signal" induced by rounding of less certain "ages", to integer values, often in multiples of 5 or 10 Ma. Despite continuing assertions, there is no compelling evidence for periodic impacts, due to cometary showers, in the terrestrial cratering record.

1.3
Recognition of Terrestrial Impact Structures

1.3.1
Morphology

With increasing diameter, impact structures become proportionately shallower and develop more complicated rims and floors, including the appearance of central peaks and interior rings. Impact craters are divided into three basic morphologic subdivisions: simple craters, complex craters, and basins (Dence 1972; Wood and Head 1976). Simple impact structures have the form of a bowl-shaped depression with an upraised rim (Fig. 1.1). At the rim, there is an overturned flap of ejected target materials, which displays inverted stratigraphy, with respect to the original target materials. Beneath the floor is a lens of allochthonous breccia that is roughly parabolic in cross-section. In places, this breccia lens may contain highly shocked, including melted, target materials. Beneath the breccia lens, parautochthonous, fractured target rocks define the walls and floor of what is known as the true crater. Shocked rocks in the true crater floor are confined to a small central volume at the base.

With increasing diameter, simple craters display increasing evidence of wall and rim collapse and evolve into complex craters. Complex impact structures on Earth

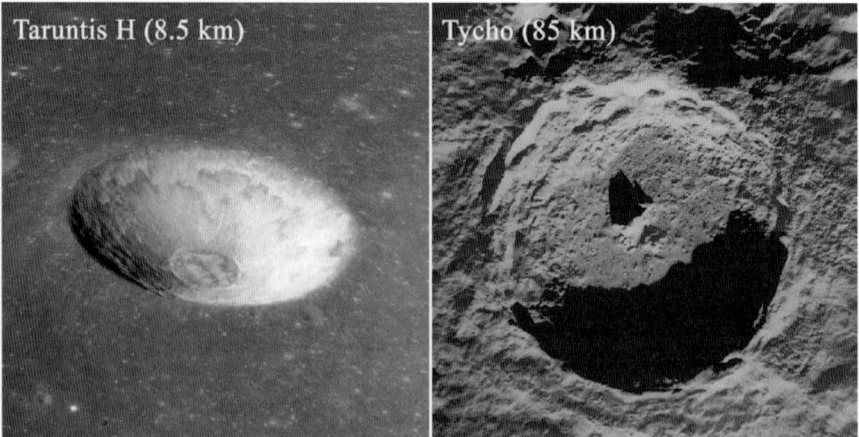

Fig. 1.1. Examples of simple and complex impact structures on the moon, where original morphologies are better preserved than on Earth. Taruntis H (*left*, Apollo 10 image H-4253) is an 8.5 km diameter simple structure and Tycho (*right*, Orbiter V image M-125) is an 85 km diameter complex structure, with central peak(s), a flat floor and structurally complex rim area

first occur at diameters greater than 2 km in layered, sedimentary rocks but not until diameters of 4 km or greater in stronger, more homogeneous, igneous or metamorphic crystalline rocks (Dence 1972). Complex impact structures are characterized by a central topographic peak or peaks, a broad, flat floor, and terraced, inwardly slumped and structurally complex rim areas (Fig. 1.1). The broad flat floor is partially filled by a sheet of impact melt rock and/or polymict allochthonous breccia. The central region is structurally complex and, in large part, occupied by a central peak, which is the topographic manifestation of a much broader and extensive area of uplifted rocks that occurs beneath the surface at the center of complex craters. Details regarding observations of terrestrial craterforms and cratering mechanics at simple and complex structures are given in Grieve and Therriault (2004) and Melosh (1989), respectively.

There have been claims that the largest known terrestrial impact structures have multi-ring forms, e.g. Chicxulub (Sharpton et al. 1993), Sudbury, Canada (Stöffler et al. 1994; Spray and Thompson 1995), and Vredefort, S. Africa (Therriault et al. 1997). Although certain of their geological and geophysical attributes form annuli, it is not clear that these correspond, or are related in origin, to the obvious topographical rings observed in lunar multi-ring basins (Spudis 1993; Grieve and Therriault 2000). Attempts to define diagnostic morphometric relations, particularly depth-diameter relations, for terrestrial impact structures have had limited success, because of the effects of erosion and, to a lesser degree, sedimentation. The most recent empirical relations can be found in Grieve and Therriault (2004).

1.3.2
Geology of Impact Structures

Although an anomalous circular feature may indicate the presence of an impact structure, there are other geological processes that can produce similar features in the terrestrial environment. The burden of proof for an impact origin for a particular structure, or lithology in the stratigraphic record, generally lies with the documentation of the occurrence of shock-metamorphic effects.

On impact, the bulk of the impacting body's kinetic energy is transferred to the target by means of a shock wave. This shock wave imparts kinetic energy to the target, which in turn leads to the ejection of target materials and the formation of a crater. It also increases the internal energy of the target materials, which leads to the formation of so-called shock-metamorphic effects. The details of the physics of shock wave behavior and shock metamorphism can be found in Melosh (1989) and Langehorst (2002), respectively. Minimum shock pressures required for the production of diagnostic shock-metamorphic effects are 5–10 GPa for most silicate minerals. Strain rates produced on impact are of the order of 10^6 s^{-1} to 10^9 s^{-1} (Stöffler and Langenhorst 1994), many orders of magnitude higher than typical tectonic strain rates (10^{-12} s^{-1} to 10^{-15} s^{-1}; e.g. Twiss and Moores 1992) and shock-pressure duration is measured in seconds, or less, in even the largest impact events (Melosh 1989). Such physical conditions are not reproduced by endogenic geologic processes. They are unique to impact and, unlike endogenic terrestrial metamorphism, disequilibrium and metastability are common phenomena in shock metamorphism. Shock-metamorphic effects are well described by Stöffler (1971, 1972, 1974), Stöffler and Langenhorst (1994), Grieve et al. (1996), French (1998), Langenhorst and Deutsch (1998), Langenhorst (2002) and others. They are discussed here only in general terms, as they relate to the recognition of impact materials in the terrestrial environment.

1.3.2.1
Impact Melting

Heating of the target rocks occurs, as not all the pressure-volume work that occurs during shock-compression is recovered upon adiabatic pressure release and the excess work is manifest as irreversible waste heat. Above 60 GPa, the residual waste heat is sufficient to cause whole-rock melting and, at higher pressures, vaporization (Melosh 1989). Impact melted lithologies occur as glass particles and bombs in crater ejecta (Engelhardt 1990), as dikes within the crater floor and walls, as glassy to crystalline pools and lenses within the breccia lenses of simple craters, or as coherent, central sheets lining the floor of complex structures (Fig. 1.2).

The final composition of impact-melt rocks depends on the wholesale melting of a mix of target rocks, as opposed to partial melting relationships for endogenous igneous rocks. The composition of impact-melt rocks is, therefore, characteristic of the target rocks and may be reproduced by a mixture of the various country rock types in their appropriate geological proportions. Such parameters as $^{87}Sr/^{86}Sr$ and $^{143}Nd/^{144}Nd$ ratios of impact melt rocks also reflect the pre-existing target rocks (Jahn et al. 1978; Faggart et al. 1985). These important characteristics of impact melt lithologies can allow impact melt material in ejecta to be traced back to specific source lithologies (e.g. Kring and Boynton 1992; Blum et al. 1993; Whitehead et al. 2000). In general, even relatively thick impact-melt sheets are chemically homogeneous over radial distances of kilometers. In large impact structure, and where the target rocks are not homogene-

Fig. 1.2.
Approximately 150 m high cliffs of impact melt rock at Manicouagan impact structure

ously distributed, this observation may not hold true in detail, such as for Manicouagan, Canada (Grieve and Floran 1978), Chicxulub (Kettrup et al. 2000), and Popigai (Kettrup et al. 2003). Differentiation is not a characteristic of relatively thick coherent impact-melt sheets, with the exception of the extremely thick ~2.5 km, Sudbury Igneous Complex, Sudbury Structure (Ariskin et al. 1999; Therriault et al. 2002).

Enrichments above target rock levels in siderophile and platinum group elements (PGEs) and Cr have been identified in some impact melt rocks and ejecta. These are due to an admixture of up to a few percent of meteoritic material from the impacting body. This attribute was critical in the initial works relating the K/T boundary material to an impact event (e.g. Alvarez et al. 1980). In some melt rocks, the relative abundances of the various siderophiles have constrained the composition of the impacting body to the level of meteorite class (Palme et al. 1979; Tagle and Claeys 2005). In other melt rocks, no geochemical anomaly has been identified. This may be due to the inhomogeneous distribution of meteoritic material within the impact melt rocks and sampling variations (Palme et al. 1981), or to differentiated impacting bodies, such as basaltic achondrites that are not relatively enriched in PGEs relative to terrestrial rocks. More recently, high precision Cr, Os and He-isotopic analyses have been used to detect meteoritic material in the terrestrial environment (e.g. Koeberl et al. 1996; Peucker-Ehrenbrink 2001; Farley 2001).

1.3.2.2
Fused Glasses and Diaplectic Glasses

Shock fused glasses are characterized morphologically by flow structures and vesiculation. Peak pressures required for shock melting of single crystals are in the order of 40 to 60 GPa (Stöffler 1972, 1974). Under these conditions, the minerals in the rock melt independently after the passage of the shock wave and melting is mineral selective. Conversion of framework silicates to isotropic, dense, but not fused, glassy phases occur at peak pressures and temperatures well below their normal melting point. These are called diaplectic glasses, requiring peak pressures of between 30 and 45 GPa for feldspar and 35 to 50 GPa for quartz in quartzo-feldspathic rocks, (e.g. Stöffler and Hornemann 1972; Stoffler 1984). The morphology of the diaplectic glass is the same as the original mineral crystal (Fig. 1.3) and they have densities lower than the crystalline form from which they are derived, but higher than thermally melted glasses of equivalent composition (e.g. Stöffler and Hornemann 1972; Langenhorst and Deutsch 1994). Maskelynite, diaplectic plagioclase glass (Fig. 1.3), is the most common example from terrestrial rocks. Diaplectic glasses of quartz (Fig. 1.3; Chao 1967) and of alkali feldspar (Bunch 1968) also occur.

1.3.2.3
High-Pressure Polymorphs

Shock can produce metastable high-pressure polymorphs, such as stishovite and coesite from quartz (Chao et al. 1962; Langenhorst 2002), and cubic and hexagonal diamond from graphite (Masaitis 1998; Langenhorst 2002). Coesite and diamond are also products of high-grade metamorphism but the paragenesis and the geological setting are

Fig. 1.3. Some shock metamorphic effects. **a** Shatter cones at Gosses Bluff impact structure, Australia. **b** Photomicrograph of planar deformation features (PDFs) in quartz in a compact sandstone from Gosses Bluff impact structure. Crossed polars, width of field of view 0.4 mm. **c** Photomicrograph of quartz (*center*, higher relief) with biotite (*darker gray, upper right*) and feldspar (*white, bottom*) in a shocked granitic rock from Mistastin impact structure, Canada. Plane light, width of field of view 1.0 mm. **d** Photomicrograph as in (**c**) but with crossed polars. The biotite is still birefringent but the quartz and feldspar are isotropic, as they have been metamorphosed to diaplectic glasses by the shock wave, while retaining their original morphology

completely different from that in impact events. Stishovite and coesite have only rarely been produced by laboratory shock recovery experiments (Stöffler and Langenhorst 1994). For terrestrial impact structures in crystalline targets, such polymorphs generally occur in small or trace amounts as very fine-grained aggregates and are formed by partial transformation of the host quartz. In porous, quartz-rich target lithologies, however, they may be more abundant. For example, coesite constitutes 35% of the mass of highly shocked Coconino sandstone at Barringer or Meteor Crater, U.S.A. (Kieffer 1971). Details on the characteristics of coesite and stishovite are given in Stöffler and Langenhorst (1994).

1.3.2.4
Planar Microstructures

The most common documented shock-metamorphic effect is the occurrence of planar microstructures in tectosilicates, particularly quartz (Fig. 1.3; Hörz 1968). The utility of planar microstructures in quartz reflects the ubiquitous nature of the mineral and the stability of the quartz and microstructures, themselves, in the terrestrial environment, and the relative ease with which they can be documented. It was the documentation of planar microstructures in quartz (Bohor et al. 1984) that provided the first physical evidence of impact involvement in K/T boundary sediments. Reviews of the nature of the shock metamorphism of quartz can be found in Stöffler and Langenhorst (1994), Grieve et al. (1996), and Langenhorst (2002). Planar deformation features (PDFs) are produced under pressures of ~10 to ~35 GPa whereas planar fractures (PFs) form under shock pressures ranging from ~5 GPa up to ~35 GPa (Stöffler 1972; Stöffler and Langenhorst 1994).

1.3.2.5
Shatter Cones

The only known diagnostic shock effect that is megascopic in scale is the occurrence of shatter cones (Dietz 1968; Sagy et al. 2002). Shatter cones are unusual, striated, and horse-tailed conical fractures (Fig. 1.3) ranging from millimeters to meters in length and are initiated most frequently in rocks that experienced moderately low shock pressures, 2–6 GPa, but have been observed in rocks that experienced up to ~25 GPa (Milton 1977). Such conical striated fracture surfaces are best developed in fine-grained, structurally isotropic lithologies, such as carbonates and quartzites. They are generally found in place as individual or composite groups of partial to complete cones in the parautochthonous rocks below the crater floor, especially in the central uplifts of complex impact structures and, more rarely, in isolated rock fragments in breccia units, indicating that the shatter cones formed before the material was set in motion by the cratering flow-field.

1.3.3
Geophysics of Impact Structures

Geophysical anomalies over terrestrial impact structures vary in their character and, in isolation, do not provide definitive evidence for an impact origin. Interpretation of a single geophysical data set over a suspected impact structure can be ambiguous (e.g. Hildebrand et al. 1991; Sharpton et al. 1993). When combined, however, with complementary geophysical methods and the existing database over other known impact structures, a more definitive assessment is possible (e.g. Ormö et al. 1999). Since potential-field data are available over large areas, with almost continuous coverage, gravity and magnetic observations have been the primary geophysical indicators used for evaluating the occurrence of possible terrestrial impact structures. Reflection seismic data, although providing much better spatial resolution of subsurface structure (e.g. Morgan et al. 2002), are generally less available. The most recent synthesis of the geophysical character of terrestrial impact structures is Grieve and Pilkington (1996).

1.4
Impacts in the Stratigraphic Record

Known occurrences of impact-related materials in the stratigraphic record are limited and are addressed most recently in Koeberl (2001). Prior to the interpretation that geochemical anomalies in K/T boundary sediments were due to impact, the known record of impact in the stratigraphic column was limited to the occurrence of the Australasian, Ivory Coast, North American as well as moldavite tektite and microtektite strewn fields. As a result of the interest generated by the K/T discoveries and potential connections between impact events and other short-term environmental events in the geological record, there have been searches for siderophile element (mostly Ir) anomalies at other major stratigraphic boundaries (Stothers 1993). In some cases, weak anomalies have been reported, as have isolated occurrences of "shocked" minerals but, at this time, there is no compelling reason or confirmatory evidence, to ascribe them to impact processes (Koeberl 2001). The majority of known impact events recorded in the stratigraphic column were recognized initially through the occurrence of physical, not geochemical, evidence of impact.

Recently, attention has focused on the Permian-Triassic boundary, with reports of extraterrestrial helium and argon, meteorites fragments and shocked quartz from Graphite Peak in Antarctica and the Sydney Basin in Australia (Becker et al. 2001; Basu et al. 2003; Retallack et al. 1998). Such discoveries have not been independently verified and evidence for potential impact-related materials at the Permian-Triassic boundary is not available from non-Gondwana sites. Most recently, Becker et al. (2004) suggested the Bedout structure, offshore northwestern Australia, as the putative Permian-Triassic impact site. The evidence they present is somewhat equivocal, with none of the "shocked" materials corresponding exactly to shocked materials at known terrestrial impact structures.

On the basis of the analysis of a Pacific deep-sea core spanning the last ~70 Ma, Kyte et al. (1993) detected only one significant siderophile (Ir) anomaly and that was at the K/T boundary. The signal to noise variation in cosmic material recorded by Ir values in the core is such that the signal of impact events even large enough to produce 100 km-sized impact structure is unlikely to be resolved (Grieve 1997). Recent high-resolution geochemical studies at Massignamo, Italy, have, however, detected elevated Ir values that have been equated with the 100 km Popigai structure in Siberia and the 80 km Chesapeake structure in the U.S.A. These two impact events are the largest known post K/T impact events on Earth and are indistinguishable in age at 35.7 ± 0.8 Ma, (Bottomley et al. 1997) and 35.3 ± 0.2 Ma (Poag and Aubrey 1995), respectively. There is some question regarding the size of the Chesapeake impact event and it was likely smaller, producing only a 40 km diameter structure (Turtle et al. 2005). Two ejecta layers identified with these impact events occur in deep-sea cores and the time separation between the events is believed to be between 20 000 and 3 000 years (Glass and Koeberl 1999). The two impact events did not lead to a mass extinction, such as at the K/T boundary, but may have resulted in global climatic perturbations (Bodiselitsch et al. 2004).

Although they appear to coincide temporally with a short period of higher delivery of interplanetary dust particles to Earth, it is not clear whether Popigai and Chesapeake represent the result of an astronomical event or a statistical sport (Tagle and Claeys 2004).

There are other calls for clusters of impacts. For example, there may have been a cluster of impacts in the Early Ordovician when an L-chondrite parent body suffered a major collisional event at ~500 Ma, which is recorded in many shock-metamorphosed L-chondrite samples. This may have produced a rain of meteorites (Schmitz et al. 2001; Heck et al. 2004). Lindstrom (2003) has equated this with a number of small impact craters, which were apparently produced at approximately the same time, in Fenno-scandia. The age estimates, however, of the individual impact structures are very poorly constrained.

1.5
Impacts and the Biosphere

1.5.1
Early Life

Only a few mineral relicts from the first 500 Ma of Earth history are known to exist. The lunar record of that time indicates that the impact rate was one to two orders of magnitude higher than today and may have been dominated by a 20 to 200 Ma pulse of bombardment at ~4 Ga (Ryder 2002; Cohen et al. 2000; Kring and Cohen 2002). By analogy and depending on the cosmic approach velocity, the number of impact structures formed on the Earth during the same time was 25 to 100 times greater, due to the Earth's larger gravitational cross-section. Such impact events could have been environmentally devastating to the Earth, the largest blowing away portions of the atmosphere and vaporizing oceans (Zahnle and Sleep 1997), potentially impeding the development of life (Maher and Stevenson 1988) or creating conditions through which only certain species may survive (Chyba 1993). Conversely, such impact events may have created subsurface hydrothermal systems suitable for prebiotic chemical reactions and possibly the origin and early evolution of life (Kring 2000, 2003). Such systems can be long-lived, persisting for > 10^5 or 10^6 yr when 200 km or larger in diameter (Abramov and Kring 2004; Daubar and Kring 2001). The impacting objects could have also delivered biogenic elements (C, S, H, N, O, P) and potentially even organic molecules like amino acids (Pierazzo and Chyba 1999; Kring and Cohen 2002), although the bulk of Earth's water had been delivered prior to the ~4 Ga bombardment (Swindle and Kring 2001; Valley et al. 2002; Campins et al. 2004).

1.5.2
Coupling through the Atmosphere and Hydrosphere

Apart from the "relatively local" formation of an impact structure, impact affects the terrestrial environment and biosphere through its interaction with the atmosphere and hydrosphere (Melosh 2004). This begins as the incoming impacting body enters the atmosphere, when a bow shock is produced and the surrounding atmosphere is heated and ionized. When the projectile nears the surface, the bow shock wave has the capacity to flatten forests and structures of comparable or less strength (Vasilyev 1998; Glasstone and Dolan 1977). A second wave radiates through the atmosphere on impact, generating an air blast. Even small impact events (e.g. at Barringer, 1.2 km in diameter) will produce surface wind velocities in excess of 2000 km hr^{-1}, shredding and uproot-

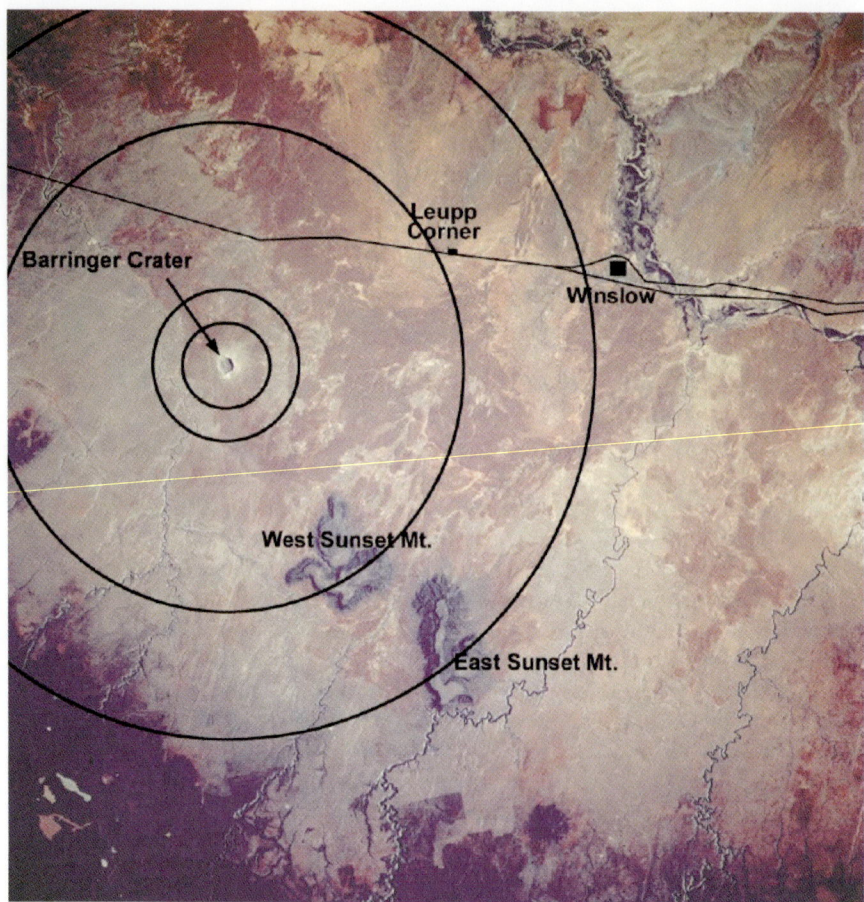

Fig. 1.4. Estimates of pressure pulse and airblast damage associated with the Barringer impact event. The blast effect was immediately lethal for human-sized animals within the inner 6 km diameter circle. Severe lung damage would occur within the next 10–12 km diameter circle due to the pressure pulse alone and animals would be severely injured and unlikely to survive. Winds would exceed 1500 km hr^{-1} within the inner circle and still exceed 100 km hr^{-1} at radial distances of 25 km (3rd circle). The outermost ~50 km circle represents the outer limit of severe to moderate damage to trees and human-structures of comparable strength. Such an event today would decimate the population of an urban area equivalent to the size of Kansas City, U.S.A. (population 425 000). See Kring (1997) for additional details

ing vegetation and severely injuring or killing local fauna (Fig. 1.4; Kring 1997). The energy of the Tunguska explosion, Siberia, in 1908 was less than the impact energy of Barringer, but it occurred at an optimum blast height. Rather than reaching the ground, the incoming body exploded 5 to 10 km above the surface, producing devastation that was similar to that of the Barringer event (Fig. 1.5; Toon et al. 1997). Twenty-one hundred km^2 of forest were damaged at Tunguska (Vasilyev 1998) and 1 000 to 2 100 km^2 are believed to have been flattened around the Barringer crater (Kring 1997). If either event occurred in the vicinity of a modern urban area, it would have been devastating.

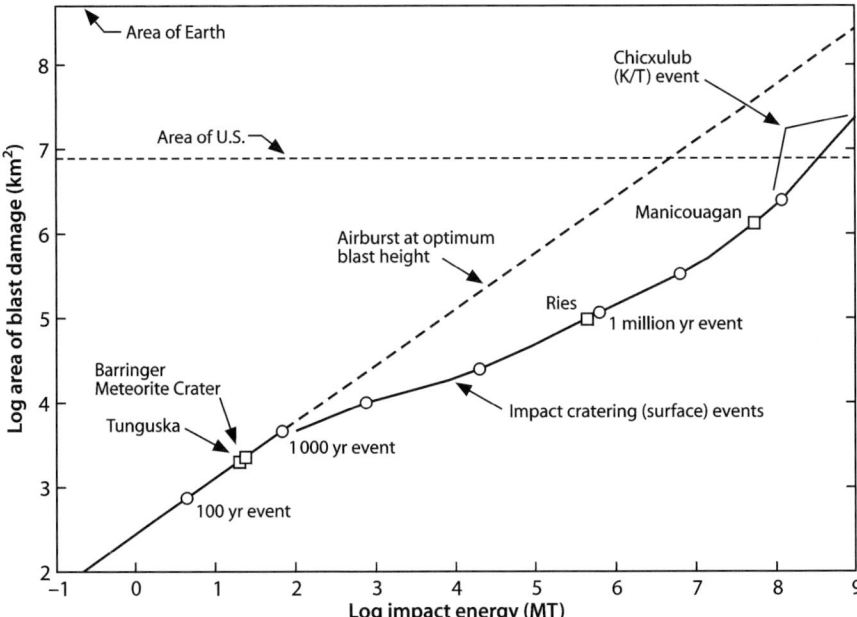

Fig. 1.5. Logarithmic plot of area damaged by overpressures in excess of 4 psi (276 hPa) as a function of impact energy in MT TNT equivalent. The smallest events will detonate in the atmosphere (like Tunguska). Large impacting bodies will impact the surface, where the efficiency of energy conversion into atmospheric shock wave is less than atmospheric explosions at optimum blast height (*dashed line*). Approximate rates of occurrence of impact events of a particular magnitude are indicated in orders of magnitude. The propagation of the air blast in the largest impact events may be affected by the curvature of the Earth, which was accommodated by assuming that the wave travels radially and does not produce over-the-horizon damage. Modified from Toon et al. (1997), which includes additional details. Approximate rates of occurrence of impact events of particular magnitudes are indicated at the level of orders of magnitude. These estimates are also from Toon et al. (1997). We note, however, that the 1 million year frequency may be too high and may be better located below the Ries event. There is also an order of magnitude uncertainty associated with derived impact energies for all events and considerable uncertainties in extrapolating to the larger events, because the finite thickness of the atmosphere. As a result, the area of airblast damage at the larger events may be an overestimate

For large impacts, the blast effects can be at sub-continental scales (Fig. 1.5), although additional uncertainties occur at larger scales due to extrapolations from smaller scale events, such as the effects of nuclear explosions that are totally contained within the atmosphere. For example, the Manicouagan impact, which resulted in a 100 km diameter impact structure, may have resulted in an air blast that affected an area in excess of a 1000 km in diameter (Fig. 1.6; Kring 2003). It could have also resulted in large-scale wildfires (Durda and Kring 2004) and could have led to an increase in the amount of S in the atmosphere on the order of 4–5 orders of magnitude (Fig. 1.7; Kring 2003).

The global extinctions that occurred at the K/T boundary were created by a different scale of atmospheric interaction, associated with the direct and indirect effects of ejected debris in the atmosphere (Melosh 2004). The Chicxulub impact created a vapor-rich plume of debris that expanded above the atmosphere and enveloped the entire

Fig. 1.6. Extent of airblast produced by the Manicouagan impact event. Near the impact site wind speeds would have exceeded 1000 km hr^{-1} and eventually decelerated to hurricane-force at the largest distances. The *white circular line* corresponds to the limit of 4 psi (27 kPa) peak overpressures derived from Toon et al. (1997) (see also Fig. 1.5), which has the capacity to severely damage and kill plants and animals (Kring 1997). The radial distance of the 4 psi limit is approximately 560 km, which is smaller than that in Kring (2003), who mis-plotted the results of Toon et al. (1997). The basemap is from NASA's Blue Marble, a true color rendering of satellite data with 1 km resolution

globe. This material and its re-entry deposited a vast amount of energy in the atmosphere, altering nitrogen chemistry, destroying ozone, producing nitric acid rain (Zahnle 1990) and heating the surface, sufficient to ignite wildfires in large areas of the world (Melosh et al. 1990; Kring and Durda 2002). Climatically active gases (greenhouse-warming H_2O and CO_2; sulfate producing SO_x; ozone-destroying Cl and Br) and dust were also injected into the atmosphere (Alvarez et al. 1980; Pope et al. 1997; Kring 2000 and references therein). In the case of Chicxulub, an unusually large amount of SO_x was liberated because the target area contained anhydrite deposits. Although, it should be noted that most impacts generate S-perturbations in the Earth's atmosphere, as it is a chemical component of asteroids and comets (Fig. 1.7; Kring et al. 1996; Kring 2003). Secondary contributions to the atmosphere (soot from fires and additional NO_x and Cl from burned vegetation) would have compounded the environmental damage.

At locations far from the Chicxulub impact, there would be an increase in temperature, as the reaccreting ejecta heated the atmosphere (Melosh et al. 1990; Kring and Durda 2002), with vegetation spontaneously igniting in the hot (several hundred degrees) air. After ~4 days, most ejecta will have reaccreted and surface temperatures would begin to decline, as a result of debris (dust, aerosols, soot) occulting sunlight. The dust may have settled to the ground in weeks to a few months, but the aerosols may have taken up to 10 years to fall as sulfuric acid rain. The aerosols would have reduced

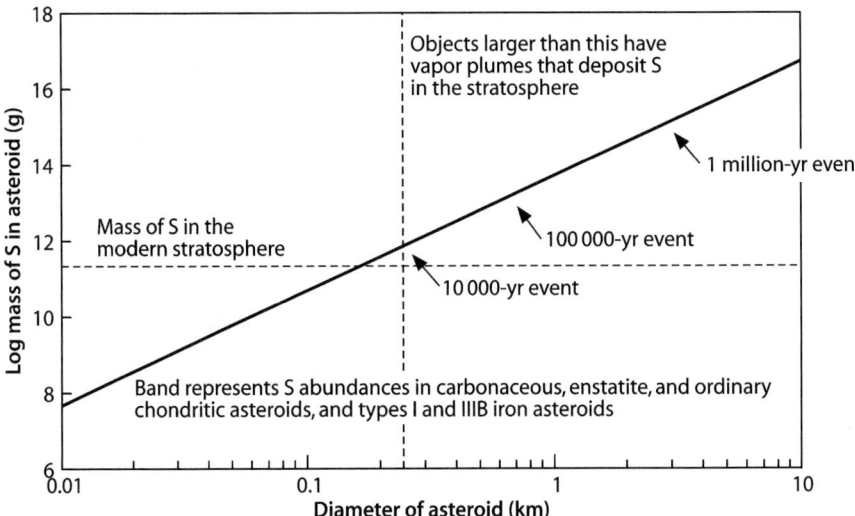

Fig. 1.7. Logarithmic plot of the mass of S in carbonaceous, enstatite and ordinary chondrites and type I and IIIB iron asteroids as a function of diameter. Impacting bodies larger than 0.3 km will produce vapor plumes that deposit S in the atmosphere. The impact of a body 3 km in diameter will release enough S from the body itself into the atmosphere to affect agricultural production on a global basis. Modified from Kring et al. (1996)

surface sunlight and resulted in cooling of the Earth' surface. This, in turn, may have induced longer term cooling in the oceans and possibly some changes in ocean circulation that lasted several thousand years (Galeotti et al. 2004). A temperature increase, due to greenhouse warming, may have followed, although the magnitude and lifetime of this effect is still unclear. Estimates of temperature increases of 1.5 to 7.5 °C have been suggested for periods up to 1 Ma (Pierazzo et al. 1998; Beerling et al. 2002). The loss to the terrestrial biosphere was tremendous, with carbon isotope studies suggesting it took ~3 Ma for the flux of organics to the deep ocean to recover (D'Hondt et al. 1998).

1.5.3
Local and Mass Extinctions

Although the shock wave, air blast, seismic activity, and simple burial beneath impact ejecta can be devastating for local flora and fauna, these are not processes that cause significant extinctions. For an impact event to cause an extinction, it must create rapid, lethal environmental changes throughout an organism's habitat and migratory range, that last longer than the organism can remain dormant (Kring 1993, 2003). Generally, an impact event creates a diverse set of environmental changes that degrade ecosystems in different ways and over different time scales. Thus, an extinction is likely to be the result of a complex series of changes, not a single environmental process. Larger impact events create greater environmental perturbations (Toon et al. 1997) but an impact event's capacity to cause extinctions will be a function of ambient conditions and will only be effective once a biologic threshold has been crossed (Kring 2002).

1.5.4
Threat to Humanity

The threshold for disrupting human civilization is much less than that needed for a significant extinction event. Relatively small events (Barringer crater) and even those involving objects too small or weak to reach the ground (Tunguska) have the capacity to have severe social and economic consequences, depending on the location of the event (Dore this vol.). Slightly larger events begin to have global effects. For example, the impact of a 300 m asteroid, which occurs approximately every 10 000 years, can significantly enhance stratospheric S by a factor similar to the 1883 Krakatau and 1982 El Chichon volcanic eruptions (Kring et al. 1996). Once per million years, an ~3 km diameter asteroid impact will produce S yields similar to the 750 000 year old Toba volcanic eruption, which is large enough to disrupt agriculture around the world.

Toon et al. (1997) found that impact events occurring on frequencies less than 60 000 years produce blast damage, earthquakes, and fires over areas (10^4–10^5 km^2) that are similar in size to those affected by recent disasters, and possibly larger areas if the impacts occur at sea (i.e. for the majority of impacts), where they can generate significant tsunamis (Bryant 2004, Melosh 2004). Serious global consequences occur on time scales of 300 000 years, when the impacts distribute water vapor and destroy ozone in the atmosphere (Birks et al. this vol.), with larger impact events creating disasters beyond anything recorded in human history.

Other natural disasters (e.g. hurricanes, earthquakes) occur more frequently than impact events (Chapman 2004). However, impact events have the capacity of creating disasters of far greater magnitude than any other natural process. They can affect much larger regions, produce several environmental perturbations simultaneously and have essentially no upper limit to their energy release and, thus, severity. The collisional evolution of asteroids and comets is an ongoing process, so impact events will continue to be a significant hazard in the future.

1.6
Concluding Remarks

Due to the highly active endogenic geologic processes on the Earth, the earth sciences were slow to recognize the evidence for the occurrence of impact events on Earth. The first terrestrial impact site (Barringer) was documented ~100 years ago, but its impact origin was highly controversial, and focused exploration efforts on terrestrial impact structures did not occur until the pre-Apollo era. During the past four to five decades, the basic physical and chemical characteristics of terrestrial impact structures and how they vary with diameter have been documented. The characteristics clearly delineate them from other geologic structures. Nevertheless, the number of known impact structures is small (~170) and it would be premature to state that the current sample is complete or that impact processes have truly entered into the mainstream knowledge base of the earth science community. Nevertheless, it is important to recognize that although endogenic processes may have destroyed much of the evidence on Earth, the lunar evidence indicates that the Earth has been the target of literally millions of impacts through geologic time.

At present, some sixteen impact structures are known with diameters greater than 20 km and ages less than 100 Ma on the Earth's land surface. These had impact energies in excess of $\sim 10^6$ MT and were dramatic events that had catastrophic regional environmental effects and moderate to severe continental and global effects (Fig. 1.5). The largest, Chicxulub, extinguished the majority of organisms living at the time and ushered in the Age of Mammals, which ultimately led to the evolution of humans. To put this in a more realistic perspective, the average impact cratering rate during the last 100 Ma of $5.6 \pm 2.8 \times 10^{-15}$ km^{-2} a^{-1} for events that produce = 20 km diameter craters indicates that ~140 events of such magnitude actually occurred on the Earth's surface, i.e. the known sample is ~10% of the actual record. Chicxulub and other impact events have begun to demonstrate the regional to global environmental effects of impacts and renewed efforts must be made to find additional impact structures and to evaluate their environmental and biologic effects, as a critical step in the assessment of the hazards of future impact events.

Although such large structures are regionally and/or globally important, smaller impact events can not be ignored on an Earth that has an ever-increasing urbanized population. Even Barringer-sized events have the potential to destroy a modern city. It is estimated that 1 km diameter cratering events occur on average once per 1600 years (Neukum and Ivanov 1994). Impact airbursts, like Tunguska, which are also capable of destroying a modern city, occur more frequently, perhaps every few hundred years. For these types of impact events, close scrutiny of small craters, crater fields, and meteorite-strewn fields are warranted, so that the strengths of the small impacting population of objects can be determined. The strengths of small asteroids will be a critical physical parameter in any effort to deflect objects that are on collisional orbits.

Impact events are not entirely a negative phenomenon with respect to the current and future human condition. They represent unusual geological events and, as such, they have resulted in local anomalous geological environments, some of which have produced significant economic deposits. About 25 per cent of known terrestrial impact structures have some form of economic deposit associated with them, and about half of these are currently exploited or have been exploited in the recent past (Grieve and Masaitis 1994). The deposits range from local and presently uneconomic (e.g. reserves of 300 000 tonnes of hydrothermal Pb-Zn ores at Siljan, Sweden) to world class (e.g. reserves of 1.6×10^9 tonnes Ni-Cu-PGE ores at Sudbury) and also include significant hydrocarbon deposits. The most recent synthesis of economic deposits related to terrestrial impact structures, which currently produce close to US$ 20 billion p.a. of resources in North America, can be found in Grieve (2005).

Although the study of terrestrial impact structures has important ramifications for understanding impact processes, their study is no longer entirely a scientific pursuit. Apart from economic considerations, there is a significant social and economic dimension (Chapman 2004). The documentation of the terrestrial impact record provides a direct measure of the cratering rate on Earth and, thus, a constraint on the hazard that impact presents to human civilization (Gehrels 1994). The K/T impact may have resulted in the demise of the dinosaurs as the dominant land-life form and, thus, permitted the ascendancy of mammals and, ultimately, humans. It is, however, inevitable that human civilization, if it persists long enough, will be subjected to an impact-induced environmental crisis of potentially extreme proportions.

Acknowledgments

We would like to thank Peter Bobrowsky and Jay Melosh, the latter who noted a computational error in the original text, for reviews of the manuscript. Geological Survey of Canada Contribution 2005157.

References

Abramov O, Kring DA (2004) Numerical modeling of an impact-induced hydrothermal system at the Sudbury crater. J Geophys Res 109(10):E10007 1–16
Alvarez W, Muller R (1984) Evidence from crater ages of periodic impact on the Earth. Nature 308:712–720
Alvarez LW, Alvarez W, Asaro F, Michel HV (1980) Extraterrestrial cause for the Cretaceous-Tertiary extinction. Science 208:1095–1108
Ariskin AA, Deutsch A, Ostermann M (1999) The Sudbury "Igneous" Complex: simulating phase equilibria and *in situ* differentiation for two proposed parental magmas. Geol Soc Amer Sp Paper 338:373–387
Baski AK (1990) Search for periodicity in global events in the geologic record: Quo vadimus? Geology 18:983–986
Basu AR, Becker L, Jacobsen SB, Petaev MI, Poreda RJ (2003) Chondritic meteorite fragments associated with the Permian-Triassic boundary in Antarctica. Science 302:1388–1392
Becker L, Bunch TE, Hunt AG, Poreda RJ, Rampino M (2001) Impact event at the Permian-Triassic boundary: Evidence from extraterrestrial noble gases in fullerenes. Science 281:1530–1533
Becker L, Basu AR, Harrison TM, Lasky R, Pope KO, Poreda RJ, Nicholson C (2004) Bedout: A possible end-Permian impact crater offshore northwestern Australia. Science 304:1469–1476
Beerling DJ, Kump LR, Lomax BH, Royer DL, Upchurch GR Jr (2002) An atmospheric pCO_2 reconstruction across the Cretaceous-Tertiary boundary from leaf megafossils. Proc Natl Acad Sci USA 99:7836–7840
Birks JW, Crutzen PJ, Roble RG (2007) Frequent ozone depletion resulting from impacts of comets and asteroids. Chapter 13 of this volume
Blum SD, Chamberlain CP, Hingston MP, Koeberl C, Marin LE, Sharpton VL, Shuraytz BC (1993) Isotopic comparison of K-T boundary impact glass with melt rock from the Chicxulub and Manson impact structures. Nature 364:325–327
Bodiselitsch B, Coccioni R, Koeberl C, Montanari A (2004) Delayed climate cooling in the Late Eocene caused by multiple impacts: high-resolution geochemical studies at Massignano, Italy. Earth Planet Sci Lett 223:283–302
Bohor B, Foord EE, Modreski PJ, Triplehorn DM (1984) Mineralogic evidence for an impact event at the Cretaceous-Tertiary boundary. Science 224:867–869
Bottomley R, Grieve RAF, Masaitis V, York D (1997) The age of the Popigai impact event and its relations to events at the Eocene/Oligocene boundary. Nature 388:365–368
Bryant E (2004) Geological and cultural evidence for cosmogenic tsunami. Paper presented at the ICSU Workshop on Comet/Asteroid Impacts and Human Society, Tenerife, Canary Islands.
Bunch TE (1968) Some characteristics of selected minerals from craters. In: French BM, Short NM (eds) Shock metamorphism of Natural Materials. Mono Book Corp, Baltimore, pp 413–432
Campins H, Swindle TD, Kring DA (2004) Evaluating comets as a source of Earth's water. In: Seckbach J (ed) Origins: genesis, evolution and diversity of life. Kluwer Academic Publishers, Dortrecht Boston New York, pp 567–590
Chao ECT (1967) Shock effects in certain rock-forming minerals. Science 156:192–202
Chao ECT, Fahey JJ, Littler J, Milton DJ (1962) Stishovite, SiO_2, a very high pressure new mineral from Meteor Crater, Arizona. J Geophys Res 67:419–421
Chapman CR (2004) The asteroid impact hazard and interdisciplinary issues. Paper presented at the ICSU Workshop on Comet/Asteroid Impacts and Human Society, Tenerife, Canary Islands.
Chyba CF (1993) The violent environment of the origin of life: progress and uncertainties. Geochim Cosmochim Acta 57:3351–3358

Cohen BA, Kring DA, Swindle TD (2000) Support for the lunar cataclysm hypothesis from lunar meteorite impact ages. Science 290:1745–1756
Daubar IJ, Kring DA (2001) Impact-induced hydrothermal systems: heat sources and lifetimes. Lunar Planet Sci XXXII, Abstract 1727
Davis M, Hut P, Muller RA (1984) Extinction of species by cometary showers. Nature 308:715–717
Dence MR (1972) The nature and significance of terrestrial impact structures. 24th Inter Geol Congr Section 15:77–89
Deutsch A, Schärer U (1994) Dating terrestrial impact events. Meteor 29:301–322
D'Hondt S, Donaghay P, Lindinger M, Luttenberg D, Zachos JC (1998) Organic carbon fluxes and ecological recovery from the Cretaceous-Tertiary mass extinction. Science 282:276–279
Dietz RS (1968) Shatter cones in cryptoexplosion structures. In: French BM, Short NM (eds) Shock metamorphism of natural materials. Mono, Baltimore, pp 267–285
Dore MHI (2004) The economic consequences of disasters due to asteroid and comet impacts, small and large. Paper presented at the ICSU Workshop on Comet/Asteroid Impacts and Human Society, Tenerife, Canary Islands
Durda DD, Kring DA (2004) Ignition threshold for impact-generated fires. J Geophys Res 109(8):E08004 1–14
Engelhardt W von (1990) Distribution, petrography and shock metamorphism of the ejecta of the Ries crater in Germany – a review. Tectonophysics 171:259–273
Faggart BE, Basu AR, Tatsumoto M (1985) Origin of the Sudbury complex by meteoritic impact: Neodymium isotopic evidence. Science 230:436–439
Farley KA (2001) Extraterrestrial helium in seafloor sediments: Identification, characteristics and accretion rate over geologic time. In: Peucker-Ehrenbrink B, Schmitz B (eds) Accretion of extraterrestrial matter throughout Earth's history. Kulwer Academic/Plenum Publishers, New York, pp 179–204
French BM (1998) Traces of catastrophe – a handbook of shock metamorphic effects in terrestrial meteorite craters. LPI Contrib 954, Lunar and Planet Inst, Houston
Galeotti S, Brinkhuis H, Huber M (2004) Records of post-Cretaceous-Tertiary boundary millennial scale cooling of western Tethys: a smoking gun for the impact-winter hypothesis. Geol 32:529–532
Gehrels T (1994) (ed) Hazards due to comets and asteroids. University of Arizona Press, Tucson
Gersonde R, Abelmann A, Bleil V, Bostwick JA, Diekmann B, Flores JA, Gohl K, Grahl G, Hagen R, Kuhn G, Kyte FT, Sierro FJ, Völker D, (1997) Geological record and reconstruction of the late Pliocene impact of the Eltanin asteroid in the Southern Ocean. Nature 390:357–363
Glass BP, Koeberl C (1999) Ocean Drilling Project Hole 689B spherules and upper Eocene microtektites and clinopyroxene-bearing spherule strewn fields. Meteor Planet Sci 34:185–196
Glasstone S, Dolan PJ (1977) The effects of nuclear weapons, 3rd edn. United States Dept Defence and United States Dept Energy, Washington DC
Grieve RAF (1997) Target Earth: evidence for large-scale impact events. Ann New York Acad Sci 822: 319–352
Grieve RAF (2005) Economic natural resource deposits at terrestrial impact structures. In: McDonald I et al. (eds) Mineral deposits and Earth evolution. Geol Soc London Spec Publ 248:1–29
Grieve RAF, Floran RJ (1978) Manicouagan impact melt, Quebec 2. Chemical interrelations with basement and formational processes. J Geophys Res 83:2761–2771
Grieve RAF, Masaitis VL (1994) The economic potential of terrestrial impact craters. Inter Geol Rev 36: 105–151
Grieve RAF, Pilkington M (1996) The signature of terrestrial impacts. AGSO J Aust Geol Geophys 16: 399–420
Grieve RAF, Shoemaker EM (1994) The record of past impacts on Earth. In: Gehrels T (ed) Hazards due to comets and asteroids. University of Arizona Press, Tucson, pp 417–462
Grieve RAF, Therriault AM (2000) Vredefort, Sudbury, Chicxulub: Three of a kind? Ann Rev Earth Planet Sci 28:305–338
Grieve RAF, Therriault AM (2004) Observations at terrestrial impact structures: Their utility in constraining crater formation. Met Planet Sci 39:199–216
Grieve RAF, Rupert JB, Goodacre AK, Sharpton VL (1988) Detecting a periodic signal in the terrestrial cratering record. Proc 18th Lunar and Planet Sci Conf, pp 375–382

Grieve RAF, Langenhorst F, Stöffler D (1996) Shock metamorphism of quartz in nature and experiment: II. Significance in geoscience. Meteor Planet Sci 31:6–35

Heck PR, Baur H, Halliday N, Schmitz B, Wieler R (2004) Fast delivery of meteorites to Earth after a major asteroid collision. Nature 430:323–325

Hesiler J, Tremaine S (1989) How dating uncertainties affect the detection of periodicity in extinctions and craters. Icarus 77:213–219

Hildebrand AR, Camargo AZ, Jacobsen SB, Boynton WV, Kring DA, Penfield GT, Pilkington M (1991) Chicxulub crater: A possible Cretaceous-Tertiary boundary impact crater on the Yucatan Peninsula, Mexico. Geol 19:867–871

Hörz F (1968) Statistical measurement of deformation structures and refractive indices in experimentally shock loaded quartz. In: French BM and Short NM (eds) Shock Metamorphism of Natural Materials. Mono, Baltimore, pp 243–253

Jahn B, Floran RJ, Simonds CH (1978) Rb-Sr isochron age of the Manicouagan melt sheet, Quebec, Canada. J Geophys Res 83:2799–2803

Jetsu L, Pelt J (2000) Spurious periods in the terrestrial impact record. Astron Astrophys 353:409–418

Kettrup B, Agrinier P, Deutsch A, Ostermann M (2000) Chicxulub impactites: geochemical clues to precursor rocks. Met Planet Sci 35:1129–1158

Kettrup B, Deutsch A, Masaitis VL (2003) Homogeneous impact melts produced by a heterogeneous target? Sr-Nd isotopic evidence from the Popigai crater, Russia. Geochim Cosmochim Acta 67:733–750

Kieffer SW (1971) Shock metamorphism of the Coconino sandstone at Meteor Crater, Arizona. J Geophy Res 76:5449–5473

Koeberl C (2001) The sedimentary record of impact events. In: Peucker-Ehrenbrink B, Schmitz B (eds) Accretion of extraterrestrial matter throughout Earth's history, Kluwer Academic/Plenum Publishers, New York, pp 333–368

Koeberl C, Reimold WV, Shirey SB (1996) Re-Os isotope and geochemical study of the Vredefort Granophyre: Clues to the origin of the Vredefort structure, South Africa. Geology 24:913–916

Kring DA (1993) The Chicxulub impact event and possible causes of K/T boundary mass extinctions. In: Boaz D, Dornan M (eds) Proceedings First Annual Symposium of Fossils in Arizona, Mesa Southwest Mus and Southwest Paleontol Soc. Mesa, Arizona, pp 63–79

Kring DA (1997) Air blast produced by the Meteor Crater impact event and reconstruction of the affected environment. Met Planet Sci 32:517–530

Kring DA (2000) Impact events and their effect on the origin, evolution, and distribution of life. GSA Today 10(8):1–7

Kring DA (2002) Reevaluating the cratering kill curve. Met Planet Sci 37:1648–1649

Kring DA (2003) Environmental consequences of impact cratering events as a function of ambient conditions on Earth. Astrobiology 3:133–152

Kring DA, Boynton WV (1992) Petrogenesis of an augite-bearing melt rock in the Chicxulub structure and its relation to K/T impact spherules in Haiti. Nature 358:141–144

Kring DA, Cohen BA (2002) Cataclysmic bombardment throughout the inner solar system 3.9–4.0 Ga. J Geophys Res 107:10.1029/2001JE001529

Kring DA, Durda DD (2002) Trajectories and distribution of material ejected from the Chicxulub impact crater: Implications for postimpact wildfires. J Geophys Res 107:10.1029/2001JE001532

Kring DA, Hunten DM, Melosh HJ (1996) Impact-induced perturbations of atmospheric sulphur. Earth Planet Sci Lett 140:201–212

Kyte FT, Heath RG, Leinen M, Zhou L (1993) Cenozoic sedimentation history of the central Pacific: Inferences from the elemental geochemistry of core LL44-GPC3. Geochim Cosmochim Acta 57:1719–1740

Kyte FT, Wasson JT, Zhou Z (1988) New evidence on the size and possible effects of a late Pliocene oceanic asteroid impact. Science 241:63–65

Langenhorst F (2002) Shock metamorphism of some minerals: Basic introduction and microstructural observations. Bull Czech Geol Surv 77:265–282

Langenhorst F, Deutsch A (1994) Shock experiments on pre-heated α and β-quartz: I. Optical and density data. Earth Planet Sci Lett 125:407–420

Langenhorst F, Deutsch A (1998) Minerals in terrestrial impact structures and their characteristic features. In: Marfunin AS (ed) Advanced Mineralogy 3. Springer-Verlag, Berlin, pp 95–119

Lindstrom M (2003) An array of offshore impact craters on mid-Ordovician Baltica. Third Inter Conf Large Meteorite Impacts, Lunar and Planetary Institute, Houston, Texas, abstract 4029
Maher KA, Stevenson DJ (1988) Impact frustration of the origin of life. Nature 331:612–614
Masaitis VL (1998) Popigai crater: origin and distribution of diamond-bearing impactites. Met Planet Sci 33:349–359
Melosh HJ (1989) Impact cratering: a geologic process. Oxford University Press, New York
Melosh HJ (2004) Indirect physical effects of comet and asteroid impacts. Paper presented at the ICSU Workshop on Comet/Asteroid Impacts and Human Society, Tenerife, Canary Islands.
Melosh HJ, Latham D, Schneider NM, Zahnle KJ (1990) Ignition of global wildfires at the K/T boundary. Nature 343:251–154
Milton DJ (1977) Shatter cones – an outstanding problem in shock mechanics. In: Pepin RO and Merrill RB, Roddy DJ (eds) Impact and Explosion Cratering. Pergamon, New York, pp 703–714
Morgan J, Warner M (1999) Chicxulub: The third dimension of a multi-ring impact basin. Geology 27: 407–410
Morgan J, Warner M, the Chicxulub Working Group (1997) Size and morphology of the Chicxulub impact crater. Nature 390:472–476
Morgan J, Grieve RAF, Warner M (2002) Geophysical constraints on the size and structure of the Chicxulub impact center. Geol Soc Amer Sp Pap 356:39–46
Neukum G, Ivanov BA (1994) Crater size distributions and impact probabilities on Earth from lunar, terrestrial-planet and asteroid cratering data. In: Gehrels T (ed) Hazards due to comets and asteroids. University of Arizona Press, Tucson, pp 359–416
Ormö J, Blomqvist G, Strukell EFF, Törnberg R (1999) Mutually constrained geophysical data for evaluating a proposed impact structure: Lake Hummeln, Sweden. Tectonophys 311:155–177
Palme H, Goebel E, Grieve RAF (1979) The distribution of volatile and siderophile elements in the impact melt of East Clearwater (Quebec). Proc 10th Lunar and Planet Sci Conference, pp 2465–2492
Palme H, Grieve RAF, Wolf R (1981) Identification of the projectile at Brent crater, and further considerations of projectile types at terrestrial craters. Geochim Cosmochim Acta 45:2417–2424
Peucker-Ehrenbrink B (2001) Iridium and osmium as tracers of extraterrestrial matter in marine sediments. In: Peucker-Ehrenbrink B, Schmitz B (eds) Accretion of extraterrestrial matter throughout Earth's history. Kulwar Academic/ Plenum Publishers, New York, pp 163–178
Pierazzo E, Chyba CF (1999) Amino acid survival in large cometary impacts. Met Planet Sci 34:909–918
Pierazzo E, Kring DA, Melosh HJ (1998) Hydrocode simulation of the Chicxulub impact event and the production of climatically active gases. J Geophys Res 103:28607–28625
Poag CW, Aubrey MP (1995) Upper Eocene impactites of the U.S. east coast: Depositional origins, biostratigraphic framework, and correlation. Palios 10:16–43
Pope KO, Baines KH, Ocampo AC, Ivanov BA (1997) Energy, volatile production and climatic effects of the Chicxulub Cretaceous/Tertiary impact. J Geophys Res 102:21, 645 - 21, 664
Rampino MR, Haggerty BM (1996) The 'Shiva hypothesis': impacts, mass extinctions and the galaxy. Earth, Moon, and Planets 72:441–460
Rampino MR, Stothers RB (1984) Geological rhythms and cometary impacts. Science 226:1427–1431
Raup DM, Sepkoski JJ (1984) Periodicity of extinctions in the geologic past. Proc Nat Acad Sci 81:801–805
Retallack GJ, Amber CP, Holser WT, Krull ES, Kyle FT, Seyedolai A (1998) Search for evidence of impact at the Permian-Triassic boundary in Antarctica and Australia. Geol 26:979–982
Ryder G (2002) Mass flux in the Earth-Moon system and benign implications for the origin of life on Earth. J Geophy Res E Planets 107:6-1-6-14
Sagy A, Fineberg J, Reches Z (2002) Dynamic fracture by large extra-terrestrial impacts as the origin of shatter cones. Nature 418:310–313
Schmitz B, Peucker-Ehrenbrink B, Tassinari M (2001) A rain of ordinary chondritic meteorites in the early Ordovician. Earth Planet Sci Lett 194:1–15
Sharpton VL, Burke K, Camargo-Zanoguera A, Hall SA, Lee DS, Marin LE, Quezaela-Muneton JM, Spudis PD and Urrita-Fucugauchi J, Suarez-Reynoso G (1993) Chicxulub multiring impact basin: Size and other characteristics derived from gravity analysis. Science 26:1564–1567
Smit J, Hertogen J (1980) An extraterrestrial event at the Cretaceous-Tertiary boundary. Nature 285: 158–200

Snyder D, Hobbs RW, the Chicxulub Working Group (1999) Ringed structural zones with deep roots formed by the Chicxulub impact. J Geophys Res 104:743–755

Spray JG, Thompson LM (1995) Friction melt distribution in terrestrial multi-ring impact basins. Nature 373:130–132

Spudis PD (1993) The geology of multi-ring impact basins. Cambridge Univ Press, Cambridge

Stöffler D (1971) Progressive metamorphism and classification of shocked and brecciated crystalline rocks in impact craters. J Geophys Res 76:5541–5551

Stöffler D (1972) Deformation and transformation of rock-forming minerals by natural and experimental shock processes. I. Behavior of minerals under shock compression. Fortsch Mineral 49:50–113

Stöffler D (1974) Deformation and transformation of rock-forming minerals by natural and experimental shock processes. II. Physical properties of shocked minerals. Fortsch Mineral 51:256–289

Stöffler D (1984) Glasses formed by hypervelocity impacts. J Non-Crystalline Solids 7:465–502

Stöffler D, Hornemann U (1972) Quartz and feldspar glasses produced by natural and experimental shock. Meteor 7:371–394

Stöffler D, Langenhorst F (1994) Shock metamorphism of quartz in nature and experiment: I. Basic observation and theory. Meteor 29:155–181

Stöffler D, Avermann M, Bischoff L, Brockmeyer P, Buhl D, Deutsch A, Lakomy R, Müller-Mohr V (1994) The formation of the Sudbury structure, Canada: Towards a unified impact model. Geol Soc Amer Sp Paper 293:303–318

Stothers RB (1993) Impact cratering at geologic stage boundaries. Geophys Res Lett 20: 887–890

Stothers RB, Rampino MR (1998) Periodicity in flood basalts, mass extinctions and impacts: A statistical view and a model. Geol Soc Amer Sp Paper 247:9–18

Swindle TD, Kring DA (2001) Cataclysm + cold comets = lots of asteroid impacts. Lunar Planet Sci XXXII, Abstract 1466

Tagle R, Claeys P (2004) Comet or asteroid shower in the Late Eocene? Science 305:492–493

Tagle R, Claeys P (2005) An ordinary chondrite impactor for Popigai crater, Siberia. Geochim Cosmochim Acta 69(11):2877–2889

Therriault AM, Grieve RAF, Reimold WU (1997) Original size of the Vredefort Structure: Implications for the geological evolution of the Witwatersrand Basin. Met Planet Sci 32:71–77

Therriault AM, Fowler AD, Grieve RAF (2002) The Sudbury Igneous Complex: A differentiated impact melt sheet. Econ Geol 97:1521–1540

Toon OB, Covey C, Morrison D, Turco RP, Zahnle K (1997) Environmental perturbations caused by the impacts of asteroids and comets. Rev Geophys 35:41–78

Turtle EP, Pierrazo E, Collins GS, Melosh HJ, Morgan JV, Osinski GR, Reimold WU (2005) Impact structures: What does crater diameter mean? Geol Soc Amer Sp Pap 384:1–24

Twiss RS, Moores EM (1992) Structural Geology. WH Freeman, New York

Valley JW, King EM, Peck WH, Wilde SA (2002) A cool early Earth. Geol 30:351–354

Vasilyev NV (1998) The Tunguska meteorite problem today. Planet Space Sci 46:129–150

Weismann P (1990) The cometary impact flux at the Earth. Geol Soc Amer Sp Pap 247:171–180

Whitehead J, Grieve RAF, Papinastassiou DA, Spray JG, Wasserburg G (2000) Late Eocene impact ejecta: Geochemical and isotopic connections with the Popigai impact structure. Earth Planet Sci Lett 181:473–487

Wood CA, Head JW (1976) Comparison of impact basins on Mercury, Mars and the Moon. Proc. 7th Lunar Sci Conf, pp 3629–3651

Yabushita S (1992) Periodicity and decay of craters over the past 600 Myr. Earth, Moon Planets 58:57–63

Yabushita S (2004) A spectral analysis of the periodicity hypothesis in cratering records. Monthly Notices-Roy Astro Soc 355: 51–56

Zahnle KS (1990) Atmospheric chemistry by large impacts. Geol Soc Amer Sp Pap 247:271–288

Zahnle KJ, Sleep NH (1997) Impacts and the early evolution of life. In: Chyba CF, McKay CP, Thomas PJ (eds) Comets and the Origin and Evolution of Life, Springer-Verlag, New York, pp 175–208

Chapter 2

The Archaeology and Anthropology of Quaternary Period Cosmic Impact

W. Bruce Masse

2.1 Introduction

Humans and cosmic impacts have had a long and intimate relationship. People live in ancient impact craters, such as at Ries and Steinheim in Germany, and use impact breccias for building material. People historically witnessed and venerated fallen meteorites, in some cases the meteorites becoming among the most sacred of objects – such as that kept in the Kaaba at Mecca. People made tools from meteoritic iron, including certain examples from the objects named the "tent," "woman," and "dog" by the Greenland Eskimos. And in one of the more peculiar ironies linking humans and cosmic impacts, people carved a portion of an ancient Ohio impact crater into the shape of a Great Serpent. This act not only created one of the more spectacular archaeological sites in North America, but also depicted a symbol used by a number of cultures to represent comets, the very source of some impact craters on the Earth.

Despite the close relationship between people and things that fall from the sky, archaeologists and anthropologists thus far have played little role in research and issues concerning cosmic impact. This situation reflects modeling by the NEO community (... those planetary scientists who study potentially-threatening near Earth objects) that "globally catastrophic" impacts – i.e. impacts capable of directly or indirectly killing a quarter of the Earth's human population (Chapman and Morrison 1994) currently estimated at an impact energy of around 10^6 megatons (MT) or slightly less – occur on the average of about once every 500 000 to a million years (Toon et al. 1997; Morrison et al. 2003; papers in this volume). A less reasonable notion by the NEO community has been that although major catastrophic impacts can occur at any time, few if any humans during the period of recorded history have ever been killed by a cosmic impact. Fortunately, at least some astrophysicists and geologists have begun to recognize the human toll (e.g. Lewis 1996).

The Quaternary period represents the interval of oscillating climatic extremes (glacial and interglacial periods) beginning about 2.6 million years ago (2.6 Ma) to the present. This encompasses the Gelasian stage of the late Pliocene geological epoch (2.6 to 1.8 Ma), the Pleistocene epoch (1.8 Ma to 10 000 years ago [10 ka]), and our present Holocene epoch of the past 10 000 years. The Quaternary contains critical developmental episodes of hominid biological and cultural evolution, including the development of urban societies during the Holocene. The Quaternary also contains a number of significant cosmic impacts that – for reasons discussed below – have yet to be identified and/or thoroughly studied. Ironically, the Quaternary may have begun with a hugely catastrophic oceanic asteroid impact (Eltanin), whereas the last sustained climatic oscillation some 4 800 to 5 000 years ago (the middle/late Holocene boundary) was possibly driven, I argue, by a sizable oceanic comet impact.

Archaeologists, paleoanthropologists, and anthropologists are largely unaware of both the nature and the potential of cosmic impact to explain much that is presently mysterious about the archaeological and paleoenvironmental record of the Quaternary, including our own Holocene period. A notable exception to this ignorance is the work of anthropologist and historian Benny J. Peiser, who not only looked closely at the topics of cosmic impact and rapid environmental change during the Holocene (e.g. Peiser et al. 1998; Peiser 2002), but who also maintains the CCNet, a scholarly electronic network servicing these broad topics throughout Earth history. Many of the contributors to the present volume have used the CCNet to help facilitate their own research and interests. Ironically, of the 1 800 subscribers to the website, only a small number, certainly less than 50, are actually professional archaeologists and anthropologists (Peiser 2004).

The paper is divided into three general parts excluding the introduction. The first (Sect. 2.2) examines the Quaternary record of known and hypothesized cosmic impact. Each subsection is presented in descending levels of relative certainty, beginning with the most concrete evidence for Quaternary period impact (Sect. 2.2.1 and 2.2.2) and working toward more uncertain and hypothetical evidence for impact (Sect. 2.2.3–2.2.5). Despite such ordering, it should not be automatically construed that the former present a clear and unequivocal picture of the impact record and the associated risks and hazards from such impacts, and that the latter are automatically suspect and should be dismissed out of hand. Rather, the only thing that is certain is that the hypothesized impacts presented in the latter sections require more study, better quality data, and a much greater effort at validation. The purpose of this paper is to provide a sample of both accepted and hypothesized impact events that serves to highlight data potentially relevant to issues of effects on human society, as well as addressing problems attendant to the recognition and validation of impact events.

Section 2.3, while more speculative builds on recent successful attempts by archaeologists, geologists, and astronomers to systematically use mythology and oral tradition to identify and productively study past major natural events (e.g., Barber and Barber 2005; Piccardi and Masse, in press). These methods are applied to Holocene cosmic impacts in South America, including some possibly responsible for regional mass fires, and for a preliminary assessment of a likely globally catastrophic mid-Holocene oceanic comet impact. Section 2.4, an epilog formulated after the ICSU workshop, presents evidence for a young potential abyssal impact structure in the Indian Ocean that may relate to the hypothesized mid-Holocene oceanic comet impact. It also highlights the dichotomy that exists between the archaeological and anthropological record of impact and current astrophysical models of the risk and effects of cosmic impact.

2.2
The Quaternary Period Cosmic Impact Record

2.2.1
Documented Impact Structures

As of June 2005, the Crater Inventory of the Earth Impact Database, maintained by the Planetary and Space Science Centre of the University of New Brunswick, contained a total of 172 identified and corroborated cosmic impact craters (University of New Brunswick 2005), although as suggested in this paper, the Argentine Rio Cuarto "craters" should require additional validation. Of this total, 27 are estimated to date to the past 2.6 Ma of the Quaternary period. A number of other potential impact locations are still undergoing study and validation. The database does not reflect airbursts nor tektite/glass melt strewn fields for which a crater has not yet been identified.

Table 2.1 depicts these 27 impacts in chronological order from most recent (Sikhote Alin) back to the beginning of the Pleistocene (Karikioselkä), and continues back to the gap between the Karikioselkä impact and those of Aouelloul and Telemzane at around 3 Ma. Several aspects of this list demand attention. Most compelling is that all listed impacts are in terrestrial settings. Because more than 70% of the earth is covered by water, including 14% terrestrial glaciers and sea ice (Dypvik et al. 2004), the table is likely missing two-thirds of the actual impacts during this time period. This situation calls into question how representative the validated terrestrial impacts are of the entire range of magnitude of all Quaternary impacts, especially given that current estimations of cratering rates model an average of three to six globally catastrophic impacts to have occurred during these three million years. Only two impact craters, Zhamanshin and Bosumtwi, approach the minimum size (10^4–10^5 MT) thought necessary for large-scale continent-wide effects (Toon et al. 1997), which would suggest that three or more larger impacts occurred in the world's oceans.

This sampling problem is compounded by the presence of large temporal gaps between cratering events. Particularly noticeable are the gaps between 100–220 ka, between 300–900 ka (one event), between 1.07–1.80 Ma (one event), and the large gap between 1.88–3.00 Ma. These gaps are the result of many different processes and do not necessarily reflect actual flux in the impact cratering rate. For example, in addition to the absence of known oceanic impacts, other perturbing forces include the scouring of land surfaces by glacial ice and the obscuration created by tropical forest canopies, shifting desert sands, and active alluvial settings. Table 2.1 illustrates the tendency for smaller craters (under about 200 m in diameter) to be more quickly obscured by the passage of time in contrast with larger craters, as can be seen both in the diameters of recorded craters and those cases including multiple small impacts.

Also, some terrestrial regions of the world have been poorly studied, whereas others such as Fennoscandia (Finland and surrounding countries) are particularly well studied. Fennoscandia has a disproportionately large number of validated craters (28 total) in the Earth Impact Database as compared with, for example, the region of China, Tibet, and Mongolia (1 total). There are at least 60 other potential craters in Fennoscandia

Table 2.1. Documented Quaternary Period cosmic impact structures on the Earth – adapted from the Earth Impact Database (University of New Brunswick 2005)

Impact structure name	Location of impact structure[a]	Diameter in km of largest crater (and number of known associated craters)		Estimated date of impact years before present (A.D. 2005)
Sikhote Alin	Russia (T)	0.027	(122)	58
Wabar	Saudi Arabia (T)	0.116	(3)	301
Sobolev	Russia (T)	0.053	(1)	<1 000
Haviland	USA (T)	0.015	(1)	<1 000
Kaalijärv	Estonia (T)	0.110	(9)	2 400 to 2 800?
Campo del Cielo	Argentina (T)	0.050	(20)	4 200 to 4 700
Henbury	Australia (T)	0.157	(11)	<4 700
Macha	Russia (T)	0.300	(1)	<7 000
Ilumetsa	Estonia (T)	0.080	(3)	7 400 to 7 700
Tenoumer	Mauritania (T)	1.900	(1)	21 400
Barringer	USA (T)	1.186	(1)	49 000
Odessa	USA (T)	0.168	(7)	<50 000
Lonar	India (T)	1.830	(1)	52 000
Rio Cuarto	Argentina (T)	Not craters?		<100 000
Morasko	Poland (T)	0.100	(8)	<100 000
Amguid	Algeria (T)	0.450	(1)	100 000
Tswaing	South Africa (T)	1.130	(1)	220 000
Dalgaranga	Australia (T)	0.024	(1)	270 000
Wolfe Creek	Australia (T)	0.080	(1)	<300 000
Boxhole	Australia (T)	0.170	(1)	540 000
Zhamanshin	Kazakhstan (T)	14.000	(1)	900 000
Veevers	Australia (T)	0.080	(1)	<1 000 000
Monturaqui	Chile (T)	0.460	(1)	<1 000 000
Bosumtwi	Ghana (T)	10.500	(1)	1 070 000
New Quebec	Canada (T)	3.440	(1)	1 400 000
Kalkkop	South Africa (T)	0.640	(1)	<1 800 000
Karikioselkä	Finland (T)	1.500	(1)	<1 880 000
Aouelloul	Mauritania (T)	0.390	(1)	<3 000 000
Telemzane	Algeria (T)	1.750	(1)	<3 000 000

[a] *T:* terrestrial; *O:* oceanic.

awaiting validation (University of Helsinki 2005), including at least two other Holocene craters in Estonia (Simuna, Tsõõrikmäe) in addition to Ilumetsa and Kaali discussed below (Veski et al. 2007, Chapter 15 of this volume).

Taking into account that the Americas likely were not occupied before about 20 ka and that Australia was not occupied before 60 ka, it should still be apparent from Table 2.1 that comet/asteroid impacts conceivably played a significant role in aspects of human history.

2.2.2
Validated Holocene Crater-Forming Impact Events

2.2.2.1
Kaali and Ilumetsa, Estonia

The Kaali meteorite impact crater field on Saaremaa Island in Estonia is the most studied impact site to date in terms of potential effects on contemporary human occupants and in the region surrounding Saaremaa Island (Veski et al. 2001, 2004, 2007 [Chapter 15 of this volume]). There is a single large lake-filled crater surrounded by eight smaller craters. The large crater is about 110 m in diameter, and collectively all the craters cover an area of about half a square kilometer.

The Kaali meteorite was a coarse octahedrite, with surviving fragments being only a few grams in weight. Despite the intensity of investigation both inside the craters and outside in nearby peat bogs, the actual date of the impact has been estimated at four widely spaced times: 6400 BC based on microspherules in peat (Raukas 2000); 5000 BC on similar evidence (Tiirmaa and Czegka 1996); 1740–1620 BC based on bulk sediment samples from the near the bottom of the crater lake, or a similar 1690–1510 BC date based on associated terrestrial macrofossils from the deepest part of the lake (Veski et al. 2004); and 800–400 BC based on peat associated with impact ejecta and iridium in nearby bogs (Veski et al. 2004). Veski and his colleagues argue for the calibrated date range of around 800–400 BC, speculating that the microspherules possibly relate to a separate earlier impact event.

The 800–400 BC date for the Kaali impact places it in a densely populated region (Veski et al. 2004), thus the estimated energy release – 20 kilotons, the magnitude of the Hiroshima and Nagasaki bombs – indicates a high probability of fatalities. It is emphasized, however, that even the earlier modeled dates coincide with known human habitation beginning by at least 5800 BC on Saaremaa Island (Veski et al. 2004). Evidence from settlement patterns and from medieval written sources suggests that Kaali was considered sacred. Myths recorded early in the 13th century describe a god that flew to Saaremaa along the reconstructed path of the impactor; likewise the Finnish national epic Kalevala has an episode where the Sun falls into a lake burning everything on its way (Veski et al. 2004). Similarly northern Estonian myths describe the time when the island of Saaremaa burned. The fortified village of Asva, about 20 km from Kaali, burned at about the same time as the date for the impact event modeled by Veski and his colleagues, although a connection has not yet been proven beyond reasonable doubt. Paleoenvironmental techniques applied in the vicinity of the impact craters suggest that farming, cultivation, and seemingly human habitation ceased in the area for several human generations after the modeled impact date of 800–400 BC (Veski et al. 2001).

The Ilumetsa crater field in southeastern Estonia contains a series of at least three and likely five or more probable impact craters (Raukas et al. 2001). The largest crater is approximately 80 m in diameter and 12.5 m deep, with the second largest crater being about 50 m in diameter and 4.5 m deep. A distance of approximately one kilometer separates the two largest craters. Fragments of the original meteorite have yet to be recovered.

Radiocarbon dating of the lowest layer of organic materials in the largest crater yielded a calibrated date range of between 4500 and 5200 BC, whereas the dating of peat layers containing glassy impact spherules in a nearby bog yielded a date range of around 5400 to 5700 BC, the latter was preferred by Raukas and his colleagues. This would place the impact at around 7500 years ago, at a time when south-eastern Estonia was known to be inhabited. Raukas et al. (2001) note that the three largest craters associated with the Ilumetsa event have names that translate as Hell's Grave, Deep Grave and Devil's Grave. They further suggest that this is consistent with an oral tradition preserving the original observation of the fall and the fact that earlier people thought of meteorites and bolides as living entities – apparently evil celestial beings who met their deaths in forming the Ilumetsa craters.

2.2.2.2
Wabar, Saudi Arabia

Situated in the dune fields of southern Saudi Arabia is Wabar, a set of three small craters in an area covering about half a square kilometer. The largest crater is 116 m across, the others much smaller, with additional craters possibly being buried by the surrounding sand dunes (Wynn and Shoemaker 1998). The craters contain bits of white shocked sandstone (impactite) created by compression of the dune sand during the impact, black melted slag and small chunks of nickel and iron from the original medium octahedrite meteorite. The energy release was estimated at 12 kilotons.

The Wabar impact is of interest not because of known harm done to humans, or work actually performed by archaeologists and anthropologists, but rather because it was witnessed from a considerable distance and appears in contemporary Arabic poems and thus can be dated to January 9, 1704 (Basurah 2003). Recent luminescence dating of the impactite and slag approximates this date (Prescott et al. 2004). The importance of this observation is to reinforce the notion that there have been a number of impact events during the past several thousand years that are undoubtedly captured in various written documents, including myths, local histories and dynastic records. For example, the famous Chinese Bamboo Annals, dating to the 3rd century BC, may contain dateable references to cosmic impacts and other natural phenomena (Masse 1998, Table 2.2).

2.2.2.3
Campo del Cielo, Argentina

Work by William Cassidy and his colleagues at the Campo del Cielo iron meteorite impact site in northern Argentina (Cassidy et al. 1965; Cassidy and Renard 1996) has been a model for the study of a low velocity impact. Campo del Cielo ("Field of the Sky") contains at least 26 known slightly elongated craters, the largest being 115 × 91 m. The crater field itself covers an area about 3 km wide and 19.2 km long, with an associated strewn field of small meteorites extending about 60 km beyond the main crater field. A number of sizeable fragments of the original octahedrite meteorite survived the impact, the largest being more than a meter in diameter and weighing about 37 tonnes. The original impactor was estimated to be minimally about 4 m in diameter (Liberman et al. 2002).

Cassidy excavated two of the craters in order to gauge the size, angle, and speed of each impactor. He also collected charcoal samples and obtained an approximate calibrated age for the impact at around 2200 to 2700 BC. It is difficult to gauge the degree to which Cassidy used the archaeological techniques of microstratigraphy, but such study on a regional scale in and around the impact site would likely be productive. Unfortunately, because of the increasing worldwide popularity of the Campo del Cielo meteorites since the 1965 report by Cassidy and his colleagues, damage to the area has occurred due to the illicit excavation and removal of meteorite fragments.

Cassidy and Renard (1996) also reported on a myth collected and reported by the medical doctor and historian Antenor Álavarez in 1926 that appears to relate to the impact:

> And there *[Campo del Cielo]* in their stories of the different tribes of their battles, passions and sacrifices, was born a beautiful, fantastic legend of the transfiguration of the meteorite on a certain day of the year into a marvellous tree, flaming up at the first rays of the sun with brilliant radiant lights and noises like one hundred bells, filling the air, the fields, and the woods with metallic sounds.

Giménez Benítez et al. (2000) have recently conducted a detailed study of the myths of the tribes of this general region of the Gran Chaco to see what relationship they may have to the Campo del Cielo impact event. They note that Álavarez, in addition to the myth noted above, was convinced that several tribes had oral historical knowledge of the impact, believing that the meteorite had detached from the Sun. Álavarez also noted that there were a number of pilgrimage paths to the crater field, covering an area of about 200 square kilometers. Giménez Benítez and his colleagues note that little archaeological work has been done around Campo del Cielo but that it needs to be done. They also note there has been little meaningful dialog between anthropologists and the astronomy community regarding the myths and the physical aspects of the impact site.

2.2.2.4
Henbury, Australia

The Henbury crater field is a series of at least 12 known craters situated in the virtual center of Australia, approximately 145 km southwest of Alice Springs (Hodge 1994, pp 67–70). The craters are scattered over an area slightly larger than 0.5 km^2, with the largest crater (possibly an eroded double crater) having dimensions of 180 m by 140 m, with the next largest crater being about half that size. The site contains shocked sandstone and impact glass melts, and more than 500 kg of nickel-iron fragments of the original medium octahedrite meteorite have been documented as being collected from the area.

The Henbury impact event has been radiocarbon dated at around 2700 BC or slightly younger, and various Aboriginal groups along the path of the impactor (coming from the southwest) would have witnessed its fall. An Aboriginal name for the crater field translates to "sun walk fire devil rock" indicative of an observed event (Grego 1998). Aboriginal myths were collected in the 1990s regarding the Henbury crater field and the sacred site therein (Parks and Wildlife Commission of the Northern Territory 1999), but such myths and sites are sensitive sacred knowledge not easily shared with the outside world.

Published Aboriginal myths about the powerful deity Rainbow Serpent are indicative of the relationship of a cosmic impactor with the great flood (see Sect. 2.3.3.5). The witnessing in the 1950s of a daylight-visible bolide near the town of Wilcannia in New South Wales was used as a teaching device to tell an ancient Aboriginal myth about a smoking "falling star" impactor that killed a number of people camped in the same vicinity and to describe ritual landscapes associated with the event (Jones 1989). The myth goes on to note details about what appears to be the great flood, but it is uncertain as to whether the myth being told is an actual great flood story (as described in Sect. 2.3.3) or is a separate witnessed impact event... or both.

2.2.3
Airbursts, Tektites, and Impact Glass Melts

The 1908 airburst event over the Tunguska region of Siberia provided unmistakable evidence of the force that can be delivered by a cosmic impact that fails to leave lasting evidence of an impact crater on the ground. Similar but smaller and less well-studied airbursts occurred in Brazil in 1930 (Bailey et al. 1995) and Guyana in 1935 (Steel 1996). Estimates for the magnitude of the Tunguska impact range between about 3 MT and 10–15 MT (Morrison et al. 2003; Longo, this volume). Given the evidence for the destruction of approximately 2 000 km^2 of Siberian forest by the Tunguska event, airbursts have received considerable attention in terms of the attempt to model their nature and frequency. Some modeling has indicated airbursts much larger than Tunguska are possible (Wasson 2003).

Several aspects of airbursts are relevant to our discussion. The first is the lack of visible evidence for impact cratering, thus airbursts are difficult to define archaeologically without recourse to signatures other than cratering. The second is the potential association of impact glass melts and other physical signatures with at least some airbursts. The third is the possibility that airbursts can cause significant ground fires (Sect. 2.3.2).

2.2.3.1
Rio Cuarto, Argentina

Schultz and Lianza (1992) published a cover-story article in *Nature* regarding a uniquely low-angle Holocene impact crater field in the Pampas of Argentina. Rather than simply remaining the stunning finding that such a discovery should engender, Rio Cuarto has turned into a case study for the difficulty of proving an extraterrestrial impact origin for a set of depressions on the Earth.

As originally defined, the Rio Cuarto "crater field" consists of a series of oblong rimmed depressions strung out over a distance of 50 km, the largest of which was 4.5 × 1.1 km. Schultz and Lianza (1992) also found highly vesicular glass melt fragments that were considered of impact origin. To their credit, they recognized that the depressions of the individual craters in their defined crater field were not so very different from aeolian depressions elsewhere on the Pampas.

Other scientists have disputed the impact origin of the Rio Cuarto structures, and have concluded that they are instead aeolian deflation features associated with predominate winds at different times within the late Quaternary period (Cione et al. 2002; Bland et al. 2002). Large numbers of similar structures exist throughout the Argentine Pampas, and the floors of some of the Rio Cuarto structures allegedly contain evidence of late Pleistocene fossils and caliche. Thus the structures themselves are of dubious origin. Nevertheless, Schultz and his colleagues (2004) presented reasonable counter arguments that appear to keep the impact crater debate alive.

Curiously, Cione et al. (2002) also had problems with an impact origin for the glass melts. These melts ("escorias") are glassy vesicular slabs found widely throughout the Pampas, and they point out that a number of researchers consider the escorias to be the product of normal anthropogenic fires created by intentional burning of fields. This topic is explored in Sect. 2.3.2 in the context of myths of mass fire from the Brazilian Highlands, and I suggest that these melts are indeed of impact origin, a position also subscribed to by Bland et al. (2002).

Schultz et al. (2004) conducted the most thorough study of the Argentine glass melts and were able to identify several separate Quaternary impact events, including one identified by Bland et al. (2002). Four of these are dated by $^{40}Ar/^{39}Ar$ ratios: 570 ± 100 ka, 445 ± 21 ka, 230 ± 30 ka, and 114 ± 26 ka. The 570 ka and 114 ka specimens are found specifically at Rio Cuarto. Of particular interest is recent glass at Rio Cuarto dated by three different techniques. By pure geological context they date to the early or middle Holocene (4–10 ka), by fission track to 2.3 ± 1.6 ka, and by $^{40}Ar/^{39}Ar$ to 6 ± 2 ka; the composite preferred date is about 1000 to 4000 BC. This is roughly similar to the previously noted age for the Campo del Cielo impact.

The extraordinary record of impact glasses in the Argentine Pampas is the result of suitable fine sandy soils (loess) high in silicates and thus suitable for the formation of impact glass, with these soils also serving to protect and to enhance the visibility of the glass layers. There are as yet no known impact structures associated with these five different glass melts. Wasson (2003) hypothesizes it may be possible to have large airbursts create immense distributions of glassy layered tektites perhaps covering areas of up to 70 000 km^2 as part of sheet melt of loess and sand from an incandescent sky. The Argentine Holocene glass melts are stated as extending at least 150 km southwest from Rio Cuarto (Schultz et al. 2004), thus covering an area considerably larger than that devastated by the Tunguska impact. It is of considerable interest that the Holocene airburst event coincides with a widespread human population replacement in the south-eastern Argentine Pampas – based on archaeological, osteological and paleoecological evidence – that took place sometime between 4000 and 1000 BC (Barrientos and Perez 2005). The airburst and the oceanic comet impact described in Sect. 2.3.3 should be given serious consideration as potential factors in this population replacement.

2.2.3.2
Australasian Tektite Strewn Field – ca. 0.8 Ma

Australasian tektites and microtektites cover more than 10% of the Earth's surface, including nearly all of Australia, island and continental Southeast Asia, the Southern Ocean below Australia, and much of the Indian Ocean as far west as Madagascar. Somewhere lurking in Thailand or perhaps off the eastern coast of Vietnam is a presently undocumented crater variously estimated at between 32 and 116 km in diameter

(Haines et al. 2004; Ma et al. 2004). Alternatively, Wasson (2003) has modeled an airburst to explain the sizeable distribution of layered tektites within the overall distribution of Australasian tektites.

There are a couple of notable aspects relating to this event, in addition to the impressive size of the impact during the middle of the Pleistocene. First, there is substantive evidence to suggest that massive flooding and other major regional environmental disturbances (such as deforestation) took place immediately after the impact (Haines et al. 2004), which together with the impact itself would have had a profound effect on ancestral human populations in Southeast Asia. Researchers suggest that the impact, believed by some to be one of the largest of the past few million years, was of such a magnitude that it "must have had serious consequences for the paleoenvironment and biogeographical history (perhaps including local hominid evolution) of Southeast Asia" (Langbroek and Roebroeks 2000).

Also intriguing is the near coincidence between the dating of the impact and the Matuyama-Brunhes boundary (MMB) that marks the last magnetic reversal in Earth history (Pillans 2003), and which has been well dated to around 780 ka. Current stratigraphic evidence suggests a separation of around 12 000 to 16 500 years between the impact and the subsequent magnetic reversal. There needs to be further study of the effects on ancestral human populations of both the impact and the MMB, including further consideration of a potential relationship between the two geophysical events, particularly if the impact crater proves to be at the larger end of the estimated size range.

2.2.4
A Sample of Current Studies of Potential Late Quaternary–Holocene Period Terrestrial Impact Sites

Potential cosmic impact site locations are proposed every year, usually based on some sort of aerial or satellite imagery. Some eventually are validated and are included in formal directories such as the University of New Brunswick's Earth Impact Database. Others can be the objects of contentious debate for years, particularly in the absence of identifiable shocked rock and other undisputed signatures of impacts. The following brief review of six interesting candidates is instructive in terms of the challenges that face field verification.

2.2.4.1
Middle East – ca. 2350 BC – the Fall of the Akkadian Empire

Archaeologists are sometimes confronted with evidence for what appears to be rapid destruction within individual archaeological sites and occasionally across large regions. The typical default conclusion is that this represents the destructive forces of a conquering army and/or some other concurrent destructive natural forces such as large-scale earthquakes and massive volcanic eruptions or perhaps rapid climate change. A prime example of such an abrupt event is that associated with the end of the Akkadian empire at around 2200 BC (commonly referred to as the "4 000 BP event"). A number of large urban cities contain evidence of widespread and apparently synchronous social collapse and destruction at around this time period (±200 years). This had been modeled as abrupt climate change (aridification) associated with volcanic ash fall as represented in a thin but widespread dust layer (Weiss et al. 1993). However, a number of researchers and archaeologists have raised serious doubts about the suggested physical causes and as well as the timing of this event (Peiser 2003).

More recent microstratigraphic examination of this dust layer and its context now suggest an impact origin together with a significant revision of chronology (Courty 1998). As reconstructed by Courty at Tell Leilan (Syria), the dust layer sits on top of an occupational surface possibly deformed by a shock wave. This surface exhibits evidence of the rapid propagation of wildfire, synchronous with the fallout of distinct black carbon associated with major forest fires in other nearby regions. The dust layer contains tiny rock fragments from various contexts (sandstones, basalts, marine limestone, gabbros), along with numerous glassy microspherules of varying mineralogical compositions and glassy grains derived from vaporized rocks. The shocked and burned occupational layer and overlying dust layer are themselves sealed with mud from a heavy rainfall. Thus what had been originally considered a tephra fall now appears to be impact ejecta. Courty notes that the occupation surface and dust layer are quite variable throughout the site as are the radiocarbon dates associated with those layers. Courty (2001) has also examined soils at Tell Brak (Syria), and has modeled a similar sequence that took place very rapidly. Although, most scholars remain sceptical of Courty's impact interpretation, her model fits well with data from surrounding regions (Masse 1998).

A problem when dealing with the study of microspherules (Raukas 2000) is that many different sources exist for such material including terrestrial (diagenic, biogenic, industrial, volcanic), extraterrestrial (interstellar and interplanetary dust, meteoritic airbursts) and melt from cosmic impacts. Thus key components of research into the origin of specific microspherules are the depositional environment and stratigraphic context as defined by the use of microstratigraphic methodologies.

The messages from this study and from the two examples in Sect. 2.2.3 are: (*1*) Airbursts and tektite strewn fields are poorly known and documented in terms of the Quaternary paleoenvironmental and cultural record; (*2*) few people, including those in the Quaternary geosciences, are trained to recognize and deal with potential tektites, impact glass melts, and microspherules – archaeologists are woefully lacking in this regard; and (*3*) the use of microstratigraphic methods and distributional studies are vital for determining the nature and context of impact glasses and other impact products.

2.2.4.2
Umm al Binni, Iraq

Umm al Binni lake is situated in the Al' Amarah marshes in southern Iraq near the junction of the Tigris and Euphrates rivers. This has been proposed by Master (2001, 2002; Master and Woldai 2004; this volume) as a 3.4 km-diameter candidate impact structure based on aerial photographic images that revealed Umm al Binni to be distinct in shape from all other marsh lakes in the region. Umm al Binni is nearly circular whereas the latter are quite irregular in shape. The sediments of this region are thought to be less that 5 000 years in age, thus suggesting that if it is an impact structure Umm al Binni could date to the Bronze Age. As such and if validated, the impact could account for some but not all of the striking devastation layers noted in the Bronze Age archaeology of Mesopotamia. Using impact modeling such as that by Marcus et al. (2005), the size of the hypothesized crater indicates a moderately small impactor, perhaps around 300 m in diameter for a stony asteroid, that would have had substantive effects only a few hundred kilometers from the impact site. Regrettably, regional politics and war have conspired to prevent the detailed physical examination of the structure, and the recent attempted draining of marshes has potentially imperiled aspects of the information value of the structure itself.

2.2.4.3
Sirente, Italy

A case was recently made for the impact origin for small depressions in a mountain plain in central Italy (Ormö et al. 2002). A single large crater, 130 m in diameter, was identified along with between 17 to 30 smaller craters. Radiocarbon samples underneath the rim of the large crater indicate a date in the 5th century AD. A correlation was made between the presumed impact and local mythology about "… a new star, brighter than the other ones came nearer and nearer, appeared and disappeared behind the top of the eastern mountain" (Santilli et al. 2003). The authors suggest that the presumed impact, which occurred during a pagan festival, inspired the observers to convert to Christianity as suggested in other documentary sources. Speranza et al. (2004) conversely claim that the larger "crater" is of anthropogenic origin, being constructed for use during historic seasonal migrations of sheep and shepherds, and the smaller "craters" are natural karstic basins. They further claim that the radiocarbon dates associated with one of the smaller "craters" are more than 2 000 years earlier than the alleged main "crater."

2.2.4.4
Iturralde, Boliva

Scientists from NASA Goddard have attempted to prove the impact origin of an 8 km wide circular depression located in an alluvial basin of the Amazonian rain forest of northern Bolivia. Based on geological context, it is likely to date between 30 000 and 11 000 years ago (Wasilewski et al. 2003), the latter date would put it coeval with humans in South America. Formal expeditions in 1987 and 1998 met with insurmountable logistical obstacles and the site was not reached and studied until 2002. Even this expedition was fraught with logistical difficulties and did not achieve all of its objectives. Definitive confirmation of the structure as an impact crater has not yet been achieved, nor has it been subjected to absolute dating techniques.

2.2.4.5
The Bavarian Crater Field, Chiemgau-Burghausen, Germany

Within the past few years, preliminary and conflicting information has appeared from two competing research groups regarding a probable impact crater field located just north of the Alps in south-eastern Germany. One group (Fehr et al. 2005) has described an area directly north of the town of Burghausen containing 12 documented and several suspected craters distributed in a south to north pattern in an area 7 km × 12 km. The craters range between 5 to 18 m in diameter and were emplaced in glacial gravels and pebbles. No meteoritic material was observed in or surrounding the observed craters. Impact breccias, shocked quartz, glass melts, and other such impact signatures were not observed, which is consonant with the small size of the craters. The main impact effect other than the cratering itself was that of the breakage and crushing of the pebbles and gravels in the bottom of the craters. Iron silicide alloys were found in the vicinity but not within the craters, and an industrial origin was suggested for their occurrence. No oral historical information was obtained regarding the craters, and radiocarbon-dated charcoal from the base of a lime kiln within one crater suggests formation of the crater before the 2nd century AD.

The second research group has offered a radically different interpretation of this crater field (Chiemgau Impact Research Group 2004; Rappenglück et al. 2004). They note the presence of 81 impact craters ranging between 3 and 370 m in diameter, encompassing an area 27 km wide and 58 km long from the southwest to the northeast. The major difference between the two crater field models is the presence of several larger craters defined by the Chiemgau Group near Lake Chiemsee – the far northeastern end of their crater field matches the previously discussed area of small craters defined by Fehr et al. (2005).

The larger craters defined by the Chiemgau group are associated with a variety of glass melts, shattered sandstone, and widespread evidence for the unusual iron-silica alloys gupeiite and xifengite stated as having been documented through microprobe analysis, polarization microscopy and X-ray diffractometry. In addition, titanium carbide is similarly present based on optical and analytical scanning electron microscopy. A date for the impact of around 200 BC has been suggested by the Chiemgau Group based on the association of cultural artifacts with the glass melts, however, an argument can be made that the impact is several hundred years more recent than modeled. A potential association between the hypothesized Chiemgau event and the AD 536–545 climatic event (e.g., Baillie 1999, 2007 [Chapter 5 of this volume]) has not been ruled out (Ernstson 2005), although a first millennium BC date is more likely. Indeed, the most intriguing aspect of the Chiemgau Group model is their hypothesis that a fragmenting comet whose original size was around 1.1 km in diameter created the crater field.

The strongly interdisciplinary nature of both research groups is noted, but the hypothesized impact cannot be fully assessed until the findings are corroborated by additional study and more fully published. In this regard, I found it curious that the Chiemgau Group chose to utilize semi-popular media (the Internet and *Astronomy Magazine*) for their initial publication. This choice allegedly (Ernstson 2005) was due to the reluctance of reputable journals to consider their submitted material and the refusal by potential reviewers to personally inspect the hypothesized impact site and its associated recovered materials. Ernstson (2005) suggests that such response to their work stems in part due to the prevalent assumption in the NEO community that a recent catastrophic comet impact on the Earth is highly unlikely based on the current modeling of hazard and impact rates. Several scientists working on other potential recent impacts have related similar experiences, thus lending credence for a claim of possible scientific bias. The Chiemgau Group has collected additional recent physical and oral tradition data supporting the impact and its cometary origin and has plans for future peer-reviewed publication.

2.2.5
Oceanic Impacts

There has been growing recognition of the importance of the study of oceanic cosmic impacts and a concerted effort to document and model such impacts (e.g. Gersonde et al. 2002; Dypvik and Jansa 2003; Dypvik et al. 2004). However, work on oceanic impacts generally has lagged far behind that of terrestrial impact studies and virtually all work to date has been performed on craters formed in water less than approximately 800 m in depth. In addition to the two oceanic impacts noted in this section, Eltanin and Mahuika, I present modeling for a hypothesized mid-Holocene globally catastrophic oceanic impact in Sect. 2.3.3, along with the physical evidence for a candidate abyssal crater in Sect. 2.4.1.

2.2.5.1
Eltanin

The best documented abyssal cosmic impact to date, but not listed in the Earth Impact Database due to lack of a detailed published and confirmed crater, is that of the early Quaternary Eltanin asteroid impact in the Bellingshausen Sea in the southeast Pacific about 1400 km west of Cape Horn (Kyte et al. 1988; Gersonde et al. 1997). Eltanin was first recognized as a tektite-strewn field on the seabed covering several hundred square kilometers, associated with high iridium counts. Initial dating placed it at around 2.15 Ma, but this has since been revised to 2.511 Ma ± 70 ka (Frederichs et al. 2002). As such the date is remarkably close to the boundary of the Quaternary period. If new data on an apparent associated impact crater (discussed below) is correct, it could be reasonably argued that the Eltanin impact is a geological boundary event.

Until recently, modeling of the size and magnitude of the Eltanin impact had a maximum of 4 km diameter for the asteroid, but most calculations placed it around 1–2 km in diameter with an impact energy of around 10^5 to 10^6 MT, the threshold for globally catastrophic impact. Recent research by Dallas Abbott and her colleagues (Glatz et al. 2002; Abbott et al. 2003a; Petreshock et al. 2004) has led to their conclusion that a putative crater 132 ± 5 km in diameter is the source crater for the tektite strewn field and iridium layer. Although not directly comparable with terrestrial craters, an abyssal oceanic crater of the size suggested by Abbott and her colleagues would rank Eltanin as the fourth largest in the current listing of the Earth Impact Database, only some 38 km smaller than the Chicxulub K-T boundary event. Although the present evidence by Abbott and her colleagues is currently poorly published, if eventually validated, the Eltanin impact not only may explain much about the erratic nature of Quaternary period warming and the abrupt cooling cycles, but also aspects of early hominid evolution.

2.2.5.2
Mahuika

At the younger end of the Quaternary period, Dallas Abbott and her colleagues have announced the discovery of an apparent sizeable oceanic impact crater on the continental shelf south of New Zealand (Abbott et al. 2003b, 2004), which they have most recently dated to around AD 1450. The crater, 20 ± 2 km in diameter, shows evidence of a widespread tektite field up to 220 km away from the crater itself (Matzen 2003). The impact is thought to be responsible for massive tsunami deposits in Australia and New Zealand, earlier documented by Bryant (2001). Assuming that the crater is real and that the impact did occur during the Maori occupancy of New Zealand, it should be possible to derive a near absolute date from myths associated with Maori royal chiefly genealogies, as dated by known astronomical events such as eclipses (see Sect. 2.3.1). In addition, it is of considerable interest to see how the impact may correlate with a period of rapid environmental degradation in New Zealand dated at around AD 1450–1550. This is seemingly part of a major Pacific-wide climatic event noted at around AD 1450 (Nunn 2000; Masse et al. 2006), thought to be associated with the onset of the Little Ice Age.

It should be noted that Steel and Snow (1992) earlier modeled an airburst in the Tapanui region of South Island as having caused the environmental degradation in New Zealand, whereas an airburst over water was originally suggested by Bryant (2001) as the source of the New Zealand mega-tsunami deposits. Goff et al. (2003) have raised a number of useful criticisms regarding the airburst model and Bryant's translations of Maori names, and most likely would have extended their argument to include the hypothesized Mahuika impact had it been available for their scrutiny. My own review of the Mahuika materials and the arguments of Goff and his colleagues indicates that not enough data have been yet marshalled to either validate or completely eliminate the claims for a cosmic impact in or near New Zealand in the 15th century. A much more detailed published treatment of Mahuika is necessary to evaluate this hypothesized impact event and putative crater. Additional research on both Eltanin and Mahuika is ongoing (Bryant et al., in press; Abbott 2005).

2.3
Oral Tradition, Myth, and Cosmic Impact

To say that science has not looked favorably upon attempts to glean meaningful historical information from oral history and mythology is to grossly understate the contempt that some physical scientists have for such endeavors. Indeed, physicists and astronomers who were active in the early 1960s understandably are still upset when confronted by anything bearing a resemblance to the infamous theories and claims of Immanuel Velikovsky (Grazia et al. 1966).

However, part of the blame for the sad state of myth as an explanatory tool must also rest on the shoulders of the ethnologists, folklorists and other scholars who most closely work with myth. The study of mythology during the last 100 years has been dominated by classical (e.g. Graves 1960), structural (e.g. Levi-Strauss 1969), and psychological (e.g. Campbell 1981) approaches. Although these approaches have produced fascinating insights into the nature and meaning of myth and have helped to highlight the critical role that myth has played in non-western and early western culture and society, they have misled generations of scholars by their assumption that myth lacks a meaningful foundation in the processes and events of real history.

Recent studies are beginning to revise our thinking with respect to the relationship between myth and history (Vitaliano 1973; Baillie 1999; Mayor 2000; Barber and Barber 2005; Piccardi and Masse, in press). The work of geologist Russell Blong (1982) with a previously undocumented 17th century Plinian style volcanic eruption in Papua New Guinea is singled out as an exquisite example of the use of mythology to complement and enhance the findings from physical geology. By collecting and analyzing the environmental details in myths about the "time of darkness" from widespread villages and tribes in and around the tephra fall, Blong documented aspects of the nature and duration of the eruption that were otherwise enigmatic in the physical record. Blong demonstrated that no one set of myths from a given village or tribe contained all of the pertinent environmental details, but rather each set had just a few details, a situation likely representing individual local circumstances and a natural response for people reacting to major natural disaster.

2.3.1
The Nature and Principles of Myth and Oral Tradition

Throughout Polynesia, myths are attached to and embedded within royal chiefly genealogies, which in Hawaii stretch back more than 95 generations prior to the reign of Kamehameha I at the end of the late 18th century. The value of this association became evident while conducting rescue archaeology in 1989 at the site of Hawaii's legendary first human sacrificial temple complex, then being overrun by lava from the ongoing eruption of Kilauea Volcano (Masse et al. 1991). In the mythology surrounding Pele, the Hawaiian volcano goddess, historically known lava flows are believed to have been created by Pele during supernatural battles attributed to the reigns of specific chiefs listed in the genealogical records. When the genealogical dates of these chiefs (based on a heuristic 20-year generation period) were compared with radiocarbon dates collected by the staff of Hawaii Volcanoes Observatory from these same named lava flows (Holcomb 1987), the close correspondence between the two sets of dates were striking (Masse et al. 1991; Masse et al., in preparation).

Hawaiian mythology contains accurate details of transient celestial events such as great comets, meteor storms, supernovae and even auroral substorms that can be exactly matched with the historic record in Asia, Europe, and the Middle East (Masse 1995). More recently, reconstructions of Polynesian solar eclipses by Fred Espenak of NASA Goddard has led to demonstrable matches with genealogically-based Polynesian eclipse stories including Hawaiian eclipses in AD 1679, 1480, 1257, 1104, and 975, a Samoan eclipse in AD 761, and perhaps a Tuamotuan eclipse in AD 605 (Masse et al. in preparation). There are dozens of exactly dated matches between natural events and Polynesian myths for a period of more than 1000 years.

Hawaiian oral tradition may support the validity of the hypothesized lunar impact witnessed on June 18, 1178 by monks in Canterbury, England, which may relate to the formation of the 20 km diameter Giordano Bruno crater on the Moon (e.g. Lewis 1996, p 50). Several Hawaiian genealogical chiefs dating to this time period (by birth or rule) have unique literal names seemingly evocative of the dusty aftermath of the lunar impact, such as *Hina-ka-i-ma-uli-awa* "Having discolored the Moon with a dark mist" and *pô'ele-i-ke-kihio-ka-Malama* "darkened in the corner of the Moon" (Masse 1995; Masse et al., in prep.). Hawaiian myth even alludes to a major meteor storm at about this time. Despite considerable scepticism from the NEO community for the lunar impact hypothesis, the Hawaiian data suggest that it may be premature to rule out the AD 1178 impact scenario for Giordano Bruno crater.

The analysis of Hawaiian myths, and similar studies in the American Southwest (Masse and Soklow 2005; Masse and Espenak 2006) and South America (Masse and Masse, in press), provide a unique window on the general nature and structure of myth that substantially differs from current anthropological characterizations. A myth is an analogical story created by highly skilled and trained cultural knowledge specialists (such as priests or historians) using supernatural images in order explain otherwise inexplicable natural events or processes. The more unusual or striking the event, the more likely the knowledge specialist will resort to using supernatural elements, such as the creation of demigods. Natural events leading to considerable loss of life for a given cultural group – such as devastating regional floods, large-scale mass fire, and massive Plinian volcanic eruptions – become part of the sacred cosmogony or creation mythology for that group. Each cataclysm typically leads to a new creation of the world and humankind and is sequenced in relative order of occurrence.

Vansina (1985) and other scholars have demonstrated that oral tradition is a particularly robust form of history, in some situations nearly matching the written word in terms of the long-term conservation of the most important details of a myth storyline. Most cultures had strict institutional mechanisms by which orally transmitted sacred knowledge could be preserved largely intact for hundreds or even thousands of years as demonstrated in Polynesia. These mechanisms included the use of highly skilled and trained narrators, typically chiefs, priests or shamans whose livelihood and sometimes their lives depended on the ability to perform their duties well as oral historians. In cultures such as China, Mesopotamia, Egypt, Mexico, Peru and Polynesia, natural events, especially great comets, meteor storms, supernovae and solar eclipses, were closely tied into the power and lineage of hereditary rulers. Typically, they were considered the property and even the euhemeristic persona of the chiefs (Masse 1995, 1998) and as such became embedded in naming chants and birthing stories attached to those chiefs.

Traditional narration of myths involved annual cycles of myth told during solstice ceremonies and other prescribed seasonal settings in which dance, chant and story repetition ensured that key details were faithfully transmitted through many generations of narrators. The unsavoury reputation currently given to oral history is largely the fault of anthropologists and historians who not only fail to understand the historical basis of the myths they collect, but who also typically record the myths in sterile settings in which the narrator has been removed from the normal highly structured and richly contextual environment of myth performance.

This is not to say that all myths and their English language translations are "literal truth." There are mechanisms that compress and distort myth storylines (Barber and Barber 2005). Adequate translations of historical documents depend on translators knowing the context of the time period in which the document was written, including cosmology and use of iconographic symbols. Few translators are trained to recognize and understand astronomical descriptions, much less cosmic impacts. Except in rare cases where myths are chronologically ordered (e.g. Polynesia), myths can only represent at best a model of past observations of the natural world. However, as defined below there are ways to systematically organize narrative data that not only strengthens the model but also provides the means by which to test and validate the encapsulated natural events (Masse et al., in press).

2.3.2
Using Myth to Identify and Model South American Cosmic Impacts

South America is both physically and culturally diverse. It was the last of the inhabited continents to be colonized, a process beginning sometime prior to 10 000 BC. Despite these recent cultural roots, there are at least 65 known language families with estimates of the numbers of individual languages ranging between 400 to as many as 3 000 (Bierhorst 1988, p 17). Prior to European contact early in the 16th century, a wide range of societies flourished throughout South America, ranging for the simple migratory hunter-gathers of Patagonia and Tierra del Fuego to well-known state-level societies of the central Andes and coastal plains of Chile and Peru. In between these two extremes were semi-sedentary and sedentary village horticulturalists occupying the tropical lowlands and highlands of Brazil and the Pampas regions of central Argentina and the Gran Chaco lying between these two areas.

South America has a rich legacy of oral traditions and mythology (c.f. Levi-Strauss 1969; Bierhorst 1988). Particularly valuable for our interests in cosmic impact are a set of 4 259 myths from 20 major cultural groups east of the Andes gathered by the University of California at Los Angeles (Wilbert and Simoneau 1992). These myths are the contribution of 111 authors, translated into English as necessary, and published over the course of 20 years as a set of 23 separate volumes. The cultural groups themselves are from widely distributed portions of South America: five from the Northwest region of Columbia and Venezuela at the northern tip of the continent, one from the Guiana Highlands along the border of Venezuela and Brazil, two from the Brazilian Highlands of central and eastern Brazil, nine from the Gran Chaco of northern Argentina, Paraguay and eastern Bolivia, one (now extinct) from Patagonia in southern Argentina, and two (also now extinct) from Tierra del Fuego at the southern tip of the continent.

Masse and Masse (in press) analyzed these 4 259 myths, concentrating on those myths that describe various local, regional or "worldwide" natural catastrophes that led to the deaths of members of a given cultural group. Events that led to the deaths of small numbers of individuals include local floods, fire, lightning – and in the case of two myths from the Brazilian Highlands, the observed thunderous fall of a meteorite into a river that killed several youths then swimming in the river. Several other myths from the Brazilian Highlands, and the Northwest talk about meteorites as being capable of causing human death – including, poignantly, the overall eventual destruction of the world – but do not describe actual impact events themselves. A single exception, not in the UCLA collection, is an Inca myth that describes a sizeable airburst in a remote mountain range near modern Cusco, which apparently did not result in any deaths.

Of greater interest is a set of 284 myths that have as their primary motif a single major cataclysm stated as having led to the deaths of most or all members of one or more cultural groups – typically referred to as having led to new creations of humanity. While one might scoff at the rational basis of such "new creations," it should be remembered that these cultural groups typically were small, a few hundred or at most a couple thousand people, and that while their overall territorial ranges may have been large the cultural group only occupied a small portion at any given time. Therefore, rare large-scale cataclysms such as Plinian eruptions, mass fires and torrential monsoons of unusual duration could indeed decimate such groups.

Table 2.2 organizes these myths by cultural group and by five defined categories of cataclysm. The stories, particularly those from the Gran Chaco, appear to be divided into relative time within the overall set of 284 myths, with certain cataclysms being stated as having occurred before or after other cataclysms. Thus myths about a lengthy time of darkness and a similar set of myths combining darkness with the sky falling or collapsing on top of people, houses, and forests are said to have occurred most recently (but still in the distant past), whereas myths of a "great" or "worldwide fire" occur in the middle of the myth cycle, and myths about a "great flood" occur at the beginning of the myth cycle, the latter sometimes coupled with a period of "great cold" stated as having occurred immediately after the flood.

The details of the "sky fell" and "darkness" myths encode ash fall from Plinian eruptions, and closely match aspects of the myths collected by Blong (1982) in his study of the Papua New Guinea Plinian eruption. Three separate ash fall events are seemingly attested in the South American myths, including one in the Northwest, one in the Guiana Highlands and a particularly convincing case in the Gran Chaco. The Gran Chaco ash fall may relate to a largely unstudied and poorly dated (1000–2000 BP?) pre-European Plinian eruption of the easternmost Holocene-active volcano, Nuevo Mundo, located in Bolivia some 500 km west of the Gran Chaco.

"World fire" myths not surprisingly are distributed in those areas most subject to devastating droughts and large-scale fires – the Gran Chaco and the Brazilian Highlands. Tribal groups of the Gran Chaco, such as the Toba, are noted for their burning of grasslands and brush as a common hunting technique, eating on the spot the charred remains of game animals (Metraux 1946, p 13). In a similar vein, the Brazilian *cerrado* is a massive mosaic of mixed grassland, planted shrub and forest occupying much of the Brazilian Highland region. The *cerrado* has been termed "the natural epicenter for Brazilian fire" (Pyne et al. 1996, p 685), whose configuration has been maintained through deliberation annual burning by tribes such as the Gé.

Given the close relationship between people and fire in the Gran Chaco and Brazilian Highland *cerrado* and the likelihood of periodic mass fires as have occurred historically due to both natural (lightning) and anthropogenic causes, it is of interest that sets of myths describing what appear to be cases of a single devastating "world fire" exist for each region. What makes the world fire distinct from all other fires is the specific *meteoritic* reason given in several of the myths for the cause of the world fire. Even in culture areas where mass fire is not common, such as that of the Bororo, the

Table 2.2. Myths of cataclysm in South America east of the Andes Mountains (from Wilbert and Simoneau 1992)

Location and culture	Great flood (earliest in myth cycle)	Great cold (after flood myth)	World fire (middle of myth cycle)	Sky fall–darkness (latest in myth cycle)	Great darkness (latest in myth cycle)
Northwest					
Cuiva	13	–	–	4	–
Guajiro	9	–	–	–	1
Sikuani	10	–	–	–	–
Warao	3	–	–	–	–
Yaruro	10	–	–	–	–
Guiana highlands					
Yanomani	17	–	–	11	2
Brazilian highlands					
Bororo	4	–	1	–	–
Ge	11	–	6	–	1
Gran Chaco					
Ayoreo	17	–	2	–	1
Caduevo	–	–	–	–	–
Chamacoco	10	–	1	1	–
Chorote	10	–	7	1	–
Makka	2	–	–	–	1
Mataco	12	–	5	–	1
Mocovi	7	–	3	–	–
Nikvale	6	1	6	9	–
Toba	24	5	27	–	12
Patagonia					
Tehuelche	2	–	–	–	–
Tierra Del Fuego					
Selknam	1	–	–	–	–
Yamana	3	1	2	–	1
Totals	171	7	60	26	20

people have an extraordinary fear of loud bolides (Masse and Masse, in press). In several stories it is pieces of the Moon or Sun breaking apart and falling, that causes the fire. This is evident in the following story from the Toba-Pilagrá of the Gran Chaco (Métraux 1946, p 33; Wilbert and Simoneau 1982, p 33):

> The people were all sound asleep. It was midnight when an Indian noticed that the moon was taking on a reddish hue. He awoke the others: "The moon is about to be eaten by an animal" *[a lunar eclipse]*. The animals preying on the moon were jaguars, but these jaguars were spirits of the dead. The people shouted and yelled. They beat their wooden mortars like drums, they thrashed

their dogs... They were making as much noise as they could to scare the jaguars and force them to let go their prey. Fragments of the moon fell down upon the earth and started a big fire. From these fragments the entire earth caught on fire. The fire was so large that the people could not escape. Men and women ran to the lagoons covered with bulrushes. Those who were late were overtaken by the fire. The water was boiling, but not where the bulrushes grew. Those who were in places not covered with bulrushes died and there most of the people were burnt alive. After everything had been destroyed the fire stopped. Decayed corpses of children floated upon the water. A big wind and a rain storm broke out. The dead were changed into birds. The large birds came out from corpses of adults, and small ones from the bodies of children.

The meteoritic cause of the fire is explicitly stated in Toba cosmology (Métraux 1946, p 19):

Moon... is a pot-bellied man whose bluish intestines can be seen through his skin. His enemy is a spirit of death, the celestial Jaguar. Now and then the Jaguar springs up to devour him. Moon defends himself with a spear tipped with a head carved of the soft wood of the bottletree..., which breaks apart at the first impact. He also has a club made of the same wood which is too light to cause any harm. The Jaguar tears at his body, pieces of which fall on the earth. These are the meteors, which three times have caused a world fire."

There has been debate as to the capacity of impacts to start ignition fires (e.g. Jones and Lim 2000; Svetsov 2002; Jones 2002; Durda and Kring 2004). Although the discussion has been geared to large impactors, it would appear – in contrast to the conclusions of Jones and Lim – that eyewitness accounts of falls, the limited archaeological record of impact sites and the myths discussed here indicate that wildfires are a common product of at least some smaller impacts. A key, of course, is the availability of fuel and suitable weather/climatic conditions, which in places such as the Gran Chaco and the Brazilian Highlands is not an issue.

The combination of ascribing the world fire to multiple meteoritic fragments (of the Moon or Sun) and that in a large percentage of stories the Toba were saved by going into "a hole many meters deep" arguably refer to the Campo del Cielo event and its multiple craters, some which would have resulted in tunnels several meters deep. The location of the majority of the Gran Chaco meteorite and mass fire stories in the general area directly north and east of the Campo del Cielo crater field is also suggestive. However, a direct link between the world fire and the Campo del Cielo impact event cannot be established without recourse to additional microstratigraphic archaeological and paleoenvironmental fieldwork in and around the Campo del Cielo impact site.

As previously noted, there also are stories of world fire in the Brazilian Highlands that appear to be linked with meteors or the fall of meteorites (Masse and Masse, in press). These include a series of elaborate myths regarding Sun and Moon in which Moon is jealous of the feather ornament that Sun has obtained from the red feathers of Woodpecker. In some stories the ornament is described as a "wheel of fire." Finally, Sun agrees to drop the ornament down to Moon, but warns Moon not to lose his grip or it will cause something bad to happen on the Earth. Sun tosses the ornament, but along with it are hot coals that prevent the Moon from holding on to the ornament. The feathers touch the ground, creating a world fire. "The sand caught fire and everything was burning. All the sand in the world, or almost all of it, was burning."

Burning sand is an unusual myth motif and is absent from Gran Chaco world fire myths. I suggest that it reflects the observation (from a safe distance) of an airburst that resulted in the creation of impact glass melt. Schultz and his colleagues (Schultz et al. 2004) have noted that temperatures in excess of 1 700 °C created the glass melts formed from the Argentine Pampas loess. A similar situation would be expected for the Brazilian Highlands loess. The maximum temperatures that can be achieved by burning of wild-land fuels are thought to be between 1 900 and 2 200 °C, but this would be an extremely rare situation not achieved in most wildfires (Pyne et al. 1996, pp 21–23). Sustained temperatures in wildfires and in the purposeful burning of fields likely would not be much greater than 1 650 °C with normal temperatures being closer to 1 000 °C. Therefore to create a large area of "burning sand" would seemingly require a meteoritic airburst. This implies that a glass melt-producing airburst has occurred in the Brazilian Highlands during the Holocene.

Two Yamana myths from Tierra del Fuego describe a "junior" and "senior" Sun in which the senior Sun creates a world fire by appearing suddenly in the east and making the ocean boil and burning down the forests. He then changes into a bright star that eventually disappears. (Masse and Masse, in press). This appears to be a somewhat confused rendering of the oceanic comet impact described in the next section. Unfortunately, since the Yamana are now extinct, there will be no future chance to clarify the details.

2.3.3
Modeling the Flood Comet Event – a Hypothesized Globally Catastrophic Mid-Holocene Abyssal Oceanic Comet Impact

The studies of myth in Polynesia, the American Southwest, and South America, coupled with Blong's (1982) work in Papua New Guinea, indicate the potential for the worldwide corpus of myth to have preserved the observation of Holocene period globally catastrophic cosmic impacts. Only one cosmogenic set of myths relate to a cataclysmic event that has universal distribution in virtually all cultures. This is the myth of the so-called "great flood."

2.3.3.1
Stilling the Waters

Two popular misconceptions exist within the scientific community regarding the flood myth. First is the belief that European missionaries and explorers diffused the myth across the world from its presumed origin in Mesopotamia or the Near East. Although there are examples of the Biblical flood having been diffused by Christian missionaries, the great majority of flood myths from more than 1 000 cultural groups worldwide demonstrate independent development of the myth within each culture (e.g. Frazer 1919; Dundes 1988). The universal nature of the great flood myth is evident in Table 2.2 where not only is this myth by far the most prevalent of all the South American catastrophe myths, but also it occurs earliest in the Gran Chaco myth cycle before the Plinian eruptions of AD 1 to AD 1000, and before the meteorite impacts at around 2700 to 2200 BC. The common claim that the myth is absent from the records of ancient Egypt and China, for example, is the result of not recognizing variant forms of the myth (Masse

1998), whereas the general lack of the myth in Sub-Saharan Africa (Dundes 1988, p 2) is possibly the product of the oceanic impact described in Sect. 2.4.1.

A second misconception is that each culture often had multiple myths of different floods, and that flood myths from each region are based on observations of local or regional floods, thus comparisons between flood events in each region would be of little consequence. In fact, in the vast majority of cultural traditions only a single worldwide flood is identified (although other more restricted local floods may also be mentioned), typically representing either the last in a cyclic sequence of global catastrophes or a unique watery disaster from which humans emerged. In either case, our modern world is seen as having evolved from a worldwide flood.

There have been a large number of attempts by reputable and well-meaning scientists to derive some kind of historical truth from the flood myth. Among the more recent are those by Ryan and Pitman (1998) regarding flooding of the Black Sea around 5600 BC and by Teller and his colleagues (Teller et al. 2000) regarding postglacial flooding of the Persian Gulf. These and similar studies invariably suffer from a biased sampling of the overall population of worldwide flood myths and by the deliberate exclusion of certain classes of environmental data – such as the presence of torrential rainfall – in those myths that they do use. To use an archaeological analogy, this is like attempting to date and interpret a stratigraphically complex archaeological site from which you have collected a total of 1 000 radiocarbon samples, but limiting actual analysis to 50 samples from but a single stratum, and then discarding half of the resultant dates because they do not fit your preconceived model.

While on this topic, I am compelled to address the interesting work of Austrian geologists Alexander and Edith Tollmann (1994) whose independent long-term study of flood mythology and geophysical evidence has resulted in findings superficially similar to my own described here. They hypothesize a major comet impact at the beginning of the Younger Dryas climatic event (ca. 9600 BC), which they claim to have resulted in seven fragments each conveniently hitting separate oceans or parts of oceans, thus creating the universal myth of the great flood. The Tollmann's particularly drew upon mythology, but also physical geology, tektites, ice cores, and other related databases.

Shortly after publication of their Flood Impact paper, a team of 13 scientists took the Tollmanns and their hypothesis to task (Deutsch et al. 1994). Their brief acerbic review highlighted a number of flaws in the Tollmann Flood Impact model. However, I suggest that the biggest flaw in the model was the failure by the Tollmanns to treat mythology with the same contextual and methodological rigor required of any scientific body of data. For example, they uncritically mix the Biblical creation myth with flood myths and make generalizations not warranted by the myths they use. Likewise, their historical illustrations are of dubious relational context to the hypothesized impact event.

I cannot overemphasize the fact that the analysis of myth requires the same stringent and systematic standards applied to all other categories of scientific data. Although my data here are admittedly preliminary, in the following discussion I attempt to provide enough details about the nature of the flood myth data and my methods of analysis so that the logic of these data and my interpretations can be understood and evaluated. This groundwork is necessary in that my conclusions about the nature of the hazards and effects of the hypothesized Flood Comet impact differ substantially from other impact models presented at the ICSU workshop and in this volume. However, I fully realize that in order to allow colleagues to satisfactorily judge my methods and inferences, it will be necessary to follow up this preliminary treatment with a subsequent detailed published analysis of the larger corpus of English language flood myths.

2.3.3.2
Preliminary Analysis of Flood Myths

Reported here is a preliminary analysis of environmental information contained in a worldwide sample of flood myths from 175 different cultural groups. The primary source for the myths is the 127 distinct myths and 46 variants contained in Frazer (1919), whereas the remaining 48 myths are from various other published sources in the initial attempt to better even out the regional distributions of the studied myths (see Masse 1998 for an earlier treatment of the Frazer myths). The myths are from the following regions: Artic Circle (5); North America (49); Central America and Mexico (11); South America (18); Africa (4); Europe (5); Middle and Near East (5); Russia (3); China/Tibet (11); Southeast Asia (31); Australia and New Guinea (22); and island Oceania (11). These 175 myths likely represent about 15% of all "great flood" myths printed in the English Language.

The premise of my comparative analysis is simple and straightforward. I hypothesize that if the universal great flood myth is based on a single worldwide natural catastrophe occurring sometime during the Holocene period, then there must be a single natural phenomenon that can logically account for the suite of all environmental information encoded in the totality of all great flood myths (Masse 1998, Table 2.1). Furthermore, these data and findings can be weighed and tested against the Holocene archaeological, geomorphological and paleoenvironmental record.

The brief analysis that follows suggests that *only* a globally catastrophic deep-water oceanic comet impact could account for all environmental information encoded in the corpus of worldwide flood myths and that my defined impact is consonant with the archaeological and paleoenvironmental record.

I identify 12 environmental variables within the corpus of great flood myths. These include: (*1*) Source and nature of the flood waters, vis-à-vis torrential rain and tsunami; (*2*) the nature of the storm, if any, associated with the flood; (*3*) earthquakes in conjunction with the flood; (*4*) time of day when the flood (or flood storm) began; (*5*) direction from which the flood storm originated; (*6*) duration of the flood storm; (*7*) unusual occurrence of light and/or darkness during the flood; (*8*) methods how survivors escaped the flood; (*9*) a rough estimate of the percentage of deaths caused by the flood; (*10*) advanced warning prior to the event that something was going to happen; (*11*) seasonal, astronomical or archaeological indicators that help to date the flood; and (*12*) descriptions of supernatural creatures associated with the flood.

Space precludes full citations and discussion of each variable and a complete distributional analysis of these preliminary data, but there are several highlights that likely speak directly to the effects of cosmic impact on human society (see also Masse 1998 for additional citations):

- *Source of floodwaters.* Some 76 (43%) of the myths do not define the nature of the "deluge" or "flood," but of the remaining 99 myths, 50 (51%) indicate the presence of torrential rainfall, 35 (35%) indicate tsunami, whereas 14 (14%) describe both rainfall and tsunami. Four of the 14 myths describing both elements indicate that the tsunami occurred before the rainfall, the others being equivocal. Nearly two-thirds of the 99 defined myths indicate the presence of torrential rainfall. Of these, 24 also indicate the presence of hurricane force winds and 23 indicate unusual darkness during the flood storm. The distribution of these elements is worldwide.
- *Duration of flood storm.* Thirty-three of the 175 myths provide a specific number of days for the flood storm. Nine are obvious outliers, several of which conflict with other myths from the same cultural group or region, including the confused dual rendering in Biblical tradition that the flood storm lasted either 40 days or 150 days (Habel 1988). The remaining 24 (73%) myths form a rough bell-shaped curve ranging between 4 and 10 days for the flood storm duration (Fig. 2.1). Intriguingly, the combined mean – 6.5 days – of these 24 worldwide myths matches exactly the duration provided in the two earliest written versions of the flood myth from Mesopotamia, the Gilgamesh (Kovacs 1989) and related Atrahasis (Lambert and Millard 1969) epics. Clay text fragments of these myths date to the 2nd and early 3rd millennium BC, and place the duration of the flood storm as six days and seven nights, and seven days, respectively. Torrential rainfall in historic hurricanes can occur at a rate of more than 10 cm per hour. In 1969, rainfall from Hurricane Camille during one six-hour period averaged more than 7.5 cm per hour throughout all of Nelson County, Virginia, leading to widespread devastating flooding and death (Clark 1982, pp 100–103). Even at the modest rate of 5.0 cm of

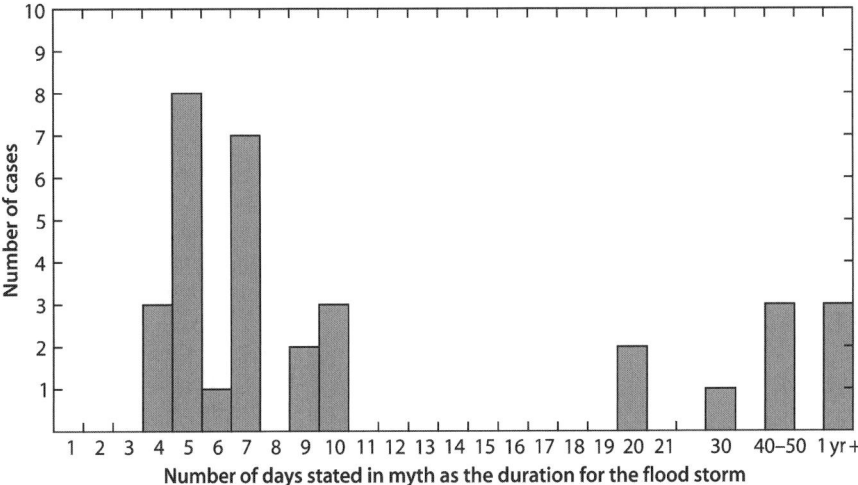

Fig. 2.1. Bar graph depicting the duration of the flood storm in days, based on 33 myths with explicit numbers out of a total sample of 175 analyzed English language flood myths. The nine cases on the right side of the graph – from 20 to more than 365 days – are considered to be non-representative outliers that encode culturally symbolic aspects of the flood event rather than the actual duration of the flood storm

rainfall per hour, the mid-Holocene flood storm, stated as being a continuous deluge throughout its duration, would yield a staggering total of 7.8 meters of water if constant for 6.5 days.

To appreciate the violence and overall duration of the flood storm, we again turn to Gilgamesh (Kovacs 1989, pp 100–101):

> Just as dawn began to glow there arose from the horizon a black cloud. Adad *[god of storms]* rumbled inside of it, before him went Shullat and Hanish *[minor storm gods]*, heralds going over mountain and land. Erragal *[Nergal – underworld god associated with forest fires and plagues]* pulled out the mooring poles, forth went Ninurta *[a warrior and farming god]* and made the dikes overflow. The Anunnaki *[Anunnakku – assistants to the sky god Anu]* lifted up the torches, setting the land ablaze with their flare. Stunned shock over Adad's deeds overtook the heavens, and turned to blackness all that had been light. The ... land shattered like a ... pot. All day long the South Wind blew ..., blowing fast, submerging the mountains in water, overwhelming the people like an attack. No one could see his fellow, they could not recognize each other in the torrent The gods were frightened by the Flood, and retreated, ascending to the heaven of Anu ... Six days and seven nights came the wind and flood, the storm flattening the land. When the seventh day arrived, the storm was pounding, the flood was a war – struggling with itself like a woman writhing (in labor). The sea calmed, fell still, the whirlwind (and) flood stopped up. I looked around all day long – quiet had set in and all the human beings had turned to clay!

- *Tsunami and storm surges.* Although a relatively small number (< 10%) of myths note that flood survivors saved themselves on the tops of high mountains such Mount Ararat (Turkey) and Mount Parnassus (Greece), most stories present more logical scenarios for surviving tsunami and cyclonic storm-induced storm surges. These tsunami locations (e.g. California; Brazil; Tierra del Fuego; Indonesia; India) are in quite believable situations, between to 15 to 100 km inland, and the hillsides or hilltops where people were stated as saving themselves typically range between 150 to 300 m above sea level. For example, near the town of Bonsall directly north of San Diego, California, the Luiseño Indians relate that the flood surrounded but did not cover the top of modern Morro (Mora) Hill (Frazer 1919, pp 288–289), a cluster of small rounded peaks that lie 16 km inland. I have visited this location. The highest elevation of Morro Hill is 280 m above mean sea level, but more pertinent is the fact that a roughly one square kilometer parcel (247 acres) lies at an elevation of more than 200 m, and is significantly higher than the directly adjacent countryside, much of which is < 100 m in elevation. That the Spanish were aware of the local flood legend for this hill is likely in their choice of the name *mora* (since corrupted to Morro), a colloquialism that means "unwatered."
- *Supernatural entities associated with the flood.* Half of the 175 myths describe supernatural entities associated with the flood, typically as undefined creators or nature deities, human-like deities, or helpful animals lacking descriptive detail. The 38 detailed descriptions include: Giant snake or water serpent (6 examples); giant bird (3); giant catfish (3); giant horned snake (3); elongated fish (3); and single examples of a giant fish with long snout; sperm whale; giant crocodile with cassowary feathers; giant horned earth dragon; monsters who grew up into the sky (perhaps a description of a debris plume); lizard thrown into the fire; battle between giant saw-fish and crocodile; dragon churning the water; battle between Sun and Moon; kite; fallen angels; blazing brand; star with fiery tail; flood begins when "dusty star-baby" is pulled apart; deity described as low-flying meteor; tongue of fire turned into flood; great light like the Sun followed by

great heat and then the flood; rain of fire associated with a serpent; battle in the sky of fiery and dark forces; and a man in a garment of lightning.

The typical elongated and celestial nature of these giant supernatural creatures, the presence of horns, their association with fire or brightness, and their presence for several days prior to the flood event are strongly suggestive of the observation of a near-Earth comet. Particularly fascinating is a composite description of a flood myth from several well-known Indian texts (*Satapatha Brahmana,* the *Mahabharata* and the *Puranas*): The progenitor of humankind, Manu, finds a tiny fish "bright as a moonbeam" in a puddle. He compassionately rescues it by putting it in a jar of water. The fish grows larger, and Manu in turn places it in a large pond, the Ganges River, and finally the ocean. By this time the fish is huge and "lotus-eyed." The fish reveals himself as a god, telling Manu that he will disappear for a period of time but will return later that year at the onset of the dissolution of the universe. The fish tells Manu that he, the fish, will be recognizable by a horn on his head. The fish later reappears, golden in color and as big as a mountain with a large single horn on his head. Manu uses a serpent rope to attach his boat to the fish and is pulled to safety across the turbulent flood-disturbed sea.

It is noted that in Indian tradition the lotus is situated in cosmic waters and is golden and radiant as the Sun (Zimmer 1946, p 90). The described characteristics of the fish well match the naked-eye visible orbital behavior of a comet observed during both the pre-perihelion and post-perihelion stages. Solar wind has the energetic velocity to blow cometary dust and ion tails away from the Sun even during the post-perihelion stage, thus creating an image visualized by naked eye observers as a headdress or horn attached to the head of the comet (Masse 1995, 1998).

- *Fire or hot water (rainfall and ocean swells) associated with the flood.* Notable is the presence of hot water, or fire/fiery rain in conjunction with the flood. At least seven myths from various parts of the world state that devastating fire or flames or rain of fiery particles occurred at some point immediately before the arrival of the flood storm. These include Arizona and Idaho in North America, the Congo in Africa, central and northern India and New Guinea. A likely related story from Egypt is the famous myth of the destruction of mankind (Pritchard 1975, pp 10–11; Ions 1968, p 106). In this myth the sky-goddess Hathor transforms into an enraged lion-headed goddess Sekmet – as the Eye of the Sun god Ra, representing the scorching, destructive power of the Sun, spitting flames at the enemies of Ra – and humankind is only saved when the land is flooded to a depth of 3 palms (ca. 25 cm) above the fields by 7 000 vats of blood-red mash brewed by the other gods and Sekmet pauses to reflect on her beauty.

  In addition are myths from Chile, Bolivia, Brazil in South America, Iraq, India and New Guinea describing hot water falling out of the sky, whereas myths in Tierra del Fuego, Taiwan and New Guinea describe hot ocean water washing up on their shores. A story about hot water bubbling out of the ground from near the Ural Mountains in Russia likely represents traditions that originated with people who migrated from India. Although not being described as "hot," the Maya Indians of Mexico describe the beginning of the flood as that of a thick resin falling out of the sky.

- *Seasonal and calendrical dating of the flood.* The seasonal and lunar data within the myths are remarkably consistent. Sixteen of the 175 myths describe seasonal indicators or name an exact month. Of these, 14 are in the northern hemisphere spring (late April–

May – early June), whereas one from the southern hemisphere is situated in the fall (equivalent to the northern hemisphere spring). In terms of described lunar phase, six of seven worldwide stories indicate that the flood began at the time of the full Moon, whereas the other story indicates a time two days later, the 17th day of the lunar cycle. In addition, there are stories in Africa and South America that place the flood at the time of a partial lunar eclipse, a phenomenon that only takes place at the time of a full Moon. The 4th century BC Babylonian historian, Berossos provides an exact day and month of the 15th of Daisios, which translates to the day of the full Moon in late April or early May (Verbrugghe and Wickersham 1996, p 49).

Equally striking are specific calendrical markers associated with these myths (Masse 1998, pp 64–65) from Chinese annals, and from well-dated archaeological contexts in Mesopotamia, Egypt, and elsewhere in the ancient Near East. China's Han Dynasty chronologists provide the date of 2810 BC for the end of the reign of first empress Nu Wa (Walters 1992). Nu Wa was a supernatural woman who at the end of her reign repaired the cosmic damage and flooding caused by the red-haired horned cosmic monster Gong Gong who knocked over a pillar of heaven, upsetting the universe. It is of some interest that Nu Wa mended the sky with melted stones of many different colors, thus matching the Biblical rainbow, as do in their own way a substantive number of traditions elsewhere in the world.

The 3rd century BC Egyptian historian, Manetho, noted that during the reign of Semerkhet, 7th of the 8 kings in Egypt's First Dynasty, *"there were many extraordinary events, and there was an immense disaster"* (Verbrugghe and Wickesham 1996, p 132). Although the nature of these events (also stated as "portents" in other renditions of Manetho) and disaster are not specified, there are several reasons to link them with the hypothesized Flood Comet impact. Semerkhet's reign is around 2800 BC, based the most recent dating of the First Dynasty between ~2920 to 2770 BC (Kitchen 1991). Not only does Semerkhet have the shortest reign of the First Dynasty kings, but he is the only one to lack an elite tomb at Saqqara (Wilkinson 1999, p 80).

Semerkhet's successor, Qa'a, the final king of the dynasty, is of interest in two respects. One is the translation of a variant of his name as *"abundance" in the sense of "flood"* (Weigall 1925, p 49). The other consists of unusual aspects of his tomb at Abydos noted by its excavator, Sir Flinders Petrie. Petrie (1900, pp 14–16) documented serious wall collapse in the lesser chambers due to insufficiently dried mud bricks; wooden timbers were unusually decayed as compared with earlier tombs; the entrance passage turned at an odd angle and was closed by rough bricks; and clean white sand was placed in and around the coffins of retainers. Re-excavation in 1992 indicated that the structure apparently was built in two or more stages over a long period of time (Wilkinson 1999, p 237). These data together suggest that the tomb of Qa'a was under construction at the time of the Flood Comet impact, suffered extensive water damage, and after a lengthy period of time was repaired and completed. This interpretation is also consonant with the fact that the succeeding kings of the 2nd dynasty abruptly shifted the location of their royal tombs at Abydos from the upper floodplain of the Nile to the nearby mesa tops, but returned to the original upper floodplain location at the end of the 2nd dynasty.

The ancient Near East exhibits a number of paleoflood deposits of various ages, typical for any region prone to flooding. Of particular interest are deposits at the ancient Mesopotamian cities of Shuruppak (modern Tell Farah), home of the legendary flood survivor, Atrahasis (Lambert and Millard 1969), and that of Kish (modern Tell Oheimer). These cities are mentioned in the famous Sumerian King list, created around 2300 BC by Enheduana, priestess of the Moon-god at Ur and daughter of King Sargon of Akkad (Postgate 1992, pp 27-28). The document lists five antediluvian cities, the last of which was Shuruppak, and then goes on to state: "*After the Flood had swept over (the earth) (and) when kingship was lowered (again) from heaven, kingship was (first) in Kish*" (Pritchard 1969, p 265). Sir Max Mallowan (1964) defined specific paleoflood deposits at both cities that he equated with Noah's Flood. The current date for these flood deposits and the establishment of Kish as a major city is estimated to be around 2800 BC (Porada et al. 1992).

This is also the time of abrupt movement of at least half of the people in Palestine from valley floors to the hill country of Galilee, Samria, and Judah, only to return to the valley floor a few generations later (Mazar 1990, pp 111-113). This unique settlement pattern is accompanied throughout much of the ancient Near East by the construction or enhancement of massive walls around most settlements, suggesting unsettled times. Many curious things from an archaeological perspective occur at around 2800 BC, including the marked dispersal and migration of five major language groups in five different parts of the world, Bantu (Africa), Indo-Aryan (Near East and Europe), Uto-Aztecan (North America), Austronesian (Southeast Asia), and Gé-Pano-Carib (South America). Significantly, this date also is roughly the boundary between the middle and late Holocene climate regimes, moving from warmer and dryer to cooler and wetter conditions.

Astrological aspects of the flood are mentioned in a number of myths. For example, Peruvian and Hindu myths mention a conjunction of planets immediately prior to the flood, whereas Hopi traditions (e.g., Mails and Evehama 1995, pp 506-509) note that the previous world ended several thousand years ago when there were violent signs in the sky and when certain "stars" (presumably planets) came together in a row. The Roman philosopher, Seneca, indicated that the 4th century BC Babylonian historian, Berossus, could date the end of the world by fire and flood by calculating when all the planets would again be positioned in a row (Verbugghe and Wickersham 1996, p 66).

Aquarius, "Water Bearer," is almost universally noted in Old World zodiacal mythology as being a source of water, with myths from China, Greece, Mesopotamia, and Egypt all specifically linking the constellation to the flood or at least some form of watery deluge (Motz and Nathanson 1988). In Greek mythology as well as in Babylonian symbolism, the asterism representing the urn carried by the Water Bear, which is located at approximately Zeta Aquarii, was the location from which the floodwaters came forth.

Pisces is of special interest due to the widespread historical astrological belief that conjunctions of planets within this sign, in particular Jupiter and Saturn, portend spectacular events and occasionally dire consequences. For example, in Biblical astrology it was predicted that another deluge would occur in the year AD 1524 when Jupiter, Saturn and Venus were in conjunction with Pisces (Allen 1963, p 341; North 1989, pp 63-68). The beginning of the modern Hindu age (*yuga*) of Kali after the flood, is stated by the 5th century AD Hindu astronomer, Âryabhata, as beginning at dawn on February 18, 3102 BC at a time when the naked-eye visible planets were in conjunction at 0° Aries, near the star Zeta Piscium (Pingree 1972; Gleadow 1968, pp 138, 147). A similar concept was expressed by the 9th century Arab astrologer, Albumasur, who predicted the destruction of the world when the five planets, Sun and Moon were in conjunction in the last degree of Pisces (Allen 1963, p 77). However, astronomy software demonstrates that such a conjunction of the five visible planets did not occur in 3102 BC or any year near that date.

This cluster of astrological details can be subjected to systematic analysis similar to that done for the environmental details in the flood myths described above to see if there is a logical explanation for these diverse statements. As reconstructed by astronomy software programs (*RedShift Multimedia Astronomy 3.0©*, *TheSky, version 5©*), it turns out that the year 2807 BC was highlighted by an extremely rare quadruple conjunction of Saturn and Jupiter at the boundary between Pisces and Aquarius (22 January, 26 April, 2 August, 10 November) with another such conjunction (including Venus) occurring on January 11, 2806 BC. On February 7, 2807 BC, the five planets were situated evenly in a row within Aquarius and Capricornus spaced about 10° apart from one another just before sunrise as seen in India, while on February 25 they were similarly situated in Aquarius in a row along with the Moon, spaced about 5° apart. During the middle of March at dawn, Venus and Mars were conjoined for several days with Saturn and Jupiter adjacent to Zeta Piscium. On April 25, 2807 BC there was a total eclipse of the Sun, and on May 10, 2807 BC there was a partial lunar eclipse.

The seasonal, calendrical and archaeological data form a compelling and logical story that well complements the rest of the environmental information in our sample of 175 flood myths. The principle of Occam's razor suggests that an oceanic comet impact on or about May 10, 2807 BC more simply and better explains the combined mythology, archaeology, paleoenvironmental record and documentary history surrounding the boundary between the middle and late Holocene (ca. 2800 BC) than do our current diverse models and theories of Holocene cultural evolution and climate change.

2.3.3.3
Modeling the Flood Comet Impact Event

Based on a reading of the preliminary set of flood myths summarized above, there are several aspects of the hypothesized impactor that can be logically elicited from these details, particularly in reference to the modeling of Toon and his colleagues (Toon et al. 1994, 1997) and the web-based impact modeling programs of Melosh and Beyer (2005) and Marcus et al. (2005).

In order to model likely impact effects, it is useful to first briefly discuss the Earth's atmosphere (Salby 1996). The atmosphere is dominated in volume by a mixture of molecular nitrogen (78%) and molecular oxygen (21%), with water vapor, carbon dioxide, ozone and other trace species comprising the remaining 1%. Although water vapor is a trace species, it plays a significant role in cloud formation, radiative processes and in energy exchanges with the oceans. About 60% of the overall water vapor is situated in the trophosphere, and then steady decreases in percentage at higher elevations. Gravity stratifies the atmosphere vertically, whereas the Earth's rotation creates meridional stratification and the development of large-scale circulation such as airflow around centers of high and low pressure. Atmospheric pressure and density decrease exponentially with increased elevation above the Earth's surface, but temperature varies in pronounced ways giving rise to the designations troposphere (lower atmosphere) from 0–10 km, the stratosphere (10–50 km) and mesosphere (50–85 km) of the middle atmosphere, and the thermosphere of the upper atmosphere (above 85 km). Upper troposphere circulation is characterized by subtropical jet streams, while the polar-night jet operates in the lower mesosphere. Collectively, the temperature-related layers below 100 km are termed the homosphere. In the heterosphere (100–500 km), molecular diffusion suppresses turbulent air motions and airflow is nearly laminar. The highest layer of the atmosphere is the exosphere, in which molecular collisions are rare and in which some molecules can achieve velocities that enable them to escape the Earth's gravity and enter deep space.

Toon et al. (1997) have noted that only limited modeling has been accomplished thus far of the potential atmospheric effects of water injection by the plume of a large abyssal oceanic impact. This was evident at the ICSU workshop in that virtually none of the presentations and papers addressed the effects and hazards of such a massive water injection.

The review and modeling of the effects of water injections in Toon et al. (1994, pp 817–821) is directly pertinent to defined effects of the hypothesized Flood Comet impact. A large comet hitting the abyssal ocean would loft an amount of water equal to about 10 times the mass of the comet into and through the middle and upper atmosphere. The latent heat of the water would cause the vapor cloud to adiabatically expand. High-altitude portions of the vapor cloud will form ice crystals that will fall downward, evaporate and humidify the lower atmosphere. Toon et al. (1994, pp 818–819) note: "Condensation after a 10^4 megaton impact may occur over several days, during which time the water will have been transported great distances from the impact site." They go on to note "a water-rich atmosphere is unstable with respect to vertical motions because any descending air parcels will have a water vapor partial pressure exceeding the vapor pressure, leading to rainout of the water, latent heat release and convective mixing." In simple terms, this means that there will be a lot of rain and very unstable atmospheric conditions. Toon et al. (1994, p 805) also note that submicron dust loading of the atmosphere caused by large terrestrial impacts may be countered by the water vapor in a large oceanic impact, and that "ice clouds formed by oceanic impacts have the potential to sweep some or all of the dust from the sky."

The environmental data in the flood myths fit remarkably well with the above modeling for a large oceanic comet impact above the threshold for global catastrophe at or greater than 10^6 MT (100 gigatons). The hypothesized Flood Comet impact is associated with six or seven days of intense atmospheric rainout, accompanied by hurricane-force winds for the duration of the period of rainout. Presumably the winds and a sizeable percentage of the rainfall are part of a system of ocean-fed worldwide cyclonic storms generated and sustained by the air pressure blast wave, the impact plume, the spread of water vapor, and its subsequent rainout. The intense darkness accompanying the flood storm is an indication of the amount of submicron and larger dust grains that accompany the water injection into the atmosphere, which is then seemingly effectively removed during the process of rainout. Intriguingly, the current myth sample suggest that torrential rainfall may have been limited to mid and low latitudes between about 55° N and 55° S. The few myths outside this range do not specifically mention rainfall.

Regardless of interpretation, the impacting comet was large enough to result in a seabed crater. Myths from Greece, Mesopotamia, India and Taiwan all indicate that the flood storm originated somewhere to their south, suggesting a possible impact location in the abyssal depths of the Atlantic-Indian Basin. Prior to the ICSU workshop, I originally modeled the impact in the general vicinity of 38° east longitude and 58° south latitude, a location reasonably close to recently discovered Burckle Crater (Sect. 2.4.1). The putative diameter of abyssal Burckle Crater at around 29 km can be modeled as the impact of a comet slightly larger than 5 km in diameter and a speed of 51 km s^{-1} entering the ocean at an act angle of 45° (Marcus et al. 2005). The energy produced by such an impact is approximately 2×10^7 MT. Of interest is the fact that such an impact would eject rocky debris to a distance of approximately 9000 km from the impact site, which is the approximate distance in which myths mention hot or fiery water falling from the sky (Fig. 2.2).

The common motif about rainbows and other similar phenomena immediately after the flood as described in a number of our sampled myths is fully consistent with

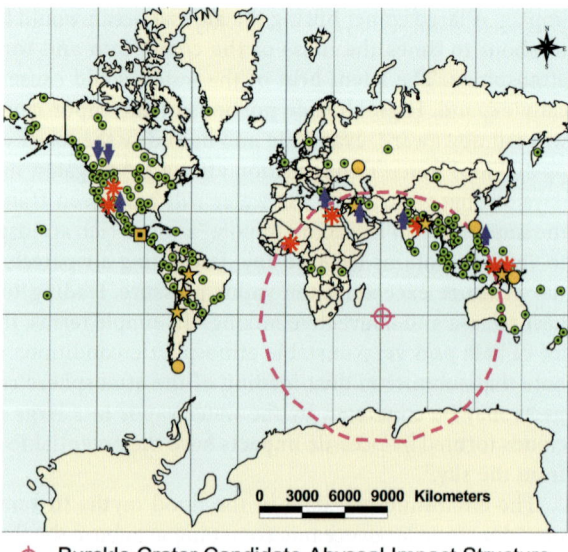

Fig. 2.2.
Map depicting the location of Burckle Crater candidate abyssal impact structure in relation to selected environmental variables as stated in a sample of 175 "Great Flood" myths. In addition to depicting the approximate locations of the sampled flood myths themselves, the variables include the apparent direction traveled by the flood storm; hot water noted as coming from the ocean; hot water and "thick resin" noted as coming from the sky; and intense heat and ignition fires at the start of the flood storm (the latter includes a story from Egypt not in the sample of 175 myths). The figure also depicts a "hypothesized ejecta re-entry splash ring" modeled as the approximate boundary between the limits of rocky ejecta and condensed water vapor from the hypothesized Burckle Crater impact

- ⊕ Burckle Crater Candidate Abyssal Impact Structure
- ▲ Apparent Direction of the Flood Storm
- ○ Hot Water from Ocean or Ground
- ☆ Hot Water from Sky
- ▪ Thick Resin from Sky
- ✱ Intense Heat and Fire Ignition at Start of Flood Storm
- ⸱ Location of Flood Myth
- ⁄ ˎ ⁄ Hypothesized Ejecta Reentry Splash Ring

the atmospheric physics of injecting a large column of water into and through the upper atmosphere. This would have led to the formation of high altitude ice clouds, that would become visible once the atmosphere had sufficiently rained out and stabilized, including the removal of obscuring dust particles. This rainbow effect, a greatly enlarged version of the common winter halo effect around the Sun and Moon, would dissipate as the ice volatilizes.

What does not fit the model of a single large Indian Ocean impact is the presence of a number of mega-tsunami myths from Brazil, the western coast of North America, the Arctic Ocean and in other locations outside the Indian Ocean basin. Likewise, the presence of hot or fiery water falling from the sky in several North and South American myths cannot have been caused by atmospheric re-entry ejecta from the Burckle Crater event. Myths from north-western North American describe the flood storm as coming from the north. And as noted in Sect. 2.4.1, Burckle Crater by itself cannot explain the large volume of rainfall indicated by worldwide mythology.

Not only was the Flood Comet likely composed of several fragments (Abbott et al. 2005), one may have considerably lagged behind the others. There are several stories from New Guinea and Australia about a flame or bright light witnessed oddly enough during the middle of the flood storm. One such Aboriginal Dreamtime story from Australia is as follows (Smith 1930):

> An old goanna *[lizard]* stuck his head out *[from the protective cave]*, but quickly withdrew it... "I have seen a wonderful sight, an awful monster with an eye as big and bright as the Moon. But wait a moment, his eye is brighter than the Moon, and nearly as bright as the Sun"... They all gathered together to discuss what they had seen, and each had a different account to give their new *Intelligence* that had arrived with the rain, the thunder, and the lightning. There was one thing, however, regarding which they were all agreed, and that was the brightness that shone from this formless being. Strange to say, whenever rays of light appeared to the vision of the watcher they were stamped upon his memory and also upon his body, and were plainly visible to those round about.

Also of interest along with these particular myths are descriptions of a second tsunami along the coast of New Guinea three days after the onset of the flood storm.

The internal consistency of these sets of myths from Australia and New Guinea are suggestive of a second smaller impact two or three days after the first, therefore indicating that the comet had calved into several separate fragments, perhaps in a prior perihelion passage of the Sun. Such a situation may help to explain the imagery of giant supernatural twins or companions that is prevalent in Mesopotamian, Egyptian and even Mesoamerican myth and iconography between the period of about 3200 BC to around 2650 BC (Masse 1998).

2.3.3.4
The "Invisible" Mid-Holocene Globally Catastrophic Comet Impact

One obvious question jumps out for anyone considering these data: How did we miss it – how did we (science) fail to recognize the signature of a globally catastrophic impact dating to less than 5000 years ago ... or even more specifically in 2807 BC? This is a disturbing question that, if the impact is real and correctly modeled, must give us great pause. There are at least four circumstances that together may extract us from this dilemma.

The first involves our current reliance on radiocarbon dating for dealing with issues of mid-Holocene archaeology and climate change. Ironically, the modeled date of 2807 BC falls into the middle of one of the largest bursts of natural radiocarbon production evident in the past 5 000 years of the calibrated radiocarbon production curve (Taylor 1997, Table 3.1). Radiocarbon production occurs in the upper atmosphere as the product of neutrons in cosmic rays interacting with nitrogen atoms to produce radiocarbon. A burst of newly formed radiocarbon has the effect of creating a "shingle" or period of a couple hundred years during which radiocarbon dating itself cannot well separate the dates for one given year from any other within that specific period. Due to such secular variation, carbon samples formed during the year of the hypothesized comet impact could be represented by radiocarbon ages anywhere between 4300 and 4080 BP.

Or perhaps it is not so ironic. We need to evaluate the possibility that the Flood Comet impact itself contributed to this radiocarbon dating shingle. The introduction of vast amounts of nitrogen into the atmosphere by the impact plume, coupled with the possibility that the plume blew off part of the atmosphere and thus would have allowed cosmic rays to more deeply penetrate and react with the nitrogen, is an ideal setting for enhanced radiocarbon production (a possibility also raised by Tollmann and Tollman 1994).

Reliance upon radiocarbon dating also masks changes in regional population size. Widespread mammalian populations such as deer typically recover rapidly from mass mortality. Even if there had been a loss of two-thirds (67%) of all people due to the Flood Comet and its aftermath, it would take the survivors only 80 years to fully recover to the previous population level, assuming a very modest average population increase of 2% per year beginning in the sixth year after the impact. Given that radiocarbon dating typically has a standard deviation of 40 to 50 years, and given that for any time period during the middle Holocene we have likely documented far fewer than 1% of all habitation sites, and given that archaeologists tend to lump population estimates into 100 or even 500-year periods for ease of data manipulation, the catastrophic loss of 67% of humanity would be hard to define in the archaeological record. Having said that, there is some indication of population decline around 3000 BC give or take a few hundred years (Masse 1998, Fig. 2.3), including the interesting extirpation of humans from sizeable Flinders Island near Australia.

The second circumstance is related to the potential environment transforming complexities of oceanic impacts. Our present models of effects for oceanic impacts at the threshold for global catastrophe (e.g. Toon et al. 1997; Marcus et al. 2004), particularly in those cases where the impact location is far from continental margins and major islands, tend to focus on tsunami rather than other effects. Based on the assumption that the hypothesized Flood Comet impact is a real event and that my preliminary modeling of magnitude is relatively accurate, I would argue that the most devastating effects on human life and infrastructure would stem from the "flood storm," that is, from the combination of atmospheric rainout and concomitant cyclonic storms.

With a globally catastrophic oceanic comet impact, there occurs – in addition to the air pressure blast wave, splash ejecta re-entry, and variable fires from ablation and ballistic re-entry of larger particulates – massive tsunamis and storm surges along coastal margins followed by even more massive water movements across the entire landscape from the flood storm. This would result in the cutting and filling of drainage systems, landslides, along with the stripping of forests and the variable destruction of vegetation communities caused by the atmospheric rainout and cyclonic storms. Ironically, the tsunamis signatures may be obscured by the surface water and sediment flows from the atmospheric rainout and cyclonic storms. Unlike the KT-boundary impact, there likely is nothing equivalent to the KT boundary iridium layer – there is no one single uniform archaeological, geomorphological or paleoenvironmental signature for the Flood Comet impact itself.

The third circumstance, related closely to the first two, is that our present field methods for studying past environmental change are ill suited for the study of abrupt large-scale catastrophe and particularly for the identification and explication of oceanic cosmic impacts. There is a need for better dating of stratigraphic columns including the securing of larger numbers of chronometric dating samples and the use of a larger range of both chronometric and relative dating techniques, including the use of microspherules as suggested by Raukas (2000) and others. There is also a need for systematic local and regional stratigraphic sampling strategies that go beyond our present research designs for looking at environmental change.

I am intrigued by the coincidence of the Flood Comet impact date with the boundary between the middle and late Holocene climatic regimes. We still have much to learn about the coupling between the atmosphere and the oceans of our world. If the Flood Comet impact were to be validated and if it could be demonstrably linked to this perceived minor climate boundary change, then what if the comet had instead crashed elsewhere into the world's oceans such as the north Pacific? Would this have created different climatic effects? Do we need to think about and to model the effects of different magnitudes and locations of oceanic impacts in relation to the El Niño-Southern Oscillation (ENSO) or some of our other climatic cycles? And what might validation of the Flood Comet impact tell us about past climate change? For example, the dramatic beginning of the Younger Dryas period at around 9600 BC, mirrors the physical signatures of the hypothesized Flood Comet impact, but is marked by an even larger burst of radiocarbon production and even greater shifts in climate and in ocean circulation, along with the destructive flooding of contemporaneous archaeological sites and the extirpation and eventual extinction of several large mammal species. Tollmann and Tollman (1994) may have been right about a hugely catastrophic comet impact at 9600 BC ... even if for the wrong reasons.

The fourth circumstance is that we are dealing with the "Flood Comet" and not the "Flood Asteroid." Not only do we lack some of the telltale clues of asteroid impact such as recognizable meteoritic fragments and possibly elevated iridium concentrations, but we also do not yet know the full range of variation in comet composition, therefore even our modeling of potential impact products for which to search becomes suspect.

2.3.3.5
Surviving the Flood Comet Impact

Settlement patterns 48 centuries ago as today favored the use of coastal margins and valley bottoms due to access to farmlands, transportation corridors, marine fisheries and river resources. Ironically, these are the areas most vulnerable to a globally catastrophic oceanic impact given the "1-2-3 punch" of tsunamis, massive flooding from storm surges and extended atmospheric rainout, as well as accompanying hurricane-force winds. In addition to the staggering loss of human life, these combined forces would destroy homes, crops, animals and plant resources, with large areas being stripped of its forest leaf cover and in many cases of the trees and shrubs themselves.

The great majority of the 175 flood myths describe in fair detail the numbers of people who survived and how they survived. Collectively, the flood myths suggest that between 50–75% of humanity died during the Flood Comet impact and its aftermath. Only about 15% of the myths indicate that more than half of a given cultural group survived, while about 35% of the myths indicate survivorship by multiple couples, families, and portions of villages. Thus half the myths indicate few or "no survivors," with the modern world being replenished by a new creation of humanity. Regions of seeming higher survivorship include Tibet, north-eastern India, portion of New Guinea and southern Australia, New Mexico in the United States, and especially portions of Alaska, northern Canada, and the North American Pacific Northwest.

About half of the myths indicate that people saved themselves on boats, canoes, makeshift rafts, or by floating on or in a log or other buoyant debris, which then typically became grounded on mountainsides or other high spots. In more than a third of the myths, survivors sought refuge by climbing tall mountains or hills near their village, in some cases occupying known caves. A few survivors found refuge at the tops of tall trees. Many refugees are stated as dying due to exposure and famine following the flood storm. The previously mentioned language dispersals around 2800 BC are an expected response to widespread destruction of habitat and the fragmentation of many societies.

The most sobering way to measure the effects on human society of the Flood Comet impact is to use the voices of the survivors and their descendents. While there are hundreds of poignant stories, I here quote two. The first is from *Metamorphoses* by Roman poet Ovid (Melville 1986):

> And out on soaking wings the south wind flew, his ghastly features veiled in deepest gloom... and when in giant hands he crushed the hanging clouds, the thunder crashed and storms of blinding rain poured down from heaven... The streams returned and freed their fountains' flow and rolled in course unbridled to the sea. Then with his trident Neptune struck the earth, which quaked and moved to give the waters way. In vast expanse across the open plains the rivers spread and swept away together crops, orchards, vineyards, cattle, houses, men, temples and shrines with all their holy things... over the whole earth all things were sea, a sea without a shore. Some gained the hilltops, others took to boats and rowed where late they ploughed... The world was drowned; those few the deluge spared for dearth of food in lingering famine died.

The second story is a composite sense of how Australian Aborigines view the coming of the powerful deity Rainbow Serpent and his role in the flood (Berndt and Berndt 1994):

> As for appearance, there is basic agreement that a great snake is involved, but other features vary. In western Arnhem Land, for instance, reference is often made to 'horns', one at each side of the snake's head, to 'whiskers' (when it is male), and to the dazzling light from the snake's eyes. But most it is the sound of the snake's approach, rather than the sight, that is mentioned in stories. The victims are so overcome by what is happening to them that they have only a vague vision of 'who' might be doing it. Apart from the sight and the feel of rising waters, trees falling and their belongings being washed

away, they hear the noise of rushing flood-streams or tides, and the roar of the wind like the combined 'voices' of many bees, or like a huge bush-fire speeding toward them. That noise is sometimes contrasted, in myths, with the stillness and quietness later on when all is over, when the bones have turned to rock. At some sites a pool of clear water reveals, deep down and unmoving, rocks that were once the domestic belongings of the people who had lived there.

2.4
Epilog and Conclusions

2.4.1
Candidate Abyssal Impact Structure

Shortly after the conclusion of the ICSU workshop, I discussed my project with geophysicist Dallas Abbott, who volunteered to perform a preliminary search for young abyssal craters in the Indian Ocean. This search resulted in the discovery of a candidate abyssal impact structure (Burckle Crater) about 1500 km southeast of Madagascar, centered at 30.87° S and 61.36° E (Fig. 2.3). The structure is approximately 29 ± 1 km in diameter, and is discernable on bathymetric topographic maps as a nearly circular feature on the edge of a fracture zone along the southeast Indian ridge at an abyssal depth of about 3800 m (Abbott et al. 2005). The rim is not continuous but rather is broken by a series of low points that likely represent resurge gulleys formed in the crater walls by water movement during the collapse of the impact water cavity. A study of pertinent seismic lines reveals that the only areas with any sediment cover are all topographic lows near Burckle Crater, while away from the crater the basement is completely bare of sediment including topographic lows.

The examination of three cores from the vicinity of Burckle Crater, but away from the ridge itself, revealed the presence of a likely ejecta layer (Abbott et al. 2005). This is represented by high levels of magnetic susceptibility in the uppermost portion of the column, along with grains of freshly broken plagioclase feldspar and other displaced mantle and fracture-related rocks such as a spinel peridotite, chrysotile asbestos and manganese-rich pyroxene. Of particular interest is a 200 micron wide grain of pure native nickel that exhibits oxidation droplets along one margin. Because pure nickel melts at 1453 °C, a higher temperature than ever occurs in mid-ocean ridge magmas, this is assumed to be evidence of impact alteration. It is presently uncertain if the nickel is extraterrestrial in origin or from the mantle. Pleistocene bedrock at the base of one of the cores and the location of the magnetic susceptibility layer at the top of each core strongly suggests a Holocene age less than 6000 years for the putative impact event.

Modeling of the injection of water vapor into the upper atmosphere from the Burckle Crater event yields a maximum rainout worldwide of around 9.2 cm (Abbott et al. 2005). As previously noted this figure is far too small to account for worldwide flood mythology rainfall. However, even with several similar-sized fragments impacting other oceanic locations as part of the overall Flood Comet impact event, this would still not produce the volume of water necessary for 6 to 7 days of worldwide torrential rainfall. I suggest that the majority of the rainfall was due to ocean-fed prolonged cyclonic storm activity stimulated by atmospheric rainout and blockage of sunlight. The termination of the cyclonic storms coincided with the return to pre-impact levels of water vapor in the atmosphere.

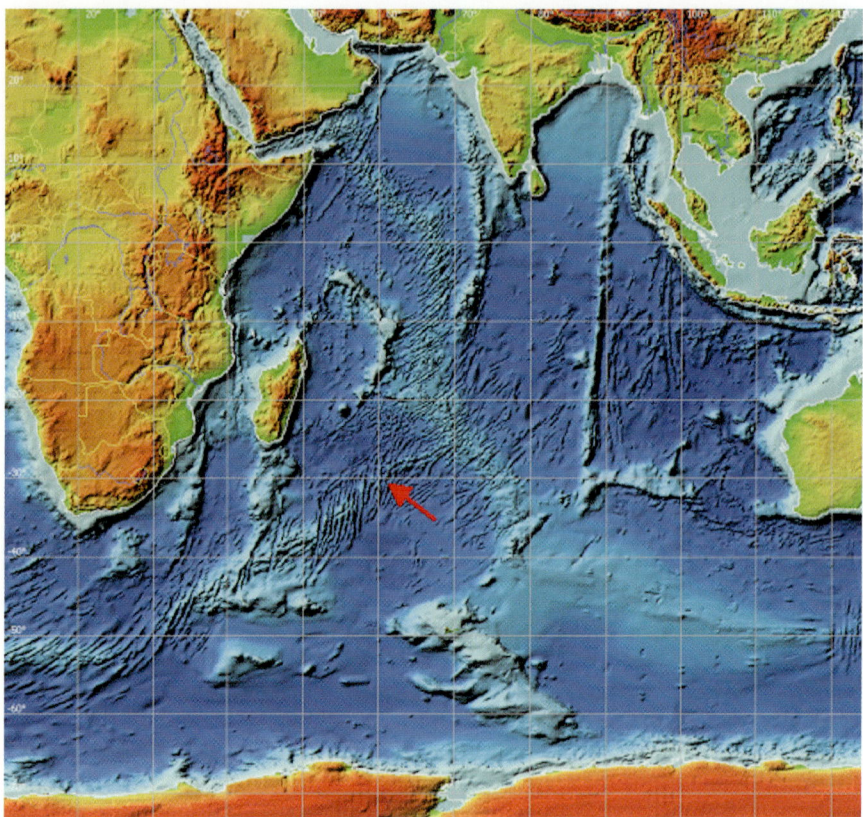

Fig. 2.3. Map of the approximate location of Burckle Crater candidate abyssal impact structure (*red arrow*) along the southeast Indian Ridge. The map is adapted from the ETOP-5 topography coverage on the *Integrated Tsunami DataBase for the Pacific* compact disc (ITDB 2004)

The location of Burckle Crater is well situated to be the source for likely Holocene mega-tsunami chevron deposits documented along the western coast of Australia (Kelletat and Scheffers 2003; Abbott et al. 2005) and for Tamil myths in which a tsunami at the time of the great flood ran inland for nearly 100 km and an elevation of 100 m only to stop at the edge of the city of Madurai in southern India (Shulman 1988). But even more compelling is the language and imagery from the Sanskrit Puranas that tell of the destruction of the world at the end of the present Kali age, but which also is paraphrased in the previously mentioned great flood myths about Manu (Dimmitt and van Buitenen 1978):

> So when Janardana in Rudra's form *[the god Visnu in the form of Siva, the destroyer god]* has consumed all creation *[with fire]*, he produces clouds from the breath of his mouth that look like a herd of elephants, emitting lightning, roaring loudly. Thus do dreadful clouds arise in the sky. Some are dark like the blossom of the blue lotus; some looking like the white water-lily; some are the color of smoke; and others are yellow. Some resemble a donkey's hue; others are like red lacquer; some have the appearance of a cat's-eye gem; and some are like sapphire. Still others are white as a conch shell

or jasmine, or similar to collyrium *[an eye lotion]*; some are like fireflies, while others resemble peacocks. Huge clouds arise resembling red or yellow arsenic, and others look like a blue-jay's wing. Some of these clouds are like fine towns, and some like mountains; others resemble houses, and still others, mounds of earth. These dense, elephantine clouds fill up the surface of the sky, roaring loudly. Pouring down rain they completely extinguish this dreadful fire which has overtaken the three worlds. And when the fire is thoroughly quenched, the clouds raining day and night overwhelm the entire world with water.

Such a description could not have been written from pure imagination. Rather, this can only be conceived as an eyewitness account of the debris plume of a cataclysmic explosion. The Sanskrit Puranas were originally written about 1000 BC, and revised during the 4th through 6th centuries AD. The contextual relation of this description to the flood and specifically to the previously noted myth of Manu and the horned, golden-colored fish suggests that the description is of the debris plume associated with the impacting Flood Comet.

2.4.2
Post-Workshop Final Thoughts

The archaeology and anthropology of cosmic impact during the Quaternary period has proven fascinating to compile and to research, but also extraordinarily complex and problematic. For example, it is clear that during the past 5 000 years – the period of recorded human history – cosmic impacts lead to significant human death and culture change. Although there can and will be debate about the scale of these effects – local, regional or global – what remains frustratingly unclear is what this record may ultimately mean toward understanding the ongoing risks of cosmic impact.

At one end of the spectrum, we have witnessed during the 20th century substantive but still local impacts such as Tunguska (1908) and Sikhote Alin (1947), each in largely uninhabited areas. The Tunguska event is typically modeled as anomalous, happening only once every few hundred or even several thousand years based on the latest modeling trends (e.g. Stuart and Binzel 2004). Such modeling seemingly fails to consider the admittedly poorly studied but very real 1930 Brazil impact (Bailey et al. 1995) and 1935 Guyana impact (Steel 1996) that also apparently exceeded one megaton in magnitude and devastated several hundred square kilometers of forest and perhaps some of its human occupants. Also, it should be remembered that the Sikhote Alin meteorite swarm impacted an area at least 4×12 km, and resulted in thousands of individual impacts, nearly 200 of which were of such size to form small craters (Gallant 2002). Had any of these four events during the 40-year period of 1908 through 1947 occurred in an urban setting, there would have been considerable property destruction and loss of life – there also would be no current need to justify research on the topic of the risks and hazards of cosmic impact.

Moving back into archaeological and anthropological time, local Holocene populations unquestionably suffered greatly from the Kaali, Campo del Cielo, Rio Cuarto (airburst), and possibly the Henbury impacts with potential regional effects (social disruption) likely for each.

There are several potential impacts toward the other end of the magnitude scale during the past 20 000 years, but for which we unfortunately are faced with preliminary data requiring varying degrees of additional research and physical validation. These include the large potential Bronze Age airburst in the Near East and the hypothesized Umm al Binni impact (ca. 10^2 MT); Iturralde in Bolivia (ca. 10^3–10^4 MT); Mahuika in the waters near New Zealand (ca. 10^4 MT); Chiemgau-Burghausen in southern Germany (ca. 10^4–10^5 MT), which may (or may not) relate to Mike Baillie's (this volume) hypothesized cometary atmospheric dust loading in the 6th century AD; Rio Cuarto, if the impact origin for the purported craters is validated (ca. 10^5–10^6 MT); the hypothesized Flood Comet impact (ca. 10^7 MT) of 2807 BC, and the putative impact associated with the Younger Dryas climate event of about 9600 BC (ca. 10^7 to 10^8 MT). There likely are other potential substantive Holocene impactor candidates that have not yet been satisfactorily identified for modeling and testing (Masse 1998). The good news is that all of these hypothesized impacts can and will be further researched and tested by all necessary physical means for eventual validation or dismissal. The length of time such research and testing will take depends partly on the degree to which the NEO community and funding agencies view this as a serious and worthwhile endeavor, and are willing to support such study.

As noted during the ICSU workshop, the validation of any one of these hypothesized larger magnitude ($> 10^3$ MT) recent events will strain existing astrophysical models of cosmic impact hazard and risk. The validation of two or more of these events, particularly if they involved comets, could not be reconciled with existing impact models. Given the nature of the information seemingly encoded in the documentary and oral historical record of humankind, and given the fact that there are several substantive widespread rapid changes in both climate and archaeological/historical culture during the past 20 000 years for which we do not yet have a satisfactory explanation, I anticipate that at least two and likely more of the hypothesized larger impacts will be validated. What this may mean in terms of the reality of assessments of the risks and hazards of cosmic impact remains to be addressed.

Virtually all past traditional knowledge keepers insisted that information about immense natural catastrophes preserved in oral traditions and myths were among the most valuable legacies that any cultural group could pass on to future generations. Perhaps they were right.

Acknowledgments

An early version of this paper was presented at the Santa Fe Institute in July 2004. I thank George Gumerman and various members of the audience, including Ted Hartwell, for comments and interest. Pete Bobrowsky, Jeff Masse and Benny Peiser provided comments on the pre-ICSU workshop draft of the present paper, and Owen Toon directed me toward pertinent literature on impact atmospheric effects. Peter Schultz provided information on his Argentine impact data. Dialog on various subjects with participants during the ICSU workshop was of great value. I particularly single out Mike Baillie, Bill Bottke, Ted Bryant, Clark Chapman, Anny-Chantal Levasseur-Regourd, Sharrad Master, Dave Morrison, Paul Slovic, and Siim Veski for information and critical comments. Discussions immediately after the ICSU workshop with Slava Gusiakov and Ted Bryant regarding oceanic impacts and satellite (radar altimeter) sea surface bathymetric data were stimulating, as have been comments and data from Dallas Abbott.

I am grateful to Slava Gusiakov for providing me with a copy of the Tsunami DataBase (ITDB 2004) compact disk. Kord Ernstson kindly provided an update on the hypothesized Chiemgau impact. Benny Peiser, Elizabeth Barber, and an anonymous reviewer provided thoughtful comments on the revised post-workshop paper. The late Alan Dundes graciously provided valuable mythology commentary in a review of the related Masse and Masse (in press) paper. None of these individuals necessarily subscribe to the opinions and conclusions presented here. Los Alamos National Laboratory assigned publication release number LA-UR-04-7204 to this study.

References

Abbott D (2005) personal communication, May 3, 2005
Abbott D, Glatz CA, Burckle L, Nunes, AA, Puchtel IS, Humayun (2003a) Multidisciplinary methods of finding and verifying abyssal impact craters: results and uncertainties. Abstract, Lunar Plan Sci XXXIV, 1858.pdf
Abbott, DH, Matzen A, Bryant EA, Pekar SF (2003b) Did a bolide impact cause catastrophic tsunamis in Australia and New Zealand? Abstract 67-7, Geol Soc Amer Annual Meeting, Seattle
Abbott D, Peckar S, and Kumar M (2004) Sand lobes on Stewart Island as probable impact-tsunami deposits. Abstract 1930.pdf, Lunar Plan Sci XXXV
Abbott, DH, Masse WB, Berger D, Burckle L, Gerard-Little P (2005) Burckle abyssal impact crater: did this impact produce a global deluge? Paper presented at Atlantis 2005 International Conf, 11–13 July 2005, Milos, Greece
Allen RH (1963) Star names: their lore and meaning (orig. 1899). Dover Publications, New York
Bailey ME, Markham DJ, Massai S, Scriven JE (1995) The 1930 August 13, Brazilian Tunguska' event. Observatory 115:250–253
Barber EW, Barber PT (2005) When they severed earth from sky: how the human mind shapes myth. Princeton, Princeton, NJ
Barrientos G, Perez SI (2005) Was there a population replacement during the Late mid-Holocene in the southeastern Pampas of Argentina? Archaeological evidence and paleoecological basis. Quat Inter 132:95–105
Basurah HM (2003) Estimating a new date for the Wabar meteorite impact. Meteor Plan Sci Supplement 38(7):A155-156
Berndt RM, Berndt CH (1994) The speaking land: myth and story in Aboriginal Australia. Inner Traditions International, Rochester VT, p 123
Bierhorst J (1988) The mythology of South America. William Morrow, New York
Bland PA, Souza Filho CR, Jull AJT, Kelley SP, Hough RM, Artemieva NA, Pierazzo E, Coniglio J, Pinotti L, Evers V, Kearsley AT (2002) A possible tektite strewn field in the Argentinian Pampa. Science 296: 1109–1111
Blong RJ (1982) The time of darkness: local legends and volcanic reality in Papua New Guinea. Washington, Seattle
Bryant EA (2001) Tsunami: the underrated hazard. Cambridge, Cambridge
Bryant EA, Walsh G, Abbott D (in press) Cosmogenic mega-tsunami in the Australia region: authenticating Aboriginal and Maori legends. In Piccardi L, Masse WB (eds) Myth and geology. Geol Soc London
Campbell J (1981) The mythic image. Princeton, Princeton, NJ
Cassidy WA, Renard ML (1996) Discovering research value in the Campo del Cielo, Argentina, research craters. Meteor Plan Sci 31:433-448
Cassidy WA, Villar LM, Bunch TE, Kohman TP, Milton DJ (1965) Meteorites and craters of Campo del Cielo, Argentina. Science 149:1055–1064
Chapman, CR, Morrisson D (1994) Impacts on the Earth by asteroids and comets: assessing the hazard. Nature 367:33–40
Chiemgau Impact Research Team (2004) Did the ancient Celts see a comet impact in 200 BC? Astronomy magazine website: *http://www.astronomy.com/default.aspx?c=a&id=2519*

Cione AL, Tonni EP, San Cristóbal J, Hernández PJ, Benítez A, Bordignon F, Perí JA (2002) Putative meteoritic craters in Río Cuarto (central Argentina) as eolian structures. Earth Moon Planets 91: 9–24

Clark C 1982) Flood. Time-Life Books, Alexandria, Virginia

Courty M-A (1998) The soil record of an exceptional event at 4000 BP in the Middle East. In: Peiser BJ, Palmer T, Bailey ME (eds) Natural catastrophes during Bronze Age civilizations: archaeological, geological, astronomical, and cultural perspectives. BAR International Series 728, Archaeopress, Oxford, pp 93–108

Courty M-A (2001) Evidence at Tell Brak for the Late EDIII/Early Akkadian Air Blast Event (4 kyr BP). In: Oates D, Oates J, McDonald H (eds) Excavations at Tell Brak. Vol. 2: Nagar in the third millennium BC. McDonald Institute for Archaeology/British School of Archaeology in Iraq, London, pp 367–372

Deutsch A, Koeberl C, Blum JD, French BM, Glass BP, Grieve R, Horn P, Jessberger EK, Kurat G, Reimold WU, Smit J, Stöffler D, Taylor SR (1994) The impact-flood connection: does it exist? Terra Nova 6: 644–650.

Dimmitt C, van Buitenen JAB (1978) Classical Hindu mythology: a reader in the Sanskrit Purânas. Temple, Philadelphia, pp 42–43, 71–74

Dundes A (ed) (1988) The flood myth. California, Berkeley

Durda DD, Kring DA (2004) Ignition threshold for impact-generated fires. Jour Geophy Res 109:E08004, pp 1–14

Dypvik H, Jansa LF (2003) Sedimentary signatures and processes during marine bolide impacts: a review. Sed Geol 161:309–337

Dypvik H, Burchell M, Claeys P (2004) Impacts into marine and icy environments – a short review. In: Dypvik H, Burchell M, Claeys P (eds) Cratering in marine environments and on ice. Springer, Heidelberg, p 1

Ernstson K (2005) personal communication, June 10, 2005

Fehr KT, Pohl J, Mayer W, Hochleitner R, Fassbinder J, Geiss E, Kerscher H (2005) A meteorite impact field in eastern Bavaria? A preliminary report. Met Plan Sci 40:187–194

Frazer JG (1919) Folk-Lore in the Old Testament, Vol 1. MacMillan, London, pp 104–361

Frederichs T, Bleil U, Gersonde R, Kuhn G (2002) Revised age of the Eltanin impact in Southern Ocean. Eos Trans. AGU, 83(47), Fall Meet Suppl, Abstract OS22C-0286

Gallant RA (2002) Meteorite hunter: the search for Siberian meteorite craters. Mc-Graw-Hill, New York

Gersonde R, Kyte FT. Bleil U, Diekmann B, Flores JA, Gohl K, Grahl G, Hagen R, Kuhn G, Sierro FJ, Völker D, Abelmann A, Bostwick JA (1997) Geological record and reconstruction of the late Pliocene impact of the Eltanin asteroid in the Southern Ocean. Nature 390:357–363

Giménez Benítez SR, López AM, Mammana LA (2000) Meteorites of Campo del Cielo: impact on the Indian culture. In: Esteban C, and Belmonte JA (eds) Oxford VI and SEAC 99: astronomy and cultural diversity. Organismo Autónomo de Museos del Cabildo de Tenerife, pp 335–341

Glatz CA, Abbott DH, Nunes, AA (2002) A possible source crater for the Eltanin impact layer. Abstract 178-7, Geol Soc Amer Annual Meeting, Denver

Gleadow R (1969) The origin of the zodiac.Atheneum, New York

Goff J, Hulme K, McFadgen B (2003) "Mystic Fires of Tamaatea": attempts to creatively rewrite New Zealand's cultural and tectonic past. Jour Royal Soc New Zealand 33:795–809.

Grego P (1998) Collision Earth! The threat from outer space: meteorite and comet impacts. Blanford, London, p 98

Graves R (1960) The Greek myths, 2 volumes, revised edition. Harmondsworth/Penguin, New York

de Grazia A, Juergens RE, Stecchini LV (1966) The Velikovsky affair. University Books, New Hyde Park, New York

Habel N (1988) The two flood stories in Genesis. In: Dundes A (ed) The flood myth. California, Berkeley, pp 13–28 (orig. 1971)

Haines PW, Howard KT, Ali JR, Burrett CF, Bunopas S (2004) Flood deposits penecontemporaneous with ~0.8 Ma tektite fall in NE Thailand: impact-induced environmental effects. Earth Plan Sci Let 225:19–28

Hodge P (1994) Meteorite craters and impact structures of the Earth. Cambridge, Cambridge

Holcomb RT (1987) Eruptive history and long-term behavior of Kilauea volcano. In: Decker RW, Wright TL, Stauffer PH (eds) Volcanism in Hawaii, Vol 1. US Geological Survey Professional Paper 1350. US Gov Print Off, Washington, DC, pp 261–350
Ions V (1968) Egyptian Mythology. Paul Hamlym, London.
ITDB (2004) Integrated Tsunami DataBase for the Pacific, version 5.11 of July 31, 2004. Intergovernmental Oceanographic Commission – U.S. National Weather Service, Pacific Region – Siberian Division Russian Academy of Sciences Institute of Computational Mathematics and Mathematical Geophysics
Jones E (1989) The story of the falling star. Aboriginal Studies Press, Canberra
Jones TP, Lim B (2000) Extraterrestrial impacts and wildfires. Palaeogeogr Palaeoclimatol Palaeoecol 164:57–66
Jones TP (2002) Reply "Extraterrestrial impacts and wildfires." Palaeogeogr Palaeoclimatol Palaeoecol 185:407–408
Kitchen KA (1991) The chronology of ancient Egypt. World Archaeo 23:201–208
Kovacs MG (1989) The epic of Gilgamesh. Stanford, Stanford CA
Kyte FT, Zhou L, Wasson JT (1988). New evidence on the size and possible effects of a late Pliocene Oceanic Asteroid Impact. Science 241:63–65
Lambert WG, Millard AR (1969) Atra-hasis: the Babylonian story of the Flood. Clarendon Press, Oxford
Langbroek M, Roebroeks W (2000) Extraterrestrial evidence on the age of the hominids from Java. Jour Human Evol 38:595–600
Legge J (1865) The annals of the bamboo books. In: The Chinese classics, vol 3, part I, Chap 4. Henry Frowde, London
Levi-Strauss C (1969) The raw and the cooked. Harper & Row, New York
Lewis JS (1996) Rain of iron and ice: the very real threat of comet and asteroid bombardment. Addison-Wesley, New York
Liberman RG, Niello F, di Tada ML, Fifield LK, Masarik J, Reedy RC (2002) Campo del Cielo iron meteorite: sample shielding and meteoroid's preatmospheric size. Meteoritics 37:295–300
Ma P, Aggrey K, Tonzola C, Schnabel C, de Nicola P, Herzog GF, Wasson JT, Glass BP, Brown L, Tera F, Middleton R, Klein J (2004) Beryllium-10 in Australasian tektites: constraints on the location of the source crater. Geo Cosmo Acta 68:3883–3896
Mallowan MEL (1964) Noah's Flood reconsidered. Iraq 26:62–82
Marcus R, Melosh HJ, Collins G (2005) Earth impact effects program. *http://www.lpl.arizona.edu/impacteffects*
Masse WB (1995) The celestial basis of civilization. Vistas in Astronomy 39:463–477
Masse WB (1998) Earth, air, fire, and water: the archaeology of Bronze Age cosmic catastrophes. In: Peiser BJ, Palmer T, Bailey ME (eds) Natural catastrophes during Bronze Age civilizations: archaeological, geological, astronomical, and cultural perspectives. BAR International Series 728, Archaeopress, Oxford, pp 53–92
Masse WB, Soklow R (2005) Black suns and dark times: the cultural response to solar eclipses in the ancient Puebloan Southwest. In: Fountain JW, Sinclair RM (eds) Current studies in archaeoastronomy: conversations across time and space. Carolina Academic Press, Durham, SC
Masse WB, Somers GF, Carter LA (1991) Waha'ula *heiau*, the regional and symbolic context of Hawai'i Island's "Red Mouth" temple. Asian Perspectives 30:19–56
Masse WB, Espenak F (2006) Sky as environment: solar eclipses and Hohokam culture change. In: Doyel DE, Dean JS (eds) Environmental change and human adaptation in the ancient Southwest. Utah, Salt Lake City, pp 228–280
Masse WB, Masse MJ (in press) Myth and catastrophic reality: using myths to identify cosmic impacts and massive plinian eruptions in Holocene South America: In: Piccardi L, Masse WB (eds) Myth and Geology. Geol Soc London, London
Masse WB, Liston J, Carucci J, Athens JS (2006) Evaluating climate, environment, resource depletion, and culture change in the Palau Islands between AD 1200 and 1600. Quat Inter national 151:106–132
Masse WB, Johnson RK, Tuggle HD (in preparation) Islands in the sky: traditional astronomy and the role of celestial phenomena in Hawaiian myth, language, religion, and chiefly power. Hawaii, Honolulu
Masse WB, Barber ET, Piccardi, L, Barber PT (in press) Exploring the nature of myth and its role in science. In: Piccardi L, Masse WB (eds) Myth and geology. Geol Soc London, London

Master S (2001) A possible Holocene impact structure in the Al 'Amarah Marshes near the Tigris-Euphrates confluence, southern Iraq. Meteor Plan Sci Suppl, 36(9):A124

Master S (2002) Umm al Binni lake, a possible Holocene impact structure in the marshes of southern Iraq: geological evidence for its age, and implications for Bronze-age Mesopotamia. Paper delivered at the conference. "Environmental Catastrophes and Recoveries in the Holocene," Brunel University, Uxbridge, UK, August 29 – September 2

Master S, Woldai T (2004). The Umm al Binni structure, in the Mesopotamian marshlands of southern Iraq, as a postulated late Holocene meteorite impact crater: geological setting and new landsat ETMt and Aster Satellite imagery. Information Circular No 382, Economic Geology Research Institute, Hugh Allsopp Laboratory, University of the Witwatersrand, Johannesburg.

Matzen AK (2003) The spatial distribution and chemical differences of tektites from a crater in the Tasman Sea. Abstract 7-10, Geol Soc Amer Annual Meeting, Seattle

Mayor A (2000) The first fossil hunters: paleontology in Greek and Roman times. Princeton, Princeton, NJ

Mazar A (1990) Archaeology of the land of the Bible: 10 000–586 BCE. Doubleday, New York

Melosh HJ, Beyer RA (2005) Crater. http://www.lpl.arizona.edu/tekton/crater.html

Melville AD (1986) Ovid: Metamorphoses. Oxford, Oxford, pp 9–10

Métraux A (1946) Myths of the Toba and Pilagá Indians of the Gran Chaco. Amer Folk Soc, Philadelphia

Morrison D, Harris AW, Sommer G, Chapman CR, Carusi A (2003) Dealing with the impact hazard. In: Bottke W, Cellino A, Paolicchi P, Binzel RP (eds) Asteroids III. Arizona, Tucson

Motz L, Nathanson C (1988) The constellations: an enthusiast's guide to the night sky. Doubleday, New York

Nunn, PD (2000) Environmental catastrophe in the Pacific Islands around AD 1300. Geoarchaeology 15: 715–740

Ormö J, Rossi AD, Komatsu G (2002) The Sirente crater field, Italy. Meteor Plan Sci 37:11

North JD (1989) Stars, minds and fate: essays in ancient and medieval cosmology. Hambeldon Press, London

Parks and Wildlife Commission of the Northern Territory (1999) Henbury Meteorites Conservation Reserve Plan of Management. Southern Regional Office, Alice Springs, Australia (http://www.nt.gvo.au/ipe/pwcnt/docs/henbury.htm)

Peiser BJ (2002) Sub-critical impacts during the Holocene. Paper delivered at the conference "Environmental Catastrophes and Recoveries in the Holocene," Brunel University, Uxbridge, UK, August 29 – September 2

Peiser BJ (2003) Climate change and civilization collapse. In Okonski K (ed) Adapt or die. Profile Books, London, pp 191–201

Peiser BJ (2004) personal communication, September 27, 2004

Peiser BJ, Palmer T, Bailey ME (eds) (1998) Natural catastrophes during Bronze Age Civilisations: archaeological, geological, astronomical, and cultural perspectives. BAR International Series 728, Archaeopress, Oxford

Petreshock K, Abbott D, Glatz C (2004) Continental impact debris in the Eltanin impact layer. Abstract, Lunar Plan Sci XXXV, 1364.pdf

Petrie WMF (1900) The Royal Tombs of the First Dynasty, Part I. Egypt Exploration Fund, London

Piccardi L, Masse WB (eds) (in press) Myth and geology. Geol Soc London, London

Pillans B (2003) Subdividing the Pleistocene using the Matuyam-Brunhes boundary (MBB): an Australasian perspective. Quat Sci Rev 22:1569–1577

Pingree D (1972) Jour. Hist. Astronom 3:27–35

Porada E, Hansen DP, Dunham S, Babcock SH (1992) The chronology of Mesopotamia, ca. 7000–1600 BC. In Ehrich RW (ed) Chronologies in Old World Archaeology. Chicago, Chicago, pp 77–121

Postgate JN (1992) Early Mesopotamia. Routledge, London

Prescott JR, Robertson GB, Shoemaker C, Shoemaker EM, Wynn J (2004) Luminescence dating of the Wabar meteorite craters, Saudi Arabia. J Geophy Res 109, no. E01, p E01008

Pritchard JB (ed)(1969) Ancient Near Eastern texts relating to the Old Testament, 3rd edn. Princeton

Pyne SJ, Andrews PL, Laven RD (1996) Introduction to wildland fire, 2nd edn. Wiley, New York

Rappenglück MA, Ernston K, Mayer W, Beer R, Benske G, Siegl C. Sporn R. Bliemetsrieder T. Schüssler U (2004) The Chiemgau impact event in the Celtic period: evidence of a crater strewnfield and a cometary impactor containing presolar matter. Website http://www.impact-structures.com/chiemgau/

Raukas A (2000) Investigation of impact spherules – a new promising method for the correlation of Quaternary deposits. Quat Inter 68(71):241–252

Raukas A, Tiirmaa R, Kaup E, Kimmel K (2001) The age of the Ilumetsa mteorite craters in southeast Estonia. Meteor Plan Sci 36:1507–1514

Ryan W, Pittman W III (1998) Noah's Flood. Simon and Schuster, New York

Salby ML (1996) Fundamentals of atmospheric physics. Academic Press, New York

Santilli R, Ormö J, Rossi AP, Komatsu G (2003) A catastrophe remembered: a meteorite impact of the 5th century AD in the Abruzzo, Central Italy. Antiquity77:313–320

Schultz PH, Lianza RE (1992) Recent grazing impacts on the Earth recorded in the Rio Cuarto crater field, Argentina. Nature 355:232–237

Schultz PH, Zárate M, Hames B, Koeberl C, Bunch T, Storzer D, Renne P, Wittke J (2004) The Quaternary impact record from the Pampas, Argentina. Earth Plan Sci Let 219:221–238

Searle R (2000) Plate tectonicsm principles. In: Hancock PL, Skinner BJ (eds) The Oxford companion to the Earth. Oxford, London, pp 827–832

Shulman D (1988) The Tamil flood myths and the Cankam legend. In: Dunde A (ed) (1988) The flood myth. California, Berkeley, pp 293–317

Smith WR (1930) Myths and legends of the Australian aboriginals. George G Harrap, London, pp 35–37

Smith WHF, Sandwell DT (1997) Global sea floor topography from satellite altimetry and ship depth soundings. Science 277:

Speranza F, Sagnotti L, Rochette P (2004) An anthropogenic origin for the "Sirente crater," Abruzzi, Italy. Meteor Plan Sci 39:635–649

Steel D (1996) A Tunguska event in British Guyana in 1935? Meteorite! February

Steel D, Snow P (1992) The Tapanui region of New Zealand: site of a "Tunguska" around 800 years ago? In Harris A, Bowell E (eds) Asteroids, comets, meteors 1991. Lunar and Planetary Institute, Houston, pp 569–572

Stuart JS, Binzel RP (2004) Bias-corrected population, size-distribution, and impact hazard for the near-Earth objects. Icarus 170:295–311

Svetsov VV (2002) Comment on "Extraterrestrial impacts and wildfires." Paleogeo Paleoclim Paleoecol 185:403–405

Taylor RE (1997) Radiocarbon daing. In: Taylor RE, Aitken MJ (eds) Chronometric Dating in Archaeology. Plenum Press, New York, pp. 65–96

Teller JT, Glennie KW, Lancaster N, Singhvi AK (2000) Calcareous dunes of the United Arab Emirates and Noah's flood: the postglacial reflooding of the Persian (Arabian) Gulf Quat Int 68-71:297–308.

Tiirma R, Czegka W (1996) The Kaali-crater field at Saaremaa (Osel), Estonia; geological investigations since 1827 and future perspectives. Meteor Plan Sci 31, A12-143

Tollmann-Kristin E, Tollman A (1994) The youngest big impact on Earth deduced from geological and historical evidence. Terra Nova 6:209–217

Toon OB, Zahnle K, Turco RP, Covey C (1994) Environmental perturbations caused by asteroid impacts. In: Gehrels T (ed) Hazards due to comets and asteroids. Arizona, Tucson, pp 791–826

Toon OB, Zahnle K, Morrison D, Turco RP, Covey C (1997) Environmental perturbations caused by the impacts of asteroids and comets. Rev Geophy 35:41–78

University of Helsinki (2005) Department of Physical Sciences (*http://wwwgeophysics.helsinki.fi/tutkimus/impacts/maps.html*)

University of New Brunswick (2005) Earth impact database. Planetary and Space Science Centre, Department of Geology. *http://www.unb.ca/passc/ImpactDatabase*

Vansina J (1985) Oral tradition as history. Wisconsin, Madison

Verbrugghe GP, Wickersham JM (1996) Berossos and Manetho: native traditions in ancient Mesopotamia and Egypt. Michigan, Ann Arbor

Veski S, Heinsalu A, Kirsimäe K, Poska A, Saarse L (2001) Ecological catastrophe in connection with the impact of the Kaali meteorite about 800–400 BC on the island of Saaremaa, Estonia. Meteor Plan Sci 36:1367–1375

Veski S, Heinsalu A, Lang V, Kestlane Ü, Possnert G (2004) The age of the Kaali meteorite craters and the effect of the impact on the environment and man: evidence from inside Kaali craters, island of Saaremaa, Estonia. Veget Hist Archaeobot 13:197–206

Vitaliano DB (1973) Legends of the earth: their geologic origins. Citadel Press, Secaucus, NJ
Walters D (1992) Chinese mythology: an encyclopedia of myth and legend. Aquarius Press, London
Wasilewski P, Kletetschka G, Tucker C, Killeen T (2003). Iturralde: a possible impact structure at the edge of the Amazon forest in northern Bolivia. Abstract, IUGG General Assembly, 30 June to 11 July, Sapporo, Japan
Wasson JT (2003) Large aerial bursts: an important class of terrestrial accretionary events. Astobiology 3:163–179
Weigall A (1925) A history of the Pharaohs: the first eleven dynasties. Thorton Butterworth Ltd, London
Weiss H, Courty M-A, Wetterstrom W, Meadow R, Guichard F, Senior L, Curnow A (1993) The origin and the collapse of Third Millennium north Mesopotamian civilization. Science 261:995-1004
Wilkinson TAH (1999) Early Dynastic Egypt. Routledge, London
Wilbert J, Simoneau K (1992) Folk literature of South American Indians: general index. UCLA Latin America Center Publications, Universitity of California, Los Angeles
Wynn JC, EM Shoemaker (1998) The day the sands caught fire. Scientific American, November, pp 64–71
Zimmer H (1946) Myths and symbols in Indian art and civilization. Princeton University Press, Princeton

Chapter 3

The Sky on the Ground: Celestial Objects and Events in Archaeology and Popular Culture

William T. Hartwell

3.1 Introduction

The celestial environment has always played a significant role in the shaping of human culture. Written records spanning thousands of years are replete with examples of the importance of the celestial constants (e.g. the Sun, moon, stars, planets) in the basic ideologies and the everyday lives of peoples around the world. Of equal or greater importance are transient celestial phenomena (e.g. eclipses, meteor storms, asteroids, comets). Because of the infrequency, unpredictability, and often fantastic manifestations that are presented by these transient events, they have been viewed as having much greater import than the much more predictable celestial constants.

Most prehistoric societies likely noted the correlation between changes in the position of the Sun, moon, and stars and cyclical seasonal changes in their environments. It is a small step from this realization to the perception that objects viewed in the sky necessarily exert some control over events occurring on the ground and, by extension, the people that live upon Earth. It is likely that this concept was a precursor to early forms of astrology. Although it is difficult to know with certainty exactly what most pre-literate societies thought celestial objects were, in those cases where documentation exists of initial historical contact with these societies it is clear that they often had richly detailed descriptions of the nature of these celestial objects and phenomena, frequently equating them to supernatural beings. The predictable nature of many celestial objects was already described at least 5000 years ago by Sumerian priests and the adoption of a systematic astrology by the Greeks by 2500 years ago was well suited to their concept of the planets and stars as divine entities (Malefijt 1968, p 217).

In order to put the near-Earth object issue in a cultural perspective, this paper will present brief examples of representations of celestial objects and events in the archaeological and ethnographic record, and then discuss their appearance in the popular culture of modern society, with specific attention to astrology and the programing of meteors, asteroids, and comets in cinematic film, video, and television productions. Finally, it will address the issue of garnering public support for near-Earth object initiatives, discussing obstacles that scientists face in increasing public awareness of the validity of the issue, and suggesting ways in which scientists can use popular cultural expressions of real-world events as educational targets of opportunity.

3.2
The Archaeological Record

The great importance of the sky and objects in it to past cultures is revealed not only in the attention they are given in written histories; it can also be seen in abundance in the archaeological record. In the structural orientation of and embellishments on architectural remains, in designs on utilitarian artifacts such as pottery, in ritual and artistic renderings on rock surfaces, and even in the oral history of specific cultural groups can be found unmistakable references to both constant celestial events and specific transient celestial phenomena.

3.2.1
Architecture

Architectural remains often offer glimpses into the working knowledge of past cultures regarding celestial objects. Displaying a range of variability in primary functionality, but often linked closely to religious or ceremonial practices, many structures and alignments built in antiquity have been shown to have orientations related to celestially significant directions. Thousands of megalithic structures that were built over a 2000-year period across western Europe beginning about 6300 years ago consisting of enormous stone-lined, chambered tombs, monuments, and circular stone rings built of large individual standing stones (Stonehenge, for example) often had orientations relating to summer and winter solstices or to other Sun and moon-related positions (Hawkins 1965; Hester and Grady 1982, pp 299–306; Mackie 1997). Other examples of structures that can show preferences for celestial orientations include Iron Age Scottish brochs, the long axis of Christian churches, Maya stone buildings, and non-circular and symmetrical stone rings (Mackie 1997). In addition to displaying significant orientations, major edifices, such as the Pyramids of the Sun and Moon in the Aztec city of Tenochtitlan in Mexico, were often built in honor of and for performing rituals related to deities associated with various celestial objects.

3.2.2
Artifacts and Rock Art

Artifacts, both utilitarian and ritual in nature, can aid in understanding the astronomical sophistication of prehistoric (pre-literate) and early historic cultures, as well as the importance with which they viewed celestial objects and events. Artifacts bearing ancient calendrical systems relying on solar and lunar observations are of great interest, although generally limited in archaeology since calendars are associated with a level of cultural attainment usually found only with literate societies, and the great majority of cultures studied by archaeologists are prehistoric. Examples of ancient civilizations with well-established calendrical systems include Greece, Rome, Egypt, Carthage, Mesopotamia, and the Maya of the Yucatan (Hester and Grady 1982, pp 53–54). The most useful artifactual information of literate societies often comes, not surprisingly, from its written records. Mayan Codices dating to 1200 to 1300 years ago are greatly concerned with celestial cycles and their relationship to other cycles of cultural interest, often concentrating on aspects of specific celestial bodies (Bricker et al. 2001). Some of China's earliest written records, inscriptions on bone artifacts dating to the Shang Dynasty (1554–1046 BC), include observations of various celestial phenomena (Xu et al. 2000). Simple tables of the length of daylight and the rising and setting of various constellations occur on Babylonian tablets dating about 5000 years ago. Later tablets known as *astronomical diaries* and dating to between 380 and 40 BC have detailed daily mathematical accounts of lunar and planetary observations, and their relationships to current events. Also included in these tablets are references to transient celestial events, including the return of comet Halley in 164 BC, the only historical record that survives of this appearance of the comet (Walker 1985).

The artifacts of prehistoric societies (those without a written language) are often more difficult to interpret, but examples of depictions of celestial objects, including many which likely relate to specific transient events, are sometimes identifiable. The cultural importance that these objects held to prehistoric societies is expressed in the treatment that several Native American groups bestowed upon meteorites. Masse and Espenak (in press) cite several examples of this, including their inclusion in medicine bundles, ritual interment, and their association with various deities. Some imagery on pottery and in rock art associated with Southwestern United States puebloan groups is highly suggestive of stylized solar eclipses (e.g. Fig. 7, Masse and Soklow, in 2005). Masse and Soklow have also shown that major cultural changes identifiable in the archaeological record during a 1300-year period appear to coincide closely with the occurrence of total solar eclipses in the region. Other Southwestern rock art, such as the well-known *Sun Dagger* petroglyph in Chaco Canyon, New Mexico, was apparently designed to identify summer and winter solstices, as well as vernal and autumnal equinoxes (Sofaer et al. 1979). Still other rock art in this area is believed to depict the supernova of 1054 and the subsequent appearance of comet Halley a few years later.

3.2.3
Oral Tradition

Most anthropologists are trained to believe that while the oral histories (stories, songs, and chants) and mythology of various cultures are very important in reflecting, maintaining, and transmitting social and cultural values, the narratives themselves are largely symbolic and the characters and events contained within them are not representative of actual individuals or events (Malefijt 1968; Masse and Espenak 2006). However, recent ethnographic and archaeological research demonstrates that oral traditions of societies that lack a written language can encode accurate accounts of specific transient celestial events that occurred at least 1000 years ago and possibly in the much more distant past. Masse (in press) presents evidence from Hawai'i that details of unmistakable, independently verifiable volcanic eruptions and transient celestial events, including eclipses and comet appearances, have been accurately transmitted through oral genealogies for at least 95 generations.

Over the past 100 000 years, humans worldwide have undoubtedly witnessed numerous visually spectacular cosmic impact or airburst events, many of which may have had significant regional, and possibly global effects. It has been proposed that oral traditions can assist in identifying cosmic impacts on the Earth in antiquity, especially those that had globally catastrophic (e.g., Masse and Masse, in press; Chapter 2, this publication) or regionally catastrophic (e.g., Bryant 2001) effects. However, in a critique of a portion of Bryant (2001), Goff et al. (2003) offer a cautionary tale on the importance of carefully considering alternative explanations for the causes of such catastrophic events (i.e. terrestrial geologic processes) and it is clear that oral traditions need to be examined on a case-by-case basis and enjoy the benefit of significant independent supporting evidence if they are to be regarded seriously by the scientific community at large.

3.3
Celestial Objects in Popular Culture

Increasingly, human populations in the Western and developing world are concentrated within and around highly urbanized areas. Significant light pollution in these areas, technological advances that have allowed individuals to control their immediate environment as never before, and access to copious amounts of information without ever having to leave one's home all have contributed to a general desensitization to the celestial environment. Although Western societies may no longer view the skies as having nearly as much direct impact on their lives as do pre-industrial peoples, popular culture in Western society is rife with examples of both constant and transient celestial symbols that linger on to remind us of the greater importance that they once held to our ancestors.

The term *popular culture* in this context refers (after Schechner 1997) to cultural attitudes and values as expressed through the artifacts and performance media of a culture. Artifacts in popular culture may include either general utilitarian or ceremonial objects imbued with meaning through decoration, their morphology, or naming; they may also include popular examples of the written word, as expressed through published works in printed or electronic form, and through the news media. Popular culture as expressed through performance may include live renditions, as with theater, storytelling, rituals, musical performance, or any recorded variants of the same, including cinematic film, recorded video, television, and music.

3.3.1
Astrology in Popular Culture

Astrology refers to the study of the relative positions and movement of various actual and construed celestial bodies in the belief that they have a direct deterministic effect on the course of human lives and events. While based chiefly on the predictable positions and movements of the celestial constants of the Sun, moon, stars, and planets as seen at the time and place of a birth or other event being studied, transient events and objects such as comets have been regarded by astrologers as particularly significant occurrences or portents that could be interpreted based on their convergence with, or proximity to, zodiac and prominent constellations (Schechner 1997, p 53).

Common varieties of astrology include Western, Jyotish, Chinese and Kabbalistic. Although all of these forms of astrology are predicated on the belief that the movement and relationships of various celestial bodies have a direct effect on human lives and events, the manner in which they arrive at an understanding or prediction of these events can differ substantially. For example, astrology may employ the use of sidereal time or tropical time, or combinations thereof; recognized constellations are usually quite different from one another; whereas celestial bodies in one astrology may be associated with deities, they will not be in another, or this attribute may change as astrology evolves.

3.3.1.1
Astrology, Medicine, and Religion

The remainder of this discussion focuses principally on Western astrology, but it is worth noting that all astrologies are intimately connected historically with both astronomy and religion, and often with medicine. Health was widely regarded as being influenced by the stars, as was the functioning of specific bodily organs such as the kidney (Ziegler 2002). The word *influenza* comes from the Medieval Latin *influencia*, thought to be a fluid or emanation given off by certain stars that controlled human affairs (Morris 1996). In Medieval and Renaissance Europe, practitioners of Islam, Judaism, and Christianity alike studied celestial relationships in the belief that it would better help them understand what God permitted to be known of the divine plan, and leading astronomers of this time were greatly involved in its study (Quinlan-McGrath 2001). Often, the dedication of important edifices, especially those related to the church, involved the selection of a specific date based on advantageous celestial relationships. There is evidence to suggest, however, that horoscopes would often be adjusted, or rectified, to match up more closely with the astrologer's personal viewpoints, as seems to be the case with the foundation horoscope for St. Peter's Basilica in Rome in 1506 (Quinlan-McGrath 2001). Later, these astrological ideologies were transported to the New World with the colonists, where they flourished both inside and outside the theological realm (Butler 1979).

3.3.1.2
Perseverance of Astrology in Modern Popular Culture

The distribution of astrology in Western culture via mass media has a long history that extends back at least as far as 17th century England (Capp 1979). Although the pseudoscientific nature of astrology cannot be denied (at least, by scientists!), neither can it be denied that astrology is also big business and pervasive in modern popular culture. Astrological advice is available in a wide variety of forms and formats. Printed and electronic newspapers, magazines, pamphlets, and face-to-face or over-the-phone horoscopes are available to anyone at virtually anytime, and the nature, extent, and complexity of the advice is limited only by the size of one's pocketbook. The majority of those who read their daily horoscope (astrological forecast) in the morning newspaper may do so only for entertainment purposes, but a significant minority treat the subject with some seriousness. There is even evidence suggesting that the recent past-President of the United States Ronald Reagan and his wife Nancy frequently consulted a professional astrologer with regards to day-to-day decisions carried out in the White House (Regan 1998). In some cases, modern popular perceptions of the efficacy of astrology result in measurable real-world consequences. For example, Yip et al. (2002)

cite the influence of the Chinese zodiac on fertility rates in Hong Kong during the *Dragon* years of 1988 and 2000. In other, somewhat more dubious cases, attempts are made to show that real-world situations are caused by actual astrological relationships (Verhulst 2000).

Horoscopes in popular publications may vary widely, and may even be contradictory across various media for a given day. Research on the content of horoscopes in popular publications have shown that they are often tailored toward consumers' socioeconomic status, and may also be responsible for reinforcing traditional gender roles and class-appropriate world views (Evans 1996). Evans also notes that from an anthropological viewpoint, they are also fulfilling some of the same social functions as religious doctrine in encouraging people in various social classes to understand their social position as part of a divine plan.

3.3.2
Art and Literature

3.3.2.1
Paintings and Printed Images

Revolutionary advances in both astronomy and naturalistic painting during the Renaissance led to an artistic interest in creating realistic depictions on canvas of various celestial objects. The convincing representations of the celestial phenomena, which were based for the most part on personal observations of these phenomena, resulted in the paintings themselves becoming more convincing to their audience (Olson and Pasachoff 2002). Prominent among the objects represented were the transient phenomena of comets, meteors and eclipses. The dominant theme of the paintings was almost always a religious one (e.g. the depiction of the Bethlehem Star as a comet, heralding the birth of Jesus), reflecting the continued association of astronomy, astrology and the church, but the placement of the paintings within a common public venue put them firmly within the realm of popular culture.

Although the popular culture of early modern Europe was primarily an oral one, a popular market for print culture had already been created in England and France by the 16th and 17th centuries (Schechner 1997, p 10). Despite the relatively high illiteracy rate, street peddlers carried chapbook, broadsheets, and prints, most of which combined text with images. Additionally, as reading was not a silent affair, the oral and printed popular cultures interacted to a great extent, allowing greater diffusion into the general populace of that which was expressed in the printed texts (Schechner 1997, p 11).

Transient celestial phenomena, particularly comets, were widely represented in printed images associated with popular culture well into the 19th century, and usually recalled the terror inspired by them for millennia. In Schechner's (1997) comprehensive study of comets and popular culture, she notes that comets, while they could occasionally herald good fortune, more often were considered to be harbingers of "war, famine, plague, ill-luck, the downfall of kings, universal suffering, and the end of the world." These ideas about comets have persisted into present-day popular culture, as discussed below.

"All I'm saying is <u>now</u> is the time to develop the technology to deflect an asteroid."

Fig. 3.1. Cartoons are a form of printed popular culture that is sometimes used to express a particular viewpoint on a current issue through the use of visual humor, as in the case of this 1998 cartoon advocating development of a technology for asteroid deflection. Used with permission, © The New Yorker Collection 1998 Frank Cotham from *cartoonbank.com*. All Rights Reserved

Cartoons are another form of printed media worthy of mention. Owing much of their origin to the broadsides of the Renaissance, cartoons can convey expressions of political and social satire that are easily understood and enjoyed by literate and non-literate segments of society alike. They enjoyed widespread use beginning in the latter half of the 19th century, and continue to provide a means of expressing opinions through visual humor (Fig. 3.1).

3.3.2.2
Modern Literature

As literacy rates have increased dramatically through the early and middle part of the 20th century, literature as a transmitter and reflection of popular culture has become more significant. The importance of celestial events to religious ideologies both great and small has remained, and is reflected in their major ideological works. Using the Bible as an example, it is possible to identify many occurrences within this tome that arguably relate to transient celestial and geophysical events. The idea that some of these events may be correlative to scientifically verifiable data for purposes of cross-dating histories described therein (e.g. Ben-Menahem 1992) is not a new one but, as with oral traditions, caution is urged with regards to drawing inferences outside the realm of the astronomical events themselves. As through antiquity, most transient celestial phenomena described in the religious literature portend ill things; for example, the Star Wormwood, heralding the Apocalypse, and often represented as a comet in images (cf. Fig. 15 in Schechner 1997).

The genre of science fiction has its modern literary origins in works such as those by Jules Verne, H.G. Wells, and Edgar Allen Poe. Although many of the elements of stories and novels by Wells and Poe especially fall more into the category of what most would today refer to as *fantasy*, the seeds were planted for works to come which would strive to incorporate recognized and theoretical scientific principals. Although literary science fiction works dealing with comets and asteroids as a central theme are too numerous to discuss here, there are some significant examples related directly to the issue at hand that are worthy of mention. They include Niven and Pournelle's (1977) *Lucifer's Hammer*, detailing the cataclysm and aftermath that follow the impact of a comet on the Earth, Arthur C. Clarke's (1993) *Hammer of God*, which concerns the discovery of a large asteroid on a collision course with Earth and the attempt to divert it and, in an original twist on the impact theme, McDevitt's (1999) *Moonfall*, wherein a large comet impacts the far side of the Moon, destroying it and sending large fragments hurtling towards the Earth. These particular novels are important in that they present fictional impact scenarios that incorporate sound scientific principals in presenting the story to a popular audience.

Finally, there are numerous popular journals and magazines targeted at laypersons that have a general interest in the sciences. These include such publications as *Popular Science, Geotimes, Discover Magazine, Scientific American,* and *New Scientist*, to name only a few. In addition, traditional newspaper media shape much of the popular view of scientific issues, including the near-Earth object problem. This topic is discussed later in this paper.

3.3.2.3
Song

Modern popular songs retain abundant references in their lyrics to celestial objects and phenomena that primarily function as metaphors. However, the way in which they are used points to the power that these objects once held. Often, there is anthropomorphizing of the object or event (e.g. *You are my sun, my moon* and *a total eclipse of the heart*), and sometimes lyrics relate directly to cultural superstition (e.g. *When you wish upon a star*). Alternatively, in a weak parallel to the encoding of significant celestial events in oral tradition, songs may commemorate specific events. There were at least three songs written concerning the impact of comet Shoemaker-Levy 9 into Jupiter in 1994, at least ten about comet Hale-Bopp (many of them related to the Heaven's Gate cult, discussed below) and dozens with references to comet Halley. Even asteroid

2004 FH, which flew by Earth in March of 2004, and 4179 Toutatis, which approached within 1.5 Mio km of the Earth on September 29, 2004 have been immortalized in music and are both available for free download on the World Wide Web.

3.3.2.4
Cinema, Video, and Television

Although literary science fiction influences popular perception and understanding of a significant niche group, programming in the form of cinema, video, and television reaches a larger audience than any of the popular media previously discussed, especially when commercial advertising and ancillary programing is taken into account. It may also be a more accurate reflection of general popular cultural views associated with potential cosmic impactors such as meteors and meteorites, asteroids and comets. In order to explore how popular views of these objects have been expressed in and influenced by these media through time, this study identified cinematic films, videos, and television programs in which one or more of these objects were a central theme of the program. The Internet Movie Database (*http://www.imdb.com/*) was the source of the majority of the data displayed in Table 3.1.

In order to be considered for inclusion in the table, a program had to have one of the aforementioned objects as a central part of the plot. In other words, it was not sufficient for the word *comet* to appear simply in the title, or for a meteor shower to occur simply as a background device in the program. The table listings are arranged in chronological order by year of release. Data in the table includes the program title, the program type, the country of origin, the year of release, the object type (as listed in the plot summary), and the program subgenre. The data in the subgenre column refer to how the object of interest in the film is treated. For example, the category *Horror* is used variously to indicate that the object brings an alien life-form to Earth that terrorizes the film's protagonists or also, for example, that the object itself has deleterious effects on the life-forms that encounter it. A total of 90 programs were considered, representing 19 countries and spanning the years 1936 through 2004. While this list is not exhaustive, it is believed to be a representative sample of general trends of the attributes discussed.

Definite trends are observable with regards to both object types as they relate to subgenres and with subgenres as they relate to the year of release. Programs dealing with meteorites and meteor showers depict them almost exclusively as objects of horror. Often associated with strange mutating radiations or energies, they frequently transform the hapless individuals they come in contact with into monsters. Meteors also are associated most often with the horror subgenre as well, and include the introduction of deadly alien life-forms to Earth as well as the other effects previously discussed. However, about 30% of the time they are associated with impact hazards. Both asteroids and comets are associated more often with impact hazards, but comets slightly less often (about 50% of the time), compared to asteroids (about 61% of programs). Another significant feature of asteroids is that they are often depicted in functional capacities as habitats or as natural resources to be mined (30%).

Probably the most striking trend has to do with changes in perception and treatment of the various potential cosmic impactors through time. From 1936 through 1993,

Table 3.1. Asteroids, comets, and meteors in cinema, television, and recorded video[a]

Program title	Program type	Country of origin	Year	Object type	Program subgenre
The Invisible Ray	Feature film	USA	1936	Meteorite	Horror
The Phantom Creeps	Feature film	USA	1939	Meteorite	Horror
The Magnetic Telescope	Animated short	USA	1942	Comet	Impact
When Worlds Collide	Feature film	USA	1951	Planetoid	Impact
It Came from Outer Space	Feature film	USA	1953	Meteor	Horror
Riders to the Stars	Feature film	USA	1954	Meteor	Research
Uchûjin Tokyo ni arawaru	Feature film	Japan	1956	Meteor	Impact
Kronos	Feature film	USA	1957	Asteroid	Horror
The Monolith Monsters	Feature film	USA	1957	Meteor	Horror
Quartermass 2	Feature film	UK	1957	Asteroid	Habitat
Teenage Monster	Feature film	USA	1958	Meteor	Horror
The Blob	Feature film	USA	1958	Meteor	Horror
Caltiki – il mostro immortale	Feature film	Italy	1959	Comet	Horror
Rymdinvasion i Lappland	Feature film	USA/Sweden	1959	meteor	Horror
Der Schweigende Stern	Feature film	E. Germany/Poland	1959	Meteor	Horror
First Man Into Space	Feature film	UK	1959	Meteor dust	Horror
Valley of the Dragons	Feature film	USA	1961	Comet	Adventure
El Barón del terror	Feature film	Mexico	1962	Comet	Horror
Yosei Gorasu	Feature film	Japan	1962	Meteor	Impact
Day of the Triffids	Feature film	UK	1962	Meteor storm	Horror
San daikaijû: Chikyu sadai no kessen	Feature film	Japan	1964	Meteor	Horror
Die, Monster, Die!	Feature film	UK/USA	1965	Meteorite	Horror
The Green Slime	Feature film	USA/Japan/Italy	1968	Asteroid	Horror
Moon Zero Two	Feature film	UK	1969	Asteroid	Mining
Na komete	Feature film	Czechoslovakia	1970	Comet	Adventure
City Beneath the Sea	TV movie	USA	1971	Asteroid	Impact
Killdozer	TV movie	USA	1974	Meteorite	Horror
Track of the Moon Beast	Feature film	USA	1976	Meteorite	Horror
The Crater Lake Monster	Feature film	USA	1977	Meteor	Horror
The Day It Came to Earth	Feature film	USA	1979	Meteor	Horror
Meteor	Feature film	USA	1979	Comet/meteor	Impact
Alien Dead	Feature film	USA	1980	Meteor	Horror
Saturn 3	Feature film	UK	1980	Asteroid	Habitat
Day of the Triffids (remake)	TV mini-series	UK	1981	Comet	Horror
Creepshow	Feature film	USA	1982	Meteor	Horror
Return of the Aliens: The Deadly Spawn	Feature film	USA	1983	Meteorite	Horror
Návstevníci	TV series	France/Czechoslovakia/ Switzerland/W. Germany	1983	Comet	Comedy-impact
Night of the Comet	Feature film	USA	1984	Comet	Horror
Lifeforce	Feature film	UK	1985	Comet	Horror
The Adventures of Mark Twain	Animated film	USA	1986	Comet	Fantasy
Chi, Cometa	Feature film	Brazil	1986	Comet	Comedy
Maximum Overdrive	Feature film	USA	1986	Comet	Horror
The Curse	Independ. film	USA	1987	Meteorite	Horror
Kidô senshi Gandamu: Gyakushû no Shaa	Animated film	Japan	1988	Meteor	Impact
The Arrival	Feature film	USA	1990	Meteor	Horror
Jetsons: The Movie	Animated film	USA	1990	Asteroid	Comedy-mining

Table 3.1. *Continued*[a]

Program title	Program type	Country of origin	Year	Object type	Program subgenre
Batoru garu	Feature film	Japan	1992	Meteor	Horror
The Meteor Man	Feature film	USA	1993	Meteor	Comedy
Without Warning	TV movie	USA	1994	Asteroids	Impact
Zombie Holocaust	Video movie	USA	1995	Asteroid	Impact/horror
The Doomsday Asteroid	TV documentary	USA	1995	Asteroid	Impact
Caged Heat 3000	Feature film	USA	1995	Asteroid	Habitat
Alien Force	Video movie	USA	1996	Meteor	Horror
I Magi randagi	Feature film	Italy	1996	Comet	Comedy
La Lengua asesina	Feature film	UK/Spain	1996	Meteorite	Horror-comedy
Gamera 2: Region shurai	Feature film	Japan	1996	Meteor	Horror
The Tomorrow Man	TV movie	USA	1996	Comet	Impact
Chicxulub la huella de un gigante	Video docum.	Mexico	1996	Meteor	Impact
3 Minutes to Impact	TV documentary	UK	1996	Comets/asteroids	Impact
Inju daikessen	Animate video	Japan	1997	Meteorite	Horror
Star Kid	Feature film	USA	1997	Meteor	Fantasy
Fire from the Sky	TV documentary	USA	1997	Asteroid/comet	Impact
Doomsday Rock	TV movie	USA	1997	Comet	Impact
Asteroids: Deadly Impact	TV documentary	USA	1997	Comets/asteroids	Impact
Asteroid	TV movie	USA	1997	Asteroid	Impact
Cosmic Traveler Series (3)	TV documentary	USA	1997	Comets/asteroids	Impact
Deep Impact	Feature film	USA	1998	Asteroid	Impact
Falling Fire	TV movie	Canada/USA	1998	Asteroid	Impact
Armageddon	Feature film	USA	1998	Asteroid	Impact
Mosura 3	Feature film	Japan	1998	Meteorite	Horror
Pete's Meteor	Feature film	Ireland	1998	Meteorite	Fantasy
Breeders	Feature film	UK	1998	Meteorite	Horror
Il Quarto re	TV movie	Italy/Germany	1998	Comet	Fantasy
Judgment Day	Video movie	USA	1999	Meteor	Impact
Murdercycle	Feature film	USA	1999	Meteor	Horror
The Last Train	Short film	USA	1999	Meteor	Impact
The Atomic Space Bug	Video movie	USA	1999	Meteorite	Horror
Tycus	Video movie	USA	2000	Comet	Impact
Supernova	Indepen. film	USA/Switzerland	2000	Comet	Mining
Dinosaur	Animated film	USA	2000	Comet	Impact
Evolution	Feature film	USA	2001	Meteorite	Comedy/horror
Pearl Harbor II: Pearlmageddon	Short film	USA	2001	Meteor	Comedy/impact
Imp, Inc.	Animated short	USA	2001	Asteroid	Habitat
Between the Moon and Montevideo	Feature film	Canada	2001	Asteroid	Habitat
Triumph of the Beasts	TV documentary	UK	2001	Asteroid	Impact
Zombie Campout	Video movie	USA	2002	Meteor	Horror
Undead	Feature film	Australia	2003	Meteorites	Horror
Asteroid! The Doomsday Rock	TV documentary	Canada	2003	Asteroids	Impact
P.I.: Post Impact	Feature film	Germany/USA	2004	Comet	Impact
Retrograde	Feature film	USA/Luxembourg	2004	Meteors	Horror

[a] The source for much of the data contained in this table is the Internet Movie Database, which can be found at http://www.imdb.com/.

a span of 57 years, a total of 48 programs were produced, with only 8 (17%) of them considering impact as the principal hazard presented by these objects. In the 10 years from 1994 to 2004, at least 42 additional programs were produced, with 22 (52%) of them emphasizing the impact hazard. The great increase both in number of programs produced and in the percentage devoted to the impact scenario is believed to be a direct result of the 1994 impact of comet Shoemaker-Levy 9 on Jupiter. A similar increase in comet-oriented programing can be observed through the mid-1980's, during the return of comet Halley.

Clearly there has been a change in popular cultural perception of the hazards of potential cosmic impactors, as evidenced by changes in the representations of these objects through time in cinematic film, video, and television. Early program representations of these objects continued to reflect traditionally held views of meteors, asteroids, and comets and discussed previously. Observable real-world events such as the Shoemaker-Levy 9 impact, coupled with resultant increases in public interest and, also importantly, contributions of information and explanation from knowledgeable scientists can have significant impact on popular culture, and the attitudes it reflects.

3.3.3
Other Examples

Other examples of celestial objects and events in modern popular culture abound. In a fairly obvious example of the symbolic power they maintain in the human psyche, sports teams of all types are often named for them (e.g. the Suns, the Comets, the Astros, the Eclipse). It is an easy exercise to invoke the names of numerous companies, products, and logos that make use of celestial objects. Dozens of countries incorporate stars, the Sun, the moon, or combinations thereof into their flags. The video game entertainment industry is well known for its science fiction themes. Interestingly, one of the earliest (1979) mass-produced coin-operated video games was *Asteroids*, the goal of which was to destroy incoming asteroids before they impacted your spaceship. Not too surprisingly, video game manufacturers, like their Hollywood counterparts, jumped on the Shoemaker-Levy 9 bandwagon following its impact on Jupiter. At least twelve video games with Earth impactors as a major theme have been produced in the decade following this event.

Another purveyor of popular culture worth mentioning is the global Internet – a relative newcomer on the scene. Originally primarily a means of communication and direct file sharing, the advent and explosive growth of the World Wide Web over the past decade has resulted in a resource whose value as a professional and personal research tool, source of news, and entertainment is equaled only by the maddening amount of misinformation to be found in its billions of pages. There are now dozens of web sites dedicated specifically to the near-Earth object issue, including personal, individual professional, and organizational sites, and literally hundreds if not thousands of sites with a more general treatment of asteroids and/or comets.

One of the more bizarre and also tragic cases involving a strange mix of the popular culture of science fiction and the internet, religious cultism, and near-Earth objects took place in the United States in spring of 1997 in Rancho Santa Fe, California. Thirty-nine men and women who were members of a religious cult known as Heaven's Gate committed mass suicide in the belief that the Earth was about to undergo a "refurbishing" and that a spaceship traveling behind comet Hale-Bopp had come to transport their disembodied souls to the "next level above human" (Wessinger 2000). The group functioned as a web design firm that also used the internet and mass media to spread the message about its ideology and its leader, Marshall Applewhite, a self-proclaimed extraterrestrial being incarnated in a human body to enlighten Earth dwellers. The views expressed by the Heaven's Gate group were significant in that they reflected millennial and apocalyptic ideas espoused by many individuals and groups toward the end of the last millennium (Stewart and Harding 1999; Wessinger 2000), in spite of the fact that this transition held absolutely no cosmic, astronomical, terrestrial or other significance (Loevinger 1997). In light of the above extreme, but not unprecedented case, future exploration of the potential range of social reactions to the occurrence or even the announcement of a potential major cosmic impacted is worthy of further attention, so that appropriate responses can be developed.

3.4
Garnering Public Support

A principal stated goal of the ICSU workshop on Comet/Asteroid Impacts and Human Society, from which this article was derived, was to produce an "unbiased consensus of the various participants regarding this type of event and will be used by ICSU and others to positively influence governments at the highest level around the world to begin to take preparatory action to deal with a possible comet/asteroid impact in the next century." Critical to the implementation of national or international policy on such an issue is both public awareness and support for funding such an effort.

3.4.1
Public Awareness and Support through Cinematic Film

Scientific principles may often be compromized in the name of Hollywood entertainment, sometimes egregiously so by filmmakers, but it is not the role of the filmmaker to ensure that science and scientists are represented in an accurate and fair manner. The filmmaker is concerned primarily with filling theater seats and making a profit by manufacturing a product which audiences will find entertaining. As long as the filmmaker feels the science in a movie has the potential for high entertainment value, they are happy to incorporate accurate representations into the film. However, once the scientific principles lose their appeal, all bets are off – time scales are compressed, sizes and outcomes are exaggerated, and the scientist intimately familiar with the subject matter cannot help but cringe at the scientific misrepresentations and outright fallacies that result.

Although scientists may cry foul regarding the general treatment of science in film, from the standpoint of raising public awareness and support for implementation of public policy on issues such as the threat posed by near-Earth objects, scientists owe a great debt of gratitude to the Hollywood blockbuster. In fact, media publicity during the impact of Shoemaker-Levy 9, and especially the resultant worldwide distribution of Hollywood blockbuster films such as *Armageddon* and *Deep Impact* have arguably produced a greater public awareness and support for this issue than almost any but the most targeted and costly public educational campaigns could have accomplished. It is principally through these media that the general public has been made aware of the key messages of importance relative to the issue, namely:

1. There are objects that have impacted the Earth in the past with devastating consequences and there are objects that will undoubtedly do so at some point in the future;
2. They are predictable – that is, we can see them coming; and
3. We can usually do something to prevent them from impacting the Earth.

Through the media of cinematic film, the NEO threat has been brought to the attention of millions of people worldwide who otherwise might have remained almost wholly ignorant of the issue. Those who are interested in learning more about the scientific validity of what they have seen will seek out documentaries, literature, and other educational resources to become better informed. Those who are not interested nevertheless have been exposed to the relevant key issues involved.

3.4.2
Public Education

Although cinematic film can be an excellent tool in raising public awareness, the task of public education on the true scientific nature of the NEO issue remains an extremely important one, albeit one that perhaps should be undertaken as a part of policy implementation rather than as a prerequisite. There are several obstacles that will need to be overcome with regards to educating the public about the near-Earth object issue. The first has to do with the general state of the public's science education, which is necessarily tied to its perception of science and scientists. The general level of science education of members of the public is quite low (speaking from a U.S.-centric position). Even students who take basic core science classes at the university level often complete those classes without a complete understanding of how the process of scientific inquiry works (Mole 2004). Instead, Mole explains, they are often introduced to classes concerning the interaction of science and society that concentrate on real and imagined deficiencies of science, while neglecting important topics such as the history of science, the role of the peer review process, and discussions on why individual scientists may have widely divergent views on a particular subject.

This last point is especially germane when dealing with members of the public who have no science background whatsoever. When addressing an issue such as near-Earth objects, the public can become easily confused by lack of consensus among scientists. The diverse range of views regarding the likelihood of catastrophic impact over a given time period and its resultant effects (e.g. Bryant 2001; Chapman 2004; Chapman and Morrison 2003; Keller 1997; Marsden 2004; Masse, Chap. 2 of this volume; Svetsov 2003; Yabushita 1997) particularly when filtered through various popular science media (e.g. Anonymous 1998; Applegate 1998; Dalton 2003; Hecht 2002; Ravilious 2002) can end with the layperson throwing up his or her hands in exasperation and walking away from the issue altogether. Peer-review, disagreement, and discourse are, after all, part of the process of conducting science, but many in the general public are unaware of this.

Finally, cinematic film may do wonders to increase public awareness of important scientific issues, but public perception of science and scientists is, at least in part, shaped by their portrayals in popular cinematic film, and video and TV programs. Such portrayals are often less than flattering, with the "mad" (e.g. Frankenstein) or bumbling scientist stereotype perpetuated and the idea that science itself is responsible for the world's ills (Haste 1997; Steinmuller 2003).

3.5
Conclusions

In industrial societies, the celestial constants and some phenomena have been relegated to the realm of scientific curiosity. However, unusual transient events can trigger significant, albeit often brief, resurgences in public interest. It is clear that public understanding of and interest in the near-Earth object issue has undergone a transformation over the last decade that was initiated by the impact of comet Shoemaker-Levy 9 on Jupiter. This real-world event and the resultant popular cultural cinematic productions helped focus the public on the actual threat that near-Earth objects present, and also greatly increased public awareness and potential support for development and implementation of public policy on the issue.

When targets-of-opportunity arise, such as feature films addressing topics of serious scientific concern, scientists should take a proactive role in initiating and participating in frank discussions that engage the public on relevant issues depicted in mass popular culture, offering correction and explanation when appropriate, and availing themselves of the opportunity to educate about the process of science at the same time. Science fiction film can also present excellent opportunities to teach students about real science and the process of critical thinking (Dubeck et al. 1988). As an additional measure, promoting good general science education at all educational levels will ensures that the future public is better equipped to independently evaluate where their support should be focused on such issues.

We may be nearing the end-life of continued popular interest in potential impact scenarios, but recent events such as the fly-by of Asteroid 4179 Toutatis in September 2004, the visit of comet c/2004, visible to the naked eye at the writing of this article in December 2004, and significant media attention focused on scientific ventures such as NASA's Deep Impact mission will ensure that the subject remains in the public eye.

Acknowledgments

The author gratefully acknowledges the Desert Research Institute's Division of Earth and Ecosystem Sciences for providing the funding that made this research possible, and also to Dr. Bruce Masse, Dr. Ruthann Knudson, and Ms. Kerry Varley for their comments on earlier versions of this manuscript. Finally, the author would like to thank the organizers of the Tenerife workshop on Asteroids/Comets and Human Society without whose efforts this research would never have taken place.

References

Anonymous (1998) Will Bruce save us? If we don't know where they are, we stand no chance against asteroid impacts. New Scientist 158:3

Applegate D (1998) Asteroid impact! Nuclear test! Why we need open discourse and data access. Geotimes 43(5):13

Ben-Menahem A (1992) Cross-dating of biblical history via singular astronomical and geophysical events over the ancient Near East. Quarterly Journal of the Royal Astronomical Society 33(3):175–190

Bricker HM, Aveni AF, Bricker VR (2001) Ancient Maya documents concerning the movements of Mars. Proceedings of the National Academy of Sciences of the United States of America 98(4):2107–2110

Bryant EA (2001) Tsunami: the underrated hazard. Cambridge University Press, London

Butler J (1979) Magic, astrology, and the early American religious heritage, 1600–1760. American Historical Review 84(2):317–346

Capp B (1979) English Almanacs 1500–1800: astrology and the popular press. Cornell University Press, Ithaca, New York

Chapman CR (2004) The hazard of near-Earth asteroid impacts on Earth. Earth and Planetary Science Letters 222(1):1–15

Chapman CR, Morrison D (2003) No reduction in risk of a massive asteroid impact. Nature 421:473

Clarke AC (1993) Hammer of God. Ballantine Dell, Los Angeles

Dalton R (2003) Long-lost wave report sinks asteroid impact theory. Nature 421:679

Dubeck LW, Moshier SE, Boss JE (1988) Science in cinema: teaching science fact through science fiction films. Teachers College Press, New York

Evans W (1996) Divining the social order: class, gender, and magazine astrology columns. Journalism and Mass Communication Quarterly 73(2):389–400

Goff J, Hulme K, McFadgen B (2003) Mystic fires of Tamaatea: attempts to creatively rewrite New Zealand's cultural and tectonic past. Journal of the Royal Society of New Zealand 33(4):795–809

Haste H (1997) Myths, monsters, and morality: understanding 'antiscience' and the media message. Interdisciplinary Science Reviews 22(2):114–120

Hawkins G (1965) Stonehenge decoded. Doubleday, New York

Hecht J (2002) It's only the really big asteroid impacts that are a threat to life on Earth. New Scientist 176:24

Hester JJ, Grady J (1982) Introduction to Archaeology. CBS College Publishing, New York

Keller G (1997) Asteroid impacts and mass extinctions – no cause for concern. Near-Earth Objects 822:399–400

Loevinger L (1997) The significance of the millennium. Interdisciplinary Science Reviews 22(4):343–351

Mackie EW (1997) Maeshowe and the winter solstice: ceremonial aspects of the Orkney Grooved Ware culture. Antiquity 71:338–359

Malefijt ADW (1968) Religion and culture: an introduction to anthropology of religion. Macmillan Publishing Co, New York

Marsden BG (2004) Comets and asteroids: searches and scares. Advances in Space Research 331514–1523

Masse BW (in press) Transient celestial events in traditional Hawaii. In: Masse WB, Johnson RK, Tuggle HD (eds) Islands in the Sky: astronomy and the role of celestial phenomena in Hawaiian myth, language, religion, and chiefly power. University of Hawaii Press, Honolulu

Masse BW, Espenak F (2006) Sky as environment: solar eclipses and Hohokam culture change. In: Doyel DE, Dean JS (eds) Environmental Change and Human Adaptation in the Ancient Southwest. University of Utah Press, Salt Lake City, pp 228–280

Masse BW, Masse MJ (in press) Myth and catastrophic reality: using myths to identify cosmic impacts and massive plinian eruptions in Holocene South America. In: Piccardi L, Masse BW (eds) Myth and geology. Geological Society of London

Masse BW, Soklow R (2005) Black suns and dark times: cultural response to solar eclipses in the ancient Puebloan southwest. In: Fountain JW, Sinclair RM (eds) Current Studies in Archaeoastronomy: Conversations Across Time and Space. Carolina Academic Press, Durham, North Carolina, pp 47–68

McDevitt J (1999) Moonfall. Eos, London

Mole P (2004) Nuturing suspicion: what college students learn about science. Skeptical Inquirer 28(3):33–37

Morris E (1996) The word detective. *http://www.word-detective.com/back-b2.html*
Niven L, Pournelle J (1977) Lucifer's Hammer. Harpercollins, New York
Olson RJM, Pasachoff JM (2002) Comets, meteors, and eclipses: art and science in early Renaissance, Italy
Quinlan-McGrath M (2001) The foundation horoscope(s) for St Peter's basilica, Rome, 1506: choosing a time, changing the storia. Isis 92(4):716-741
Ravilious K (2002) Did lava cover traces of asteroid impacts? New Scientist 176:16-17
Regan D (1998) For the record. Harcounrt, New York
Schechner, SJ (1997) Comets, popular culture, and the birth of modern cosmology. Princeton University Press, Princeton, New Jersey
Sofaer A, Zinser V, Sinclair RM (1979) A unique solar marking construct. Science 206:283-291
Steinmuller K (2003) The uses and abuses of science fiction. Interdisciplinary Science Reviews 28(3): 175-178
Stewart K, Harding S (1999) Bad endings: American apocalysis. Annual Review of Anthropology 28: 285-310
Svetsov VV (2003) Numerical modelling of large asteroidal impacts on the Earth. International Journal of Impact Engineering 29:671-682
Verhulst J (2000) World cup soccer players tend to be born with sun and moon in adjacent zodiacal signs. British Journal of Sports Medicine 34(6):465-466
Walker CBF (1985) Archaeoastronomy - Halley's comet in Babylonia. Nature 314:576-577
Wessinger CL (2000) How the millennium comes violently: from Jonestown to Heaven's Gate
Xu Z, Jiang Y, Pankenier DW (2000) East-Asian archaeoastronomy: Historical Records of Astronomical Observations of China, Japan, and Korea. Gordon and Breach, Paris
Yabushita S (1997) On the possible hazard on the major cities caused by asteroid impact in the Pacific Ocean II. Earth, Moon and Planets 76:117-121
Yip PSF, Lee J, Cheung YB (2002) The influence of the Chinese zodiac on fertility in Hong Kong SAR. Social Sciences and Medicine 55:1803-1812
Ziegler J (2002) The medieval kidney. American Journal of Nephrology 22:152-159

Chapter 4

Umm al Binni Structure, Southern Iraq, as a Postulated Late Holocene Meteorite Impact Crater

S. Master · T. Woldai

4.1 Introduction

Master (2001) discovered a ca. 3.4 km diameter circular structure, in the marshes of southern Iraq, on published satellite imagery (Fig. 4.1, after North 1993a), and interpreted it to be a possible meteorite impact crater, based on its morphology (its roughly polygonal outline, an apparent raised rim and a surrounding annulus), which differed greatly from the highly irregular outlines of surrounding lakes. The structure, which is situated in the Al 'Amarah Marshes, near the confluence of the Tigris and the Euphrates Rivers (at 47° 4' 44.4" E, 31° 8' 58.2" N), was identified by Master (2002) as the Umm al Binni lake, based on a detailed map of the marshes published by Wilfred Thesiger (1964). Following the Gulf War of 1991, Saddam Hussein's regime embarked on a massive program to drain the Al 'Amarah marshes, by building a huge canal named the "Glory River" parallel to the Tigris River (Fig. 4.3) (North 1993a, b; Wood 1993; Pearce 1993, 2001; Partow 2001a; Naff and Hanna 2002). After the almost complete draining of the marshes since 1993 (Munro and Touron 1997; Partow 2001a, b; Nicholson and Clark 2002) the Umm al Binni Lake has disappeared and in recent Landsat TM and ASTER satellite imagery, it appears as a light colored area, due to surface salt encrustations (Fig. 4.4). Following the Iraq War of 2003, there are moves afoot to re-flood the marshes in an attempt to restore its devastated ecology (Brookings Institution 2003; Jacobsen 2003; Lubick 2003; Martin 2003; Sultan et al. 2003; Richardson et al. 2005; Lawler 2005).

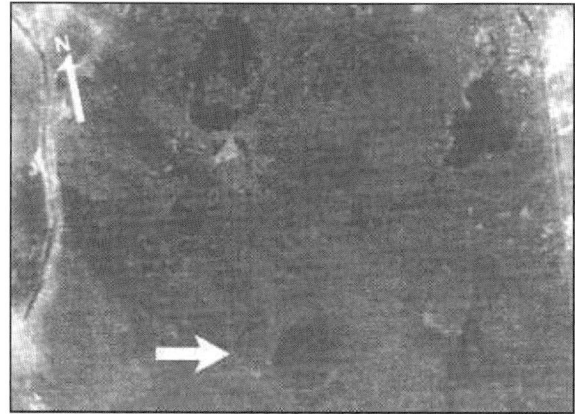

Fig. 4.1.
Detail of published Landsat image (from Master 2001; enlarged from an image published by North 1993a), showing the ca. 3.4 km diameter Umm al Binni Lake (*arrow*), and other marsh lakes with highly irregular outlines, in the Al 'Amarah Marshes of southern Iraq

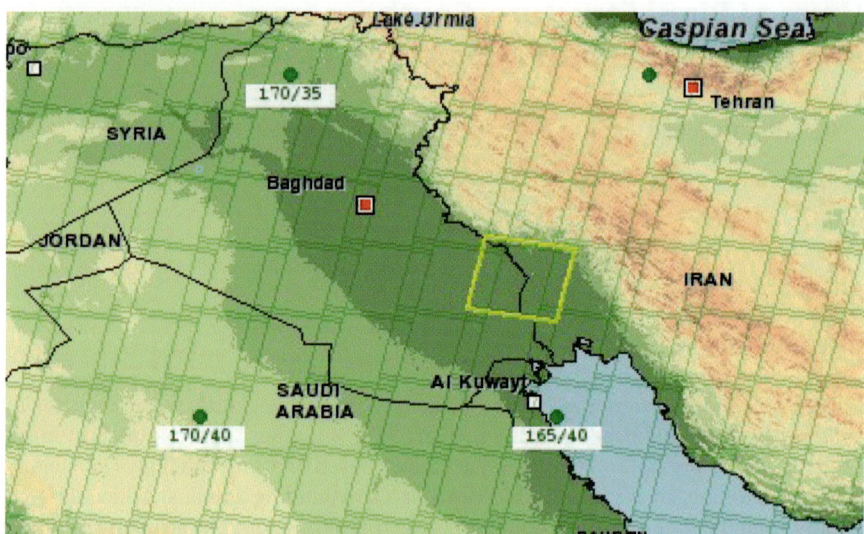

Fig. 4.2. Study area location map. The *green stripes* correspond to satellite flight paths in a N-S direction. The study area is shown in *yellow*. *170/40* indicates the path and row corresponding to the Landsat TM and ETM+ images. The Mesopotamian Basin, at a low elevation, is shown in *dark green color*. Higher elevations of the Zagros Mountains in Iran and NE Iraq are shown in *brown and yellow colors*

4.2
Geological Setting

The alluvial plains of Iraq occupy a structural trough, known as the Mesopotamian Basin (Fig. 4.2), which is linked to active subduction-related orogenic processes in the Zagros Mountains of Iran and northeast Iraq. The Mesopotamian basin is part of the larger Zagros foreland basin associated with the closure of the Neotethys Ocean and the collision of the Arabian passive margin and Eurasian plate (Beydoun et al. 1992). Convergence in the Zagros collision zone still continues and the region is tectonically active today (Lees 1955; Mitchell 1957, 1958b). The Mesopotamian basin is floored by Neoproterozoic crystalline basement rocks of the Arabian shield (Bahroudi and Talbot 2003). Overlying this basement, there is a thick pile of Phanerozoic sedimentary rocks, consisting of an attenuated Paleozoic succession of Cambro-Ordovician, Devonian-Lower Carboniferous and Upper Permian rocks; a well-developed Mesozoic succession of Triassic, Jurassic and Cretaceous rocks, and a Cenozoic succession of Eocene to Pliocene rocks, overlain by Pleistocene to Holocene alluvium (Beydoun et al. 1992). The alluvium, consisting of clay, silt, sand and gravel, is related to the floodplain of the Euphrates and Tigris rivers and associated swamps, as well as to marine incursions (Loftus 1855; Baghdadi 1957). The Tigris and Euphrates rivers and their tributaries arise in the mountains of Syria, Turkey, northern Iraq and Iran, and they meet, after traversing through marshlands, at Al Qurna, north of Basra, to form the Shatt al Arab estuary, which extends for 140 km from Basra to the Gulf (Al Ghunaim et al. 1994). The Karun River, rising in the Iranian Zagros, joins the Shatt al Arab at Khorramshahr, about 40 km

Fig. 4.3. Map of south-eastern Iraq showing extent of former marshlands, and water diversion projects. Image from *http://geography.about.com/ library/maps/ Iraq_marshes_1994.jpg*

ESE of Basra. A mineralogical study of the sediments of the Tigris and Euphrates Rivers, the Shatt al Arab, and some older terraces, shows similar source areas with the main light mineral fraction made up of quartz, cryptocrystalline silica, carbonates, biotite, muscovite, chlorite and plagioclase feldspars, while 32 heavy mineral species were identified (Philip 1968). The suspended loads of the Tigris and Euphrates show marked differences, with the Euphrates richer in both chlorite and expandable lattice clays (Berry et al. 1970).

The Mesopotamian region has the world's oldest examples of large-scale water engineering for irrigation purposes, and the Euphrates and Tigris river systems have been extensively canalized for more than four millennia (Adams 1958; Adams and Nissen 1972; Lees and Falcon 1952; Lees 1955; Harris and Adams 1957; Nelson 1962; Wagstaff 1985; Naff and Hanna 2002). Smith (1872) mentioned waterworks on the Tigris River undertaken during the reign of Hammuragas in the mid-second Millenium BCE.

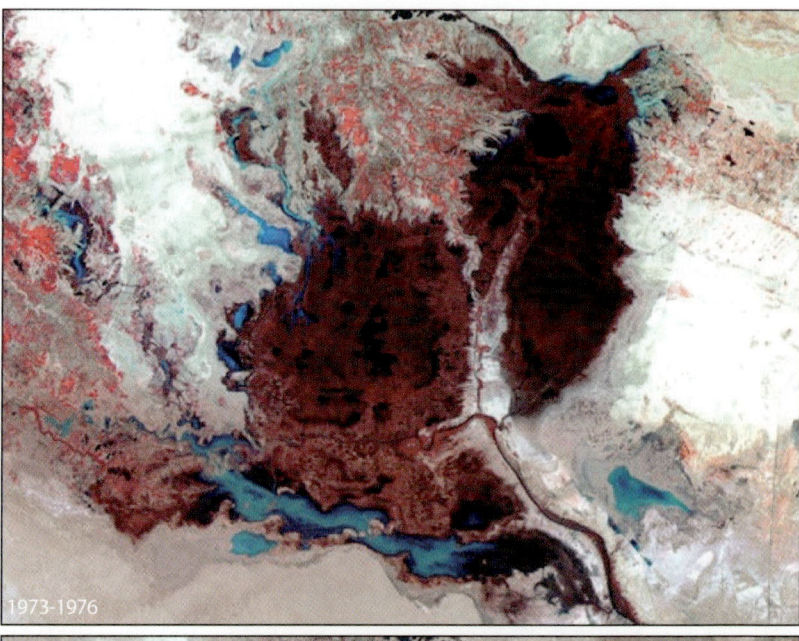
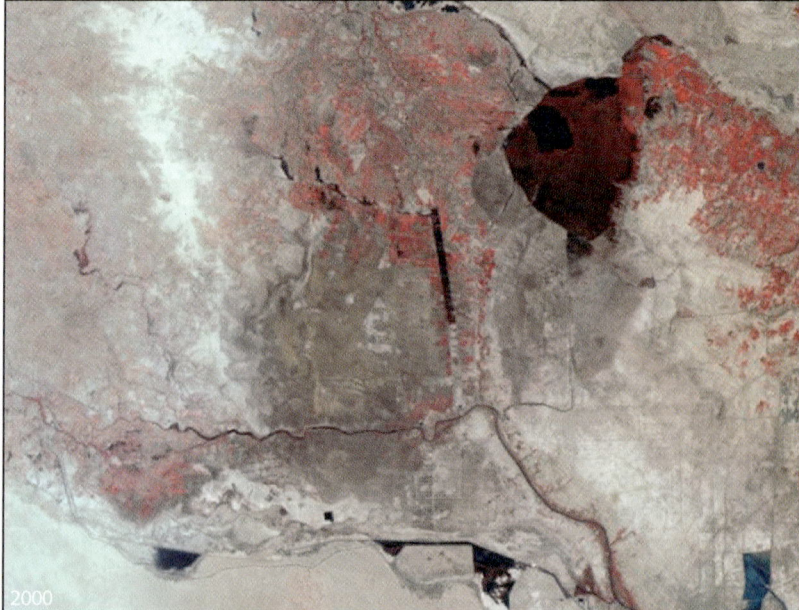

Fig. 4.4. Landsat MSS false-color composite images showing the destruction of the marshlands of southern Iraq between 1976 and 2000. The *red areas* show vegetated marshland. The lakes that appear as *black areas* within the marshlands in the earlier images, appear as *white areas* in the 2000, because of desiccation and encrustation with white salt. Most of the destruction took place in the period from 1992 to 2000. Images from Partow (2001b)

Herodotus, who flourished ca. 490–425 BCE, refers to waterworks in Babylon, and the confluence of the Tigris and Euphrates rivers (Herodotus 1972). Nearchus, in his voyage of 325 BC, mentions that the Euphrates and Tigris had separate entrances into the sea or an estuarine gulf (de Morgan 1900; Hansman 1978). Le Strange (1905), citing the Islamic geographer Baladuri, indicated that the large Khawr al Hammar lake south of the Euphrates and west of Basra was formed during the reign of the Sassanian king Kubadh I in the fifth century CE by breaching of levees on the Tigris; these being repaired in the following reign, the waters of both rivers rose again in flood in 636 CE, and laid the surrounding country under water. Modern changes in the morphology of the delta region have been recorded on Admiralty charts dating from as early as 1826, with various updates (Lees and Falcon 1952). In the last few decades, the Shatt-al-Arab (on the Iraq/Iran border) and the Khawr as Sabiyah (a possible former mouth of the Euphrates north of Kuwait) have been extensively dredged to keep the channels open for large ships such as oil tankers (Al Ghunaim et al. 1994). Several new large canals built in the past decade have drained the Al Amarah marshes and the Khawr al Hammar, and devasted their ecology (Partow 2001a, b). The marshlands of southern Mesopotamia have been the home of the Marsh Arabs or Ma'adan for millennia, and their way of life, described by Thesiger (1964) and Young (1977) has been severely disrupted by the draining of the marshes (Brookings Institution 2003). The shifting watercourses of the Mesopotamian floodplain thus represents a dynamic system in which there is an interplay of natural processes including neotectonic subsidence, fluvial (and aeolian) aggradation, eustatic marine incursions, and human-induced canalization, draining and dredging (Nicholson and Clark 2002).

The bedrock in the region close to the Tigris-Euphrates confluence consists of marine clastics of the Miocene-Pleistocene Dibdibba Formation (Macfayden 1938; Baghdadi 1957; Larsen and Evans 1978). These rocks consist mainly of sandstones, granulestones and conglomerates with rounded igneous clasts and white quartz pebbles, in places with calcareous cements (Baghdadi 1957). The overlying Holocene marine sediments (fine silt and silty clay) of the Hammar Formation contain a Recent fauna consisting of gastropods, lamellibranchs, scaphopods, bryozoa, crab and echinoid fragments (Loftus 1855; Hudson et al. 1957; Eames and Wilkins 1957; Mitchell 1958a; Dance and Eames 1966; Macfayden and Vita-Finzi 1978). The Hammar Formation in turn is overlain by Recent delta plain and delta front deposits of the Mesopotamian Plains, in which there were numerous marshes and permanent lakes until the recent destruction of the marshlands (Lees and Falcon 1952; Larsen and Evans 1978; Partow 2001a,b).

The geological and geographical history of the Tigris-Euphrates-Karun delta region and the head of the Persian/Arabian Gulf have been debated since the 1830's. Beke (1834, 1835) argued from historical evidence that the former head of the Gulf was situated much farther inland in Mesopotamia, based on the voyage of Nearchus in 325 BC, under instruction from Alexander the Great, as recounted by Arrian in his Indica (e.g. Arrian 1983) and by the geographer Strabo (Larsen and Evans 1978; Hansman 1978). As a result of the Euphrates Expedition of 1835–1837, the first geological mapping of Mesopotamia was carried out by Ainsworth (1838) and was followed by the work of Loftus (1855) along the current Iraq/Iran frontier. De Morgan (1900) published very influential diagrams showing the reconstructed paleogeography of the Mesopotamian delta region, utilizing information from the Assyrian king Sennacherib's expedition against the Elamites in ca. 696 BC, and Nearchus' voyage (Larsen and Evans 1978; Hansman 1978). Lees and Evans (1952) questioned the model of a simple outbuilding of the Mesopotamian delta, as argued by de Morgan (1900) and presented evidence for a much more complex interplay of tectonically induced subsidence and fluvial (and aeolian) aggradations in the delta region. This was supported by the observations of Ionides (1954), Smith (1954), Hudson et al. (1957), Mitchell (1957, 1958a, b) and Hansman (1978). Roux (1960) discovered Neo-Babylonian and Kassite (last half of Second Millenium BCE) sites on the southern part of the Khawr al Hammar, an area that was supposed to have been submerged beneath the waters of the Gulf at this time, according to de Morgan (1900). Many authors have presented evidence for the presence of Recent marine or estuarine fauna far inland from the current head of the Gulf, especially in the vicinity of Basra (Loftus 1855; Hudson et al. 1957; Mitchell 1958a), but also at Qurmat Ali (Al Qurna) and Amara (Macfayden and Vita-Finzi 1978), and as far inland as the Abu Dibbis depression southwest of Baghdad (Voûte 1957). Ai-Adili (2004) studied clay minerals from the West Qurna Field, and found mainly mixed layer illite-smectite clays and chlorite, suggesting a marine depositional environment. While such evidence was explained as the result of marine incursions due to tectonic subsidence (Lees and Falcon 1952; Mitchell 1957), Larsen (1975) and Larsen and Evans (1978) invoked eustatic sea-level changes, and attributed the marine sediments to transgressions during Holocene highstands. Larsen and Evans (1978) estimated that the Recent sediments of the Tigris-Euphrates plains were deposited in the last 5000 years, during which time about 130–150 km of seaward progradation has taken place.

4.3
Origin of the Umm al Binni Structure

Because of the extremely young nature of the sediments in the marshlands of the Tigris-Euphrates confluence area (< 5000 years), it is difficult to find a geological explanation for the shape of the Umm al Binni structure. Salt diapirs are common in the Makran coast of Iran and in the Persian Gulf, but are absent from the Mesopotamian Basin (Edgell 1996). Sinkholes are present in Eocene and Miocene limestones (Damman Formation) of the Southern Desert in western Iraq (Baghdadi 1957), but they are two orders of magnitude smaller, as seen on X-SAR Shuttle Radar imagery. The only possible large sinkhole (Al Naqib 1967), is the Al Umchaimin structure, 2.75 km in diameter, in western Iraq, which, however, from its circular crateriform morphology, has been postulated to be a meteorite impact crater (Merriam and Holwerda 1957; Underwood 1994). The sediments of the Mesopotamian plain are un-deformed, whereas their substrate is only very gently folded (Lees and Falcon 1952; Lees 1955). There is no Recent igneous activity in the Mesopotamian basin (Buday et al. 1980; Weiss et al. 1993). The presence of extensive young volcanic fields in adjacent areas of Jordan, Saudi Arabia and Syria prompted Mitchell (1958c) to propose that the Al Umchaimin structure in western Iraq was produced by surface collapse following magma withdrawal in a volcanic intrusion. However, there is a complete absence of igneous rocks at this structure, which is regarded as of meteorite impact origin (Underwood 1994). Thus an origin of the Umm al Binni structure by salt doming, karst dissolution, interference folding or igneous intrusion can be effectively ruled out.

The postulate that the structure was formed by a Recent bolide impact can account for the simple bowl-shaped geometry with markedly polygonal outline, and the apparent rim and annulus around the structure in pre-1993 imagery. For a crater of 3.4 km diameter, scaling equations given by Shoemaker (1983) can be used to calculate the size of an impacting body. For an impactor made of iron with a density of 7 860 kg m^{-3}, and using a range of densities of the target of 1 500 to 2 000 kg m^{-3}, one derives the diameter of a spherical impactor to be between 90 and 108 m, or roughly 100 m. An iron impactor of this diameter, traveling with a velocity of 20 km s^{-1}, would have an energy of 7.86×10^{17} J, or the energy equivalent of 9 400 Hiroshima atomic bombs (20 kT TNT equivalent). A similar calculation determined for an impactor made of typical asteroidal material with density of 2 380 kg m^{-3} yields a diameter of about 355 m. If the postulated impact site was under water, the water column would have absorbed some of the energy, resulting in a smaller crater than if the impact had been on dry land (Ormö et al. 2001). Hence estimates of the bolide diameter are only a minimum, and the bolide could have been larger and more energetic. A wet impact would also have generated huge tsunamis.

Master (2001, 2002) speculated on the possible consequences of this structure, if it was indeed of impact origin, for Bronze-Age Mesopotamia, and suggested that it might possibly be linked with an ~2350 BC "ash" layer found at Tell Leilan, Syria (Weiss et al. 1993), and in sea-sediment core off Oman (Kerr 1998), re-interpreted by Courty (1998) to be an impact fallout layer. Master (2001, 2002) also suggested that an impact-generated tsunami could have been responsible for the Babylonian and Sumerian "flood" legends of Atra-Hasis, Utnapishtim and Ziusudra, as recounted in the appendix to the Epic of Gilgamesh, and other accounts (Smith 1876; Speiser 1958; Sandars 1960; Civil 1969; Lambert and Millard 1969; George 2003). Following these suggestionss, a host of commentators in the popular press and on the Internet rushed to print in sensational articles about meteorite impacts causing the end of Mesopotamian civilizations. It was pointed out by Lyon (2001) and by Master (2002), that the proposed impact structure has not yet been investigated on the ground, and has not been proven to be of impact origin. Until it has been properly studied, and dated, it is difficult to speculate about the possible role of impactors in ancient Mesopotamian history.

4.4
New Satellite Imagery

We have obtained recent Landsat TM and high-resolution ASTER satellite imagery over the Al 'Amarah marshes. Figure 4.2 shows the paths and rows for the Landsat images obtained. The new Landsat TM and ETM+ images are shown in Fig. 4.5. Figure 4.5a is a false-color image showing the marshland (red) surrounding the Umm al Binni and other lakes (black), in an image acquired in 1990. The same area is shown in Fig. 4.5b, in an image acquired 10 years later, and it shows the almost total destruction of the marshland vegetation through the draining of the marshes, and the drying of all the former lakes and wetlands, which are now light-colored because of encrustation with salt. Recent investigations of these former lake-beds has revealed that some of the salt crusts are up to 60 cm deep (Sultan et al. 2003). The salt crusts are probably formed from the evaporation of the brackish marsh waters, which are known to be quite saline

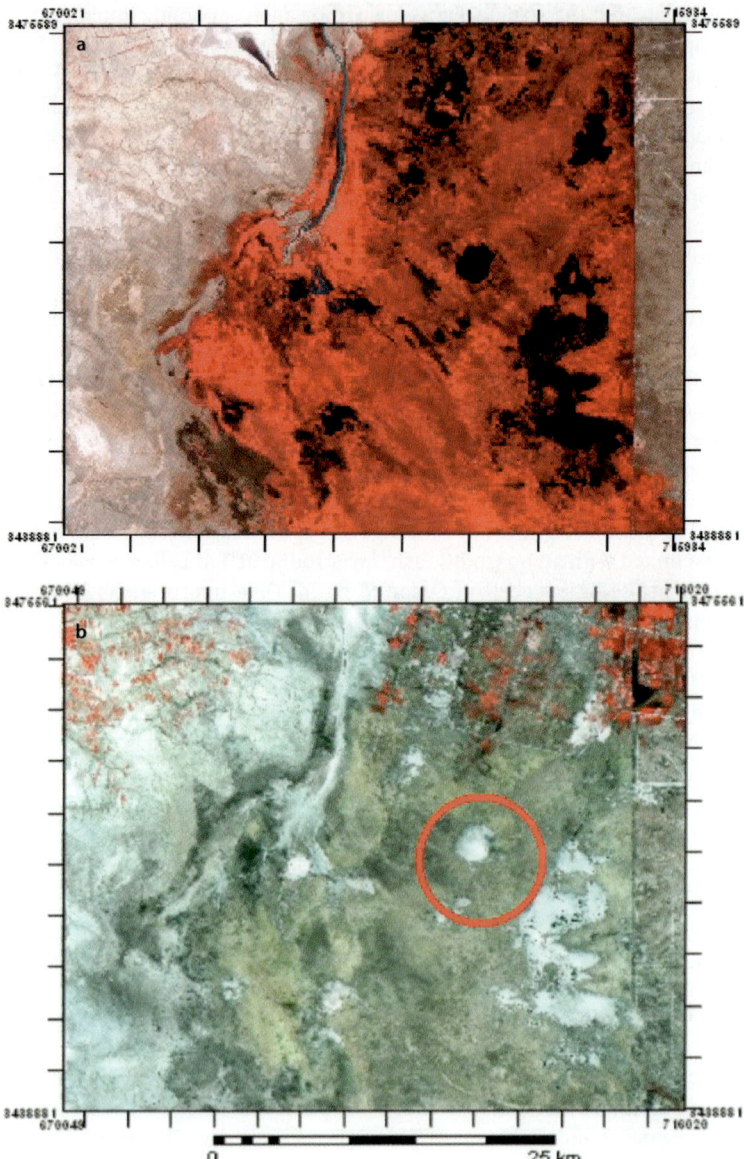

Fig. 4.5. Landsat TM (**a**) and Landsat ETM+ (**b**) bands 4, 3, 2 in RGB order of the study area acquired on the 7th September 1990 and 26th March 2000. Images (**a**) and (**b**) are sub-windows of larger Landsat scenes. Note the changes in marshland as denoted by *reddish color* (marshy area covered by vegetation designating high chlorophyll content) and *dark color* (designated as water bodies) in (**a**) as compared to *light yellowish gray* (no vegetation) and *light tones* (due to surface salt encrustations) in (**b**). Most of the water bodies in the area have disappeared and have become encrusted with salt (shown by their high reflectance signatures in all bands). As a result, the Umm al Binni structure (shown by *red circle*), which was filled by fresh water (*dark* in **a**) is seen with *light tones* in (**b**).

Fig. 4.6. ASTER VNIR (Visible Near Infrared) image of the confluence between the Euphrates (flowing from *left* to *right*) and the Tigris (*top* to *bottom*) rivers – showing also the canal parallel to the Tigris which was used to drain the marshes, Former marsh lakes appear white. The Umm al Binni structure is shown outlined by the *red rectangle*. AST_L1B_00304142001074934_01232004124911) bands 1, 2, 3 in RGB order of the study area (coordinates: ULX = 679992.550102, LRX = 725003.687500, ULY = 3471760.447664, LRY = 3434939.750000) acquired on the 14th April 2001

(Russel 1956), and from evapotranspiration of subsurface waters, which are also saline (e.g. in the Dibdibba Formation aquifers, Hassan and Al-Kubaisi 2002).

A high resolution ASTER image (acquired in April 2001) of the Tigris-Euphrates confluence area has been studied in the Visible-Near-Infra-Red (VNIR) bands (Fig. 4.6). In this image it can be seen that the marshlands have been completely destroyed, and the only vegetation is present in irrigated fields along the Euphrates and along the new canal parallel to the Tigris. The Umm al Binni lake is now a dry lake whose bed is encrusted with white salt deposits (Figs. 4.7 and 4.8). The high resolution ASTER imagery clearly shows a strikingly polygonal outline of the lake, which is in maximal contrast to the highly irregular outlines of most of the other former marshland lakes within the region. The new images of the dry lake show a highly asymmetrical aspect to the lake: the southern half has smooth straight edges to the polygon sides, whereas

Fig. 4.7. ASTER VNIR (Visible Near Infrared) bands 3, 2, 1 (in RGB order) of the Umm al Binni structure (enlarged from Fig. 4.6). The morphology of the crater (roughly polygonal outline and raised rim) is clearly different from the surrounding lakes viewed in all images shown above. The southern part of the crater is surrounded by a series of scalloped concentric zones, which in appearance are similar to ejecta blankets from young terrestrial and non-terrestrial impact structures. The structure is quite asymmetrical – only the southern half has straight, polygonal outline, a scalloped "ejecta-blanket" type zone, and a pure white salt crust. The northern part of the structure is characterised by irregular outline, absence of "ejecta-type" scalloped material, and the presence of dark-reflecting material lining the crater rim. These "ejecta"-type material are totally absent from the northern half of the structure

in the northern half these edges are quite irregular and neither smooth nor polygonal. The southern part of the crater is surrounded by a series of convex-outward scalloped concentric zones, which in appearance are similar to fluidized ejecta blankets from young terrestrial and non-terrestrial impact structures (Melosh 1989). However, this "ejecta"-type material is totally absent from the northern half of the structure. If the structure is of impact origin, then it should normally have a symmetrical ejecta blanket surrounding it on all sides, unless it was the result of a very low angle oblique impact, or if part of the ejecta blanket was eroded away (Melosh 1989).

In the false color composite of Fig. 4.8, which shows the Thermal Infra-Red (TIR) bands 12, 13, 10 in RGB order, long blue streaks trending southwards (diagonally to the bottom right in the image) from the edge of the structure are interpreted as flow lines showing the former position of channels in the marshlands. In these images, higher thermal reflectance shows up as red (warm) colors, lower thermal reflectance shows up as blue (cool) colors. In the inset in Fig. 4.8, which shows a close up of the Umm al Binni structure, a north to south gradation is observed which corresponds to a decrease in thermal reflectance, from areas that were pure white (i.e. salt encrusted) in

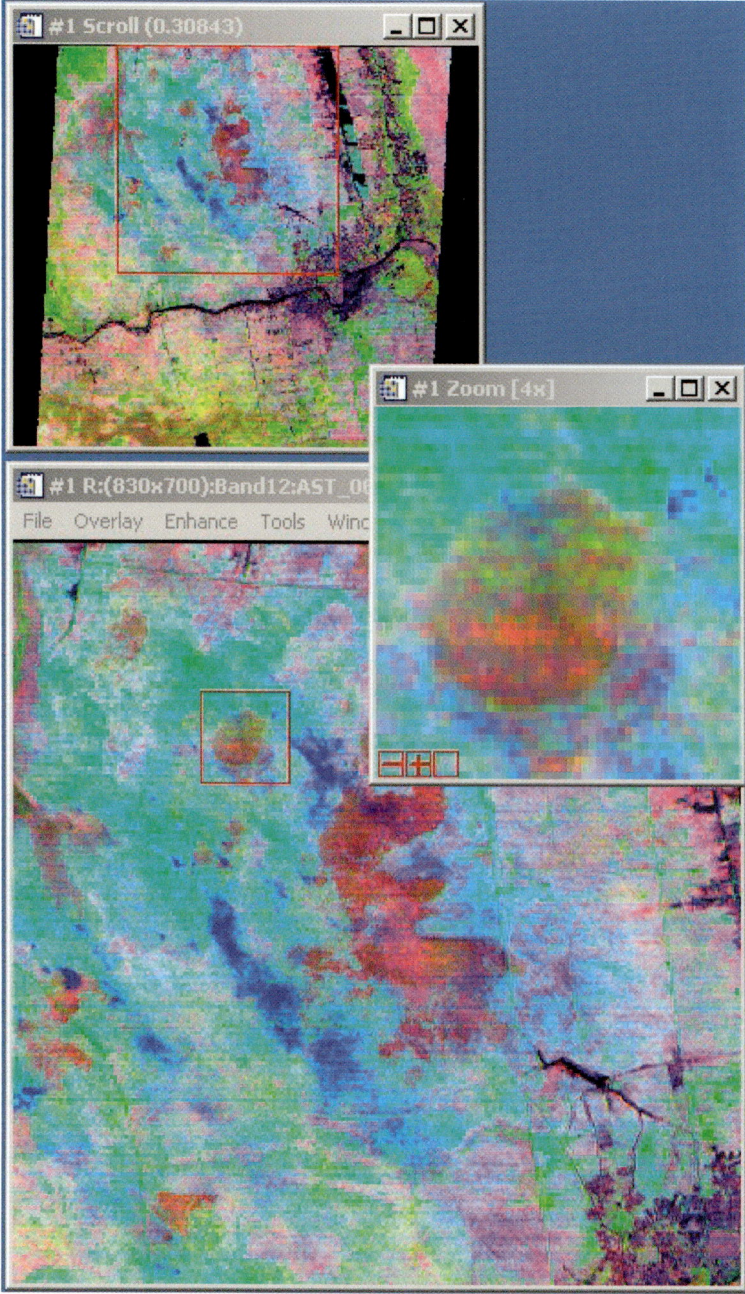

Fig. 4.8. ASTER image with Thermal Infrared (TIR) false color composite bands 12, 13, 10 in RGB order. The upper image covers the whole ASTER scene of Fig. 4.6. An enlargement of the *inset above* is shown below with the Umm al Binni structure enlarged 4 times in the *middle right part*

the image of Fig. 4.7; to the northern part of the structure, where there is a dark band adjacent to the rim (probably corresponding to an increase in the clay content and a decrease in salt). In the area surrounding the Umm al Binni structure, there is an opposite effect: the "ejecta-like" scalloped material to the south has a lower reflectance than the smooth "ejecta-free" area to the north of the structure. We interpret the blue streaks extending past the edge of the Umm al Binni structure as representing former channels where increased clay fractions were deposited, in contrast to the areas to the north of the structure, where erosion took place. We also infer that some deposition of clay minerals took place in the northern rim of the structure. We explain the marked north-south asymmetry of the Umm al Binni structure (in terms of smoothness and polygonality of outline; presence or absence of "ejecta-like" material, and the differing TIR and VNIR spectra), by invoking a north to south water flow within the marshes which eroded an originally continuous "ejecta" blanket, and which was obstructed by the presence of the crater with an uplifted rim, in the northern part of which there was more deposition of clay. There is at least one example of a terrestrial impact structure (the Tsenkher structure in Mongolia) with only a partially preserved ejecta blanket, due to the removal by erosion of the rest of the ejecta around it (Komatsu et al. 1999).

Finally, our high-resolution imagery shows the presence, in an area that was marshland just a decade ago, of a settlement about 4 km ENE of the Umm al Binni structure, from which paths radiate in all directions, possibly caused by domestic animal tracks. This settlement corresponds to the position of the former island village of Ishan abu Shajar, which was visited by Thesiger (1964) in 1951, and which then had thirty or forty houses erected close together on a black Earth island 100 m across and about 3.3 m high at its highest point.

A road leads from Abu Shajar to the NE, towards the larger settlement of Qubur. This shows that the area is currently accessible overland. It is imperative that the structure is studied on the ground in order to determine its origin, and that this work is completed soon, before the proposed re-flooding of the marshes makes the area inaccessible again. All previous attempts to study the structure on the ground have been frustrated by the extremely dangerous political and military situation that has prevailed in Iraq in the past 3 years. A major improvement in the security situation is necessary before the structure may be investigated scientifically.

If the security situation does improve, we propose the following lines of research on the Umm al Binni structure: The structure needs to be examined all along its rim, where a search should be made for deformation features such as overturned sediments, and breccias. The scalloped terrain to the south of the structure must be given special attention. Gravity and magnetic profiles should be made in a north-south direction. A gravity survey will be especially useful in delineating the shape of the crater bottom, and in deciphering the nature of its fill (e.g. Wong et al. 2001). A magnetic survey will aid in detecting any igneous rocks or subsurface magnetic rocks that may have moved upwards in a central uplift (Pilkington and Grieve, 1992). We also propose to implement a series of auger holes, in a north-south profile, extending from well beyond the structure (in order to obtain "background" readings), through the "ejecta" layer in the south, through the crater and out onto the northern flanks. The auger cores from outside the structure should be examined petrographically and geochemically, in order to detect any "fallout" layers related to a possible impact event. The cores from inside the structure must be examined in great detail petrographically, in order to detect macroscopic and microscopic evidence for shock deformation (planar deformation lamellae, diaplectic glasses, impact melts, microbreccias, pseudotachylites, shatter cones) (French 1998). If the structure shows evidence for an impact origin, then it needs to be dated and this can most likely be accomplished using any number of Quaternary dating methods, because of the young age of the country rocks. Only once all of the above has been accomplished, will it be feasible to evaluate the possible role this structure may have played in the history of Mesopotamia.

References

Adams RM (1958) Survey of ancient water courses and settlements in central Iraq. Sumer 14(1–2):101–103
Adams RM, Nissen HJ (1972) The Uruk countryside: the natural setting of urban societies. The University of Chicago Press, Chicago and London
Ai-Adili AS (2004) The study of clay minerals and their applications in petroleum projects – West Qurna Field, Iraq. In: Agenda and Abstr. 7th Int. Conf. Geology Arab World, Cairo Univ., Giza, Egypt, February 2004, p 20
Ainsworth W (1838) Researches in Assyria, Babylonia, and Chaldea. John W. Parker, London
Al Ghunaim AY, Ghunemi ZEDA, Abd al Razzaq FHA, Al Mayyal AY, Al Aryan JY, Moati YA (1994) Iraq navigational outlets. Centre for Research and Studies on Kuwait, Almansoria, Kuwait
Al Naqib KM (1967) Geology of the Arabian Peninsula: Southwestern Iraq. U.S.G.S. Prof. Paper, 560-G, p. G7
Arrian (Flavius Arrianus) (1983) Anabasis Alexandri, Books V–VII (with an English translation by P. A. Brunt). Loeb Classical Library, Harvard University Press, Cambridge, Massachusetts, USA
Baghdadi AI (1957) Ground-water in Iraq, its domestic use, supply and planned utilization of underground reservoirs. In: Seccion IV: Geohidrologia de regiones aridas y sub-aridas. Congreso Geologico Internacional, XX^a Sesión, Ciudad de México, pp 231–246
Bahroudi A, Talbot CJ (2003) The configuration of the basement beneath the Zagros Basin. J Petrol Geology 26(3):257–282
Beke CT (1834) On the former extent of the Persian Gulf and on the comparatively recent Union of the Tigris and Euphrates. London and Edinburgh Phil. Mag. and J. Sci., Ser. 3, IV, 107–112
Beke CT (1835) On the historical evidence of the advance of the land upon the sea at the head of the Persian Gulf. London and Edinburgh Phil Mag and J Sci, Ser. 3, VI, 401–408
Berry RW, Brophy GP, Naqash A (1970) Mineralogy of the suspended sediment in the Tigris, Euphrates, and Shatt-al-Arab rivers of Iraq and the Recent history of the Mesopotamian plain. J Sedim Petrol 40:131–139
Beydoun ZR, Hughes-Clarke MW, Stoneley R (1992). Petroleum in the Zagros Basin: A Late Tertiary foreland basin overprinted onto the outer edge of a vast hydrocarbon-rich Paleozoic-Mesozoic passive-margin shelf. In: Macqueen RW, Leckie DA (eds) Foreland basins and fold belts. AAPG Memoir 55, Tulsa, Oklahoma, USA, pp 309–339
Brookings Institution (2003) The Iraqi Marshlands: can they be saved? Assessing the human and ecological damage. Brookings Institution, Washington, D.C., USA, *http://www.brook.edu/dybdocroot/ comm/ events/20030507.pdf*
Buday T, Kassab IIM, Jassim SZ (1980) The regional geology of Iraq, vol 1: Stratigraphy and paleogeography. State Organisation for Minerals, Baghdad
Civil M (1969) The Sumerian flood story. In: Lambert WG, Millard AR, Atra-Hasîs (eds) The Babylonian story of the flood. Clarendon Press, Oxford
Courty M-A (1998) Causes and effects of the 2350 BC Middle East anomaly evidenced by micro-debris fallout, surface combustion and soil explosion. In: Peiser BJ, Palmer T, Bailey ME (eds) Natural catastrophes during Bronze Age civilisations: archaeological, geological, astronomical and cultural Perspectives. British Archaeol Reports S728, Archaeopress, Oxford
Dance SP, Eames FE (1966) New molluscs from the Recent Hammar Formation of south-east Iraq. Proc Malacological Soc 37:35–53

de Morgan J (1900) La délégation en Perse, Mémoires I, Leroux, Paris, pp 4–48
Eames FE, Wilkins GL (1957) Six new molluscan species from the alluvium of Lake Hamar, near Basrah, Iraq. Proc Malacological Soc 32(5):198–203
Edgell HS (1996) Salt tectonics in the Persian Gulf basin. In: Alsop GL, Blundell DL, Davison I (eds) Salt tectonics. Spec Publ Geol Soc London 100:129–151
French BM (1998) Traces of catastrophe: a handbook of shock-metamorphic effects in terrestrial meteorite impact structures. LPI Contribution No. 954, Lunar and Planetary Institute, Houston, Texas, USA
George AR (2003) The Babylonian Gilgamesh epic: introduction, critical edition and cuneiform text, 2 volumes. Oxford University Press
Hansman JF (1978) The Mesopotamian delta in the first millenium BC. Geogr J 14:49–61
Harris SA, Adams RM (1957) A note on canal and marsh stratigraphy near Zubediyah. Sumer 13(102):157–162
Hassan HA, Al-Kubaisi QY (2002) Pliocene groundwater evolution of the Dibdiba aquifers, Iraq. In: Youssef E-SAA (ed) Agenda and Abstr. 6th Int. Conf. Geology Arab World, Cairo University, Giza, Egypt, February 2002, p. 67
Herodotus (1972) The histories (translated by Aubrey de Sélincourt). Revised, with an Introduction by A. R. Burn. Penguin Books, Harmondsworth
Hudson RGS, Eames FE, Wilkins GL (1957) The fauna of some Recent marine deposits near Basra, Iraq. Geol Mag 94(5):393–401
Ionides MG (1954) The geographical history of the Mesopotamian Plains. Geogr J 120(3):394–395
Jacobsen L (2003) Scientists hope to restore historic Iraqi marshlands. Milwaukee Journal Sentinel, Milwaukee, Wisconsin, USA, 4 May, 2003
Kerr RA (1998) Sea-floor dust shows drought felled Akkadian Empire. Science 279:325–326
Komatsu G, Olsen JW, Baker VR (1999) Field observation of a possible impact structure (Tsenkher structure) in southern Mongolia. Lunar Planet Sci XXX, Abstr. No. 1041, LPI, Houston, Texas, USA (CD-ROM)
Lambert WG, Millard AR (1969) Atra-Hasîs: The Babylonian story of the flood. Clarendon Press, Oxford
Larsen CE (1975) The Mesopotamian delta region: a reconsideration of Lees and Falcon. J Amer Oriental Soc 95:43–57
Larsen CE, Evans G (1978) The Holocene geological history of the Tigris-Euphrates-Karun delta. In: Brice WC (ed) The environmental history of the Near and Middle East since the last Ice Age. Academic Press, London, pp 227–244
Lawler A (2005) Reviving Iraq's wetlands. Science 307:1186–1189
Lees GM (1955) Recent Earth movements in the Middle East. Geol Rund 43:221–226
Lees GM Falcon NL (1952) The geographical history of the Mesopotamian Plains. Geogr J 118:24–39
Le Strange G (1905) The Lands of the Eastern Caliphate. Oxford University Press, Cambridge, pp 26–27
Loftus WK (1855) On the Geology of portions of the Turko-Persian Frontier, and of the districts adjoining. Quart J Geol Soc London 11:247–344
Lubick N (2003) Iraq's marshes renewed. Geotimes, October 2003, pp 25–27
Lyon I (2001) The importance of peer review. Meteor Planet Sci 36(12):1569
Macfayden WA (1938) Water supplies in Iraq. Iraq Geol. Dept., Publ. No. 1, Baghdad
Macfayden WA, Vita-Finzi C (1978) Mesopotamia: the Tigris-Euphrates delta and its Holocene Hammar fauna. Geol Mag 115:287–300.
Martin G (2003) A dream of restoring Iraq's great marshes – wetlands destroyed by Hussein could thrive again. San Francisco Chronicle, 7 April 2003
Master S (2001) A possible Holocene impact structure in the Al 'Amarah Marshes, near the Tigris-Euphrates confluence, southern Iraq. Meteor Planet Sci 36(9), Suppl., p. A 124
Master S (2002) Umm al Binni lake, a possible Holocene impact structure in the marshes of southern Iraq: geological evidence for its age, and implications for Bronze-age Mesopotamia. In: Leroy S, Stewart IS (eds) Environmental catastrophes and recovery in the Holocene, Abstr. Vol., Dept Geogr., Brunel Univ., Uxbridge, West London, U.K., 29 August – 2 September 2002, pp 56–57. http://atlas-conferences.com/cgi-bin/abstract/caiq-15
Melosh HJ (1989) Impact cratering: a geologic process. Oxford University Press, New York
Merriam R, Holwerda JG (1957) Al Umchaimin, a crater of possible meteoritic origin in western Iraq. Geogr J 123:231–233
Mitchell RC (1957) Recent tectonic movements in the Mesopotamian Plains. Geogr J 123(4):569–571

Mitchell RC (1958a) Recent marine deposits near Basrah. Geol Mag 95(1):84–85
Mitchell RC (1958b) Instability of the Mesopotamian Plains. Bull Soc Géogr Egypte 3:127–139
Mitchell RC (1958c) The Al Umchaimin crater, western Iraq. Geogr J 124:578–580
Munro DC, Touron H (1997) The estimation of marshland degradation in southern Iraq using multitemporal Landsat TM images. Int J Remote Sensing 18(7):1597–1606
Naff T, Hanna G (2002) The Marshes of Southern Iraq: a hydro-engineering and political profile. In: Nicholson E, Clark P (eds) The Iraqi Marshlands: a human and environmental study. The Amar International Charitable Foundation, AMAR Publications, London
Nelson HS (1962) An abandoned irrigation system in Southern Iraq. Sumer 18:67–72
Nicholson E, Clark P (2002) The Iraqi Marshlands: a human and environmental study. The Amar International Charitable Foundation, AMAR Publications, London
North A (1993a) New evidence shows marshlands draining away. The Middle East London 227:22–23
North A (1993b) Saddam's water war. Geogr Mag, July 1993, pp 10–14
Örmo J, Shuvalov V, Lindström M (2001) A model for target water depth estimation at marine impact craters. Meteor Planet Sci 36(9), Suppl., p. A154
Partow H (2001a) The Mesopotamian Marshlands: demise of an ecosystem. Early Warning and Assessment Technical Report, UNEP/DEWA/TR.01.2.Rev.1. UNEP, Nairobi, Kenya. www.grid.unep.ch/activities/sustainable/ tigris/marshland
Partow H (2001b) Landsat witnesses the destruction of Mesopotamian ecosystem. NASA Goddard Spaceflight Center, Scientific Visualization Studio. http:// svs.gsfc.nasa.gov/vis/a000000/a002200/ a002210/Mesopotamia_v2.html
Pearce F (1993) Draining life from Iraq's marshes. New Scientist 1869:11–12
Pearce F (2001) Iraqi wetlands face total destruction. New Scientist 2291:4–5
Philip G (1968) Mineralogy of Recent sediments of Tigris and Euphrates rivers and some of the older detrital deposits. J Sediment Petrol 38:35–44
Pilkington M, Grieve RA (1992) The geophysical signature of terrestrial craters. Rev Geophysics 30:161–181
Richardson CJ, Reiss P, Hussain NA, Alwash AJ, Pool DJ (2005) The restoration potential of the Mesopotamian marshes of Iraq. Science 307:1307–1311
Roux G (1960) Recently discovered sites in the Hammar Lake District. Sumer 16:20–31
Russel JC (1956) Historical aspects of soil salinity in Iraq. Majallatu-'Zzira'ati'l-Iraqiyan, Ministry of Agriculture Iraq, XI(2–3), pp 204–215
Sandars NK (1960) The Epic of Gilgamesh. Penguin Books, Harmondsworth
Shoemaker EM (1983) Asteroid and comet bombardment of the Earth. Ann Rev Earth Planet Sci 11:461–494
Smith G (1872) Early history of Babylonia. Trans Bibl Archaeol Soc I:55–62
Smith G (1876) The Chaldean account of Genesis. Sampson Low, Marston, Searle & Rivington, London
Smith S (1954) The geographical history of the Mesopotamian Plains. Geogr J 120(3):395–396
Speiser EA (1958). The Epic of Gilgamesh. In: Pritchard JB (ed) The ancient Near East, vol I: An anthology of text and pictures. Princeton University Press, Princeton, pp 40–75
Sultan M, Becker R, Al-Dousari A, Al-Ghadban AN, Bufano E (2003) Water, agriculture and land cover: lessons for the post-war era. Geotimes, October 2003, pp 22–24
Thesiger W (1964) The Marsh Arabs. Longmans, Green, London
Underwood JR (1994) Al Umchaimin depression, western Iraq: an impact structure? In: Dressler BO, Grieve RAF, Sharpton VL (eds) Large meteorite impacts and planetary evolution. GSA Spec Paper 293:259–263
Voûte C (1957) A prehistoric find near Razzaza (Karbala Liwa). Sumer 13:135–148
Wagstaff JM (1985) The evolution of Middle Eastern landscapes: an outline to AD 1840. Croom Helm, London and Sydney
Weiss H, Courty M-A, Wetterstrom W, Guichard F, Senior L, Meadow R, Curnow A (1993) The genesis and collapse of Third Millenium North Mesopotamian civilization. Science 261:995–1004
Wong AM, Reid AM, Hall SA, Sharpton VL (2001) Reconstruction of the subsurface structure of the Marquez impact crater in Leon County, Texas, USA, based on well-log and gravity data. Meteor Planet Sci 36:1443–1455
Wood M (1993) Saddam drains the life of the Marsh Arabs. The Independent, Saturday 28 August 1993
Young G (1977) Return to the marshes: life with the Marsh Arabs of Iraq. Collins, London

Chapter 5

Tree-Rings Indicate Global Environmental Downturns that could have been Caused by Comet Debris

M. G. L. Baillie

5.1
Introduction

The dates of a series of *narrowest ring events* (dates where numbers of long-lived oaks showed catastrophically narrow growth rings at the same time) have been identified in a long Irish oak tree-ring chronology (Baillie and Munro 1988). The dates were christened 'marker dates' because they were immediately noted to fall in clusters of information relating to traumatic happenings in widely separated areas around the world. For example, one of the Irish oak dates was 207 BC. In China events in 208 BC, and the years following, included a dim Sun, crop failures, famine and high death rates; and a new dynasty, the Han, is believed to have started in 206 (Pang et al. 1987). Meanwhile, in Europe, problems in Rome called for consultation of the Sibylline Books resulting in the return of the Goddess Cybele from Asia Minor; Cybele was manifest as a 'small black meteorite.' This latter occurrence made sense of a series of references by Livy to 'stones falling from the sky' and strange lights in the sky, 'prodigies of Jupiter', et cetera (Forsyth 1990). Clearly, dates around 207 BC might be expected to show up in other records.

Earlier potential marker horizons are at 2345 BC, 1628 BC and 1159 BC, all fixed in time by tree rings. However, understanding these earlier events is hampered by the poor dating control in such ancient times. This threw the spotlight on the only narrowest-ring event in the present era, that at AD 540. As more tree-ring chronologies became available it was discovered that this Irish tree-ring event was duplicated in oak chronologies across Europe. The event was there in pines from Finland (Zetterberg et al. 1994) and Sweden (Briffa et al. 1992); it was there in trees from Siberia and Mongolia (D'Arrigo et al. 2001) and from North and South America (Scuderi 1990; Boninsegna and Holmes 1985), see Figs. 5.1, 5.2 and 5.3. Thus, by the mid-1990s it was realized that, around AD 540, there was a *global* environmental downturn that had affected tree growth in widely separated regions around the world (Baillie 1994, 1995). Moreover, it was almost immediately apparent that the event was two-stage. It appears that the initial effects were in 536 and that these were followed by a second pulse somewhere in the window 538–543; thus it became sensible to refer to the '540 event' as something spanning 536–545.

Clearly, from the period around AD 540 there should have been enough historical information to define the nature of the global event – just what did people record? A preliminary excursion into history indicated that whereas conditions were very bad in China in the late 530s (Weisburd 1985), and while Justinian's attempt to re-establish the Roman Empire was going into reverse around 540, there was actually notably little

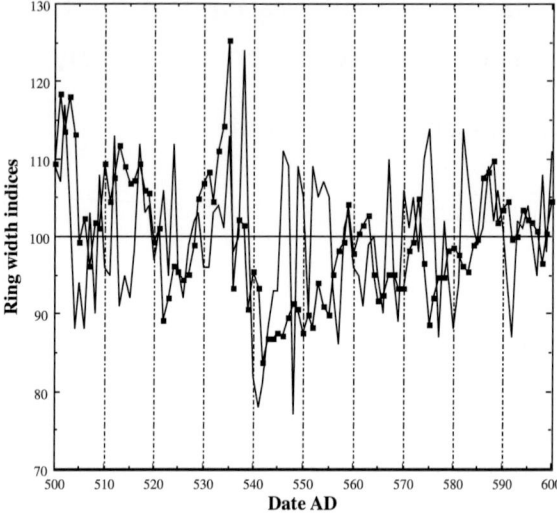

Fig. 5.1.
Plots of annual growth indices (raw ring widths normalised to values around 100) for Irish oak [*solid line*] (Baillie 1995) and Finnish pine [*black dots*] (Zetterberg et al. 1994), showing a notable simultaneous growth reduction in the early 540s

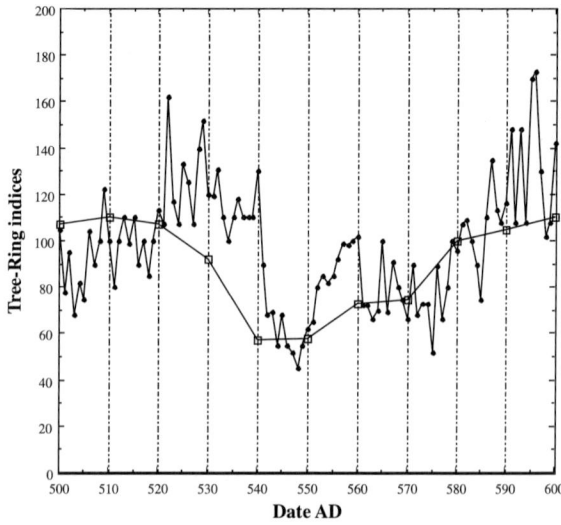

Fig. 5.2.
Plots of annual growth indices (raw ring widths normalised to values around 100) for Argentinian *Fitzroya* [*dots*] (Boninsegna and Holmes 1985), and 20-year averages of ring widths for Nevadan foxtail pine [*open squares*] (Scuderi 1990), showing the synchronous growth reduction across the 540s

information available about the years immediately around this precise date. To make it absolutely clear, according to *world* tree-rings this is the *worst* environmental downturn in the last two millennia. This made it all the more strange that the environmental event was not referred to in conventional history.

So, what can cause a global environmental downturn? Several things were known from the historical record. There was a severe 'dry fog' in 536–537, assumed by volcanologists to be the dust-veil associated with a large volcanic eruption (Stothers and Rampino 1983; Stothers 1984). There were famines in China and in the Mediterranean region in the later 530s. A major plague, named after the Emperor Justinian, broke out around 540 and arrived into Constantinople in 542, thereafter killing perhaps one third

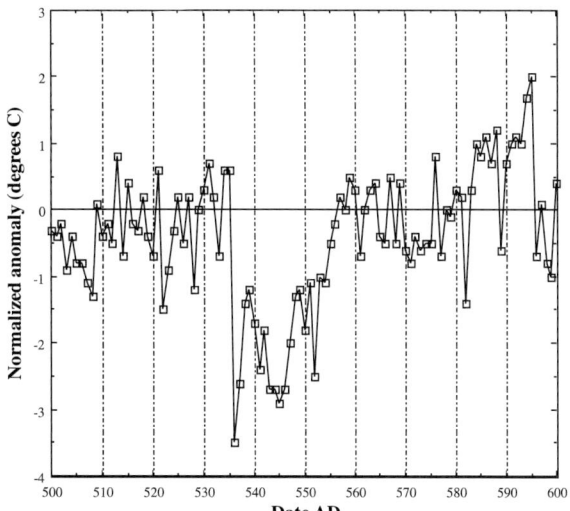

Fig. 5.3.
Average series of temperature anomalies (normalised to values around 0 °C) constructed using three Eurasian chronologies from Tornetrask (Sweden), Yamal and Taimyr (Russia). Figure re-drawn from data provided by Keith Briffa (pers. comm., 9 January 2004), see also Briffa (1999). Note the dramatic reduction in 536 and the more prolonged departure in the 540s

of Europe's population. In terms of cause, all the initial thinking, following Stothers and Rampino, involved volcanoes. Was the event the result of an exceptional volcanic eruption that produced unusual levels of atmospheric aerosol? Was there more than one large volcano involved? Here it is necessary to turn from tree-rings and history to the ice-core record from Greenland. A preliminary analysis of the ice records raised questions about linking a volcano to the event (Baillie 1994). It is now known, on the basis of three replicated ice cores (Dye3, GRIP and NGRIP), that there is no significant volcanic-acid signal in the time window 536–545 (Clausen et al. 1997). The latest statement states specifically:

> With the chemistry and the isotope data it is possible to do a very precise dating for the eruption. The volcanic eruption is dated to AD 527 ± 1 year. The AD 527 volcanic eruption is the only eruption in the period (Larsen et al. 2002).

The authors go on to say that this volcano is the only likely candidate to have caused the 536–545 global event, but that the dating 'suggest(s) that the event is not the same one described by other sources' (Larsen et al. 2002). There are two ways to deal with this observation. One option is to disregard the dating by the ice-core workers and simply assume that 527 ± 1 really *means* 536 or 540 – there are currently no compelling arguments for moving the date derived from three replicated ice-cores in this manner. The other is to make the more logical jump, namely that the global environmental downturn was not volcanic in origin, but rather was caused by loading of the atmosphere from another source, presumably from space. Such a suggestion immediately reduces to the idea that around 536–545 we most probably had a brush with a comet or its debris. This is the logical step that this author made after 1994. Instead of asking the historical record what happened around 540 – a question that produces almost no answer – the question was re-worded as 'we suspect that the Earth had a brush with a comet – what do the records say'? Let us look at what the records do say and be prepared to 'read between the lines' of the only relevant historical records.

5.2
The Historical Record

It transpires that the historical record is un-naturally thin around 540. For example, in Britain there is only one writer believed to be contemporary, Gildas. Morris, in the foreword to the 1978 edition, says "Gildas wrote his main work, the 'Ruin of Britain', about 540 AD or just before..." (Winterbotham 1978). Note that trees suggest a worldwide environmental downturn just at this date, and the *only* known British text is entitled the 'Ruin of Britain.' Gildas' writings are, moreover, essentially apocalyptic.

In the Mediterranean area there is a strange pattern. One major writer, Cassiodorus, stops writing in 538 (Barnish 1992). Another, Malalas, produces a record so thin as to be useless across the relevant period (Jeffreys et al. 1986). Zachariah of Mythilene, whose 12 volume history, compiled in the sixth century, originally covered the 460s to 560s, is complete only to the end of volume nine which *ends* in AD 536 (Hamilton and Brooks 1899). Of significance for this article, Zachariah's key volume 10 is missing and much of the rest is fragmentary. Procopius, who is a major source for the Justinian period does mention the sun being dim in 536, and the major plague, but provides no really useful record of any 'global' event. It is fair to say that there is little in the mainstream historical record that would have led anyone to believe, *pre tree-rings*, that there had been a global environmental event around 540; however, there are historical hints.

Cassiodorus in one of his last letters – that describes the dry-fog, or dim-sun, event of 536–537 – does make a note that people may be worried about "what is coming on us from the stars." Gibbon (1832) mentions a "great comet" in 539 that caused worry of calamitous things to come, and whose "prognostications were abundantly fulfilled". The medieval historian, Roger of Wendover, writing in the thirteenth century, makes the following statement concerning 540/541:

> 540 Battles in the Air
> The reference is probably to aurorae seen in France [Britton's suggestion]. Roger of Wendover has an account of this: In the year of grace 541, there appeared a comet in Gaul, so vast that the whole sky seemed on fire. In the same year there dropped real blood from the clouds ... and a dreadful mortality ensued ... (Britton 1937).

Britton did not know that there was a global environmental event around 540, so, in keeping with the prevailing paradigm that there is no threat from space, he interpreted this statement as the probable appearance of an aurora. We now know about the 536–545 global tree-ring event, *and* the dry-fog references, *and* the plague. Wendover's record would therefore appear to be accurate; especially as the ice-core evidence now suggests that the event did not involve a volcano. This Wendover statement, in keeping with most early isolated references has been dismissed elsewhere as "These entries are almost certainly purely fictional" (James 1999) telling us that we cannot expect historians to do much "reading between the lines." However, there is another ancient record from Britain apparently relating to this period. In AD 542, according to Hector Boetius:

> The sun appeared about noondays, all wholly of a bloody colour. The element appeared full of bright stars to every man's sight, continually, for the space of two days together (Chatfield 2002).

This seems to fit well with the other available comments. Returning to Gildas, writing around 540, he uses passages from the Bible to illustrate what may happen to contemporary sinners, effectively making a collage of quotations all of an apocalyptic nature. Here is an example:

> Behold, the day of the Lord shall come … to make a wilderness of the land … the brilliant stars in the sky shall cease to spread their light, and the sun shall be shadowed at its rising … The moon will grow red, the sun will be confounded … (Winterbotham 1978; 44:1).

Gildas seems to be drawing together references that relate to a dust veil that affects the light of the Sun, the Moon, and the stars, and which in turn produces a "wilderness". This seems to be a purposeful use of biblical quotation to describe something affecting Britain that we already know had affected the Mediterranean in 536–537. Overall, these British sources appear to confirm the idea of a dust-veil, with material dropping from the sky, a dim Sun and a plague, combined with a close comet. It could be asked why Gildas does not mention a comet overtly? The answer may be that the word *comet* does not appear in the Bible and hence no relevant quotation was available. Gildas does however make an interesting statement based on an early version of the Book of Zechariah:

> And the angel said to me: What do you see? I replied: I see a flying sickle, twenty cubits long. It is a curse that goes over the face of the whole land … and I shall cast it forth, says the almighty Lord … (Winterbotham 1978; 57:2).

By the seventeenth century the King James Bible version of this text says "Then I turned, and lifted up mine eyes, and looked, and behold, a flying roll…" (Old Testament, Zechariah 5,1) rendering it essentially incomprehensible. Historically, of course, comets with their curved tails are often described as *sickles*; the later translation as *roll* appears to have no meaning at all. So although Gildas did not mention a comet directly, he came as close as was possible using biblical quotes. His selection of quotations also included "When the overflowing scourge passes over" (Winterbotham 1978; 79:1) and assorted allusions to "fire from heaven," "famine" and "the land will be scattered and laid waste".

Everett (2001), who went through Gildas' apocalyptic choices in detail, points out a pattern wherein Gildas uses these Old Testament quotations but makes them contemporary.

> Gildas continues ad nauseam to hammer home his apocalyptic message, and it could only have carried conviction if his contemporaries had had some sort of apocalyptic experience (Everett 2001).

For example here is Gildas making a "telling aside"

> After a while he [Isaiah] discusses the day of judgment and the unspeakable fear of sinners: 'Howl! The day of the Lord is near' (and if it was near then, what are we to suppose today?) 'for destruction is on the way from the Lord.' (Winterbotham 1978; 44:1; Everett 2001).

The comment in brackets is one of several where Gildas relates his quotations to contemporary sixth century happenings that would be recognisable to his readers. Again, here is Everett making such a point:

Gildas refers to the immoral actions of Constantine as 'poisoned showers of rain' (28:4), a curious phrase to use unless the population had recently experienced such a downpour. He (also) urges Aurelius Caninus to shake himself 'free of your stinking dusts' (30:3) (Everett 2001).

Given Gibbon's strange reference:

> Such was the universal corruption of the air, that the pestilence which burst forth in the fifteenth year of Justinian [AD542] was not checked or alleviated by any difference of the seasons… but it was not until the end of a calamitous period of fifty-two years that mankind recovered their health, or the air resumed its pure and salubrious quality (Gibbon 1832).

We can see that Gildas' remarks appear to be part of a pattern. Taken all together, these scattered pieces of information raise the specter that the air around the globe was somehow corrupted in the immediate vicinity of AD 540. What is particularly interesting is that Gibbon's text actually reinforces the assertion: there was "universal corruption of the air," and later "the air resumed its pure and salubrious quality." This is hardly a slip of the pen; Gibbon seems to have been quite confident that the air was "corrupted." Moreover, Gibbon could not have known that the time period he chose to specify corresponds very closely to the "Maya Hiatus" of AD 534–593 (Robichaux 2000). Is this just a coincidence?

5.3
Mythology

Obviously scientists should not involve themselves with myth, unless there is a good reason to do so. However, the irony is that, in its own way, myth seems to contain a better description of what happened around 540 – and its causes – than any history book. There is not space here to go into myth in detail. The salient facts have been published elsewhere (Baillie 1999, 2002; McCafferty and Baillie 2005). What follows is a précis of a complex story.

In Britain King Arthur, probably the most famous Briton of the first Millennium, is said to have died around 540 (variously 537, 539 or 542). He is, without doubt, a Celtic god (scholars have known this for a long time but it is ignored by those who wish Arthur to be a flesh and blood hero). Arthur is cognate with a range of Celtic deities that include Cúchulainn, Mongan and Lugh. Of interest is the fact that Lugh is described in one text as "coming up in the west, as bright as the sun, with a long arm", he is also known for his "terrible blows" (Loomis 1927). As these descriptive elements only befit a comet (what else can be as bright as the Sun, can come up in the west, has a long arm and can deliver terrible blows?), myth is telling us that a 'comet god' died at the time of a global environmental disaster. Another major aspect of Arthurian romance is the "Wasteland" wherein three kingdoms are destroyed. In the stories, this destruction was caused by a "Dolorous Blow" that was delivered by Balin with a bleeding spear. Scholars have traced this Arthurian bleeding spear back to Lugh's spear (Christianized to the spear of Longinus) (Loomis 1927). Thus, mythology, by linking the death of Arthur to the period immediately around 540, tells us what conventional history does not, i.e. *a comet god caused a wasteland around 540*. To repeat, Arthur is cognate with Lugh who is described as a comet, and it is Lugh's spear that causes the Wasteland in Arthurian romance.

I can imagine readers not familiar with such literature saying things like "he lost me when he jumped from King Arthur to Lugh; Arthur might have been around 540 but there is no evidence that Lugh was". It may concentrate the mind, therefore, to know that a document, *Vita s. Mochtaei De Hibernia*, relates to AD 535. Its content is described as follows:

> Mochteus; or Mochta Lugh, a Briton, is said to have been a disciple of St. Patrick, and became the first Bishop of Louth. He died in 535. The piece is, to a great extent, quite fabulous (Hardy 1862, p 117)

Now, Mochta in Old Irish means *great* or *mighty*. So, Mochta Lugh could, at its most simple, mean 'Great Lugh'. An entry to the same effect occurs in the Annals of Ulster and recent scholarship suggests that this basic statement about his death was written before 700 (Sharpe 1990). Sharpe also points out the exceptional nature of the quotation about Mochteus (Mauchteus or Mochta Lugh) in the Annals

> … the quotation from a document … to no annalistic purpose is without parallel in the Annals of Ulster (Sharpe 1990, p 88)

So the compiler of the Annals, sometime before 700, takes an unparalleled step in introducing a reference that could be to Lugh, with a date in the 530s. It might be quite reasonable to interpret this as another metaphor especially as County Louth is named after Lugh, thus, the first Bishop of Louth could also be a cryptic reference to Lugh. As we will see later, there are other uses of metaphor in the period around 540, including another link to Lugh.

Of necessity this is an extreme compression of an enormous amount of information. The simplest way to show that this Arthur/Lugh/540 idea has some substance is to show that there is another version of the myth that tells essentially the same story. Another deity cognate with Lugh is Mongan; in the stories he is Lugh's 'son' or more strictly Lugh's 're-birth', i.e. Lugh back again (MacKillop 1998). In the story *Mongan's Frenzy* (Stephens 1920) we read how Mongan is at a week-long festival at the Navel of Ireland in the year 538. Suddenly the skies go dark, with clouds coming from both east and west, and there is a horrendous shower of hail stones. In order to get away from this unusual phenomenon Mongan has to enter the Otherworld. Therefore, mythology not only tells us that one version of a comet god, Arthur, 'dies' around 540, but another aspect of the same deity 'goes to the Otherworld' at the same time associated with a darkened sky and an unusual hail-storm. Given the belief that Arthur did not die, but actually went to the Otherworld – Avalon – the similarities between the stories is striking. Both come with long-attributed dating within the environmental (tree-ring) window 536–545, and, if we imagine that the Otherworld is in all probability the sky, then both stories have the god going away into the sky to eventually return. Thus myth, when considered with the arguable 'comet' paradigm can be made to make sense where no viable interpretation existed before. The critical point is the placing of these sky-god myths in time, *precisely* at a global environmental downturn defined by dendrochronology.

5.4
What Actually Happened – the Global Consequences?

Let us, for the sake of argument, accept that the cause of the global environmental downturn in the window 536–545 (and running on even later if we accept Gibbon's comments) was a brush, or brushes, with a comet or its debris. We already know that the consequences were reduced tree-growth around the world and widespread famines implying reduced cereal production; should we be imagining reduced plant growth generally? We know that there was a serious plague after 540. We have hints that the primary vector – the cause of the dim-Sun condition – was dust loading of the atmosphere, through some combination of dust, gas and in all probability Tunguska-class impactors. We have direct written testimony that there was the 'dry fog' in 536–37. We also have Zachariah telling us that the "stars were dancing" from 533 to 540 (Hamilton and Brooks 1899); something that might imply atmospheric disturbance. As noted, we have Cassiodorus telling us that "something is coming on us from the stars" (Barnish 1992).

All of this must raise some concern about the plague at the time of Justinian. It has long been assumed that the Plague of Justinian was bubonic plague. But was it? The descriptions of the phenomenon reaching the British Isles in the 540s do not sound much like bubonic plague. There is the Yellow Plague recorded in Wales (Senior 1979), and there is a plague simply called 'Blefed' in Ireland (O'Donovan 1848); neither of these descriptions argue persuasively for the disease having been bubonic plague. Then we have Gibbon's comment about the "universal corruption of the air." Given the allusions to material – dust and showers (variously of stones and blood) – falling from the sky in the period immediately around 540, it has to be asked whether the plague might have included some sort of atmospheric pollution in addition to bubonic plague. We are at liberty to imagine that one of the two most widespread and severe 'plagues' of the last two millennia might have more than a single killing vector – why not bubonic plague and corrupted atmosphere? I want to look a bit further at this aspect of the devastation in the sixth century.

5.5
The Dust and Corrupted Air

There is another source that bears on this issue. Zachariah, the later volumes of whose history are largely missing, at least preserves Book 12, Chapter 5 (Hamilton and Brooks 1899). This section is entitled *The fifth chapter treats of the powder, consisting of ashes, which fell from heaven,* and dates to 556. In this bizarrely entitled chapter Zachariah tells us that:

> In addition to all the evil and fearful things described above and recorded below [mostly lost!], the earthquakes and famines and wars in divers places … there has also been fulfilled against us and against this last generation the curse of Moses in Deuteronomy … (Hamilton and Brooks 1899).

For someone writing in 556, "the last generation" would include those who had lived through the 536–545 events. So, the things that had been fulfilled against those who had lived across 540 could be identified in the writer's mind with Moses' curse from soon after the Exodus. Might the curses in Deuteronomy give us a clue as to what may have been fulfilled against that past generation. What do they include? The key items are that you shall be cursed in the city and in the field and in your store.

> Cursed shall be the fruit of thy body and the fruit of thy land... The Lord shall make the pestilence cleave onto thee...The Lord shall smite thee with a consumption and with a fever, and with an inflammation, and with an extreme burning, and with the sword (or drought), and with blasting, and with mildew... The Lord shall make the rain of your land powder and dust: from heaven shall it come down upon thee... The Lord will smite you with the botch of Egypt, and with the emerods, and with the scab, and with the itch, whereof thou canst not be healed. The Lord shall smite you with madness, and blindness, and astonishment of heart: And you shall grope at noon-days as the blind gropeth in darkness. (Old Testament, Deuteronomy 28,18–28)

Here we see another contemporary writer, like Gildas (but almost certainly independent of Gildas), indulging in biblical metaphor to try to describe the happenings around 540. Again, it would seem that Zachariah is being quite accurate in his choice of metaphor; he is essentially using the Plagues of Egypt as an analogy. He is suggesting darkness at mid-day, dust from heaven, famine and pestilence, and he is talking *specifically* about the period around AD 540. While in this same time window, 536–545, we have historical evidence for a dry fog that renders the Sun dim, for famine, and for plague. A nice twist in Zachariah's metaphor is that there is one last key element in Moses' curse, as follows:

> ... and the stranger that shall come from a far land, shall say, when they see the plagues of that land, and the sickness which the Lord has laid upon it; And that the whole land thereof is brimstone, and salt, and burning, that it is not sown, nor beareth, nor any grass groweth therein, like the overthrow of Sodom and Gomorrah, Admah and Zeboim, which the Lord overthrew in his anger and in his wrath. (Old Testament, Deuteronomy 29,22–23)

By using the Curse of Moses to describe the happenings around 540, Zachariah is incorporating Sodom and Gomorrah into the description, and, of course, those cities were destroyed by "brimstone and fire from the Lord out of heaven" (Old Testament; Genesis 19,24). So, Zachariah's choice of metaphor – with fire and brimstone, darkness at mid-day, famine and pestilence – would seem to be confirming the 540 scenario given by Gildas and Roger of Wendover.

It is apparent that there is a sub-text here. Gildas did not spell out what was happening around 540, nor did Zachariah; both used Old Testament extracts as metaphor. This is interesting in itself, and may well give rise to a whole field of study. But once sensitized to this concept, it seemed relevant to look for other examples. It turns out there is yet another, again in the Irish Annals, bearing the date 539. Here is the entry:

> The Age of Christ, 539. The decapitation of Abacuc at the fair of Tailltin [Teltown], through the miracles of God and Ciaran; that is, a false oath he took upon the hand of Ciaran, so that a gangrene took him in his neck (i.e. St. Ciaran put his hand upon his neck), so that it cut off his head (O'Donovan 1848).

In a sense it doesn't matter what the entry itself says (it reads at first sight like medieval gobbledegook). The important point is that Abacuc is not an Irish name, but here is someone called Abacuc being killed at *Lugh's Fair* at Tailltin in 539 (it was Lugh who traditionally founded the fair at Teltown).

Who, then, is Abacuc? The answer is that he is Habakkuk of the Old Testament. Hence, what the 'complier' of the Annals was doing by saying that Abacuc lost his head in 539 is that, embedded in Chapter 3 of the Book of Habakkuk is what we need to know. (It is in Chapter 3 that Habakkuk mentions 'Thou smotest down the head in the house of the ungodly, and discovered the foundations, even onto the neck of him.') In this case, again, we have an anonymous monk using Old Testament metaphor to describe what was going on around 540. So what does Habakkuk, Chapter 3 tell us? It includes:

> Before him went the pestilence and burning coals (or burning diseases): he ... drove asunder the nations; and the everlasting mountains were scattered ... The sun and moon stood still in their habitation ... the fields shall yield no meat ... and there shall be no herd in the stalls (Old Testament, Habakkuk 3:5–17)

Again this seems like a consistent description of what was going on around 540, with pestilence and burning coals from the sky and famine on the ground. In this case, the writer also provides the strong secondary link to Lugh's Fair – the Festival of the comet god Lugh. However, the links do not end there. In Habakkuk 3, the entity causing the havoc is described as: '... *his* brightness was as the light; he had horns *coming* out of his hand;' It is widely accepted that an alternative to the 'horns coming out of his hand' is 'bright beams out of his side' (Old Testament, Habakkuk 3,4). Given that consideration is being given here to a possible brush with a comet, around 540, how strange that an Irish monk would use an Old Testament metaphor for the happenings at 539 that could be interpreted as a description of some aspect of the dust/gas/ion tail(s) of a comet.

5.6
The Scientific Prior Hypothesis

On the basis of this accumulated evidence, and in the absence of any evidence for a volcano, it now seems reasonable to suggest that around AD 540 – in the window 536–545 – Earth had a brush with a close comet that dumped material into the atmosphere and caused a global environmental downturn. We have the scenario deduced scientifically from dendrochronology and ice-core work. We have several British and Irish recorders telling us essentially the same story, and we have an independent Mediterranean source repeating the same catastrophic elements – all with pre-existing dates.

But there is a scientific surprise. It turns out that there is a pre-existing scientific hypothesis, dating from the 1980s, wherein Clube and Napier (1990) elaborate their 'cosmic swarm' scenario. In this scenario, in a short period of months to years the Earth encounters a range of comet debris. The essential point is that Clube and Napier estimate, because of the more active sky, that running into a 'cosmic swarm' of small objects may have been likely and they make the following statement:

> Overall, it seems likely that during a period of a few thousand years, there is an expectation of an impact, possibly occurring as part of a swarm of material, sufficiently powerful to plunge us into a Dark Age.

Indeed they were even more specific:

If large boulders do form in swarms, then during close encounters with the comet or its degassed remnant there is a risk of occasional bombardment on a scale comparable with that of a nuclear war…The occurrence of Tunguska-like swarms in recorded history is therefore expected … Thus we expect a Dark Age within the last two thousand years.

They reviewed the evidence and, in collaboration with Mark Bailey, went on to suggest:

… it seems probable that the least biased measure of relative meteor activity during the Dark Age is now provided by the recorded incidence of meteor showers… There have probably been at least two significant surges in meteor shower activity [in the last two millennia], namely 400–600AD and 800–1000AD (Bailey et al. 1990).

Thus, scientists are confronted with the scientific case, from tree rings and ice cores, for a brush with a comet around 540. They are confronted with several independent suggestions, from history and mythology, that such a thing *did take place* around 540. Now it seems that there was even a prior hypothesis that a closely-related event involving comet debris might have occurred in the time window AD 400–600.

5.7
The AD 540 Symptoms

Tree-ring chronologies from around the world show that we had a global environmental event, involving reduced growth, in the time window 536–545. Mainstream history does not record the event in any thorough way. However, accumulation of marginal references, annals and mythology indicates that the events are recorded quite widely in non-conventional ways. These records hint strongly that a comet god was involved.

Given that the environmental event, coupled with plague, directly or indirectly killed one third of the population of Europe (there are reasons to assume that the rest of the world may have suffered similarly) it is surprising that the whole issue does not have a more conspicuous place in history. However, from the non-conventional records we see a consistent pattern of references that the atmosphere may have been corrupted – a situation stated by Gibbon but normally ignored for lack of context or corroboration. So, for the purpose of this discussion I am going to suggest that in the mid-sixth century the Earth's atmosphere may have been loaded with cosmic material to a level that was harmful to humans. Moreover, if this dust were indeed debris from a comet we could reasonably expect that it might include a volatile fraction, particularly an organic component.

With that in mind, and given the preceding interpretation of the use of biblical metaphor in the sixth century, Hoyle and Wickramasinghe make the following statement:

By about the sixth century AD, Christian beliefs included the dogma that nothing that happens in the heavens could have any conceivable effect on the Earth (1993; 2–3).

Perhaps this is the reason why early medieval churchmen felt that they could only express themselves metaphorically; to talk about goings on in the sky overtly would have been to go against Church dogma. It would appear, however, that some felt sufficiently motivated by events to circumvent the dogma and to leave clues for anyone who, for whatever reason, might recognize the significance of the biblical quotations. Thus, when our interpretation of the tree-ring data indicated a sixth-century, global, environmental event, and the ice cores indicated, by default, that it might have been extraterrestrial in origin, the metaphors finally make sense.

We can now reasonably re-ask the question prompted by Gibbon, Zacharaih, Gildas, Roger of Wendover and Cassiodorus – was the atmosphere compromized by extraterrestrial pathogens, 'dust' of some sort in the mid-sixth century? The answer is that people writing at the time seem to have been trying to tell us that it was so corrupted.

However, we have access to records they could not have dreamed of. If we go to the Greenland ice cores and look at the ammonium record (Fuhrer et al. 1993) we find that the two highest values in the last two millennia are 46 ppb and 35 ppb ammonium at depths of 238 and 336 meters respectively. These depths correspond to calendar years at or close to AD 1014 and AD539 (see Fig. 5.4). So, an unusual ammonium layer in the GRIP ice core coincides with records of a corrupted atmosphere.

Obviously, given the thrust of this discussion, this is a quite remarkable observation. However, from the point of view of the reliability of some of these ancient records it is hard to improve on the 1014 ammonium signal. If we go to Britton's (1937) meteorological compilation we find the following:

1014
Short refers to a remarkable calamity in this year. He says 'a heap of cloud fell and smothered thousands'. He adduces the Anglo-Saxon Chronicle as authority for this phenomenon, a work in which there is certainly no mention of it. It might conceivably be a poetic distortion for a heavy rainstorm in which many people were drowned (Britton 1937, p 39).

We now have evidence that this reference to a smothering heap of cloud coincides with the largest atmospheric concentration of ammonium in this era (Fuhrer et al. 1996). This raises the question as to the source of such unusual – once in a thousand years – concentrations of ammonium. Conventional wisdom *suggests* that ammonium may be attributable to forest fires (Legrand et al. 1992), however, the authors of that paper display their uncertainty by the insertion of a question mark in the title. Am-

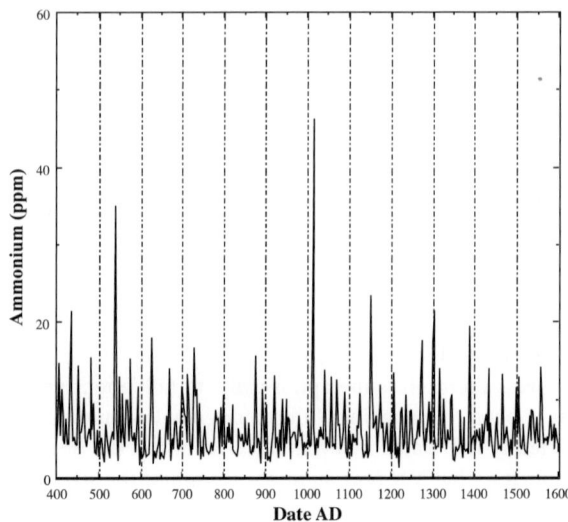

Fig. 5.4.
The ammonium record from AD 400 to 1600 derived from the Greenland GRIP ice core (Furher et al. 1993) [Data provided by the National Snow and Ice Data Center, University of Colorado at Boulder, and the WDC-A for Paleoclimatology, National Geophysical Data Center, Boulder, Colorado]

monium could also come from ocean-bed clathrates, for example, and it is surmised to occur in comets (Sagan and Druyan 1997). What makes 1014 particularly interesting is that it is listed by Sekanina and Yeomans (1984) as a year when a comet made a relatively close approach of the Earth. Thus, the two highest ammonium spikes in the last two millennia both have some comet association.

With respect to the 540 event, we can also ask – how long did its effects endure? The answer in this case is that we don't know, but Gibbon's corrupted air, when combined with the duration of the Maya Hiatus suggests that it could have been prolonged. In Fig. 5.5 we see a notable depression in the envelope of Irish oak growth that lasts from 540 to 590. Could this be a symptom?

The trouble is that more and more pieces of information can be added to this story. Once we accept the duration of corrupted air in the sixth century we find that there is another relevant Irish story. The story involves a sea monster called the Rosault. It washes up on the Atlantic coast in the mid-sixth century. The dating is determined by the story being set at the time of St Columcille (traditional dates 518–597). The monster is described as follows:

> ... he was able to vomit in three different ways three years in succession. One year he turned up his tail, and with his head buried deep down, he spewed the contents of his stomach into the water, in consequence of which all the fish died in that part of the sea... Next year he sank his tail into the water, and rearing his head high up in the air, belched out such noisome fumes that all the birds fell dead. In the third year he turned his head shoreward and vomited towards the land, causing a pestilential vapour to creep over the country that killed men and four-footed animals (Joyce 1913).

Yet again, we see a story set in the sixth century that specifically refers to noisome fumes and pestilential vapour creeping over the country. This ancient Irish story parallels Gibbon's comment on corruption, as does Gildas' metaphorical note 'For from the prophets of Jerusalem pollution has gone out over all the Earth' (Winterbotham 1978, 82:3).

Fig. 5.5.
Widest and narrowest growth rings in each year of the sixth century in Irish oak (widest rings plotted as ring widths; narrowest rings plotted as indices [normalised departures around a value of 100] for clarity) showing the systematic reduction in the envelope of oak growth from AD 540 to 590. *Black bars* represent average values for the periods indicated

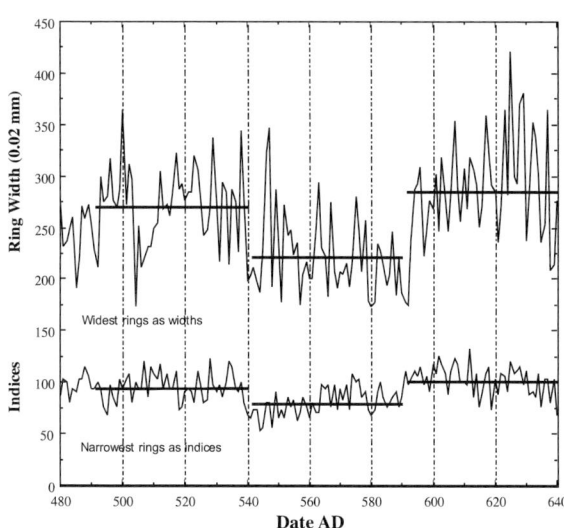

Here is another story:

> In 550 [sic] the Yellow Plague was said to be roaming through the land in the guise of a loathly monster. This was in Wales but in Ireland too the plague was regarded as a living thing that roamed the land. The power of prayer against this creature was amply demonstrated when, at the prayer of St MacCreiche in Kerry, a fiery bolt from heaven fell upon it and reduced it to dust and ashes in the presence of the people (Twigg 1984).

A yellow plague monster roaming the land, with fire from heaven reducing it to dust and ashes! Overall, there is enough information out there to allow the suggestion that the event(s) in the sixth century that triggered the Plague of Justinian (or Yellow plague, or, in Ireland, the plague called Belfed) and the Maya Hiatus included corruption of the atmosphere due to a close brush with a comet. Finally, it is important to realize that there appear to be no equivalent clusters of dated information on atmospheric corruption in the other centuries of the first millennium. The mid-sixth century stands out in this respect; just as the global tree-ring event stands out; just as the cluster of dated myths stands out.

5.8 Linkages to Other Events

Having highlighted the 'strangeness' of the literature relating to the period around the AD 540 global tree-ring event, it seems reasonable to look briefly at the other events high-lighted by the Irish oaks. We have already touched on 207 BC with its Chinese environmental trauma and dynastic change, and Livy's references to 'stones from the sky'. Below are some brief coincidences involving other dated environmental events wherein Irish trees exhibited catastrophic growth reductions. The tree-ring dates are given in bold.

- **2354–2345 BC**. This date marks the transition from the Neolithic to the Bronze Age in the British Isles. This occurred at about the same time as a widespread societal collapse in the Near East (Weiss 1996; Courty 1998; Peiser 1998). It coincides uneasily with Archbishop Ussher's date for the Biblical Flood (2349 BC). Curiously, Isaac Newton, no less, suggested that the biblical Flood of 2349 BC might have been due to a comet (Schechner Genuth 1997).
- **1159–1141 BC**. This event in the middle of the 12th century BC falls close (there being no precisely-dated history at this time) to both the traditional date for the fall of Troy and the end of the Chinese Shang Dynasty. In both cases, the Trojan war and the mythical battle of Mu, it is observed that the battles involve humans *and* sky gods. In the case of Troy it is the god Apollo who brings plague. The 12th century BC sees the start of the four century long Greek Dark Age.

So, these dates, derived purely from tree-rings, provide curious resonances to two of the major events in ancient history, namely the Flood and the Fall of Troy. Now let us rehearse the rest of the Chinese story.

In the 12th year of his reign (trad. 2346 BC) the first Chinese emperor, Yao, meets the Divine Archer Shên I (clearly a version of Apollo). At the time there are terrible catastrophes including *ten suns in the sky*, famines, floods etc. The Divine Archer, having shot down nine of the ten suns, sets out to seek the cause of these catastrophic events and finds that they are due to the activities of one Fei Lien (a wind spirit) (Werner 1995). Now, noting the tree-ring dates, let us look at the associations of this story. In the Chinese story Fei Lien who was responsible for the calamities in the 24th century BC was later a minister of King Chòu, the last emperor of the Shang dynasty who was defeated at the battle of Mu. The Shang dynasty ends by tradition in the 12th century BC. Hence, preserved in a Chinese story is *a link* from the 24th to the 12th century BC, something that implies that some observers in China recognized the similar causes of the two events; such recognition might best be explained by people having seen things in the sky.

By tradition it is at the Fall of Troy that the Greek god Apollo shoots plague arrows, while in China at the time of Chòu (also 12th century BC) a Zeus-like character, No-cha, finds a wonderful bow and three magic arrows. No-Cha shoots an arrow towards the south-west "a red trail indicated the path of the arrow, which hissed as it flew". Subsequently it was observed that the arrow bore the inscription 'Arrow which shakes the heavens' (Werner 1995). Again, in case this seems far fetched, there is an accepted reference to a comet at the fall of the Shang, *viz*:

> When King Wu-wang waged a punitive war against King Chòu [the last king of the Shang dynasty], a comet appeared with its tail pointing towards the people of Yin ... (Sagan and Druyan 1997, p 15)

So, close to two of the early tree-ring dated environmental events (2350 BC and 1150 BC) we have associations with Apollo-like gods. Then, with Arthur's death (542), Mongán's frenzy (538), and possibly with the death of Mochta Lugh (535), we have characters cognate with Lugh, the Celtic Apollo, recorded just around the time that plague breaks out, arriving into Constantinople in AD 542.

5.9
Conclusion

Given that the tree-ring dates are derived scientifically and are well replicated, they cannot easily be moved in time. Thus dated growth departures in these tree-ring records are fully equivalent with any other precisely dated records. It is therefore interesting that both mythical stories and normally disregarded historical records, with dates, should sit so comfortably with the tree-ring dates. In each case there appears to be some reason to invoke a link to comets, or comet debris, or meteorites, whether it is a reference to ten suns in the sky, or Cybele – a goddess manifest as a meteorite – or a suggestion of Isaac Newton, or direct historical references as noted from Roger of Wendover, Gibbon, etc. To these can be added the consistent appearance of comet-associated sky gods, be it the Divine Archer or Apollo or Lugh at the events described in dated myths. More surprising is the consistent use of biblical metaphor to describe happenings in the sixth century. None of this ancient information need exist; but, not only does it exist, it has mostly been treasured from antiquity.

The historian Gibbon tells us that there was a comet in 539 and that the atmosphere was corrupted from 542 to 594, but, while his comet record is accepted, his assertion about corruption has previously been disregarded for lack of corroboration. As shown, there are clear indications that some other ancient writers – writing in the 6th century, and using biblical metaphor – were attempting to convey this same concept of a corrupted atmosphere. Now, relevant to the tree-ring event bracketing AD 540, there is direct scientific information, from the Greenland ice record, showing unusually elevated levels of atmospheric ammonium at the time. It is hard to imagine that this extended package – including information from the written record, from tree rings and from ice cores – is just a coincidence. Rather, it should be an important clue as to the true nature of the 540 event, stated by the normally disregarded medieval historian, Roger of Wendover, as involving "a comet seen from Gaul so vast that the whole sky appeared to be on fire" in 540/541. It is time for a more concerted look to be taken at what is hidden in ancient records and indeed what else is present in the ice cores.

Note: A question to atmospheric scientists. Pentti Zetterberg informs me (pers. comm., 6 April 1999) that the AD 535 growth ring in Finnish pine was the *widest* in his whole 7000-year record. It is suggested that in Japan the year 535 was 'perfectly wonderful' (Aston 1956), while Cassiodorus (Barnish 1992) notes that the year before the dry-fog 'such was last year's [presumably 535] fortunate abundance.' What mechanism might induce a *widespread fertilization effect* in the run up to a dry-fog catastrophe? Could it have been something involving nitrogen fertilization?

Acknowledgments

The author would like to thank Dr Duncan Steel for many helpful comments on the draft text and numerous colleagues in dendrochronology who have willingly shared tree-ring data. Funding towards establishing links between the tree-ring and ice-core records has been provided by the 14-CHRONO initiative at Queen's University, Belfast.

References

Aston WG (1956) Nihongi: chronicles of Japan from the earliest times to AD 697. George Allen and Unwin, London
Bailey ME, Clube SVM, Napier WM (1990) The origin of comets. Pergamon Press, London
Baillie MGL (1994) Dendrochronology raises questions about the nature of the AD 536 dust-veil event. The Holocene 4:212–217
Baillie MGL (1995) A slice through time: dendrochronology and precision dating. Routledge, London
Baillie MGL (1999) Exodus to Arthur: catastrophic encounters with comets. Batsford, London
Baillie MGL (2002) Dublin Institute for Advanced Studies website (*www.celt.dias.ie/publications/tionol/baillie02.pdf*)
Baillie MGL, Munro MAR (1988) Irish tree-rings, Santorini and volcanic dust veils. Nature 332:344–346
Barnish SJB (1992) The variae of Magnus Aurelius Cassiodorus Senator. University Press, Liverpool
Boninsegna JA, Holmes RL (1985) *Fitzroya cupressoides* yields 1534-year long South American chronology. Tree-Ring Bulletin 45:37–42
Briffa KR (1999) Analysis of dendrochronological variability and associated natural climates in Eurasia – the last 10 000 years (ADVANCE-10K). PAGES vol 7(1):6–8
Briffa KR, Jones PD, Bartholin TS, Eckstein D, Schweingruber FH, Karlen W, Zetterberg P, Eronen M (1992) Fennoscandian summers from AD 500: temperature changes on short and long timescales. Climate Dynamics 7:111–119
Britton CE (1937) A meteorological chronology to AD 1450. Geophysical Memoirs No 70, HMSO, London
Chatfield C (2002) Dark days. In: *http://www.phenomena.org.uk/siteguide.htm*

Clausen HB, Hammer CU, Hvidberg CS, Dahl-Jensen D, Steffensen JP (1997) A comparison of the volcanic records over the past 4000 years from the Greenland Ice Core Project and Dye 3 Greenland ice cores. Journal of Geophysical Research 102(C12):26707–26723

Clube SVM, Napier B (1990) The cosmic winter. Blackwell, Oxford

Courty M-A 1998 The soil record of an exceptional event at 4000 B.P. in the Middle East. In: Peiser BJ, Palmer J and Bailey ME (eds) Natural catastrophes during Bronze Age civilizations. BAR International Series 728: 93–108

D'Arrigo R, Frank D, Jacoby G, Pederson N (2001) Spatial response to major volcanic events in or about AD 536, 934 and 1258: frost rings and other dendrochronological evidence from Mongolia and Northern Siberia. Climatic Change 49:239–246

Everett D (2001) Gildas and the plague: sixth-century apocalyptic and global catastrophe. Medieval Life 15:13–18

Forsyth PY (1990) Call for Cybele. The Ancient History Bulletin 4.4:75–78

Fuhrer K, Neftel A, Anklin M and Maggi V (1993) Continuous measurements of hydrogen peroxide, formaldehyde, calcium and ammonium concentrations along the new GRIP ice core from Summit, Central Greenland. Atm Environ 12:1873–1880

Gibbon E (1832) The history of the decline and fall of the Roman empire. Nelson and Brown, Edinburgh

Hamilton FJ, Brooks EW (eds) (1899) The Syriac chronicle known as that of Zacharias of Mytilene. Methuen and Co, London

Hardy TD (1862) Materials relating to the history of Great Britain and Ireland, vol 1. Longman, Green, Longman and Roberts, London

Hoyle F, Wickramasinghe C (1993) Our place in the cosmos. Phoenix, London

James E (1999) Review of David Keys' *Catastrophe*. Medieval Life 12:3–6

Jeffreys E, Jeffreys M, Scott R (1986) The chronicle of John Malalas, Byzantina Australiensia. Australian Assoc, Byzantine Studies 4, Melbourne

Joyce PW (1913) A social history of ancient Ireland. Gill, Dublin

Larsen LB, Siggaard-Andersen M-L and Clausen HB (2002) The sixth century climatic catastrophe told by ice cores. Abstract from the 2002 Brunel University Conference Environmental Catastrophes and Recoveries in the Holocene 29 Aug – 2 Sept 2002 (available at Atlas Conferences Inc. Document #caiq-21)

Legrand MR, De Angelis M, Staffelbach T, Neftel A, Stauffer B(1992) Large perturbations of ammonium and organic acids content in the Summit-Greenland ice core – Fingerprint from forest fires? Geophysical Research Letters 19:473–475

Loomis RS (1927) Celtic myth and Arthurian romance. Columbia University Press, New York

MacKillop J 1998 Dictionary of Celtic mythology. Oxford University Press

McCafferty P, Baillie MGL (2005) Celtic gods: comets in Celtic myth. Tempus, London (in press)

O'Donovan J (1848) Annals of the Kingdom of Ireland by the four masters. Hodges and Smith, Dublin

Pang KD, Slavin JA, Chou H-H (1987) Climatic anomalies of late third century BC: correlation with volcanism, solar activity and planetary alignment. EOS Transactions American Geophysical Union 68:1234

Peiser B 1998 Comparative analysis of late Holocene environmental and social upheaval: evidence for a global disaster around 4000 BP. In: Peiser BJ, Palmer J and Bailey ME (eds) Natural catastrophes during Bronze Age civilizations. BAR International Series 728:117–139

Robichaux HR (2000) The Maya Hiatus and the AD 536 atmospheric event. British Archaeological Reports (International Series) 872:45–53

Sagan C, Druyan A (1997) Comet. Headline, London

Scuderi LA (1990) Tree-ring evidence for climatically effective volcanic eruptions. Quaternary Research 34:67–85

Schechner Genuth S (1997) Comets, popular culture, and the birth of modern cosmology. Princeton University Press, Princeton

Sekanina Z and Yeomans DK (1984) Close encounters and collisions of comets with the Earth. Astronomical Journal 89(1):154–161

Senior M (1979) Myths of Britain. Book Club Associates, London

Sharpe R (1990) Saint Mauchteus, *discipulus* Patricii. In: Bammesberger A, Wollmann A (eds) Britain 400-600: language and history. Carl Winter, Heidelberg, pp 85-93

Stephens J (1920) Irish fairy tales. McMillan, New York

Stothers RB (1984) Mystery cloud of AD 536. Nature 307:344-345

Stothers RB, Rampino MR (1983) Volcanic eruptions in the Mediterranean before AD 630 from written and archaeological sources. Journal of Geophysical Research 88:6357-6371

Twigg G (1984) The black death: a biological reappraisal. Batsford, London

Weisburd S (1985) Excavating words: a geological tool. Science News 127:91-96

Weiss H (1996) Late third millennium abrupt climate change and social collapse in West Asia and Egypt. In: Dalfes HN, Kukla G and Weiss H (eds) Third millennium BC climate change and Old World collapse. Springer-Verlag, Berlin, pp 711-723

Werner ETC (1995) Ancient tales and folklore of China. Senate, London (originally published 1922)

Winterbotham M (1978) Gildas: The ruin of Britain and other works. Phillimore, London

Zetterberg P, Eronen M, Briffa KR (1994) Evidence on climatic variability and prehistoric human activities between 165 BC and AD 1400 derived from subfossil Scots Pines (*Pinus sylvestris*) found in a lake in Utsjoki, Northernmost Finland. Bulletin of the Geological Society of Finland 66:107-124

Chapter 6

The GGE Threat:
Facing and Coping with Global Geophysical Events

W. J. McGuire

6.1 Introduction

The threat posed to our planet and our civilization by future comet and asteroid impacts (CAIs) is now widely recognized and is becoming increasingly well constrained. Recent studies have provided tighter estimates of the numbers of potentially-threatening objects, particularly within the near-Earth space (Near Earth Object Science Definition Team 2003), better approximations of likely frequencies of collision with objects of various diameters (e.g. Chapman 2004), and a more realistic appreciation of the effects of CAIs on society and the environment (e.g. Toon et al. 1997; Morrison et al. 2004). In this regard, the hazard and risk associated with CAIs are now far better comprehended than those linked with other geological and geophysical phenomena capable of affecting the entire planet or impinging in some detrimental way upon the global community. Such *global geophysical events (GGEs)* form a compendium of low frequency-high magnitude phenomena of which CAIs are just a single element. While far less well understood, and therefore scientifically much more controversial, *terrestrial GGEs* currently appear at least as hazardous as impacts of kilometer-sized and larger bolides, and to have frequencies that are considerably shorter than CAIs capable of comparable levels of destruction and disruption (Tables 6.1 and 6.2). A miniscule glimpse of this capability was provided by the December 26, 2004 Asian earthquake and tsunami, which claimed an estimated 250 000 lives (including 100 000 children), destroyed close to half a million buildings, and led to eight million people being made homeless, impoverished, displaced or unemployed.

At the top end of the potential damage scale are so-called volcanic super-eruptions (e.g. Rampino and Self 1992, 1993a), with return periods that may be as short as 5×10^4 yr, and giant (mega) tsunamis of ocean-basin extent (e.g. Ward and Day 2001) arising as a consequence of the catastrophic collapse of the flanks of ocean-island volcanoes, with time-averaged frequencies estimated at 10^4 yr. Volcanic super-eruptions have been charged with having the potential to severely impact upon society and the environment through triggering a period of severe global cooling (*Volcanic Winter; the terrestrial equivalent of Cosmic Winter*) (Rampino et al. 1988) lasting on the order of 10^3 days. Although the damage and disruption 'footprint' of an ocean-wide giant tsunami is likely to be, at most, sub-hemispherical rather than global, the level of physical destruction may be considerably greater than for a super-eruption. It is worth noting here that while the formation of so-called mega-tsunami due to large-scale structural failure of island volcanoes forms a focus here, the triggering of such phenomena has also been

Table 6.1. Summary of terrestrial, rapid-onset GGEs (age and frequency data from Bryan et al. 2003; Chesner 1998; Christiansen 2001; Decker et al. 1990; Maslin et al. 2004; Mason et al. 2004; McMurtry et al. 2004; Oppenheimer 2003a)

GGE type	Selected past occurrences (ages: years before present except where indicated)	Identified future threats location/region (selected)	Event frequency (yr)
Volcanic super-eruption	Yellowstone, U.S. (0.64×10^6 and 2.1×10^6) Toba, Indonesia (7.35×10^5)	Pacific rim and SE Asia	$\geq 5 \times 10^4$
Ocean-island volcano collapse and tsunami	Mauna Loa, Hawaii (1.20×10^5)	Hawaii, Canary and Cape Verde archipelagos	$\sim 10^4$
Continental margin sediment failure and tsunami	Storegga slide, Norway ($\sim 8.15 \times 10^3$)	Atlantic margin; Puerto Rico Trench	27 slides identified in North Atlantic region over last 4.5×10^4 yr
Climate-purturbing volcanic eruptions	Baitoushan, China A.D. 1030 A.D. 1259 (location unknown) Laki, Iceland A.D. 1783 Tambora, Indonesia A.D. 1815	?	250 – 1 000
Global economy-disrupting earthquake	Tokyo (Japan) A.D. 1923	Tokyo	200 – 500

Table 6.2. Impact frequencies of near Earth objects proposed by Stuart and Binzel (2004)

Impator size (m)	Impact frequency (10^3 yr)
40–50	2–3
~200	56 ±6
1 000	600

recognised as a consequence of gigantic sediment slides along submarine continental margins (e.g. Bugge et al. 1988; Ward 2001) and postulated by some to result from CAIs occurring in the ocean environment. A sub-group of smaller-scale or lower intensity *GGEs* have destruction and damage potential orders of magnitude smaller than those of super-eruptions and mega-tsunami, but remain capable of severe impacts on global society as a consequence of progressive effects on the world's climate or economy. In the former category are climate-perturbing volcanic eruptions (e.g. Oppenheimer 2003a, 2003b) that, by dint of ejected volume or mass, fall short of super-eruption status and catastrophic earthquakes that strike at the heart of a G8 economy (Rikitake 1991; Bendimerad 1995).

All *GGEs* so far addressed have in common the property of spontaneity. In other words, whether forecast in advance or not and although the consequences arising may be long-lasting, they are sudden-onset events. This is not, however, a required diagnostic attribute of a *GGE*. Contemporary climate change, although slower-acting should, in fact, be considered the most threatening *GGE* of all, partly because its true scale and ramifications remain – as yet – cryptic, and partly because it is the only *GGE* whose effects are already becoming apparent. Discrete, large-scale, geological and geophysical phenomena that may arise as a consequence of climate change may also qualify for GGE status in their own right, in particular major changes in the behavior of North Atlantic currents (e.g. Dickson et al. 2003; Häkkinen and Rhines 2004; Hansen et al. 2001; Bryden et al. 2005), capable of leading to severe regional cooling, large-scale sea-level rise linked to wholesale melting of the Greenland or Antarctic ice sheets (e.g. Gregory et al. 2004) and the increased persistence of ENSO (El Niño-Southern Oscillation) conditions in the eastern Pacific Ocean (e.g. IPCC 2001).

Coping with climate change – through a combination of greenhouse-gas reduction, increased emphasis on more sustainable energy production and lifestyles, and adaptation to the warmer and more hazardous world that is now inevitable – is a huge issue in its own right and not one I address here. Instead I focus on summarising our current knowledge of rapid-onset terrestrial GGEs, addressing pertinent matters of debate, discussion and controversy, and considering ways in which the threat they pose might initially be approached.

6.2
Volcanic Super-Eruptions

The term *super-eruption* has, in the last few years, become synonymous with volcanic events registering a score of 8 on the Volcanic Explosivity Index (VEI). Introduced in 1982, the VEI (Newhall and Self 1982) uses a number of parameters, including height of the eruption column, volume of material ejected and eruption rate, to determine the scale of an eruption. The index starts at 0 and is open-ended, although nothing larger than a VEI 8 has yet been identified in the geological record. This may, perhaps, reflect the fact that crustal properties will not support a magma chamber large enough to supply greater volumes of magma to the surface in a single eruptive episode. Like the Richter Scale of earthquake magnitude, the index is semi-quantitative logarithmic so that each value on the scale represents an eruption ten times larger in volume than the previous value. The lowest value on the VEI is reserved for non-explosive eruptions that involve the gentle effusion of low-viscosity basaltic magmas that characterise eruptions of the Hawaiian volcanoes such as Kilauea and Mauna Loa. VEI 1 and 2 eruptions are described as small to moderate explosive eruptions that eject less than 10^7 m^3 of debris. VEI values 3 to 7 designate progressively more violent explosive eruptions of andesitic (containing 52–63 percent silica or SiO_2) and dacitic (63–68 percent SiO_2) magma, capable of ejecting greater and greater volumes of debris and gas to higher levels in the atmosphere. Eruptions registering 8 on the VEI scale are very rare and involve the explosive ejection of 10^3 km^3 or more of high-viscosity rhyolitic magma. Such enormous explosions can deposit tephra (volcanic debris – mainly ash – that has traveled through the atmosphere) across millions of square kilometers; for example, the Toba (Sumatra, Indonesia) super-eruption, which occurred around 7.35×10^4 yr BP (Chesner at al. 1991), covered one percent of the Earth's surface with more than 10 cm of ash (Rose and Chesner 1987). In a similar manner to CAIs, such events are held capable of triggering rapid and dramatic changes in the Earth's physical environment through the emplacement of enormous volumes of debris and gas into the stratosphere and its distribution across the planet.

Although eruptions in the 0–4 range are commonplace, the larger events have progressively lower frequencies. VEI 5 eruptions, of which the 1980 Mount St. Helens blast is an example, occur – on average – every decade or so, whereas VEI 6 events (e.g. the 1991 eruption of Pinatubo, The Philippines) have return periods approximating a century. The only historic eruption to merit a 7 occurred at Tambora (Indonesia) in 1815, and the frequency of such potentially climate-perturbing events is likely to lie somewhere between 500 and 1000 years (Oppenheimer 2003a).

Over the last 2 million years, Decker (1990) estimates that there have been some 40 eruptions of a size deserving of VEI 8 super-eruption status, yielding an average return period of around 5×10^4 yr. Little is known about the majority of these cataclysmic eruptions, and the most closely studied are those that occurred at Yellowstone (Wyoming, USA) 2.1×10^6 and 6.40×10^5 yr BP (Smith and Braile 1994; Christiansen 2001) (Fig. 6.1), the 7.35×10^4 yr BP event at Toba, and the 2.65×10^4 yr BP Oruanui eruption (Taupo, New Zealand) (Wilson 2001), the most recent VEI 8 eruption. Invariably, explosive eruptions on this scale involve siliceous, high-viscosity rhyolitic magma and result in the formation of large calderas. At both Yellowstone and Toba these caldera systems remain active and *restless* and are characterized by continuing hydrothermal activity, seismicity and surface deformation. The formation of the large volumes of rhyolitic (SiO_2 > 68 percent) magma required to feed a super-eruption necessitates the involvement of similarly silica-rich continental crust in magma formation and limits such volcanic systems, therefore, to ocean-continent destructive plate margins (e.g. Toba) or continental mantle plume settings (e.g. Yellowstone). During super-eruptions, magma is commonly ejected from ring fractures that during the later stages of

Fig. 6.1. The Yellowstone region (Wyoming, USA) has hosted two VEI 8 super-eruptions, 2.1×10^6 and 6.40×10^5 yr BP. The Yellowstone caldera remains restless and future cataclysmic events cannot be ruled out

the eruptions act as faults along which a central crustal block subsides to form a caldera. Magma is expelled in the form of curtain-like eruption columns that may last for up to two weeks (Ledbetter and Sparks 1979), and which can loft tephra to altitudes of 40 km or more. Column collapse typically results in the formation of extensive pyroclastic flows that deposit *ignimbrite* (pumice-rich pyroclastic flow material) over areas of 10^4 km^2 or more. While the coarser tephra component falls to Earth locally, progressively finer fractions are deposited over an area the size of a continent, with the finest material being distributed globally by stratospheric winds. On the VEI, super-eruptions are classified as those that eject volumes of debris on the order of 10^3 km^3. In actual fact, however, volumes may be considerably greater, with the 2.1×10^6 years BP eruption at Yellowstone ejecting around 2.45×10^3 km^3 of debris and the Toba event expelling at least 2.8×10^3 km^3 (Rose and Chesner 1990) and possibly as much as 6×10^3 km^3 (Bühring et al. 2000) of material.

6.3
The Toba Super-Eruption

In terms of its environmental impact, the latest and greatest eruption of Toba is the most closely studied (e.g. Rampino and Self 1992, 1993a; Bekki et al. 1996; Yang et al. 1996; Zielinski et al. 1996). According to Rampino and Self (1992) the eruption lofted 10^{12} kg of fine ash and 10^{13} kg of sulfur gases to altitudes of between 27 and 37 km, creating dense stratospheric clouds of dust and aerosols. Zielinski et al. (1996) estimate, on the basis of volcanic sulfate recorded in the GISP2 Greenland ice core that the total stratospheric sulfate aerosol loading due to the combination of SO$_2$ and atmospheric water may have been as high as 4.40×10^9 kg. This is perhaps 20 times greater than that caused by the 1815 eruption of Tambora, which resulted in a Northern Hemisphere temperature fall of 0.7 °C. The resulting global aerosol optical depth (a measure of the opacity of the atmosphere) following the Toba eruption is estimated to have been 10, in comparison to 1.3 following the Tambora blast, and is sufficient to have caused a northern hemisphere temperature fall of 3–5 °C (Rampino and Self 1992, 1993a).

The length of the *Volcanic Winter* triggered by the Toba event is not well constrained, but assuming an *e-folding* stratospheric residence time for the Toba aerosols of about 1 year, Rampino and Self (1992) suggest that it could have lasted for several years. In support of this, a ~6 year long period of volcanic sulfate recorded in the GISP2 ice core at about the time of the Toba eruption suggests that the residence time of the Toba aerosols may have been on this order (Zielinski et al. 1996). This is supported by modeling undertaken by Bekki et al. (1996), which suggests that SO$_2$ aerosol levels in the stratosphere would have been above background for nearly a decade. Zielinski et al. (1996) also recognize a 1000 year long cooling episode – prior to Dansgaard-Oeschger event 19 (a warm *interstadial around* 7×10^4 yr BP) – and immediately following the deposition in the ice of the Toba sulfate, and suggest that the longevity of the Toba stratospheric loading may account at least for the first two centuries of this event.

The impact on our human ancestors of such an extended period of volcanogenic cooling remains largely a matter for speculation, although Rampino and Self (1993b) have made a link with a putative late Pleistocene human population crash that may have reduced the race to a few thousand individuals. More recently, Rampino and Ambrose (2000) have invoked the Toba eruption to explain a severe culling of the human population, from the survivors of which the modern human races differentiated around 7×10^4 yr BP. Although appreciating the ability of a single volcanic eruption, however large, to dramatically influence the evolutionary development of the human race, a similar impact has also been proposed for the much smaller (~200 km^3) Campanian Ignimbrite eruption in the Bay of Naples region of Italy (Fedele et al. 2002).

6.4
Reassessment of the Super-Eruption Threat

Recently, the sizes and frequencies of the largest explosive eruptions have been revisited and reassessed, along with the degree to which they are capable of significantly affecting the global climate. Mason et al. (2004), in particular, draw attention to problems with the VEI in the context of providing a means of comparing the mass of material ejected by different eruptive events. As the allocation of a VEI value is primarily dependent on the bulk volume of ejected material, it takes no account of the density of the material deposited. In consequence, the eruption of 0.5×10^3 km^3 of rhyolitic magma may result in the deposition of more than 10^3 km^3 of poorly consolidated tephra or just 0.6×10^3 km^3 of dense ignimbrite. On the VEI scale, the former would qualify as a VEI super-eruption, and the latter as just a VEI 7 event. To ensure better comparability between eruptions, Mason et al. (2004) apply a logarithmic *magnitude* scale based upon erupted mass, which they use to define the largest known eruptions. Forty-seven events are identified with masses of 10^{15} kg or more, 42 of which occurred in the last 3.6×10^7 yr, and six in the last 2×10^6 yr. The latter include the two VEI 8 eruptions at Yellowstone, along with those of Toba and Oruanui. Frequency figures determined by Mason et al. (2004) for a Magnitude 8 event are wide ranging, from 1.4 to 22 every 10^6 yr, the latter close to the figure proposed by Decker (1990) on the basis of a far less complete data set. Assuming homogeneous Poisson behavior, this translates into at least a 75 percent probability of a Magnitude 8 eruption within the next 10^6 yr and a one percent chance of a Magnitude 8 eruption in the next 4.6×10^2 to 7.2×10^3 years (Table 6.3).

The received wisdom on the environmental effects of VEI 8/Magnitude 8–9 volcanic events has also been readdressed, most notably in Oppenheimer (2002), where a note of caution is expressed regarding the scale and consequences of the resulting episode of global cooling. Oppenheimer (2002) draws attention to the fact that whereas the Toba event represents the largest known Quaternary eruption, estimates of its sulfur yield – the primary determinant of resulting cooling – vary by two orders of magnitude, from 3.5×10^{10} to 330×10^{10} kg (Becki et al. 1996; Zielinski et al. 1996; Scaillet et al. 1998). As a consequence, Oppenheimer (2002) suggests that previous estimates of globally averaged surface cooling due to the eruption of 3–5 °C may be too high, and proposes a more realistic figure of one degree Centigrade. A role for the Toba eruption in triggering the millennium of colder climate prior to Dansgaard-Oeschger event 19 is also questioned; indeed Oppenheimer (2002) points out that a similar cool stadial preceding Dansgaard-Oeschger event 20 was not associated with an eruption. Consensus still holds that the Toba eruption had a major effect on the planet's cli-

Table 6.3. Likelihood of future large magnitude eruptions, assuming homogeneous Poisson behaviour (from Mason et al. 2004). A frequency of 1.4 events per 10^6 yr corresponds to the known (minimum) large eruption rate since 1.35×10^7 yr B.P.; 2 events per 10^6 is approximately the known large eruption rate from 2.5–3.6×10^7 yr B.P., and for the period 0–6×10^6 yr. A frequency of 22 events per 10^6 yr is the upper bound determined by extreme value analysis

Eruptions of magnitude 8 and larger	Frequency		
	1.4 events/Ma	2 events/Ma	22 events/Ma
Probability of 1 event in the next 100 yr (%)	0.014	0.02	0.2
Probability of 1 event in the next 10^6 yr (%)	75	86	100
Time for 1% chance of an eruption (yr)	7 200	5 000	460
Time to 95% probability of an eruption (10^6 yr)	2.1	1.5	0.14

mate, environment, and perhaps even human demography, it is also clear that the true scale and extent of that impact will not be resolved until the details of the eruption are better understood. It is highly likely, however, that even based upon a best-case scenario, a future eruption on this scale would have a major impact on our civilization as a consequence of, at the very least, a global fall in temperatures lasting for several years.

6.5
Collapsing Ocean-Island Volcanoes and Mega-Tsunami Formation

Landslides from ocean-island volcanoes (Keating and McGuire 2000, 2004) are among the biggest catastrophic mass movements on the planet. Around 70 major landslides have been identified around the Hawaiian Island archipelago, the largest having volumes in excess of 5 000 km³ and lengths of over 200 km (e.g. Moore et al. 1994). Such volcanic landslides are now proving to be widespread in the marine environment (Holcomb and Searle, 1991; McGuire 1996, 2006) and have been identified around other island groups, such as the Canary and Cape Verde islands, and around individual island volcanoes including Stromboli, Piton des Neiges and Piton de la Fournaise (Réunion Island, Indian Ocean), Tristan de Cunha, the Galapagos Islands, Augustine Island (Alaska) and Ritter Island (Papua New Guinea).

6.6
Volcano Instability and Structural Failure

Serious attention became focused on the unstable nature of volcanic edifices, and their tendency to experience structural failure, following the spectacular landslide that triggered the climactic eruption of Mount St. Helens during May 1980 (Lipman and Mullineaux 1981). Such behavior is now recognized as ubiquitous, and evidence for collapsing volcanoes has been recognized both within the geological record and at many of the world's currently active volcanoes (e.g. Ui 1983; Siebert 1984). Although Siebert (1992) estimated that structural failure of volcanic edifices had occurred roughly four times a century over the past 500 years, this may be an underestimate. Belousov (1994), for example, points out that there were three major collapses in the last century occurring in the Kurile-Kamchatka region of Russia alone.

Active volcanoes are dynamically evolving structures, the growth and development of which are typically punctuated by episodes of edifice instability, structural failure and ultimately collapse (McGuire 1996, 2006). Growing volcanoes may become unstable and experience collapse at any scale, ranging from minor rock falls with volumes of the order of a few hundred to a few thousand cubic meters to the giant 'Hawaiian-type' megaslides involving in excess of 10^3 km^3. Low volume collapses occur at some volcano on an almost daily basis whereas the largest events have frequencies of 10^4–10^5 years. The causes of volcano instability are manifold and some volcanoes clearly have a greater propensity to become destabilized than others. Despite low slope angles and an essentially homogeneous structure, instability development is common at large basaltic volcanoes, where persistent dyke-related rifting is implicated in large-scale failure of the flanks. In the marine environment, instability at large basaltic volcanoes may be increased by edifice spreading along weak horizons of oceanic sediment. Strato-volcanoes composed of a mixture of lavas and pyroclastic materials are also easily destabilized, partly due to their unsound mechanical structure and partly due to their characteristic steep slopes and the high precipitation rates that often accompany their elevation. The development of instability and the potential for failure are enhanced at all types of volcano by the fact that actively-growing edifices experience continuous changes in morphology, with the *endogenetic* (by intrusion) and *exogenetic* (by extrusion) addition of material often leading to over-steepening and overloading at the surface. Once a volcanic edifice has become sufficiently destabilized, structural failure and collapse may be induced by any one of a number of triggers. These include earthquakes, elevated mechanical stress or pore-water pressurization resulting from the emplacement of fresh magma, or environmental factors such as changes in sea level or variations in the prevailing climate.

6.7
Environmental Triggers of Ocean-Island Volcano Collapse

At volcanic oceanic islands and coastal volcanoes large, rapid changes in sea level are likely to play a significant role in contributing to edifice destabilization and collapse. By linking sea-level change and the incidence of explosive volcanism in the Mediterranean, McGuire et al. (1997) proposed that some of the eruptions might be triggered by structural failure and collapse. The seaward-facing flank(s) of any volcano is inevitably the least buttressed. This applies both to coastal volcanoes such as Mount Etna (Sicily), where the topography becomes increasingly elevated inland, and to island volcanoes such as Hawaii, where younger centers (such as Kilauea) are buttressed on the landward side by older edifices (such as Mauna Loa). This morphological asymmetry leads to a preferential release, in a seaward direction, of accumulated intra-edifice stresses due, for example, to surface over-loading or to persistent dyke emplacement. Stress release may take the form of co-seismic down-faulting towards the sea, the slow displacement of large sectors of the edifice in the form of giant slumps, the episodic generation of catastrophic landslides, or a combination of all three. Inevitably, the relatively unstable nature of the seaward-facing flanks of any marine volcano is further enforced by the dynamic nature of the land-sea contact. McGuire et al. (1997) demonstrated that large sea level changes are implicated in significant internal stress variations at coastal and island volcanoes, which may contribute towards eruption, collapse or both. More directly, peripheral erosion associated with rapid sea-level rise and the removal of lateral buttressing forces due to a large sea-level fall might also be expected to promote collapse of the flanks of island and coastal volcanoes (McGuire 1996).

An alternative model is proposed by Day et al. (2000), who advocate a correlation between the timing of prehistoric giant lateral collapses on low latitude volcanic archipelagos, such as the Canaries and Hawaiian Islands, and the precession-forced sea-surface temperature (SST). Day and co-authors note that as sea levels rise following glacial terminations so does the low latitude SST. This sea-surface warming is in turn accompanied by changes in the pattern and characteristics of the trade winds so that they bring increased humidity to low-latitude volcanic islands and increased precipitation on their mid-flanks and summit regions. This, the authors propose, leads to a rise in the water table on the order of several hundred meters and so to an increased opportunity for collapse as a result of intruded magma pressurising groundwater in the core of the volcano. Day et al. (2000) also point out that, at least over the past 200 000 years, giant collapses of ocean-island volcanoes appear to be clustered, with the clusters having periodicities that reflect the ca. 20 ka Milankovitch precessional forcing of sea-surface temperature maxima at low latitudes. In proposing that the ocean volcano collapse hazard is greatest during warm periods such as the present, they also tentatively suggest that contemporary global warming might further exacerbate the situation.

6.8
Tsunami Generation from Ocean-Island Volcano Collapses

There is increasing evidence in the geological and geomorphological records that major collapses at ocean-island volcanoes trigger ocean-wide giant tsunami. Around five percent of all tsunami are related to volcanic activity, and at least a fifth of these are the result of volcanic landslides entering the ocean (Smith and Shepherd 1996). Due to an often greater vertical drop and to the high velocities attained, the tsunami-producing potential of a large body of debris entering the sea is much greater than that of a similar-sized submarine landslide, and even small subaerial volcanic landslides can generate highly destructive waves if they enter a large body of water. In 1792 at Mount Unzen (Japan), for example, a landslide with a volume of only $\sim 0.33 \times 10^9$ m^3 – which was not connected with an eruption – entered Ariake Bay and triggered a series of tsunami that caused 14 500 deaths. More recently, many deaths are thought to have resulted from the collapse of the Ritter Island volcano (Papua New Guinea) in 1888, which generated tsunami with wave run-up heights of 12–15 m (Johnson 1987).

Tsunami associated with giant collapses at oceanic-island volcanoes can, however, have run-up heights an order of magnitude greater. For example, a wave train associated with collapse of part of Mauna Loa (Hawaii) – the so-called Alika 2 Slide – around 120 000 years ago has been implicated in the deposition of coral and other debris to an altitude of up to 400 m above current sea level on the neighboring island of Kohala (McMurtry et al. 2004). Giant waves generated by ancient collapses in the Hawaiian Islands appear to have been of Pacific-wide extent, and Young and Bryant (1992) explain signs of catastrophic wave erosion up to 15 m above current sea level along the New South Wales coast of Australia – 14000 km distant – in terms of impact by tsunami associated with a major collapse in the archipelago around 1.05×10^7 yr BP. These phenomena have also, however, been interpreted in terms of tsunami generated by marine impacts.

Potential giant-tsunami deposits continue to be identified at increasing numbers of locations. On Gran Canaria (Pérez-Torrado 2006) and Fuerteventura in the Canary Islands, for example, deposits consisting of rounded cobbles and broken marine shells have been recognized at elevations of up to 100 m above present sea level (S. J. Day, pers. comm.) and may have been emplaced by waves associated with ancient collapses in the archipelago. Similarly, large coral boulders weighing up to 2000 tonnes on the Rangiroa reef (French Polynesia) have been linked by Talandier and Bourrouilh-le-Jan (1988) with giant tsunami formed by the early 19th century collapse of the Fatu Hiva volcano (Marquesas Islands, Southeast Pacific). Most spectacularly of all, boulders measuring up to 10^3 m^3, scattered along the north-east coast of the Bahamiam island of Eleuthera, and associated geomorphological features (Hearty 1997; Hearty et al. 1998) may provide evidence of the impact of giant tsunami from a major collapse event at the Canary Island of El Hierro that occurred around 120 000 years ago (S. J. Day, pers. comm.).

Fig. 6.2. The Cumbre Vieja volcano on the Canary Island of La Palma is the most active in the archipelago. During an eruption in 1949, the western flank of the Cumbre Vieja detached itself from the remainder of the edifice and dropped 4 m

6.9
Contemporary North Atlantic Mega-Tsunami Risk

Much attention has been focused recently, both within the world media and the tsunami community, upon the incipient giant landslide on the Canary island of La Palma and its potential for triggering devastating mega-tsunami. During an eruption in 1949, the western flank of the Cumbre Vieja volcano (Fig. 6.2), which occupies the southern half of the island, appears to have detached itself from the remainder of the edifice and spontaneously dropped 4 m. Geodetic monitoring during the mid-1990s (Moss et al. 1999), hinted that the landslide – which may have a volume approaching 5×10^2 km^3 – continues to creep downslope at the rate of 1 cm or less a year. As is a common feature of ocean-island volcanoes, the eventual entry of the detached mass into the North Atlantic Ocean is likely to occur catastrophically, with a velocity on the order of 100 m s^{-1}. Ward and Day (2001) have modeled the consequences, predicting the formation of an initial dome of water 900 m in height that subsides to a series of devastating waves hundreds of meters high. For collapse scenarios involving a range of masses from $1.5-5 \times 10^2$ km^3, Ward and Day predict a wave train that transits the entire Atlantic Ocean (Fig. 6.3), with wave heights along the coast of the Americas ranging from 10–25 m (for a 5×10^2 km^3 slide) to 3–8 m (for a 1.5×10^2 km^3 slide). The timing of future collapse is completely unconstrained, but the event is most likely to occur during a future eruption when elevated seismic shaking, the pressure of intruded magma and additional impetus provided by magma-heated ground water, will provide optimum conditions for the slide to complete its journey to the sea floor. This can only happen, however, when sufficient strain has accumulated along the future slide plane.

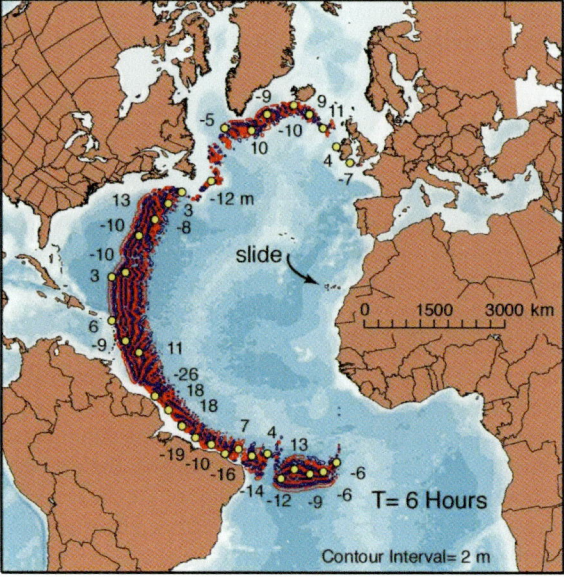

Fig. 6.3.
Time-slice from the Ward and Day (2001) La Palma tsunami model, showing the location of the wave train 6 h after collapse. *Positive numbers* = wave crest heights (m); *negative numbers* = wave trough heights (m)

Areas of controversy with respect to the future failure of the Cumbre Vieja's western flank center upon: (*i*) whether or not entry into the ocean will be catastrophic or sufficiently slow as to minimize or negate the tsunami threat, and (*ii*) if catastrophic, whether resulting tsunami will retain sufficient energy to be destructive at remote locations. With regard to the former, there is overwhelming evidence – both observational and in the geological record – for catastrophic failure and high transport velocities being the norm in respect of the lateral collapse of steep-sided volcanoes. In May 1980, failure of the north flank of the Mount St. Helens volcano (Washington State, US) occurred in less than a minute, with the landslide achieving velocities in excess of 80 m s^{-1} (Lipman and Mullineaux 1981). In the Canaries archipelago itself, an aborted prehistoric landslide on the neighboring island of El Hierro is evidenced by a 300 m slip-surface upon which friction-melted rock known as *pseudotachylite* testifies to very rapid transport (Carracedo et al. 1999). In the Hawaiian archipelago, the recently identified tsunami deposits at an elevation of 400 m on the flanks of Kohala volcano (McMurtry et al. 2004) also require the very high velocity entry into the ocean of the Alika 2 landslide. Similarly, large subaerial landslides (Sturzstroms), such as that responsible for the 1963 Vajont (NE Italy) dam disaster (Kilburn and Petley 2003), are also catastrophic phenomena. A number of authors (e.g. Mader 2001; Pararas-Carayannis 2002) have argued that even if giant volcanic landslides do enter the ocean environment catastrophically, because they are essentially point sources, the tsunami they generate will lose energy sufficiently rapidly to prevent destruction of coastal venues on an ocean-basin scale. Arguments over volcano collapse tsunami propagation and dispersion mechanisms in relation to the far-field effects are likely to continue. In support of destructive wave persistence from landslides in the ocean environment, however, Ward (2001) and Ward and Day (2003) have been successful in modeling known far-field parameters of tsunami generated – respectively – by the giant Storegga submarine sediment slide off the Norwegian coast around 7.2 × 10^3 yr BP, and the collapse of part of the Ritter Island (Papua New Guinea) volcano in 1888.

6.10
High-frequency GGEs

Geophysical phenomena capable of global impact are not required to have average return periods measured in millennia or tens of millennia. Most notably, a number of large volcanic eruptions during the last thousand years, although falling far short of super-eruption status, have had a significant effect on regional or global weather and climate. By far the best known and studied is the 1815 eruption of Tambora (Sumbawa island, Indonesia), which is held responsible for 1816 being a *year without a summer* in Europe and North America. The Tambora blast scores a seven on the VEI scale and is estimated to have ejected ~1.4 × 10^{14} kg of debris. The eruption's climatic impact arose from the injection of perhaps 60 megatonnes of sulfur into the stratosphere, some six times more than was released by the 1991 Pinatubo (Philippines) eruption (Oppenheimer 2003a). The resulting sulfate aerosol veil led to significant climate perturbations, with unusually cold weather affecting Europe, eastern Canada and the northeast US the following year. Among other effects, the aftermath of the eruption is blamed for widespread crop failures, livestock deaths, and major typhus epidemics and the events of 1816 have been described as the western world's *last great subsistence crisis* (Post 1977). Estimates of the frequency of such climatically disruptive volcanic eruptions vary from 250–500 years (Decker 1990) to 500–1000 (Oppenheimer 2003a). Certainly an event comparable in magnitude occurred in AD 1030 at Baitoushan volcano on the border between North Korea and China. A major volcanic eruption is postulated as the cause of a serious climate perturbation around AD 540 (Keys 2000), and the most prominent sulfate layer (four times the magnitude of the Tambora signal) in the Greenland GISP2 ice core record for the last 7×10^3 yr (Zielinski 1995) testifies to a huge, but as yet unidentified, eruption in AD 1259. On a smaller scale, the massive Laki (Iceland) effusive eruption of 1783 resulted a serious perturbation of the European climate, alongside severe atmospheric pollution and noticeably elevated mortality rates in (e.g. Grattan and Pyatt 1999; Grattan et al. 2003). Clustered volcanic events that together elevate stratospheric sulfate aerosol loading may also conspire to perturb the global climate. Both Free and Robock (1999) and Crowley (2000) propose that the medieval cold period known as the *Little Ice Age* can be explained in terms of multiple volcanic eruptions significantly raising the mean optical depth of the atmosphere over a period long enough to cause decadal-scale cooling. Highly energetic explosive eruptions, that eject insufficient mass to qualify as super-eruptions, may also have global consequences through the triggering of potentially damaging worldwide meteorological tsunami (due to the atmospheric shock-waves generated), and this has been proposed for the (150 ± 50 megatonne explosive yield) eruption of Taupo (North Island, New Zealand) in 181 AD (Lowe and de Lange 2000).

More than one million earthquakes are recorded each year, of which perhaps a hundred or so have the potential to be severely destructive should they coincide with an urban center where anti-seismic building codes are absent or inadequately enforced. In low to medium income countries, a major earthquake disaster can have a drastic impact on the national economy. The cost of the 1999 Kocaeli earthquake, for example, amounted to 10 percent of Turkey's GDP, whereas economic losses due to the 2003 Bam earthquake totalled around 12 percent of Iran's GDP. Notwithstanding this, no earthquake has resulted in consequences that are sufficiently wide-ranging to have a serious and deleterious impact on the global economy. The conditions for such an event may, however, now exist in the Japanese capital. In 1923, Tokyo and the neighburing city of Yokohama were virtually obliterated by the *Great Kanto Earthquake* – a Magnitude 7.9 event that destroyed 360 000 buildings including 20 000 factories and 1 500 schools. In Tokyo, 71 percent of the population lost their homes, with this figure rising to over 85 percent in Yokohama. Out of a population of 11.7 million, 104 000 were killed and a further 52 000 injured, with 3.2 million people left homeless. The worst natural disaster in the country is estimated to have cost around US$ 50 billion, at today's prices, and proved an unsustainable drain on the national economy. Together, the earthquake and the global economic crash that followed six years later, triggered economic collapse and plunged the country into deep depression. The ensuing climate of despair and misery is held as leading ultimately to the rise of fascism and a thirst for empire and war. The cities of Yokohama and Tokyo have now largely merged to form the Greater Tokyo Metropolitan Region; a gigantic agglomeration of 33 million people – some 26 percent of the nation's population – and the largest urban concentration on the planet. Despite improved building construction and a better understanding of the hazard, a three-fold rise in the population of the region is predicted to see up to 60 000 lives lost when the next major quake strikes. The economic cost of the event is forecast to reach a staggering US$ 4.3 trillion (IIE, 2004) – up to 43 times more than the 1995 Kobe earthquake that occurred 400 km south west of the capital – at US$ 100 billion, the most expensive natural catastrophe to date. After more than a decade of stagnation and the accumulation of a gigantic government debt one and a half times the country's GDP, serious concerns are already being voiced about the possibility of a future collapse of the Japanese financial system and resulting global economic turmoil. Prior to 1923, the last major quake to strike the capital region occurred in 1703, suggesting that such events may have recurrence intervals of just a few centuries. Rikitake (1991) calculates a 40 percent 10-year probability of a Magnitude 6 or greater earthquake and a 5 percent probability for a shock of Magnitude 7 or greater. Such is the complexity of the tectonic situation in the Tokyo region, however, and the potential for interaction between different seismogenic faults, that making an accurate probabilistic forecast of the timing of the next *big one* remains beyond current capabilities. Nevertheless, it would not be unreasonable to make an assumption that it will arrive sometime within the next 100–500 years, heralding – depending upon the precise states of the Japanese and the global economies – a period of serious economic difficulties on a planetary scale.

6.11
Addressing the GGE Threat

Although the CAI threat has been firmly planted in the minds of many national governments, international agencies and much of the developed world public, other terrestrial geological and geophysical phenomena with the potential to exact major loss of life, cause unprecedented levels of physical destruction, or impinge drastically upon the social and economic fabric of our society, have remained relatively obscure. The Asian tsunami has begun to change this, as has increasing appreciation of climate change and its plethora of hazardous consequences and ramifications. Nevertheless, sudden-onset GGEs, such as super-eruptions and megatsunami, continue to be treated as scientific curiosities rather than true threats. As was the case for CAIs prior to the extraordinarily high-profile media coverage of the impacts of the Comet Shoemaker-Levy 9 fragments on Jupiter, there is a general tendency to accept the occurrence of such terrestrial geophysical phenomena in the geological record, but to blot out the thought that they are certain to occur again. Such denial is rarely conscious, but arises primarily because in the past modern human civilization has never experienced a volcanic super-eruption or an ocean-wide giant tsunami; *ergo* they will not happen in our future either. Something needs to be done to combat such perception.

In the early years of the new millennium, those of us fortunate enough to live in the developed world, are strongly risk aware and highly risk averse. Within a culture that is increasingly built around compensation, individuals, groups, organizations and governments insure themselves against almost every eventuality. Unfortunately, there is a problem. Often, risk awareness and perception are highly flawed with excessive credence given to risks that may be infinitesimally small, such as the probability of contracting CJD (Creutzfeld-Jacob Disease) from eating beef-on-the-bone, while being withheld from those risks that are high, for example driving while talking into a handheld mobile phone. On the one-hand, therefore, the UK is under threat from an epidemic of childhood diseases such as mumps, measles and rubella, because parents are failing to have their babies appropriately vaccinated due to a wrongly perceived risk that the combined MMR vaccine might cause autism. On the other, very few citizens are even considering making their lifestyles more sustainable in order to reduce their 'carbon footprint' and help to slow climate change, by far the greatest current threat to our planet and our society. Within this climate of skewed risk awareness, the so-called war on terror is paramount while greenhouse gases continue to rise at unprecedented rates and the far from adequate Kyoto Protocol only now looks like coming into force. At the same time, sudden-onset gee-gees such as super-eruptions and mega-tsunami, and the smaller events, such as climate-perturbing volcanic eruptions and the next major Tokyo earthquake, only appear on the radar screens of small numbers of academics, an even smaller number of concerned politicians and – inevitably – the insurance community.

In the light of the devastating Asian tsunami of December 2004, the time is clearly ripe for taking stock of the global risk portfolio, from terror to tsunami and from climate change to CAI. All risks need to be identified and – where the data are available – quantified. Gaps in our knowledge must be recognized and initiatives begun to help plug these gaps. With respect to the volcanic threat, for example, 1500 volcanoes have erupted since the start of the Holocene 10^4 yr BP, and at least that number can probably still be classed as active despite being dormant for the last ten millennia. Of the resulting 3000 active and potentially active volcanoes, only a few hundred are being monitored to any serious extent. This number has to be increased considerably if we are to have any possibility of advance warning of either a super-eruption or a smaller event capable of significant climate perturbation. Specific locations already identified as presenting a credible threat need to be monitored closely. La Palma's Cumbre Vieja volcano, for example, at present hosts an inadequate seismic monitoring network designed to provide some warning of the rise of fresh magma, and the unstable flank remains completely unmonitored. Almost inevitably, satellite sensors provide the key to improving volcano monitoring worldwide, and in geophysical hazard identification and quantification in general. The PS InSAR (Permanent Scatterer Interferometric Synthetic Aperture Radar), for example, is able to monitor crustal movements in volcanic, seismic and landslide-prone terrains at the sub-centimeter level.

Terrestrial GGE awareness needs to be raised dramatically in the public domain, not in order to terrify but in order to inform. Given the tendencies for hyperbolae and economy of truth that permeate media coverage of the GGE threat – and the blanket coverage of the La Palma situation three times between the years 2000 and 2004 provides an excellent example – this will not prove easy. National governments have a role and a duty to play to inform their electorates of all and any threats to the state, and it would seem that multi-national blocks, such as the EU, or global organizations/initiatives with a scientific interest, such as UNESCO or UNISDR (UN International Strategy for Disaster Reduction), might be better placed to develop and promote a more effective awareness campaign, perhaps incorporating under a single banner attention to the risks we face from climate change and its implications, CAIs and terrestrial gee-gees. It may well turn out that the Asian tsunami catastrophe provides the driving force for progress along this path. In January 2005, UK Prime Minister, Tony Blair, instituted a Natural Hazard Working Group, charged with examining geological and geophysical hazards of high global or regional impact and looking at the feasibility of an effective global early warning system. In June 2005, the group published a report (NHWG 2005) that recommended the establishment of an International Science Panel for Natural Hazard Assessment to identify, evaluate and warn of future major geological and geophysical events. Tacit support was provided for the initiative at the G8 meeting in Gleneagles (Scotland) in July 2005, but only time will tell if the plan becomes reality.

Alongside awareness raising, it is time for the development of an internationally agreed framework that maximizes the chances of successfully mitigating and managing a future GGE. In particular, no protocols currently exist to provide guidance for scientific researchers who uncover evidence of a potential global geophysical threat about when, where and how this information is best presented. The incident of Near Earth Asteroid AL00667 (Later designated 2004AS1), which for several hours on the night of January 13th 2004, looked as if it might be on a collision course with our planet, dramatically highlighted the requirement for some agreed and consistent mechanism for disseminating warnings. In combination with establishing an effective and coherent means of communicating the GGE threat, there is also a need at a range of levels, from national to multilateral, for civil defense and contingency planning studies that address a wide range of pertinent issues such as managing food and energy supplies following a CAI or super-eruption, establishing plans for wholesale evacuation of coastal environments in mega-tsunami prone states, and minimising the economic fallout of the next major Tokyo earthquake scenario. Detailed plans to cope with these and other GGE related scenarios may be too much to expect within the next few years, but recognition of the threats and preliminary planning exercises would constitute a useful start. Tackling climate change is forcing individuals, national governments and multilateral organizations, to take a longer-term view of the future. This new perspective can and should be utilized to ensure that, at last, the full range of GGEs take their place as a key element of an all-encompassing risk-response portfolio.

References

Bekki S, Pyle JA, Pyle DM (1996) The role of microphysical and chemical processes in prolonging climate forcing of the Toba eruption. Geophysical Research Letters 23:2669–2672
Belousov AB (1994) Large-scale sector collapses at Kurile-Kamchatka volcanoes in the 20th century. In: Abstracts of the International Conference on Volcano Instability on the Earth and Other Planets. The Geological Society of London
Bendimerad F (1995) What if the 1923 earthquake strikes again? A five-prefecture Tokyo region scenario. Topical Issue Series, Risk Management Solutions, Fremont
Bryden HC, Longworth HR, Cunningham SA (2005) Slowing of the Atlantic meridional overturning circulation at 25° N. Nature 438:665–657
Bryn P, Solheim A, Berg K, Lien R, Forsberg CF, Haflidason H, Ottesen D, Rise L (2003) The Storegga complex; repeated large scale sliding in response to climatic cyclicity. In: Locat J, Mienert J (eds) Submarine mass movements and their consequences. Advances in natural and technological research series. Kluwer, Dordrecht, pp 215–222
Bühring C, Sarnthein, M, Leg 184 Shipboard Scientific party (2000) Toba ash layers in the South China Sea: evidence of contrasting wind directions during eruption ca. 74 ka. Geology 28:275–278

Bugge T, Belderson RH, Kenyon NH (1988) The Storegga Slide. Philosophical Transactions of the Royal Society, Series A 325:357–388
Carracedo JC, Day SJ, Guillou H, Pérez Torrado FJ (1999) Quaternary collapse structures and the evolution of the western Canaries (La Palma and El Hierro). Journal of Volcanolohy and Geothermal Research 94:169–190
Chapman CR (2004) The hazard of near-Earth asteroid impacts on Earth. Earth and Planetary Science Letters 222:1–15
Chesner CA (1998) Petrogenesis of the Toba tuffs, Sumatra, Indonesia. Journal of Petrology 39:397–348
Chesner CA, Rose WI, Deino A, Drake R, Westgate JA (1991) Eruptive history of Earth's largest Quaternary caldera (Toba, Indonesia) clarified. Geology 19:200–203
Christiansen RL (2001) The Quaternary and Pliocene Yellowstone Plateau volcanic field of Wyoming, Idaho and Montana. U.S. Geological Survey Professional Paper 729
Crowley TJ (2000) Causes of climate change over the past 1000 years. Science 289:270–277
Day SJ, Elsworth D, Maslin M (2000) A possible connection between sea surface temperature variations, orographic rainfall patterns, water-table fluctuations and giant lateral collapse of ocean-island volcanoes. Abstract volume, Western Pacific Geophysics Meeting, Tokyo, WP251
Decker RW (1990) How often does a Minoan eruption occur? In: Hardy DA et al. (eds) Thera and the Aegean World III, pp 444–452
Dickson Robert R et al. (2003) Recent changes in the North Atlantic. Philosophical Transactions of the Royal Society A 361:1917–1934
Fedele FG, Giaccio B, Isaia R, Orsi G (2002) Ecosystem impact of the Campanian Ignimbrite eruption in Late Pleistocene Europe Quaternary Research 57:420–424
Free M, Robock A (1999) Global warming in the context of The Little Ice Age. Journal of Geophysical Research 104:19057–19070
Grattan JP, Pyatt FB (1999) Volcanic eruptions, dust veils, dry fogs and the European Palaeoenvironmental record: localised phenomena or hemispheric impacts? Global and Planetary Change 21:173–179
Grattan JP, Durand M, Taylor S (2003) Illness and elevated human mortality coincident with volcanic eruptions. Geological Society Special Publication 213:401–414
Gregory JM, Huybrechts P, Raper SCB (2004) Climatology – threatened loss of the Greenland ice sheet. Nature 428:616
Häkkinen S, Rhines PB (2004) Decline of sub-polar North Atlantic circulation during the 1990s. Science 304:555–559
Hansen B et al. (2001) Decreasing overflow from the Nordic Seas into the Atlantic Ocean through the Faroe Bank Channel since 1950. Nature 411:927–930
Hearty PJ (1997) Boulder deposits from large waves during the last interglaciation on north Eleuthera Island, Bahamas. Quaternary Research 48:326–338
Hearty PJ, Conrad Neumann A, Kaufman DS (1998) Chevron ridges and runup deposits in the Bahamas from storms late in oxygen-isotope substage 5e. Quaternary Research 50:309–322
Hills JG, Mader CL (1997) Tsunami produced by the impacts of small asteroids. In: Remo JL (ed) Near-Earth Objects: The United Nations International Conference. Annals of the New York Academy of Sciences 822:381–394
Holcomb RT, Searle RC (1991) Large landslides from oceanic volcanoes. Marine Geotechnology 10:19–32
Insurance Information Institute (2004) Catastrophes: insurance issues. Hot topics and Issues Updates, November. *http://www.iii.org/media*
Inter-governmental Panel on Climate Change (IPCC) (2001) Climate Change 2001: the scientific basis. Third Assessment Report, Cambridge University Press
Johnson RW (1987) Large-scale volcanic cone collapse: the 1888 slope failure of Ritter volcano, and other examples from Papua New Guinea. Bulletin of Volcanology 49:669–679
Keating BH, McGuire WJ (2000) Island edifice failures and associated tsunami hazards. Pure and Applied Geophysics 157:899–955
Keating BH, McGuire WJ (2004) Instability and structural failure at volcanic ocean islands and continental margins and the climate change dimension. Advances in Geophysics 47
Keys D (2000) Catastrophe: an investigation into the origins of the modern world. Arrow, London

Kilburn CRJ, Petley DN (2003) Forecasting giant, catastrophic slope collapse: lessons from Vajont, Northern Italy. Geomorphology 54(1–2): 21–32

Ledbetter M, Sparks RSJ (1979) Duration of large magnitude explosive eruptions deduced from graded bedding in deep-sea ash layers. Geology 7:240–244

Lipman PW, Mullineaux D (eds) (1981) The 1980 eruptions of Mount St. Helens. U.S. Geological Survey Professional Paper 1250

Lowe DJ, de Lange WP (2000) Volcano-meteorological tsunamis, the c. AD 200 Taupe eruption (New Zealand) and the possibility of a global tsunami. The Holocene 10:401–407

Mader CL (2001) Modelling the La Palma landslide tsunami. Science of Tsunami Hazards 19, 160–180

Maslin M, Owen M, Day S, Long D (2004) Linking continental slope failures and climate change: testing the clathrate gun hypothesis. Geology 32:53–56

Mason BG, Pyle DM, Oppenheimer C (2004) The size and frequency of the largest explosive eruptions on Earth. Bulletin of Volcanology

McGuire WJ (1996) Volcano instability: a review of contemporary themes. In: Volcano instability on the Earth and other planets. In: McGuire W J, Jones AP, Neuberg J (eds) Geological Society of London Special Publication 110, pp 1–23

McGuire WJ (2006) Lateral collapse and tsunamigenic potential of marine volcanoes. In: Troise C, De Natale G, Kilburn CRJ (eds) Mechanisms of activity and unrest at large calderas. Geological Society, london, spec. pub. 269, pp 121–140

McGuire WJ, Howarth RJ, Firth CR, Solow AR, Pullen AD, Saunders SJ, Stewart IS, Vita-Finzi C (1997) Correlation between rate of sea-level change and frequency of explosive volcanism in the Mediterranean. Nature 389:473–476

McMurtry GM, Fryer GJ, Tappin DR, Wilkinson IP, Williams M, Fietzke J, Garbe-Schoenberg D, Watts P (2004) Megatsunami deposits on Kohala volcano, Hawaii, from flank collapse of Mauna Loa. Geology 32:741–744

Moore JG, Normark WR, Holcomb RT (1994) Giant Hawaiian landslides. Annual Review of Earth and Planetary Sciences 22:119–144

Morrison DM, Chapman CR, Steel D, Binzel R (2004) Impacts and the public: communicating the nature of the impact hazard. In: Belton MJS (ed) Mitigation of hazardous comets and asteroids. Cambridge University Press, Cambridge

Moss JL, McGuire WJ, Page D (1999) Ground deformation monitoring of a potential landslide at La Palma, Canary Islands. Journal of Volcanology and Geothermal Research 94:251–265

NHWG (2005) The role of science in physical natural hazard assessment. UK Government Natural Hazard Working Group. Office of Science & Technology. London

Near-Earth Object Science Definition Team (2003) Study to Determine the feasibility of extending the search for Near-Earth Objects to smaller limiting diameters. NASA Office of Space Science, Solar System Exploration Division. Washington, DC

Oppenheimer C (2002) Limited global change due to largest known Quaternary eruption, Toba ~74 kyr BP? Quaternary Science Reviews 21:1593–1609

Oppenheimer C (2003a) Climatic, environmental and human consequences of the largest known historical eruption: Tambora volcano (Indonesia) 1815. Progress in Physical Geography 27:230–259

Oppenheimer C (2003b) Ice core and palaeo-climatic evidence for the timing and nature of the great mid-13th century volcanic eruption. International Journal of Climatology 23:417–426

Pararas-Carayannis G (2002) Evaluation of the threat of megatsunami generation from postulated massive slope failures of island stratovolcanoes on La Palma, Canary Islands, and on the island of Hawaii. Science of Tsunami Hazards 20:251–277

Pérez-Torrado FJ, Paris R, Cabrera MC, Schneider J-L, Wassmer P, Carracedo J-C, Rodriguez-Santana A, Santana F (2006) Tsunami deposits related to flank collapse in oceanic volcanoes : The Agaete Valley evidence, Gran Canaria, Canary Islands. Mar Geol 227(1–2):135–149

Post JD (1977) The last great subsistence crisis in the Western World. John Hopkins University Press, Baltimore, MD

Rampino MR, Ambrose SH (2000) Volcanic winter in the Garden of Eden: the Toba super-eruption and the late Pleistocene human population crash. In: McCoy FW, Heiken G (eds) Volcanic hazards and disasters in human antiquity. Geological Society of America Special Paper 345, pp 71–82

Rampino MR, Self S (1992) Volcanic winter and accelerated glaciation following the Toba super-eruption. Nature 359:50–52

Rampino MR, Self S (1993a) Climate-volcanism feedback and the Toba eruption of ~74000 years ago. Quaternary Research 40:269–280

Rampino MR, Self S (1993b) Bottleneck in human evolution and the Toba eruption (correspondence). Nature 262:1954

Rampino MR, Self S, Stothers RB (1988) Volcanic winters. Annual Review of Earth and Planetary Sciences 16:73–99

Rikitake T (1991) Assessment of earthquake hazard in the Tokyo area, Japan. Tectonophysics 199:121–131

Rose WI, Chesner CA (1990) Worldwide dispersal of ash and gases from Earth's largest known eruption. Palaeontology 89:269–275

Scaillet B, Clemente B, Evans BW, Pichavant M (1998) Redox control of sulphur degassing in silicic magmas. Journal of Geophysical Research 103(23):937–949

Siebert L (1984) Large volcanic debris avalanches: characteristics of source areas, deposits, and associated eruptions. Journal of Volcanology and Geothermal Research 22:163–197

Siebert L (1992) Threats from debris avalanches. Nature 356:658–659

Smith RB, Braile LW (1994) The Yellowstone hotspot. Journal of Volcanology and Geothermal Research 61:121–187

Smith MS, Shepherd JB (1996) Tsunamigenic landslides at Kick 'em Jenny. In: McGuire WJ, Jones AP, Neuberg J (eds) Volcano instability on the Earth and other planets. Geological Society of London Special Publication 110:293–306

Stuart JS, Binzel RP (2004) Bias-corrected population, size distribution and impact hazard for the near-Earth objects. Icarus 170, 295–311

Talandier J, Bourrouilh-le-Jan F (1988) High energy sedimentation in French Polynesia: cyclone or tsunami? In: El-Sabh MI, Murty TS (eds) Natural and man-made hazards. Department of Oceanography, University of Quebec, Rimiuski PQ, Canada, pp 193–199

Toon OB, Zahnle K, Morrison D, Turco RP, Covey C (1997) Environmental perturbations caused by the impacts of asteroids and comets. Reviews in Geophysics 35:41–78

Ui T (1983) Volcanic dry avalanches deposits – identification and comparison with non-volcanic debris stream deposits. Journal of Volcanology and Geothermal Research 18:135–150

Ward SN (2001) Landslide tsunami. Journal of Geophysical Research 106(11):201–211, 216

Ward SN, Day SJ (2001) Cumbre Vieja volcano – potential collapse and tsunami at La Palma, Canary Islands. Geophysical Research Letters 28:397–400

Ward SN, Day SJ (2003) Ritter Island volcano – lateral collapse and the tsunami of 1888. Geophysical Journal International 154:891–902

Wilson CJN (2001) The 26.5 ka Oruanui eruption, New Zealand: an introduction and overview. Journal of Volcanology and Geothermal Research 112:133–174

Yang Q, Mayewski PA, Zielinski GA, Twickler M (1996) Depletion of atmospheric nitrate and chloride as a consequence of the Toba eruption. Geophysical Research Letters 23:2513–2516

Young RW, Bryant EA (1992) Catastrophic wave erosion on the southeastern coast of Australia: impact of Lanai tsunami c 105 ka? Geology 20:199–202

Zielinski GA (1995) Stratospheric loading and optical depth estimates of explosive volcanism over the last 2100 years derived from the Greenland Ice Sheet Project 2 ice core. Journal of Geophysical Research 100(20):937–955

Zielinski GA, Mayeswki PA, Meeker LD, Whitlow S, Twickler MS (1996) Potential atmospheric impact of the Toba mega-eruption ~71000 years ago. Geophysical Research Letters 23:837–840

Part II Astronomy and Physical Implications

Chapter 7 The Asteroid Impact Hazard and Interdisciplinary Issues

Chapter 8 The Impact Hazard: Advanced NEO Surveys and Societal Responses

Chapter 9 Understanding the Near-Earth Object Population: the 2004 Perspective

Chapter 10 Physical Properties of NEOs and Risks of an Impact: Current Knowledge and Future Challenges

Chapter 11 Evaluating the Risk of Impacts and the Efficiency of Risk Reduction

Chapter 12 Physical Effects of Comet and Asteroid Impacts: Beyond the Crater Rim

Chapter 13 Frequent Ozone Depletion Resulting from Impacts of Asteroids and Comets

Chapter 14 Tsunami as a Destructive Aftermath of Oceanic Impacts

Chapter 15 The Physical and Social Effects of the Kaali Meteorite Impact – a Review

Chapter 16 The Climatic Effects of Asteroid and Comet Impacts: Consequences for an Increasingly Interconnected Society

Chapter 17 Nature of the Tunguska Impactor Based on Peat Material from the Explosion Area

Chapter 18 The Tunguska Event

Chapter 19 Tunguska (1908) and Its Relevance for Comet/Asteroid Impact Statistics

Chapter 20 Atmospheric Megacryometeor Events versus Small Meteorite Impacts: Scientific and Human Perspective of a Potential Natural Hazard

Chapter 7

The Asteroid Impact Hazard and Interdisciplinary Issues

Clark R. Chapman

7.1 Introduction

Sometime in the foreseeable future, perhaps during this decade or maybe not until our great-great-grandchildren are adults, an asteroid the size of a large building will crash into the Earth's atmosphere, exploding in an air-burst with the force of megatons or more of TNT. Most likely, such an event will happen over an ocean or sparsely populated desert; but, if it occurs over an urban area, the consequences could be very destructive and deadly. Actually, small strikes by cosmic grains of sand happen all the time (witness meteors or "shooting stars", visible in a dark, clear sky several times an hour) and every year many large rocks, called "meteorites", survive their atmospheric plunge to be collected and exhibited in museums.

The unique threat from the skies, however, is the very small but finite chance that a large asteroid or comet, 2 km or more across, will slam into the Earth at 100 times the speed of a jetliner, instantly producing a global environmental crisis unprecedented in human history and threatening the future of civilization, as we know it. Half-a-dozen times since the beginning of the Cambrian Period half-a-billion years ago, when large, life forms evolved on our planet, giant asteroids or comets 10 or 20 km across have struck Earth, producing a global holocaust that killed almost everything alive and hence transformed the biosphere. Human civilization is one subsequent result of such a mass-extinction, which ended the Cretaceous Period (when dinosaurs reigned) 65 million years ago. Such a mass-extinction could conceivably happen again, although the chance of it happening during our lives is extraordinarily small. In this sense, the impact hazard exceeds any other known natural or man-made threat to civilization's or even our species' future. It is the ultimate low-probability high-consequence hazard.

In this paper, I begin by outlining the facts, and associated uncertainties, concerning the impact hazard. I consider the astronomical data on asteroids and comets; the physics of impact into and through Earth's atmosphere and subsequent explosive cratering of the land or ocean; and what is known or speculated about the resulting environmental effects. I have recently reviewed these issues at some length, in a fashion accessible by non-physicists:

- My recent review of the impact hazard, emphasizing the physical-scientific features of the hazard, is "The hazard of near-Earth asteroid impacts on Earth", by Clark R. Chapman, *Earth & Planetary Science Letters*, vol. 222, pp 1–15, 2004, downloadable from: *http://www.boulder.swri.edu/clark/crcepsl.pdf*. This is referred to below as CRC04.

- My 2003 report to the Organisation for Economic Cooperation and Development (OECD) on the potential consequences of asteroid impacts of various sizes: "How a Near-Earth Object Impact Might Affect Society", by Clark R. Chapman, commissioned by the OECD Global Science Forum for "Workshop on Near Earth Objects: Risks, Policies, and Actions" (Frascati, Italy, January 2003), downloadable from: *http://www.oecd.org/dataoecd/18/40/2493218.pdf* or *http://www.boulder.swri.edu/clark/oecdjanf.doc*. This is referred to below as CRC03.

Accordingly, I will avoid redundancy with those publications and keep this part of my paper succinct.

What is practically relevant to society are the less certain extrapolations of the effects of impacts (or predictions of future impacts) on elements of society, including human mortality and the physical infrastructure (which I will assume roughly follows mortality), but more interestingly on the psychology and sociology of different societal institutions (e.g. the military, science, religion, government, business/economy and media). Whether or not modern civilization can survive a large impact depends on whether these institutions are robust or fragile in the face of unprecedented disaster that threatens sustainability. What does it take to cause collapse of the global economy and disintegration of human communities? But we also need to understand even the potential responses of different societal sectors to predictions of rather modest-scale impacts, because they are most likely to happen "on our watch".

The central thrust of this paper is to translate what is known, and what is not known, about the impact hazard into familiar frameworks that can enable non-astronomers to appreciate this unusual, newly recognized hazard in the context of other more familiar threats with which society is wrestling. In some ways, the impact hazard is as objectively threatening (in a statistical sense) as many other natural hazards that command much news coverage and expenditures by disaster management and recovery agencies. The impact hazard has features that can cause people of different temperaments either (*a*) to wholly ignore it (after all, almost nobody has been killed by a cosmic object during the past century) or (*b*) to respond disproportionately to the objective death and damage (e.g. as has been happening in the United States following terrorist killings of ~3 000 people on 11 September 2001). Despite the low probabilities of a major impact catastrophe occurring in our lifetimes, the chances are rising – due to the advancing technology of telescopic searches for threatening asteroids – that "near misses", misinterpretations of actual "small" impacts, or mistaken/hyped media reports will accelerate during the next few years. Such scares already have posed serious issues for officials and policy-makers and will continue to do so.

Despite its low-probability high-consequence character, the impact hazard has many features in common with other natural hazards, including the specific effects that cause death and destruction (fire, shaking, hurricane-force winds, flying objects, flooding, etc.). But it also has some other distinct differences, which I review. A crucial one is that it is within the capability of space agencies to devise missions that could divert an oncoming asteroid, causing it to miss the Earth, thus totally preventing the disaster. Given the possibility of 100% protection, unusual in mitigation of natural hazards, policy issues are raised about the degree to which the impact hazard should be treated seriously by individual nations and/or international entities. To date, the impact hazard has achieved scant governmental recognition and almost zero funding, relative to its actuarial cost. Is this implicit down-weighing of the impact hazard (despite its recent prominence in the news, in science education and in entertainment) the correct decision or not? My prime purpose here is to foster further thinking about the impact hazard in order to inspire a serious evaluation of how, and to what degree if at all, society should become proactive about this threat.

7.2
Near-Earth Asteroids (NEAs)

The Earth resides in a cosmic "shooting gallery". Despite the great emptiness of interplanetary space, objects ranging from dust to cosmic bodies many tens of kilometers in size move around the Sun in paths that can intersect the Earth's orbit. Relative velocities (hence impact velocities) are tens of kilometers per second. The largest objects are called asteroids and comets. Very much smaller ones, generated by the disintegration or collisional fragmentation of the larger ones, are called meteoroids while in space, meteors (or bolides) while passing through the Earth's upper atmosphere, and meteorites if parts of them make it to the ground. Asteroids and comets are remnants of bodies that gathered together to form the planets 4.5 billion years ago. While there are technically interesting patterns to their orbital behavior, the statistical probabilities of Earth being impacted by objects of various sizes can be thought of as a constant, random process that has changed little for at least 3 billion years.

Near Earth Asteroids (NEAs) are defined as those whose perihelia (closest orbital distances to the Sun) are < 1.3 Astronomical Units (1 AU = the mean distance of Earth from the Sun). About 20% of NEAs are currently in orbits that can approach the Earth's orbit to within < 0.05 AU; these are termed Potentially Hazardous Objects (PHOs). In terms of their origin and physical nature, PHOs are no different from other NEAs; they just happen to come close enough to Earth at the present time so that close planetary encounters could conceivably perturb their orbits so as to permit an actual near-term collision, hence they warrant careful tracking. The Spaceguard search programs (chiefly LINEAR in New Mexico; LONEOS in Flagstaff, Arizona; NEAT in Maui and southern California; Spacewatch on Kitt Peak, Arizona, and the Catalina Sky Survey near Tucson, Arizona and in Siding Spring, Australia) continue to discover a new NEA every few days. As of October 2006, over 4200 NEAs are known (of which about 1/5 are PHOs). The census is probably complete for NEAs > 3 km diameter. The estimated number of NEAs > 1 km in diameter (the size for which NASA established the Spaceguard Goal of 90% completeness by 2008) is ~1100 ± 200, of which about 75% have been found; since those are now known not to be dangerous during the next century, any near-term danger from > 1 km sized NEAs can come only from the remaining 25% not yet discovered. In this way, the Survey is actually helping to reduce the danger from a global asteroid catastrophe. But since current searches are not optimized for discovering smaller-but-still-dangerous NEAs, and are virtually useless for discovering a large comet headed for Earth from the outer Solar System, Spaceguard does not significantly lessen those dangers. Plans to survey

NEAs down to a couple of hundred meters are in various degrees of development. Pan-STARRS (*pan-starrs.ifa.hawaii.edu/*) is being built in Hawaii and may begin searching by 2008. The Large Synoptic Survey Telescope (LSST) is bigger, but farther in the future, and less certain to be built. Other ground and space-based approaches to NEA-searching have been evaluated by NASA's Science Definition Team (SDT 2003) and by the European Space Agency Near-Earth Object Mission Advisory Team (Harris et al. 2004).

Figure 7.1 shows the rapid increase in estimated numbers of NEAs with decreasing size, down to the billion-or-so NEAs ≥ 4 m diameter; 4 m is the size of NEA that impacts Earth about once per year, exploding harmlessly but frighteningly very high in the atmosphere with an energy equivalent to ~5 kT of TNT. The numbers are least secure (at least a factor of several) for NEAs too rare to be witnessed as bolides (brilliant meteors) but too small to be readily discovered telescopically, e.g. ~10–200 m diameter. This includes objects of the size (~60 m) that produced the dramatic 15 MT Tunguska lower atmospheric explosion in Siberia in 1908. Though uncertain, the

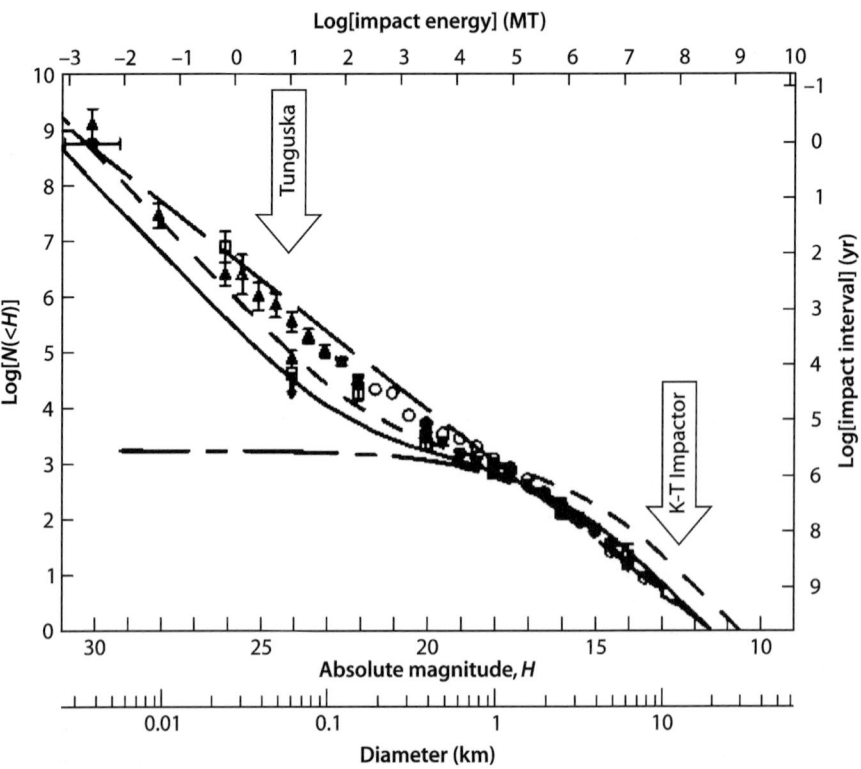

Fig. 7.1. The size-frequency relationship for NEAs, for the cumulative number larger than a particular size, based chiefly on telescopic search programs. There are two reference curves: The *straight, long-dashed line* is a power-law; the *curve that flattens out to the lower left* is the number of NEAs discovered as of early 2002. Courtesy A. Harris

expected frequency of Tunguska-like events is less than once per thousand years; possibly the destruction of thousands of square kilometers of Siberian forest was accomplished by a blast much less energetic than 15 MT, due to a more common, smaller object (Boslough and Crawford 1997).

Most NEAs originate as fragments of colliding asteroids in the inner half of the main asteroid belt, between Mars and Jupiter. Chaotic processes, associated with the gravitational forces of Mars, Jupiter and Saturn, combined with heating by sunlight, bring many such fragments into Earth-crossing orbits over some millions of years. As NEAs strike the Sun or terrestrial planets, or are tossed back out of the inner solar system, fresh fragments from the inner asteroid belt provide replenishment. Probably a small fraction of NEAs (5–10%) originate as comets in the outer solar system. NEAs that were originally comets, as well as some from the asteroid belt, are believed to be composed of rather fluffy, structurally weak materials (e.g. ices and carbon-rich mud-like substances); the Deep Impact experiment on Comet Tempel 1 found its surface to be unexpectedly weak. Most NEAs are made of harder rocks, like the common ordinary chondrite meteorites. A few are composed of solid metal (nickel-iron alloy), also represented in meteorite collections. NEAs smaller than about 200 m diameter are mostly solid, monolithic rocks – like a big meteorite. However, most NEAs > 200 m are likely to be "rubble piles" – collections of smaller objects, weakly held together by gravity. Nearly 20% of NEAs are actually double bodies; they often take the form of a larger central body with a smaller satellite revolving about it. When a double NEA strikes a planetary surface, two side-by-side craters may result.

Particularly important as we contemplate the practicalities of deflecting an oncoming NEA away from Earth impact are (*a*) the nature of the NEA's surface and (*b*) the structural integrity of the body. The first is important because we may have to grab onto the surface of the body, or otherwise interact with its surface. The second is important because we need to have confidence in the outcome of our deflection attempt (whether we are pushing on it or blasting it with a bomb). It is expected that surface and interior attributes of NEAs vary widely from body to body. But we currently have very little information about either trait, even for one body. The NEAR-Shoemaker spacecraft landed on one of the largest NEAs, named Eros, in 2001, but such a landing was not in the mission plan and the spacecraft lacked instruments to study the detailed physical nature of the surface; in any case, Eros has a million times the mass of a 200 m NEA, which is a size we are much more likely to have to deal with, and it is doubtful that Eros' structure is a good analog for such small, nearly gravity-free bodies. The Japanese Hyabusa mission studied the 300 m NEA Itokawa in late 1995, revealing it to be a rocky, nearly crater-free rubble-pile, with several flat regions of coarse gravel. The B612 Foundation (Schweickart et al. 2003) has proposed a demonstration mission, using a space tugboat or gravity tractor, to measurably move a ~200 m NEA in a controlled fashion. There is recent interest in the NEA Apophis, which will undergo major tidal forces during its exceptionally close pass to Earth in 2029; it potentially could strike the Earth during the subsequent decade if it happens to pass through a small "keyhole" in 2029, without being deflected from the keyhole beforehand. Scientific exploration of Apophis before, during, and after its 2029 pass is under consideration.

7.3
Consequences of NEA Impact

If a rocky object strikes the Earth, consequences vary depending on its size (and where it hits). Dust and sand grains burn up in the upper reaches of the Earth's atmosphere (as meteors). Larger objects lose their energy more brilliantly (as bolides), but still high in the atmosphere; small portions may reach the Earth's surface (as meteorites), falling at terminal velocity through the atmosphere. Objects 30–150 m in diameter explode in the lower atmosphere, as dangerous air-bursts. Somewhat larger objects (150–250 m explode at the bottom of the atmosphere but may also excavate the surface (whether land or ocean). Still larger objects explode beneath the surface, like a buried nuclear bomb, forming a crater perhaps 20 times the projectile's diameter, ejecting material to re-impact at great distances. If the ocean is excavated, the transient crater immediately collapses, generating a tsunami that propagates to the edges of the ocean, and runs up onto the shore with potentially devastating effects. In the case of a land impact, the cratering event greatly exceeds even the largest nuclear bomb test; examination of comparatively recent craters on the Moon and other planets provides evidence about the scale of destruction. Impacting NEAs larger than 1 or 2 km approach the threshold for truly global effects, such as pollution of the stratosphere with dust, which could induce global cooling with disastrous consequences for agriculture. The explosive impacts of fragments of Comet Shoemaker-Levy 9 into Jupiter's atmosphere in 1994 had energies in this range, and they resulted in dark patches in Jupiter's atmosphere the size of planet Earth, which lasted for months. The very largest impacts that could conceivably happen, and which have happened several times since larger plants and animals evolved on Earth, generate larger and additional global consequences (e.g. global firestorms, poisoning of the oceans, etc.), which can result in the permanent extinction of numerous species.

Projectiles made of strong metals are not so readily broken up when they penetrate the atmosphere (although smaller ones are greatly slowed down); but only a few percent of NEAs are metallic. Projectiles made of fluffy, icy, and/or under-dense materials (e.g. comets), penetrate the atmosphere less readily. They explode at higher altitudes, or it takes a larger one to reach the ground at cosmic velocities. Actual live comets (as distinct from dead ones described above as constituting a small fraction of NEAs) contribute to the impact hazard at the level of ~1% (SDT 2003), although the very largest objects that could threaten Earth (> 3 km diameter) would be comets; all NEAs of such sizes have been discovered and are not an immediate threat.

The energetic interactions of an impacting NEA with the atmosphere, ocean, and land generate various immediate, secondary, and perhaps long-term effects – physical, chemical, and perhaps biological. The most thorough evaluation of the environmental physical and chemical consequences of impacts is by Toon et al. (1997); more recent research, summarized by the SDT (2003), has begun to elucidate the previously poorly understood phenomena of impact-generated tsunami. I now briefly describe the chief environmental effects, for impacts of NEAs > 300 m diameter:

- *Total destruction in the crater zone:* No structure or macroscopic life form would survive being in or adjacent to the explosion crater, a region roughly 30 times the size of the projectile (falling ejecta could be lethal over far greater distances).

- *Tsunami:* Flooding of historic proportions along proximate ocean shores would be caused by a > 300 m impact, but run-up is highly variable depending on shore topography. An extinction-level impact (by a 10–15 km NEA) could inundate low-lying regions adjacent to oceans worldwide. (There is considerable debate and uncertainty about the scale and character of impact-caused tsunami.)
- *Stratospheric dust obscures sunlight:* 300 m impacts would cause noticeable but relatively minor effects similar to those of the largest volcanic explosions (e.g. the "year without summer" caused by the 1815 explosion of Tambora). For a > 2 km NEA impact, sunlight would drop to "very cloudy days" nearly worldwide, threatening global food supplies by cessation of agriculture due to prolonged summertime freezing temperatures. Severe immediate effects (permanent "night" globally) and possible catastrophic long-term climate oscillations result from an extinction-level impact.
- *Fires ignited by fireball and/or re-entering ejecta:* Even the Tunguska impact, which did not reach the ground, caused trees to burn in the center of the zone where trees were toppled. But fires are of only local-to-regional importance even for a > 2 km impact that would have global climate effects due to dust. In an extinction-level event, the broiling of the entire surface of our planet by re-entering ejecta – and the resulting global firestorm – would be the chief immediate cause of general death of plants and animals on land.
- *Poisoning of the biosphere:* Immediate atmospheric effects (sulfate production, injection of water into the stratosphere, destruction of the ozone layer, production of nitric acid, etc.) and subsequent poisoning of lakes and oceans augment the effects of stratospheric dust for a > 2 km impact and dramatically worsen the already hellish conditions created by an extinction-level impact. (Birks, Chap. 13 of this volume, suggests that an NEA as small as 0.5 km might cause destruction of the ozone layer.)
- *Earthquakes:* Although local-to-global earthquakes (in response to the cratering explosion of various sized NEAs) would be serious if considered in isolation, they are minor compared with other more damaging and lethal consequences listed above.

These effects are most securely understood for the 300 m case and for even smaller impacts, where man-made and natural explosions provide relevant analogs with only modest extrapolations. The larger impacts not only have never been witnessed (fortunately), but they involve enormous extrapolations from existing knowledge. Of course, there are logical constraints dictated by the laws of physics, and some evidence can be gleaned from actual past impacts (e.g. the Cretaceous-Tertiary [K/T] boundary impact on Earth, giant craters on other worlds). But synergies between the multiple effects are poorly understood. Nevertheless, for the larger impacts, the magnitude of energy released in virtually an instant is so enormous compared with the scale of the biosphere that catastrophic effects are assured.

Table 7.1, modified from a similar table in CRC03, is an attempt to characterize impacts in terms of their practical consequences. It lists three relevant attributes for impacts by bodies ranging from a 10–15 km extinction-level NEA down to the size of a basketball: (*a*) the TNT-equivalent energy of the explosion, (*b*) the chances of such an impact happening this century (or, for frequent events, how many will happen this century, or per year), and (*c*) a qualitative description of the consequences and potential for mitigation. I amplify on issues related to mitigation below.

Table 7.1. Frequency of cosmic impacts of various magnitudes

Asteroid/comet diameter	Energy and where deposited	Chance this century (world)	Potential damage and required response
>10 km	100 million MT global	<1 in a million[a]	Mass extinction, potential eradication of human species; little can be done about this extraordinarily unlikely eventuality
>3 km	1.5 million MT global	<1 in 50 000[a]	Worldwide, multi-year climate/ecological disaster; civilization destroyed (a new Dark Age), most people killed in aftermath; chances of having to deal with such a comet impact are extremely remote; mitigation extremely challenging
>1 km	80 000 MT major regional destruction; some global atmospheric effects	0.02%	Destruction of region or ocean rim; potential worldwide climate shock – approaches global civilization-destruction level; consider mitigation measures (deflection or planning for unprecedented world catastrophe)
>300 m	2 000 MT local crater, regional destruction	0.2%	Crater ~5 km across and devastation of region the size of a small nation *or* unprecedented tsunami; advance warning or no notice equally likely; deflect, if possible; internationally coordinated disaster management required
>100 m	80 MT lower atmosphere or surface explosion affecting small region	1%	Low-altitude or ground burst larger than biggest-ever thermonuclear weapon, regionally devastating, shallow crater ~1 km across; after-the-fact national crisis management (advance warning unlikely)
>30 m	2 MT stratosphere	40%	Devastating stratospheric explosion; shock wave may topple trees, weak wooden houses, ignite fires within 10 km; deaths likely if in populated region (1908 Tunguska explosion was several times bigger); advance warning very unlikely, all-hazards advanced planning would apply
>10 m	100 kT upper atmosphere	6 per century	Extraordinary explosion in sky; broken windows, but little damage on ground; no warning
>3 m	2 kT upper atmosphere	2 per year	Blinding explosion in sky; could be mistaken for atomic bomb
>1 m	100 t TNT upper atmosphere	40 per year	Bolide explosion approaching brilliance of the Sun for a second or so; harmless, may yield meteorites
>0.3 m	2 t TNT upper atmosphere	1 000 per year	Dazzling, memorable bolide or "fireball" seen; harmless

[a] Frequency from Morrison et al. (2002); but no asteroid of this size is in an Earth-intersecting orbit; only comets (a fraction of the cited frequency) contribute to the hazard, hence "<."

Which of the cases in Table 7.1 are of greatest consequence? Obviously, cases involving house-sized or smaller NEAs (< 20 m) are of little practical concern, notwithstanding the fact that meteorites have struck several people and one killed a dog. Education about the causes of brilliant explosions in the sky is useful to prevent inappropriate military responses to misinterpreted natural phenomena. It is also difficult to regard extinction-level impacts as meriting any concern; not only is such an impact extremely unlikely to happen in our lifetimes, but there is little to do but wait for the apocalypse (comets cause such events and provide only months of warning time). For NEAs > 30 m but less than a few km in size, however, the chances of such an impact happening are within the range regarded as serious for hazards in other walks of life; moreover, there is at least a chance of averting the impact in the first place or at least mounting effective disaster management activities before and after the impact.

Within this range of possible NEA impacts meriting practical concern, which scenario is objectively the most threatening? Although giant impacts are very rare, the potential mortality is unprecedented large once the threshold for global disaster is exceeded (NEAs > 1.5–3 km diameter); such impacts dominate mortality, perhaps 1 000 deaths per year worldwide. This threat is comparable with mortality from other significant natural and accidental causes (e.g. fatalities in airliner crashes). Of course, this is a statistically averaged mortality; in almost any year there are zero deaths, but a tiny chance that one year a billion people will be killed. This threat motivated implementation of the Spaceguard Survey. Since most of that mortality has been eliminated by discovery of over $3/4^{ths}$ of NEAs > 1 km diameter and demonstration that none of them will strike the Earth in the next century, the remaining global threat is from the $1/4^{th}$ of yet-undiscovered large NEAs plus the minor threat from comets. Once the Spaceguard Survey is complete, the residual threat from a globally destructive impact will be < 100 annual fatalities worldwide (see Fig. 7.2).

Two sources of mortality are due to smaller NEAs: (*a*) impacts onto land, with local and regional consequences analogous to the explosion of a huge nuclear bomb and (*b*) impacts into an ocean, resulting in inundation of shores by tsunamis. My own interpretation of an analysis by the SDT (2003; see CRC04) is shown in Fig. 7.2. The post-Spaceguard residual hazard for land impacts is ~50 deaths per year and for tsunami-

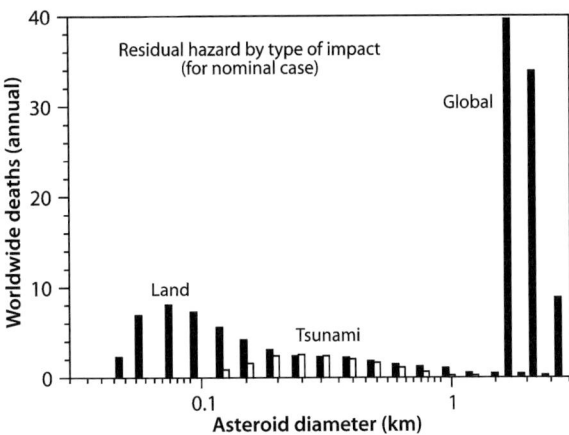

Fig. 7.2.
Annualized global mortality for NEA impacts of three different types (nominal, SDT 2003, CRC04), applicable after the Spaceguard Survey is completed. The residual global threat from NEOs > 2 km is being reduced, leaving primarily the local or regional threats from land impacts by bodies of order 100 m in size. The tsunami threat is very uncertain (it pertains to deaths rather than the SDT's "persons affected")

producing impacts ~15 per year. These are very modest mortality rates compared with most natural hazards (not to mention disease, war, and famine), but exceed the ~5 deaths each year worldwide from shark attacks, which merit much popular and official concern. But the chances are better than 1% that a land-impact by a 70–200 m NEA will kill ~100 000 people during the 21st century. I argue below that various factors may magnify public concern about frequent, modest-scale impacts in ways that may demand greater official attention to the impact hazard than might seem warranted from the modest mortality statistics.

7.4
Mitigation: Deflection and/or Disaster Management and Response

What can and cannot be done about this threat? First, I describe what is currently being done. Since the scientific community first seriously considered the impact hazard 25 years ago, astronomers worldwide have increased NEA searches with ever-better instrumentation. In 1998, NASA committed to the Spaceguard Goal and roughly doubled its funding of NEA searches to several million US$ per year. Funding of NEA research in other countries has been minimal, but widespread interest of amateur astronomers in many nations has enabled most professional discoveries (dominated by the LINEAR telescopes in New Mexico) to be followed up, in order to determine orbits and any possible future impact with Earth. As I showed above, discovering NEAs and demonstrating that they are not hazardous reduces the pool of NEAs that might strike soon and thus reduces the hazard. The surveys also stand some chance (a good chance for NEAs > 1 km diameter) of identifying an NEA years or decades before it will strike, which would enable various mitigation options. The popular concept, depicted in movies, that a large NEA would be detected only hours or months before impact, is exceedingly unlikely; false alarms of such a scenario have actually happened, however, and may happen again.

Besides these modest telescopic efforts, little serious research has been devoted to mitigation of an NEA impact. In a series of conferences during the past dozen years, aerospace engineers and physicists have addressed approaches to modifying the path of an NEA, years or decades before a predicted impact, so that it would miss rather than hit the Earth. The latest meeting (the AIAA Planetary Defense Conference, held in Garden Grove CA in February 2004) is thoroughly documented (with video and associated PowerPoint charts for all presentations, *http://www.planetarydefense.info/*). The proceedings of a late-2002 mitigation conference were published in late 2004 (Belton et al. 2004). Since funding of NEA deflection research has been minimal, mission designs are immature. Even fundamental issues like how much warning is needed to mount a successful deflection, or how soon can we tell whether an NEA will surely hit and where, are only beginning to be studied. The main point is that there are a variety of scenarios – involving relatively modest-sized NEAs with warning times of > 5 years, preferably much longer – in which it is plausible that a combination of existing technologies could be used to gently, and controllably, move a threatening NEA into a path that would miss rather than hit the Earth by a comfortable margin. In other cases, typically involving very large NEAs or comets in which there is inadequate warning for controlled deflection, there is the possibility of altering the object's path with a nuclear bomb or other violent means; the outcomes of such interventions are less readily predictable and even the development of some of these concepts threatens treaty obligations prohibiting use of nuclear weapons in space.

In the event that an oncoming NEA is discovered, but deflection is either impossible or unreliable, conventional disaster management approaches could be employed (or modified) to mitigate the consequences of a major impact. In some cases, regions around ground-zero or shorelines could be evacuated, food reserves augmented, and so on. If the impact were actually to happen, with or without warning, conventional approaches to rescue and recovery could be implemented to reduce casualties.

Although conceptually similar to normal disaster management, on-the-ground mitigation of an asteroid impact necessarily has features that differ from conventional practices. Consider evacuation, for example. Through numerous events, public officials have gradually learned who should be evacuated and when, and who should not, during the days before landfall of an approaching typhoon or hurricane. The evolution of predictions of where an asteroid might strike would be very different, and there would be large uncertainties about how large a region would need to be evacuated. Public reactions are much less readily predictable concerning a never-before-experienced event (fewer cases of "I will ride this one out"; some may regard it as the coming of the Apocalypse). Impacts have some features in common with more familiar disaster scenarios (flying objects, fire, smoke), they differ from others (no harmful radiation, no willful perpetrators), and they are unique in still other ways (e.g. likely very long lead times, different tsunami behavior).

The most salient fact about integration of asteroid impact disaster planning into the broader responsibilities of public disaster management agencies is that there has been none. Despite publication of a few papers on the topic (e.g. Garshnek et al. 2000), I am aware of no consideration at all of the impact hazard by United States or international agencies responsible for managing a broad spectrum of other disasters. Theoretically, one might expect that an "all-hazards" approach would suffice for the impact hazard, because of some of the similarities. But I expect that there are sufficient differences between this particular never-before-witnessed kind of disaster and others that a specific focus on the unusual or unique features of the impact hazard is also essential.

Indeed, even as NASA tries to formalize procedures for communications within that agency if the cognizant official is notified by astronomers of an impact prediction, it remains uncertain who the NASA Administrator should notify within the Federal Emergency Management Agency (a part of the U.S. Dept. for Homeland Security) or whether anyone is prepared to receive such information and would know what to do with it. Although Britain has established an NEO Information Centre (*http://www.nearearthobjects.co.uk*), I am unaware that the British government, any other national agency, or the United Nations has even a rudimentary plan for responding to announcement of an impending impact. The only significant steps that have been taken have been by astronomers: (*a*) formulation of an impact prediction evaluation process by the Working Group on Near Earth Objects of the International Astronomical Union (a member of ICSU), (*b*) the development and promulgation of the Torino Scale (Binzel 2000) for articulating the significance of an impact prediction to the public through the news media, and (*c*) the maintenance of several web sites where up-to-date information is available on NEAs (*http://neo.jpl.nasa.gov/, http://newton.dm.unipi.it/cgi-bin/*

neodys/neoibo?, and *http://spaceguard.rm.iasf.cnr.it/*; background information is maintained at *http://www.nearearthobjects.co.uk* and *http://impact.arc.nasa.gov/index.html*, among other sites. But for an end-to-end disaster management plan to be effective, astronomers constitute only the first link in a lengthy, so-far-undefined chain of communications and responsibilities.

7.5
Perceptions of the Impact Hazard

Above, I outline what is objectively known about asteroids and the direct consequences of an impact with Earth. But there is an enormous leap between such "facts" and what policy makers require to address the issue. I now touch briefly on one of the most important factors, although it is outside the realm of my professional expertise: risk perception. About a decade ago, Paul Slovic (Morrison et al. 1994) polled a small sample of the American public about their perceptions of the impact hazard. This was at a fairly early stage of public awareness of the hazard, before the blockbuster movies "Armageddon" and "Deep Impact". At the time, about a quarter of respondents had some familiarity with the hazard; surely awareness has increased considerably since then. An interesting aspect of the responses by those who were aware of the hazard and had an opinion is that roughly half considered the impact hazard to be "serious" while the other half considered it to be a "silly" thing to worry about.

Probably the chief reasons a person would consider the impact hazard as "silly" are (*a*) the extremely low probability of a catastrophic impact happening and (*b*) lack of personal (or even historical) familiarity with asteroid impacts. There are also several reasons that motivate concern about the impact hazard that are probably responsible for the tendency of some to consider it to be "serious". In particular, as Slovic (1987) has demonstrated for perceptions of other hazards with which people are especially concerned, NEA impacts have enormous catastrophic potential and are "dreadful". Moreover, many people probably accept the contention of scientists and engineers that something practical can be done to avert an impact catastrophe; many other natural hazards (e.g. earthquakes) are difficult or impossible to predict in advance or to prevent.

In my own discussions of the impact hazard in public forums during the past two decades, I have learned several things (not all surprising) about perceptions of this issue, both by lay people and scientists:

- There is a common tendency for people to think of long "waiting times" before the next impact rather than in terms of "chances" of a disaster in the near-term. For the same reason people will build in a hundred-year floodplain, thinking (especially in the aftermath of an actual flood) that a flood won't happen for a hundred years, many people believe that an urgent response to the NEA threat isn't required: we can let the next generation deal with it. Yet many people buy lottery tickets (or avoid very low-probability hazards) with odds of winning (or dying) that are much lower than the chances of a large NEA impact happening this decade.
- People have enormous difficulty judging consequences of different degrees. It is very difficult for me to communicate the differences between a civilization-killing im-

pact and a mass-extinction event (although it would take a thousand of the former impacts to equal one of the latter). Should/will people consider 100 deaths per year (roughly the statistically averaged threat from NEAs) to be serious or not? We live in a society that can become very concerned about the life of a single individual highlighted by the news, yet remain oblivious to the plight of millions in a different context. At the peak of the Rwanda genocide killings, newspaper headlines were instead dominated for a week by the impact hazard (when Comet Shoemaker-Levy 9 fragments were crashing into Jupiter). American society felt that "the world had changed" when ~3000 people died on 11 Sept. 2001, yet the ~3000 American traffic fatalities in Sept. 2001 (and every month) go unnoticed. Since a large NEA impact has never been witnessed, it is difficult to predict how seriously even properly informed people might react to such a predicted impact.

- People are inclined to visualize the problem as involving an NEA that is on its way in and the way to deal with it is to "blow it up" shortly before it hits. The picture of an NEA orbiting the Sun countless times (and for decades, centuries, or longer) before it hits – all the while remaining in our cosmic neighborhood, where it is accessible by spacecraft – is difficult to get across.

- The process of NEA discovery and orbit determination is an arcane art, not even well appreciated by non-specialist astronomers, let alone journalists or the public. The reality is that when an NEA is discovered, its track is extremely poorly determined in the first hours and days, so it is likely to have a "chance" (although a very low one) of colliding with Earth. It might take days or even months before additional observations of the NEA's movements in the sky permit refinement of the orbit to the degree that the chances of collision go to zero (which happens for almost every NEA). The NEAs of interest, of course, are those very rare cases where refinement of the orbit results in the chances of impact going *up*: the body is likely to pass very close to the Earth at some point during the next decades or century. How the uncertainties of impact (e.g. the "error ellipse") behaves as still more observations are made is complicated and non-intuitive. Although simulations of such behavior have been run, it is plausible that if and when the first real prediction of an actual impact (or very near miss) is made, a complete understanding of the uncertainties will elude astronomers, as well as the public officials who will have to make decisions. For accounts of some past examples of misunderstood impact predictions, see accounts by Chapman (2000), Morrison et al. (2004) and Chapman (2004b).

7.6
Societal Impacts

I now address, from an astronomer's perspective, the theme of this volume. I attempt to provide a bridge between the "facts" about the impact hazard I have described and what I perceive to be issues about the impact hazard that affect different sectors of society. First I examine the aspect of the impact hazard that is most likely to affect society during the near future. Then I briefly discuss four institutions, the news media, religion, the military and science; then I conclude by focusing on hazards research and disaster management.

I outlined above the most serious, objective threats of mortality (and damage) from NEA impacts. It must be re-emphasized that the major, unique element of the NEA impact hazard (end of civilization, or even extinction of our species) comes from NEAs > 1.5 km in size, and that threat is rapidly being reduced by the Spaceguard Survey (though it will never go to zero, because current technology cannot handle the threat from large comets). Although impact of a small NEA, say 100 to 300 m across, could be as devastating and lethal as one of the largest natural disasters of the past century, such impacts are rare: hundreds of such floods, earthquakes, and other natural disasters will occur for each NEA impact of comparable seriousness. Certainly the issue that will most likely face society and public officials in the near future are the most frequent, smallest events – indeed those that directly kill nobody, but which make the news and could result in public response (panic [though Clarke, 2002, argues that such a response is unlikely], political calls for action, etc.). I describe below some recent news stories about "near-misses", actual impacts by harmlessly small bodies, or predictions that a dangerous NEA *might* impact in the future. Such events in the past have illuminated difficulties in communication between astronomers and the public. Clearly we must improve communications channels if we are to avoid ever-increasing "scares" and *ad hoc* responses by officials as NEA discoveries accelerate due to Pan-STARRS and other searches coming on-line. I now focus briefly on four societal institutions.

7.6.1
The News Media

The news media have been a prime route whereby people have learned that the impact hazard exists. Most commonly the stories have concerned three kinds of events: (1) A "near-miss" generates interest, when an NEA is about to pass close to the Earth, or is found to have just passed by (and that fact was discovered only afterwards, incorrectly implying that astronomers weren't looking carefully enough beforehand). The distance may be genuinely close (e.g. the 10-m NEA 2004 FU_{162} was reported in August 2004 to have passed just 6500 km above the Earth's surface on 31 March 2004) or many times farther away than the Moon (e.g. news stories in September 2004 before the large NEA Toutatis missed the Earth by 1.5 million km, an event which had been accurately predicted years beforehand). (2) An actual bolide (or "fireball") is reported, an actual meteorite strikes someone's house, or some other phenomenon related to very tiny cosmic objects that are essentially harmless draws attention. (3) A prediction of a *very small chance* of a catastrophic impact at some time (typically decades) in the future.

One or another such media report, meriting a CNN "crawler" or at least "page-two" coverage, has happened every couple of months during the last five years. The media reports are often inaccurate about the details or use hyped or even wholly fallacious language; for example, a 2002 BBC on-line report that an NEA was "on a collision course with Earth" misrepresented the truth that the astronomers' press release had estimated that it would miss by tens of millions of kilometers and never had an estimated chance of Earth impact higher than 1-in-100 000. There are several science journalists associated with major media who have become well-informed about the topic and whose reports are generally reliable. But most casual viewers/readers hear reports passed along by TV weather forecasters and other reporters who often magnify misunderstandings. Of course, concerns about news media affect risk communication generally, and most other aspects of life. But fears are augmented when the issue involves a relatively new, difficult-to-comprehend hazard and predictions of a frightening catastrophe. I have heard of cases where misleading headlines about an NEA impact have caused citizens to "run into the streets" or schoolchildren to run home crying. How to achieve better, more accurate communications between NEA experts, journalists and citizens should be part of broader dialogs addressing larger issues affecting science journalism and science literacy generally. (For lengthier discussion of past NEA scares, see Morrison et al. 2004.)

7.6.2
Religion

Some researchers believe that the sacred Kaaba stone in Mecca may be a meteorite. In 1910, some people thought that the approach of Halley's Comet signaled the Apocalypse. Perusal of the web with Google uncovers a surprising number of religious sites that are fascinated by asteroid impacts. Surely, many people on the fringe keep asteroid researchers (those who choose to do so) busy answering questions on late-night radio shows. What I cannot predict is the degree to which mainstream religions would become interested in an actual predicted future impact, and what they might do about it.

7.6.3
The Military

There are numerous past and potential connections between the impact hazard and the military. Much of our present knowledge about the frequency of impacts by objects roughly 1 m in size comes (often reported rather belatedly) from military assets deployed in space for other purposes. The U.S. Air Force has partially supported several elements of the Spaceguard Survey, especially the LINEAR project, which currently is the leading NEA detection survey. The U.S. Department of Defense (DoD) has shown other interests in NEAs on occasion, such as development of the Clementine space mission to the Moon and the NEA Geographos (the spacecraft was lost before the asteroid phase of the mission). On the other hand, the DoD has never taken ownership of "planetary defense". NASA, on the other hand, has pointedly never taken ownership of the topic, either, except for telescopic discoveries of NEAs and space missions to comets and asteroids (specifically motivated by planetary science objectives, not by the impact hazard). Most interest in the U.S. concerning military options for deflecting asteroids (using bombs and other technologies developed for the "Star Wars" Strategic Defense Initiative) has emerged from the Dept. of Energy national laboratories (e.g. Los Alamos and Livermore) or occasional individuals within the DoD; such individual interest has never been translated into serious programs. There have been parallel interests on the part of Russian scientists with military backgrounds.

Of course the potential use of bombs in space has important ramifications. This may be one motivation for the occasional interest in the impact hazard by the U.N. Office for Outer Space Affairs (Committee on the Peaceful Uses of Outer Space), which has co-sponsored at least two conferences on the impact hazard, including one held at U.N. Headquarters in New York in 1995 (Remo et al. 1997). But some U.N. involvement has attempted to foster scientific programs in underdeveloped countries, including possible construction of telescopes for NEA follow-up observations. It was reported in the early 1990s that the NEO hazard had been used as a bargaining chip by China with regards to nuclear disarmament (the argument was that nuclear weapons had to be maintained for potential use to protect Earth from an NEA strike).

One hazard posed by smaller NEA impacts mentioned above is the possible misinterpretation of the upper atmospheric explosion of an NEA as an offensive military action. This possibility has been recognized for decades, and we must hope and assume that there has been adequate promulgation of information about bolides to preclude inappropriate military responses to bolides in areas of conflict in the world.

All of these minor involvements of military institutions with the impact hazard could sharply crystallize if a specific impact threat were to develop. We would quickly focus on such questions as civilian-versus-military responsibilities for mitigation and national-versus-international approaches to deflection and disaster management. I think it would be prudent to think about these issues in advance.

7.6.4
Science

There is an uneasy relationship between basic scientific research and the impact hazard. Normally, in the past century at least, astronomy has had little direct, practical applications to the Earth. Until recognition of the impact hazard, solar flares were the only aspect of the heavens with practical effects. Unlike geology, which deals with earthquakes, landslides and oil reserves, astronomy has been a bastion of "pure science". Thus, when the NEA hazard arose, the question was asked about whether it is "real science". The only recent American National Academy of Sciences/National Research Council evaluation of NEA research priorities explicitly set aside hazard issues and focused on strict science issues. Hemmed in by flat budgets, NASA's Office of Space Science (recently transformed into the Science Mission Directorate), took the "high road" and declared that funds would not be carved out from "real astronomy" for practical matters like planetary defense; thus NASA-funded NEA research in the 1990s addressed questions involving the origin and evolution of the solar system. NASA's only forays into the NEA hazard arena have been under pressure from Congress and usually in the narrow endeavour of telescopic searches for NEAs. NASA spacecraft missions like NEAR Shoemaker and Deep Impact have some obvious relevance to NEA hazard mitigation issues, but they were funded to meet pure scientific objectives. There has been more willingness, in principle, to address the NEA hazard within the European Space Agency. But there has been little direct funding of NEA hazard research in Europe or by any other national science agency, presumably in part because the budgetary pie has already been sliced up for existing scientific constituencies. The scientific establishment is as conservative as any other human institution and, barring an actual NEA impact; it may prove difficult to shift priorities in order to accommodate the impact hazard.

In another arena of science, the NEA impact hazard has had enormous influence: so-called *informal* science education (i.e., in TV documentaries, planetarium presentations, etc.). Well over a dozen widely distributed documentaries on the impact hazard have been produced by Nova, National Geographic, NBC, BBC, CBC, the National Film Board of Canada, national television networks in Germany and Japan, and so on. Cosmic impacts have been themes of planetarium shows far out of proportion to research funding of the topic. Thus interest in asteroid catastrophes (and related topics of popular interest, like dinosaurs) has provided a focus for educating the public about a wider range of scientific issues, such as primordial accretion processes, planetary cratering and climate change.

7.7
Hazards Research/Disaster Management

In the last few decades, people have become much more aware of hazards, both in their personal lives and in the news about hazards facing communities and humanity in general. These include natural hazards (whose locally catastrophic effects are compellingly broadcast on 24/7 TV news channels), technological hazards (like Chernobyl and Bhopal), and the threats of war and terrorism (e.g., 9/11 and weapons of mass destruction). Natural hazards research, risk assessment and risk communication, disaster management and recovery, and issues of insuring against unpredictable catastrophes have been growing topics in recent years. There has been a trend, in the last decade, to use an "all-hazards" approach to emergency preparedness and crisis management, in order to take advantage of the many common elements of disasters, to simplify warning systems, and coordinate other elements of mitigation and response. In the United States, Homeland Security Presidential Directive #5, issued in February 2003, orders that such an all-hazards approach be developed to manage all "domestic incidences." But the NEA hazard has notably been missing from most discussions of "all-hazards" and NEAs have not explicitly been incorporated into discussions of implementing HSPD-5.

One of the chief challenges and opportunities of this multi-disciplinary volume is to develop an understanding of how NEA impacts might fit within the larger umbrella of risk perception and hazard mitigation. Surely consideration of this extreme low-probability high-consequence hazard may prepare us to deal with other analogous, but less extreme, disasters that face us. And NEA researchers may also learn from the broader hazards fields the ways to more effectively approach implementation – at whatever level of priority seems to be appropriate – of the end-to-end processes from discovery of a threatening NEA ... through management of mitigation efforts ... to response to any NEA disaster that may be required. I believe that a thorough evaluation of the NEA threat, in the context of other hazards, by one or more authoritative national or international scientific advisory bodies, is essential to establish the appropriate priorities for researching the NEA hazard, for extending searches, for developing deflection options, and for treating this hazard within the context of other hazards.

References

Belton M, Morgan TH, Samarasinha N, Yeomans DK (eds) (2004) Mitigation of hazardous comets and asteroids. Cambridge Univ Press

Binzel RP (2000) The Torino impact hazard scale. Planet. Space Science 48:297–303

Boslough MBE, Crawford DA (1997) Shoemaker-Levy 9 and plume-forming collisions on Earth. In: Remo (1997), p 236

Chapman CR (2000) The asteroid/comet impact hazard: *Homo sapiens* as dinosaur? In: Sarewitz D et al. (eds) Prediction: science, decision making, and the future of nature. Island Press, Washington DC, pp 107–134

Chapman CR (2003) (CRC03) How a near-Earth object impact might affect society, commissioned by the OECD Global Science Forum for Workshop on Near-Earth Objects: risks, policies, and actions. Frascati, Italy, January 2003. Downloadable from: *http://www.oecd.org/dataoecd/18/40/2493218.pdf* or *http://www.boulder.swri.edu/clark/oecdjanf.doc*

Chapman CR (2004a) (CRC04) The hazard of near-Earth asteroid impacts on Earth. Earth and Planetary Science Letters 222:1–15. *http://www.boulder.swri.edu/clark/crcepsl.pdf*

Chapman CR (2004b) NEO impact scenarios, Amer. Inst of Aeronautics and Astronautics, 2004 Planetary Defense Conference: Protecting Earth from Asteroids, Garden Grove, CA

Clarke L (2002) Panic: myth or reality, contexts. Univ Calif Press 1(3)

Garshnek V, Morrison D, Burkle FM (2000) The mitigation, management, and survivability of asteroid/comet impact with the Earth. Space Policy 16:213–222

Harris AW, Benz W, Fitzsimmons A, Green S, Michel P, Valsecchi G (2004) Recommendations to ESA by the near-Earth object advisory panel, July 2004. *www.esa.int/gsp/NEO/other/NEOMAP_report_June23_wCover.pdf*

Morrison D, Chapman CR, Slovic P (1994) The impact hazard. In: Gehrels T (ed) Hazards due to comets and asteroids. Univ Arizona Press, Tucson, pp 59–91

Morrison D, Harris AW, Sommer G, Chapman CR, Carusi A (2002) Dealing with the impact hazard. In: Bottke WF Jr, Cellino A, Paolicchi P, Binzel RP (eds), Asteroids III. Univ Arizona Press, Tucson, pp 739–754

Morrison D, Chapman CR, Steel D, Binzel RP (2004) Impacts and the public: communicating the nature of the impact hazard. In: Belton et al. loc cit, Chapter 16

Remo JL (ed) (1997) Near-Earth objects: the United Nations international conference. Annals of the New York Academy of Sciences 822:632

Schweickart RL, Lu ET, Hut P, Chapman CR (2003) The asteroid tugboat. Scientific American 289:54–61

SDT (Near-Earth Object Science Definition Team) (2003) Study to determine the feasibility of extending the search for near-Earth objects to smaller limiting diameters. NASA Office of Space Science, Solar System Exploration Div, Washington DC. *http://neo.jpl.nasa.gov/neo/neoreport030825.pdf*

Slovic P (1987) Perception of risk. Science 236:280–285

Toon OB, Zahnle K, Morrison D, Turco RP, Covey C (1997) Environmental perturbations caused by the impacts of asteroids and comets. Revs Geophysics 35:41–78

Chapter 8

The Impact Hazard:
Advanced NEO Surveys and Societal Responses

David Morrison

8.1 Background

The Earth is immersed in a swarm of Near Earth Asteroids (NEAs) capable of colliding with our planet, a fact that has become widely recognized within the past decade. The first comprehensive modern analysis of the impact hazard resulted from a NASA study requested by the United States Congress. This *Spaceguard Survey Report* (Morrison 1992) provided a quantitative estimate of the impact hazard as a function of impactor size (or energy) and advocated a strategy to deal with such a threat.

Impacts represent the most extreme example of a hazard of very low probability but exceedingly grave consequences. Chapman and Morrison (1994) concluded that the greatest hazard was associated with events large enough to risk a global environmental disaster, with loss of crops and mass starvation worldwide – an event that happens on average once or twice per million years (see also Morrison et al. 1994, Toon et al. 1997). The NASA Spaceguard study (Morrison 1992) advocated focusing on these global-scale events, a result of asteroids larger than 1–2 km striking the Earth. The Spaceguard Survey that was proposed in that report and formally initiated in 1998 would discover such asteroids and determine their orbits well in advance of any actual impact. The relative orbital stability of even the Earth-crossing asteroids makes such discovery and cataloguing a practical task.

A follow-up NASA study (Shoemaker 1995) described a practical way to implement such a Spaceguard Survey using modest-sized ground-based telescopes equipped with modern electronic detectors and computer systems. The Shoemaker team suggested a goal to discover and track 90 percent of the NEAs larger than 1 km within ten years. A government-sponsored study in the United Kingdom (Atkinson et al. 2000) confirmed the NASA conclusions and also advocated extending the survey to smaller NEAs, down to 500 m diameter, as a first step toward dealing with impacts below the threshold for global disaster. Recent comprehensive reviews of the impact hazard and ways to deal with it include Morrison et al. (2003) and Chapman (2004).

A handful of telescopes, primarily located in the United States, are now used in the Spaceguard Survey. This survey has already found nearly 75 percent of the NEAs with diameter greater than 1 km (a total of 800 in August 2005, out of an estimated population of 1100). The surveys are deemed worthwhile because we have the technology, at least in principle, to deflect a threatening asteroid, given sufficient (decades) warning. The impact hazard is unique in that it is possible to avoid the damage entirely. In most natural hazard areas, "mitigation" consists of ways to plan for a disaster or to deal with the disaster after it happens. Cosmic impacts represent an opportunity to take steps to avoid the disaster itself.

Recently NASA sponsored a NEO study (Stokes 2003) that focused on the role of impacts by sub-kilometer asteroids, below the global hazard threshold. Such impacts are much more frequent, since there are many more small asteroids than larger ones, but the damage would be local or at most regional in scale. As we retire the risk from the global threats, it seems prudent to also examine the options for defending ourselves against smaller impacts. This study raises (but does not settle) the issue of how much society should invest in protecting against impacts across the full range of energy and risk. As in so many other cases, we question how much protection we need and seek to strike the balance between cost and mitigation.

Comets as well as asteroids can strike the Earth. We do not know if the impact that killed the dinosaurs, for example, was from a comet or an asteroid. Statistically, however, asteroid hits are more frequent than comet hits. This disparity increases as the size declines, to the point where comets are virtually absent below 1 km diameter (Yeomans 2003). Therefore, the discussions in this paper refer only to asteroids, which account for 99 percent or more of the risk in the sizes of primary interest.

This paper discusses the impact hazard from the 2005 perspective, in which the Spaceguard Survey is steadily reducing the threat from global-scale impacts. The issues for asteroids larger than 1 km are how far to push the survey toward completeness and what plans should be made to develop technology to deflect an asteroid in the absence of a clear and present threat. For the smaller (sub-kilometer) asteroids, the immediate question is how much should be invested in reducing the risk of the smaller impacts. There are broad international implications in dealing with both the globally threatening impacts and smaller impacts, which might target one country while leaving its neighbors relatively unscathed. Finally, there are issues of public perception (and misperception) that cut across all of these issues.

While the level of hazard is sufficient to warrant public concern and justify possible government action, its nature places it in a category by itself. Unlike more familiar hazards, the impact risk is primarily from extremely rare events, essentially without precedent in human history. Although there is a chance of the order of one in a million that each individual will die in any one year from an impact, it is not the case that one out of each million people dies each year from an impact. The expectation value for impact casualties within any single lifetime is nearly zero. The most important consideration for society is not, therefore, the average fatalities per year, but rather the question of *when and where the next impact will take place*. Surveys must find each asteroid, one at a time, and calculate its orbit, in order to determine whether any are actually on a collision course with Earth.

8.2
The Spaceguard Survey

A decade ago, when we were trying to establish the credibility of the impact hazard in the face of widespread scepticism, it was important to calculate probabilities of impact of various sizes. It was also essential to make some estimate, however crude, of the expected losses in lives and property (e.g. Chapman and Morrison 1994). We are now beyond that stage in evaluating the hazard. Any estimates made of expected casualties are rendered extremely uncertain by lack of knowledge of societal responses to such an unprecedented calamity, as well as by the unknowns associated with the nature of the impactor (composition, density, etc.) and with the wide range of possible target conditions (land vs. ocean vs. ice etc.). Newspapers sometimes run headlines indicating that the statistical risk from impacts has risen or declined based on some new astronomical data, but such a conclusion is not very meaningful. We know that impacts in any size range are unlikely within a human lifetime. But we also know that if, against the odds, there is an impactor on a collision course for Earth, people and governments want to know about it – hence the survey approach, directed at identifying the next impactor and providing decades of warning before it hits.

Although they are quite faint, asteroids down to one kilometer diameter can be detected by their motion using modest-sized ground-based telescopes (aperture about 1 m) equipped with state-of-the-art electronic detectors. Moving objects are identified automatically by the search software and a preliminary orbit can be obtained with data from even a single night. Lists of new NEAs are posted every day on public websites, and this information is used to guide both the ongoing surveys and the follow-up support.

Although asteroid searches had been underway for more than two decades, the formal beginnings of the NASA Spaceguard Survey were in 1998, the same year that the highly successful LINEAR survey became fully operational (Stokes et al. 2000). The Spaceguard objective is to find 90 percent of the Near Earth Asteroids (NEAs) larger than one kilometer within ten years, or by the end of 2008 (Pilcher 1998). Halfway into this survey decade, more than 60 percent of the estimated 1100 ± 100 of these NEAs had already been found. This is not as positive a result as might seem, however, since the rate of new discoveries falls off as the survey nears completeness. As expected, the discovery rate of one kilometer NEAs has been dropping since 2002, after a decade of steady increases. Estimates of when the 90-percent level will be met vary from 2008 to beyond 2010. This survey is being carried out with approximately $ 4 million per year from NASA, plus voluntary and in-kind contributions – a tiny sum compared to the ongoing cost of mitigation for numerically comparable but better-known hazards such as earthquakes, severe storms, airplane crashes and terrorist activities.

If we focus on asteroids larger than two kilometers in size, which is the nominal threshold size for a global catastrophe, then we are already (in 2005) approaching 90 percent completeness. For five kilometers diameter, which may be near the threshold for an extinction event, we are complete today for asteroids (but at this size long period comets may represent a significant contribution to the hazard). Thus astronomers have already assured us that we are not due for an extinction level impact from an asteroid within the next century. Barring an unlikely strike by a large comet, we are not about to go the way of the dinosaurs (status summarized by Morrison et al. 2003).

The field of impact studies is still too young to determine what society (and representative governments) seeks in the way of protection (Chapman 2000). For those who mainly fear an extinction event that might end human life forever, we have already achieved a considerable level of reassurance. For those whose concern is a global, civilization-threatening disaster, we are more than halfway complete. But for those who are primarily concerned about the smaller but more frequent impacts by sub-kilometer asteroids (100 m to 1000 m diameter), the astronomers have not achieved even 1 percent completeness in our surveys. The cost of the present Spaceguard Survey is much lower than the estimates of expected equivalent annual loses in lives and property for the U.S. alone, justifying the effort even if it is supported solely by the U.S. taxpayer. It is not equally obvious that the survey should be extended to smaller impacts.

8.3
Sub-Kilometer Impacts

The term "sub-kilometer impacts" is intended to include all potentially destructive impacts from asteroids with diameters between the threshold for global disaster (nominally 1–2 km) and the sizes where the Earth's atmosphere offers protection (nominally 50–100 m). Below this size range, atmospheric friction and shear stress on a stony or icy projectile cause it to decelerate and disintegrate at high altitudes, with little blast damage on the ground.

Members of the 2003 NASA Science Definition Team (SDT) (Stokes 2003) focused on two classes of sub-kilometer impacts by stony asteroids that do pose a substantial hazard: land impacts yielding massive ground or air-burst explosions, and ocean impacts that produce tsunami waves that endanger exposed coastlines.

The effects of land impacts can be derived by extrapolation of our knowledge of large nuclear explosions. The SDT analysis uses estimates of blast damage as a function of impactor size by Hills and Goda (1993). From about 50 to 150 m diameter, these are primarily airbursts, and the impactor disintegrates explosively before reaching the ground. Impactors larger than 150 m produce craters. At 300 m diameter, the area of severe damage is as large as a U.S. state or small European country. Because of the highly uneven distribution of population on the Earth, most of these sub-kilometer impacts, which are near the lower size limit, will produce few if any casualties, but much rarer impacts over heavily populated areas could kill tens of millions. Combining their explosion models with frequency-of-impact estimates and a model population distribution, the SDT concluded that the greatest hazard in the sub-kilometer realm is from NEAs 50–200 m diameter, with total expected equivalent annual deaths from sub-kilometer impacts at a few dozen – roughly two orders of magnitude less than the similar metric for larger (global-hazard) impacts at the start of the Spaceguard Survey, and still more than an order of magnitude less than that from the residual of undiscovered NEAs larger than one kilometer remaining in 2005.

Ocean impacts are less well understood, since we do not have any examples of impact tsunamis to provide "ground truth." Chesley and Ward (2005) have analyzed the risk from impact tsunamis as a function of impactor size, based in part on an earlier study by Ward and Asphaug (2000). They modeled the production and propagation of the waves and, with greater uncertainty, the run-up and run-in of the waves as they reach the coast. The impact tsunamis have an intermediate wavelength between seismic tsunamis (tens of kilometers scale) and familiar storm waves (tens of meters scale), leading to intermediate run-in. Even large impact tsunamis, with open ocean waves many meters high, are unlikely to flood more than a few kilometers inland.

These wave penetration predictions have been convolved with the distribution of coastal populations on the Earth. Chesley and Ward find that impact tsunamis constitute less of a hazard than one might guess based on the greater run-in of seismic tsunamis. They conclude that the highest risk comes from smaller but more frequent events, as was the case with land impacts. However, since airbursts over water do not generate tsunamis, the peak hazard is shifted to impactor sizes from about 200–500 m. The total impact tsunami hazard is larger than that of land impacts by roughly factor of 5. However, since it should be possible to provide warning of an approaching wave in time to evacuate coastal populations, the actual casualties might be much smaller. Therefore the tsunami at-risk estimates are properly understood as a surrogate for property damage rather than human fatalities.

Chesley and Ward (2005) and the NASA SDT (Stokes 2003) provide the data to assemble a ranked estimate of the impact hazards remaining after the present Spaceguard Survey achieves its 90 percent goal. The largest hazard in terms of fatalities remains the residual 10 percent of undiscovered NEAs larger than one kilometer, with an equivalent annual fatality rate of roughly 100, as well as the potential to destabilize global civilization. Even larger is the risk to property from impact tsunamis by sub-kilometer NEAs (down to about 200 m diameter), but the fatalities can be easily reduced by the application of tsunami warning systems. Third in rank in terms of both property damage and fatalities are the land impacts from sub-kilometer NEAs (down to about 100 m diameter).

The present Spaceguard Survey will, if continued, eventually deal with the residual of undiscovered NEAs larger than one kilometer, but it will require several decades of additional work to do so. However, if society desires to make serious progress within the next decade or two in retiring the risk from sub-kilometer NEAs, we will need a much more ambitious survey using telescopes larger than the current one meter systems. Such surveys have been supported by two panels of the U.S. National Academy of Sciences / National Research Council under the general name of LSST, or Large Synoptic Survey Telescope (NRC 2001). One wide-field telescope of approximately eight meter aperture at a superior observing site could carry out an asteroid survey that is 90 percent complete down to 200 m diameter within a decade while also accomplishing several other high-priority astronomy objectives that require all-sky surveys (Strauss 2004). Alternatively, the NASA SDT note that this task could also be accomplished with two or more four meter telescopes, or with a combination of ground-based and space-based survey telescopes.

No decision has been made on construction of the full-scale LSST or on the option of searching from space. Meanwhile, however, a similar survey instrument using smaller telescopes, called Pan-STARRS, is under construction at the University of Hawaii with U.S. Air Force support and will begin tests in 2006. Deeper asteroid surveys are also part of the program planned for a new four meter telescope planned for Lowell Observatory. These surveys should push the detection size limit down to 300 m. It is not clear whether any of these new instruments, including a full-up LSST, can extend the survey to 100 m NEAs, but they can certainly retire at least 80% of the risk that remains in 2008.

8.4
Communication and Miscommunication

There are numerous challenges in communicating the nature of the impact hazard to both decision-makers and the public. NEO impacts are qualitatively different from any other hazard, in that the numbers of people killed could be far larger than in any natural disaster that has occurred during historical times, and may approach the whole population of the planet. Because of their rarity, people do not have direct knowledge of the destructive potential of an impact. Many political leaders feel they can ignore this problem, since it is unlikely that anything bad will happen "on their watch." At the opposite extreme, however, there is a tendency in some quarters to exaggerate the risk and to issue repeated warnings of impacts that never happen.

One other catastrophe that might provide a similar outcome to an impact by an NEA with energy greater than one million megatons would be a global nuclear war, a concern of obvious interest to people and governments. The question might therefore be asked: why is the impact hazard not taken more seriously? One reason, of course, is a lack of awareness of the nature and level of the hazard, but another is the fact that blame might be attached to various people/countries in the case of a nuclear war, whereas an impact might be regarded as an Act of God.

The NEO community has taken several actions to facilitate communications with the media and the public, as discussed by Morrison et al. (2003). First, the nature of the hazard itself has been explained in a variety of public forums (for example, hearings in the United States Congress and documentaries produced for television broadcast). Second, the Internet has been widely used to explain the hazard and to provide up-to-date information on asteroid discoveries and orbits. Third, the International Astronomical Union (IAU) has attempted to provide authoritative information on NEOs and possible future impacts.

Whether testifying before Congress, providing sound bites on television, or writing messages to post on the Internet, it is vital for scientists involved in disaster prediction to communicate their calculations in simple ways that can be understood by the news media, public officials, and disaster relief agencies, so that news of a potential disaster evokes consistent and appropriate responses. By analogy with other hazard scales, such as those associated with forest fire danger or homeland security alerts, a simple one-dimensional color-coded scale seems useful to characterize potential impact events. The discovery of asteroid 1997 XF11, the first asteroid for which an orbit calculation apparently gave a non-negligible probability for impact of globally catastrophic consequences, provided a baptism by fire for the NEO community. This experience precipitated the adoption the 'Torino Impact Hazard Scale' (Binzel 2000), which was announced simultaneously by the IAU and by NASA in July 1999.

The Torino Scale values range from 0 (no threat) to 10 (certain threat of bad global consequences), along with essential information such as the name of the object and the date(s) of its close approach. It provides a simple vehicle for allowing the public to appreciate whether the object merits their concern. Further education and public familiarity are necessary to understand the scale, but even sound-bite news reporting "The December 2037 encounter by object XYZ ranks only a 1 on the 10-point Torino hazard scale" correctly conveys a very low level of concern on a distant date without knowing anything else about the scale or going into details of probability calculations, error ellipses, orbital nodes, etc.

In the absence of national or intergovernmental agencies to deal with the NEO impact issues, the International Astronomical Union (IAU) has assumed some of the responsibility by default. The IAU formed a Working Group on NEOs in the early 1990s to advise on coordination of NEO activities worldwide, on reporting of NEO hazards, and on research relevant to NEOs. When someone predicts a close approach to Earth by an asteroid, a committee of the IAU Working Group can be convened to advise the IAU on the reliability of the prediction. This IAU Technical Review Committee of international specialists offers prompt, expert review of the scientific data, computations, and results on NEOs that might present a significant danger of an impact on Earth in the foreseeable future. The use of this review process is voluntary, and researchers worldwide remain free to publish whatever results they wish in whichever way they wish, at their own responsibility.

The role of the IAU is limited: it deals only with the discovery of NEOs, not with mitigation, and it has limited ability to respond rapidly to new discoveries. From the IAU perspective, it remains the responsibility of the individual science teams who discover NEAs or make orbital predictions to decide whether to release information to the public.

The basic challenges of communication with the public and the journalistic media remain with us today. It is difficult to understand low-probability events, especially by a largely innumerate public. Probabilities enter into the dialog whether we are discussing the apriori risks of impact by an unknown object or the accuracy of the orbital predictions for a newly discovered object. Most of the media miscommunications of the past seven years have arisen from exaggerated concern or outright misrepresentations of the very low probabilities of impact assigned to objects before accurate orbits are determined (see detailed discussion by Morrison et al. 2004). Yet it seems essential to make timely information on newly discovered NEAs available to both the scientific and public communities. In this way we at least protect ourselves against concerns about governments or groups of scientists suppressing information on possible threats that the public justifiably feels it has a right to know.

8.5
Public Policy Issues

The preceding sections of this paper hint at a number of policy issues that are summarized in this concluding section. The following questions are all addressed to what steps we should undertake beyond the current Spaceguard Survey.

1. Is it important to extend asteroid surveys to sub-kilometer impactors, perhaps down to the limit of penetration of the Earth's atmosphere? Such an undertaking is consistent with a legal imperative for governments to make an effort to identify and protect their populations from preventable disasters (Gerrard 1997; Seamone 2002). It may or may not be cost effective, depending on accounting assumptions. This

effort would be considerably less cost-effective than the current Spaceguard Survey, since we would need to spend at least an order of magnitude more funds to protect against a risk that is at least an order of magnitude smaller than that of NEAs larger than one kilometer.

2. Should we begin to develop technologies for deflecting asteroids? To date, essentially no funds have been spent for this purpose. Many would argue that it is prudent to begin such research before an actual threat is identified. Others argue that since these technologies are unlikely to be needed within the next few decades, it is a waste of resources to do any work at present. The most compelling case is probably to accelerate our study of NEAs, including visits by spacecraft (Belton 2004). The knowledge gained by such scientific exploration is also needed to make plans for future deflection efforts, if they are required.

3. Should we test asteroid deflection technologies? Edward Teller was an advocate during the final decade of his life for conducting such experiments. He argued not only that such experiments were needed to test deflection schemes, but also that the experience gained in planning such an international test project would be invaluable if and when we faced the real thing – especially if the options for defense included nuclear explosives (Morrison and Teller 1994). The recent proposal by the B612 Foundation for test of a space tug represents such an experimental approach (Shweickart et al. 2003). Both the NASA Deep Impact mission and the ESA Don Quijote mission explore the technology for high-speed impacts, with asteroids and comets respectively, as an alternative option for deflection.

4. Who should be in charge of these efforts, from possible extensions of the Spaceguard Survey to potential testing of defensive systems? Is NASA the correct agency within the U.S. government? For that matter, are these topics the responsibility of the U.S. government? Why have other nations not contributed funding to these efforts to defend our planet from possible cosmic catastrophe?

5. Should civil defense and disaster relief agencies be planning to deal with the aftermath of an impact explosion that occurs without warning? Today, no warning would be expected for most sub-kilometer impacts. Who should assume responsibility in planning for mitigation if such a disaster should occur (cf. Garshnek et al. 2000)?

6. How important is international cooperation? While the impact hazard has been discussed internationally by the United Nations, the Council of Europe, the Organization for Economic Co-operation and Development, the International Astronomical Union, and the International Council for Science, no concrete action has been taken. The most comprehensive study of the problem outside the U.S. was carried out in the U. K. However, of the 14 recommendations in the UK NEO Task Group Report (Atkinson et al. 2000), only one has been fully implemented – the establishment of a British National Center for public education on the impact hazard.

7. Which impacts (if any) do not require mitigation, and who will make the decision? Suppose the astronomers discover a 100 m asteroid that will impact in the ocean – even if the science community concludes that there is no danger from tsunami, will that satisfy the public? Or suppose that a land impact is predicted; if the target area is deserted it may be easy to decide to let it hit, but suppose there are cities or other major infrastructure such as dams in the target area. Who will decide whether

a multi-tens-of-billions of dollars effort should be undertaken to deflect the asteroid? Who will pay for it?
8. If a sub-kilometer impactor is identified and a decision is made to change the orbit, there are a number of scenarios that could be complex and divisive. Suppose the initial target is identified as being in Country A. To change the asteroid orbit we must supply continuous thrust that gradually moves the impact point off the planet. But in this process the impact point crosses Nations B, C, and D, which were originally not at risk. Who will the nations trust to carry out the deflection maneuver? And what if the maneuver is only partially successful and the asteroid ends up striking Nation C rather than missing the Earth? Who is responsible? (for example, see Harris et al. 1994; Sagan and Ostro 1994).
9. In any of these examples, will the public trust either scientific judgments or the decisions of public officials? If an asteroid is discovered with an initial well-publicized non-zero chance of collision, and subsequent observations ultimately convince the scientific community that it will miss by a very small margin, will the public believe them? Or suppose an asteroid is found that is indeed on a collision course but the scientists estimate that it is only 30 m in diameter and is predicted to disintegrate harmlessly at high altitude. Will the people who live at ground zero trust this conclusion? What level of proof (or acceptance of responsibility) will be required?
10. Is the public likely to support continued and perhaps accelerated government spending to protect the Earth from asteroids? It is difficult to sustain interest and support in the absence of known threats, and there has never been an asteroid impact in a populated area in all of recorded history (Chapman 2000). In recent years, there have been a number of media-inspired scare stories, mostly based on very preliminary orbits, with the "threat" disappearing within a day or two. Such stories may sustain public interest, but they can also backfire if the public or the media conclude either that the astronomers don't know what they are doing or that they are "crying wolf" to attract public attention. Communicating the nature of this hazard, with no historical examples but potential fatalities of a billion or more people, is challenging. Yet if we are to create and sustain international programs for planetary defense, public understanding and support is required (Park et al. 1994).

We cannot today answer the above questions. All would profit by a wider dialog and the participation of individuals and groups who may never have been exposed to this unique natural hazard.

Note. The question of what government agencies should take responsibility for asteroid impact mitigation was resolved within the United States by Congressional action in January 2006. The Congress changed the NASA Charter to give responsibility to NASA and to establish an expanded NEO survey program to detect, track, catalogue, and characterize the physical characteristics of NEAs greater than 140 m diameter. This survey is to achieve 90 percent completeness by 2020. The NASA Administrator is asked to provide Congress with a plan by December 2006 to carry out this mandate, and also to provide an analysis of possible alternatives that NASA could employ to divert an object on a likely collision course with Earth.

This action has focused attention on a new Spaceguard Deep Survey designed to discover NEAs at approximately 100 times the current rate. Such a survey could be carried out with either large ground-based telescopes or with optical or infrared telescopes in space. The first element of the new survey is Pan-STARRS, with an initial 2 m telescope operational on Haleakala in Hawaii in 2007. This should be followed by a 4-telescope Pan-STARRS system on Mauna Kea (funded by the U.S. Air Force), a 4 m Discovery Telescope at Lowell Observatory in Arizona, and ultimately by the 8 m Large Synoptic Survey Telescope (LSST) in Chili, currently being designed with support from the U.S. National Science Foundation. There may also be search telescopes in space as well as increased space missions to characterize NEAs.

Increase of the NEA discovery rate by a factor of 100 will require that the calculation of orbits and archiving of data grow by the same factor, and it will place new burdens on astronomers carrying out follow-up observations. Whereas today one or two NEAs are found each year that receive publicity, we can expect an "interesting" asteroid at least once a week. It is not clear what these new NASA programs will mean for public understanding and support for NEA defense, or their international implications.

Acknowledgments

I am grateful to many colleagues who have helped sharpen my understanding of the NEO impact hazard, especially Richard Binzel (MIT), Clark Chapman (SWRI, Boulder), Alan Harris (Space Science Institute, Boulder), Don Yeomans (JPL), and Kevin Zahnle (NASA ARC).

References

Atkinson, H, Tickell C, Williams D (2000) Report of the Task Force on potentially hazardous Near Earth Objects, British National Space Center, London. http//:www.nearthearthobject.co.uk
Belton MJS (2004) Towards a national program to remove the threat of hazardous NEOs. In: Belton M, Morgan T, Samarasinha N, Yeomans D (eds) Mitigation of hazardous comets and asteroids. Cambridge University Press, Cambridge, pp 391–410
Binzel, RP (2000) The Torino Impact Hazard Scale. Planetary and Space Science 48:297–303
Chapman CR (2000) The asteroid/comet impact hazard: Homo sapiens as dinosaur? In: Sarewitz D, Pielke Jr. RA, Byerly R (eds) Prediction: science, decision making, and the future of nature. Island Press, Washington DC, pp 107–134
Chapman CR (2004) The hazard of near-Earth asteroid impacts on Earth. Earth Planetary Science Letters 222:1–15
Chapman CR, Morrison D (1994) Impacts on the Earth by asteroids and comets: assessing the hazard. Nature 367:33–39
Chesley SR, Ward SN (2005) A quantitative assessment of the human and economic hazard from impact-generated tsunami. J Natural Hazards (in press)
Garshnek V, Morrison D, Burkle FM (2000) The mitigation, management, and survivability of asteroid/ comet impact with the Earth. Space Policy 16:213–222
Gerrard MB (1997) Asteroids and comets: U.S. and international law and the lowest-probability, highest consequence risk. New York University Environmental Law Journal 6:1
Harris AW, Canavan GH, Sagan C, Ostro SJ (1994) The deflection dilemma: Use versus misuse of technologies for avoiding interplanetary collision hazards. In: Gehrels T (ed) Hazards due to comets and asteroids. University of Arizona Press, pp 1145–1156

Hills JG, Goda MP (1993) The fragmentation of small asteroids in the atmosphere. Astronomical J 105: 1114–1144

Morrison D (1992) The Spaceguard Survey Report of the NASA International Near-Earth-Object Detection Workshop. NASA Publication: *http://impact.arc.nasa.gov*

Morrison D, Teller E (1994) The impact hazard: Issues for the future. In: Gehrels T (ed) Hazards due to comets and asteroids. University of Arizona Press, pp 1135–1144. Some of Teller's other remarks are summarized in a news entry for October 9, 2003: *http://impact.arc.nasa.gov*

Morrison D, Chapman CR, Slovic P (1994) The impact hazard. In: Gehrels T (ed) Hazards due to comets and asteroids. University of Arizona Press, pp 59–92

Morrison D, Harris AW, Sommer G, Chapman CR, Carusi A (2003) Dealing with the impact hazard. In: Bottke W, Cellino A, Paolicchi P, Binzel RP (eds) Asteroids III. University of Arizona Press, Tucson, pp 739–754

Morrison D, Chapman CR, Steel D, Binzel R (2004) Impacts and the public: Communicating the nature of the impact hazard. In: Belton M, Morgan T, Samarasinha N, Yeomans D (eds) Mitigation of hazardous comets and asteroids. Cambridge University Press, Cambridge, pp 353–390

National Research Council (2001) Astronomy and astrophysics in the new millennium. National Academy Press, Washington. *http://www.nas.edu*

Park RL, Garver LB, Dawson T (1994) The lesson of Grand Forks: Can defense against asteroids be sustained? In: Gehrels T (ed) Hazards due to comets and asteroids. University of Arizona Press, pp 1225–1232

Pilcher C (1998) Statement at hearing on "Asteroids: Perils and Opportunities", Subcommittee on Space and Aeronautics, Committee on Science, May 21, 1998 (transcript available at *http://impact.arc.nasa.gov*)

Sagan C, Ostro S (1994) Dangers of asteroid deflection. Nature 369: 501

Schweickart, RL, Lu ET, Hut P, Chapman CR (2003) The asteroid tugboat. Scientific American, November 2003, pp 54–61

Seamone ER (2002) When wishing on a star just won't do: The legal basis for international mitigation of asteroid impacts and similar transboundary disasters. Iowa Law Review 87: 1091–1139

Shoemaker G (1995) Report of the Near Earth Objects Survey Working Group (unpublished NASA report, June 1995)

Stokes GH (2003) Study to determine the feasibility of extending the search for Near Earth Objects to smaller limiting diameters. Report of the NASA NEO Science Definition Team. *http://neo.jpl.nasa.gov/neo/report.html*

Stokes GH, Evans JB, Viggh HEM, Shelly FC, Pearce EC (2000). Lincoln Near-Earth Asteroid Program (LINEAR). Icarus 148:21–28. See also *www.ll.mit.edu/LINEAR/*

Strauss M (2004) Candidate specifications and observing protocols for the LSST. Report of the NOAO-LSST Science Working Group. *http://www.noao.edu/lsst/*

Toon OB, Zahnle K, Morrison D, Turco RP, Covey C (1997) Environmental perturbations caused by the impacts of asteroids and comets. Reviews of Geophysics 35: 41–78

Ward SN, Asphaug E (2000) Asteroid impact tsunami: A probabilistic hazard assessment. Icarus 145: 64–78

Yeomans D (2003) Population Estimates: In: Stokes GH (ed) Study to determine the feasibility of extending the search for Near Earth Objects to smaller limiting diameters. Report of the NASA NEO Science Definition Team. *http://neo.jpl.nasa.gov/neo/report.html*, Chapter 2

Chapter 9

Understanding the Near-Earth Object Population: the 2004 Perspective

William F. Bottke, Jr.

9.1 Introduction

Over the last several decades, evidence has steadily mounted that asteroids and comets have impacted the Earth over solar system history. This population is commonly referred to as "near-Earth objects" (NEOs). By convention, NEOs have perihelion distances $q \leq 1.3$ AU and aphelion distances $Q \geq 0.983$ AU (e.g. Rabinowitz et al. 1994). Subcategories of the NEO population include the Apollos ($a \geq 1.0$ AU; $q \leq 1.0167$ AU) and Atens ($a < 1.0$ AU; $Q \geq 0.983$ AU), which are on Earth-crossing orbits, and the Amors (1.0167 AU $< q \leq 1.3$ AU) that are on nearly-Earth-crossing orbits and can become Earth-crossers over relatively short timescales. Another group of related objects that have not yet been considered part of the "formal" NEO population are the IEOs, or those objects located inside Earth's orbit ($Q < 0.983$ AU). To avoid confusion with standard conventions, I treat the IEOs here as a population distinct from the NEOs. The combined NEO and IEO populations are comprised of bodies ranging in size from dust-sized fragments to objects tens of kilometers in diameter (Shoemaker 1983).

It is now generally accepted that impacts of large NEOs represent a hazard to human civilization. This issue was brought into focus by the pioneering work of Alvarez et al. (1980), who showed that the extinction of numerous species at the Cretaceous-Tertiary geologic boundary was almost certainly caused by the impact of a massive asteroid (at a site later identified with the Chicxulub crater in the Yucatan peninsula). Today, the United Nations, the U.S. Congress, the European Council, the UK Parliament, the IAU, OECD, NASA, and ESA have all made official statements that describe the importance of studying and understanding the NEO population. In fact, among all world-wide dangers that threaten humanity, the NEO hazard may be the easiest to cope with, provided adequate resources are allocated to identify all NEOs of relevant size. Once we can forecast potential collisions between dangerous NEOs and Earth, action can be taken to mitigate the potential consequences.

In this paper, I review the progress that has been made over the last several years to understand the NEO population. As such, I employ theoretical and numerical models that can be used to estimate the NEO orbital and size distributions. The model results are constrained by the observational efforts of numerous NEO surveys that constantly scan the skies for as of yet unknown objects. The work presented here is based on several papers (Bottke et al. 2002a; Morbidelli et al. 2002a; Morbidelli et al. 2002b; Jedicke et al. 2003) as well as a recent report prepared for NASA entitled "Study to Determine the Feasibility of Extending the Search for Near-Earth Objects to Smaller Limiting Diameters" by Stokes et al. (2003).

9.2
Dynamical Origin of NEOs

9.2.1
Near-Earth Asteroids

The dynamics of bodies in NEO space are strongly influenced by a complicated interplay between close encounters with the planets and resonant dynamics. Encounters provide an impulse velocity to the body's trajectory, causing the semimajor axis, eccentricity, and inclination to change by an amount that depends on both the speed/geometry of the encounter and the mass of the planet. Resonances, on the other hand, keep the semimajor axis constant while changing a body's eccentricity and/or inclination.

Dynamical studies over the last several decades have shown that asteroids located in the main belt between the orbits of Mars and Jupiter can reach planet-crossing orbits by increasing their orbital eccentricity under the action of a variety of resonant phenomena (e.g. J.G. Williams, see Wetherill 1979; Wisdom 1983). Most asteroidal NEOs, or near-Earth asteroids (NEAs) for short, are believed to be collisional fragments that were driven out of the main belt by a combination of Yarkovsky thermal forces (i.e. see Bottke et al. 2002b for a review) and secular/mean motion resonances (e.g. J. G. Williams, see Wetherill 1979; Wisdom 1983). In a scenario favored by many scientists, main belt asteroids with diameter $D < 20$–30 km slowly spiral inward and outward via the Yarkovsky effect until being captured by a dynamical resonance capable of increasing their orbital eccentricity enough to reach planet-crossing orbits. Hence, by understanding the populations of asteroids entering and exiting the most important main belt resonances, we can compute the true orbital distribution of the NEAs as a function of semimajor axis a, eccentricity e, and inclination i.

Here I classify resonances according to two categories: "powerful resonances" and "diffusive resonances", with the former distinguished from the latter by the existence of associated gaps in the main belt asteroid semimajor axis, eccentricity, and inclination (a, e, i) distribution. A gap is formed when the timescale over which a resonance is replenished with asteroidal material is far longer than the timescale over which resonant asteroids are transported to the NEO region. The most notable resonances in the "powerful" class are the v_6 secular resonance at the inner edge of the asteroid belt and several mean motion resonances with Jupiter (e.g. $3:1, 5:2$ and $2:1$ at 2.5, 2.8 and 3.2 AU respectively). Because the $5:2$ and $2:1$ resonances push material onto Jupiter-crossing orbits, where they are quickly ejected from the inner solar system by a close encounter with Jupiter, numerical results suggest that only the first two resonances are important delivery pathways for NEOs (e.g. Bottke et al. 2000, 2002a). For this reason, I focus my attention here on the properties of the v_6 and $3:1$ resonances.

9.2.1.1
The v_6 Resonance

The v_6 secular resonance occurs when the precession frequency of the asteroid's longitude of perihelion is equal to the sixth secular frequency of the planetary system. The latter can be identified with the mean precession frequency of Saturn's longitude of perihelion, but it is also relevant in the secular oscillation of Jupiter's eccentricity (see Chap. 7 of Morbidelli 2002). The v_6 resonance marks the inner edge of the main belt. In this region, asteroids have their eccentricity increased enough to reach planet-crossing orbits. The median time required to become Earth-crosser, starting from a quasi-circular orbit, is about 0.5 Myr. Accounting for their subsequent evolution in the NEO region, the median lifetime of bodies started in the v_6 resonance is ~2 Myr, with typical end-states being collision with the Sun (80% of the cases) and ejection onto hyperbolic orbit via a close encounter with Jupiter (12%) (Gladman et al. 1997). The mean time spent in the NEO region is 6.5 Myr, longer than the median time because v_6 bodies often reach $a < 2$ AU orbits where they often reside for tens of Myr (Bottke et al. 2002a). The mean collision probability of objects from the v_6 resonance with Earth, integrated over their lifetime in the Earth-crossing region, is ~1% (Morbidelli and Gladman 1998).

9.2.1.2
The 3:1 Resonance

The 3:1 mean motion resonance with Jupiter occurs at ~2.5 AU, where the orbital period of the asteroid is one third of that of the giant planet. The resonance width is an increasing function of the eccentricity (about 0.02 AU at $e = 0.1$ and 0.04 AU at $e = 0.2$), while it does not vary appreciably with the inclination. Inside the resonance, one can distinguish two regions: a narrow central region where the asteroid eccentricity has regular oscillations that bring them to periodically cross the orbit of Mars, and a larger border region where the evolution of the eccentricity is wildly chaotic and unbounded, so that the bodies can rapidly reach Earth-crossing and even Sun-grazing orbits. Under the effect of Martian encounters, bodies in the central region can easily transit to the border region and be rapidly boosted into the NEO space (see Chap. 11 of Morbidelli 2002). For a population initially uniformly distributed inside the resonance, the median time required to cross the orbit of the Earth is ~1 Myr, whereas the median lifetime is ~2 Myr. Typical end-states for test bodies include colliding with the Sun (70%) and being ejected onto hyperbolic orbits (28%) (Gladman et al. 1997). The mean time spent in the NEO region is 2.2 Myr (Bottke et al. 2002a), and the mean collision probability with the Earth is ~0.2% (Morbidelli and Gladman 1998).

9.2.1.3
Diffusive Resonances

In addition to the few wide mean motion resonances with Jupiter described above, the main belt is also crisscrossed by hundreds of thin resonances: high order mean motion resonances with Jupiter (where the orbital frequencies are in a ratio of large integer numbers), three-body resonances with Jupiter and Saturn (where an integer combination of the orbital frequencies of the asteroid, Jupiter and Saturn is equal to zero; Nesvorny et al. 2002), and mean motion resonances with Mars (Morbidelli and Nesvorny 1999). The typical width of each of these resonances is of order of a few 10^{-4}–10^{-3} AU.

Because of these resonances, many, if not most, main belt asteroids are chaotic (e.g. Nesvorny et al. 2002). The effect of this chaoticity is very weak, with an asteroid's eccentricity and inclination slowly changing in a secular fashion over time. The time required to reach a planet-crossing orbit (Mars-crossing in the inner belt, Jupiter-crossing in the outer belt) ranges from several 10^7 years to billions of years, depending on the resonances and the starting eccentricity. Integrating real objects in the inner belt ($2 < a < 2.5$ AU) for 100 Myr, Morbidelli and Nesvorny (1999) found that chaotic diffusion drives many main belt asteroids into the Mars-crossing region. The flux of escaping asteroids is particularly high in the region adjacent to the v_6 resonance, where effects from this resonance combine with the effects from numerous Martian mean motion resonances.

It has been shown that the population of asteroids solely on Mars-crossing orbits, which is roughly 4 times the size of the NEO population, is predominately resupplied by diffusive resonances in the main belt (Migliorini et al. 1998; Morbidelli and Nesvorny 1999; Michel et al. 2000; Bottke et al. 2002a). We call this region the "intermediate-source Mars-crossing region", or IMC for short. To reach an Earth-crossing orbit, Mars-crossing asteroids random walk in semimajor axis under the effect of Martian encounters until they enter a resonance that is strong enough to further decrease their perihelion distance below 1.3 AU. The mean time spent in the NEO region is 3.75 Myr (Bottke et al. 2002a).

The paucity of observed Mars-crossing asteroids with $a > 2.8$ AU is not due to the inefficiency of chaotic diffusion in the outer asteroid belt, but is rather a consequence of shorter dynamical lifetimes within the vicinity of Jupiter. For example, Nesvorny and Morbidelli (1999) showed that the outer asteroid belt – more specifically the region between 3.1 and 3.25 AU – contains numerous high-order mean motion resonances with Jupiter and three body resonances with Jupiter and Saturn, such that the dynamics are chaotic for $e > 0.25$. To investigate this, Bottke et al. (2002a) integrated nearly 2000 observed main belt asteroids with $2.8 < a < 3.5$ AU, $i < 15°$, and $q < 2.6$ AU for 100 Myr. They found that ~20% of them entered the NEO region. Accordingly, they predicted that, in a steady state scenario, the outer main belt region could provide ~600 new NEOs per Myr, but the mean time that these bodies spend in the NEO region was only ~0.15 Myr.

9.2.2
Near-Earth Comets

Numerical simulations suggest that comets residing in particular parts of the Transneptunian region are dynamically unstable over the lifetime of the solar system (e.g. Levison and Duncan 1997; Duncan and Levison 1997). Comets also contribute to the NEO population. Comets can be divided into two groups: those coming from the Transneptunian region (the Kuiper belt or, more likely, the scattered disk; Levison and Duncan 1994; Levison and Duncan 1997; Duncan and Levison 1997) and those coming from the Oort cloud (e.g. Weissman et al. 2002). Some NEOs with comet-like properties may come from the Trojan population as well, though it is believed their contribution is small compared to those coming from the Transneptunian region and Oort cloud (Levison and Duncan 1997). The Tisserand parameter T, the pseudo-energy of the Jacobi integral that must be conserved in the restricted circular three-body problem, has been used in the past to classify different comet populations (e.g.

Carusi et al. 1987). Writing T with respect to Jupiter, the Tisserand parameter becomes (Kresak 1979):

$$T = \frac{a_{Jup}}{a} + 2\cos i \sqrt{\frac{a}{a_{Jup}}(1-e^2)}$$

where a_{Jup} is the semimajor axis of Jupiter. Adopting the nomenclature provided by Levison (1996), we refer to $T > 2$ bodies as ecliptic comets, since they tend to have small inclinations, and $T < 2$ bodies as nearly-isotropic comets, since they tend to have high inclinations.

Those ecliptic comets that fall under the gravitational sway of Jupiter ($2 < T < 3$) are called Jupiter-family comets (JFCs). These bodies frequently experience low-velocity encounters with Jupiter. Though most model-JFCs are readily thrown out of the inner solar system via a close encounter with Jupiter (i.e. over a timescale of ~0.1 Myr), a small component of this population achieves NEO status (Levison and Duncan 1997). The orbital distribution of the ecliptic comets has been well characterized using numerical integrations by Levison and Duncan (1997), who find that most JFCs are confined to a region above $a = 2.5$ AU. Comets that are gravitationally decoupled from Jupiter ($T > 3$), like 2P/Encke, are thought to be rare. It is believed that comets reach these orbits via a combination of non-gravitational forces and close encounters with the terrestrial planets.

Nearly isotropic comets, comprized of the long-period comets and the Halley-type comets, come from the Oort cloud (Weissman et al. 2002) and possibly the Transneptunian region (Levison and Duncan 1997; Duncan and Levison 1997). Numerical work has shown that nearly isotropic comets can be thrown into the inner solar system by a combination of stellar and galactic perturbations (Duncan et al. 1987). At this time, however, a complete understanding of their dynamical source region (e.g. Levison et al. 2001) is lacking.

To understand the population of ecliptic comets and nearly isotropic comets, an understanding of more than cometary dynamics is needed. Comets undergo physical evolution as they orbit close to the Sun. In some cases, active comets evolve into dormant, asteroidal-appearing objects, with their icy surfaces covered by a lag deposit of non-volatile dust grains, organics, and/or radiation processed material that prevents volatiles from sputtering away (e.g. Weissman et al. 2002). Accordingly, if a $T < 3$ object shows no signs of cometary activity, it is often assumed to be a dormant, or possibly extinct, comet. In other cases, comets self-destruct and totally disintegrate (e.g. comet C/1999 S4 (LINEAR)). The fraction of comets that become dormant or disintegrate amidst the ecliptic and nearly isotropic comet populations must be understood to gauge the absolute impact hazard to the Earth. We return to this issue in Sect. 9.5.

9.2.3
Evolution in NEO Space

In general, NEOs with $a < 2.5$ AU do not approach Jupiter even at $e \sim 1$, so that they end their evolution preferentially by an impact with the Sun. Particles that are transported to low semimajor axes ($a < 2\,\mathrm{AU}$) and eccentricities have dynamical lifetimes that are tens of Myr long (Gladman et al. 1997) because there are no statistically significant dynamical mechanisms to pump up eccentricities to Sun-grazing values. To be dynamically eliminated, the bodies in the evolved region must either collide with a terrestrial planet (rare), or be driven back to $a > 2\,\mathrm{AU}$, where powerful resonances can push them into the Sun. Bodies that become NEOs with $a > 2.5\,\mathrm{AU}$, on the other hand, are preferentially transported to the outer solar system or are ejected onto hyperbolic orbit by close encounters with Jupiter. This shorter lifetime is compensated by the fact that these objects are constantly re-supplied by fresh main belt material and newly-arriving Jupiter-family comets.

9.3
Quantitative Modeling of the NEO Population

Although there is currently a good working understanding of NEO dynamics, it is still challenging to deduce the true orbital distribution of the NEOs. There are two main reasons for this: (*i*) it is not obvious which source regions provide the greatest contributions to the steady state NEO population, and (*ii*) the observed orbital distribution of the NEOs, which could be used to constrain the contribution from each NEO source, is biased against the discovery of objects on some types of orbits. Given the pointing history of a NEO survey, however, the observational bias for a body with a given orbit and absolute magnitude can be computed as the probability of being in the field of view of the survey with an apparent magnitude brighter than the limit of detection (Jedicke 1996; Jedicke and Metcalfe 1998, see review in Jedicke et al. 2002). Assuming random angular orbital elements of NEOs, the bias is a function $B(a, e, i, H)$, dependent on semimajor axis, eccentricity, inclination and the absolute magnitude H. Each NEO survey has its own bias. Once the bias is known, in principle the real number of objects N can be estimated as:

$$N(a,e,i,H) = \frac{n(a,e,i,H)}{B(a,e,i,H)}$$

where n is the number of objects detected by the survey. The problem, however, is that there are rarely enough observations to obtain more than a coarse understanding of the debiased NEO population (i.e. the number of bins in a 4-dimensional orbital-magnitude space can grow quite large), though such modeling efforts can lead to useful insights (Rabinowitz 1994; Rabinowitz et al. 1994; Stuart 2001).

An alternative way to construct a model of the real distribution of NEOs relies on dynamics (Bottke et al. 2000; 2002a). Using numerical integration results, it is possible to estimate the steady state orbital distribution of NEOs coming from each of the main source regions defined above. The method used by Bottke et al. (2002a) is described below. First, a statistically significant number of particles, initially placed in each source region, is tracked across a network of (a,e,i) cells in NEO space until they are dynamically eliminated. The mean time spent by these particles in those cells, called their residence time, is then computed. The resultant residence time distribution shows where the bodies from the source statistically spend their time in the NEO region. As it is well known in statistical mechanics, in a steady state scenario, the residence time distribution is equal to the relative orbital distribution of the NEOs that originated from the source. This allowed Bottke et al. (2002a) to obtain steady state orbital distributions for NEOs coming from all the prominent NEO sources: the v_6 resonance, the 3:1 resonance, the population coming from numerous diffusive resonances in the main belt, and the Jupiter family comets. The overall NEO orbital distribution was then constructed as a linear combination of these distributions, with the contribution of each source dependent on a weighting function. (Note that the nearly isotropic comet population was excluded in this model, but its contribution is discussed in Sect. 9.5).

The NEO magnitude distribution, assumed to be source-independent, was constructed so its shape could be manipulated using an additional parameter. Combining the resulting NEO orbital-magnitude distribution with the observational biases associated with the Spacewatch survey (Jedicke 1996), Bottke et al. (2002a) obtained a model distribution that could be fit to the orbits and magnitudes of the NEOs discovered or accidentally re-discovered by Spacewatch. A visual comparison showed that the best-fit model adequately matched the orbital-magnitude distribution of the observed NEOs. The resulting best-fit model nicely matches the distribution of the NEOs observed by Spacewatch (see Fig. 10 of Bottke et al. 2002a).

Once the values of the parameters of the model are computed by fitting the observations of *one* survey, the steady state orbital-magnitude distribution of the *entire* NEO population is determined. This distribution is also valid in regions of orbital space that have never been sampled by any survey because of extreme observational biases. This underlines the power of the dynamical approach for debiasing the NEO population.

9.4
The Debiased NEO Population

Bottke et al. (2002a) predict as a function of absolute magnitude H that 37 ± 8% of the NEOs come from the v_6 resonance, 23 ± 9% from the 3:1 resonance, 33 ± 3% from the numerous diffusive resonances stretched across the main belt, and 6 ± 4% come from the Jupiter-family comet region. Their model results were constrained in the JFC region by several objects that are almost certainly dormant comets. For this reason, factors that have complicated the discussions of previous JFC population estimates (e.g. issues of converting cometary magnitude to nucleus diameters, etc.) are avoided. Note, however, that the Bottke et al. (2002a) model does not account for the contribution of comets of Oort cloud origin. This issue will be discussed in Sect. 9.5.

Figure 9.1 displays the debiased (a, e, i) NEO population as a residence time probability distribution plot. To display as much of the full (a, e, i) distribution as possible in two dimensions, the i bins were summed before plotting the distribution in (a, e), while the e bins were summed before plotting the distribution in (a, i). The color scale depicts the expected density of NEOs in a scenario of steady state replenishment from the main belt and transneptunian region. Red colors indicate where NEOs are statistically most likely to spend their time. Bins whose centers have perihelia $q > 1.3$ AU are not used and are colored white. The gold curved lines that meet at 1 AU divide the NEO region into Amor (1.0167 AU $< q < 1.3$ AU), Apollo ($a > 1.0$ AU; $q < 1.0167$ AU) and Aten ($a < 1.0$ AU; $Q > 0.983$ AU) components. IEOs ($Q < 0.983$ AU) are inside Earth's orbit.

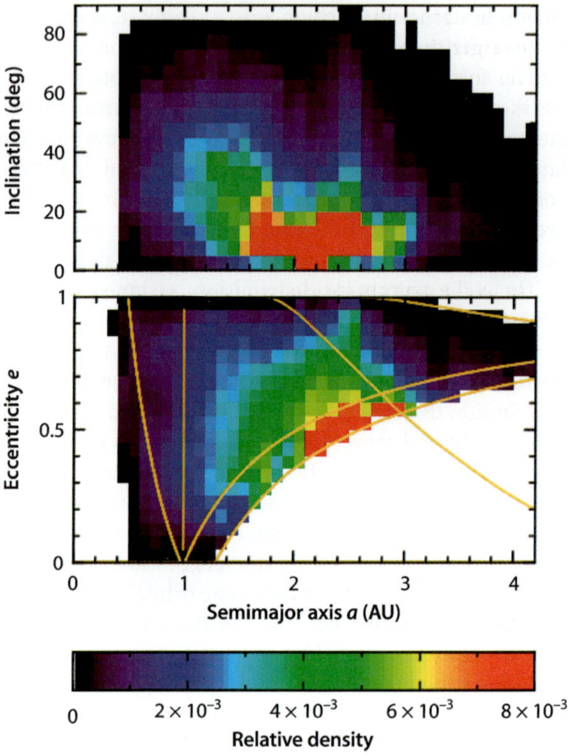

Fig. 9.1.
A representation of the probability distribution of residence time for the debiased near-Earth object (NEO) population

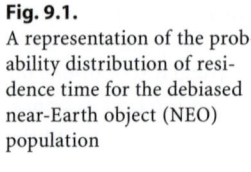

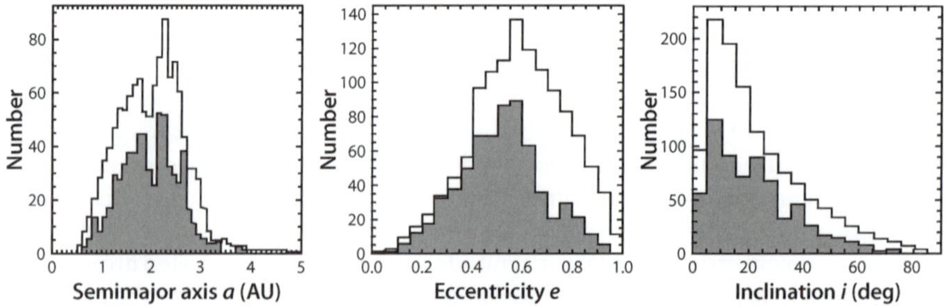

Fig. 9.2. The debiased orbital distribution for NEOs with absolute magnitude $H < 18$. The predicted NEO distribution (*dark solid line*) is normalized to 1200 NEOs. It is compared with the 645 known NEOs (as of April 2003) from all surveys (*shaded*).

The Jupiter-family comet region is defined using two lines of constant Tisserand parameter $2 < T < 3$. The curves in the upper right show where $T = 2$ and $T = 3$ for $i = 0$ deg.

Figure 9.2 displays the debiased distribution of the NEOs with absolute magnitude $H < 18$ as a series of three one-dimensional plots (see Bottke et al. 2002a for other representations of these data). For comparison, the figure also reports the distribution of the objects discovered up to $H < 18$, all surveys combined, as of 2003. For objects with an absolute magnitude brighter than about 18, the object's diameter would be expected to be larger than one kilometer.

The absolute magnitude and size-frequency distributions of the NEO population are discussed in the next section. Most of the NEOs that are still undiscovered have H larger than 16, e larger than 0.4, a in the range 1–3 AU and i between 5–40°. The populations with $i > 40°$, $a < 1$ AU or $a > 3$ AU have a larger relative incompleteness, but contain a much more limited number of undiscovered bodies. Of the total NEOs, $32 \pm 1\%$ are Amors, $62 \pm 1\%$ are Apollos, and $6 \pm 1\%$ are Atens. Some $49 \pm 4\%$ of the NEOs should be in the evolved region ($a < 2$ AU), where the dynamical lifetime is strongly enhanced. As far as the objects inside Earth's orbit, or IEOs, the ratio between the IEO and the NEO populations is about 2%. Thus, there are only about 20 IEOs with $H < 18$.

With this orbital distribution, and assuming random values for the argument of perihelion and the longitude of node, about 21% of the NEOs turn out to have a Minimal Orbital Intersection Distance (MOID) with the Earth smaller than 0.05 AU. The MOID is defined as the closest possible approach distance between the osculating orbits of two objects. NEOs with MOID < 0.05 AU are defined as Potentially Hazardous Objects (PHOs), and their accurate orbital determination is considered top priority. About 1% of the NEOs have a MOID smaller than the Moon's distance from the Earth; the probability of having a MOID smaller than the Earth's radius is 0.025%. This result does not necessarily imply that a collision with Earth is imminent since both the Earth and the NEO still need to rendezvous at the same location, which is unlikely.

9.5
Nearly Isotropic Comets

I now address the issue of the contribution of nearly isotropic comets (NICs) to the NEO population (and the terrestrial impact hazard). Dynamical explorations of the orbital distribution of the nearly isotropic comets (Wiegert and Tremaine 1999; Levison et al. 2001) indicate that, in order to explain the orbital distribution of the observed population, nearly-isotropic comets (NIC) need to rapidly "fade" (i.e. become essentially unobservable). In other words, physical processes are needed to hide some fraction of the returning NICs from view. One possible solution to this so-called "fading problem" would be to turn bright active comets into dormant, asteroidal-appearing objects with low albedos. If most NICs become dormant, the potential hazard from these objects could be significant. An alternative solution would be for cometary splitting events to break comets into smaller (and harder-to-see) components. If most returning NICs disrupt, the hazard to the Earth from the NIC population would almost certainly be smaller than that from the NEA population.

To explore this issue, Levison et al. (2002) took several established comet dynamical evolution models of the NIC population (Wiegert and Tremaine 1999; Levison et al. 2001), created artificial populations of dormant NICs from these models, and ran these artificial objects through a NEO survey simulator that accurately mimics the performance of various NEO surveys (e.g. LINEAR, NEAT) over a time period stretching from 1996–2001 (Jedicke et al. 2003). Levison et al. (2002) then compared their model results to the observed population of dormant comets found over the same time period. For example, the survey simulator discovered 1 out of every 22 000 dormant NICs with orbital periods > 200 years, $H < 18$, and perihelion $q < 3$ AU. This result, combined with the fact that only 2 dormant objects with comparable parameters had been discovered between 1996–2001, led them to predict that there are a total of 44 000 ± 31 000 dormant nearly-isotropic comets with orbital periods $P > 200$ years, $H < 18$, and perihelion $q < 3$ AU.

Levison et al. (2002) then used these values to address the fading problem by comparing the total number of artificial dormant nearly isotropic comets discovered between 1996–2001 to the observed number. The results indicated that dynamical models that fail to destroy comets over time produce ~100 times more dormant NICs than can be explained by current NEO survey observations. Hence, to resolve this paradox, Levison et al. (2002) concluded that, as comets evolve inward from the Oort cloud, the vast majority of them must physically disrupt.

Assuming there are 44 000 dormant comets with $P > 200$ years, $H < 18$, and perihelion $q < 3$ AU, Levison et al. (2002) estimated that they should strike the Earth once per 370 Myr. In contrast, the rate that active comets with $P > 200$ years strike the Earth (both new and returning) is roughly once per 32 Myr (Weissman 1990; Morbidelli 2002). For NICs with $P < 200$ years, commonly called Halley-type comets (HTCs), Levison et al. (2002) estimate there are 780 ± 260 dormant objects with $H < 18$ and $q < 2.5$ AU. This corresponds to an Earth impact rate of once per 840 Myr. Active HTCs strike even less frequently, with a rate corresponding to once per 3500 Myr (Levison et al. 2001, 2002). Hence, since all of these impact rates are much smaller than that estimated for $H < 18$ NEOs (one impact per 0.5 Myr; Bottke et al. 2002a; Morbidelli et al. 2002a), we conclude that nearly-isotropic comets currently represent a tiny fraction of the total impact hazard.

Another way to look at the issue is as follows. If we assume the bulk densities for a cometary nucleus and an S-type NEA are 0.6 and 2.6 g cm^{-3}, respectively, and the mean Earth impact velocities for long-period comets and NEAs are 55 and 23 km s^{-1}, respectively, then the average impact energy of a long-period comet impact would be only 30% more than a similarly-sized NEA that impacts the Earth. Stokes et al. (2003), using these results as well as methods described in Sekanina and Yeomans (1984) and Marsden (1992), showed that the threat of long-period comets is only about 1% the threat from NEAs. Thus, asteroids rather than comets provide most of the present-day impact hazard.

9.6
NEA Size-Frequency Distribution

Many groups have made estimates of the NEO population in the recent literature (Stuart 2001; D'Abramo et al. 2001; Bottke et al. 2002; Brown et al. 2002; Stuart and Binzel 2004; Bottke et al. 2004). Despite using a wide variety of techniques, all tend to yield comparable results. To keep things simple, it is useful to adopt in this paper the estimate made by Stokes et al. (2003), that, within limits of reasonable uncertainty, fits the NEO absolute magnitude H distribution to a constant power law in logarithmic units:

$$\log[N(<H)] = -5.414 + 0.4708\,H$$

In units of diameter, taking an equivalence of $H = 18$ to be equal to $D = 1$ km (i.e. Morbidelli et al. 2002a) and Stuart and Binzel (2004) estimate that the mean NEO albedo should be ~0.13–0.14, which would implying an equivalence of $H = 17.75$–17.85 to $D = 1$ km), we obtain the relationship:

$$N(D) = 1148\, D^{-2.354}$$

This population model lies slightly above the number currently estimated for the population of NEOs larger than 1 km (1 000–1 100). Its main advantage is that it lies within about a factor of 2 (on the high side) of numerous NEO small body population estimates for $D > 1$ m. This estimate is used in computing the NEO hazard studies described below.

9.7 Conclusion

The question of how to deal with the threat represented by comets and asteroids was recently reviewed by Near-Earth Object Science Definition Team (Stokes et al. 2003). They found that searching for potential Earth-impacting objects could help eliminate the statistical risk associated with the hazard of impacts. Even though the impact rate of hazardous objects on Earth is low, the "average" rate of destruction due to impacts was deemed large enough to merit additional interest.

Stokes et al. argued that the cost/benefit ratio for finding such objects was favorable enough to warrant the construction of a new NEO search survey. This goal of this new survey would be to discover and catalog the potentially hazardous population enough to eliminate 90% of the remaining hazard (i.e., 90% of the $D > 170$ m objects). This same survey program would also find essentially all of the undiscovered $D > 1$ km objects remaining in the NEO population, thus eliminating the global risk from these larger objects. Once the above goal was met, the average casualty rate from impacts would be reduced from about 300 per year to less than 30 per year. Systems capable of meeting this goal over a period of 7–20 years would likely cost between $ 236 million and $ 397 million, comparable to NASA Discovery-class missions.

The costs of a new survey system, which are tiny relative to the costs of proposed missions to deflect NEOs, could be considered a form of term life insurance taken out by humanity against the hazard represented by infrequent but potentially dangerous impacts. It seems prudent to approach the problem from this direction before taking additional steps that could be both costly and dangerous.

References

Alvarez LW, Alvarez W, Asaro F, Michel HV (1980) Extraterrestrial cause for the Cretaceous Tertiary extinction. Science 208:1095–1099

Bland PA, Artemieva NA (2003) Efficient disruption of small asteroids by Earth's atmosphere. Nature 424:288–291

Boslough MBE, Crawford DA (1997) Shoemaker-Levy 9 and plume-forming collisions on Earth. In: Remo JL (ed) Near-Earth Objects. The United Nations International Conference: Proceedings of the International Conference held April 24–26, 1995, in New York, NY. Annals of the New York Academy of Sciences 822:236–282

Bottke WF, Jedicke R, Morbidelli A, Petit JM, Gladman B (2000) Understanding the distribution of near-Earth asteroids. Science 288:2190–2194

Bottke WF, Morbidelli A, Jedicke R, Petit J, Levison HF, Michel P, Metcalfe TS (2002a) Debiased orbital and absolute magnitude distribution of the near-Earth objects. Icarus 156:399–433

Bottke WF, Vokrouhlicky D, Rubincam DP, Broz M (2002b) The effect of Yarkovsky thermal forces on the dynamical evolution of asteroids and meteoroids. In: Bottke WF, Cellino A, Paolicchi P, Binzel RP (eds) Asteroids III. Univ of Arizona Press, Tucson, pp 395–408

Brown P, Spalding RE, ReVelle DO, Tagliaferri E, Worden SP (2002) The flux of small near-Earth object colliding with the Earth. Nature 420:294–296

Carusi A, Kresak L, Perozzi E, Valsecchi GB (1987) High order librations of Halley-type comets. Astron Astrophys 187:899

Chesley SR, Ward SM (2003) A quantitative assessment of the human and economic hazard from impact-generated tsunami. J Environmental Hazards

Chyba CF, Thomas JP, Zahnle KJ (1993) The 1908 Tunguska explosion – atmospheric disruption of a stony asteroid. Nature 361:40–44

D'Abramo G, Harris AW, Boattini A, Werner SC, Valsecchi JB (2001) A simple probabilistic model to estimate the population of near-Earth asteroids. Icarus 153:214–217

Duncan MJ, Levison HF (1997) A scattered comet disk and the origin of Jupiter family comets. Science 276:1670–1672

Duncan M, Quinn T, Tremaine S (1987) The formation and extent of the solar system comet cloud. Astron J 94:1330–1338

Duncan M, Quinn T, Tremaine S (1988) The origin of short-period comets. Astrophysical Journal Letters 328:L69–L73

Gladman BJ, Migliorini F, Morbidelli A, Zappala V, Michel P, Cellino A, Froeschle C, Levison HF, Bailey M, Duncan M (1997) Dynamical lifetimes of objects injected into asteroid belt resonances. Science 277:197–201

Harris AW (2002) A new estimate of the population of small NEAs. Bulletin of the American Astronomical Society 34:835

Hills JG, Goda MP (1993) The fragmentation of small asteroids in the atmosphere. Astron J 105:1114–1144

Ivezic Z, 32 colleagues (2001) Solar system objects observed in the Sloan Digital Sky survey commissioning data. Astron J 122:2749–2784

Jedicke R (1996) Detection of near-Earth asteroids based upon their rates of motion. Astron J 111:970

Jedicke R, Metcalfe TS (1998) The orbital and absolute magnitude distributions of main belt asteroids. Icarus 131:245–260

Jedicke R, Larsen J, Spahr T (2002) Observational selection effects in asteroid surveys. In: Bottke WF, Cellino A, Paolicchi P, Binzel RP (eds) Asteroids III. Univ of Arizona Press, Tucson, pp 71–87

Jedicke R, Morbidelli A, Petit J-M, Spahr T, Bottke WF (2003) Earth and space-based NEO survey simulations: prospects for achieving the Spaceguard goal. Icarus 161:17–33

Kresak L (1979) Dynamical interrelations among comets and asteroids. In: Gehrels T (ed) Asteroids. Univ of Arizona Press, Tucson, pp 289–309

Levison HF (1996) Comet taxonomy. In: Rettig TW, Hahn JM (eds) Completing the inventory of the solar system. ASP Conf Series 107:173–191

Levison HF, Duncan MJ (1994) The long-term dynamical behavior of short-period comets. Icarus 108:18–36

Levison HF, Duncan MJ (1997) From the Kuiper belt to Jupiter-family comets: the spatial distribution of ecliptic comets. Icarus 127:13–32

Levison HF, Dones L, Duncan MJ (2001) The origin of Halley-type comets: probing the inner Oort cloud. Astron J 121:2253–2267

Levison HF, Morbidelli A, Dones L, Jedicke R, Wiegert PA, Bottke WF (2002) The mass disruption of Oort cloud comets. Science 296:2212–2215 [for a detailed treatment, see *http://www.boulder.swri.edu/~hal/PDF/disrupt.pdf*]

Marsden BG (1992) To hit or not to hit. In: Canavan GH, Solem JC, Rather JDG (eds) Proceedings, Near-Earth Objects Interception Workshop. Los Alamos National Laboratory, Los Alamos, NM, pp 67–71

Melosh HJ (2003) Impact-generated tsunamis: an over-rated hazard. Lunar and Planetary Science XXXIV:2013
Michel P, Migliorini F, Morbidelli A, Zappala V (2000) The population of Mars-crossers: classification and dynamical evolution. Icarus 145:332–347
Migliorini F, Michel P, Morbidelli A, Nesvorny D, Zappala V (1998) Origin of Earth-crossing asteroids: a quantitative simulation. Science 281:2022–2024
Morbidelli A (2002) Modern celestial mechanics: aspects of solar system dynamics. Taylor & Francis, London
Morbidelli A, Nesvorny D (1999) Numerous weak resonances drive asteroids toward terrestrial planets orbits. Icarus 139:295–308
Morbidelli A, Jedicke R, Bottke WF, Michel P, Tedesco EF (2002a) From magnitudes to diameters: the albedo distribution of near Earth objects and the Earth collision hazard. Icarus 158:329–342
Morbidelli A, Bottke WF, Froeschle CH, Michel P (2002b) Origin and evolution of near-Earth objects. In: Bottke WF, Cellino A, Paolicchi P, Binzel RP (eds) Asteroids III. Univ of Arizona Press, Tucson, pp 409–422
Morrison D (1992) The Spaceguard survey: report of the NASA international near-Earth-object detection workshop. NASA, Washington, DC
Nesvorny D, Ferraz-Mello S, Holman M, Morbidelli A (2002) Regular and chaotic dynamics in the mean motion resonances: implications for the structure and evolution of the main belt. In: Bottke WF, Cellino A, Paolicchi P, Binzel RP (eds) Asteroids III. Univ of Arizona Press, Tucson, pp 379–394
Rabinowitz DL, Bowell E, Shoemaker EM, Muinonen K (1994) The population of Earth-crossing asteroids. In: Gehrels T (ed) Hazards due to comets and asteroids. Univ of Arizona Press, Tucson, pp 285–312
Rabinowitz DL, Helin E, Lawrence K, Pravdo S (2000) A reduced estimate of the number of kilometre-sized near-Earth asteroids. Nature 403:165–166
Sekanina Z, Yeomans DK (1984) Close encounters and collisions of comets with the Earth. Astron J 89: 154–161
Shoemaker EM (1983) Asteroid and comet bombardment of the Earth. Annual Review of Earth and Planetary Sciences 11:461–494
Stokes GH, Yeomans DK, Bottke WF, Chesley SR, Evans JB, Gold RE, Harris AW, Jewitt D, Kelso TS, McMillan RS, Spahr TB, Worden SP (2003) Report of the Near-Earth Object Science Definition Team: a study to determine the feasibility of extending the search for near-Earth objects to smaller limiting diameters. NASA-OSS-Solar System Exploration Division (*http://neo.jpl.nasa.gov/report.html*)
Stuart JS, (2001) A near-Earth asteroid population estimate from the LINEAR survey. Science 294: 1691–1693
Stuart JS (2003) Observation constraints on the number, albedos, sizes, and impact hazards of the near-Earth asteroids. MIT PhD thesis
Stuart JS, Binzel RP (2004) Bias-corrected population, size distribution, and impact hazard for the near-Earth objects. Icarus 170:295–311
Toon OB, Zahnle K, Morrison D, Turco RP, Covey C (1997) Environmental perturbations caused by the impacts of asteroids and comets. Reviews of Geophysics 35:41–78
Van Dorn WG, LeMehaute B, Hwant L-S (1968) Handbook of explosion-generated water waves, vol 1: state of the art. Tetra Tech, Pasadena, CA
Ward SN, Aspaugh E (2000) Asteroid impact tsunami: a probabilistic hazard assessment. Icarus 145: 64–78
Weissman PR, Bottke WF, Levison H, (2002) Evolution of comets into asteroids. In: Bottke WF, Cellino A, Paolicchi P, Binzel R (eds) Asteroids III. Univ of Arizona Press, Tucson, pp 669–686
Wetherill, GW (1979) Steady state populations of Apollo-Amor objects. Icarus 37: 96–112.
Wiegert P, Tremaine S (1999) The evolution of long-period comets. Icarus 137:84–121
Wisdom J (1983) Chaotic behavior and the origin of the 3/1 Kirkwood Gap. Icarus 56:51–74

Chapter 10

Physical Properties of NEOs and Risks of an Impact: Current Knowledge and Future Challenges

A. Chantal Levasseur-Regourd

10.1 Introduction

10.1.1 Key Questions before Impact

Someday, in a not too far away future ... A potentially hazardous astronomical object, with an estimated size significantly above 10 meters, is just detected. Quite soon, the probability of its impact with the Earth in, again, a not too far away future, is found to be close to 1. We certainly want to predict with a decent accuracy the effects of the impact and, even better, to tentatively initiate a mitigation strategy.

We need to estimate the mass of the object, since the energy released at impact is proportional to it. We also want to have some ideas about the structure of the object, which could explode or break into fragments in the lower layers of the atmosphere. Finally, for any mitigation technique, we have to know the surface properties, in order to use efficient tools for impacting, landing or anchoring on it.

10.1.2 The True Nature of NEOs

The near Earth objects (hereafter NEOs) population consists of asteroids (or fragments thereof), which are rocky objects; it also includes cometary nuclei, consisting of ice and dust, which happen to eject gases and dust whenever they are sufficiently heated by the solar radiation, and of so-called defunct or dormant comets, which have lost all their ice or are coated by an insulating dust mantle. Asteroids most likely represent the main population. However, dormant and defunct comets could represent up to 18% of the total population, and active comets about 1% of the total population (Binzel et al. 2004).

It is thus necessary to consider the physical properties of both asteroids and comets, especially taking into account the fact that cometary orbits may be quite elongated and inclined (with respect to the Earth orbital plane), leading to high relative velocities and impact energies. All these objects, as we will see now, present a wide diversity in their properties. The reason is that their parent bodies have been formed over a large range of solar distances in the early solar system, with different temperatures, compositions and concentrations; besides, they have been going through various evolutionary processes, in relation to their collision and evaporation history.

Numerous astrophysical observations, together with unique data from space probes, already provide significant information about some key characteristics. Below I summarize our understanding of the masses and densities of NEOs. Then questions about their structure and porosity are addressed. Observations relevant to their surface properties are also discussed. Finally, I provide a review of the knowledge expected from scientific projects under development, with emphasis on future space missions.

10.2
Densities: from Feather to Lead?

10.2.1
Determining Mass and Density

The determination of the mass of small bodies in the solar system is seldom possible. It can be estimated from gravitational interactions between asteroids or from the gravitational perturbations undergone by a nearby spacecraft; it can also be derived from the orbital motion of a potential satellite in orbit around the object. As an example, the mass of the NEO 433 Eros has been inferred from the movement of the orbiting space probe NEAR-Shoemaker, whereas the mass of the asteroid 45 Eugenia has been derived from the motion of its satellite Petit Prince. On the opposite side, the masses of the five comet nuclei up to now encountered by a space probe (1P/Halley, 26P/Grigg-Skjellerup, 19P/Borrelly, 81P/Wild 2 and 9P/Tempel 1) are so low that mostly upper limits have been obtained. The fact that masses (and densities) are very low has also been inferred by modeling the sublimation induced gravitational forces for quite a few nuclei (see e.g. Rickman et al. 1987).

Once the mass is estimated, the bulk density (ratio of mass to volume) is derived from the estimation of the equivalent radius, or from the observed dimensions and shape. The bulk density, when expressed in g cm^{-3}, is directly comparable to the density of water (equal to 1 g cm^{-3}). It provides information about the composition and structure of the object.

10.2.2
Typical Results on Densities

Table 10.1 (from Hilton 2002 and Britt et al. 2002 reviews) presents the densities of some main belt and near Earth asteroids. The taxonomic type is also mentioned, with C and P for dark asteroids (with spectra suggestive of carbon-rich material) that could be analogous to carbonaceous chondrites, S for brighter asteroids (with spectra typical of iron and magnesium bearing silicates) that could be analogous to ordinary chondrites or stony-iron meteorites, and M for asteroids that exhibit the characteristics of metallic iron-nickel.

A correlation seems to exist between the taxonomic type (as derived from spectroscopic observations) and the density, with S and M types denser than C or P types. It is of interest to notice that (except for the largest asteroids) the bulk density is usually smaller than the density of the corresponding meteorite analog, suggesting, as developed below, the existence of some porosity.

Table 10.1. Estimated bulk density and porosity of some main belt asteroids and near Earth asteroids (from Hilton 2002, Britt et al. 2002, and references within).

Asteroid name or number (type)	Density (g cm^{-3})	Porosity (%) (macro-porosity)
Main Belt		
1 Ceres (G)	2.12 ±0.04	<10 (0)
2 Pallas (B)	2.7 ±0.1	<10 (0)
4 Vesta (V)	3.4 ±0.1	<10 (0)
16 Psyche (M)	2.0 ±0.6	72 (≈60)
22 Kalliope (M)	2.5 ±0.3	65 (≈60)
45 Eugenia (C)	1.2 (+0.6/−0.2)	55 (≈40)
253 Mathilde (C)	1.3 ±0.2	27 (≈40)
87 Sylvia (P)	1.6 ±0.3	52 (≈20)
Near Earth		
433 Eros (S)	2.67 ±0.03	27 (18)
1999 KW4	2.4 ±0.9	
2000 DP107 (C)	1.6 (+1.2/−0.9)	
2000 UG11	1.5 (+0.6/−1.3)	
25143 Itokawa (S)	≈2.5	

10.2.3
Open Questions

The mass and density of cometary nuclei have, up to now, been impossible to estimate. The above-mentioned modeling studies (Rickman et al. 1987; Davidsson and Gutiérrez 2004) mostly lead to densities in the 0.1 to 0.6 g cm^{-3} range for cometary nuclei. Such results agree with the very low values (about 0.1 g cm^{-3}) derived in comet Halley coma for dust particles, which seem to consist of dark and fluffy aggregates of smaller grains (Levasseur-Regourd et al. 1999; Fulle et al. 2000).

From the few values already derived for near Earth asteroids, it may be assumed that the densities are in the 1 to 3 g cm^{-3} range. However, for some metallic monolithic fragments, the density might possibly reach a value of the order of 8 g cm^{-3}.

10.3
Structure: from Monoliths to Rubble Piles?

10.3.1
Determining the Structure

Determining the interior structure of a solar system body is certainly one of the most difficult tasks for planetary scientists, since it requires active space experiments (e.g. radar tomography, blast experiments, drilling). The internal structure of small bodies, including Eros, is still unknown. In the absence of any direct information, some clues about the interior structure are obtained from the outer shape and from the porosity, as derived from the bulk density.

10.3.2
Outer Shape and Structure

Optical or radar imaging of near Earth asteroids immediately shows that they are highly irregularly shaped. A wide variety of shapes has been observed: 216 Kleopatra (possibly a binary object) is somewhat dog-bone shaped, 1620 Geographos is mostly elongated, 6489 Golevka presents an extraordinarily angular shape, and 2100 Ra-Shalom is rather spheroidal (Ostro et al. 2002). Besides, some NEOs (typically 16 % of those with a size larger than 200 m, Margot et al. 2002) could be binary or multiple objects. Evidence for numerous binary objects is also given by the detection of double impact craters on Earth (as illustrated by e.g. Clearwater Lakes in Canada). Although some near Earth asteroids could be monolithic, it may thus be estimated that quite a few of them are second or multi-generation collision fragments from larger bodies. They may then be significantly fractured, and some fragments may form "rubble piles", i.e. gravitational aggregates that remain close to one another under the effect of their mutual gravity.

Optical imaging of cometary nuclei requires in-situ missions, since these small bodies are either point sources when they are far from the Sun (and the Earth) or are hidden by their bright gas and dust comae when they get closer to the Sun. Four nuclei have up to now been imaged: Halley, Borrelly, Wild 2 and Tempel 1, respectively by Giotto, Deep Space 1, Stardust and Deep Impact space probes (Fig. 10.1). A comparison immediately reveals a significant diversity, and suggests that some cometary nuclei could be gravitational aggregates of smaller bodies, whereas others could be more compact (Weaver 2004).

10.3.3
Porosity and Structure

The porosity is defined as the ratio of the bulk density to the building grains density, i.e. as the percentage of volume with empty space (Britt et al. 2002). A very porous object could be more likely to disintegrate while traveling through the atmosphere than a compact object, although it may better resist an impact.

The determination of the porosity requires the estimation of the density of the object (as previously discussed), of the meteoritic analog (from reflectance spectra) and of the average porosity thereof. The values presented in Table 10.1 suggest that most asteroids have a significant porosity. These values actually represent the sum of the micro-porosity (from micro-pores and voids) and of the macro-porosity (from large-scale fractures and voids).

The macro-porosity, which determines the asteroid internal structure, is estimated to be about 18% for 433 Eros, indicating an internally fractured consolidated body with coherent strength. It could be above 40% for 45 Eugenia and 253 Mathilde, two C-type objects that seem to be most porous and robust. Interestingly, Mathilde has obviously suffered energetic cratering (providing permanent compaction of the target material) without breaking-off. Assuming their densities have been accurately estimated, the M

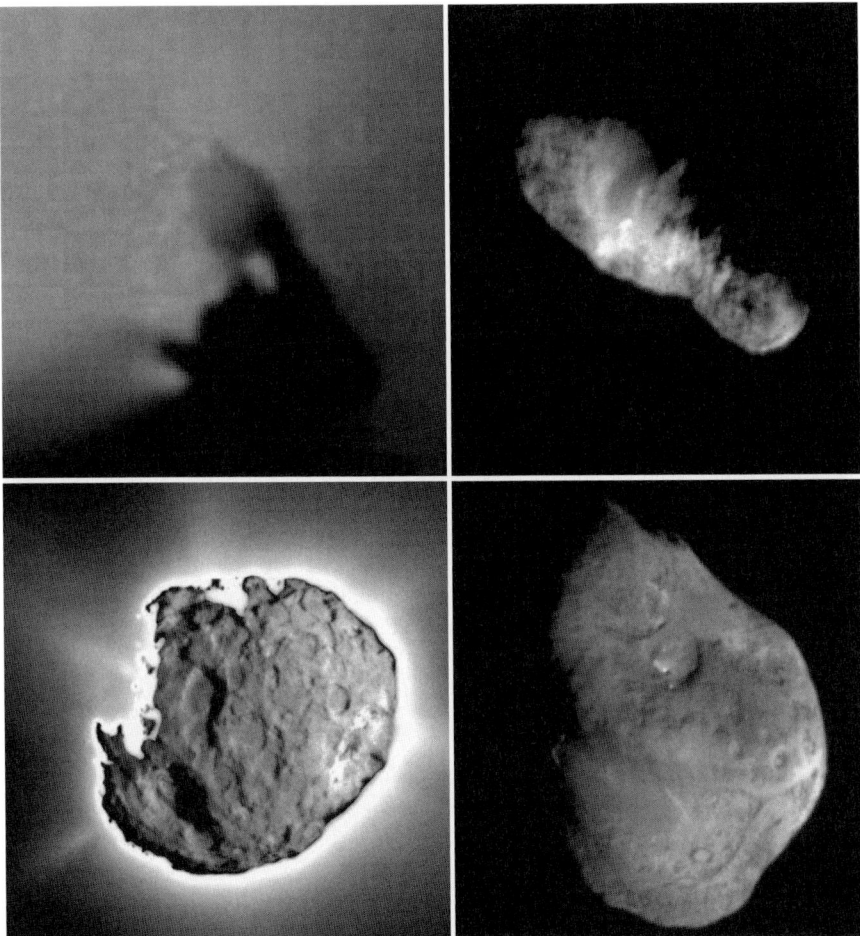

Fig. 10.1. Images of cometary nuclei obtained during spacecraft flybys (Giotto, MPIA/ESA; Deep Space 1, Stardust and Deep Impact, NASA). From *left* to *right* and *top* to *bottom*: Halley in 1986 (length about 16 km), Borrelly in 2001 (length about 8 km), Wild 2 in 2004 (size about 4 km, coma numerically enhanced) and Tempel 1 in 2005 (size about 6 km)

type asteroids 16 Psyche and 22 Calliope could even have higher macro-porosities (above 60%), and would then be likely to be disrupted objects, loosely reassembled, with fragments held together by mutual gravitation.

10.3.4
Comets Disruption and Fragmentation

The fact that, with instrumental refinement, an increasing number of comets have been observed to suffer complete disruption or partial fragmentation gives us some clues about the internal structure of cometary nuclei.

Complete disruption (i.e. breaking-up) has been observed quite a few times. The most famous example, illustrated in Fig. 10.2a, is that of the nucleus of Shoemaker-Levy 9 (D/1993 F2). It suffered a (tidal) fragmentation while passing close to Jupiter in 1992, and the multiple fragments (with sizes up to 1 or 2 km for the larger ones) later impacted the giant planet in July 1994. Comet LINEAR (C/1999 S4) also suffered a complete disruption in July 2000 (Fig. 10.2b). Large telescopes (i.e. HST and VLT) could observe at least 16 bright condensations around icy fragments (Weaver et al. 2001), and the light scattering observations suggested the presence of rather large (hundred of microns) particles fragmenting within these condensations (Hadamcik and Levasseur-Regourd 2003). Such nuclei could actually consist of gravitational aggregates of a few tens or hundreds of meters cometesimals.

Partial fragmentation (i.e. peeling-off of the cometary nucleus) is even more frequent a process. It was observed on Hyakutake (C/1996 B2), while the comet was passing not too far from Earth in March 1996. The size of the biggest icy fragment was in a 50 to 250 m range (Desvoivres et al. 2000). It is also likely that the huge cloud of dust particles encountered by Stardust over 4000 km after closest approach resulted from the progressive disintegration of a fragment of Wild 2 (Sekanina et al. 2004; Levasseur-Regourd 2004). A similar event might have been observed during the flyby of Grigg-Skjellerup (McBride et al. 1997), and such "crumbles" might actually be often present inside the coma of some comets that would be fragile bodies with some internal cohesiveness (McDonnell, pers. comm., 2004).

10.3.5
Open Questions

Although their porosity seems to be significant, it is likely that some asteroids are monolithic, whereas some others are fractured, or may even be gravitational aggregates of smaller objects. Similarly, cometary nuclei could be gravitational aggregates or more compact (but nevertheless fragile) objects.

The whole question of the internal structure (and its diversity) of NEOs is still an open one, which certainly requires further investigation. This topic is all the more important since it appears that some low density objects resist quite well to impacts, whereas fractured objects could be most fragile (see e.g. Michel et al. 2003).

Fig. 10.2. Illustration of cometary nuclei fragmentation, through the detection of bright condensations (i.e. mini comae) around the fragments. **a** Shoemaker-Levy 9 (before its impact on Jupiter in 1994; **b** LINEAR (1999 S4) before its complete disruption in 2000 (HST, NASA)

10.4
Surface Properties: from Sand Dunes to Concrete?

Precise information about the surface physical properties, including the texture, is obtained through in-situ studies by space probes. However, after more than 40 years of space exploration, in-situ studies of asteroids are mostly restricted to the flybys of Gaspra, Ida and Mathilde in the main belt, and to the rendezvous with the NEO Eros, whereas in-situ studies of cometary nuclei are restricted to the above-mentioned flybys of Halley, Grigg-Skjellerup, Borrelly, Wild 2 and Tempel 1.

Impact craters are conspicuous on the above-mentioned asteroids (Fig. 10.3). The surfaces seem to be mostly covered by a regolith, that is, a layer of fragmentary incoherent rocky debris, which nearly everywhere forms the surface terrain. Such a loose material may even be found on very low gravity objects, with e.g. evidence for downslope motion and flat ponds of smaller debris detected inside Eros craters (Thomas et al. 2002).

As far as cometary nuclei are concerned (see Fig. 10.1), the images obtained during Wild 2 flyby indicate a large variety of landforms (and physical processes taking place),

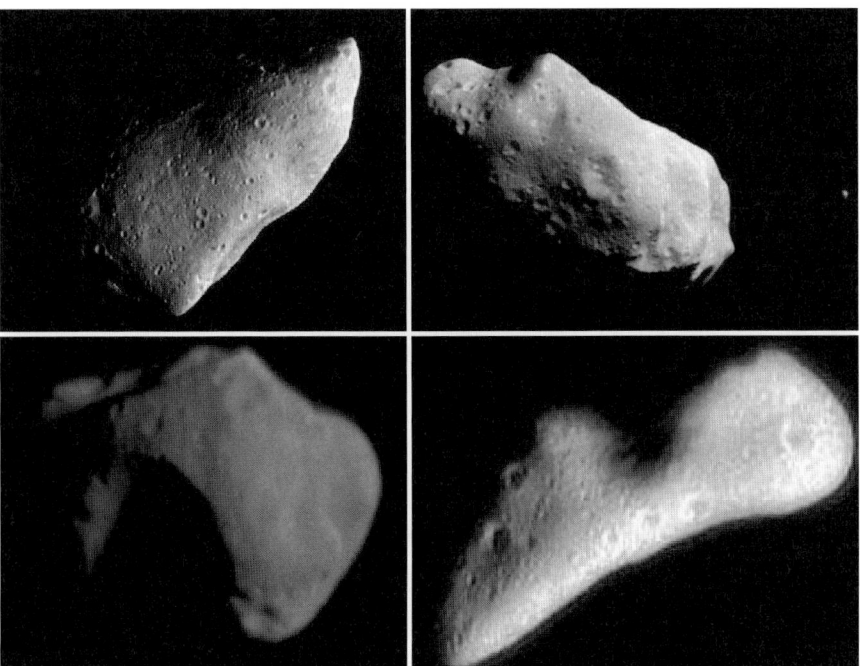

Fig. 10.3. Images of asteroids obtained during flyby and rendezvous missions (Galileo and NEAR, NASA). From *left* to *right* and *top* to *bottom*: Gaspra (length about 17 km), Ida (length about 57 km) with its satellite Dactyl on the right, dark and porous Mathilde (length about 59 km), and NEO Eros (length about 33 km)

with some areas likely to consist of cohesive porous material (Brownlee et al. 2004). The images obtained by the impactor on Tempel 1 reveal topographic features quite different from those seen on Borelly and Wild 2, with scarps, smooth terrains and impact craters; the impact event was controlled by gravity and the outer layer consisted of very fine (possibly organic rich) particles (A'Hearn et al. 2005a, 2005b).

10.4.1
Estimating the Surface Properties

Information on the physical properties of the surfaces may be derived from the properties of the light they scatter in the visible and near infrared domains. Solar light scattered by such surfaces is essentially partially linearly polarized. The linear polarization is defined by the ratio of the difference to the sum of two polarized components of the intensity (with the electric field vector respectively parallel and perpendicular to the scattering plane components, see e.g. Hapke 1993). It only varies with the phase angle (between the direction of the Sun and of the observer, as seen from the object), with the wavelength, and with the physical properties of the surface. Polarization can thus be used to compare data obtained at different times and on different objects. It may be added that the temporal modulation of the intensity provides information about the period of rotation of asteroids. Fast rotating objects (with a period smaller than about 1h) are small and necessarily monolithic. On the opposite end, slow rotators may be bigger gravitational aggregates.

10.4.2
Typical Results on Surface Properties

The changing geometry of the scattering NEO and of the observer with respect to the Sun is used to define for a given object a (disk integrated) polarization phase curve, tentatively between 0° (backscattering) and 180° (forward scattering).

Asteroidal polarization phase curves are similar to those of numerous particulate media in the solar system, such as the Moon or cometary dust (see e.g. Muinonen et al. 2002; Levasseur-Regourd and Hadamcik 2003). As illustrated in Fig. 10.4, they are smooth, with a small negative branch (electric field vector parallel to the scattering plane predominating over electric field vector perpendicular to it) near backscattering, an inversion region near 20°, and a wide positive branch with a near 90° maximum for larger phase angles. Such curves have been estimated by various authors to be typical of the interaction of light with irregular particles media, with a size larger than the wavelength.

An enhancement of the intensity near backscattering may be observed, together with a sharp increase of the negative polarization. It is attributed to optical effects within a porous regolitic surface (mutual shadowing and coherent backscattering). Different slopes at inversion (or different maxima in polarization for NEOs observed at large phase angles) are easily noticed; the slope at inversion, as well as the maximum in polarization, actually increases with decreasing albedo of the asteroid. It is

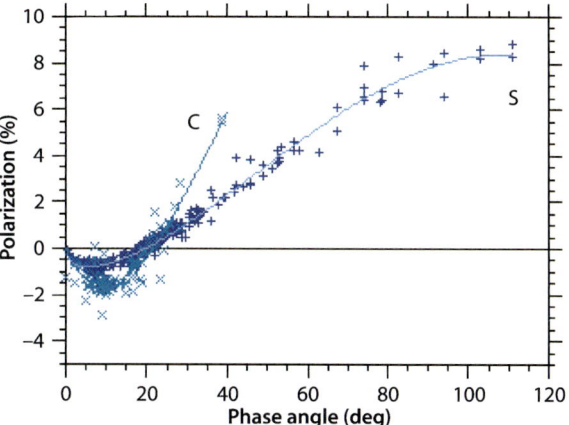

Fig. 10.4.
Polarization versus phase angle (from backscattering to forward scattering) for asteroidal surfaces. The curves are typical of irregular particles with a size greater than about one micrometer. (+) bright S-type asteroids; (×) dark C-type asteroids (adapted from Levasseur-Regourd et al. 2005)

related to the existence of multiple scattering (in the sub-surface) and can be used to derive an asteroidal taxonomy, with two main classes, corresponding respectively to bright S-type (also M and E) asteroids and dark C-type (also G and P) asteroids (see e.g. Goidet-Devel et al. 1995). The light scattering properties of asteroidal surfaces should thus also provide clues, through a precise classification, about their composition and possibly density.

Finally, the analysis of the variation of the polarization with the wavelength suggests that, for a fixed phase angle (above about 30°, where it is high enough), the polarization varies linearly with the wavelength, at least in the visible domain (Levasseur-Regourd and Hadamcik, 2003). Numerous observations have been obtained for 4179 Toutatis (e.g. Ishiguro et al. 1997), an S-type NEO actually named from a Celtic god that was supposedly prayed to prevent the sky from falling down. A decrease of the polarization with increasing wavelength (getting steeper from 20° to 90°) is observed for Toutatis while, on the opposite, the polarization hardly varies with the wavelength for C-type asteroids. Such different behaviors correspond to different physical properties.

10.4.3
Open Questions

NEOs surfaces are likely to be covered by porous and rough regolitic layers. However, they are certainly far from being homogeneous, and the presence of harder consolidated areas cannot be ruled out, as emphasized by the detection of local variations in the physical characteristics (e.g. albedo).

Some parameters in the light scattering properties (e.g. maximum in polarization phase curve, polarization wavelength dependence) differ significantly from one object to the next. They need to be exactly translated in terms of morphological properties (e.g. size distribution, porosity) of the particulate media building up the surfaces.

10.5
Knowledge Expected from Future Science

10.5.1
Remote Observations and Simulations under Development

More observations of (newly detected and already known) NEOs will certainly take place in the near future, and provide more information, not only on their size, albedo, spectral type and rotation (which may give indirect information on their monolithic or aggregated structure), but also on their light scattering properties. The differences noticed in the parameters characterizing the scattered light provide numerous constraints, to be taken into account to infer the physical properties from numerical and experimental simulations.

Numerical simulations are essential in the understanding of the physical properties of NEOs. Collision simulations may be used to understand how the response to an impact varies with the internal structure (see e.g. Michel et al. 2003) and thus to develop strategies to deflect an hazardous object. Light scattering simulations provide insights on the physical properties of the regolith (see e.g. Muinonen et al. 2002) and thus on the techniques to be used for landing or anchoring on the surface.

Experimental simulations in the laboratory are under development to simulate the formation of low-density bodies and to study their response to impacts and their light scattering properties with the ICAPS project on board the international Space Station (see e.g. Levasseur-Regourd and Hadamcik 2003), and thus to better understand the physical properties of regoliths.

10.5.2
Future Space Missions

Numerous missions to NEOs are in their cruise phase, under development, or under consideration for future developments. These missions originate from various space agencies (e.g. JAXA in Japan, NASA in the USA, ESA in Europe) and are summarized below.

After the successful NASA Deep Impact mission in July 2005, the next step is JAXA Hayabusa mission, launched in 2003, to reach asteroid 25143 Itokawa in September 2005. Similar to Eros, Itokawa is a S type asteroid, but its smaller size (about 500 m) could provide new results. Besides, the probe (equipped with electric propulsion and autonomous navigation) will not only study the asteroid from orbit, but launch a micro-rover on it and collect some dust samples that should reach Earth in 2007.

Information about cometary nuclei structure is expected from the ESA Rosetta probe, launched in March 2004 to rendezvous comet 67P/Churyumov-Gerasimenko in 2014. It will allow an accurate determination of the density and surface properties. Besides, the CONSERT experiment should provide unique information about the interior structure through the radar tomography technique (Kofman et al. 1998).

A similar technique has been proposed for a mission called ISTHAR, in response to an ESA call for ideas for NEO exploration and discovery (Barucci et al. 2005); the probe should also study the density and the surface properties. Also, the Deep Interior mission concept had been proposed to NASA to determine the geophysical properties of NEOs through radio reflection tomography and blast experiments (Asphaug et al. 2003).

Among other projects of missions to NEOs, it is worth mentioning two other proposals to ESA, SIMONE that should provide information about density and surface properties of a series of targets, and Don Quijote that should, together with its penetrator, determine the density and surface properties, while gathering information for the design of an effective mitigation mission (Harris et al. 2004).

10.6
Conclusion

An extreme diversity is already noticed among the few NEOs that have been tentatively studied. Some of them may be monolithic, whereas other ones are likely to be shattered collisional fragments or gravitational aggregates. They seem to be mostly covered with regolith layers, the thickness of which may vary significantly over the surface. More information about the surface properties of NEOs is within reach, with numerous remote observations of newly discovered objects, as well as with numerical and experimental simulations.

Huge uncertainties will nevertheless remain in the estimation of densities, interior structures and mechanical properties of potentially hazardous objects, leading to failed estimations of the effect of an impact that would be both economically and psychologically unacceptable. Programing quite a few space missions (rendezvous and landing), to objects belonging to the different classes already suspected, would provide a unique knowledge on the physical properties, with relevant statistics. Such missions, within the range of modern technologies, would then allow a relevant estimation of the effect of the impact of a newly discovered (thus only documented by remote observations) potentially hazardous object.

Finally, future missions to near Earth asteroids and cometary nuclei should be co-ordinated between various agencies and countries to improve significantly the public awareness and education on NEO issues.

Note added in proof. Recent results from Hayabusa have confirmed the huge diversity already noticed between NEOs. Images of asteroid Itakawa reveal some rough terrains with boulders; its density is about 1.95 g cm^{-3} and it is likely to be a gravitational aggregate.

References

A'Hearn MF, Belton MJS, Delamere A, Blume WH (2005a) Deep Impact, a large-scale active experiment on a cometary nucleus. Space Sci Rev 117:1–21
A'Hearn MF, Belton MJS, Delamere WA et al. (2005b) Deep Impact: Excavating Comet Tempel 1. Science 310:258–264
Asphaug E, Belton MJS, Cangahuala A, Keith L, Klaasen K, McFadden L, Neumann G, Ostro SJ, Reinert R, Safaeinili A, Scheeres DJ, Yeomans DK (2003) Exploring asteroid interiors: the Deep Interior mission concept. Lunar Planet Sci XXXIV:1906

Barucci MA, D'Arrigo P, Ball AJ, Doressoundiram A, Dotto E, Kofman W, Orosei R, Pätzold M, Perozzi E (2005) The ISHTAR Mission: Probing the Internal Structure of NEOs. In: Engvold O (ed) Highlights of astronomy 13 (B), Astron Soc Pacific, San Francisco, pp 796-800

Binzel RP, Rivkin AS, Stuart JS, Harris AW, Bus SJ, Burbine TH (2004) Observed spectral properties on near-Earth objects: results for population distribution, source regions, and space weathering processes. Icarus 170:259-294

Britt DT, Yeomans D, Housen K, Consolmagno G (2002) Asteroids density, porosity, and structure. In: Bottke WF, Cellino A, Paolicchi P, Binzel RP (eds) Asteroids III, U of Arizona Press, Tucson, pp 485-500

Brownlee DE, Horz F, Newburn RL, Zolensky M, Duxbury TC, Sandford S, Sekanina Z, Tsou P, Hanner MS, Clark BC, Green SF, Kissel J (2004) Surface of young Jupiter family comet 81P/Wild 2: view from the Stardust spacecraft. Science 1764-1769

Davidsson BJR, Gutiérrez PJ (2004) Estimating the nucleus density of comet 19P/Borrelly. Icarus 168: 392-408

Desvoivres E, Klinger J, Levasseur-Regourd AC (2000) Modeling the dynamics of fragments of cometary nuclei: Application to comet C/1996 B2 Hyakutake. Icarus 144:72-181

Fulle M, Levasseur-Regourd AC, McBride N, Hadamcik E (2000) In-situ dust measurements from within the coma of 1P/Halley: first order approximation with a dust dynamical model. Astron J 119: 1968-1977

Goidet-Devel B. Renard JB, Levasseur-Regourd AC (1995) Polarization of asteroids: Synthetic curves and characteristic parameters. Planet Space Sci 43:779-786

Hadamcik E, Levasseur-Regourd AC (2003) Dust coma of comet C/1999 S4 (LINEAR): Imaging polarimetry during the nucleus disruption. Icarus 166:188-194

Hapke B (1993) Theory of reflectance and emittance spectroscopy. Cambridge University Press, Cambridge

Harris AW, Benz W, Fitzsimmons A, Green SF, Michel P, Valchecchi G (2004) Report from the NEOs mission advisory panel. ESA advanced concepts team, ESA

Hilton JL (2002) Asteroid masses and densities. In: Bottke WF, Cellino A, Paolicchi P, Binzel RP (eds) Asteroids III, U of Arizona Press, Tucson, pp 103-112

Kofman W, Barbin Y, Klinger J, Levasseur-Regourd AC, Barriot JP, Herique A, Hagfors T, Nielsen E, Grün E, Edenhofer P, Kochan H, Picardi G, Seu R, van Zyl J, Elachi C, Melosh J, Veverka J, Weissman P, Svedhem LH, Hamran SE, Williams IP (1998) Comet nucleus sounding experiment by radiowave transmission. Adv Space Res 21:1589-1598

Ishiguro M, Nakayama H, Kogachi M, Mukai T, Nakamura R, Hirata R, Okazaki A (1997) Maximum visible polarization of 4179 Toutatis in apparition of 1996. Publ Astron Soc Japan, 49: L31-L34

Levasseur-Regourd AC (2004) Cometary dust unveiled. Science 304:1762-1763

Levasseur-Regourd AC, Hadamcik E (2003) Light scattering by irregular dust particles in the solar system: observations and interpretation by laboratory measurements. J Quant Spectros Radiat Transfer 79: 903-910

Levasseur-Regourd AC, McBride N, Hadamcik E, Fulle M (1999) Similarities between in situ measurements of local dust scattering and dust flux impact data within the coma of 1P/Halley. Astron Astrophys 348:636-641

Levasseur-Regourd AC, Hadamcik E, Lasue J (2006) Interior structure and surface properties of NEOs. Adv Space Res 37:161-168

Margot JL, Nolan MC, Benner LAM, Ostro SJ, Jurgens RF, Giorgini JD, Slade MA, Campbell DB (2002) Binary asteroids in the near Earth objects population. Science 296:1445-1448

McBride N, Green S, Levasseur-Regourd AC, Goidet-Devel B, Renard JB (1997) The inner dust coma of comet 26P/Grigg-Skjellerup: multiple jets and nucleus fragments? Mon Not R Astron Soc 289: 535-553

Michel P, Benz W, Richardson D (2003) Disruption of fragmented parent bodies as the origin of asteroid families. Nature 421: 608-611

Muinonen K, Piironen J, Shkuratov YG, Ovcharenko A, Clark BE (2002) Asteroid photometric and polarimetric phase effects. In: Bottke WF, Cellino A, Paolicchi P, Binzel RP (eds) Asteroids III, U of Arizona Press, Tucson, pp 123-138

Ostro SJ, Hudson RS, Benner LAM, Giorgini JD, Magri C, Margot JL, Nolan MC (2002) Asteroid radar astronomy. In: Bottke WF, Cellino A, Paolicchi P, Binzel RP (eds) Asteroids III, U of Arizona Press, Tucson, pp 151–168

Rickman H, Kamel L, Festou MC, Froeschlé C (1987) Estimates of masses, volumes and densities of short period comets. In: Rolfe EJ, Battrick B (eds) Diversity and similarity of comets, ESA, Noordwijk, pp 471–481

Sekanina Z, Brownlee DE, Economou TE, Tuzzolino AJ, Green SF (2004) Modelling the nucleus and jets of comet 81P/Wild 2 based on the Stardust encounter data. Science 304:1769-1774

Thomas PC, Joseph J, Carcich B, Veverka J, Clark BE, Bell III JF, Byrd AW, Chomko R, Robinson M, Murchie S, Prockter L, Cheng A, Izenberg N, Malin M, Chapman C, McFadden LA, Kirk R, Gaffey M, Lucey PG (2002) Eros: shape, topography, and slope processes. Icarus 155:18-37

Weaver HA (2004) Not a rubble pile. Science 304:1760-1762

Weaver HA, Sekanina Z, Toth I, Delahodde CE, Hainaut OR, Lamy PL, Bauer JM, A'Hearn MF, Arpigny C, Combi MR, Davies JK, Feldman PD, Festou MC, Hook R, Jorda L, Keesey MSW, Lisse C.M, Marsden BG, Meech KJ, Tozzi GP, West R (2001) HST and VLT investigations of the fragments of comet C/1999 S4 (LINEAR). Science 292:1329-1334

Chapter 11

Evaluating the Risk of Impacts and the Efficiency of Risk Reduction

G. B. Valsecchi · A. Milani Comparetti

11.1
Introduction

The space missions of the past decades have shown that impacts represent an ubiquitous phenomenon in the Solar System, and occur at all scales, from dust particles up to planetary bodies. In fact, a clue to the importance of this phenomenon also for our planet has always been available on the heavily cratered surface of the Moon, that testifies to the present and past fluxes of bodies on Earth crossing orbits.

The impacts of interplanetary bodies above a critical size (40–50 meters diameter) are an environmental threat. Unlike most other types of natural hazards, the risk of asteroid/comet impact is perfectly deterministic, that is, given sufficient information gathered with existing technology it is possible to decide whether a given catastrophic event will or will not take place. Thus gathering information on the population of potentially impacting bodies implies an immediate effect of risk reduction: the known objects for which enough information is available to exclude the possibility of an impact (within a given time span) can be removed from the estimate of the risk.

It is only relatively recently that astronomers have started to aim at an effective risk reduction by astronomical means, that is by observations and computations. In this paper we summarize what is currently done in this field, and discuss what could be done to improve the situation and to achieve a very significant risk reduction.

To simplify the discussion, we subdivide in five steps the overall process of risk reduction seen from the astronomical point of view; these steps are:

1. the early detection of near-Earth objects, that is a prerequisite for all further action;
2. the accurate determination of their orbits;
3. the computation, for each near-Earth object, of all the possibilities of collisions with our planet within a reasonable time span in the future;
4. the acquisition of further observations for the objects that have the possibility of colliding with the Earth, in order to be able to exclude (or confirm) these collisions in the given time span;
5. the measures that have to be put in place to prevent a collision for an object for which the previous steps have led to ascertain that it will impact on our planet.

In the following sections we discuss each step in turn.

11.2
Near-Earth Objects Surveys

Asteroids and comets are small Solar System bodies thought to be the remnants of the processes that led to the accretion of the planets in the early phases of the evolution of the Solar System. In particular, if a small body observed telescopically appears as a point-like light source, it is called "minor planet" or "asteroid"; if it does not appear point-like, it is then called a "comet".

The number of asteroids known is much larger than the number of known comets (according to the current theories, there is a much larger number of comets with orbits so large that their presence in the planetary region is very rare). All the populations of small bodies have a size distribution, typically with the number of bodies increasing with decreasing size; for asteroids the cumulative size distribution is known over a large range of sizes, with numbers growing with the inverse of the diameter to the power k (k is between 2 and 3 in most of the size range of interest for this discussion).

An asteroid or a comet is considered a Near-Earth Object (NEO) if the perihelion distance of its orbit is smaller than 1.3 AU; the acronym AU stands for Astronomical Unit, and is equal to the mean distance of the Earth from the Sun (about 150 million kilometers). The perihelion is the location in the orbit where the distance from the Sun is minimum. Another acronym frequently used in the rest of this paper is NEA, that stands for Near-Earth Asteroid.

An asteroid (or a comet) is detected because it moves with respect to the fixed stars; the information made available by the first detection amounts to four measured quantities, two angular positions and two rates of change of the same (at a given observation time). Since the information needed to place an object in its orbit consists of six quantities at a given time, it is clear that the intial detection never allows to compute an orbit. Thus, it is not possible to deduce from the initial detection that the object is a NEO, with the exception of the few cases in which the rate of motion is very large: in these cases the object must be moving close to the Earth and is a NEO, but it also needs to be comparatively small (for a given apparent brightness).

This argument on the detection procedure implies that it is not possible to design a survey to discover all the NEOs and nothing else: unavoidably, the large majority of the objects detected will be run of the mill asteroids, belonging to the so-called Main Belt (with orbits between those of Mars and Jupiter). The proportion of NEOs is typically 1 in 1000 detections. To decide whether a given detection really corresponds to a NEO we need much more information than the one contained in the detection itself.

Moreover, it is clear that not all NEOs represent an immediate threat to the Earth, since the simple fact that the perihelion distance is smaller than 1.3 AU does not imply that a collision with our planet is possible. In fact, a collision is possible if, at a given time:

1. the minimum distance between the orbits (often called MOID, an acronym that stands for Minimum Orbital Intersection Distance) is of the order of one Earth radius;
2. the Earth and the NEO arrive at the MOID points along the respective orbits at the same time.

Since the MOID changes only slowly with time, for a collision in the near future (say, within the next century) to be at all possible, the MOID must be currently of the order of a few tens of Earth radii (one Earth radius is about 0.000043 AU). Thus, the potentially dangerous NEOs that must be detected early in order to allow preventive actions is a subset of the entire population, and a good knowledge of the orbit is required to catalog a discovery among the potentially hazardous objects.

It is a wise policy to try and get good orbital data on all asteroids discovered, irrespective of whether they can actually approach the Earth very closely, for various reasons. First, these data have scientific value independently from the risk reduction goal. Second, all asteroids for which we are not able to obtain a good knowledge of the orbit will sooner or later be "rediscovered", and at that time we will have to waste telescopic, computational and possibly human resources just to establish that we had already seen it; being able to predict a second apparition is more economical than to have to make a rediscovery.

The so called NEO surveys in fact aim at detecting anything moving against the fixed stars background. There are various techniques to accomplish this task, reviewed in (Carusi et al. 1994; Stokes et al. 2002). Essentially the same techniques detect also objects with variable luminosity, corresponding to other astronomically interesting phenomena. If the object is indeed an asteroid (or comet) the observations need to be pursued until an orbit can be computed: this process is called "follow-up".

There are a number of NEO surveys currently ongoing (Stokes et al. 2002). All of the most successful ones are carried out by the U.S.A., and all but one of them from the Northern Hemisphere. Although this Northern/Southern asymmetry does not prevent the achievement of a NEO catalog complete up to a given size, it makes the process slower and less efficient.

11.2.1
The Problem of Orbit Determination

Since, as discussed in the previous section, a single detection does not allow to compute an orbit, two or more belonging to the same object must be available to achieve this. Targeted follow-up deliberately accumulates observations of the same object until an orbit can be computed and the nature of the object determined (e.g., NEA, Main Belt).

Large surveys currently detect thousands of objects each night; in the near future, the next generation surveys will detect hundreds of thousands objects per night. The surveys aim at covering as much sky area as possible, thus they cannot perform the targeted follow up of all their detections. The method used by most surveys is to revisit the same general area in the sky several times over a time span of few days/weeks, so that most objects are detected several times.

This creates the problem of identification: among the detections obtained in a given time span (e.g., one month) how to find which ones belong to the same physical object. This is a mathematically interesting and computationally challenging problem, for which there has been significant progress in the last few years (Bowell et al. 2002; Milani 2005).

For the typical MB asteroid, orbit determination takes time but is achieved in the long run because geometric and illumination conditions favorable to detection repeat with a more or less regular pattern. Orbit determination can be more difficult for a NEA, especially if it is small, because it can easily move to a region where it becomes practically unobservable: by the time of the next apparition its position in the sky could become so badly predictable that it has in fact to be serendipitously rediscovered.

With automated surveys currently operating (LINEAR, LONEOS, NEAT, Catalina, Spacewatch) there has been rapid progress; as of November 2004 more than 700 NEAs with estimated diameter > 1 km have been discovered (and followed up until a reasonably good orbit could be determined). The estimation of the total population is tricky, but about 2/3 of the 1 km NEAs have been discovered. The remaining ones, however, will take long to discover, because their orbits are such that they are less often visible than the ones already discovered (Bottke et al. 2002).

11.3
Checking for Impact Possibilities

Before 1998 the problem of computing all possible impact solutions for objects with a given set of observations had not been solved. However, since orbital evolution is deterministic and is computable with the required accuracy, it is not clear at first sight why this should be a difficult task. Moreover, it is also not immediately apparent that probabilities would have anything to do with a deterministic problem like this one.

However, it must be taken into account that there is no such thing as the orbit of an asteroid determined from the observations. Actually, there is always a range of possible orbits, all compatible with the observations; this range may be very small, but anyway of finite size. Probability then enters the picture as a measure of our ignorance: we just know the region containing all the possible orbits for our asteroid.

One way of describing our knowledge of the orbit of a specific asteroid is to introduce the concept of Virtual Asteroid (VA). The orbits compatible with the observations of an asteroid can be described as a swarm of VAs: only one of them is real, but we don't know which one. Thus, we can compute the orbital evolution of each individual VA, as if it were a real body. The purpose of the computation would be to check whether the VA has an impact with the Earth, in which case we call it a Virtual Impactor (VI), with an associated Impact Probability (IP) depending upon the statistics of the observational errors (see Milani et al. 2000b and 2002 for the technical details).

Let us now consider a NEA that has an IP of 1/1000; if we computed the orbital evolution of 1000 VAs chosen at random within the region containing all orbits compatible with the observations of that NEA, we can expect to find one VI among the 1000 VAs. However, if the IP is 1/1 000 000, to find a VI with such a brute force approach we need to compute ~1 000 000 VAs: this is too much to be done on a daily basis, even for current computers.

The strategy to detect efficiently VIs with low IP consists of arranging the VAs along a string. As the VAs proceed on their separate orbits, the string stretches, mostly along track, until it wraps around a large portion of the orbit. If there is a point where the orbits are close to the Earth's orbit, some VAs have close approaches to the Earth. We can then interpolate along the string. If two consecutive VAs straddle the Earth, an intermediate VA can be constructed to find the minimum possible approach distance, and the check of whether a collision with the Earth is possible requires only a relatively short additional computation. The efficiency gain with this computational strategy is more than 1000.

In March 1999, with the first application of this strategy, we could detect a VI with IP 1/1 000 000 000 with only a few thousand VAs (Milani et al. 1999). Later that same year, in November 1999, the software robot CLOMON begun operations at the University of Pisa, monitoring each new NEA for possible impacts in the next 80 years (Milani et al. 2000a). The results of the computations, i.e. the VIs that have since then been found, have been posted on the Risk Page of the NEODyS web site (*http://newton.dm.unipi.it/neodys/*).

In 2002 the 2nd generation impact monitoring robots CLOMON2 and Sentry (this second one at the Jet Propulsion Laboratory in Pasadena, with results posted at the URL *http://neo.jpl.nasa.gov/risk/*) became operational (Chamberlin et al. 2001; Milani et al. 2005). The VIs found by the two monitoring systems are routinely cross checked by the two teams operating them. The experience accumulated over five years of operations has led to significant improvements in the reliability of the computations.

11.4
Eliminating Virtual Impactors

The fact that a NEA has some VIs can change only as a result of observations. In these cases, the astronomical community has to provide further observations, and in fact the prime usefulness of the impact monitoring software robots consists in their ability to highlight the NEAs for which additional observations are needed in order to eliminate the VIs associated with them.

After an initial turbulent period, in which unnecessary attention of the media was called by people aiming at the sensationalization of the issue, the publication of VIs on the WWW has become a well established procedure that does not lead anymore to frequent (and counterproductive) media storms. In fact, nowadays this procedure makes sure that the essential information (i.e. the need for further observations) reaches all the interested parties.

The consequence of posting VIs on the NEODyS and Sentry risk pages is thus to alert observers, that in general react quickly, most often within 24 hours, providing new observations. These are then processed together with the already available ones, and as a consequence the probability of each VI changes. Actually, the new observations can push the probability both up and down (in the end, an IP can only go to 0 or 1), and in general the result is that all VIs are eliminated in a matter of weeks.

It can happen that a NEA becomes unobservable while it still has one or more VIs; in these cases, the IP cannot change until the NEA is recovered, deliberately or by chance. This currently happens only for small asteroids, i.e. for objects with estimated diameter well under 1 km. Of course, if the surveys will become in the near future able to detect and accurately track fainter objects, the size range of the objects that become lost before losing all their VIs will be reduced.

11.4.1
Decrease of the Risk Estimate

Since most of the risk can be shown to come from 1 km objects (Chapman and Morrison 1994), if all the known objects in this size range cannot impact in the next 100 yrs, and the fraction discovered with respect to the presumed total population is 2/3, then the risk has been decreased, as result of the work of the astronomers, by about 2/3 with respect to the background risk (by definition, the one present before human intervention, measured by the count of craters on the lunar surface).

However, not all NEOs that can be shown to be harmless over the next century will remain harmless forever: due to the changes that take place over time of their orbits, their MOID can become dangerously low at some future epoch, and the sooner we will be able to establish that, the easier will be any preventive action. Thus observations and orbit computations will have to be continued also for objects known not to have VIs over the time span so far monitored.

11.5
Deflection

We now discuss the problem of what to do in the unlikely, but possible, case in which a NEA were discovered, going to impact our planet with reasonable certainty at a specific time in the coming decades. In such a case, the preventive actions to be taken will of course depend on the likely level of damage expected, and such an estimation involves aspects that are outside the context of astronomy and space sciences. However, if as a result of the damage estimation it would be decided that the only sensible action would be to prevent the collision from happening, then space activities aimed at the deflection or at the destruction of the potential impactor would be necessary. It is obvious that an adequate level of preparedness should be in place beforehand, and hereafter we describe one specific example of space mission, aimed at the deflection of a NEA in the half kilometer diameter range, that is currently being studied by the European Space Agency.

Necessary conditions for such a mission to be meaningful are at least the following:

1. the potential impactor has to be discovered several decades before impact, so that the impulse needed to deflect the impactor is within technologically feasible bounds,
2. we must be able to control the amount of deflection imparted, in order to transfer the impactor onto a safe orbit.

11.5.1
Kinetic Energy Deflection

A quantitative analysis of the problem shows that, if the two conditions just mentioned are met, then a relatively small mass spacecraft (about 500 kg), impacting at a speed of the order of 10 km s^{-1}, could transfer enough linear momentum to deflect a NEA of

300–500 m diameter away from its Earth-colliding orbit. In fact, the largest unknown in this scenario is the amount of linear momentum transferred, that depends not only on the mass and speed of the impacting spacecraft, but also on the detailed physics of the formation of a crater on the NEA, with the ensuing ejection of material in the direction opposite to that from which the spacecraft arrives.

Thus, a precursor mission, in which one aims at determining the "reaction" of the NEA in question to a spacecraft impact, is needed before setting up the "real" mission, i.e., the mission aiming at the accomplishment of the full deflection.

The ESA study Don Quijote aims at acquiring the know-how to do such a deflection (*http://www.esa.int/gsp/completed/neo/donquijote_execsum.pdf*). It envisions two spacecrafts, named after the two main characters of Cervantes' masterpiece. The first one, named Sancho (also because it does not take risks …) is put in orbit about the asteroid several months before the arrival of the other spacecraft. During its permanence in orbit, Sancho carries out a number of investigations aimed, among other things, at knowing with a precision much greater than that achievable from ground the states of motion and of proper rotation of the asteroid, as well as some investigation of its internal structure, performed through the implantation on the NEA surface of seismometers that are later used to measure the seismic waves excited by some pyrotechnic.

The second spacecraft, named Hidalgo (this is the daring one!) arrives at the asteroid after an interplanetary journey completely different from that of Sancho, and impacts the surface of the asteroid at > 10 km s^{-1}. At the time of Hidalgo's arrival, Sancho retreats at a safe distance from the NEA, and continues to carry out its observations. Afterwards, Sancho continues to carry out measurements of the states of motion and of proper rotation of the asteroid; these will have changed, due to the impact of Hidalgo, and their precise values are crucial to assess the effectiveness of the deflection.

It is clear that the results of a mission like Don Quijote will be valid only for the NEA that will be its objective, and cannot be easily generalized. This is a problem that we will have to face anyway, given the number and the variety of the population of potential Earth impactors. On the other hand, a lot can be learned through a space mission of this type, both on the target NEA and, perhaps more importantly, on our real degree of mastering the technology of an actual asteroid deflection.

11.6
Conclusions

As we have seen, NEO impacts are fully predictable, and the possibility of actually predicting the next one only depends on our will to invest the necessary telescopic, computational and manpower resources needed to collect the necessary information.

In absolute terms, the data gathering allowing prediction is doable, with available know-how, at a cost that is not larger than that of many other scientific endeavors in fields like particle physics or medical research. Thus, the risk of NEO impact is not in the hands of fate: preventive actions mean zero damage, provided that we acquire the know-how needed for deflection.

References

Bottke WF, Morbidelli A, Jedicke R, Petit J-M, Levison HF, Michel P, Metcalfe TS (2002) Debiased orbital and absolute magnitude distribution of the Near-Earth Objects. Icarus 156:399–433

Bowell E, Virtanen J, Muinonen K, Boattini A (2002) Asteroid orbit computation. In: Bottke WF, Cellino A, Paolicchi P, Binzel RP (eds) Asteroids III. University of Arizona Press, Tucson, pp 27–43

Carusi A, Gehrels T, Helin EF, Marsden BG, Russell KS, Shoemaker CS, Shoemaker EM, Steel DI (1994) Near-Earth objects: present search programs. In: Gehrels T, Matthews MS, Schumann AM (eds) Hazards due to comets and asteroids. University of Arizona Press, Tucson, pp 127–147

Chamberlin AB, Chesley SR, Chodas PW, Giorgini JD, Keesey MS, Wimberly RN, Yeomans DK (2001) Sentry: an automated close approach monitoring system for Near-Earth Objects. Bulletin of American Astronomical Society 33:1116

Chapman CR, Morrison D (1994) Impacts on the Earth by asteroids and comets: assessing the hazard. Nature 367:33–40

Milani A (2005) Virtual asteroids and virtual impactors. In: Knezevic Z, Milani A (eds) Dynamics of populations of planetary systems. IAU Coll. 197, Cambridge University Press, pp 219–228

Milani A, Chesley SR, Valsecchi GB (1999) Close approaches of asteroid 1999 AN10: resonant and non-resonant returns. Astronomy and Astrophysics 346:L65–L68

Milani A, Chesley SR, Valsecchi GB (2000a) Asteroid close encounters with the Earth: risk assessment. Planetary and Space Science 48:945–954

Milani A, Chesley SR, Boattini A, Valsecchi GB (2000b) Virtual impactors: search and destroy. Icarus 145:12–24

Milani A, Chesley SR, Chodas PW, Valsecchi GB (2002) Asteroid close approaches: analysis and potential impact detection. In: Bottke WF, Cellino A, Paolicchi P and Binzel RP (eds), Asteroids III. University of Arizona Press, Tucson, pp 55–69

Milani A, Chesley SR, Sansaturio ME, Tommei G, Valsecchi GB (2005) Nonlinear impact monitoring: line of variation searches for impactors. Icarus 173:362–384

Stokes GH, Evans JB, Larson SM (2002) Near-Earth asteroids search programs. In: Bottke WF, Cellino A, Paolicchi P, Binzel RP (eds) Asteroids III. University of Arizona Press, Tucson, pp 45–54

Chapter 12

Physical Effects of Comet and Asteroid Impacts: Beyond the Crater Rim

H. J. Melosh

12.1
Introduction: the Impact Hazard

Astronomical and geological investigations initiated in the past century have revealed that the Earth is continually subjected to the infall of a variety of solid solar system debris. Most of this debris is so small that it evaporates harmlessly, as it enters the Earth's upper atmosphere at high speed. However, an occasional larger object survives atmosphere entry. Small examples of such objects result in meteorites on the surface of the Earth, with harmful consequences only for the rare individuals, who happen to be struck by them. More infrequent, but larger, objects can cause local or even global devastation. A recent report on the number and consequences of such impacts (Team 2003) proposes that the impact frequency can be computed as a function of the energy release, equal to the kinetic energy of the object before it strikes the Earth:

$$T_{RE}(\text{years}) = 110\, E_{MT}^{0.77}$$

where T_{RE} is the recurrence interval (in years) and E_{MT} is the energy release in megatons of TNT equivalent (1 MT = 10^{15} cal ≈ 4.2×10^{15} J).

The impact of a large meteoroid on the Earth initiates a rapid series of events that, for sufficiently large impactors, may cause a large number of human deaths (Team 2003). These deadly effects are principally a function of the kinetic energy of the impact. The impactor's type (comet or asteroid, stony or iron), shape and angle of impact are all secondary compared to the energy released. The speed of an impacting body is relatively well constrained for different types of objects. Most asteroids strike between 15 and 23 km s^{-1}, whereas short period comets average about 50 km s^{-1}, so for a given mass object, about 4 times more energy is delivered by comets (which, however, only form about 1% of the total impact risk [Team 2003]). Although higher energy impacts are more devastating, there are, fortunately, fewer of them.

Small asteroids may be stopped or dispersed by the atmosphere (Bland and Artemieva 2003). This is the fate of most impactors delivering up to about 20 MT to the Earth. Such objects may explode in the air, as did the 1908 Siberian Tunguska object (Chyba et al. 1993) and create substantial local damage. The 1908 explosion felled meter-diameter trees over an area about 20 km in diameter and vaporized a herd of reindeer (Krinov 1966). Iron asteroids of equivalent energy reach the ground intact and form small meteorite craters, such as the famous 1.2 km diameter Barringer or Meteor Crater in Arizona (Shoemaker 1963). The very largest impact for which we have good evidence is the 15 km diameter Chicxulub impactor that delivered about 100 000 000 MT to the Earth, created a 170 km diameter crater in Yucatan, and initiated the greatest biological extinction in the past 250 Myr (Grieve and Therriault 2000). Somewhere between these two extremes lies an energy release that, while too small to initiate a profound biological extinction, is nevertheless large enough to devastate global civilization. It is generally supposed that such an object releases about 30 000 MT and has a recurrence interval of about 1 Myr (Chapman and Morrison 1994).

Most of the immediate effects of an impact of any size are a strong function of distance from the event. These effects can be classified as either local or regional in extent (from kilometers to thousands of kilometers from the impact site. There is no exact demarcation of these designations – the actual physical phenomena are the same, but they extend farther from the impact site for larger impacts) or global. Since the Alvarez's proposal that the K/T extinction was initiated by an impact (Alvarez et al. 1980), most research has focused on the global effects of very large impacts (Alvarez 1986). However, a number of authors (e.g., Toon et al. 1997) have recognized that the local and regional effects of smaller impacts may have serious consequences for our delicately balanced global civilization, even though no biological extinctions are known to be associated with smaller impact events. The following paper, thus, concentrates mainly on the "Local and Regional" effects and only mentions the more serious "Global" effects in a short final section.

12.2
Local and Regional Devastation by Impacts

Although humans have never recorded a crater-forming hypervelocity impact, experience with nuclear explosions (Glasstone and Dolan 1977) informs us that it is extremely hazardous to be within a few tens of kilometers of even a small impact. Meteor Crater, Arizona, which formed about 50 000 years ago, created a shock wave in the air (usually referred to as an "airblast") that probably killed most large animals within a radius of about 20 km (Kring 1997). The eminent meteorite researcher H. H. Nininger summarized his years of thinking about the effects accompanying the formation of Meteor Crater with the following vivid passages from the preface to his book on the crater (Nininger 1956):

> The grazing bands of deer, elk, and antelope face southwest into the roaring wind as twilight deepens across the grassy plain. Suddenly the fields are lighted with the brilliance of noonday. A deafening swishing roar from out of the northern sky brings each head erect and frightened eyes watch, as 20 miles away a giant blazing sun screams downward, spewing an exploding train of fiery sparks as from a raging blast furnace.
>
> A blinding flash, a billowing fountain of flame, and a swirling, blazing, mushroom cloud shoots skyward into the stratosphere. Five, ten, fifteen miles, and up it goes while a deadly pall of smoke and dust covers the spot where the blazing sun dived to its doom.
>
> The wide eyes stare, their terror-stricken owners frozen into statues. Sharp ears strain forward to catch the faintest sound on the momentarily quiet air.
>
> A searing blast of heat and wind. The straining ears are deaf, the sharp eyes sightless bulges on the crushed and roasted heads. The herds have vanished in a stench of burning hair and flesh, and on the charred grass, so lately green, lie twisted, blackened hulks, insensible to roaring wind and to the warm drizzle of tiny metal droplets which are blanketing the land.
>
> And 20 miles away, steam and smoke swirl from the gaping mile-wide hole and from the mountain of shattered rocks and twisted bits of metal that now strew the land.

Nininger probably overestimated the distance to which incendiary effects are important, but he touched on nearly all of the phenomena that we presently believe to be important in the vicinity of an impact (the only one he omitted entirely is seismic shaking, which would have knocked the antelope off their feet).

The recurrence time for Meteor Crater size events is about 20 000 years for the entire Earth surface, after factoring in the 5% abundance of iron meteorites that are uniquely capable of penetrating the atmosphere at the relatively small energy, 20 MT, of the Meteor Crater event. A similar energy event occurs about once per 1 000 years, by more abundant stony asteroids, but they disintegrate in the atmosphere, producing Tunguska-like explosions (Chyba et al. 1993).

Aside from the creation of an impact crater and obliteration of anything actually inside the crater, there are four major effects (five, if tsunamis from oceanic impacts are included). In the order of arrival at some point distant from the impact, they are: (1) Thermal radiation from the fireball and incandescent ejecta, (2) seismic shaking from the force of the impact, (3) burial by ejecta from the crater, and (4) airblast from the sudden expansion of the impact plume. Of these four effects, the one that extends to greatest distances is probably seismic shaking.

Large impacts can affect the Earth to quite large distances: A Chicxulub-scale impact today on San Francisco would bury Los Angeles (distance 650 km) with ejecta, ignite fires in San Diego (distance 850 km) and level Denver (distance 1 700 km) from the seismic shaking. Even in New York City (distance 4 500 km) a few buildings would collapse. (For detailed information on the effects of any size impact, visit our website, *www.lpl.arizona.edu/impacteffects*. We describe the algorithms used on this website in a recent paper [Collins et al. 2005]). Serious as these effects may be, they are all classified as local or regional. The last section of this paper will describe the unique global phenomena associated with very large impacts. The present section considers each of the local and regional effects described above in more detail.

12.2.1
Thermal Radiation

When a rapidly moving object collides with the Earth's surface, approximately one-half of its kinetic energy is immediately converted to heat (Melosh 1989). At velocities above about 15 km s^{-1}, peak temperatures on impact exceed 10 000 K and the formerly solid projectile and a roughly equivalent mass of target material is converted to incandescent gas or plasma. Lower velocity projectiles melt a few times their own mass of rock, which also emits heat as thermal radiation. This initial heat is lost rapidly, as the gas and hot melt rock expand away from the impact site. Although most of this energy is converted back to kinetic energy again during this phase, a small but important fraction is emitted as electromagnetic radiation; both as visible light and radiant heat. When sufficiently intense, this radiation can ignite fires over the entire region from which the fireball is visible (Nemchinov and Svetsov 1991).

Numerical modeling (Nemtchinov et al. 1998) indicates that the conversion efficiency of total kinetic energy to light and heat is in the range of 1×10^{-4} to 5×10^{-4}. This radiant energy is emitted from a hot fireball of expanding gas or plasma as it cools through a critical temperature T_*, known as the transparency temperature (Zel'dovich and Raizer

1967). This is the temperature at which the hot gases in the fireball become transparent and permit the electromagnetic radiation formerly bottled up in the hot plasma to escape to the surrounding air. For the Earth's atmosphere, T_* is about 3000 K, so at this point the fireball would appear as a "second sun" in the sky. Although radiant heat travels at the speed of light, the time at which irradiation begins depends on how long the fireball takes to develop and for the transparency temperature to be reached. This is typically a few seconds or less for impacts ranging between the size of Meteor Crater and Chicxulub.

Conversion of the heat lost from the fireball to the consequences for humans and structures at a given distance from the impact depends on many factors, such as the size of the fireball, curvature of the Earth (the fireball must be above the horizon for an observer at the selected distance), cloud cover, atmospheric transparency, duration of the exposure, and the nature of the materials affected. Such factors are discussed in detail by Collins et al. (2005) for impacts and Glasstone and Dolan (1977) from the nuclear weapons perspective. However, although this factor is unlikely to be very important for small cratering events, an impact the size of Chicxulub could have ignited fires up to a thousand kilometers from the impact site (Nemchinov and Svetsov 1991).

12.2.2
Seismic Shaking

The impact of a meteorite with the Earth's surface produces ground shaking analogous to that created by an earthquake. Unfortunately, the efficiency of conversion of impact energy to seismic energy is not well known. Values in the literature range from 10^{-5} to 10^{-3}, with a generally accepted mean of 10^{-4} (Schultz and Gault 1975). Adopting this mean and using the standard Gutenberg-Richter relation between earthquake energy and surface wave magnitude M, produces a relation between cratering energy E_c (note that this is not the same as the total impact energy if a significant fraction of the energy is dissipated in the atmosphere during entry) and equivalent Richter magnitude M:

$$M = 0.67 \log_{10} E_c (\text{Megatons}) + 4.44$$

The most damaging seismic waves emitted in a strong earthquake or an impact are surface waves, which travel at about 5 km s^{-1} over the Earth's surface. The more rapidly moving P and S body waves are generally much smaller in amplitude than the surface waves and can be ignored from the hazard point of view. The arrival time of these waves at a distance r_{km} in kilometers from the impact is, thus, about $r_{km}/5$ seconds after the impact. The amount of devastation at a given distance from the impact can be estimated by computing the intensity of shaking I as defined on the Modified Mercalli Intensity Scale (Richter 1958). A somewhat complex procedure can be constructed to estimate the intensity at any given distance from the impact and then used to predict the consequences for human habitations. This is described in detail in Collins et al. (2005), where the well-known saturation of the Richter scale at large magnitudes is taken into account by the use of the modern Moment-Magnitude scale at large magnitudes.

Although humans have not directly observed any large impact, there are many indications that the Chicxulub impact caused massive landslides both on land (Busby et al. 2002), in the Caribbean (Bralower et al. 1998), and off the eastern North American continental shelves from Florida (Klaus et al. 2000) to the Grand Banks of Canada (Norris et al. 2000). Chicxulub's seismic shaking apparently fluidized near shore marine sediments of the Fox Hills formation in South Dakota, some 2000 km from the impact site (Terry et al. 2001). A recent set of observations indicate that even relatively small amounts of seismic shaking can induce the eruption of geysers (Husen et al. 2004), trigger small earthquakes (Gomberg et al. 2004) and disturb hydrothermal systems at great distances from the earthquake epicenter (Stark and Davis 1996), so the seismic consequences of even a relatively small impact might be very widespread. It has been suggested that seismic shaking created by the putative Upheaval Dome impact crater in Utah caused a massive petroleum injection into the nearby Roberts Rift (Huntoon and Shoemaker 1995).

12.2.3
Ejecta Deposition

Even a casual observation of fresh lunar craters through a small telescope reveals that they are surrounded by a raised rim and that preexisting surface features are blanketed by a sheet of material (ejecta) that extends about 1 crater diameter from the crater's rim. The volume of this material is roughly equal to the volume of the crater bowl (Schröter's 1802 "Rule": see Melosh 1989, p 90). Careful observations of crater topography, coupled with data from explosions and small impacts indicate that the thickness of the ejecta blanket t_e, although highly variable around the circumference of the crater, is approximately given by:

$$t_e = \frac{h_{tr}}{8}\left(\frac{D_{tr}}{r}\right)^3$$

where h_{tr} is the height of the rim of the transient cavity (the cavity that opens just after the impact, before it collapses to become either a simple crater or a complex crater), D_{tr} is the diameter of the transient cavity crater at the pre-impact ground surface and r is the distance from the center. The power law in this equation can vary between 2.5 and 3.5: 3 is an average that gives roughly correct results in most cases. For further information on crater type see Melosh (1989, p 90).

The visible ejecta deposit around lunar craters is called the "continuous ejecta blanket". Beyond the edge of this blanket are often seen fields of small secondary craters that indicate some material is launched at higher speed and lands still farther from the impact site. A surprising observation is that a lunar-like continuous ejecta blanket is apparent around impact craters even on the planet Venus, which possesses an atmosphere 100 times denser than the Earth's. The reason for this apparent indifference to the presence of an atmosphere was revealed by numerical simulations of impacts on the surface of Venus (Ivanov et al. 1992). These simulation show that the hot fireball of vaporized rock that expands out of the crater rapidly pushes back the ambient atmosphere and, for a time sufficient for the deposition of the nearby ejecta, the impact site is surrounded by an attenuated atmosphere of very hot, low density gas that permits the nearby ejecta to travel as it if were moving in a vacuum. The time required for this process is relatively short, a few times

$$\sqrt{\frac{D_{tr}}{g}}$$

where g is the surface acceleration of gravity on the Earth. This is about 10 seconds for a 1 km diameter crater and 100 seconds for a Chicxulub-sized event.

The fate of ejecta traveling fast enough to fly beyond the low-density fireball region depends on the energy release in the impact. The fireball that forms near impacts that release less than about 200 MT rises buoyantly, after it equilibrates with the surrounding atmosphere, blocking the flight of fast-moving ejecta and drawing most of the ejecta particles and dust upward with it. This material later rains out downwind of the impact site, similar to the observed deposition of the ash from volcanic eruptions. Although distal ejecta from small craters is, thus, blocked by the atmosphere, impacts that release more than about 200 MT create a fireball that pushes out of the atmosphere (Jones and Kodis 1982), accelerating as it rises and eventually ejecting debris well above the atmosphere itself. Although never observed directly on the Earth, this process must have acted to permit tektites to travel the observed thousands of kilometers from their source craters (Taylor 1973). Moreover, a model based on this process likely accounts of the formation of radar-dark parabolas surrounding impact craters on Venus (Vervack and Melosh 1992; Schaller and Melosh 1998). These Venusian studies also permit estimates of the mean fragment size of ejecta, as a function of the distance from the crater and crater diameter. These estimates can also be used for terrestrial craters as well (Collins et al. 2005).

Assuming that ejecta deposited at ranges of more than a few tens of kilometers (equivalent to a few atmospheric scale heights, which on the Earth is about 8 km) from an impact travels ballistically over most of its path, the time of arrival of ejecta T_{fl} at a range r from an impact is approximately

$$T_{fl} = \sqrt{\frac{r}{2g}\tan\Theta}$$

where Θ is the angle of ejection, approximately 45° for most solid ejecta. This equation ignores the curvature of the Earth. At great ranges, where this is important, a much more complex form of this equation must be used (Collins et al. 2005).

In the case of large impacts, even beyond the range of continuous ejecta, the ejecta deposit thickness may be sufficient to cause damage to human beings and structures. Note that at large distances, and small ejecta fragments, the atmosphere plays an important role in the ultimate deposition of the particles, a role that will be discussed under the heading of Global effects.

12.2.4
Airblast

The atmosphere in the neighborhood of a large impact is greatly disturbed by the expansion of the fireball and ejecta plume. The sudden displacement of the air near the impact produces strong shock waves that compress and heat the air. As these shocks expand away from the impact site, they eventually decay to sound waves that continue to weaken with distance from the impact. These waves are analogous to a sonic boom or thunder produced by similar rapid disturbances of the air by supersonic aircraft or lightening. At short ranges such airblasts can be very destructive, collapsing buildings, bridges and overturning cars and trucks. The strength of such waves is measured by the overpressure, the excess of pressure in the wave compared to the ambient atmosphere. Buildings and glass windows, in particular, are surprisingly vulnerable to small overpressures. An overpressure of only 0.3 atm is sufficient to collapse a steel-framed structure and 0.004 atm to shatter glass windows. In contrast, an unprotected human being can withstand overpressures of 1 atm without serious harm and 5 atm before death is likely. Most injuries due to airblast are from flying debris.

Since the airblast is considered one of the most destructive aspects of nuclear explosions, it has been studied in great detail in the nuclear weapons effects literature (Glasstone and Dolan 1977). Extensive tables have been published giving the rate at which the airblast declines with distance from a nuclear explosion, as a function of explosion energy and height of burst (generally, impacts correspond to a surface burst, unless the impacting object is small enough that it disintegrates in the atmosphere, as did the 1908 Tunguska object). Although these tables probably exaggerate the strength of the airblast for large energy releases, because they ignore the finite thickness of the atmosphere and curvature of the Earth, they probably give a good first estimate of the airblast effect of impacts as well as nuclear explosions.

The effects of a given overpressure on a variety of structures was measured directly in above ground nuclear testing before 1962, and for high explosive tests in subsequent years. Extensive tables of structural response to various overpressures exist and can be directly applied to the impact hazard (Glasstone and Dolan 1977). Depending on the strength of the wave, this response ranges from a barely perceived sound to total collapse of reinforced concrete structures. This information is incorporated on our Web-based impact effects calculator (Collins et al. 2005).

A strong airblast travels faster than the speed of sound. However, such strong waves weaken rapidly with distance from the impact site and travel close to the speed of sound, 300 m s^{-1}, over most of their path outside the near vicinity of the fireball itself. The airblast is, thus, usually the last of the destructive consequences of an impact to arrive at a given distance from the crater.

12.2.5
Tsunamis from Oceanic Impacts

Since 3/4 of the Earth's surface is underwater, oceanic impacts are 3 times more likely than impacts on land and the subject of impact-generated tsunami from oceanic impacts invariably arises in any discussion of impact effects. Explosion-generated waves were clearly observed breaking on the beaches of Bikini Atoll in the aftermath of the 20 kiloton BAKER nuclear test (Glasstone and Dolan 1977). Because of its possible strategic value, waves generated by explosions at or below the sea surface have received considerable attention, some of which appears in the unclassified literature (LeMéhauté 1971). A much-cited paper by Hills (Hills et al. 1994) and more recent papers by Ward and Asphaug (Ward and Asphaug 2000, 2003) emphasize the potential importance of impact tsunami.

Unfortunately, the true importance of impact tsunami has been obscured by an apparent oversight in these latter papers (Melosh 2003). The most spectacular effects emphasized in all three papers involve waves that, at the outset of their propagation, have amplitudes that exceed the depth of the ocean itself, which is impossible. These gigantic waves arise from the uncritical use of a linear approximation to the true tsunami propagation equations. In reality, such waves would break near the impact site and dissipate an unknown (but probably large) amount of their energy in turbulence. In addition, a cold-war report that recently surfaced (Van Dorn et al. 1968) suggests that explosion generated (and impact) tsunami have periods such that they would break on the continental shelves and may, thus, pose little threat to shoreline installations, although coastal shipping would be at risk. This is often referred to as the "Van Dorn Effect" and played an important role in the Congressional decision to base MX missiles on land rather than in offshore mini-submarines (Van Dorn's report was actually read into the Congressional record during the debate). Unfortunately, the unclassified portion of the Van Dorn report does not contain enough information to fully support his claim, and the situation on the hazard from impact-generated tsunami remains murky, although very recent work seems to vindicate Van Dorn's analysis (Korycansky and Lynett 2005).

12.3
Global Devastation?

Of the local and regional effects described in the last section, the most far ranging is probably seismic shaking: Ejecta deposit thickness and airblast decay much more rapidly with distance than seismic ground motion. Most localities are shielded from direct thermal radiation from the fireball by the Earth's curvature. Although the 15 km diameter asteroid impact that created the Chicxulub crater 65 million years ago evidently initiated huge landslides over a large fraction of the North American continent, these effects were probably not themselves capable of initiating the major biological extinctions observed in the geologic record.

At the scale of the Chicxulub impact a new set of phenomena with global consequences becomes important. These effects are, in order of their operation after an impact: (1) The thermal pulse from ejecta rain back, (2) dust loading of the atmosphere, (3) injection of climatically active gases, and (4) indirect effects of biological extinctions. All of these have been discussed extensively in Toon et al. (1997), so only a short description and updates are presented here. This section focuses almost exclusively on the Chicxulub impact because it is the only large impact known to have affected the world's biota. Although other large impacts have occurred since the Cretaceous-Tertiary event, such as the 35 Myr old, 100 km diameter, Popagai impact crater in Siberia, none are associated with extinctions.

12.3.1
The Thermal Pulse from Ejecta Rain Back

The most immediate global consequence of a very large impact is ejecta rain back, lasting for a few hours after the impact. Ejecta particles condensed from the melt and vapor plume reenter the atmosphere all over the Earth and release a vast amount of energy as heat in the upper atmosphere. This heat was, in the case of Chicxulub, intense enough to ignite global wildfires (Wolbach et al. 1988) and directly scorch unprotected animals (Melosh et al. 1990). Indeed, the pattern of survival of land animal populations is in good agreement with the supposition that intense thermal radiation was the first lethal punch from this impact (Robertson et al. 2004). Aside from the theoretical computations cited above, an analogous thermal pulse was directly observed during the impact of SL/9 comet fragments on Jupiter in July 1994. Current estimates suggest that 30 to 50% of the total kinetic energy of these fragments was later emitted as thermal infrared radiation over an area comparable in size to the area of the Earth (Zahnle and MacLow 1994; Zahnle and MacLow 1995). Although this incendiary effect was important for the Chicxulub impact, it appears that Chicxulub is just at the threshold at which it becomes important. Smaller impacts are probably not capable of igniting global wildfires.

An important aspect of this mechanism is the amount of mass ejected at a given velocity. Most of the very fast ejecta that travels on Earth-spanning ballistic trajectories is generated by vaporization of the projectile and a comparable mass of the target, as attested by the high concentration of the siderophile element iridium in the "fireball layer" component of the ejecta deposit (Smit 1999). To date, no good computations of the mass-velocity relation for this portion of the ejecta have been made. Melosh et al. (1990) assumed that the fireball expands as a sphere of hot gas, following a model of Zel'dovich and Raiser (1967). Alvarez et al. (1995) supposed the ejecta was emplaced by "hot" and "warm" ejecta plumes, but did not present detailed computations. Kring and Durda (2002) proposed a model in which the mass of the ejected material increases as the 3rd or even 5th power of the ejection velocity. This relation is not supported by any observed distribution or computation. Although many existing numerical hydrocodes are capable of estimating the mass and velocity of this fraction of the impact ejecta, the problem in the past has been the lack of a reliable equation of state for rock materials (Melosh and Pierazzo 1997). This lack has recently been addressed (Melosh 2000) and new computations can be expected soon.

12.3.2
Dust Loading of the Atmosphere

Since the Alvarez's first paper on the K-T extinction (1980), dust loading of the atmosphere in the wake of a large impact has been a favorite extinction mechanism. Indeed, in the popular press one usually hears about no other mechanism. Although large dust particles quickly settle out of the atmosphere, submicrometer dust can remain suspended for years. This dust may block solar radiation from reaching the surface, leading to extended periods of sub-freezing temperatures and the death of photosynthetic plants. Toon et al. (1997) made a fine state-of-the-art attempt at estimating the mass of submicrometer dust raised by the Chicxulub impact, and their results seem to confirm that, at a Chicxulub scale, dust may cause serious climatic changes for decades, subsequent to the impact (Luder et al. 2002). However, there are few data on which these estimates can be based and, indeed, there is no direct evidence for any dust at all in the K-T ejecta deposits (Pope 2002).

A further great uncertainty is the rate at which upper atmospheric dust coagulates and is rained out.

Much of the support for the "darkness at noon" extinction scenario comes from consideration of the "nuclear winter" scenario (Turco et al. 1983) that was, in fact, inspired by the Alvarez et al. discovery. Although this scenario is still, thankfully, mostly hypothetical, a small-scale equivalent in the form of the Kuwait oil fires demonstrated that the soot rapidly rained out of the lower troposphere and did not produce the global climatic effects expected (Pilewskie and Valero 1992). The amount and effect of submicrometer dust raised by an impact is thus highly uncertain at present. This is clearly an area needing clarification and further research.

12.3.3
Injection of Climatically Active Gases

In addition to lofting a putative dust cloud, the Chicxulub impact vaporized large masses of sulfur-rich sediments (Brett 1992; Sigurdsson et al. 1992), which may subsequently have condensed as H_2SO_4 aerosols in the upper atmosphere (Toon et al. 1997). This would have caused surface temperatures to plummet for several years (Pierazzo et al. 2003) and initiated an episode of intense acid rain as the aerosol filtered down and washed out in tropospheric rains (Retallack 1996; Sigurdsson et al. 1992). Acidification of the upper ocean waters presently seems to be the only agent capable of explaining the extensive marine, as opposed to terrestrial, extinctions.

Sulfur-rich target rocks are not common on Earth. Only about 5% of the Earth's surface is underlain by large accumulations of sulfur-bearing sediments. It may be that the Chicxulub impact event caused such widespread extinctions because it struck an unusually lethal type of target rock. A similar size impact striking anywhere else might not have had the same profound influence on the biosphere.

Carbon dioxide has also been frequently blamed for the climatic excursions at the end of the Cretaceous era (Pierazzo et al. 1998). However, the warm temperatures prevalent at that time suggest that the carbon dioxide content of the atmosphere was approximately four times that at present, so the amount released by vaporization of the carbonate-rich target rocks would not have greatly enhanced the eisting background abundance. There is also great present uncertainty in how much carbon dioxide is released during a shock event (Ivanov et al. 2002). At the moment, carbon dioxide release no longer seems like a good candidate for major impact-induced climatic changes.

Several other noxious additions to the atmosphere have been considered (Toon et al. 1997). Impact heating of our N_2 and O_2 rich atmosphere produces NO_x gases that may destroy the ozone layer and increase the amount of UV radiation reaching the surface. Furthermore, upon reaction with water vapor, NO_x creates nitric acid and leads to acid rain. Water vapor deposited directly into the otherwise dry stratosphere may also have climatic consequences, because water is an excellent greenhouse gas. Fires create pyrotoxins that act as poisons. Finally, it appears that impacts may cause ozone depletion, opening the atmosphere to enhanced UV radiation (Birks et al. 2006).

A new contender for a global impact disturbance is methane. Methane clathrates underlie a large area of the sediments on continental shelves. If disturbed by large submarine landslides (themselves initiated by seismic shaking), a large amount of methane might be suddenly released into the atmosphere, perhaps explaining the large excursions in carbon isotope ratios across the K-T boundary (Day and Maslin 2005).

12.3.4
Indirect Effects of Biological Extinctions

Although a large number of possible effects of a large impact have been investigated to date, none of the physical or chemical consequences of a large impact has been able to explain the much longer-term perturbations apparent in the geologic record (Smit 1999). Carbon and oxygen isotopic excursions apparently persisted for millennia after the impact event. The only plausible cause for these long-lasting effects is the biological extinctions themselves. The flourishing and diverse Cretaceous planktonic populations were replaced with an impoverished Paleogene population consisting of a comparatively homogeneous group of small, simple foraminifera. This population may have been unable to match the ability of the Cretaceous planktonic community in recycling carbon dioxide and other nutrients, leading to the observed long-lasting isotopic excursions. Only after evolution had time to fill the vacated ecological niches could the upper ocean return to its previous state of efficient recycling.

12.4
Conclusion

Large impacts occasionally disturb the course of Earth history. They have occurred in the past and will continue to occur at a low, but predictable, rate in the future. Analysis of the effects of a large impact shows that, although the consequences are frightful close to the event itself, they decline rapidly with distance. The most widespread harmful effect of an impact is probably seismic shaking. However, a major uncertainty is the role of dust in extending the deleterious effects from a regional to a global scale. Volcanic eruptions, such as that of the 1883 Krakatau or the 1783 Icelandic Laki fissure eruption, are known to have caused acid hazes and year-long drops in temperature, resulting in crop failures and human starvation (Francis 1993). Most of these effects seem to be caused by sulfur-rich aerosols, with long atmospheric residence times. Unless an impact, by bad chance, strikes a sulfur-rich target rock, the global effects might not be comparable to those of large volcanic eruptions.

It has often been assumed that the impact of a kilometer-scale asteroid or comet will cause global disruption of our delicately balanced modern civilization (Toon et al. 1997). This figure relies heavily on the assumption that dust ejected from an impact will cause global climatic changes leading to global crop failures. This may or may not be true. The role of dust in impacts is one of the most poorly constrained of all impact effects. Not only is the amount of dust raised by an impact uncertain, the residence time in the atmosphere is poorly constrained, especially for dust injected to very high altitudes. This is clearly an area needing further research, if we are to fully understand the consequences of large impacts on the Earth.

References

Alvarez LW, Alvarez W, Asaro F, Michel HV (1980) Extraterrestrial cause for the Cretaceous-Tertiary extinction. Science 208:1095–1108

Alvarez W (1986) Toward a theory of impact crises. EOS 67:653–655

Alvarez W, Claeys P, Kieffer SW (1995) Emplacement of Cretaceous-Tertiary boundary shocked quartz from Chicxulub crater. Science 269:930–935

Birks JW, Crutzen PJ, Roble GG (2007) Frequent ozone depletion resulting from impacts of asteroids and comets. Chapter 13 of this volume

Bland PA, Artemieva NA (2003) Efficient disruption of small asteroids by Earth's atmosphere. Nature 424:288–291

Bralower TJ, Paull CK, Leckie RM (1998) The Cretaceous-Tertiary boundary cocktail: Chicxulub impact triggers margin collapse and extensive sediment gravity flows. Geology 26:331–334

Brett R (1992) The Cretaceous-Tertiary extinction: a lethal mechanism involving anhydrite target rocks. Geochem Cosmochim Acta 56:3603–3606

Busby CJ, Yip G, Blikra L, Renne P (2002) Coastal landsliding and catastrophic sedimentation triggered by Cretaceous-Tertiary bolide impact: a Pacific margin example? Geology 30:687–690

Chyba CF, Thomas PJ, Zahnle KJ (1993) The 1908 Tunguska explosion: atmospheric disruption of a stony asteroid. Nature 361:40–44

Collins GS, Melosh HJ, Marcus RA (2005) Earth Impact Effects Program: a web-based computer program for calculating the regional envionmental consequences of a meteoroid impact on Earth. Meteoritics and Planet Sci 40: 817–840

Chapman CR, Morrison D (1994) Impacts on the Earth by asteroids and comets: assessing the hazard. Nature 367:33–39

Day S, Maslin M (2005), Widespread sediment liquefaction and continental slope failure at the K-T boundary: the link between large impacts, gas hydrates and carbon isotope excursions. In: Kenkmann T, Hörz F, Deutsch A (eds) Large meteorite impacts III. Geol Soc Amer Special Paper 384, pp 239–258

Francis P (1993) Volcanoes: a planetary perspective. Oxford Univ Press, Oxford

Glasstone S, Dolan PJ (ed) (1977) Effects of nuclear weapons. United States Departments of Defense and Energy

Gomberg J, Bodin P, Larson K, Dragert H (2004) Earthquake nucleation by transient deformations caused by the $M = 7.9$ Denali, Alaska, earthquake. Natg 427:621–624

Grieve R, Therriault A (2000) Vredefort, Sudbury, Chicxulub: three of a kind? Ann Rev Earth Planet Sci 28:305–338

Hills JG, Nemchinov IV, Popov SP, Teterev AV (1994) Tsuanmi generated by small asteroid impacts. In: Geherls T (ed) Hazards from comets and asteroids. Univ of Arizona Press, Tucson, AZ, pp 779–789

Huntoon PW, Shoemaker EM (1995) Roberts Rift, Canyonlands, Utah, a natural hydraulic fracture caused by comet or asteroid impact. Ground Water 33:561–569

Husen S, Taylor R, Smith RB, Healser H (2004) Changes in geyser eruption behavior and remotely triggered seismicity in Yellowstone National Park produced by the 2002 M 7.9 Denali fault earthquake, Alaska. Geology 32:537–540

Ivanov BA, Langenhorst F, Deutsch A, Hornemann U (2002) How strong was impact-induced CO_2 degassing in the Cretaceous-Tertiary event? Numerical modeling of shock recovery experiments. In: Koeberl C, MacLeod KG (ed) Catastrophic events and mass extinctions: impacts and beyond. Geological Society of America Special Paper 356, pp 587–594

Ivanov BA, Nemchinov IV, Svetsov VA, Provalov AA, Khazins VM, Phillips RJ (1992) Impact cratering on Venus: physical and mechanical models. J Geophys Res 97:16167–16181

Jones EM, Kodis JW (1982) Atmospheric effects of large body impacts: the first few minutes. In: Silver LT, Schultz PH (ed) Geological implications of impacts of large asteroids and comets on the Earth. Geol Soc Amer Sp Pap 190:175–186

Klaus A, Norris RD, Kroon D, Smit J (2000) Impact-induced mass wasting at the K-T boundary: Blake Nose, western North Atlantic. Geology 28:319–322

Korycansky DG, Lynett PJ (2005) Offshore breaking of impact tsunami: the van Dorn effect revisited. Geophys Res Lett 33: DOI:10.1029/2004GL021918

Kring DA (1997) Air blast produced by the Meteor Crater impact event and a reconstruction of the affected environment. Meteoritics and Planet Sci 32:517–530

Kring DA, Durda DD (2002) Trajectories and distribution of material ejected from the Chicxulub impact crater: implications for postimpact wildfires. J Geophy Res 107(6):1–22

Krinov EL (1966) Giant Meteorites. Pergamon Press

LeMéhauté B (1971) Theory of explosion-generated water waves. In: Chow VT (ed) Advances in Hydroscience 7:1–79. Academic Press, New York and London

Luder T, Benz W, Stocker TF (2002) Modeling long-term climatic effects of impacts: First results. In: Koeberl C, MacLeod KG (ed) Catastrophic Events and Mass Extinctions: Impacts and Beyond, vol Special Paper 356:717–729. Geological Society of America, Boulder

Melosh HJ (1989) Impact Cratering: A Geologic Process. Oxford University Press, New York

Melosh HJ (2000) A new and improved equation of state for impact studies. In: 31st LPSC, Abstract #1903, Lunar and Planetary Institute, Houston (CD-ROM)

Melosh HJ (2003) Impact tsunami: an over-rated hazard, LPSC XXXIV, Abstract #1338

Melosh HJ, Pierazzo E (1997) Impact vapor plume expansion with realistic geometry and equation of state, LPSC XXVIII, pp 935–936

Melosh HJ, Schneider NM, Zahnle KJ, Latham D (1990) Ignition of global wildfires at the Cretaceous/Tertiary boundary. Nature 6255:251–254

Nemchinov IV, Svetsov VV (1991) Global consequences of radiation impulse caused by comet impact. Adv Space Res 112:95–97

Nemtchinov IV, Shuvalov VV, Artem'eva NA, Ivanov BA, Kosarev IB, Trubetskaya IA (1998) Light flashes caused by meteoroid impacts on the lunar surface. Solar System Research 32:99–114

Nininger HH (1956) Arizona's Meteorite Crater. American Meteorite Laboratory, Denver, Colo

Norris RD, Firth J, Blusztajn JS, Ravizza G (2000) Mass failure of the North Atlantic margin triggered by the Cretaceous-Paleogene bolide impact. Geology 28:1119–1122

Pierazzo E, Hahmann AN, Sloan LC (2003) Chicxulub and climate: radiative perturbations of impact-produced S-bearing gases. Astrobiology 3:99–118

Pierazzo E, Kring DA, Melosh HJ (1998) Hydrocode simulation of the Chicxulub impact event and the production of climatically active gases. J Geophys Res 103:28607–28625

Pilewskie P, Valero FPJ (1992) Radiative effects of the smoke clouds from the Kuwait oil fires. J Geophy Res 97:14541–14544

Pope KO (2002) Impact dust not the cause of the Cretaceous-Tertiary mass extinction. Geology 30:99–102

Retallack GJ (1996) Acid trauma at the Cretaceous-Tertiary boundary in eastern Montana. GSA Today 6:1–7

Richter CF (1958) Elementary Seismology. Freeman, San Francisco and London

Robertson DS, McKenna MC, Toon OB, Hope S, Lillegraven JA (2004) Survival in the first hours of the Cenozoic. Geol Soc Amer Bull 116:760–768

Schaller CJ, Melosh HJ (1998) Venusian ejecta parabolas: comparing theory with observation. Icarus 131:123–137

Schultz P, Gault DE (1975) Seismic effects from major basin formation on the Moon and Mercury. The Moon 12:159–177

Shoemaker EM (1963) Impact mechanics at Meteor Crater, Arizona. In: Middlehurst BM, Kuiper GP (ed) The Moon, meteorites and comets. The Solar System, 4:301–336. University of Chicago Press, Chicago, Ill

Sigurdsson H, D'Hondt S, Carey S (1992) The impact of the Cretaceous/Tertiary bolide on evaporite terrane and generation of major sulfuric acid aerosol. Earth and Planetary Science Letters 109:543–559

Smit J (1999) The global stratigraphy of the Cretaceous-Tertiary boundary impact ejecta Ann Rev Earth Planet Sci 27:75–113

Stark MA, Davis SD (1996) Remotely triggered microearthquakes at The Geysers geothermal field, California. Geophys Res Lett 23:945–948

Taylor SR (1973) Tektites: a post-Apollo view. Earth Sci Rev 9:101–123

Team N-EOSD (2003) Study to determine the feasibility of extending the search for near-Earth objects to smaller limiting diameters, NASA

Terry DO, Chamberlain JA, Stoffer PW, Messina P, Jannett PA (2001) Marine Cretaceous-Tertiary boundary section in southwestern South Dakota. Geology 29:1055–1058

Toon OB, Zahnle K, Morrison D, Turco RP, Covey C (1997) Environmental perturbations caused by the impacts of asteroids and comets. Rev Geophys 35:41–78

Turco RP, Toon OB, Ackerman TP, Pollack JB, Sagan C (1983) Nuclear winter: global consequences of multiple nuclear explosions. Science 222:1283–1292.

Van Dorn WG, LeMéhauté B, Hwang L-S (1968) Handbook of explosion-generated water waves, vol I: state of the art. TC-130, Final Report, 1968, Tetra Tech

Vervack R, Melosh HJ (1992) Wind interaction with falling ejecta: origin of the parabolic features on Venus. Geophys Res Lett 19:525–528

Ward SN, Asphaug E (2000) Asteroid impact tsunami: a probabilistic hazard assessment. Icarus 145: 64–78

Ward SN, Asphaug E (2003) Asteroid impact tsunami of 2880 March 16. Geophys J. Int. 153:F6–F10.

Wolbach WS, Gilmour I, Anders E, Orth CJ, Brooks RR (1988) Global fire at the Cretaceous-Tertiary boundary. Nature 334:665–669

Zahnle K, MacLow M-M (1994) The collision of Jupiter and comet Shoemaker-Levy 9. Icarus 108:1–17

Zahnle K, MacLow M-M (1995) A simple model for the light curve generated by a Shoemaker-Levy 9 impact. J Geophy Res 100(16):885–16, 894

Zel'dovich YB, Raizer YP (1967) The physics of shock waves and high temperature hydrodynamic phenomena. Academic, New York

Chapter 13

Frequent Ozone Depletion Resulting from Impacts of Asteroids and Comets

John W. Birks · Paul J. Crutzen · Raymond G. Roble

13.1
Introduction

The fossil record reveals that the evolution of life on Earth has been punctuated by a number of catastrophic events, of which one of the most devastating occurred at the end of the Cretaceous, approximately 66 million years ago. The postulate introduced in 1980 by Alvarez et al. (1980) that the collision of an approximately 10 km diameter asteroid with the Earth caused the extinction of the dinosaurs along with more than half of all plant and animal species has resulted in a greatly expanded research efforts in the area of catastrophic events (Alvarez et al. 1980).

Large events such as the K-T impact, which may have baked plants and animals at the surface of the Earth with thermal radiation from re-entering ejecta (Robertson et al. 2004), injected enough dust (Toon et al. 1982) and/or sulfate aerosol (Pope et al. 1997) into the atmosphere to block most of the incoming solar radiation for months to years, produced enough nitric acid (Lewis et al. 1982; Prinn and Fegley 1987) and sulfuric acid (Pope et al. 1997) to reduce the pH of rainfall to phytotoxic levels, and/or injected enough carbon dioxide into the atmosphere to cause a global warming (O'Keefe and Ahrens 1989) are estimated to occur with extremely low frequency – perhaps once every hundred million years or longer. Although not as devastating as the K-T impact, impacts of much smaller objects, which occur much more frequently, could have serious consequences for humanity and the ecosystems on which we depend. Here we postulate that the method of harming the global biosphere with the least amount of impact energy is to deplete the stratospheric ozone layer, thereby allowing enhanced levels of UV radiation to reach the Earth's surface. Both terrestrial and aquatic plants, at the base of the food chain, are highly sensitive to UV radiation (Nachtway et al. 1975). Model calculations presented here indicate large stratospheric ozone depletions occurring as often as once every few tens of thousands of years.

13.2
Physical Interactions with the Atmosphere

Here we consider the impacts of asteroids having diameters ≥ 150 m. Objects of this size pass through the atmosphere with only minimal loss of energy (approximately

5.8% for a 150 m stony meteoroid) to the atmospheric shock wave that they produce (Melosh 1989). The energy of impact, $1/2\,mv^2$, is given by

$$E = \frac{1}{6}\pi\rho D^3 v^2 \tag{E1}$$

where D is the diameter of the asteroid and ρ is its density. The minimum impact velocity, v, is the Earth's escape velocity, 11.2 km s^{-1}, whereas the maximum velocity is 72.8 km s^{-1}, corresponding to the sum of the Earth's escape velocity, orbital velocity about the Sun and the velocity of an object just barely bound to the sun at the Earth's orbital position (Melosh 1989). For all calculations made below, the impact velocity is assumed to be 17.8 km s^{-1}, the mean impact velocity with the Earth, and the density is assumed to be 2500 kg m^{-3}, characteristic of stony meteoroids. The amounts of energy released in the impact, $1/2\,mv^2$, are 400 MT, 2100 MT and 6200 MT for 200-m, 350-m and 500-m diameter asteroids, respectively. Upon impact, this energy will be released in the form of a strong shock wave that heats the surrounding medium to temperatures of a few tens of thousands of degrees, producing in the atmosphere what in military parlance is termed a "fireball."

The hemispherical fireball expands until its internal pressure matches that of the surrounding atmosphere. For relatively small impacts, the radius of the fireball, R_f, may be calculated assuming adiabatic expansion as (Melosh 1989)

$$R_f = \sqrt[3]{\frac{\left(\frac{3V_i}{2\gamma}\right)}{\left(\frac{P_i}{P_0}\right)^{1/\gamma}}} \tag{E2}$$

where P_i and V_i are the initial pressure and volume of the gas, P_0 is the pressure after expansion, and γ is the ratio of heat capacities, C_p/C_v, and is approximately 1.4 for air. This simplifies to

$$R_f = 0.009\sqrt[3]{E_d} \tag{E3}$$

when E_d, the energy deposited in the fireball, is known (Melosh 1989). Calculations using this equation with the assumption that one-half the impact energy is deposited in the fireball predicts radii of 8.5 km, 14.8 km and 21.1 km for 200 m, 350 m and 500 m diameter meteoroids, respectively. For meteoroids slightly larger than 165 m in diameter, the calculated radius of the initial fireball is greater than the scale height (≈ 7 km) of the atmosphere. Under such conditions, Eqs. E2 and E3 no longer apply; instead, detailed computer models indicate that the phenomenon of "blowout" or "backfire" will occur in which the hot fireball of vaporized material rises rapidly, partially funneled by the "vacuum straw" formed during passage through the atmosphere, and spills into the near vacuum of the mesosphere and above (Melosh 1982, 1989; Jones and Kodis 1982). Under such conditions, the contents of the fireball is expected to be distributed nearly uniformly across the globe.

13.3
Chemical Perturbations of the Upper Atmosphere

Upon impact of a meteoroid with the solid Earth or its oceans, stratospheric ozone depletion could result from the injection of large quantities of (*1*) nitric oxide produced in the shock-heated air, (*2*) water vaporized and injected high into the upper atmosphere, and (*3*) halogens, especially chlorine, chemically activated from sea salt contained in vaporized seawater. Our calculations show that all three effects may be important on time scales of ≈ 60 000–100 000 years, although the effect of halogens, which are the most effective catalysts for ozone destruction, is speculative and requires further laboratory data and modeling to be confirmed. The idea that halogens, especially chlorine and bromine produced in the vaporization of the bolide and the lithosphere (seawater, sediments and the granitic crust) might be produced in sufficient quantity to cause ozone stratospheric ozone depletion was suggested earlier by Kring et al. (1995) and Kring (1999).

13.3.1
Nitric Oxide Production

In air heated to high temperatures, nitric oxide is in equilibrium with N_2 and O_2:

$$N_2 + O_2 = 2NO \tag{R1}$$

This equilibrium is rapidly established at the initially high temperature of a few thousand degrees in the shock wave as the bolide enters the atmosphere, in the shock-heated air produced by the high-velocity ejecta plume, and within the fireball itself. The equilibrium given by the stoichiometric reaction R1 is maintained by the following forward and reverse elementary chemical reactions:

$$O_2 + M = O + O + M \tag{R2}$$

$$O + N_2 = NO + N \tag{R3}$$

$$N + O_2 = NO + O \tag{R4}$$

$$NO + M = N + O + M \tag{R5}$$

where *M* represents any air molecule. These reactions have large activation energies, with the result that as air cools a "freeze out" temperature is reached where the time constant associated with maintaining equilibrium becomes long in comparison to the cooling rate. In the production of NO within the fireballs of nuclear explosions, for example, the cooling time is of the order of a few seconds, and the freeze out temperature is about 2000 K (Foley and Ruderman 1973; Johnston et al. 1973; Gilmore 1975). At this temperature approximately 0.7% of the air molecules are present as NO. For lightning discharges, the cooling time of the shock wave is of the order of 2.5 ms, and the freeze out temperature is about 2660 K where the equilibrium mole fraction of NO is about 2.9% (Chameides 1986).

The problem of nitric oxide production upon impact of a meteoroid or comet with the Earth has been treated by Prinn and Fegley (1987). They estimate NO production from two terms as follows:

$$P = [\varepsilon_1 Y_1 + (1 - \varepsilon_1)\varepsilon_2 Y_2] E_k \tag{E4}$$

where ε_1 is the fraction of kinetic energy of the bolide, E_k, transferred to the atmospheric shock wave during passage through the atmosphere and ε_2 is the fraction of energy coupled to the atmosphere by the impact fireball. Y_1 and Y_2 are yields of NO for the two processes in molecules per Joule. The value of ε_1 is estimated from the energy required to accelerate laterally the intercepted mass of air to the incoming bolide velocity and is given by

$$\varepsilon_1 = \frac{\pi D^2 P_s}{4 m g \cos\phi} \tag{E5}$$

where m is the bolide mass, P_s is the surface pressure of one atmosphere, and ϕ is the angle of impact, assumed here to be 45°. The calculated value of ε_1 varies from 0.058 to 0.018 for 150 m and 500 m diameter meteoroids, respectively. A production factor, $Y_1 = 1 \times 10^{17}$ molec J^{-1}, characteristic of NO production by lightning (Chameides 1986), was adopted by Prinn and Fegley (1987) for NO production during passage through the atmosphere. Based partly on the work of Emiliani et al. (1981), Prinn and Fegley (1987) estimate ε_2, the fraction of impact energy coupled to the atmosphere, to be 0.125. For Y_2, they assumed a value of 2×10^{16} molec J^{-1} as characteristic of NO production in nuclear explosions. In a careful analysis, Gilmore (1975) estimates a central value of 2.4×10^{16} molec J^{-1} for NO production in air bursts of nuclear weapons with an uncertainty of ±50%, and this is the production factor most commonly adopted for studies of the effects of nuclear warfare on the ozone layer (NRC 1975; Whitten et al. 1975; Crutzen and Birks 1975; Turco et al. 1983; NRC 1985; Pittock et al. 1985). An important difference, however, is that nuclear fireballs lose about one-third of their energy to radiative emission whereas impact fireballs do not. For this reason, we assume a higher but still conservative value of 3.6×10^{16} molec J^{-1} for Y_2.

For the range of bolide diameters and energies considered here, the two terms in Eq. E4 make comparable contributions to the amount of NO produced. However, as discussed above, the radius of the impact fireball is larger than the scale height of the atmosphere, and as a result nearly all of the NO produced during entry through the atmosphere will be raised to high temperature again and brought back into thermal equilibrium with N_2 and O_2. Therefore, the amount of NO ultimately produced is solely determined by shock-wave heating following the impact. For this reason, we consider only the second term in Eq. E4 for production of NO (i.e. we set $Y_1 = 0$), while adopting the higher emission factor for Y_2 with its associated uncertainty of ±50%. Thus, our NO emission factor is 3.6×10^{16} molec J^{-1} or 1.5×10^{32} molec MT^{-1} with $\varepsilon_2 = 0.125$. In past work, no differentiation has been made for NO production during land vs. ocean impacts. There is strong shock heating of the atmosphere in both cases, and we assume identical emission factors. However, additional theoretical studies are required to better estimate NO emissions for both types of impacts.

13.3.2
Lofting of Water

An enormous amount of liquid water is ejected into the atmosphere upon collision of a meteoroid with the oceans. However, the altitude of injection of liquid water is well below the tropopause and will be removed rapidly from the atmosphere as rain (Emiliani et al. 1981). Of great significance, however, is water vaporized by the impact. Croft (1982) estimated the amount of water vaporized in an ocean impact by means of the Gamma model of shock wave heating and vaporization. The Gamma model is semi-empirical in that some features are derived from fundamental physics, while others are generalizations of the results obtained by computer code calculations of shock wave propagation. Croft (1982) calculated the number of projectile volumes of liquid water that would be vaporized for impacts of gabbroic anorthosite ($\rho = 2936$ kg m^{-3}, a mineral simulating stony meteoroids) with the ocean in the velocity range 5–80 km s^{-1}. For an average impact velocity of 17.8 km s^{-1}, a third-order polynomial interpolation of their results predicts a total water vapor volume of 27.6–44.2 projectile volumes (for values of the semi-empirical parameter γ in the range 2.4–2.0). Since their assumed meteoroid density (2936 kg m^{-3}) is slightly larger than ours (2500 kg m^{-3}), we reduce the volume of vapor to 23.5–37.6 projectile volumes, which amounts to assuming that the mass of water vaporized varies linearly with impact energy for a given velocity. For impact velocities between 15 and 20 km s^{-1}, the Gamma model predicts that 36.5% of the water vapor is present in the isobaric core produced by the shock wave and the remaining 63.5% is intimately mixed with an approximately equal volume of liquid water at its boiling point in a region of "incipient" vaporisation. As discussed below, it is likely that all of the water contained in the region of incipient vaporization, both vapor and liquid, will be injected to altitudes well above the stratosphere, and we therefore take 38.4–61.5 projectile volumes as the amount of water vapor ultimately reaching the stratosphere and affecting the chemistry there.

The sea salt contained in the region of incipient vaporization will all be partitioned into the liquid phase. When this water evaporates in the upper atmosphere, salt particles having diameters of 100 µm or larger will be formed and will rapidly settle out of the atmosphere, as discussed below. Thus, for calculations of the amount of sea salt contributing to perturbations in stratospheric chemistry, we consider only the water vapor in the isobaric core, which amounts to 8.6–13.7 projectile volumes.

A small downward correction ($\approx$ 6% for 150 m meteoroids and decreasing with increasing bolide diameter) is made for all injection volumes to account for loss of kinetic energy during transit of the meteoroid or comet through the atmosphere. Gamma Model calculations indicate that ocean impacts of comets (ice of density 0.917) vaporize approximately the same amount of seawater as meteoroids of the same mass (Croft 1982).

As discussed above, for meteoroids having diameters greater than about 150 m (impact energies $\cong$ 165 MT) the atmosphere offers little resistance to the expanding vapor plume. The asymptotic limit ($t \rightarrow \infty$ for the vertical component of the root-mean-square velocity, v_∞, of the supersonic expansion of the impact "fireball" into the near vacuum above the stratosphere is given by (Zel'dovich and Raizer 1968)

$$v_\infty = \sqrt{\frac{2E_k}{m}} \cos\theta \qquad (E6)$$

where E_k is the kinetic energy of the material, m is its mass, and θ is the ejection angle with respect to normal. The mean altitude reached is given by

$$h = \frac{E_k}{mg} \cos\theta \qquad (E7)$$

For an average ejection angle of 45°, it requires partitioning of only 2% of the total impact energy into kinetic energy in order to loft the water vapor within the strongly shocked isobaric zone to an altitude of 66 km. Besides the initial kinetic energy, additional lofting can result from conversion of latent heat of vaporization to kinetic energy as the water vapor condenses. The latent heat within this strongly shocked isobaric zone amounts to $\approx 7\%$ of the total impact energy. The much less strongly shocked region of incipient vaporization is probably lofted to high altitudes as well. The mass of vapor and liquid is ≈ 4.5 times greater, so the maximum altitude reached is ≈ 4.5 times less for the same amount of kinetic energy. There is sufficient latent heat alone, however, to loft this mass of water to an altitude of ≈ 90 km. Numerical solutions of the shock wave equations for a wide range of kinetic energies and projectile and target compositions show that the vaporized bolide is typically injected well above the stratosphere and to altitudes of up to several hundred kilometers (O'Keefe and Ahrens 1982; Jones and Kodis 1982).

13.3.3
Fate of Salt Particles

Ozone depletion caused by halogens critically depends on the fate of the salt contained in the vaporized seawater. Sea salt and meteoroid vapor would be the first material to begin to condense as adiabatic expansion and work done against the Earth's gravitational field causes the temperature to decline in the rising fireball. The size distribution of salt particles produced is not possible to calculate but may be estimated from results of high-temperature combustion of materials containing salts or other low-vapor pressure substances. Condensation from the hot vapor results in primary particles, or Aitken nuclei, that is typically in the 0.005 to 0.1 µm diameter range (Finlayson-Pitts and Pitts 1986). Once the gas-phase molecules are consumed, these particles may grow further by agglomeration. If the particles grow to sizes of a few microns, they will rapidly settle out of the atmosphere and have no effect on stratospheric ozone. Rapid agglomeration of salt particles most likely are prevented, however, since those particles serve as condensation nuclei for water vapor in the near vacuum of the mesosphere and above where within a few minutes from the time of impact water will condense, probably directly to ice, onto the salt particles. The size distribution of the salt/ice particles is determined by the size distribution of the salt particles initially formed. Given 3.5 wt-% salt in seawater, and taking into account the relative densities of salt and ice, the salt/ice particles will have diameters ≈ 5 times as large as the salt condensation nuclei, i.e. diameters = 0.5 µm. The diameters may actually be less due to free fall of particles through the vapor cloud before condensation is complete. The time to free fall, $(2\Delta z/g)^{1/2}$, from 300 km to 150 km, for example, is only 175 s, at which point the velocity, $(2g\Delta z)^{1/2}$, of the particle is 1.7 km s^{-1}. At about 150 km, the particles will reach terminal velocity where the gravitational force is balanced by frictional drag, and as these salt/ice particles settle through denser regions of the atmosphere they will begin to evaporate due to frictional heating. Any water that remains will be completely removed as the particles pass through the ≈ 44–66 km region where the static pressure is less than the vapor pressure of water (List 1984). Considering the measured rate of sublimation of ice as a function of temperature (Haynes et al. 1992), the time for complete removal of water from the salt/ice particles is only a fraction of a second as particles settle through the ≈ 44–66 km region. These relatively dry, sub-micron salt particles will be transported primarily by vertical mixing through the 65–50 km region and into the upper stratosphere within a few days to weeks. Because of the "backfire" nature of the impact event, the particles will be fairly uniformly distributed in both zonal and meridional directions.

Water vapor also will be slowly transported downward to the stratosphere. A small fraction of the injected water will be photolyzed to form H and OH (and ultimately H_2 and O_2 through subsequent thermal reactions), but this fraction will be small since the characteristic time for vertical transport by eddy diffusion is shorter than the photolysis lifetime by about an order of magnitude at 65 km (Brasseur and Solomon 1984).

13.3.4
Activation of Halogens from Sea Salt Particles

Chloride, bromide and iodide ions contained in the sub-micron-sized salt particles settling through the stratosphere may be oxidized to form catalytically active species by several processes. Considering their relative abundances in seawater and relative catalytic efficiencies, only chlorine and bromine species will have a significant impact, with chlorine being by far the most important. Therefore, the following discussion will focus on chlorine even though bromide and iodide are much more easily oxidized. The amount of chloride converted to catalytically active species is limited by the NO_y concentration in the stratosphere, where NO_y is defined here as the sum $NO + NO_2 + NO_3 + 2 N_2O_5 + HNO_2 + HNO_3 + HNO_4 + ClONO_2$. The presence of these species, either directly or indirectly, will result in the oxidation of chloride, with some of the most important heterogeneous reactions being (Finlayson-Pitts et al. 1989; Timonen et al. 1994; Livingston and Finlayson-Pitts 1991; Zangmeister and Pemberton 1998):

$$ClONO_2 + NaCl(s) \longrightarrow Cl_2 + NaNO_3(s) \tag{R6}$$

$$N_2O_5 + NaCl(s) \longrightarrow ClNO_2 + NaNO_3(s) \tag{R7}$$

$$HNO_3 + NaCl(s) \longrightarrow HCl + NaNO_3(s) \tag{R8}$$

The key to "activation" of halide ions is replacement of the Cl$^-$ or other halide ions in the salt particle with another anion; for NO_y activation that anion is nitrate, NO_3^-.

Sufuric acid also activates chloride by substituting sulfate for the chloride ion (Finlayson-Pitts and Pitts, 1986):

$$H_2SO_4(l) + 2\,NaCl(s) \longrightarrow 2\,HCl + Na_2SO_4(s) \tag{R9}$$

In the absence of acids or acid gases (e.g., NO_y species), the halogens will remain in inactive salt forms. However, the natural stratosphere contains approximately 10 ppbv of NO_y (Brasseur and Solomon 1984) available for halogen activation. Small amounts of sulfuric acid aerosol are naturally present in the lower stratosphere with highly elevated levels during periods of volcanic activity (Junge et al. 1961; Brasseur et al. 1999). Because chlorine and other halogens catalyze ozone depletion at a much faster rate than oxides of nitrogen, replacement of the naturally occurring NO_y species with halogen species is expected to result in large ozone depletion. Thus, ozone depletion resulting from sea salt injection does not necessarily require additional input of NO_y from the asteroid impact. In fact, higher levels of ozone depletion might occur in the case where the amount of sea salt exceeds the NO_y content so that chlorine sequestered in the form of $ClONO_2$ becomes activated via reaction R6; a situation analogous to what occurs during formation of the Antarctic "ozone hole" where the polar stratosphere becomes "denitrified" (Brasseur et al. 1999).

13.3.5
Catalytic Cycles for Ozone Depletion

Catalytic destruction of ozone based on reactions involving hydrogen oxides (HO_x) derived from water (Bates and Nicolet 1950), nitrogen oxides (NO_x) derived from naturally occurring N_2O (Crutzen 1970) and potentially from aircraft engine exhaust (Crutzen 1970; Johnston 1971), and chlorine oxides (ClO_x) derived from chlorofluorocarbons (Molina and Rowland 1974) is well known. At mid-latitudes the most important reaction cycles for catalytic ozone destruction based on HO_x, NO_x and ClO_x are:

$$NO + O_3 \rightarrow NO_2 + O_2 \quad (R12) \qquad Cl + O_3 \rightarrow ClO + O_2 \quad (R15)$$

$$OH + O_3 \rightarrow HO_2 + O_2 \quad (R10) \qquad O_3 + h\nu \rightarrow O_2 + O \quad (R13) \qquad O_3 + h\nu \rightarrow O_2 + O \quad (R13)$$

$$HO_2 + O_3 \rightarrow OH + O_2 \quad (R11) \qquad NO_2 + O \rightarrow NO + O_2 \quad (R14) \qquad ClO + O \rightarrow Cl + O_2 \quad (R16)$$

Net: $2\,O_3 \rightarrow 3\,O_2$ Net: $2\,O_3 \rightarrow +3\,O_2$ Net: $2\,O_3 \rightarrow +3\,O_2$

Other catalytic cycles also contribute, especially in polar regions where it is well established that chlorine derived from CFCs currently results in a seasonal "ozone hole" with typically more than half of the ozone column depleted over Antarctica each austral spring. Because the catalytic species are regenerated in the cycles of reactions, part-per-billion levels of HO_x, NO_x and ClO_x can destroy ozone at three orders of magnitude higher concentration.

13.3.6
Estimates of Asteroid Impact and Ozone Depletion Frequency

As summarized by Chapman (2004), recent observations of near Earth objects has allowed improved estimates of their impact frequency as a function of size, with the size distribution and impact frequency roughly following a power law. Expressed as the period, the average time between impacts for asteroids having a diameter greater than or equal to D is estimated to be

$$t_{\text{impact}} = 3.71 \times 10^{-2} D^{2.377} \tag{E6}$$

This power law expression is a fit to the estimates that a $\geq$ 1-km diameter asteroid strikes the Earth on average once every 500 000 years and that a $\geq$ 4-m meteoroid strikes the Earth once annually (Chapman 2004). At small sizes, Eq. E6 is consistent with the observation by Nemtchinov et al. (1997) that approximately 25 exploding meteors strike the Earth each year in the energy range 0.25 to 4 kT TNT (0.75 m < R < 1.19 m for V = 20 km s^{-1} and density of 3 g cm^{-3}). Equation E6 predicts objects having diameters larger than 200, 350 and 500 m impact once every 10 900, 41 200 and 96 300 years, respectively. It should be noted that for a $\geq$ 1-km asteroid, this estimate is a factor of approximately five less frequent than estimated earlier by Shoemaker et al. (1990) and approximately three times less frequent than estimated by Toon et al. (1994). For estimates of ozone depletion requiring impacts with the oceans (i.e., depletions resulting from injection of water vapor and sea salt into the upper atmosphere), the time between impacts is increased by a factor of 1.41 to account for the fact that oceans cover only 71% of Earth's surface.

The impact frequency implied by Eq. E6 includes the short-period Jupiter family comet population, which comprises about 6 ± 4% of the NEOs. The contribution from long-period (e.g., from the Ort Cloud) or new comets to the NEO population was recently assessed to be only about 1% (Chapman, 2004).

Table 13.1 summarizes the estimated lower and upper limits for contributions to the stratospheric mixing ratios of NO_y, chlorine, bromine and water vapor for impacts of meteoroids having diameters in the range 150–1 000 m. Also given is the mean time between Earth (ocean + land) impacts and ocean impacts. Mixing ratios are calculated for illustrative purposes assuming a constant mixing ratio for each species throughout the entire stratosphere containing 1.3×10^{43} molecules.

Estimates of the diameters of asteroids required to cause stratospheric ozone depletion are shown in Fig. 13.1. Here we assume that an approximate doubling of the natural level of water vapor ($\approx$ 4 ppmv) or NO_y mixing ratio ($\approx$ 10 ppbv) in the stratosphere will result in a major ozone depletion at mid latitudes. For NO_y this requires impact of a 450–660 m asteroid with either land or ocean, which occurs about once every 75 000 to 190 000 years. Doubling of the natural water vapor mixing ratio in the stratosphere could occur as often as every 65 000–95 000 years as the result of the impact of an asteroid having a diameter of 370–430 m. Similarly, an increase in the chlorine content of the stratosphere to 10 ppbv requires the impact of a 390–450 m asteroid, which is predicted to occur about once every 75 000–110 000 years. Bromine, which is a much bet-

Table 13.1. Impact-induced changes in stratospheric composition for asteroid impacts with the Earth's oceans[a]

Asteroid diameter (m)	Earth impact interval (yr)	Ocean impact interval (yr)	NO (ppbv)	Chlorine (ppbv)	Bromine (pptv)	Water (ppmv)
150	5 500	7 800	0.11 – 0.33	0.35 – 0.55	0.53 – 0.84	0.16 – 0.26
200	10 900	15 000	0.27 – 0.81	0.84 – 1.3	1.3 – 2.0	0.40 – 0.64
250	19 000	26 000	0.53 – 1.6	1.7 – 2.7	2.5 – 4.0	0.79 – 1.3
300	29 000	40 000	0.93 – 2.8	2.9 – 4.6	4.4 – 7.0	1.4 – 2.2
350	41 000	58 000	1.5 – 4.5	4.6 – 7.4	7.0 – 11.2	2.2 – 3.5
400	57 000	80 000	2.2 – 6.7	6.9 – 11	11 – 17	3.1 – 4.9
450	75 000	110 000	3.2 – 9.6	9.09 – 16	15 – 24	4.7 – 7.5
500	96 000	140 000	4.4 – 13	14 – 22	21 – 33	6.5 – 10
600	150 000	210 000	7.6 – 23	24 – 38	36 – 57	11 – 18
700	210 000	300 000	12 – 36	37 – 60	57 – 91	18 – 29
800	290 000	410 000	18 – 54	56 – 90	85 – 140	27 – 43
900	390 000	550 000	26 – 77	80 – 130	120 – 190	38 – 61
1 000	500 000	700 000	35 – 106	110 – 180	170 – 270	52 – 84

[a] Calculated mixing ratios assume a uniform, global distribution everywhere above the tropopause for illustrative purposes. No account is taken of loss of NO_y due to cannibalistic reactions in the thermosphere and above. "Earth impact intervals" apply to NO_y mixing ratios while "ocean impact intervals" apply to chlorine, bromine and water intervals.

ter catalyst for ozone depletion than chlorine, would contribute to ozone depletion for the larger ocean impacts when its mixing ratio reaches a few tens of parts-per-trillion (see Table 13.1). Within error, the frequency of ozone depletion by any one of these mechanisms is about the same (although we assign the NO_y mechanism a much larger uncertainty). Any of these ozone-depletion catalysts (NO_y, water vapor, chlorine, bromine) acting alone could have a large effect on stratospheric ozone; however, because of the complexity of the chemistry of the stratosphere, the effects are not additive, and detailed model calculations, such as those discussed below, are required to account for the various synergisms involved. Overall, we estimate that large stratospheric ozone depletions probably occur about once every 60 000 to 120 000 years as a result of perturbations of stratospheric chemistry resulting from impacts of asteroids and comets.

Antarctic ozone holes may occur much more frequently. Assuming that the sea salt vaporized by ocean impacts becomes activated in the stratosphere to release chlorine and bromine, levels of ≈ 3 ppbv of chlorine (and 2.9 pptv of bromine), which we now know is sufficient to cause an Antarctic ozone hole with loss of greater than 50% the ozone column in the austral spring, will occur for impacts of asteroids as small as 260–300 m with an average time between impacts of about 30 000–40 000 years.

Fig. 13.1.
Estimated upper and lower limits to diameters of asteroids that result in large ozone depletions due to increases in the stratospheric mixing ratio of NO_y of 10 ppbv (*upper*), H_2O by 4 ppbv (*center*) and ClO_x by 10 ppbv (*lower*). The lower graph also shows the threshold for forming an Antarctic ozone hole. These estimates do not take into account any loss of NO to cannibalistic reactions in the mesosphere and above

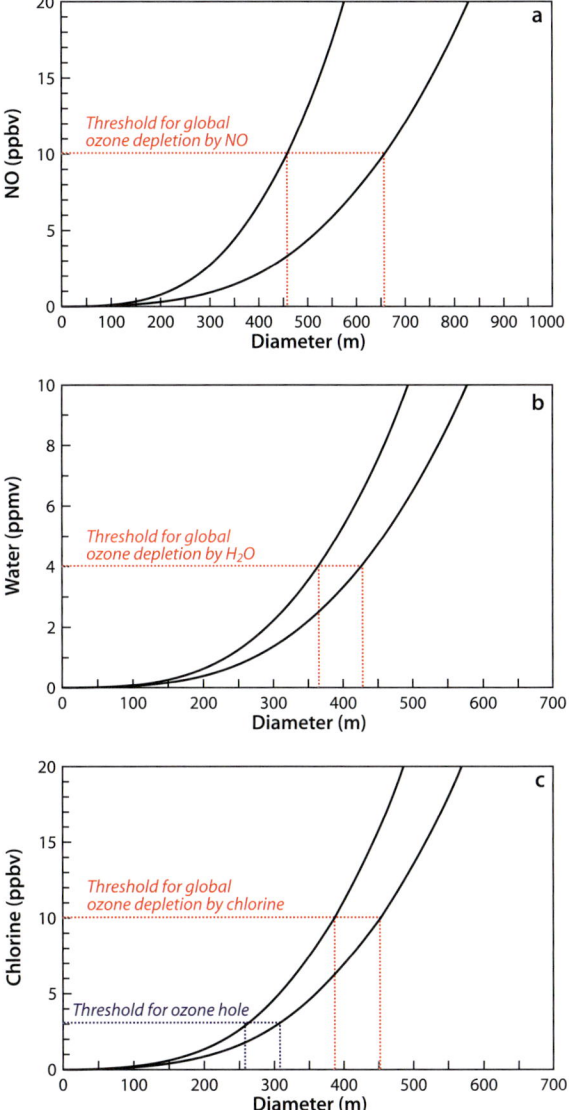

13.3.7
Model of Coupled Chemistry and Dynamics of the Upper Atmosphere

The Thermosphere-Ionosphere-Mesosphere-Electrodynamics General Circulation Model (TIME-GCM) was used to explore the effects on the upper atmosphere of nitric oxide and water vapor injected by an asteroid impact. This model has been described in detail by Roble and Ridley (1994) and Roble (2000). The model extends between 30 and 500 km altitude with a vertical resolution of four grid points per scale height and a horizontal resolution of 5° latitude and longitude. The model includes interactive chemistry and dynamics at each time step and is self-consistent, requiring only the specification of solar flux, auroral heat and momentum forcing from the solar wind parameters and dynamical and chemical forcing at the 30 km lower boundary. The model is hydrostatic and thus cannot model the initial phase of the impact, but for an assumed initial distribution of constituents, it can calculate the transport of chemical constituents and their transformations as a function of time and determine the effects of those species on stratospheric ozone down to about 30 km. This model was chosen for our initial work because it allows us to evaluate the effects of the chemical inputs on atmospheric dynamics and chemistry immediately after the impact event. Results from this model can be used as inputs to more detailed models of the stratosphere to more accurately evaluate levels of ozone depletion. As discussed in the model results below, it is critical to determine what fraction of the NO produced in the impact event survives chemical reactions in the mesosphere and thermosphere that act to destroy NO.

13.3.8
Model Results for Injections of Nitric Oxide and Water Vapor

We have simulated the effects on the upper atmosphere for three different cases corresponding to impacts of ≈ 1200 m (large), ≈ 560 m (medium), and ≈ 260 m (small) diameter asteroids having impact energies of $\approx 87\,000$, ≈ 8700 and ≈ 870 MT, respectively. Based on the previous discussion, it is estimated that the largest impact injects 1.3×10^{39} molecules of H_2O and 1.3×10^{36} molecules of NO. The medium-sized impact injects 10% as much NO and H_2O, and the small impact injects 1% as much. For the large impact, this corresponds to about 100 ppbv of NO and 100 ppmv of H_2O if uniformly distributed over the globe and above the tropopause at constant mixing ratio. However, the injection was made within the boundaries of a single grid point and at constant mixing ratio at and above 70 km, well above the stratosphere. It was assumed that the impact occurred in the vicinity of the Yucatan Peninsula of Mexico on January 10 of a hypothetical year with present day chemical and dynamical structure. Goals of the simulation were to determine (1) how fast the H_2O and NO are redistributed, (2) how the injection affects the temperature structure and dynamics of the upper atmosphere, (3) what fraction of the NO survives photochemical decomposition at such high altitudes and is transported downward to the stratosphere and (4) the combined effects of NO and H_2O on ozone in the upper stratosphere.

The results of these simulations are shown as global average differences from a perturbed minus unperturbed case. The model behavior for the unperturbed case for a year simulation has been discussed in detail in Roble (2000). The perturbed simulation is similar except that an impact occurred on day 10 (January 10). The parallel runs recorded histories daily, and difference fields were constructed for the entire year. The mean circulation redistributed the chemical perturbations seasonally so that for the January 10 impact most of the chemicals were first transported toward the northern hemisphere polar region and then downward to the stratosphere. The circulation shifts during equinox so that the injected species are then transported toward the southern hemisphere. The latitudinal distributions are not shown, but globally averaged distributions are presented to show that a significant fraction of the

Chapter 13 · Frequent Ozone Depletion Resulting from Impacts of Asteroids and Comets

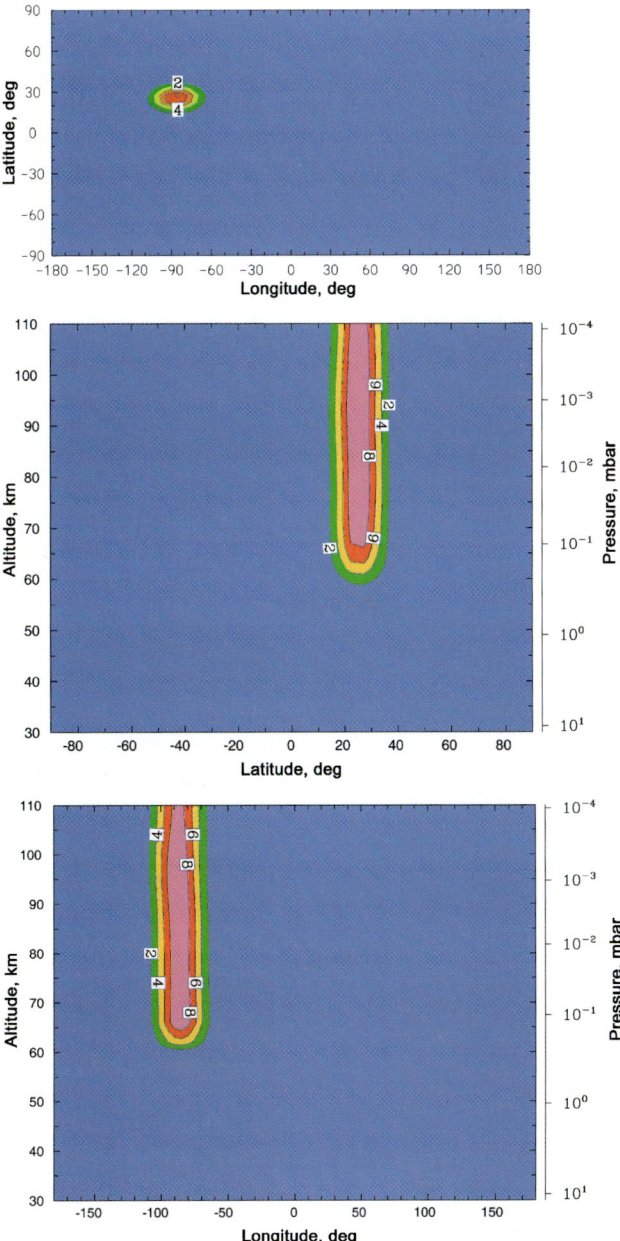

Fig. 13.2. Horizontal and vertical distributions of H_2O one hour following the large simulated impact on January 10. Impact injects a total of 1.3×10^{39} molecules of H_2O. Contour units are number density mixing ratio, i.e., the ratio of water molecules to air molecules ($N_2 + O_2 + O$) per unit volume. For all of the contours shown the concentration of injected water exceeds the concentration of air molecules. The distribution of NO at this time is virtually identical but at 1 000 times lower concentration

chemical species injected are transported to the lower atmosphere and not destroyed in the upper atmosphere.

Figure 13.2 shows the horizontal and vertical distributions of injected H_2O one hour after the large impact. Nitric oxide has essentially the same spatial distribution but with a factor of 1000 lower concentration. By this time the plume already has spread approximately 30° in latitude and 60° in longitude, and its lower boundary has descended nearly 10 km. This rapid dispersion continues over the next weeks to months,

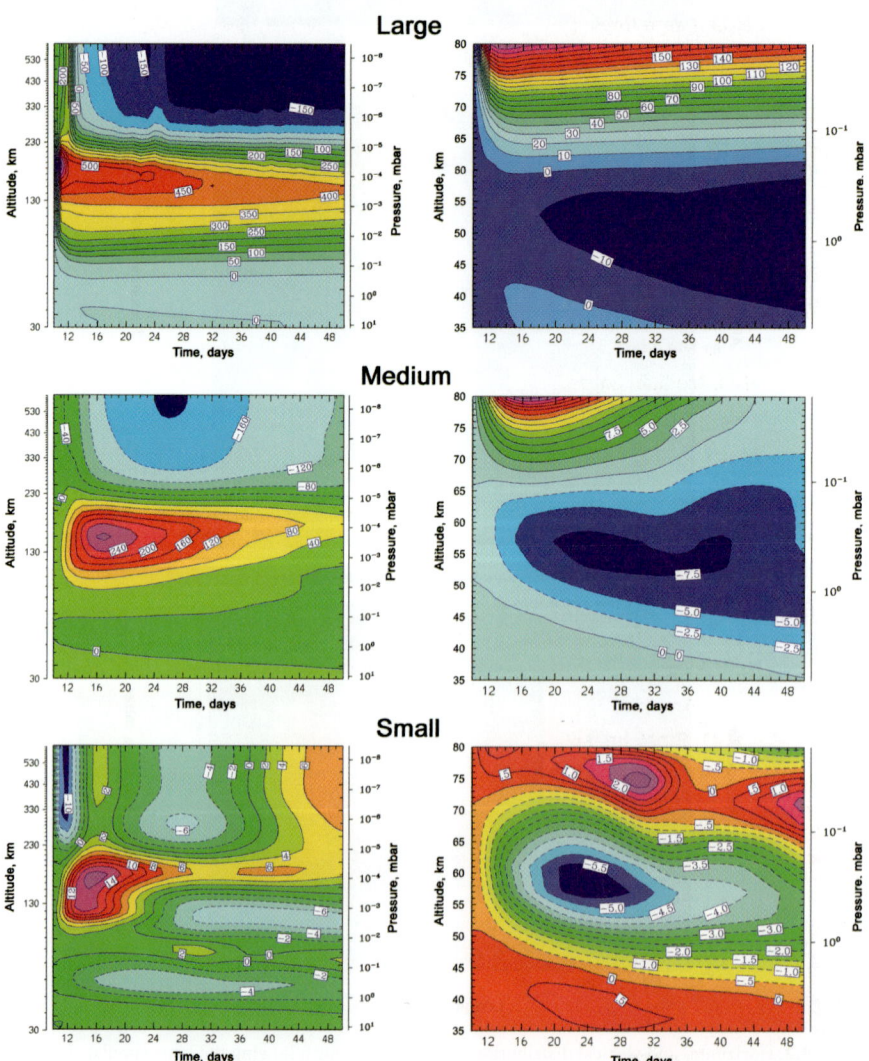

Fig. 13.3. Globally averaged atmospheric temperature difference fields (perturbed minus unperturbed cases) as a function of altitude and time following the simulated large, medium and small impacts. Contour units are degrees Kelvin

driven by the seasonal winds and by the intense chemical heating and decreased ozone heating from absorption of solar radiation. Figure 13.3 shows the global average temperature changes that occur as a function of altitude during the first year following the impact for the high, medium and small impact cases. For the large case, increases in temperature by up to several hundred °C in the region 100–200 km destroys the mesosphere as a separate atmospheric layer, with temperature increasing rather than decreasing above what is normally the stratopause. Decreased absorption of incoming

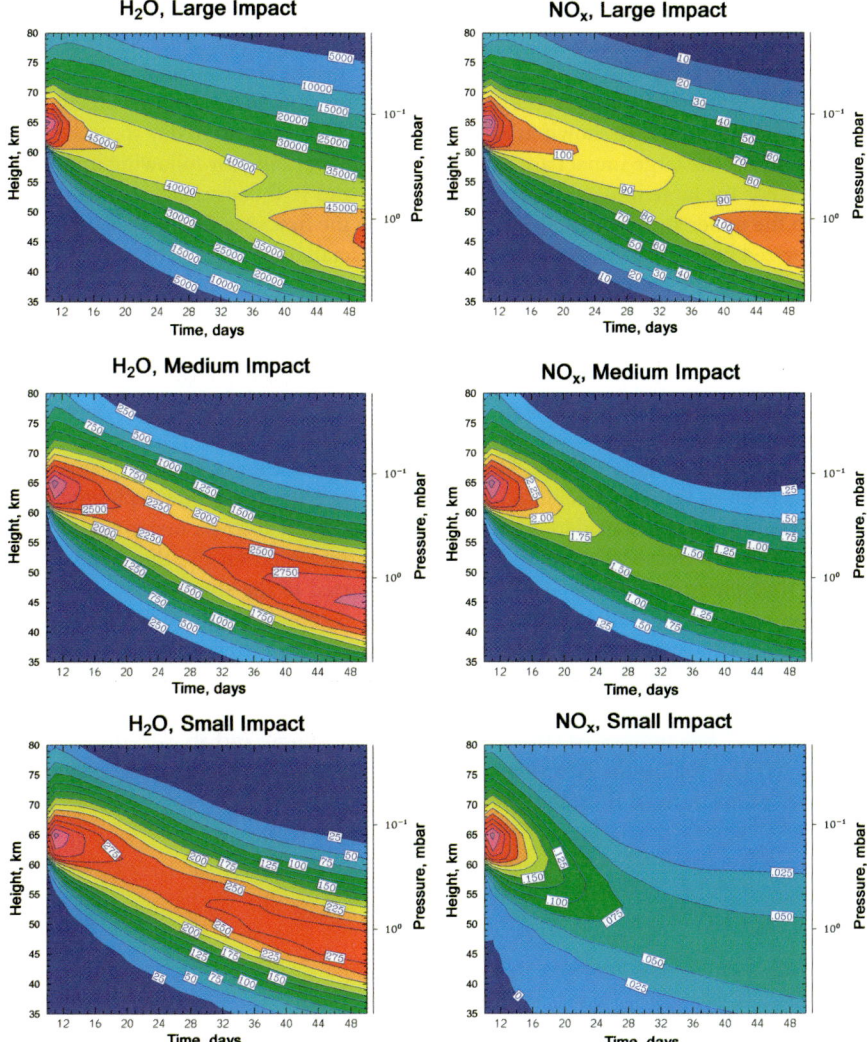

Fig. 13.4. Simulated globally averaged increases (perturbed-unperturbed cases) in water vapor (*left*) and NO_x ($NO_x = NO + NO_2$, *right*) concentration as a function of altitude and time following the simulated large, medium and small impacts. Contour units are 10^{10} molec cm^{-3}.

radiation causes the upper stratosphere to cool by as much as 40 °C for the large impact case. Heating of the lower thermosphere by several hundred degrees occurs in the medium case as well, with a globally averaged cooling in the stratosphere by as much as 10 °C. Temperature effects on the upper atmosphere are relatively minor for the small impact case, with a maximum heating near 110 km of 10–20 °C and cooling of the stratosphere by only about 5 °C.

Both water vapor and NO_x are rapidly transported downward to the stratosphere for all three impact cases. Figure 13.4 shows changes in the vertical distribution of these species during the first 50 days following large, medium and small impacts; the results for NO_x and water vapor are extended to one year in Fig. 13.5. Large fractions of both water vapor and NO_x have descended below 50 km (height of the normal stratopause), with concentrations peaking in the stratosphere after only three months. Changes in ozone concentrations are shown in Fig. 13.6 for all three cases out to day 50, and ozone depletions for the large impact are simulated for the first full year following impact in Fig. 13.5. For the large impact case, injection of NO and H_2O causes large ozone depletions in the upper stratosphere that persist through the first year. Ozone depletions are summarized in Fig. 13.7 for the large, medium and small impact cases. By day 50 ozone depletion of the globally integrated ozone column (above 30 km) has been depleted by 58%, 9% and 1% for the large, medium and small impact cases. These depletions continue to increase beyond day 50 for the large and medium impact cases. Local depletions within the hemisphere of impact are much larger. Stratospheric ozone levels are expected to recover over a period of 2–3 years as water vapor and NO_y are slowly removed to the troposphere.

It is important to note that the TIME-GCM model has a lower boundary at 30 km, whereas most of the ozone column lies below 30 km. Thus, the ozone depletions predicted should only be considered qualitative. As discussed above, a major goal of this study was to determine what fraction of the NO_x input would survive the chemistry of the thermosphere and mesosphere and be transported to the stratosphere. In future work, we will use the model results described here as inputs to a more detailed model of the lower atmosphere that extends to the ground in order to better quantify ozone depletion as a function of meteoroid size.

Of particular interest to this study is the degree to which NO is destroyed by the "cannibalistic" reactions of NO that occur in the mesosphere and above. This destruction of NO can reduce the effect on stratospheric ozone. Nitric oxide is destroyed by nitrogen atoms derived from the photolysis of NO:

$$NO + h\nu \longrightarrow N + O \quad (R17)$$

$$N + NO \longrightarrow N_2 + O \quad (R18)$$

Net: $2 NO \longrightarrow N_2 + 2 O$

At high NO concentrations the rate-limiting step for this cycle of reactions is the photolysis of NO (reaction R17), which has a lifetime of only about 3 days. However, at

Fig. 13.5.
Simulated globally averaged difference fields of water vapor (*upper*), NO_x (*middle*) and ozone (*lower*) as a function of altitude and time for the first year following the large impact. Contour units are 10^{10} molec cm^{-3}

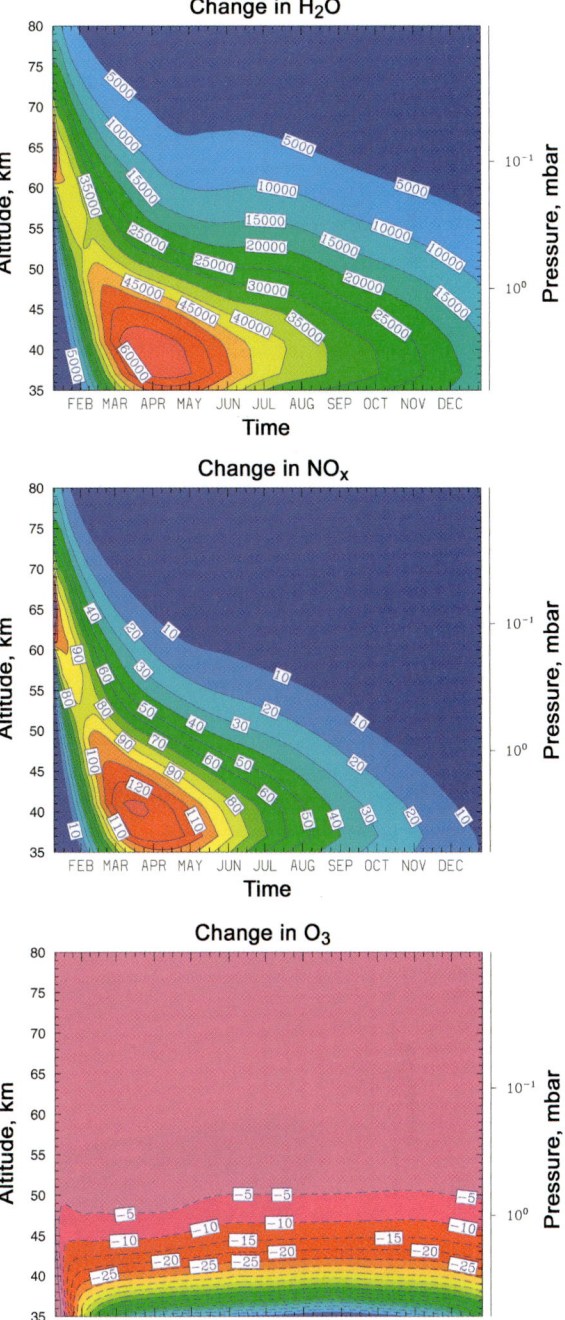

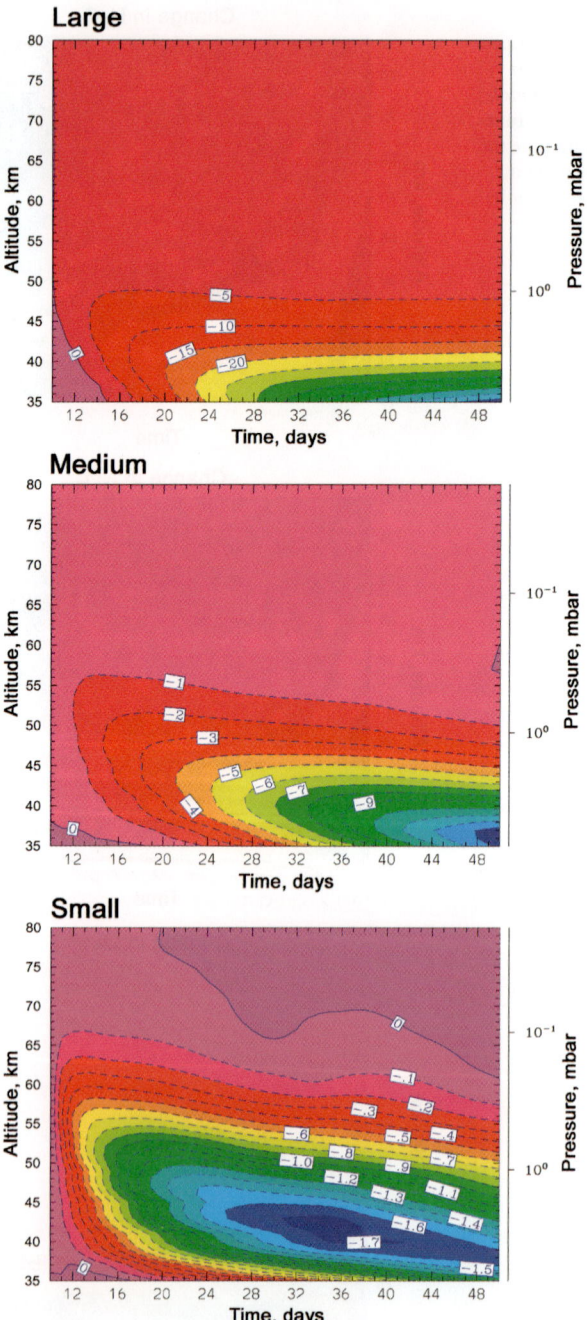

Fig. 13.6. Changes in globally averaged ozone concentration as a function of altitude and time following the simulated large, medium and small impacts. Contour units are 10^{10} molec cm^{-3}.

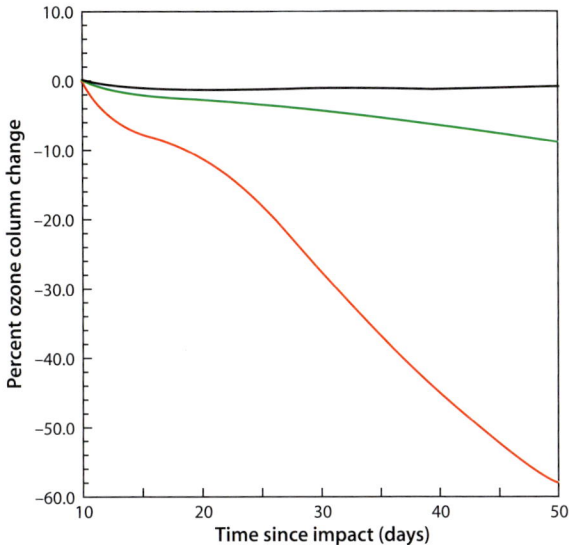

Fig. 13.7.
Calculated depletions of the globally integrated ozone column above 30 km for the large (*red*), medium (*green*) and small (*blue*) impacts as a function of time following impact; it is important to note, however, that most of the ozone column lies below 30 km

high concentrations of NO, the upper atmosphere becomes optically thick in NO, so that the rate of NO destruction is limited by the solar flux, and, as a result, the lifetime of NO is greatly increased. Thus, we expect the fraction of NO that survives destruction in the mesosphere and thermosphere to increase with the impact size. Consistent with this expectation, for the high, medium, and low impact cases we found that 63, 41 and 18% of the injected NO is still present as NO_x at 50 days following the impact event.

In conclusion, we find that the medium case, corresponding to the impact of a ≈ 560 m asteroid with the oceans, would result in a very significant ozone depletion that would have serious implications for the biosphere. These calculations do not take into account the additional ozone-destruction chemistry made possible by activation of chlorine and bromine in the sea salt vaporized and deposited at high altitudes, which could further reduce the size of asteroid required to produce a major ozone depletion. Despite the many uncertainties, it appears likely that the size threshold for asteroids producing a major ozone depletion that would seriously damage the global biosphere is in the range 390–660 m for an average density and impact velocity.

13.3.9
Possible Test of the Impact-Induced Ozone Depletion Hypothesis

Ice cores, such as the Antarctic Dome C core dating back to 740 000 BP, contain pollen grains that may serve as tiny UV-B dosimeters. When exposed to UV-B radiation, adjacent pyrimidine bases of the pollen DNA photodimerize to produce thymine-thymine (T-T), cytosine-thymine (K-T) and cytosine-cytosine (C-C) photodimers (Setlow and Carrier 1966; Witkin 1969; Hall and Mount 1981). Thus, the degree of pyrimidine photodimer formation may serve as a proxy for UV-B radiation. Monoclonal antibodies have been developed against T-T dimers (Wani et al. 1987; Mori et al. 1991; Lee and Yeung 1992), and capillary electrophoresis has been used in conjunction with laser-induced fluorescence detection to analyze for biomolecules in single cells (50). Many ultrasensitive bioanalytical techniques for DNA analysis have been developed in the past decade, largely driven by the successful Human Genome Project. Thus, it should be possible to test the hypothesis presented here that asteroid and comet impacts result in large ozone depletions every few tens of thousands of years.

Acknowledgments

The National Center for Atmospheric Research is sponsored by the National Science Foundation.

References

Alvarez LW, Alvarez W, Asaro F, Michel HV (1980) Extraterrestrial cause for the Cretaceous-Tertiary extinction. Science 208:1095–1108
Bates DR, Nicolet M (1950) J Geophys Res 55:301
Brasseur G, Solomon S (1984) Aerononomy of the middle atmosphere. Reidel, Dordrecht
Brasseur GP, Orlando JJ, Tyndall GS (1999) Atmospheric chemistry and global change. Oxford University Press, Oxford
Chameides WL (1986) The role of lightning in the chemistry of the atmosphere. In: The Earth's electrical environment. National Academy Press, Washington, pp 70–77
Chapman CR (2004) The hazard of near-Earth asteroid impacts on Earth. Earth Planet Sci Lett 222:1–15
Croft SK (1982) A first-order estimate of shock heating and vaporization in oceanic impacts. Geol Soc Amer Spec Paper 190:143–152
Crutzen PJ (1970) Quart J Roy Meteorol Soc 96:320
Crutzen PJ, Birks JW (1982) The atmosphere after a nuclear war: twilight at noon. Ambio 11:114–125
Emiliani C, Kraus EB, Shoemaker EM (1981) Sudden death at the end of the Mesozoic. Earth Planet Sci Lett 55:317–334
Finlayson-Pitts BJ, Pitts JN, Jr (1986) Atmospheric chemistry: fundamentals and experimental techniques. Wiley, New York
Finlayson-Pitts BJ, Ezell MJ, Pitts J (1989) Formation of chemically active chlorine compounds by reactions of atmospheric NaCl particles with gaseous N_2O_5 and $ClONO_2$. Nature, 337:241–244
Foley HM, Ruderman MA (1973) Stratospheric NO production from past nuclear explosions. J Geophys Res 78:4441–4451
Gilmore FR (1975) J Geophys Res 80:4553
Hall JD, Mount, DW (1981) Prog Nucleic Acid Res Mol Biol 25:53–126
Haynes DR Tro NJ, George SM (1992) Condensation and evaporation of H_2O on ice surfaces. J Phys Chem 96:8502–8509
Johnston HS (1971) Reduction of stratospheric ozone by nitrogen oxide catalysts from supersonic transport exhaust. Science 173:517–522
Johnston H, Whitten G, Birks J (1973) Effect of nuclear explosions on stratospheric nitric oxide and ozone. J Geophys Res 78:6107–6135
Jones EM, Kodis JW (1982) Atmospheric effects of large body impacts: the first few minutes. Geol Soc Amer Spec Paper 190:175–186
Junge CE, Chagnon CW, Manson JE (1961) J Meteor 18:81
Kring DA (1999) Meteor Planet Sci 34, A67–A68
Kring DA, Melosh HJ, Hunten DM (1995) Meteoritics 30:530
Keller G (1989) Paleonoceanography 4:287–332
Lee TT, Yeung ES (1992) Anal Chem 64:3045–3051
Lewis JS, Watkins GH, Hartman H, Prinn RG (1982) Chemical consequences of major impact events on Earth. Geol Soc Amer Spec Paper 190:215–221

List RJ (1984) Smithsonian meteorological tables. Smithsonian Institution Press, Washington
Livingston FE, Finlayson-Pitts BJ (1991) Geophys Res Lett 18:17–20
Melosh HJ (1982) The mechanics of large meteoroid impacts in the Earth's oceans. Geol Soc Amer Spec Paper 190:121–127
Melosh HJ (1989) Impact cratering. Oxford University Press, New York
Molina MJ, Rowland FS (1974) Nature 249:810
Mori T et al. (1991) Photochem Photobiol 54:225–232
Nachtway DF, Caldwell MM, Biggs RH, eds (1995) CIAP, monograph 5, US Department of Transportation, Impacts of Climatic Change on the Biosphere
National Research Council (1975) Long term world-wide effects of multiple nuclear-weapon detonations. National Academy Press, Washington
National Research Council (1985) The effects on the atmosphere of a major nuclear exchange. National Academy Press, Washington
Nemtchinov IV, Svetsov VV, Kosarev IB, Golub AP, Popova OP, Shuvalov VV, Spalding RE, Jacobs C, Tagliaferri E (1997) Icarus 130:259–274
O'Keefe JD, Ahrens TJ (1982) The interaction of the Cretaceous/Tertiary extinction bolide with the atmosphere, ocean, and solid Earth. Geol Soc Amer Spec Pap 190:103–120
O'Keefe JD, Ahrens TJ (1989) Impact production of CO_2 by the Cretaceous/Tertiary extinction bolide and the resultant heating of the Earth. Nature 338:247–248
Pittock AB et al. (1985) Environmental consequences of nuclear war, vol I: Physical and atmospheric effects. Scientific Committee on Problems in the Environment, SCOPE 28, Wiley
Pope, KO, Baines KH, Ocampo AD, Ivanov B. (1997) Energy, volatile production, and climatic effects of the Chicxulub Cretaceous/Tertiary impact. J Geophs Res 102:21645–21664.
Prinn RJ, Fegley JB (1987) Bolide impacts, acid rain, and biospheric traumas at the Cretaceous-Tertiary boundary. Earth Planet Sci Lett 83:1–15
Robertson DS, McKenna MC, Toon OB, Hope S, Lillegraven JA (2004) Survival in the first hours of the Cenozoic. GSA Bulletin 116:760–768.
Roble RG (2000) Geophysical Monograph 123:53–67
Roble RG, Ridley EC (1994) Geophys Res Lett 21:417–420
Setlow RB, Carrier WL (1966) J Mol Biol 17:237–254
Shoemaker EM, Wolfe RF, Shoemaker CS (1990) In: Sharpton VL, Ward PD (ed) Global catastrophes in Earth history. GSA Special Paper 247, Geological Society of America, Boulder, CO, pp 155–170
Timonen RS, Chu LT, Leu M, Keyser LF (1994) Heterogeneous reaction of $ClONO_2(g) + NaCl(s) \rightarrow Cl_2(g) + NaNO_3(s)$. J Phys Chem 98:9509–9517
Toon OB et al. (1982) Evolution of an impact-generated dust-cloud and its effects on the atmosphere. Geol Soc Amer Spec Pap 190:187–200
Toon OB, Zahnle K, Turco RP, Covey C (1994) Environmental perturbations caused by impacts. In: Gehrels T (ed) Hazards due to comets and asteroids. Univ of Arizona Press, Tucson, pp 791–826
Turco RP, Toon OB, Ackerman TP, Pollack JB, Sagan C (1983) Science 222:1283–1292
Wani AA, D'Ambrosio SM, Alvi NK (1987) Photchem Photobiol 46: 477–482
Whitten RC, Borucki WJ, Turco RP (1975) Nature 257:38
Witkin EM (1969) Annu Rev Microbiol 23:487–514
Zangmeister CD, Pemberton JE (1998) In situ monitoring of the $NaCl + HNO_3$ surface reaction: the observation of mobile surface strings. J Phys Chem 102:8950–8953
Zel'dovich YB, Raizer YP (1968) In: Elements of gas dynamics and the classical theory of shock waves. Academic Press, New York, pp 101–106

Chapter 14

Tsunami as a Destructive Aftermath of Oceanic Impacts

V. K. Gusiakov

14.1
Introduction

Tsunamis belong to the long-period oceanic waves generated by underwater earthquakes, submarine or subaerial landslides or volcanic eruptions. They are among the most dangerous and complex natural phenomena, being responsible for great losses of life and extensive destruction of property in many coastal areas of the World's ocean. The tsunami phenomenon includes three overlapping but quite distinct physical stages: the generation by any external force that disturbs a water column, the propagation with a high speed in the open ocean and, finally, the run-up in the shallow coastal water and inundation of dry land (Gonzalez, 1999). Most tsunamis occur in the Pacific, but they are known in all other areas of the World including the Atlantic and the Indian oceans, the Mediterranean and many marginal seas. Tsunami-like phenomena can occur even in lakes, large man-made water reservoirs and large rivers.

In terms of total damage and loss of lives, tsunamis are not the first among other natural hazards. Actually, they rank fifth after earthquakes, floods, typhoons and volcanic eruptions. However, because of their high destructive potential, tsunamis have an extremely adverse impact on the socioeconomic infrastructure of society, which is further strengthened by their suddenness, terrifying rapidity, heavy destruction of property and high percentage of fatalities among the population exposed. The feature that differs tsunamis from other natural disasters is their ability to produce a destructive impact far away from the area of their origin (up to 10 000 km). In the ocean where the bottom is flat, their far-field amplitude decreases as $1/\sqrt{r}$ because of cylindrical divergence that is a minimum possible attenuation allowed by the energy conservation law. One of the largest Pacific tsunamis historically known was generated by a strong (magnitude 8.6) earthquake which occurred on May 22, 1960 near Chiloe Island (southern Chile), and some 22 hours later reached Japan still generating waves 6–7 meters high, producing extensive damage (nearly 10 000 houses were destroyed) and claiming some 229 lives (Fig. 14.1).

In the open ocean, tsunamis travel at a speed ranging from 400 to 700 km per hour depending on the water depth. The velocity is controlled by the ocean depth as described by the formula

$$v = \sqrt{gH}$$

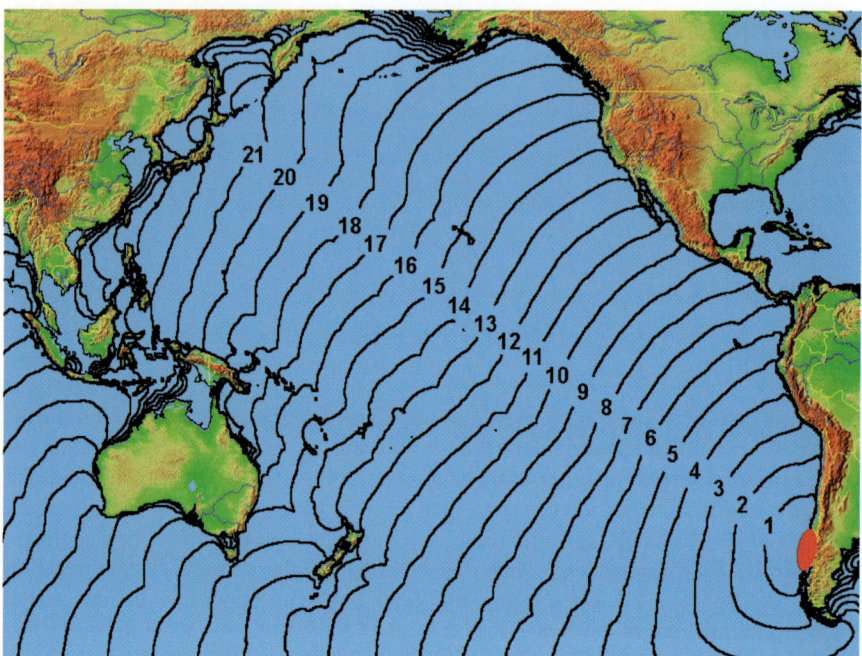

Fig. 14.1. Tsunami travel-time chart for the 1960 Chilean tsunami. *Digits* near the isochrones – propagation time in hours. The *solid ellipse* shows the position of the earthquake source

where g is acceleration due to gravity, and H is depth of water. The typical periods of oscillations in these waves cover the range from 5–6 minutes to 1 hour. Due to their great wavelength, reaching 500–700 km in the deep ocean and 50–100 km on the continental shelf, tsunamis rarely approach the coast as breaking waves, rather, they appear as a quick succession of floods and ebbs producing strong, up to 10 m s^{-1}, currents.

Destruction from tsunamis results from the three main factors: inundation of salt water, dynamic impact of water current and erosion. Considerable damage is also caused by floating debris that enhance the destructive force of the water flood. Flotation and drag force can destroy frame buildings, overturn railroad cars and move large ships far in-land. Ships in harbours and port facilities can be damaged by the strong current and surge caused by even a weak tsunami.

A typical height of tsunami, generated by an earthquake with magnitude of 7.0 to 8.0 (the range where most of tsunamigenic earthquakes occur) is from 5 to 10 meters at the nearest coast where run-ups are typically observed along 100 to 300 km of the coastline. This height is still within the range of the largest possible storm surges for many coastal locations. However, having a longer wavelength, tsunami can penetrate in-land to much greater distances reaching in many places several hundreds of meters and sometimes several kilometers. The highest run-up of tectonically induced tsunamis can reach 20–30 meters (1952 Kamchatka, 1960 Chile, 1964 Alaska tsunamis). Even a higher water splash (up to 50–70 meters) can be produced by submarine or coastal landslides when compared to those triggered by earthquakes. Such strong tsunamis have an enormous destructive power and sweep everything on land lying in their way, removing soil, vegetation and all traces of existing settlements.

All destructive tsunamis can be divided into two categories: *local* or *regional* and *trans-oceanic*. For local tsunamis, the destructive effect is confined to the nearest coast located within one hour of the propagation time (from 100 to 500 km). In all tsunamigenic regions of the World oceans, most of damage and casualties come from local tsunamis. Far less frequent but potentially much more hazardous are trans-oceanic tsunamis capable of widespread distribution. Formally, this category includes the events that have run-ups higher than 5 meters at a distance of more than 5000 km. Historically, all trans-oceanic tsunamis are known in the Pacific with only one case recorded in the Atlantic (the 1755 Lisbon tsunami, that reached the Caribbean with 5–7 m waves).

The overall physical size of a tsunami is measured on several scales. Among most common is the *Soloviev-Imamura* scale denoted by *I*. This is an intensity scale, first proposed by A. Imamura (Imamura, 1942) and then slightly modified by S. Soloviev (Soloviev 1978). It is based on the average run-up height of waves h_{av} along the nearest coast according to the formula

$$I = \frac{1}{2} + \log_2 h_{av}$$

In this scale, the largest trans-Pacific tsunamis have an intensity of 4–5, destructive regional tsunamis – an intensity of 2–3, not damaging but still visually observable and finally local tsunamis – an intensity of 0–1. Tsunamis detectable only on instrumental records have a negative intensity (from –1 to –4).

14.2
Geographical and Temporal Distribution of Tsunamis

The world-wide catalog and database on tsunamis and tsunami-like events that is being developed under the GTDB (Global Tsunami DataBase) Project (Gusiakov 2003), covers the period from 1628 BC to the present and currently contains nearly 2250 historical events with 1206 of these for the Pacific, 263 for the Atlantic, 125 from the Indian ocean and 545 from the Mediterranean region. The geographical distribution of tsunami is shown in Fig. 14.2 as a map of seismic, volcanic and landslide sources of historical tsunamigenic events. When analyzing this map, one should take into account that it reflects not only the level of tsunami activity, but also the regional historical and cultural conditions that strongly influence the availability of the historical data. From geographical distribution of tsunamigenic sources, we can see that most of tsunamis were generated along subduction zones and the major plate boundaries in the Pacific, the Atlantic and the Mediterranean regions. Very few historical events occurred in the Deep Ocean and central parts of the marginal seas, except several cases of small tsunamis that originated along the middle-ocean ridges and some major transform faults.

The temporal distribution of historical tsunamis is shown in Fig. 14.3 for the last 1000 years. From this graph we can see that the historical data have a highly non-uniform distribution in time with three quarters of all events reported within the last two hundred years. The most complete data exist for the 20th century, when the instrumental measurements of weak tsunamis became available. In all tsunamigenic regions (except possibly Japan) there are obvious gaps in reporting even large destructive events for the period preceding the 20th century. Thus, any estimates of tsunami recurrence should be considered with this fact (data incompleteness) in mind.

In 1901–2000, a total of 943 tsunamis were observed in the World Ocean that results in about ten events per year. Most of these events were weak, observable only on tide

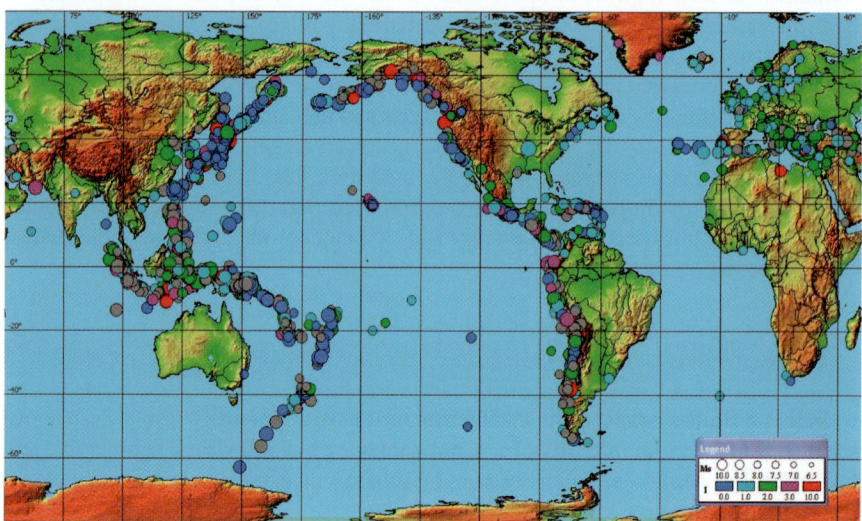

Fig. 14.2. Geographical distribution of tsunami sources in the World Ocean. The size of *circles* is proportional to the earthquake magnitude, density of *gray tone* – to the tsunami intensity on the Soloviev-Imamura scale

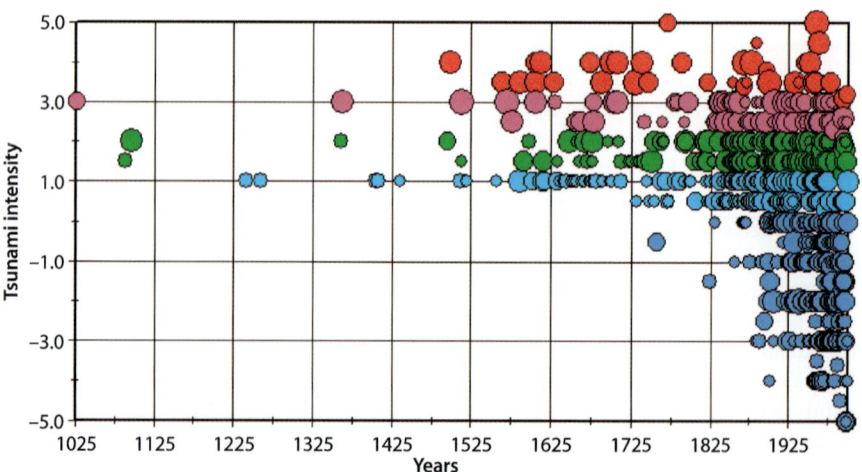

Fig. 14.3. Tsunami occurrence versus time for the last 1000 years. Events are shown as *circles* with the color depending on tsunami intensity and the size proportional to the earthquake magnitude

gauge records. About 260 tsunamis were "perceptible", having a run-up height exceeding one meter. Among these, in 33 cases the run-up was greater than one meter and it was observed at a distance of more than 1000 km from the source. During the last 100 years, five destructive trans-oceanic tsunamis, all in the Pacific are known to have occurred (1946 Aleutians, 1952 Kamchatka, 1957 Aleutians, 1960 Chile, 1964 Alaska).

14.3
Basic Types of Tsunami Sources

Most of oceanic tsunamis (up to 75% of all historical cases) reported in historical catalogs are generated by shallow-focus earthquakes capable of transferring sufficient energy to the overlying water column. The rest are divided between landslide (7%), volcanic (5%), meteorological (3%) tsunamis and water waves from explosions (less than 1%). Up to 10% of all the reported coastal run-ups still have unidentified sources.

Seismotectonic tsunamis. Tectonic tsunamis are generated by submarine earthquakes due to the large-scale co-seismic deformation of the ocean bottom and the dynamic impulse transferred to a water column by compression waves. Tsunamigenic earthquakes occur along subduction zones, middle ocean ridges and main transform faults, i.e. within the areas with a large vertical variation of the bottom relief. The size of tsunami generated by an earthquake relates to the energy released (earthquake magnitude), source mechanism, hypocentral depth and the water depth at the epicenter.

Concerning the spatial distribution of tsunami damage, the rule of thumb is that in all but largest seismically induced tsunamis, their damage is limited to an area within one hour of the propagation time. The typical distribution of tsunami run-up heights along the coast is shown in Fig. 14.4. This is a modification of the figure from (Chubarov and Gusiakov 1985) obtained as a result of calculations of tsunami genera-

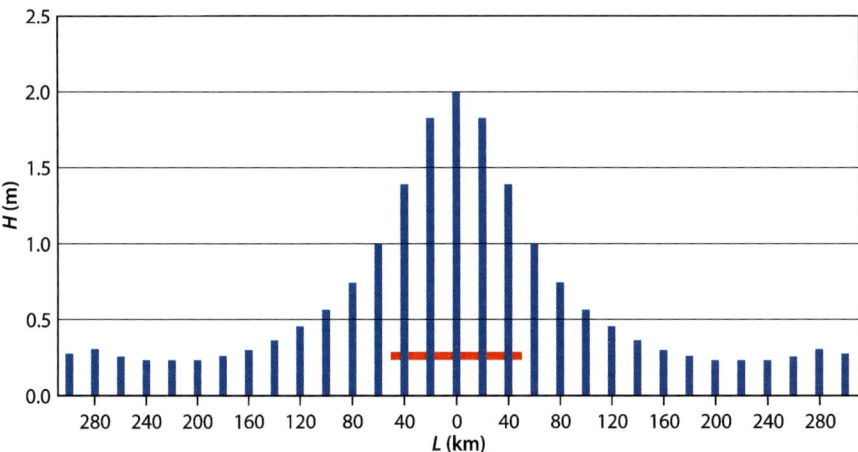

Fig. 14.4. Typical distribution of tsunami run-up heights along the coast calculated for a model source equivalent to a magnitude 7.5 submarine earthquake. Section of the *solid line* shows the position and size of the seismic source

tion by a model source having some basic features of a real earthquake with moment-magnitude 7.5 and wave propagation over the inclined bottom modeling the continental slope and shallow-water shelf. One can see that the area of the dominating heights is roughly limited to twice the size of the earthquake source (100–200 km for an earthquake of magnitude 7.0–7.5). Outside of this area, the run-up heights rapidly decrease. Such a strong directivity results from three main factors: (1) initial directivity of energy radiation by a seismic source, (2) ellipticity of a source, and (3) wave refraction on the inclined bottom. Among these factors, the most important is the third – refraction on the inclined bottom – and this effect dominates in the coastal run-up distribution in all regional tsunamis, having their sources on the continental slope and shelf.

Note, that initial wave height in the source area as derived from the co-seismic displacement produced by an earthquake source is about 0.7 m, thus giving the magnification factor (ratio of the maximum coastal height to the wave height in the deep water) of about 3. Further wave amplification during run-up on to dry land can give another factor of 2, thus resulting in total magnification from 5 to 6, that is significantly less than 10 to 40 as postulated in (Morrison et al. 1994). Such great amplification is possible just under very specific combination of the near-shore bathymetry, configuration of the coastline and the coastal relief. Indeed, the 30.8 m run-up measured after the 1993 Okushiri tsunami in the Japan Sea, was a rare feature above the average 8–10 m run-up along the rest of the Okushiri coast.

Slide-generated tsunamis. Not as frequent as tectonic generation, but still very common world-wide, slide-generated tsunamis result from rock and ice falling into the water, or sudden submarine landslides. Typically, they produce an extremely high water splash (up to 50–70 m, with the highest historical record of 525 m noted in Lituya Bay, Alaska in 1958) but not widely extended along the coast. In general, the energy of landslide tsunamis rapidly dissipates as they travel away from the source, but in some cases (e.g., if the landslide covers a large depth range), a long duration of slide movement can focus the tsunami energy along a narrower beam than the equivalent seismic source (Iwasaki 1997). One of the most recent cases where the involvement of slide mechanism in tsunami generation was definitely confirmed is the 1998 Papua New Guinea tsunami when 15-m waves were observed after the Mw 7.0 earthquake (Okal and Synolakis 2001; Synolakis et al. 2002; Tappin et al. 2002). The slide-generated water waves occur not only in the oceans and seas, but pose a clearly recognized hazard to reservoirs, harbors, lakes and even large rivers where they may endanger lives, overtop dams, or destroy the waterside property.

In the case of large earthquakes, the accompanying landslides, locally triggered by strong shaking, can produce waves greatly exceeding the height of the main tectonic tsunami. They are particularly dangerous as they arrive within a few minutes after the earthquake, leaving no time for a warning. One of the primary causes of death in the 1964 Alaska earthquake was the secondary tsunami generated by slides from the fronts of the numerous deltas at the Alaska coast (Lander 1996). These locally-triggered landslide tsunami can be an important factor even for a land impact especially in the case where it happens within a coastal area particularly vulnerable to landslides given the existence of numerous fiords, narrow bays and steep submarine canyons having large potential for slumping (e.g. Norway, Kamchatka, Alaska, west coast of Canada and US) (Rabinovich et al. 2003).

Volcanic tsunamis. Although relatively infrequent, explosive volcanic eruptions on small islands can generate extremely destructive water waves in the immediate source area. The 1883 Krakatau eruption with 150–200 MT of TNT equivalent and 18–20 km^3 of the estimated volume of the eruptive material resulted in 25-meter tsunami that flooded the coast of the Sunda Strait and killed 36 000 people (Yokoyama 1981). The catastrophic tsunami that devastated the northern coast of the island of Crete, was generated by an explosion of the Santorini volcano in 1628 BC with the estimated volume of eruptive material 50–60 km^3 (McCoy and Heiken 2000; Minoura et al. 2000). Smaller eruptions can generate a significant tsunami if they are accompanied by a volcanic slope failure (e.g. 1792 Unzen volcano collapse in Japan) or a large lahar or a pyroclastic flow (e.g. 1902 Mount Pele eruption in Martinique). As compared to tectonically-induced tsunamis, volcanic tsunamis can be extremely destructive locally, but rarely transport their energy far from the area of origin. It is widely known that the 1883 Krakatau tsunami was globally observed and recorded by 35 remote tide stations including several in the northern Atlantic, but it is rarely recognized that most of the damage and all deaths actually occurred in the very limited area along the coast of the Sunda Strait within the distance of 300 km from the site of explosion.

Meteorological tsunamis. Tsunami-like waves can be generated by a rapidly moving atmospheric pressure front moving over a shallow sea at approximately the same speed as a tsunami could allow them to couple. The resulting run-up can be increased by the hydrostatic water rise due to the low pressure zone in the cyclone center and the dynamic surge resulting from a strong wind pressure. In fact, after the 1883 Krakatau eruption at some remote tide stations the recorded sea level disturbance was a result of the water response to the air pressure waves traveling in the atmosphere from the site of explosion (Ewing and Press 1955; Press and Harkrider 1966).

Explosion-generated tsunamis. The world-wide historical tsunami catalog contains several cases of tsunamis generated by large explosions. In December of 1917, large waves were generated by the greatest man-made explosion before the nuclear era – this happened in the Halifax Harbour (Nova Scotia, Canada) after a collision of the munitions ship *Mont Blanc*, having 3 000 tons of TNT on board, with the relief ship *Imo*. At the coast near to the explosion site, the waves were over 10 meters high, but their amplitude diminished greatly with distance (Murty 2003). An extensive study of water waves generated by submarine nuclear explosions, both on and under the sea surface and up to 10 MT yield, and also on a series of smaller-scale tests carried out in Mono Lake (California) was made by W. Van Dorn (Van Dorn 1968) for the US Navy. The main conclusions from his study were that tsunamis from explosions have a shorter wavelength as compared to the size of the resulting cavity (a few km in diameter), in near-field the tsunami height can be very large, but rapidly decays as the waves travel outside the source area. He also indicated (however, without any proof and details presented) to the effect of breaking of short-length waves when they cross the continental shelf, generating large-scale turbulence, but leaving the coast without dampable run-up (Morrison 2003).

14.4
Tsunamigenic Potential of Oceanic Impacts

Since evidence for asteroid impact on Earth exists, we have to conclude that there is a four-to-one chance that they hit oceans, seas or even large internal water reservoirs and therefore tsunami or tsunami-like water waves can be generated by an extra-terrestrial impact. There has been a general concern that the tsunami from a deep-water impact of a 1-km asteroid could contribute substantially to its overall hazard for the people living near coasts and would wash out all coastal cities of the entire ocean (Chapman 2003; Morrison 2003). However, a 1-km asteroid is quite close to the global disaster threshold (impact of a 2–3 km object) and tsunami could therefore contribute somewhat to other hazardous aftermaths of this natural catastrophe that would have a large enough potential to end our modern civilization era. Fortunately for humankind, it is indeed a very rare event, available estimates of its return period vary in the range of 100 000 to 1 000 000 years. Much more frequent are the Tunguska-class impacts (the size of an object being 100 m or less) with the return period being more relevant to the human time scale and spanning from several hundred to one thousand years. Unless the small asteroid is made of solid metal (iron or nickel), it would likely explode in the upper atmosphere with a TNT equivalent in the first tens of megatons. Available estimates, based mainly on nuclear tests results, show that tsunami from such an airblast should be from several tens of centimeters to one meter (Glasstone and Doland 1977), so the water impact of such a small, once-per-century asteroid could be in general less hazardous than an equivalent explosion above land.

A practical concern related to impact tsunamis is that the risk they impose can be significant for asteroids with a diameter between 200 m and 1 km (Hills et al. 1994). Possible effects of tsunamis are mentioned in numerous publications devoted to the estimation of impact aftermaths (Hills et al. 1994; Hills and Mader 1997; Hills and Goda 1998; Mader 1998; Ward and Asphaug 2000, 2003). The resulting deep-water wave height and expected run-up distribution along the coast depends on many factors – the size of an impactor and its composition, velocity and angle of collision, finally, the particular site of an impact. Even for a concrete set of these parameters, researchers are very uncertain about the expected height of an impact generated tsunami. The main reason for that is, of course, that the problem of modeling of the generation stage and, especially, the first initial 10 seconds of the impact process is extremely complicated. The full-scale modeling of this high-speed process requires the solution of 3D equations describing the non-linear dynamics of compressible multi-substance fluid (model of the ocean) overlying the layered elastic half-space (model of the Earth's crust) and allowing for hyper-velocity shock waves and large deformations. This is still a very challenging task for the modern hydrodynamics and computational mathematics required and the application sophisticated numerical algorithms, like LPIC (Lagrangian Particles In Cell) method, and supercomputing.

One of the most fascinating examples of this kind of computations was made by D. Crawford of Sandia National Laboratory (Crawford, 1998) during the initial testing of the Intel Teraflop supercomputer and with additional purpose of generating unclassified data to test innovative visualization techniques. The CTH Shock Physics Hydrocode was used to model the impact of a 1 km diameter comet (with 300 GT

Fig. 14.5. Snapshot of the Crawford's numerical model of 1-km ice comet into the ocean. The comet and large quantities of ocean water are vaporized and ejected onto suborbital ballistic trajectories. Picture is downloadable from *http://sherpa.sandia.gov/planet-impact/comet/*

TNT equivalent) into a 4-km depth ocean. A large tsunami initially several kilometers high was generated and radiated from the point of impact (Fig. 14.5). However, it rapidly decayed having just 50–100 m high crests in the open ocean at the distance of 1 000 km from the impact site.

In this paper, I refer to the estimates of possible wave heights from the water impacts of a solid asteroid as function of its basic parameters (diameter, density and velocity) that were obtained by V. Petrenko (Petrenko, 2000), based on the available experimental data on underwater nuclear explosions (Glasstone and Dolan 1977), the rules of similarity for hydrodynamic processes and application of models developed for simulation of dynamics of compressible multi-substance fluids with large deformations (Petrenko, 1970). These estimates are shown in Table 14.1 for the "deep water" case (the size of the resultant cavity being smaller than the average water depth) and in Table 14.2 for a "shallow water" case (the size of the resultant cavity being comparable to or larger than the average water depth).

From the data in the tables, we can see that in the "deep water" case, a 200-meter stone asteroid (density 3 g cm^{-3}) falling into the water at 20 km s^{-1} speed is capable of generating 5-meter waves at a distance of 1 000 km. Similar waves are generated in the "shallow water" case. However, for a 500-meter asteroid the resulting wave height in the "deep water" case is almost double the size compared to the "shallow-water" case (21.9 m and 10.1 m, respectively).

Table 14.1. The estimated wave height h (m) and the impact kinetic energy W (GT TNT) in *deep water* at a distance of 1 000 km from the impact site as function of impactor diameter for iron-nickel ($\rho = 8.0$ g cm^{-3}) and stony ($\rho = 3.0$ g cm^{-3}) asteroids and ice comet ($\rho = 0.5$ g cm^{-3}) falling into water with the velocity 20 km s^{-1}

Diameter (m)	Density (g cm^{-3})					
	8.0		3.0		0.5	
	W (GT)	h (m)	W (GT)	h (m)	W (GT)	h (m)
200	1.60	8.4	0.60	5.0	0.10	1.9
300	5.39	16.2	2.02	9.6	0.34	3.6
400	12.77	25.9	4.79	15.2	0.80	5.8
500	24.93	37.1	9.35	21.9	1.56	8.3
1 000	199.47	114.1	74.80	67.2	12.47	25.5

Table 14.2. The estimated wave height h (m) and the impact kinetic energy W (GT TNT) in *shallow water* at a distance of 1 000 km from the impact site as function of impactor diameter for iron-nickel ($\rho = 8.0$ g cm^{-3}) and stony ($\rho = 3.0$ g cm^{-3}) asteroids and ice comet ($\rho = 0.5$ g cm^{-3}) falling into water with the velocity 20 km s^{-1}

Diameter (m)	Density (g cm^{-3})					
	8.0		3.0		0.5	
	W (GT)	h (m)	W (GT)	h (m)	W (GT)	h (m)
200	1.60	6.5	0.60	5.1	0.10	3.3
300	5.39	8.8	2.02	6.9	0.34	4.4
400	12.77	11.0	4.79	8.6	0.80	5.5
500	24.93	13.0	9.35	10.1	1.56	6.5
1 000	199.47	21.8	74.80	17.1	12.47	10.9

The further evolution of the initial water displacement strongly depends on the particular site conditions – whether it is a deep ocean, a marginal sea, an island archipelago or shallow-water coastal areas. Scattering in the final run-up and run-in distribution along the coast can be well above the factor of 10, thus making any scenario estimates of potential damage or human loss due to an oceanic asteroid tsunami very doubtful or even misleading. My personal feeling, based on long-term involvement in the study of historical and contemporary tsunamis and the analysis of available scenarios is that the total risk of asteroid tsunamis is somewhat overestimated in the literature, in particular, in the papers published in the 1990s (see for instance, Hills et al. 1994; Morrison et al. 1994). Under no conditions will 1% of the total population (that is more than 60 million people) be killed by tsunami from a single oceanic impact if it is below the global threshold.

14.5
Operational Tsunami Warning

The tectonically generated tsunamis can be predicted shortly before their arrival to the coast based on seismic observations and deep-water measurements. This is the task of the international Tsunami Warning System (TWS) that is in operation in the Pacific since the beginning of 60s (Master Plan 2000). The main operational center of this system is located in Ewa Beach, Hawaii, and provides 26 Pacific countries with operational warnings in about half to one hour after occurrence of an earthquake with magnitude above the threshold value (7.8 for most of the Pacific). Unfortunately, due to complexity and statistical nature of the tsunami generation process, these warnings quite often turn out to be false and, at the same time, several dangerous events in the last decade were not provided with timely warnings (1992 Nicaragua, 1994 Mindoro, Philippines, 1995 Jalisco, Mexico, and 1998 Papua New Guinea).

The International Co-ordination Group for the Tsunami Warning System in the Pacific (ICG/ITSU) was established by the Intergovernmental Oceanographic Commission (IOC) of UNESCO in 1965 for promoting the international cooperation and coordination of tsunami mitigation activities. It consists of national representatives from 26 Member States in the Pacific region and conducts biannual meetings to review progress and to coordinate the activity in improvement of the Tsunami Warning System. The IOC/UNESCO also supports the International Tsunami Information Center (ITIC) in Honolulu, Hawaii, whose mandate is to collect and distribute the data and information on tsunamis, to monitor and recommend improvements to the TWS, to assist in establishing national and regional TWSs in the Pacific and other tsunamigenic regions.

As mentioned above, on Earth the probability of an asteroid impact into a water basin is essentially higher that onto the land. Whereas available quantitative estimates of resulted run-up heights vary greatly, it is clear that a sub-kilometer asteroid can generate the significant tsunami that can be devastating locally or regionally. Such an impact will also produce a seismic waves that will be almost immediately detected by the global seismic network and, after routine processing, will be identified as a submarine earthquake with the very shallow focus depth. Even for a 10 MT impact, the estimated equivalent Richter magnitude is about 5.1, that is still well below the threshold ($Ms = 6.5$) for in-depth investigation adopted in the Pacific TWS, and such an event will be routinely placed on the list of current earthquakes. However, tsunami from the oceanic impact can be considerably higher (at least, locally) as compared to a submarine earthquake with the equivalent seismic magnitude.

Because of relative slowness of tsunami propagation on the continental slope and shelf, there will be a limited time interval, spanning from tens of minutes to several hours, to warn the population of coastal areas at risk and to implement the Tsunami Response and Mitigation Plan existing in many countries faced with the threat of tectonic tsunamis. However, being exceptionally oriented to seismically-induced tsunamis, the Pacific TWS may not recognize the signature of an asteroid impact if it occurred in an unusual place (i.e. in abyssal oceanic plate or aseismic marginal sea) and may not timely implement the standard tsunami evaluation procedure based on the analysis of telemetric tide gauge records and start the warning dissemination as prescribed by the TWS Communication Plan. As a result, the essential part of a possible warning time may be lost before non-standard warning situation is resolved and a potentially dangerous asteroid tsunami is identified and evaluated.

14.6
Detection of Impact Tsunamis by Tide Gauge Network

For the last one hundred years, we are sure that we did not miss any damageable impact-generated tsunami if it happened to occur in the World oceans. All the considerable coastal run-ups were associated with identified seismic, volcanic or landslide sources. However, we cannot be so confident in relation to numerous weak events that are identified only on tide gauge records.

Instrumentally, tsunamis are recorded by the world-wide network of tide gauge stations that has almost a 200-year history starting from the first tide gauge installed in Brest, France in 1807. In 1883 a distant tsunami resulting from the catastrophic Krakatau eruption was recorded by 35 instruments situated along the coast of the Pacific, the Atlantic and the Indian oceans (Simons 1888). By the beginning of the 20th century, there were nearly 100 tide stations in operation. Presently, the sea level recording system includes almost 1500 instruments installed all over the world (Fig. 14.6), some of them having real-time or near real-time telemetry to the data processing centers.

Normally the search for instrumental records starts from a report about the "event occurrence" (that usually comes from seismologists) or from a local account about unusual wave activity or coastal run-up. After that, the examination of records of the nearest tide stations is made in the time windows corresponding to the expected arrival times of tsunami, and the parts of records, containing the tsunami signal, stored

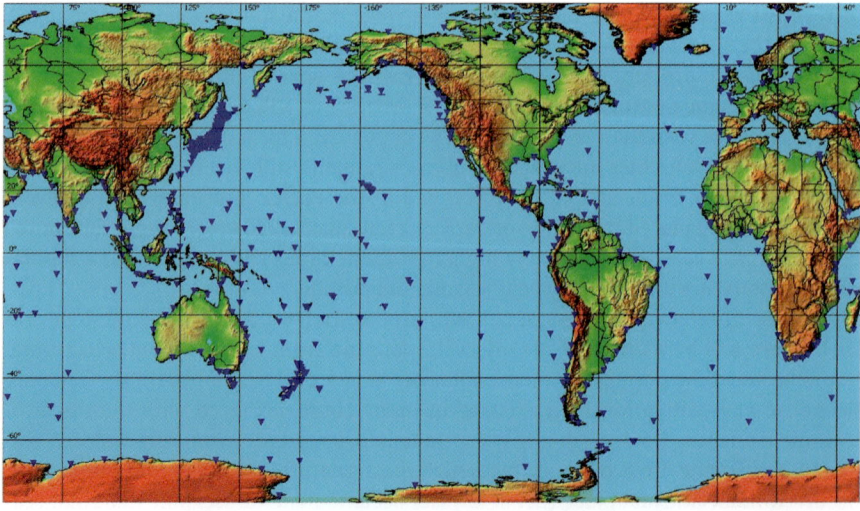

Fig. 14.6. Geographical distribution of the world-wide tide gauge network. The stations are shown as *blue triangles*

in the local archives. Agencies, that are responsible for operation of tide gauges, are interested in recording of long-term sea level changes, tides and storm surges and usually ignore small-amplitude impulsive signals appearing on tide gauge records. Thus, the signature of a small, "non-seismic" tsunami has little chance to be discovered and reported.

As distinct from seismologists, tsunami researchers do not have a routine system for systematic examination of tide gauge records with the purpose of identification of a "signal onset" on the tide records, their association between different stations and search for a possible source. The only exception is the work of S. Wigen (Wigen 1983) who systematically studied records from Tofino tide station (Canada) from 1906 to 1980 in order to identify potential tsunami arrivals and locate their sources. There is no doubt that a wealth of data on non-identified tsunami-like signals exists and these data are lying dormant in archives of tide gauge stations. Being found and reported, most of them would be later identified as records of small tsunamis from weak earthquakes, submarine slides or underwater eruptions. However, all such events originate in well-defined zones of seismic, volcanic or slumping activity. Discovering the signals with an estimated source outside of the well-known tsunamigenic areas would be a strong indication to a possible trace of oceanic impact. The absence of known reports of such an impact into the ocean doesn't mean that these events did not actually occur during the last one hundred years. It is worth noting that the 1908 Tunguska explosion remained almost unknown for scientists and general public for another 20 years (the first expedition to the area of explosion was conducted only in 1927).

14.7
Geological Traces of Tsunamis

The time coverage of historical tsunami catalogs that for many regions does not exceed 300–400 years can be greatly extended by application of geological methods of studying the paleotsunami traces preserved in coastal sediments or erosion features left by water impact on the coastal bedrocks. As was demonstrated by K. Minoura and S. Nakaya in Japan (Minoura, Nakaya, 1991) and later confirmed in numerous field studies in other countries, the invasion of a large volume of salt water onto land can produce a serious disturbance of the normal sedimentary process and leave unique deposits which could remain in the intertidal environment for a long time and, being investigated and interpreted in the correct manner, may represent a "geological chronicle" of a local tsunami history. In fact, tsunami deposits found in Texas was one of most essential evidences in favor of the reality of K/T boundary global catastrophe resulted from the Chicxulub impact (Bourgeois et al. 1988).

At present, paleotsunami studies are in progress in many countries providing a significant amount of new information about past tsunamis (Atwater 1987; Atwater et al. 1995; Bourgeois and Reinhardt 1989; Buckman et al. 1992; Darienzo and Peterson 1990; Dawson et al. 1988, 1991). At the Kamchatka Peninsula, the paleotsunamis studies made in 1993–1996 discovered up to 50 previously unknown pre-historical events that flooded the Kamchatka east coast during the last two thousand years (Pinegina and Bourgeois 2001). Thus, the historical catalog, covering in Kamchatka only the last 250 years, was extended more than ten times.

The search and investigation of geological traces of paleotsunami can be one of the primary methods to confirm the occurrence of impact-generated tsunamis in the past. Since 1990 numerous geological evidence of destructive water wave impact have been found on south-east and north-west coasts of Australia suggesting a mega-tsunami impacts at these aseismic coastlines that have not been flooded any historically known considerable tsunami (Bryant 2001; Bryant and Young 1996; Bryant et al. 1996; Bryant and Nott 2001; Nott and Bryant 2003). Characteristics of these tsunami (vertical flooding of 30–40 meters covering more then 1500 km of coastline, maximum vertical run-up reaching 130 m, horizontal flooding up to 5 kilometers inland) so dramatically differentiate them from the largest tectonically-induced tsunamis that their cosmogenic origin becomes almost obvious. Recent discoveries of the Mahuika crater in the shallow water near the South Island of New Zealand along with numerous Aboriginal and Maori legends about comets, smog, dust, fire and flood gives complementary evidence for a mega-tsunami resulting from a comet or asteroid impact in the Tasman Sea around 1500 AD (E. Bryant, pers. comm. December 2004).

Geological evidence of mega-tsunami exists in other regions. R. Paskoff (Paskoff 1991) describes large boulders in the Herradura Bay scattered over the shelly beach deposits associated with an abrasion platform at an elevation of 30–40 m above the modern beach southwest of Coquimbo, Chile. Similar boulders exist in the Guanaquero Bay about 25 km southwest of the Herradura Bay. By his opinion, these boulders were displaced by powerful waves coming from the northwest. At that time of publication his paper, ideas about cosmogenic tsunami were rather unusual, perhaps it was the reason why R. Pascoff talked about "an earthquakes of exceptional magnitude that happens only once in Plio-Quaternary time, probably around 300 000 years ago". Now we cannot completely ruled out the hypothesis that those boulders were deposited by a tsunami resulted from an asteroid impact somewhere in the southeast Pacific.

14.8
Conclusions

1. Both distantly and locally generated tsunamis are a typical example of "low probability – high consequence" hazard. Having, as a rule, a long recurrence interval (from 10–20 to 100–150 years) for a particular coastal location, they produce an extremely adverse impact on the coastal communities resulting in heavy property damage, a high rate of fatalities, disruption of commerce and social life.
2. In most of historical tsunamis, the major damage is confined to the nearby coast, but in some cases, waves may cross the entire ocean and devastate distant shorelines. Locally highly destructive tsunamis are generated after earthquake-triggered subaerial or submarine landslides. The size of tsunami generated by an earthquake relates to the energy released (earthquake magnitude), source mechanism, hypocentral depth and the water depth at the epicenter. The size of tsunami generated by landslide relates mainly to its volume as well as to the angle of inclined bottom (that controls the maximum slide velocity), initial water depth (for submarine slides) or relative slide body height (for subaerial slides) and the type of material involved in mass movement.

3. The expected tsunami height from an oceanic impact of sub-kilometer asteroid remains highly controversial. Estimates available in the literature vary by more than factor of ten. In the near-field zone, impact tsunamis can be much higher than any tectonically induced tsunamis. However, they have essentially shorter wave length that results in a faster energy decay (as $1/r$ against $1//\sqrt{r}$ for tectonic tsunamis) as they travel across the ocean. Therefore, the total area of perceptible damage for asteroid tsunamis cannot be very large (e.g., essentially more than 1000 km in radius) for all but the largest possible impacts approaching the threshold for a global catastrophe.
4. Being exceptionally oriented to seismically-induced tsunamis, the international Tsunami Warning System, established in the Pacific since 1965, might not recognize the signature of an asteroid impact and might not implement in a timely matter a standard tsunami evaluation procedure based on the analysis of tide gauge records available in real-time from many locations in the Pacific. As a result, an essential part of a possible warning time may be lost before an asteroid-induced tsunami can be identified.
5. The small tsunamis generated by Tunguska-class middle-oceanic impacts can hardly be visually observed even at the nearest coast. However, they can be recorded by the world-wide tide gauge network as small amplitude (from first centimeters to several tens of centimeters) short-period (from one to several minutes) impulsive wave trains. The careful search for "tsunami-like" signals of unknown origin on sea-level records available for the last one hundred years for many coastal locations can reveal the "signature" of oceanic impacts of extra-terrestrial bodies that otherwise could pass unnoticed.

Acknowledgments

The author is very grateful to Prof A. S. Alekseev, former director of the ICMMG SD RAS, who stimulated his interest to the study of the tsunami problem and, especially, its cosmogenic aspects. He also wishes to thank Mrs. T. Kalashnikova for her assistance with graphics for this paper. The studies mentioned in the paper were supported by the SD RAS grant 2006-113, RFBR grants 05-05-64460 and 04-07-90069.

References

Atwater BF (1987) Evidence for great Holocene earthquakes along the outer coast of Washington state. Science 236:942–944
Atwater BF, Nelson AR, Clague JJ et al. (1995) Summary of coastal geologic evidence about past great earthquakes at the Cascadia subduction zone. Earthquake Spectra 11:1–18
Bourgeois J, Hansen TA, Wilberg PL, Kauffman EG (1988) A tsunami deposits at the Cretaceous-Tertiary boundary in Texas. Science 241:576–570
Bourgeois J, Reinhardt MA (1989) Onshore erosion and deposition by the 1960 tsunami at the Rio Lingue estuary, South-Central Chile. EOS Trans Am Geoph Union 70:1331
Bryant E (2001) Tsunamis. The underrated hazard. Cambridge University Press, Cambridge
Bryant E, Nott J (2001) Geological indicators of large tsunami in Australia. Natural Hazards 24:231–249
Bryant E, Young RW (1996) Bedrock-sculpturing by tsunami, South Coast New South Wales, Australia. J Geology 104:565–582

Bryant E, Young RW, Price DM (1996) Tsunami as a major control on coastal evolution, Southeastern Australia. Journal of Coastal Research 12:831–840

Buckman RC, Hempton-Haley E, Leopold EB (1992) Abrupt uplift within the past 1700 years at souther Pudget Sound, Washington. Science 258:1611–1614

Chapman CR (2003) How a near-Earth object impact might affect society, commissioned by the OECD Global Science Forum for Workshop on Near Earth Objects: Risks, Policies, and Actions. Frascati, Italy, January 2003, downloadable from: *http://www.oecd.org/dataoecd/18/40/2493218.pdf*

Chubarov LB, Gusiakov VK (1985) Tsunamis and earthquake mechanism in the island arc region. Sci Tsunami Hazard 3(1):3–21

Crawford DA (1998) Modeling asteroid impact and tsunami. Sci Tsunami Hazards 16:21–30

Darienzo ME, Peterson CD (1990) Episodic tectonic subsidence of the late Holocene salt marshes, northern Oregon coast, central Cascadia margin. Tectonics 9:1–22

Dawson AG, Long D, Smith DE (1988) The Storegga slides: evidence from eastern Scotland for a possible tsunami. Marine Geology 82:271–276

Dawson AG, Foster ID, Shi S, Smith DE, Long D (1991) The identification of tsunami deposits in coastal sediment sequences. Science of Tsunami Hazard 9:73–82

Ewing M, Press F (1955) Tide-gauge disturbances from the Great Eruption of Krakatoa. Trans AGU 36:53–60

Glasstone S, Dolan PJ (1977) The effects of nuclear weapons. US Government Printing Office, Washington, DC

Gonzalez F (1999) Tsunami!, Scientific American 280(5):56–65

Gusiakov VK (2003) NGDC/HTDB meeting on the historical tsunami database proposal. Tsunami Newsletter 35(4):9–10

Hills JC, Goda P (1998) Tsunami from asteroid and comet impacts: the vulnerability of Europe. Sci Tsunami Hazards 16:3–10

Hills JG, Mader CL (1997) Tsunami produced by the impacts of small asteroids. Annals of the New York Academy of Science 822:381–394

Hills, JG, Nemchinov IV, Popov SP, Teterev AV (1994) Tsunami generated by small asteroid impacts. In: Gehrels T (ed) Hazards due to comets and asteroids. University of Arizona Press, pp 779–790

Imamura A (1942) History of Japanese tsunamis. Kayo-No-Kagaku (Oceanography) 2(2):74–80

Iwasaki SI (1997) The wave forms and directivity of a tsunami generated by an earthquake and a landslide. Sci Tsunami Hazard 15:23–40

Lander JF (1996) Tsunamis affecting Alaska, 1737–1996. National Geophysical Data Center, Boulder, Colorado

Mader CL (1998) Asteroid tsunami inundation of Japan. Sci Tsunami Hazards 16:11–16

Master Plan for the Tsunami Warning System in the Pacific (2000) Second Edition Intergovernmental Oceanographic Commission, IOC/INF-1124, Paris

McCoy, FW, Heiken, G (2000) Tsunami generated by the Late Bronge Age eruption of Thera (Santorini), Greece. Pure Appl Geophys 157:1227–1256

Minoura K, Nakaya S (1991) Traces of tsunami preserved in inter-tidal lakustrine and marsh deposits: some examples from northeast Japan. J of Geology 99(2):265–287

Minoura, K, Imamura, F, Kuran U, Nakamura T, Papadopoulos GA, Takahashi T, Yalciner AC (2000) Discovery of Minoan tsunami deposits. Geology 28:59–62

Morrison D (2003) Tsunami hazard from sub-kilometer impacts In: Summary of Impact Tsunami Hazard Workshop in Houston, March 16, 2003, downloadable from *http://128.102.38.40/impact/news_detail.cfm?ID=123*

Morrison D, Chapman C, Slovic P (1994) The impact hazard. In: Gehrels T (ed) Hazards due to comets and asteroids. University of Arizona Press, pp 60–91

Murty TS (2003) A review of some tsunamis in Canada In: Yalciner A, Pelinovsky E, Synolakis C, Oka E (eds),Submarine landslides and tsunamis. NATO Science Series: IV, Earth and Environmental Sciences: vol. 21, Kluwer Academic Publishers, Dordrecht, pp 173–183

Nott J, Bryant E (2003) Extreme marine inundation (tsunamis?) of coastal Western Australia. Journal of Geology 111:691–706

Okal EA, Synolakis CE (2001) Identification of the sources, NATO Advanced Research Workshop "Underwater Ground Failures on Tsunami Generation, Modeling, Risk and Mitigation", Istanbul, 2001, pp 36–37

Paskoff R (1991) Likely occurrence of a mega-tsunami in the Middle Pleistocene, near Coquimbo, Chile. Revista Geologica de Chile 18(1):87–91

Petrenko VE (1970) The Lagrangian particle-in-cell method for the calculation of the dynamics of compressible multi-substance fluids with large deformations. Computing Center, Novosibirsk, Report no. 58-A (in Russian)

Petrenko VE (2000) Numerical modeling of asteroid impact. In: Tsunami hazard for the Mediterranean region. Report to the INTAS-RFBR-95-1000 Project, Novosibirsk, ICMMG SD RAS (in Russian)

Pinegina TK, Bourgeois J (2001) Historical and paleotsunami deposits on Kamchatka, Russia: long-term chronology and long-distance correlation. Natural Hazards and Earth System Sciences 1:177–185

Press F, Harkrider D (1966) Air-sea waves from the explosion of Krakatoa. Science 154:1325–1327

Rabinovich AB, Thompson RE, Bornhold BD, Fine IE, Kulikov EA (2003) Numerical modeling of tsunamis generated by hypothetical landslides in the Strait of Georgia, British Columbia. Pure Appl Geoph 160:1273–1313

Soloviev SL (1978) Tsunamis. In: The assessment and mitigation of earthquake risk. UNESCO, Paris, pp 118–139

Symons GJ (1888), The eruption of Krakatoa and subsequent phenomena. Report of the Krakatoa committee of the Royal Society. Trubner & Co, London

Synolakis CE, Bardet JP, Borrero JC, Davies HL, Okal EA, Silver EA, Sweet S, Tappin DR (2002) The slump origin of the 1998 Papua New Guinea tsunami. Proc Royal Soc London 458:763–790

Tappin DR, Watts P, McMurtry GM, Lafoy Y, Matsumoto T (2002) Prediction of slump generated tsunamis: The 17 July 1998 Papua New Guinea event. Sci Tsunami Hazards 20(4):222–238

Van Dorn WG (1968) Handbook of explosion-generated water waves, vol 1: state of the art. TTR Report TC-130

Ward SN, Asphaug E (2000) Asteroid impact tsunami: a probabilistic hazard assessment. Icarus 145: 64–78

Ward SN, Asphaug E (2003) Asteroid impact tsunami of 2880 March 16. Geophys J Int 153:F6–F10

Wigen SO (1983) Historical study of tsunamis at Tofino, Canada. In: Iida K, Iwasaki T (eds) Tsunami – their sciences and engineering. Terra Scientific Publ. Co., Tokyo, pp 105–119

Yokoyama I (1981) A geophysical interpretation of the 1883 Krakatau eruption. J Volcanology and Geothermal Research 9:359–378

Chapter 15

The Physical and Social Effects of the Kaali Meteorite Impact – a Review

Siim Veski · Atko Heinsalu · Anneli Poska · Leili Saarse · Jüri Vassiljev

15.1
Introduction

There is a concern that the world we know today will end in a global ecological disaster and mass extinction of species caused by a meteorite impact (Chapman and Morrison 1994; Chapman 2004). We are aware that rare large impacts have changed the face of our planet as reflected by extinctions at the Permian/Triassic (~251 Ma; Becker et al. 2001), Triassic/Jurassic (~200 Ma; Olsen et al. 2002) and Cretaceous/Tertiary (~65 Ma; Alvarez et al. 1980) boundaries. Today astronomers can detect and predict the orbits of the asteroids/comets that can cause similar impacts. Yet, Tunguska, Meteor Crater-size and smaller meteorites that could cause local disasters are unforeseeable. However, while planning to avoid the next bombardment by cosmic bodies we can look at past interactions of human societies, environment and meteorite impacts to understand to what extent human cultures were influenced by meteorite impacts. The question is whether the past examples are relevant in the modern situation, but they are certainly useful. The Kaali crater field in Estonia, in that respect, is an excellent case study area for past human–meteorite interactions. Moreover, Kaali is not the only Holocene crater field in this region: in fact, during the last 10 000 years Estonia has been targeted at least by four crater forming impacts and there are five registered meteorite falls (Fig. 15.1). The two large craters, Neugrund and Kärdla, originate from 535 and 455 Ma, respectively (Suuroja and Suuroja 2000). The role of earth sciences combined with other natural sciences and archaeology in meteorite impact research is mainly to study the physical record of (pre)historic impacts (cratering), the evidence for past effects on biological organisms (extinction and disturbance events), causal effects of the impact such as the impact created tsunami damages and human cultures. The latter issue in Estonia is somewhat difficult to evaluate directly as unfortunately the possible people witnessing the Kaali impact were illiterate and there is no direct written record of the impact event, although there is a variety of indirect archaeological and oral material present. Considering that the Kaali meteorite impact had a wider reverberation in the contemporary world we may argue that some of the much-quoted early European written records possibly describe the event.

There are many ways a meteorite impact can influence societies, including changes in climate, tsunamis, earthquakes, wildfires, acid rain, greenhouse effects, the intensity of which depends on the size and target of the impact and the distance from it. But in the long run there are basically two options: (1) by extermination, and (2) when the impact is smaller by utilization and worship. From the past, though, we seek the signal

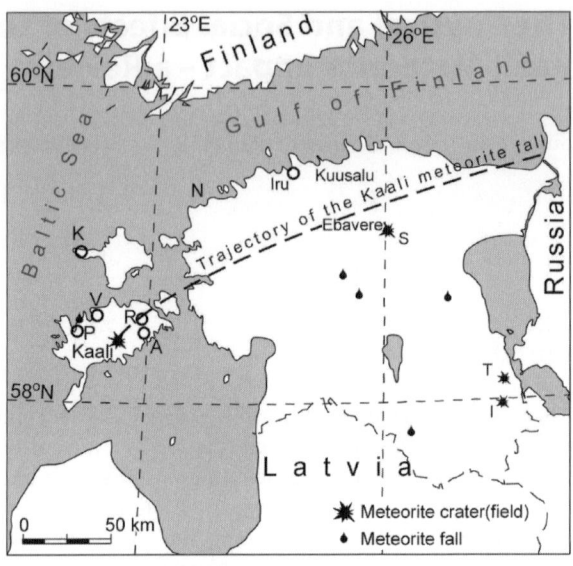

Fig. 15.1.
Map of Estonia showing Holocene impact craters (Kaali on the Island of Saaremaa, S – Simuna, T – Tsõõrikmäe, I – Ilumetsa), registered meteorite falls and places mentioned in the text (A – Asva, P – Pidula, R – Ridala, V – Võhma, K – Kõivasoo bog, N – Neugrund)

of the meteorite in meteorite utilization, worship and cultic activity, but today we would assess the impact damage and put it all into an economic framework. There are dozens of examples from all over the world of meteorite utilization, worship and legends (e.g. Blomqvist 1994, Hartmann 2001, and references within; Santilli et al. 2003). One of the legends is the voyage of Pytheas, a Greek explorer, who between 350–325 BC visited the island Ultima Thule far in the north, where the barbarians showed him "the grave where the Sun fell dead". According to the interpretation of Meri (1976) the place was the Kaali crater on the Island of Saaremaa. The reason he suggested that Lake Kaali and the meteorite impact were known among the geographers and philosophers before Cornelius Tacitus, who in his book *De Origine et Situ Germanorum Liber* wrote "Upon the right of the Suevian Sea [the Baltic] the Aestyan nations [Estonians] reside, who use the same customs and attire with the Suevians [Swedes]. They worship the Mother of the Gods" (Tacitus 1942). The Mother of Gods, Cybele (Rhea), is associated with meteorites (Burke 1986). Also Phaeton is connected with celestial bodies and more precisely with Kaali (Blomqvist 1994). The Argonautica of Apollonius of Rhodes (295–215 BC) may describe the Kaali crater lake: "… where once, smitten on the breast by the blazing bolt, Phaethon half-consumed fell from the chariot of Helios into the opening of that deep lake; and even now it belcheth up heavy steam clouds from the smouldering wound" (Seaton 1912). Apart from classical literature the Kaali phenomenon may be reflected in the Estonian and Finnish folklore and eposes (Jaakola 1988).

15.2
The Meteorite

The Kaali meteorite impact site (main crater 58° 22' 22" N, 22° 40' 08" E) with nine identified craters is located on the Island of Saaremaa, Estonia (Fig. 15.2). Geological and chemical studies in the area suggest that an iron meteoroid of type IAB, weighing

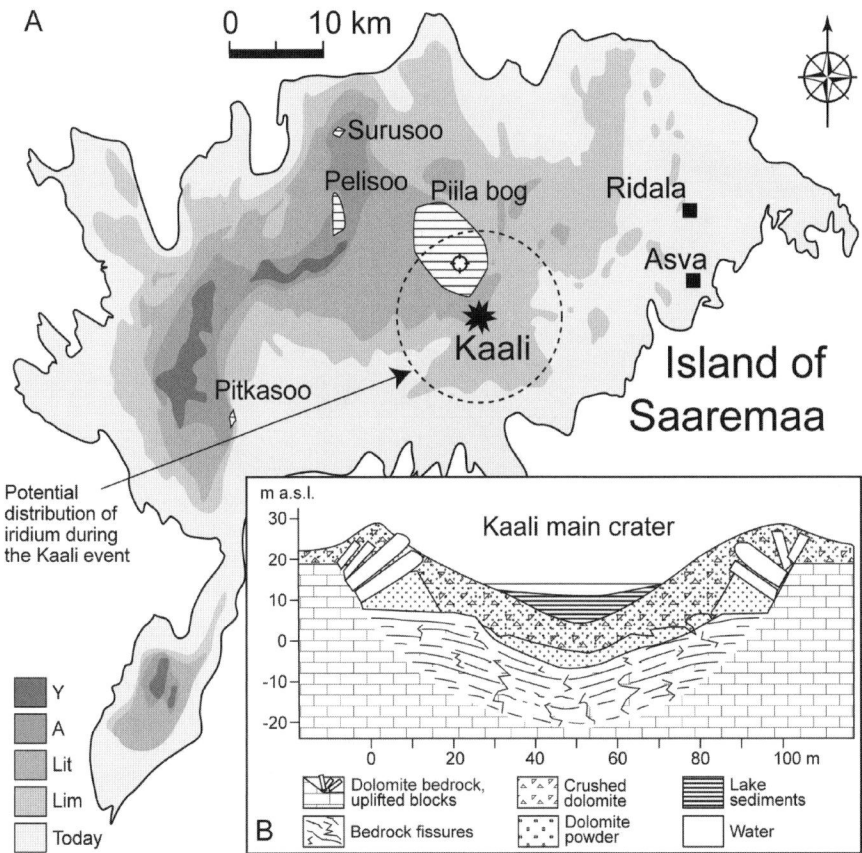

Fig. 15.2. a Map showing the Island of Saaremaa, Estonia, location of the investigated sites, Kaali meteorite crater field, and Piila, Surusoo, Pelisoo and Pitkasoo bogs. The map displays the location of the shoreline during the different stages of the Baltic Sea (*Y* – Yoldia Sea 9000 BC, *A* – Ancylus Lake 8000 BC, *Lit* – Litorina Sea 6000 BC, *Lim* – Limnea Sea 2000 BC). **b** Geological cross-section of the main crater at Kaali (modified from Aaloe 1968)

~1000 tons (estimations range from 400 to 10 000 t) fell at an angle estimated to be ~35° from the northeast (Bronshten 1962; Aaloe 1968). Others suggest that the meteorite fell from the southeast (Reinwaldt 1937) or south (Krinov 1961). The target rock consisted of Silurian dolomites covered by a thin layer of Quaternary till (Fig. 15.2b). Altogether 3.5 kg of meteorite iron of coarse octahedrite class (Buchwald 1975) has been collected in Kaali (Raukas 2004). The largest piece weighs ~30 g (Saarse et al. 1991), but the bulk consists of small, less than a gram, particles (Marini et al. 2004). The iron contains 7.25% of Ni, 2.8 µg g^{-1} of Ir, 75 µg g^{-1} of Ga and 293 µg g^{-1} of Ge (Yavnel 1976). While penetrating the atmosphere, the meteoroid heated and broke into pieces. It is estimated that the largest fragment was ~450 tons, and struck the ground surface with an energy of ca. 4×10^{12} J, corresponding to an impact velocity of ~15 km s^{-1} (Bronshten and Stanyukovich 1963). The resulting crater is 16 m deep (rim-to-floor depth is

Fig. 15.3. Kaali main crater from air. Photo by Ants Kraut (Estonian National Heritage Board)

22 m) and has a diameter of 105–110 m (Fig. 15.3). The depression is today filled with water and at least 5–6 m of lake and bog deposits (Fig. 15.2b). The cluster of smaller meteoroids produced eight satellite craters with diameters ranging from 12 to 40 m and up to 4 m deep scattered over an area of 1 km². The total energy of all nine impacts was ~4.7×10^{12} J, which is equivalent to about 5–20 kilotons of TNT (calculated from Bronshten and Stanyukovich 1963). Raukas (2004) estimated the energy forming the main crater at Kaali to be 1–2 kT TNT.

15.3
Age of the Impact

Kaali craters were not always regarded to be of cosmic origin. Rauch (1794) suggested that the crater was a fossil volcano (Raukas 2002). Subsequently, the peculiar Kaali landform was also considered as created by eruption, karst phenomena, gypsum or salt tectonics (Raukas et al. 1995). Since the 1920s, when the craters were first described as potentially of meteoritic origin (Kalkun 1922; Kraus et al. 1928; Reinwaldt 1928), discussions about their age were initiated. First, Linstow (1919) estimated the age of the craters to be 8000–4000 BC. Reinwald (1938) proved the meteoritic origin of the craters by collecting 30 fragments of meteoritic iron from satellite craters and further concluded, based on the presence of land snails, that the craters were young, possibly of postglacial time. He also understood the need to perform pollen analysis on the lake sediments inside the crater to estimate the age of impact (Reinwaldt 1933). Aaloe (1958) took into account the speed of the glacio-isostatic land uplift on the Island of Saaremaa and suggested that the craters formed around 3000–2500 BC, when the area had emerged during the Litorina Sea stage from the Baltic basin. First conventional radiocarbon dating of charcoal, wood and peat from the satellite craters suggested that they may have formed about 1100–600 BC (Aaloe et al. 1963). By interpolating the pollen evidence from a sedimentary core in the main crater, Kessel (1981) estimated the age of the impact as around 1800 BC. Saarse et al. (1991) radiocarbon dated the near-bottom lake sediments of the Kaali main crater from a bulk sample of calcareous gyttja overlying the dolomite debris and proposed an age of 1740–1620 BC. Raukas et al. (2003) acquired several wood samples from excavations on the crater slope obtaining ages ranging from 760 to 390 BC. Veski et al. (2004) re-dated terrestrial macrofossils in the bottom sediments of the water-filled Kaali main crater by AMS radiocarbon method by sampling the deepest minerogenic layers of the Kaali main crater (representing the meteorite impact fallback ejecta consisting of crushed dolomite debris and dolomite powder) indicating the initial post-impact filling of the crater. The obtained age of the crater is 1690–1510 BC using traditional paleolimnological approaches, though Rasmussen et al. (2000) and Veski et al. (2004) put forward some doubts on the possibility of ^{14}C dating inside the Kaali craters. Sediment disturbance by falling trees, mixing with in washed old humus and/or hard-water effects could have influenced the radiocarbon determinations from sediments of shallow hard-water lakes such as Kaali.

Apart from estimating the age of the meteorite impact from inside the crater one can detect the signature of the meteorite shower and impact ejecta in peat bogs near the crater. Different groups of researchers have used ^{14}C dating of peat layers with extraterrestrial material or particles supposedly formed by melting and vaporization of impactor and target material during the impact. A horizon with glassy siliceous microspherules in the peat of Piila bog (6 km northwest from Kaali craters and 310–300 cm below bog surface; Fig. 15.2a) is dated radiometrically back to ~6400 BC (Raukas et al. 1995; Raukas 2000, Raukas 2004). Similar microspherules have been found in the Early Atlantic layers of peat at the Pelisoo and Pitkasoo mires some 18 km NW and 30 km SW from Kaali, respectively (Fig. 15.2a) as well as in the peat of Kõivasoo bog on Hiiumaa Island (Fig. 15.1) ~70 km NW from Kaali (Raukas 2004). However, the early Holocene age of the impact can be ruled out given the local Quaternary geology and history of the Baltic Sea. At 6400 BC (see Raukas et al. 1995), the water level of the Baltic Sea basin, whose shore was situated about 2 km away from Kaali at that time, was approximately 16 m above the present sea level and was still rising (Fig. 15.4). This means that the bottom of the Kaali main crater had to be at least 9 m below the contemporary sea-level and consequently filled with water as the groundwater level cannot be lower than the sea-level (Veski et al. 2002, 2004). Moreover, the initial sediments of the crater contain pollen of spruce (Veski et al. 2004) that immigrated and established on Saaremaa Island starting only about 3800 BC (Saarse et al. 1999).

In the abovementioned Piila bog Rasmussen et al. (2000) found a peat layer at 172–177 cm below peat surface that has an elevated Ir content (up to 0.53 ppb) and is dated to about 800–400 BC (Fig. 15.5). This marker horizon has been considered to represent the signal of the Kaali iron meteorite outside the crater area (Rasmussen et al. 2000; Veski et al. 2001). Thus, currently there are three contradicting hypotheses about the age of the Kaali meteorite impact. Two of them rely on ^{14}C dating of peat

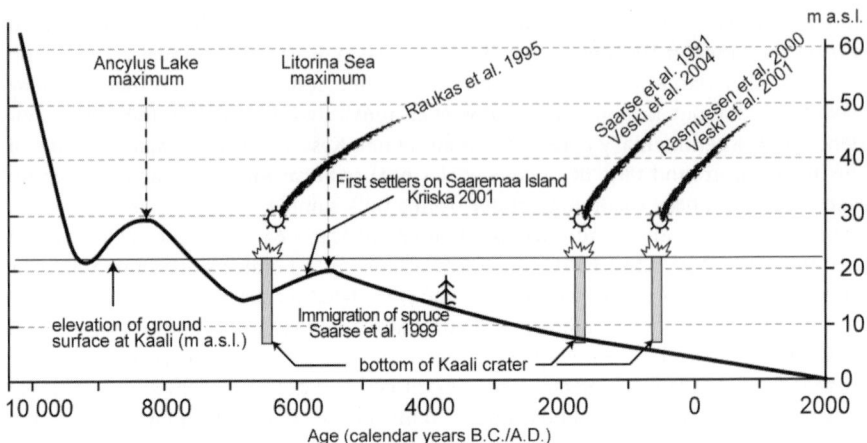

Fig. 15.4. The altitude of the Kaali meteorite target area (22 m a.s.l.) projected to the shore displacement curve for the Kaali area (after the database of Saarse et al. 2002) and three hypotheses of the age of the Kaali meteorite fall

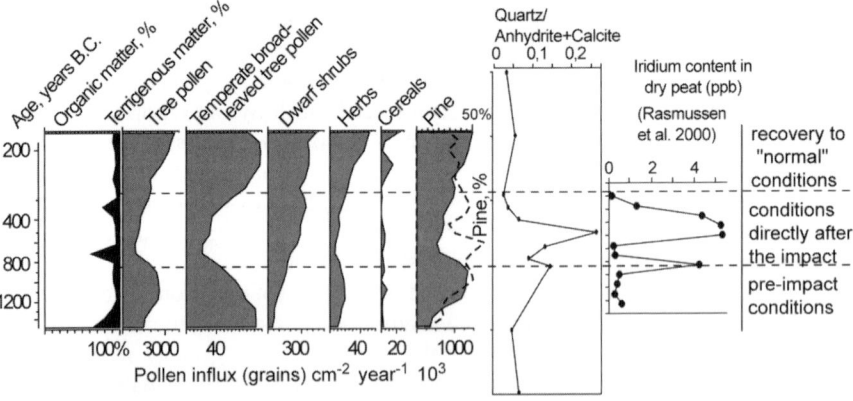

Fig. 15.5. Loss-on-ignition (LOI) and selected pollen accumulation rates (pollen grains cm^{-2} yr^{-1}) diagram from the investigated section at Piila bog. Percentage values of *Pinus* (*dotted line*) are given together with influx of *Pinus* pollen. Generalized mineral composition of peat ash, the ratio of quartz / anhydrite + calcite and the iridium concentration (ppb) in the peat ash (Rasmussen et al. 2000)

layers with impact ejecta found in nearby bogs and the third, more classic approach, on radiocarbon dated terrestrial macrofossils from the near-bottom lake sediments of the Kaali main crater. The relevance of these dates is more thoroughly assessed in Veski et al. (2004). Briefly, the age of the impact estimated inside the crater is 1690–1510 BC which is about 1000 years older from that revealed from the Ir-rich marker-horizon in a contemporaneous peat sequence. The microspherules discovered by Raukas et al. (1995) could indicate another much older event not connected with the Kaali impact.

15.4
Effects of the Meteorite Impact

The statistical frequency of impacts by bodies of various sizes is fairly well known, less well understood are the physical and environmental consequences of impacts of various sizes (Chapman 2004). Impact hazard studies tend to focus on Tunguska-size and larger bodies (Chapman and Morrison 1994), but as far as we know the only crater forming impact that fell into a relatively densely inhabited region was Kaali, which is a magnitude smaller than Tunguska. Nevertheless, even smaller meteorites disturb the local environment for a long time period. Effects of the impact vary from the initial deposition of meteoritic matter during the entrance of the meteoroid, the impact explosion, deformation of the target rock, blast and heat wave, and cratering. After the impact the crater and its nearest vicinity is a classic primary succession habitat (Cockell and Blaustein 2002), at sufficient distances from the crater the blast wave may fell trees and destroy vegetation (Kring 1997) leaving a secondary succession habitat with damages similar of tornadoes and hurricanes. Cockell and Lee (2002) divide the post-impact biology of craters into three phases: phase of thermal biology, phase of impact succession and climax, and phase of ecological assimilation.

The physical effects of the impact, apart from the crater field itself, may be recorded in the surrounding sedimentary archives – the bogs. The above-mentioned peat horizon with elevated Ir contents in the Piila bog combines additional multiple evidence that may be connected to the Kaali event. Significant changes occur in loss-on-ignition (LOI), pollen accumulation, composition of pollen and mineral matter in the 8-cm peat layer, that contains iridium (Fig. 15.5). Enriched Ir values in the peat are possibly primarily formed as a result of atmospheric dispersion of Ir during the entrance and break-up of the meteoroid. The Ir signal is present at Piila bog, but was not found at Surusoo bog 25 km NW from Kaali (Fig. 15.2a), which sets a limit to the effect of the meteorite impact. Associated with the Ir-enriched horizon in Piila is a marked charred layer of peat spread over the entire bog basin and indicating that the whole bog probably suffered from a severe burn. LOI and X-ray diffraction analysis of the same peat layer show increased input of inorganic allochthonous material (up to 20% of quartz and feldspars). Above-mentioned mineral matter accumulated in the peat as impact ejecta during the explosion and/or later, as post-impact aeolian dust during the period of increased erosion of the fire-destroyed topsoil in the surroundings. Pollen evidence reveals that the impact swept the surroundings clean of forest, which is shown by the threefold decrease in pollen influx (especially tree pollen influx), increase in influx and diversity of herb taxa and the relative dominance of pine (Fig. 15.5). Over representation of *Pinus* percentages is a common feature for barren landscapes. The temperate broad-leaved trees on fertile soils outside the bog were most affected, which indicates that the disruptions in vegetation were not just local features around the sampling site in the bog. Pollen evidence indicates a gradual recovery of vegetation from the impact, thus, the effect of the Kaali impact on landscape is hidden by new generations of vegetation.

Indicators of cultivated land, such as the pollen of cereals *Triticum*, *Hordeum* and *Secale*, which were continuously present in pre-impact conditions, disappear after the impact. The disappearance of cereals suggests that farming, cultivation and possibly human habitation in the region was disturbed for a period. However, archaeological evidence from the ring-wall of the main crater at Kaali displays signs of habitation in the Late Bronze Age–Pre-Roman Iron Age (approximately 500–700 years BC; Lõugas 1980), indicating that people did not abandon the area, but, on the contrary, soon after used the rim of the crater as part of their fortification and/or ceremonial purposes. We do not possess many examples of human societies interacting with crater forming meteorites and these very few should be studied thoroughly. Currently, archaeological evidence does not tell much about the imprint of meteorite impacts nor may we with certainty associate the legends and folk songs with these impacts. One, however, provides a relatively clear picture. The Island of Saaremaa was inhabited since the Mesolithic period, around 5800 BC (Kriiska 2000). During the Neolithic and Bronze Ages, Saaremaa was densely populated, and half of the bronze artefacts of Estonia originate from this island (Ligi 1992). Three late Bronze Age fortified settlements, Asva, Ridala and Kaali, are known from Saaremaa (Fig. 15.2a). The main economy was cattle rearing and agriculture. Continuous signs of crop cultivation (cereal pollen grains in sediments) on the island of Saaremaa appear at approximately 2300 BC (Poska and Saarse 2002). Archaeological evidence around, inside, and on the Kaali crater slopes suggests human habitation since about 700–200 BC. The impact must have been witnessed and most probably worshiped in some way as the archaeological record at Kaali suggests primarily a ceremonial purpose for the complex (Veski et al. 2004). Although there is a record that small meteorites have caused human casualties (Yau et al. 1994 and references within) we may say nothing of the kind in the case of Kaali. Some archaeological sites on Saaremaa seem to mirror the shape of the Kaali complex, consisting of two concentric circles built of stones. The best examples come from Võhma and Pidula (Fig. 15.6). There is evidence of impact craters utilization by humans in other different parts of the world. For instance, the Tswaing impact crater in South Africa appears to have been visited by Stone Age people to collect salt (Reimold et al. 1999). Several other craters have been preferentially used as agricultural land (Cockell and Lee 2002).

The ethnographic material that has been related to the Kaali impact is wide and outside the expertise of environmental scientists. The Kaali phenomenon supposedly had a major impact on Estonian-Finnish mythology, folklore, involvement in ironmaking and trade (Meri 1976; Jaakola 1988; Raukas 2002; Haas et al. 2003). Particularly

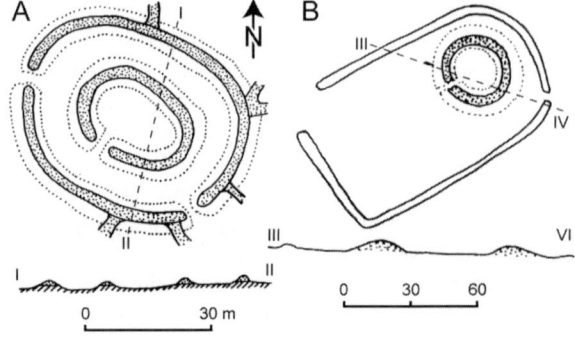

Fig. 15.6.
Archaeological sites on the island of Saaremaa that seem to mirror the Kaali complex, having two concentric circles built of stones. A – Võhma. B – Pidula. Both from northwest Saaremaa, see Fig. 15.1

the natural phenomena such as the birth of fire, the giant figures, the blind archer and the role of iron is believed to originate from the Kaali event (Jaakola 1988). The epics refer to the birth of iron in a lake. The legend written down in the chronicle of Henry the Livonian (*Heinrici Chronicon Livoniae*, early 13th century, in Tarvel and Kleis 1982) about the god Tharapita (Taara), who was born on the hill of Ebavere (located in northeast Estonia, on the trajectory of the meteorite, see Fig. 15.1) and flew to the island of Saaremaa from there may be a reflection of this event. Songs in north Estonian (Kuusalu) folklore describe the burning of the Island of Saaremaa, etc. Outside the Baltic area, the legend of Phaeton is connected with celestial bodies and more precisely with Kaali (Blomqvist 1994; Raukas 2002). There can be no scientific verification that all these tales are reflections of the Kaali meteorite explosion, but we cannot exclude the possibility. Even today the craters at Kaali are a major tourist attraction and part of ceremonial traditions.

Acknowledgments

The research was supported by ESF 4963. This is a contribution to the ICSU Workshop: Comet/Asteroid Impacts and Human Society. We thank Ants Kraut, Estonian National Heritage Board for Kaali images. The paper was much improved by reviewers W. B. Masse, U. Miller and J. Plado.

References

Aaloe A (1958) Kaalijärve meteoriidikraatri nr. 5 uurimisest 1955. aastal. ENSV TA Geoloogia Instituudi uurimused 2:105–117 (in Estonian)
Aaloe A (1968) Kaali meteoriidikraatrid. Eesti Raamat, Tallinn (in Estonian)
Aaloe A, Liiva A, Ilves E (1963) Kaali kraatrite vanusest. Eesti Loodus 6:262–265 (in Estonian)
Aaloe A, Andra H, Andra V (1982) Direction of the Kaali meteorite shower according to geophysical data. Proceedings of the Estonian Academy of Science Chemistry-Geology 31:56–61 (in Russian)
Alvarez LW, Alvarez W, Asaro F, Michel HV (1980) Extraterrestrial cause for the Cretaceous–Tertiary extinction: experimental results and theoretical interpretation. Science 208:1095–1108
Becker L, Poreda RJ, Hunt AG, Bunch TE, Rampino M (2001) Impact event at the Permian–Triassic boundary: evidence from extraterrestrial noble gases in Fullerenes. Science 291:1530–1533
Blomqvist J (1994) The fall of Phaethon and the Kaalijärv meteorite crater: is there a connexion? Eranos. Acta philologica Suecana 92:1–16
Bronshten VA (1962) Ob obstojatelstvah padenija Kaalijarvskogo meteorita [On the fall of Kaali Meteorite]. Meteoritika 22:42–46 (in Russian)
Bronshten VA, Stanyukovich K (1963) On the fall of the Kaalijärv meteorite. ENSV TA Geoloogia Instituudi Uurimused 11:73–83 (in Russian)
Buchwald VF (1975) Handbook of iron meteorites. University of California Press, Berkeley
Burke JG (1986) Cosmic debris: meteorites in history. University of California Press, Berkeley
Chapman CR (2004) The hazard of near-Earth asteroid impacts on Earth. Earth and Planetary Science Letters 222:1–15
Chapman CR, Morrison D (1994) Impacts on the Earth by asteroids and comets – assessing the hazard. Nature 367:33–40
Cockell CS, Blaustein AR (2002) Biological recovery after impact events – effects of scaling. Astrobiology in Russia, March 25–29, 2002, St. Petersburg, Russia, pp 185–203
Cockell CS, Lee P (2002) The biology of impact craters – a review. Biological Reviews 77:279–310
Haas A, Peekna A, Walker RE (2003) Echoes of Stone Age cataclysms in the Baltic Sea. Folklore 23:49–85

Hartmann WK (2001) Sociometeoritics. Meteoritics and Planetary Science 35:1294–1295
Jaakkola T (1988) The Kaali giant meteorite fall in the Finnish–Estonian folklore. In: Hänni U, Tuominen I (eds) Proceedings of the 6th Soviet-Finnish Astronomical Meeting. W. Struve Astrophysical Observatory of Tartu, Tallinn, 10–16 Nov 1986, pp 203–216
Kalkun JO (1922) Üldine geoloogia. Pihlakas, Tallinn (in Estonian)
Kessel H (1981) Kui vanad on Kaali järviku põhjasetted. Eesti Loodus 24:231–235 (in Estonian)
Kraus E, Meyer R, Wegener A (1928) Untersuchungen über den Krater von Sall auf Ösel. Kurlands Beiträge zur Geophysik 20:312–378
Kriiska A (2000) Settlements of coastal Estonia and maritime huntergatherer economy. Lietuvos Archeologija 19:153–166
Kring DA (1997) Air blast produced by the Meteor Crater impact event and a reconstruction of the affected environment. Meteoritics and Planetary Science 32:517–530
Krinov EL (1961) The Kaalijärv meteorite craters on Saaremaa Island, Estonian SSR. American Journal of Science 259:430–440
Ligi P (1992) The prehistory of Saaremaa. PACT 37:163–173
Linstow O (1919) Der Krater von Sall auf Oesel. Zentralblatt für Mineralogie, Geologie und Paläntologie 21/22:326–329
Lõugas V (1980) Archaeological research at Kaali meteorite crater. Proceedings of the Estonian Academy of Sciences, Humanities 29:357–360 (in Estonian)
Marini F, Raukas A, Tiirmaa R (2004) Magnetic fines from the Kaali impact-site (Holocene, Estonia): preliminary SEM investigation. Geochemical Journal 38:107–120
Meri L (1976) Hõbevalge. Eesti Raamat, Tallinn (in Estonian)
Olsen PE, Kent DV, Sues H-D, Koeberl C, Huber H, Montanari A, Rainforth EC, Fowell SJ, Szajna MJ, Hartline BW (2002) Ascent of dinosaurs linked to an iridium anomaly at the Triassic–Jurassic boundary. Science 296:1305–1307
Poska A, Saarse L (2002) Vegetation development and introduction of agriculture to Saaremaa Island, Estonia: the human response to shore displacement. The Holocene 12:555–568
Rasmussen KL, Aaby B, Gwozdz R (2000) The age of the Kaalijärv meteorite craters. Meteoritics and Planetary Science 35:1067–1071
Rauch JE (1794) Nachricht von der alten lettischen Burg Pilliskaln, und von mehreren ehemaligen festen Plätzen der Letten und Ehsten; auch von etlichen andern lief- und ehstländischen Merkwürdigkeiten. Neue Nordische Micellaneen 9/10:540–541
Raukas A (2000) Investigation of impact spherules – a new promising method for correlation of Quaternary deposits. Quaternary International 68–71:241–252
Raukas A (2002) Postglacial impact events in Estonia and their influence on people and the environment. In: Koeberl C, MacLeod KG (eds) Catastrophic events and mass extinctions: impacts and beyond. Geological Society of America Special Paper 356:563–569
Raukas A (2004) Distribution and composition of impact and extraterrestrial spherules in the Kaali area (Island of Saaremaa, Estonia). Geochemical Journal 38:101–106
Raukas A, Pirrus R, Rajamäe R, Tiirmaa R (1995) On the age of the meteorite craters at Kaali (Saaremaa Island, Estonia). Proceedings of Estonian Academy of Sciences, Geology 44:177–183
Raukas A, Laigna K, Moora T (2003) Olematu looduskatastroof Saaremaal 800–400 aastat enne Kristust. Eesti Loodus 54:12–15 (in Estonian)
Reimold WU, Barndt D, De Jong R, Hancox J (1999) The Tswaing Meteorite Crater. Council for Geoscience, Geological Survey of South Africa, Pretoria
Reinwaldt I (1928) Bericht über geologische Untersuchungen am Kaalijärv (Krater von Sall) auf Ösel. Loodusuurijate seltsi aruanded 35:30–70
Reinwaldt I (1933) Kaali järv – the meteorite craters on the Island of Ösel (Estonia). Loodusuurijate Seltsi Aruanded 39:183–202
Reinwaldt I (1937) Kaali järve meteoorkraatrite väli. Loodusvaatleja 4:97–102 (in Estonian)
Reinwaldt I (1938) The finding of meteorite iron in Estonian craters – A long search richly rewarded. The Sky Magazine of Cosmic News 2:28–29

Saarse L, Rajamäe R, Heinsalu A, Vassiljev J (1991) The biostratigraphy of sediments depoisted in the Lake Kaali meteorite impact structure, Saaremaa Island, Estonia. Bulletin of the Geological Society of Finland 63:129-139

Saarse L, Poska A, Veski S (1999) Spread of *Alnus* and *Picea* in Estonia. Proceedings of the Estonian Academy of Sciences, Geology 48:170-186

Saarse L, Vassiljev J, Miidel A (2002) Simulation of the Baltic Sea shorelines in Estonia and neighbouring areas. Journal of Coastal Research 18:639-650

Santilli R, Ormö J, Rossi AP, Komatsu G (2003) A catastrophe remembered: a meteorite impact of the 5th century AD in the Abruzzo, central Italy. Antiquity 77:313-320

Seaton RC (1912) Apollonius Rhodius: Argonautica. Harvard University Press, Cambridge MA

Suuroja K, Suuroja S (2000) Neugrund structure – the newly discovered submarine early Cambrian impact crater. Lecture Notes in Earth Sciences 91: 389-416

Tacitus PC (1942) De origine et situ Germanorum liber. In: Hadas M (ed) The complete works of Tacitus. The Modern Library, New York

Tarvel E (ed), Kleis R (transl.) (1982) Heinrici Chronicon Livoniae (early 13th century). Eesti Raamat, Tallinn

Veski S, Heinsalu A, Kirsimäe K, Poska A, Saarse L (2001) Ecological catastrophe in connection with the impact of the Kaali Meteorite about 800-400 BC on the island of Saaremaa, Estonia. Meteoritics and Planetary Science 36:1367-1376

Veski S, Heinsalu A, Kirsimäe K (2002) Kaali meteoriidi vanus ja mõju looduskeskkonnale Saaremaa Piila raba turbaläbilõike uuringu põhjal. Eesti Arheoloogia ajakiri 6:91-108 (in Estonian, English summary)

Veski S, Heinsalu A, Lang V, Kestlane Ü, Possnert G (2004) The age of the Kaali meteorite craters and the effect of the impact on the environment and man: evidence from inside the Kaali craters, island of Saaremaa, Estonia. Vegetation History and Archaeobotany 13:197-206

Yau K, Weissman P, Yeomans D (1994) Meteorite Falls in China and some related human casualty events. Meteoritics 29:864-871

Yavnel AA (1976) On the composition of meteorite Kaalijärv. Astronomicheskii Vestnik 10:122-123

Chapter 16

The Climatic Effects of Asteroid and Comet Impacts: Consequences for an Increasingly Interconnected Society

Michael C. MacCracken

16.1 Introduction

The Earth's atmosphere, ocean and land surface interact together to provide the environmental conditions to which life and society have become accustomed. Society has come to depend on these components working together to provide relatively stable (or at least regularly varying) and livable conditions that are conducive to growing and gathering necessary food, providing sufficient freshwater, limiting the domains and viability of disease vectors and, except on rare occasions, providing safe habitat for living and reproducing.

Early peoples, initially nomads and later agriculturalists, had to learn to work hard to survive the natural variations of the climate. For nomads, surviving environmental threats such as drought was accomplished by moving around, their survival depending on the large, sparsely populated areas that were then available. Learning to grow crops provided more food, allowing for a larger population, but also introduced the need for storage of sufficient reserves to survive extremes in climate and other environmental stresses that occurred where they then lived. Not all of these various groups and early societies survived the various environmental stresses they faced; the best prepared, however, survived and occasionally flourished.

Continuing societal development has occurred now over many centuries, most intensely over the past two centuries. During this period, and particularly over the last several decades, society has developed the capabilities and infrastructure that have allowed a separation of more and more people from the challenges and experiences of living off the land, leading to an ever increasing fraction of the global population living in cities.

As this has occurred, and as succeeding generations have lived mainly in urban areas, the impression has grown that societies have become more and more resilient to variations and perturbations in the natural environment. For a greater and greater fraction of the world's people, the international market economy that has developed seems able to draw forth resources from across the world, ensuring that supplies of necessary food and life-facilitating medicines and machines are continuously available. Life now seems to be proceeding apace, the main challenge being to make cities more livable.

Four trends, however, are likely actually increasing the vulnerability of society to environmental stresses. First, the world population is increasing, with projections being that over the 21st century the number of people spread across the Earth will rise from about 6 billion to 8 to 12 billion. A result of this is that there is less and less unoccupied arable land, thereby restricting relocation as an adaptation option. Native Americans used to adapt to climate fluctuations by following the buffalo to non-impacted regions and large numbers of north eastern Brazilians relocated to avoid El Niño-induced droughts. Now the areas to which they formerly moved hold other people or are being used to provide resources to take care of others, and many regions are now so populated that transportation routes are not adequate for full evacuation. For example, even with a few days warning, there is no way that Long Island, New Orleans, or Haiti can be evacuated when a major hurricane is imminent. And the situation is no better for non-human species, which have become increasingly isolated in smaller and smaller domains that can more and more easily be disrupted, meaning that smaller and smaller stresses could adversely impact biodiversity and ecologically provided resources.

Second, the push to optimize the global market economy has led to lower and lower reserves of food, seeds, medicine, and other necessary resources. Global grain reserves have been continuing to decline and now amount to less than a two month supply. This amount is less than the amount produced during a typical growing season, and is certainly much less than the amount by which production could be increased in the next growing season to replace the loss of a season or year's production. While the standard-of-living globally is rising for many people, this is often a result of dependence upon a continuing and growing stream of "necessities;" we have all become dependent on the routine functioning of more and more nodes and channels, and so the range of possibilities that could lead to disruption seems to be actually increasing.[1] No longer is it really small changes in the multi-year statistical-average climate that is the main concern; with the economic system so tightly interconnected, disruption of the weather over a month or season is all that is needed to create significant economic disruption. Although survival might not be immediately threatened, extreme events such as El Niño episodes can cause not only regional environmental problems (e.g. the drought in Indonesia and Southeast Asia several years ago caused both local and international economic impacts), but also can affect nations around the world.

Third, as the market economy has developed, there has been a tendency for particular locations to each become specialized in particular economic activities. For example, virtually all of the grain traded internationally (and so the supplies needed to sustain peoples in many nations around the world) comes from only a few regions (i.e. U.S., Russia, China, Australia, Argentina and India), and failures in even one region can cause a significant disturbance in world prices; clearly, simultaneous crop failures in more than one region could significantly increase prices, exposing large populations to reduced food supplies. The specialized seeds that underpin the green revolution and the needed fertilizers also come from a relatively few locations. As pursuit of economic efficiency has driven the international economic system to become more concentrated, interdependencies have increased and few regions remain independent of vital services and supplies from other regions. Just as wider groups of people in a community

---

[1] That many areas have largely avoided such problems has been due in part to decreasing the number of non-climatically induced and non-geophysically related breakdowns. Improving reliability further, however, seems likely to require overcoming more difficult challenges.

became more vulnerable to disruption due to floods when modern sewage treatment plants near rivers and coastlines replaced personal outhouses, society has become more vulnerable because activities are now much more concentrated in our interdependent world.

Fourth, due to an unusual geological roll of the dice, the natural environment to which we have become adapted has been unusually stable over recent centuries. As a result, society is not particularly well prepared for the wider range of possible conditions that have occurred naturally or for the increasing intensity of extreme events being brought on by anthropogenic climate change. Natural oscillations, such as the El Niño/Southern Oscillation and the North Atlantic Oscillation, already demonstrate that relatively limited changes in climate in particular regions can influence seasonal to decadal weather patterns in significant ways over large regions. While the Holocene as a whole has been a time of unusually low climatic variability, records for the last 100 000 years (and longer) from the Greenland ice core (NRC 2002) provide indications of large shifts in temperature that were comparable in their mid-latitude effects to going from interglacial to glacial conditions (or at least for the atmospheric and oceanic circulations that over time have created the climate). It is thought that these interglacial to glacial transitions, which occurred over a few years and typically lasted several centuries, resulted from an outbreak of glacial meltwater into the North Atlantic, but whatever the cause, the lesson is that a stable climate cannot be taken for granted, and that relatively modest events have the potential, if they occur in the right location, to prompt rather large, persistent shifts in climatic conditions over very large regions.

What is particularly disturbing is that all of these factors are increasing societal vulnerability at the same time. More and more people are crowding into vulnerable coastal areas, and more and more people are dependent on the well-timed, long-distance transmission of critical resources (e.g. water, food, fuel, electricity, etc.). In addition, the range of climatic extremes is increasing as the world warms, with, for example, more intense precipitation events already documented and more intense tropical cyclones projected (e.g. see Sect. 2.7 in IPCC 2001).

With the primary purpose of trying to understand if a large asteroid impact could explain the end of the age of dinosaurs, initial studies of the likely environmental consequences of the impact of a comet or asteroid have focused on the wide range of very disastrous impacts that could blot out sunlight and dramatically alter temperature, precipitation, and other climatic variables over the planet as a whole. A brief overview of the results of these analyses is included in the next section (also see Melosh 2007). Were such an event to occur today, even with, and maybe especially because of, our advanced technological capabilities, the result would likely wipe out the international economic support system on which virtually all societies depend, leaving probably only the few millions who could survive off the devastated natural environment. Clearly, such a situation needs to be avoided – and the effort to identify all potential threats is of critical importance.

Far fewer studies have examined what would happen in the event of impacts of modest or even small asteroids and comets, considering in a probabilistic way the expected damage from various levels of blast from atmospheric explosions, of dust injection and ground shaking from various levels of impacts on land, and of tsunamis from various sized impacts into the ocean. While it is true that the actual areal extent of cities is quite small, making the odds of a direct impact quite small, the world is not randomly covered with ocean or with developed and undeveloped land areas – there are many vital links and nodes. Section 16.3 is an initial exploration of some of the types of larger-than-expected consequences that could result if even a relatively modest sized impact were to strike a sensitive location. Because such events could lead to unusually significant consequences, this may suggest that even more effort than has been planned is needed to at least detect, if not deflect, such objects.

16.2
The Global Climatic Effects of Large Asteroid or Comet Impacts

The climate system is fundamentally a heat engine, being driven by the incoming solar radiation and modulated by the loss of energy from thermal (infrared) radiation, all moderated by the storage and redistribution of energy, which are determined by the various motions, composition and thermal capacities of the atmosphere, oceans, land and stored waters (e.g. as ice, clouds, vapor, etc.) and the biological activity on land and in the oceans. The state of each of these systems is often dependent on variations in the distribution of a key variable (e.g. temperature) that is in turn often closely tied to the characteristics and distribution of radiatively active constituents (e.g. gases, aerosols). Disturbing any of these processes activates couplings to many others, although disturbances smaller than those created by natural internal oscillations may be hard to distinguish and so be relatively unimportant.

The impacts of asteroids and comets of a diameter of roughly one kilometer or more have the potential to loft large quantities of substances that would significantly alter the absorption, transmission and emission of radiation by the atmosphere. The magnitude of these influences depends strongly on the amount and characteristics of what materials are created and lofted into the atmosphere by the impact, on the areal extent and the altitude of the injection, and on the combined effects of materials being injected, as there can be amplifying or compensating influences.

Toon et al. (1997) provide the most thorough review of the many types of atmospheric impacts that have been suggested, considering the potential injection of materials for impacts of a range of possible sizes; earlier important reviews of such effects can be found in Morrison (1992), Gehrels (1994), Chapman (2003) and Shoemaker (1995). The Toon et al. (1997) analysis indicates that only objects having a diameter larger than about 1 km (so energy level of nearly 10^5 MT)[2] cause global scale perturbations to the atmosphere [Birks et al. (2007), however, now calculate that an asteroid of less than half this diameter could virtually destroy the global ozone layer; see below]. Whereas the Toon et al. (1997) paper also covers the potential impacts from shock waves, earthquakes, tsunamis and heavy metals, the large, global-scale climatic impacts are only likely to result from materials that are injected into the atmosphere, including material

---

[2] For comparison, the Earth system absorbs about 2.5×10^6 MT of solar radiation each day and re-radiates the same amount in the long-wave. Thus, a 1 km asteroid strikes the Earth with about as much energy as contained in 1-hour of absorbed sunlight. The Chicxulub impact 65 million years ago, for which the diameter is estimated to have been about 10 km with an energy of as much as 10^9 MT, would be roughly equivalent to the solar radiation absorbed by the Earth over a year.

from the asteroid or comet itself; then dust, water, sulfur dioxide and nitrogen oxides from the impact process; and smoke and carbon dioxide from the burning of affected vegetation.

16.2.1
Injection of Asteroidal and Cometary Material

For large asteroids or comets, although some of its mass is burned off during its passage through the atmosphere, most of the asteroidal or cometary material that ends up in the atmosphere would be there as a result of lofting after the object impacts the surface. For the material that is lofted to be climatically important (in comparison to other material injected), the lofted debris must end up as submicron sized particles at stratospheric altitudes. If particles are larger than submicron in size, they have only limited influence on atmospheric radiation. Such particles also tend to fall out of the atmosphere over a few days and, if injected only into the troposphere (i.e. to less than 10 to 15 km in altitude), the particles tend to be rained out or filtered out (i.e. dry deposited) when air contacts the surface. If lofted too powerfully, the particles can be launched ballistically above the atmosphere; while they would be dispersed more widely over the Earth, the particles might well be so large (due to the heating and then condensation) that they would tend to fall through the atmosphere relatively rapidly. Compared to the other likely materials injected, debris from the asteroid or comet itself is not considered to be a major factor in directly inducing changes in climate.

16.2.2
Injection of Dust

The impact of a large asteroid with the land surface, or even with the ocean bottom, will create a large crater, with the potential for much of the material to be lofted into the atmosphere. Toon et al. (1997) estimated that impacts of 10^5 to 10^8 MT lead to stratospheric loadings of submicron sized dust particles of from 10^{-4} to 10^{-1} g cm^{-2}, respectively, based on an assumption that the submicron mass loading will be roughly 0.1% of the pulverized rock from the impact [a value suggested from analyses by O'Keefe and Ahrens (1982)]. For such large impacts, the amount injected from an oceanic impact is suggested to be similar.

The initial optical depth of such a loading would be of the order 10 to 1000, meaning that the light level would range from the equivalent of a very cloudy day to pitch darkness[3] (Toon et al. 1997). Because such particles would collide and coagulate, they calculate that such a dust injection would be unlikely to persist for more than 6 months. A model simulation of the climatic effects of a dust injection toward the lower end of this range (so only somewhat larger than the Tambora volcanic eruption of 1815) by Covey et al. (1994) calculated that the average temperature over land areas would drop by about 8° C in the weeks following the injection but would gradually recover over about a year; changes would be larger inland and smaller in coastal regions buffered

---

[3] Based on these optical parameters, and considering the full range of injected materials, global nightfall would be expected to occur for an impact having an energy between 10^6 and 10^7 MT (Toon et al. 1997).

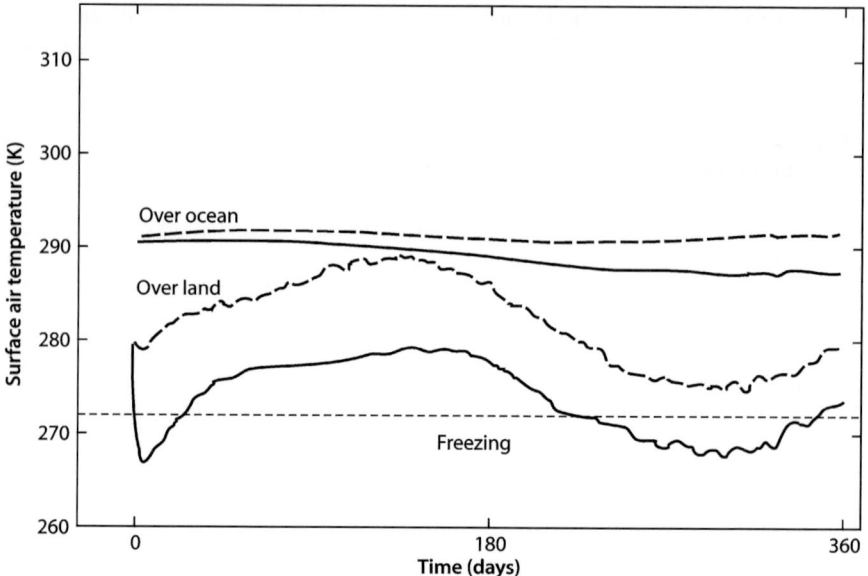

Fig. 16.1. Model-simulated near-surface air temperature, averaged separately over all ocean and over all land areas, as a function of time beginning on May 20, for an unperturbed climate (*dashed line*) and following an impact on day zero of an impactor that creates a global dust loading of 2.5×10^5 kg of fine dust particles. This amount of dust particles is one-half the amount deduced by Alvarez et al. (1980) from the average worldwide thickness of the Cretaceous-Tertiary (K-T) boundary deposit, based on the assumption that the global deposition is a measure of the fine particle loading that would spread globally (alternatively, if the dust loading is assumed to be injected ballistically, then this amount would reflect the total injected mass, and the fine particle loading for the K-T impact would be somewhat less). Based on nuclear explosions at ground level, this injection would result from roughly a 10^8 megaton impact. This figure is reprinted from Covey et al. (1994) with kind permission of Elsevier Science-NL, Sara Burgerhartstraat 25, 1055 KV Amsterdam, Netherlands

by oceanic warmth (see Fig. 16.1). Because the surface cooling and warming aloft would tend to stabilize the atmosphere, the model also calculated a substantial decrease in global precipitation, with near total shutoff for several months and only a slow recovery in the ensuing year. The combination of low light levels, cooling and reduced precipitation, each of which would be amplified as the size of the impactor increased above 1 km, makes clear that, even though not dissimilar to the impacts of really large volcanic eruptions of the past, the food, fuel, and water resources of the modern world would be very sorely stressed should such an event occur today (or tomorrow).

16.2.3
Injections from Fires

Both the heat of the impact itself and the fallback of heated debris over a much wider region would make the occurrence of fires quite likely (Toon et al. 1997); indeed there is evidence that widespread fire occurred following the K-T impact (e.g. Wolbach et al. 1985, 1990a, 1990b). For that very large event, Toon et al. (1997) summarized observations which suggest that a large fraction of the world's aboveground biomass must have burned and quite efficiently created and lofted soot into the atmosphere, where it would have absorbed virtually all of the incoming solar radiation. Based on simulations done in assessing the effects of fire-injected smoke following a nuclear war, the smoke layer would have sharply cooled the surface and dramatically reduced precipitation for an extended period (Ghan et al. 1988).

The death of the vegetation in the dark environment and the occurrence of fires would also lead to the injection of various gases. Among these would be carbon dioxide and methane, both of which are radiatively active and would tend to induce warming once the atmosphere cleared enough for solar radiation to be absorbed at the surface and in the lower troposphere (absorption of solar radiation at higher layers would not lead to amplification of the natural greenhouse effect). How long the levels of these gases would stay elevated is unclear, as, for example, the cooling of the ocean and the regrowth of vegetation would likely pull excess carbon dioxide from the atmosphere, while changes in atmospheric chemistry could modify the average residence time of methane in the atmosphere.

16.2.4
Injection of Water

An asteroid or comet impact into the ocean would be likely to loft and inject large amounts of water. Because of its effectiveness in absorbing upward directed infrared radiation and re-radiating roughly half back downward toward the surface, the warming influence of water vapor in the atmosphere is largest in the upper troposphere and stratosphere, assuming that the solar radiation reaches to below this level. How these processes would work with a mixture of dust and soot and other substances has not been calculated, although one would nominally expect the water vapor to exert a cooling influence in the layer where it is located (because the ability of the layer to radiate thermal energy away would be increased) and to exert a warming influence at the surface (because the stratosphere is radiating downward more energy than before). Toon et al. (1997) suggested that the amount of water vapor injected by a large asteroid or comet could be roughly 1 to 100 times the amount of water vapor that is currently present. The amount that remains in the atmosphere, however, could not be so large because condensation that leads to ice clouds and possible precipitation would remove any excess over saturation. While large injections of water seem quite likely as a result of ocean impacts of large asteroids or comets, how this effect would play out climatically has not been simulated, although a relatively strong warming influence for a few years may essentially counterbalance at least part of the relatively large cooling influence from injected dust and soot.

16.2.5
Injection of Sulfur Dioxide

If the impacted materials contain significant sulfur, there is the possibility of lofting significant amounts of sulfur dioxide into the stratosphere, much as an explosive volcanic eruption can do. In addition, asteroids and comets may also contain significant amounts of sulfur. Toon et al. (1997) review studies of potential injection amounts and conclude that the injected amounts could range from that of a Pinatubo-like eruption (which caused global temperatures to decrease about 0.5 °C for a year or more) up to 10^4 times this much or even more (which would exert a very large cooling influence). Lofted as sulfur dioxide, there would be a relatively rapid conversion of the sulfur dioxide to sulfate aerosols, assuming there is sufficient water vapor. Toon et al. (1997) suggested that, due to limitations imposed by atmospheric chemistry, the optical depth of the sulfate could be as much as 10 or more for a decade; the resulting enhancement of the Earth albedo would cause very significant cooling for this period, and then over a longer period due to the cooling of the ocean that would result.

16.2.6
Injection of Nitrogen Oxides

During and following an impact, nitric oxide could be created in several ways (Toon et al. 1997), including rapid cooling following dissociation of O_2 and N_2 by shock waves, in the plume following impact with the surface, as debris ejected ballistically re-entered the atmosphere, and possibly as a result of induced fires. Model simulations and observations following nuclear weapons tests indicate that injection of substantial amounts of nitric oxide into the stratosphere would lead to large-scale ozone depletion (NRC 1975). Considering the effect of an asteroid impact, Birks et al. (2007) calculate that an object of about 0.5 km or greater would lead to significant depletion of the global ozone layer; this estimated diameter is about half that cited by Toon et al. (1997), suggesting that asteroid impacts with only about one-eighth the energy have the potential for devastating global-scale impacts and substantially raising the frequency that such events have likely occurred in the past and should be expected in the future.

In that ozone is the primary absorber of solar radiation in the stratosphere (balanced by loss of thermal radiation due mainly to carbon dioxide), the stable vertical structure of the stratosphere could also be disturbed, causing changes in atmospheric circulation and in the concentrations of ozone and other substances. Such depletion would allow UV radiation to pass downward toward the surface, where it could do biological harm if not otherwise absorbed by other substances. Nitric oxide injected into the lower atmosphere would also interact chemically, although the lack of light would diminish its role in ozone formation and the suppression of precipitation might slow its removal in rain. Ultimately, however, the nitric oxide (probably in the form of nitric acid) would be likely to be removed from the atmosphere and would tend to acidify water bodies and harm remaining vegetation.

Despite many limitations in our understanding, the Earth's climatic history and model simulations indicate that the impact of a large asteroid or comet would have global-scale effects on climate, which would, in turn, very adversely impact the environment and many key societal activities. While an individual might well have a chance of surviving, the impacts would be such as to make it very difficult for societies to function, leading to very large numbers of secondary deaths (even aside from direct effects of blasts, tsunamis, etc.). That mass extinctions would result from the largest impacts appears quite plausible.

16.3
Potential Weather and Climate-Related Impacts of Small to Modest-Sized Asteroids and Comets

As it has developed, society has expanded the range of conditions that it can endure. Thus, much of society has become reasonably well adapted to the prevailing atmospheric and oceanic means and variations (what we often call the climate, which is typically defined as the state of the atmosphere as represented by a statistical amalgamation of the weather over a 30-year period). Even with these adaptive efforts, however, disasters result when extreme conditions occur; for example, when intense tropical cyclones (i.e. typhoons and hurricanes) strike, monsoons persistently fail, or ice storms disable power grids. Whereas society as a whole will survive, many in vulnerable situations could die and devastation could be quite widespread.

Toon et al. (1997) suggest that an "energy of 10^5 MT [i.e., an asteroid with a diameter of roughly 1 km] is a conservative lower limit at which damage might occur beyond the experience of human history" and they offer World War II (in which many tens of millions died over the course of about 7 years) as an example of an event that civilization survived. For our purposes, we are not interested only in what society could possibly survive. For World War II, there was time to prepare, there were large regions not directly involved in the conflict from which resources could be drawn in order to assist the combatant regions, and deaths and damage were spread out over time and space; for even a modest asteroid impact, even with warning, the damage would be concentrated in time and space,[4] and impacts would then be likely to spread not only because of the atmospheric disruption, but also because of disruptions of key economic lifelines and health-preserving services. Given how tightly society is interconnected (e.g. see Dore 2007 and Carusi et al. 2007), it seems worth considering what sort of special vulnerabilities exist for impact energies considerably lower than have previously been thought to have the potential for significant widespread damage.

16.3.1
Asteroid and Comet Impacts that do not Involve a Surface Impact

Depending on an impactor's composition and the speed and angle of impact, objects with diameters of up to roughly 50 to 100 m typically create blast effects of up to of order 50 to 100 MT without impacting the surface. The Tunguska event of 30 June 1908, for example, created a shock wave that leveled the forest over an area of roughly 2000 km^2 (Kolesnikov et al. 2007). This event is variously estimated to have resulted from an atmospheric explosion of 10 to 100 MT from an object estimated to be between about 60 and 100 m in diameter (Covey 2002; Yazev 2002). Based on astronomical evidence, such events are statistically expected to occur approximately every few

---

[4] And this aspect of the concentration of impacts in time and space is critical, at all scales. For example, while a human could readily donate 6 pints of blood spread over a year, an instantaneous loss would lead to a quite different outcome; and while a society might be able to survive on 25% fewer calories over a year, a 100% loss for a season would be devastating.

centuries, with an event occurring over land roughly every 500 to 1000 years, which is not inconsistent with historical records.

Fortunately, the Siberian forest area where the Tunguska object exploded was relatively moist and only very lightly populated. If such an impact had occurred in the summer over a drought-stressed forested area in the western US, fires surely would have been started and been very hard to contain. Although even much larger forest fires have been experienced in relatively rural areas (e.g. during the summer of 2004, about 20 000 km² of forest were consumed by fire in Alaska), the occurrence of so much smoke over so short a time in as highly an inhabited region as the western United States would be very disruptive and damaging. Major injections of smoke would also occur if facilities such as oil wells or refineries were in the impact zone. Similarly, blast impacts over populated areas would be expected to, among many other effects, ignite fires that could, under certain conditions, readily spread, potentially displacing many tens of thousands of people. The resulting fire-induced smoke could then lead to a range of additional impacts on a region's weather, including suppressing diurnal temperature variations and regional precipitation.

Drawing by analogy from the atmospheric explosion of nuclear weapons (e.g. see Luther, 1983), the injection of nitric oxides from a multi-megaton explosion due to a relatively small impactor would likely lead to regional diminution of stratospheric ozone. Not only would such changes allow transmission of harmful levels of UV radiation, but model simulations also indicate that, for example, substantial changes in the stratospheric ozone concentration in high latitudes can affect atmospheric circulation (Rind et al. 2005). While the circulation changes are within the range of natural variability already being experienced, the timing and patterns of apparent climatic oscillations could be affected, potentially leading to changes in the occurrence and intensity of extreme conditions.

16.3.2
Modest-Sized Asteroid and Comet Impacts that do Involve a Surface Impact

For impacts having energies of less than roughly 10^4 MT (so a diameter of less than about half a kilometer and a frequency of only roughly once per 10^4 to 10^5 years)[5], the effects felt at the surface would include blast, cratering, earthquakes, fires and, for ocean impacts, tsunamis.[6] While Toon et al. (1997) described the likelihood of regional and global climatic consequences imagining no special location on Earth, the potential for significant impact on society would seem to be quite dependent on the location of the impact and the time of year that it occurs. Impacts affecting dry forests or dry urban or suburban areas could create massive, likely uncontrollable fires that would

---

[5] Several papers drawing upon archeological and geological evidence presented at the ICSU Workshop on Comet/Asteroid Impacts and Human Society, however, suggested that such impacts appear to have occurred more frequently during the Holocene than would be expected based on analysis of astronomical data (e.g., see Masse 2007 and Bryant 2004).

[6] While attention has been given to tsunami run-up onto coastal plains, there are a number of especially vulnerable areas near or below sea level that are protected by levees (e.g., The Netherlands, New Orleans). That earthquake shaking could also have an exaggerated influence on levees is suggested by Reisner (2004), though he was not considering earthquakes from impact events.

cause much destruction and many deaths. In addition, the induced fires would lead to smoke injections that could limit light and precipitation over wide regions, thereby impacting agricultural production, water supply systems, etc. The light reduction, cooling, and precipitation reduction caused by such smoke could also lead to the death of vegetation over even wider areas, opening up the potential for later fires to inject more smoke. With the land cleared and charred, there would be the potential for further regional influences on the climate. Impacts into an ocean could not only create tsunamis that would damage coastal areas (see Gusiakov 2007), but could have amplified consequences if the impact itself or the tsunami were to cause major loss to the Greenland or Antarctic ice sheets or even to simply fracture Arctic sea ice.

The effect of the impact on the stratospheric ozone layer would increase with the size of the impactor, with an object of roughly 0.5 km in diameter or greater causing significant depletion of the global ozone layer (Birks et al. 2007). At what level new weather extremes or an abrupt change in atmospheric or oceanic circulation patterns might be triggered by the impacts on the ozone layer, by the induced climatic effects, or by the direct impacts of the impact is less clear. Both geological evidence and modeling studies indicate that the climate system does seem to have, in some sense, transition thresholds that it might be possible to exceed, and sudden transitions do seem to have been triggered, presumably by other factors, in the past (NRC 2002). Even modest depletion of the stratospheric ozone layer would be expected to lead to significant increases in UV radiation at the surface.

Not only natural systems are vulnerable to such impacts. The international economic system has a number of crucial nodes, as made clear from the far-flung consequences from the comparatively localized devastation (as compared to an asteroid impact) that terrorists wreaked on 11 September 2001. As another example, the near failure of the Soviet grain harvest in the 1970s led to economic repercussions around the world; imagine the impact of simultaneous failures in two or more regions. The Tambora eruption and the year without a summer similarly indicate how a relatively small cooling led to unusual frosts that caused relatively widespread crop losses due to their occurrence early in the growing season, leading to regionally important food limitations.

Even though the likelihood is very low that especially vulnerable sites might be affected, there are actually quite a large number of them and the potential consequences to the global social system could be very large if such an impact occurred (see Dore 2007 and Carusi et al. 2007).[7] Just this result arose in the analysis of the "nuclear winter" situation (see Harwell and Hutchinson 1989), suggesting that analysts of damage from nuclear war were underestimating the potential effects by not considering how highly interconnected the international economic system has become.

16.4
Discussion

Paleoclimatic records and the recent history of human influences indicate that the surface climate is quite dependent on variations in factors that control the Earth's energy

---

[7] A similar argument has been made regarding the ramifications of impacts or atmospheric explosions in locations of political tensions and international conflict.

balance. As one example, even though orbital variations cause virtually no change in the annual integral of incoming solar radiation at the top of the atmosphere, these changes in the latitudinal and seasonal influx of solar radiation, amplified by accompanying changes in atmospheric composition and other feedback processes, appear to be the drivers of the cycling of the Earth's climate between glacial and interglacial conditions (see Berger 2002). As another example, the relatively sudden release of glacial meltwater that spread over the North Atlantic Ocean appears to have been the cause of a rapid slowing of the ocean's thermohaline circulation, which in turn disrupted mid-latitude weather patterns in the Northern Hemisphere for hundreds of years, to an extent that near-glacial conditions prevailed over much of Europe and eastern North America during the Younger Dryas some 11 000 years ago (NRC 2002).

Impacts by a comet or asteroid, depending on when and where the impact occurs, have the potential to loft substantial amounts of materials, both directly and indirectly that could substantially alter the climate. For large impactors, the directly induced consequences would be global in extent and very seriously disruptive; for modest sized impactors, amplification of the direct consequences by various indirect feedbacks (e.g. smoke from fire) have the potential to loft radiatively active gases and aerosols into the atmosphere that may in turn cause further modification of atmospheric and/or oceanic conditions that in turn would affect the climate and/or atmospheric composition. Model and analysis studies to date (e.g. SDT 2003) seem to have focused on the atmospheric and near surface response to the directly injected materials from small to modest sized impacts and seem to date to have not fully accounted for emissions and feedbacks that might result as a consequence of disturbing the tightly interconnected global social and economic system. Policy makers need to understand that whereas probabilistic analyses can lead to seemingly low estimates of potential damage (e.g. as described in Chapman 2004), the range of possible outcomes is very large and some cases could be much, much more disruptive.

If the climate is indeed as sensitive to radiative forcings as is being indicated by current research and paleoclimatic indications of abrupt change in the past, simulations covering a wider range of possible scenarios are needed to further fill out the set of possible consequences over the near to long-term versus the size, latitude, and season of the impact itself.

References

Alvarez LW, Alvarez W, Asaro F, Michel HV (1980) Extraterrestrial cause for the Cretaceous-Tertiary extinction. Science 208:1095–1108
Berger A (2002) The role of CO_2, sea-level and vegetation during the Milankovitch forced glacial-interglacial cycles. In: Bengtsson LO, Hammer CU (eds) Geosphere-biosphere interactions and climate. Cambridge University Press
Birks, JW, Crutzen, PJ, Roble, RG (2007) Frequent ozone depletion resulting from impacts of asteroids and comets. Chapter 13 of this volume
Bryant, E (2004) Geological and cultural evidence for cosmogenic tsunami. Paper presented at the Comet/Asteroid Impacts and Human Society Workshop, Tenerife, Canary Islands
Carusi A, Carusi A, Pozio P (2007) May land impacts induce a catastrophic collapse of society? Chapter 25 of this volume
Chapman CR (2003) How a near-Earth object might affect society. Workshop on Near Earth Objects: Risks, Policies, and Actions, Global Science Forum, Frascati, Italy, OECD

Chapman CR (2004) The hazard of near-Earth asteroid impacts on Earth. Earth and Planetary Science Letters 222:1–15

Covey C (2002) Asteroids and comets, effects on Earth, pp 205–210. In: MacCracken MC, Perry JS (eds) Encyclopedia of global and environmental change, vol 1: the Earth system: physical and chemical dimensions of global environmental change. John Wiley and Sons, London

Covey C, Thompson SL, Weissman PR, MacCracken MC (1994) Global climatic effects of atmospheric dust from an asteroid or comet impact on Earth. Global and Planetary Change 9:263–273

Dore, MHI (2007) The economic consequences of disasters due to asteroid and comet impacts, small and large. Chapter 29 of this volume

Gehrels T (ed) (1994) Hazards due to Comets and Asteroids. Univ. of Arizona Press

Ghan SJ, MacCracken MC, Walton JJ (1988) The climatic response to large atmospheric smoke injections: sensitivity studies with a tropospheric general circulation model. Journal of Geophysical Research 93:8315–8337

Gusiakov, VK (2007) Tsunami as a destructive aftermath of oceanic impacts. Chapter 14 of this volume

Harwell MA, Hutchinson TC (1989) Environmental consequences of Nuclear War, vol II: ecological and agricultural effects, SCOPE 28. John Wiley, New York

Intergovernmental Panel on Climate Change (IPCC) (2001) Climate change 2001: the scientific basis. Houghton J et al. (eds), Cambridge Univ. Press, 881 pp (available at *http://www.grida.no/climate/ipcc_tar/wg1/index.htm*)

Kolesnikov, EM, Rasmussen, KL, Hou, Q, Xie, L, Kolesnikova, NV (2007) Nature of the Tunguska impactor based on peat material from the explosion area. Chapter 17 of this volume

Luther FM (1983) Nuclear war: short-term chemical and radiative effects of stratospheric injections. Proceedings of the International Seminar on Nuclear War 3rd Session: The Technical Basis for Peace, Erice, Italy, 19–24 August 1983. Servizio Documentazione dei Laboratori Frascati dell'INFN, pp 108–128

Masse WB (2007) The archaeology and anthropology of Quaternary period cosmic impacts. Chapter 2 of this volume

Melosh HJ (2007) Indirect physical effects of comet and asteroid impacts. Chapter 12 of this volume

Morrison D (ed) (1992) The Spaceguard Survey: report of the NASA international Near-Earth orbit detection workshop. NASA, Washington, DC. (available at *http://impact.arc.nasa.gov/gov_nasastudies.cfm*)

National Research Council (1975) Long-term worldwide effects of multiple nuclear weapons detonations. National Academy Press

National Research Council (NRC) (2002) Abrupt climate change: inevitable surprises. National Academy Press

O'Keefe JD, Ahrens TJ (1982) Impact mechanisms of large bolides interacting with Earth and their implication to extinction mechanisms. In: Silver LT, Schultz PH (eds) Geological implications of impacts of large asteroids and comets on the Earth. Spec Pap Geol Soc Am 190:103–120

Reisner, M (2004) A dangerous place: California's unsettling fate. Penguin Books

Rind D, Perlwitz J, Lonergan P (2005) AO/NAO response to climate change, part I: The respective influences of stratospheric and tropospheric climate change. J Geophys Res 110:D12107, doi:10.1029/2004JD005103

SDT, Near-Earth Object Science Definition Team (2003) Study to determine the feasibility of extending the search for near-Earth objects to smaller limiting diameters. NASA Office of Space Science, Solar System Exploration division, Washington, DC. *http://neo.jpl.nasa.gov/neo/neoreport030825.pdf*

Shoemaker EM (ed) (1995) Report of the Near-Earth Objects Survey Working Group. NASA, Washington DC 85 pp. (available at *http://impact.arc.nasa.gov/gov_nasastudies.cfm*)

Toon OB, Turco RP, Covey C (1997) Environmental perturbations caused by the impacts of asteroids and comets. Reviews of Geophysics 35:41–78

Wolbach WS, Lewis RS, Anders E (1985) Cretaceous extinctions: evidence for wildfires and search for meteoritic material. Science 230:167–170

Wolbach WS, Gilmour I, Anders E (1990a) Major wildfires at the Cretaceous/Tertiary boundary. In: Sharpton V, Ward P (eds) Global catastrophes in Earth history. Spec Pap Geol Soc Am 247:391–400

Wolbach WS, Anders E, Nazarov M (1990b) Fires at the K-T boundary: carbon at the Sumbar, Turkmenia, site. Geochim Cosmochim Acta 54:1133–1146

Yazev S (2002) Tunguska phenomenon, pp 730–731. In: MacCracken MC, Perry JS (eds) Encyclopedia of global and environmental change, vol 1: the Earth system: physical and chemical dimensions of global environmental change. John Wiley and Sons, London

Chapter 17

Nature of the Tunguska Impactor Based on Peat Material from the Explosion Area

Evgeniy M. Kolesnikov · Kaare L. Rasmussen · Quanlin Hou · Liewen Xie
Natal'ya V. Kolesnikova

17.1
Introduction

The nature of the bright bolide and the giant explosion that took place on June 30, 1908, in the Podkamennaya Tunguska river basin, Central Siberia, is still being discussed. The area with fallen trees is in excess of 2000 square km (Fast et al. 1967), whereas the kinetic energy deposited by the impactor has been estimated to be ca. 15 million tons of TNT equivalent (or 1500 Hiroshima bombs; Vasiljev 1998). Nevertheless, Kolesnikov et al. (1973) have shown that the explosion could not be of nuclear nature. Its energy release was, in fact, too big to be a nuclear explosion. Two other nuclear hypotheses, one of annihilation and one of thermonuclear origin, have been tested by measuring ^{39}Ar activity in rocks and soil at the explosion epicenter. No excess ^{39}Ar was detected, and this method is much more sensitive than the method of measuring radiocarbon in tree rings (Cowan et al. 1965). Likewise no excess beta activity was observed in 1908, or the following years, in two ice cores from Camp Century nor in an ice core from DYE-3, all three on the Greenland ice sheet (Rasmussen et al. 1984).

The turbidity of the atmosphere after the explosion was observed by the Mount Wilson Observatory in California. The increased turbidity was probably due to dispersed cosmic material (about 1 million tons; Fesenkov 1978) which is in accord with the recent estimations (Vasiljev 1998; Bronshten 2000). However, not even a gram of the Tunguska Cosmic Body (TCB) material has ever been discovered.

Among other more than 100 hypotheses put forward in order to explain the Tunguska event, the hypotheses of a large meteorite (Kulik 1939; Krinov 1966; Chyba et al. 1993) and of a small cometary core (Whipple 1930; Fesenkov 1969; Petrov and Stulov 1975; Kolesnikov 1988; Bronshten and Zotkin 1995; Grigoryan 1998; Kolesnikov et al. 1995a,b, 1998a,b, 1999, 2003; Rasmussen et al. 1984, 1995, 1999) are still being debated. In determining the nature of the TCB, the most important problem is locating and studying some of its material.

17.2
Search for the TCB Remnants in the Epicenter Area

During the 1961–1962 expeditions of the USSR Academy of Sciences, cosmic magnetic spherules 20–100 μm in diameter were found in soil from the Tunguska explosion area (Florenskij 1963; Florenskij et al. 1968). Their cosmic origin has been confirmed by Ganapathy (1983) and Nazarov et al. (1990). However, it is difficult to prove that these spherules belong to the TCB material because such spherules can be found almost everywhere.

Peat, *Sphagnum fuscum*, from the event layer, containing material correlative to 1908, can be isolated (L'vov 1984). This appears to be more promising material in the search for the TCB remnants as compared to soil. Since peat lives only on aerosol nutrition, it thus, could have incorporated extraterrestrial fall-out from the Tunguska event.

In order to determine the presence of the TCB material, layer-by-layer chemical analyses of bulk peat samples have been made by several research teams. In the event layers of several peat columns, increased abundances of Fe, Co, Al, Si, and several volatile elements, Zn, Br, Pb, and Au, werer observed and are probably due to the entrapment and conservation in the peat of the TCB material (Golenetskiy et al. 1977; Kolesnikov et al. 1977). Small particles in tree resin formed in 1908 have a similar composition (Longo et al. 1994). Most element concentrations, i.e. of Fe, Al, Si, Au, Cu, Zn, Cr, Ba, Ti, and Ni, are almost the same as for the anomalous elements in the peat column sampled at the Northern peat bog located near the explosion epicenter (Golenetskiy et al. 1977).

It has been suggested that the sharp increase of volatile element concentrations in the 1908 peat layers is a consequence of the cometary nature of the TCB (Kolesnikov 1980; Kolesnikov et al. 1998b, 2003). In addition, it has been shown that Pb in the event layer has an isotopic composition different from that in other peat layers and of typical Pb in this area (Kolesnikov and Shestakov 1979).

17.3
Platinum Group Elements (PGE) Investigation

It is generally accepted that the presence of dispersed cosmic material in terrestrial sediments can be detected by measuring the Ir (or other PGE elements) because, for example, the content of Ir in chondrites is about 25 000 times more abundant than in average rocks of the Earth's crust. This approach is widely used to identify large meteorite impacts (e.g. Alvarez et al. 1980; Rasmussen et al. 2000). In an Antarctic ice core at the depth corresponding to the Tunguska event, Rocchia et al. (1990) did not detect an increase in the Ir content.

In two ice cores from North Greenland and one ice core from South Greenland Rasmussen et al. (1984) found no excess in nitrate fall-out related to the Tunguska event. This is inconsistent with the predictions made by Turko et al. (1982) based on model calculations. Later, Rasmussen et al. (1995) measured increased concentrations of Ir, Ni, Cr, Au, Zn, Sb and As compared to terrigenic dust in a Greenlandic ice core and have in this way shown the presence of a cosmic dust component in the Greenlandic ice sheet. However, in the 1905–1914 layers, the concentrations were within the limits of typical variations, and no excess input of cosmic material as a result of the Tunguska event in 1908 was detected. These data seem to be inconsistent with the stony meteorite hypothesis of the TCB, but do not contradict the cometary hypothesis because the solid dust component of a cometary core carrying the Ir may only be a small part of the mass of the comet. Rasmussen et al. (1995) regarded the fraction of chondritic material in the TCB to be less than 5%.

Geochemical data show that the fall-out of the TCB material at the explosion area was not homogeneous (Golenetskiy et al. 1977; Kolesnikov 1980; Serra et al. 1994). That is probably the reason why Rocchia et al. (1996) did not find Ir in two peat columns from the explosion area. However, in other peat columns taken near the explosion epicenter, Nazarov et al. (1990), Hou et al. (1998) and Rasmussen et al. (1999) have proven the presence of cosmic dust by measuring an increase in the Ir content. Therefore, at the explosion epicenter there are a number of sites enriched with the TCB material although there seemingly are also places without enrichment. In fact, there are data on a number of the smaller TCB explosions at lower altitudes in addition to the main high-altitude giant explosion (Krinov 1966; Serra et al. 1994). These are in agreement with eyewitness accounts concerning the occurrence of several TCB explosions. These smaller explosions are consistent with a cometary scenario, and could be due to the explosive atmospheric entries of several fragments of the icy core. Many eyewitnesses of the Tunguska bolide reported crushing during its motion (L'vov 1984; Epiktetova 1998).

An increased concentration of iridium, i.e. an Ir anomaly, was for the first time revealed in the Southern swamp peat column by Nazarov et al. (1990). The maximum Ir content in the event layer was 17.2 ppt, corresponding to 735 ppt in the ash, i.e. in the mineral fraction of the peat. This content is significantly higher than the average of 20 ppt Ir typical for upper crustal rocks (Taylor and McLennan 1985). Therefore, the Ir anomaly in the peat is not likely to be explained by terrestrial sources.

Hou et al. (1998) discovered the sharp Ir anomaly (0.24–0.54 ppb) at the event layer extending as well into the lower layers of the Northen peat bog column. This is about 10–20 times larger than the Ir-concentration reported by Nazarov et al. (1990). In addition, anomalies in the contents of Ni, Fe, Co and rear earth elements (REE) in the event layer indicate that the mineral, or dust, fraction of the TCB must have been composed of material similar to CI carboneaceous chondrites (or a comet), rather than ordinary chondrites.

Rasmussen et al. (1999) found the Ir anomaly (39.9 ppt) and a ^{14}C depletion in the event layer of a column taken in the Near Khushma peat bog. This may imply that in the explosion area the distribution of the TCB fallout is indeed not homogeneous. Unfortunately, very few investigations have been made of the PGE (including Ir) in other Siberian peat bogs. Several PGEs have been analyzed for in the Northern peat bog column from the Tunguska explosion area by Hou et al. (2000).

In Fig. 17.1 we show the distribution of elements in another peat column from the Northern peat bog analyzed by Xie et al. (2001). The concentrations of the PGE: Ru, Rh, Pd and other elements in the event and lower layers are higher than the background values for the upper layers. The Pd concentration in the event layer (317.4 ppb) is ten times as high as its background value. And the concentrations of other elements are eight times as high for Ni, ten times for Co and fifteen times for REE, as their background values, respectively.

There is a good correlation between Rh, Pd, Ru, and Co concentrations in all the works of Hou et al. (1998, 2000, 2004), which points to the same source of the anomalies of these elements, thus indicating the presence of the TCB material.

Golenetskiy et al. (1977) showed that the mineral component of the soil had a composition similar to that of nearby basaltic rocks. In the peat columns from the explosion area analyzed for PGEs, anomalies in the event layers have been demonstrated quite clearly, but no PGEs have been detected in any of the nearby basaltic rocks (Xie

Fig. 17.1.
Elemental abundances in the peat column from the epicenter of the Tunguska event

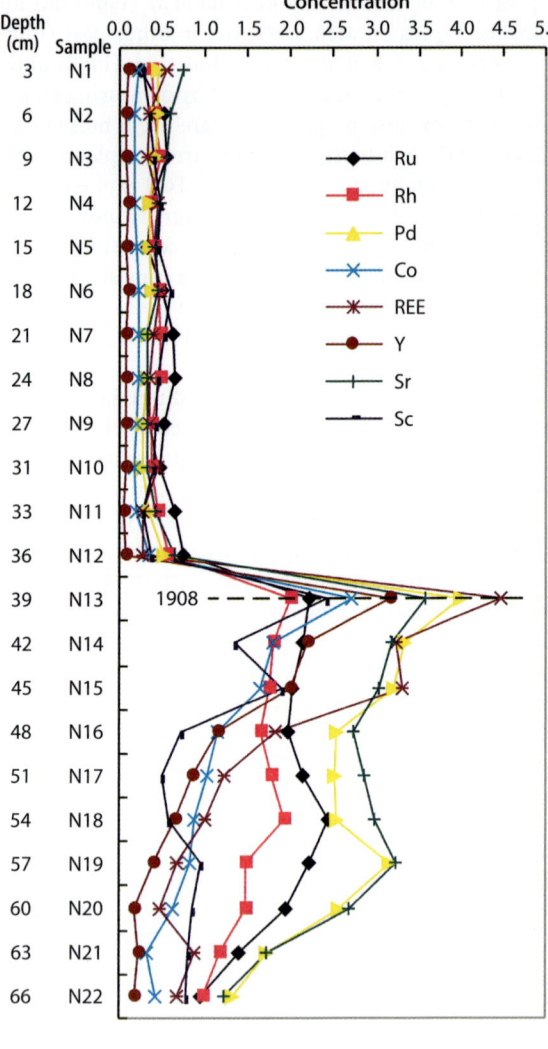

Fig. 17.2.
Patterns of CI-chondrite-normalized REE in the peat samples in the 1908 Tunguska explosion area. It is shown that the patterns in the event layers (corresponding to C40 – C50 samples) are different from that in the normal layers

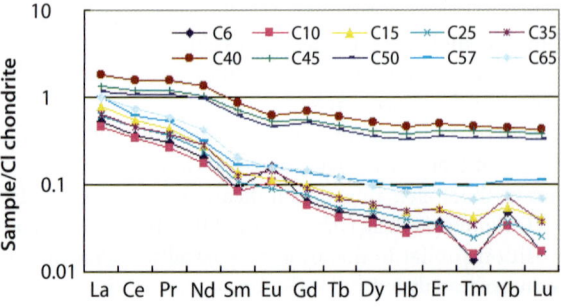

et al. 2001; Hou et al. 2004). Therefore, the increase of the concentration of PGEs in the event peat layers cannot be attributed to an extra input from the event of terrigenic or basaltic dust, but is more probably caused by the fall-out of the TCB material.

It was found that the REE concentrations in the event and lower layers are much lower than those in the nearby basaltic rocks and clearly higher than those in the normal peat layers (Fig. 17.2). The pattern of CI-chondrite-normalized REE in the event layers are different from those of the basaltic rocks and of the normal peat layers. These characteristics indicate that the peat, especially in the event layers, is unlikely to be contaminated by terrestrial dust. The slight slope of the distribution of the REE for the event layers points to the chemical composition similar to CI chondrites.

17.4
Isotopic Investigations of Light Elements in the Peat

The TCB material in the peat can also be diagnosed by isotopic methods. If the TCB was a small cometary core it most likely contained substantial amounts of H_2O, CO_2, NH_3, various hydrocarbons, and organic fractions in the form of bitumen frozen together (Churyumov 1980). Hydrogen, C, N, and S, being the main cometary elements, are biologically available and essential as well and, when these were deposited as a result of the explosion became (or may well have been) incorporated into organic molecules of the growing peat biomass, or alternatively adsorbed by the peat.

In meteorites and lunar material isotopic composition of the light elements are distinctly different from terrestrial materials. Attempts have been made to use the isotopic composition as an indicator of the presence of the cometary material in the peat (Kolesnikov 1988). In five peat columns sampled at the explosion area the isotopic shifts have been observed relative to the terrestrial values, namely for carbon, nitrogen, and hydrogen (Kolesnikov et al. 1995a, 1995b, 1996, 1999, 2003; Rasmussen et al. 1999). The positive sign of the isotopic shifts for carbon ($\Delta^{13}C$ reaches +4.3‰) and the negative for hydrogen (ΔD reaches –22‰) cannot be explained by ordinary terrestrial processes like fall-out of terrestrial dust, soot deposition, emission from the Earth of natural gas, climate changes, or other terrestrial processes. Moreover, the isotopic effects are closely connected with the area and the time of the TCB explosion and are absent in the upper and lowest peat layers and in the control columns sampled at the other locations (Fig. 17.3). These effects cannot be explained by contamination of the peat by ordinary chondrite materials as well. Rasmussen et al. (1999) and Kolesnikov et al. (1999) have shown that to explain the isotopic effect for carbon from $\delta^{13}C_{PDB} = +2‰$ to +4‰ in the peat it is necessary to assume about 2–3% of exogenic carbon with a very heavy isotopic composition ($\delta^{13}C_{PDB}$ from +40‰ to +60‰). Such heavy carbon does not normally occur on Earth (Galimov 1968), nor in ordinary chondrites or achondrites (Faure 1986). Rasmussen et al. (1999) have shown that this carbon is of an abiogenic origin, because of its depletion in radiocarbon, ^{14}C, ("^{14}C-dead"), which is otherwise present in all biological systems on Earth (Fig. 17.4). The isotopically heavy carbon is typical only of some mineral fractions of the CI carbonaceous chondrites (Halbout et al. 1986). It is known from investigations of Halley's Comet that the composition of cometary dust is very close to that of carbonaceous chondrites (Jessberger et al. 1986).

Fig. 17.3.
Variations of the content and isotopic composition of carbon in the peat column from the epicenter of the explosion area (●) and from Vanavara 65 km of the epicenter (+), in burnt peat samples (*points F and d*), in bush and in roots of birch (*P*), and in moss *Polytrichum* (*K*)

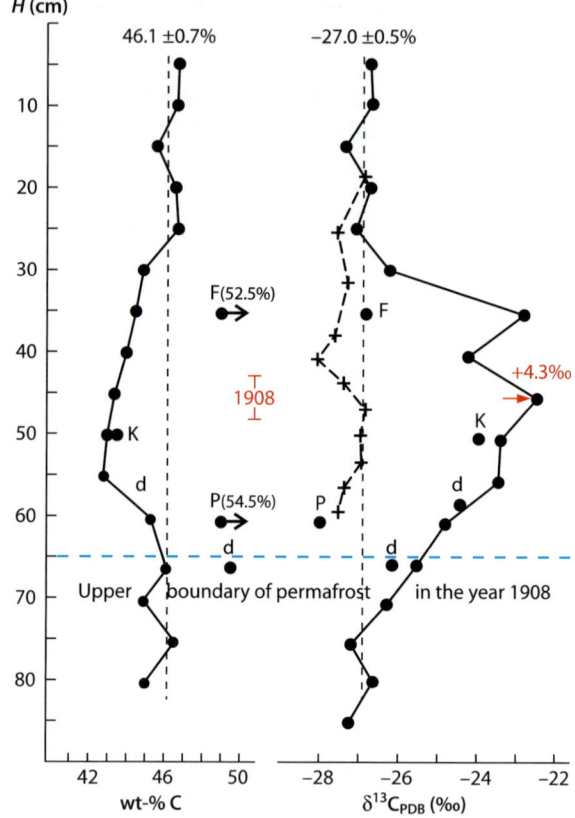

Thus, we suggest that in this area peat was contaminated by extraterrestrial material compositionally similar to cometary dust. Therefore isotopic evidence points towards a cometary nature of the TCB.

The present data speak against the hypothesis that suggest the Tunguska explosion was due to the explosion of methane outbursts from the Earth (Kundt 2001). Terrestrial methane is known to have light isotopic composition of its carbon with $\delta^{13}C_{PDB}$ from −30 to −50‰ (Galimov 1968). We analyzed Dyulyushma oil sampled from the well located near the Tunguska explosion area and obtained a value for $\delta^{13}C_{PDB} = -33.7‰$, that is close to above-mentioned values (Kolesnikov et al. 1995b). On the contrary, an admixture to peat of abiogenic cosmic carbon has heavy isotopic composition from +40‰ to +60‰. In addition, many eyewitnesses undoubtedly saw the passage of the Tunguska bolide, thus further rejecting Kundt's hypothesis.

The first data on the N-content and its isotopic composition in the peat columns from the explosion area are consistent with the assumption of acid rain fall-out after the passage and explosion of the TCB (Kolesnikov et al. 1998a, 2003), quite similar to the K/T boundary sediments (Gardner et al. 1992). In the event and lower layers, one can observe shifts in the isotopic composition of nitrogen (up to $\Delta^{15}N = +7.2‰$) and carbon (up to $\Delta^{13}C = +2‰$) and also an increase in the nitrogen concentration com-

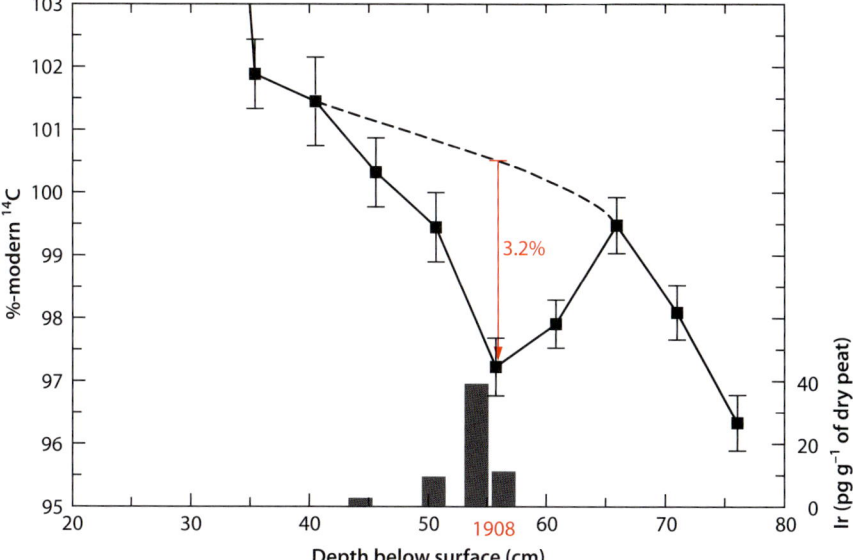

Fig. 17.4. Results of the ^{14}C radioactive measurements in a peat as a percent of the modern (1950) value. The samples above 35 cm are influenced by bomb-produced ^{14}C. The dashed curve is an extrapolation of the points below and is the expected ^{14}C content. The excess caused by cosmic influx of nonradioactive C is seen between the *dashed curve* and the measurements. Also shown the Ir values that are above the detection limit

pared to those in the normal upper layers, unaffected by the Tunguska event (Kolesnikov et al. 2003). One possible explanation for these effects could be the presence in the event and lower peat layers of nitrogen and carbon from the TCB material and from acid rains, followed the TCB explosion. We found that the highest quantity of isotopically heavy nitrogen fell down near the explosion epicenter and along the TCB trajectory. The quantity of the nitrogen fallen down above the forest devastation area is 200 000 tons that is only about 30% of the value calculated by Rasmussen et al. (1984).

17.5
Discussion

Rasmussen et al. (1999) have measured exceptionally high C/Ir ratio of $12 \pm 3 \times 10^8$ in the dry peat, which is at least a factor 10^4 higher than that in the meteorites. During the atmospheric entry, the loss of C is a much more likely process than the loss of PGEs, but the loss of C will only make the initial C/PGE ratio of the TCB more impressive. So, we are forced to conclude that the high C/PGE ratio is not well understood and not very well in accordance with any chondritic or achondritic type of the explosive body. However, these data are better explained by a cometary type of the TCB.

In comets PGEs are mostly localized in the dust. Therefore, the high C/PGE-ratio points to a rather low content of the dust in the TCB. This means that if the TCB was a comet, its core would have been almost pure ice with admixtures of soot, hydrocarbons and other "organic" compounds. Such a pristine core, with a very low content of dust, is very different from the rather mature core of Halley's Comet, which had a high fraction of dust of approximately 40% (Gruen and Jessberger 1990). This is in good agreement with Halley's Comet having experienced many solar approaches during each of which the core has lost many volatiles.

The relatively high volatile content and low dust content of the TCB is also indicated by the eyewitness accounts. Among the more than 700 reports, not a single individual reported an intense smoky trail after the TCB passage, which is otherwise typical of stony or iron meteorites passages through the Earth's atmosphere. This is in accordance with a low content of the dust in the TCB (L'vov 1984; Plekhanov 1997). This is also in good agreement with the negative results of searching for the traces of a global deposition of iridium in 1908 in both Antarctic (Rocchia et al. 1990) and Greenland ice fields (Rasmussen et al. 1995).

Golenetskiy et al. (1977) and Kolesnikov et al. (1977) found positive anomalies of several volatile elements (Zn, Br, Pb etc.) in the event peat layers, which were probably due to the conservation in the peat of the TCB material. In addition, Kolesnikov and Shestakov (1979) have shown that Pb in the event layer has a different isotopic composition compared to other peat layers in the Tunguska area. The sharp increase of many volatile elements, Li, Na, Rb, Cs, Cu, Zn, Ga, Br, Ag, Sn, Sb, Pb, and Bi, in the peat below the event layer is most likely caused by the presence of these elements in cometary material (Kolesnikov 1980; Kolesnikov et al. 1998b).

Hou et al.(2004) have reported Pd and Rh depositions of 46.0 ng cm^{-2} and 2.6 ng cm^{-2} in the peat column from the Northern peat bog. If we assume as a rough estimate that the whole mass of the TCB was spread out over the ~2000 km^2 of the devastated forest area (Fast et al. 1967), we can take this column site to be representative for the deposition of the entire area, and if, as discussed above, we assume the chemical composition of the TCB's solid part to be similar to that of CI chondrites, we can estimate the mass of chondritic material (the solid, or dust, component) of the explosive body to be ~1.6×10^6 tons by Pd, and 0.4×10^6 tons by Rh.

17.6
Conclusions

The results of several studies show the presence of the cosmic material in the peat of the Tunguska explosion area, distinct from terrestrial material by its chemical and isotopic composition. The various indicators of the presence of the cosmic material, i.e. the shifts in the isotopic composition of the light elements, which is not likely to be attributed to terrestrial processes, the presence of abiogenic carbon not carrying ^{14}C, and the sharp increase of Ir and other PGE are in good accordance.

The chemical composition of the dust fraction of the TCB seems to be close to that of the CI chondrites, which is also in agreement with data on Halley's Comet. All this gives credence to the hypothesis that the TCB was a cometary core. However, compared to Halley's Comet, the Tunguska comet had a very low dust content and was very rich in carbon and volatile elements.

The results of recent theoretical calculations support the hypothesis of the cometary origin (Grigoryan 1998; Bronshten 2000) whereas others support the asteroid hypothesis (Farinella et al. 2001). This distinction has, however, lost its scientific sharpness after the recent discovery of asteroids that behave like comets, and comets that behave like asteroids (Yeomans 2000). Furthermore, we feel that theoretical calculations must have less weight than measurements.

If the TCB was an asteroid, it might resemble Mathilde 253, a C-type asteroid, whose density, measured directly by the NEAR-Shoemaker space probe, is about 1.3 g cm^{-3}. Mathilde 253 is enriched in carbon and seems to be an ex-comet (Basilevsky 1987). If the TCB was a cometary core with very high C/Ir ratio (Rasmussen et al. 1999) then it could be similar to the core of comet Borelly which, unlike Halley's Comet, has a tar-like surface recently explored by NASA Deep Space-1 probe (Soderblom et al. 2002).

Acknowledgments

We are grateful to M.E. Kolesnikov, I.K. Doroshin, D.F. Anfinogenov, R. Serra, H.F.Olsen and the other participants of the Tunguska expeditions for their help in peat sampling, Dr. T. Boettger (UFZ Centre for Environmental Research, Halle, Germany) and Dr. P. Gioacchini (Bologna University, Italy) for helping in isotope measurements, and Prof. Giuseppe Longo (Bologna University, Italy) for the very useful discussion. Thanks are also given to the National Natural Science Foundation of China (No. 40072046), and the Russian Foundation of Fundamental Investigations (No. 02-05-39015) for their financial support.

References

Alvarez LW, Alvarez W, Asaro F, Michel HV (1980) Extraterrestrial cause for the Cretaceous-Tertiary extinction. Science 208:1095–1108
Basilevsky AT (1987) Images of asteroid 253 Mathilde (in Russian). Astronomicheskiy Vestnik 31:571–574, or (in English) Sol System Res 31:514–517
Bronshten VA (2000) Nature and destruction of the Tunguska cosmical body. Planet Space Sci 48: 855–870
Bronshten VA, Zotkin IT (1995) Tunguska meteorite: fragment of a comet or an asteroid. Solar Syst Res 29:241–245
Churyumov KI (1980) Comets and their observation (in Russian). Nauka, Moscow.
Chyba CF, Thomas PJ, Zahnle KJ (1993) The 1908 Tunguska explosion: atmospheric disruption of a stony asteroid. Nature 361:40–44
Cowan C, Atluri CR, Libby WF (1965) Possible anti-matter content of the Tunguska meteor of 1908. Nature 206:861–865
Epiktetova LE (1998) Crushing of the Tunguska Body during its motion through the atmosphere according to eyewitness evidences (in Russian). Internat. Conf. "90 years of the Tunguska Problem (TKT-90)", Krasnoyarsk, Abstracts
Farinella P, Foschini L, Froeschlé Ch, Gonczi R, Jopek TJ, Longo G, Michel P (2001) Probable asteroidal origin of the Tunguska Cosmic Body. Astron Astrophys 377:1081–1097
Fast VG, Bojarkina AP, Baklanov MV (1967) Destruction caused by blast wave of the Tunguska meteorite (in Russian). In: Problema Tungusskogo Meteorita, Part 2, Izdatelstvo Tomskogo Universiteta, Tomsk, pp 62–104
Faure G (1986) Principles of isotope geology. John Wiley and Sons, New Work
Fesenkov VG (1969) Nature of comets and the Tunguska phenomenon. Solar System Res 3:177–179
Fesenkov VG (1978) Meteorites and Meteor Matter (in Russian). Nauka, Moscow.
Florenskij KP (1963) Preliminary results from the 1961 combined Tunguska meteorite expedition (in Russian). Meteoritika 23:3–37

Florenskij KP, Ivanov AV, Iljin NP, Petrikova MN, Loseva LE (1968) The chemical composition of the cosmic spherules from the Tunguska explosion area and some problems of differentiation of cosmic body material (in Russian). Geokhimija (10):1163–1173

Galimov EM (1968) Geochemistry of stable isotopes of carbon (in Russian). Nedra, Moscow.

Ganapathy R (1983) The Tunguska explosion of 1908: discovery of meteoritic debris near the explosion site and the South Pole. Science 220:1158–1161

Gardner A, Hildebrand A, Gilmour I (1992) Isotopic composition and organic geochemistry of nitrogen at the Cretaceous-Tertiary boundary. Meteoritics 27:222–223

Golenetskiy SP, Stepanok VV, Kolesnikov EM (1977) Signs of cosmochemical anomaly in the area of Tunguska Catastrophe 1908 (in Russian). Geochimiya (11):1635–1645

Grigoryan SS (1998) The cometary nature of the Tunguska meteorite. On the predictive possibilities of mathematical models. Planet Space Sci 46:213–217

Gruen E, Jessberger EK(1990) Dust. In: Huebner WF (ed) Physics and chemistry of comets. Springer Verlag, Berlin, pp 113–176

Halbout J, Mayeda TK, Clayton RN (1986) Carbon isotopes and light element abundances in carbonaceous chondrites. Earth and Planetary Science Letters 80:1–18

Hou QL, Ma PX, Kolesnikov EM (1998) Discovery of iridium and other element anomalies near the 1908 Tunguska explosion site. Planet Space Sci 46:179–188

Hou QL, Kolesnikov EM, Xie LW, Zhou MF, Sun M, Kolesnikova NV (2000) Discovery of probable Tunguska Cosmic Body material: anomalies of platinum group elements and REE in peat near explosion site (1908). Planet Space Sci 48:1447–1455

Hou QL, Kolesnikov EM, Xie LW, Kolesnikova NV, Zhou MF, Sun M (2004) Platinum group element abundances in a peat layer associated with the Tunguska event, further evidence for a cosmic origin. Planet Space Sci 52:331–340

Jessberger EK, Kissel J, Fechtig H, Krueger FR (1986) On the average chemical composition of cometary dust. Comet Nucl Sample Return Mission Eur Space Agency Proc Workshop, Canterbury, pp 27–30

Kolesnikov EM (1980) On some probable features of chemical composition of the Tunguska Cosmic Body (in Russian). In: Vzaimodeystviye Meteoritnogo Veshchestva s Zemley. Nauka, Novosibirsk, pp 87–102

Kolesnikov EM (1982) Isotopic anomalies in H and C in peat from the Tunguska meteorite explosion area (in Russian) Doklady Akad Nauk SSSR 266:993–995

Kolesnikov EM (1984) Isotopic anomalies in peat from the Tunguska meteorite explosion area (in Russian). In: Meteoritnye Issledovaniya v Sibiri. Nauka, Novosibirsk, pp 49–63

Kolesnikov EM (1988) Isotopic investigations in the area of the Tunguska Catastrophe in 1908 year. Conference "Global Catastrophes in Earth History", Abstracts, Snowbird, pp 97–98

Kolesnikov EM (1989) Search for traces of Tunguska Cosmic Body dispersed material. Meteoritics 24:288

Kolesnikov EM, Shestakov GI (1979) Isotopic composition of lead from peat of the area of the 1908 Tunguska explosion (in Russian). Geochimiya (8):1202–1211

Kolesnikov EM, Lavrukhina AK, Fisenko AV (1973) Experimental check of hypothesis about an annihilation and thermonuclear nature of the Tunguska explosion of 1908 (in Russian). Geochimiya (8):1115–1121

Kolesnikov EM, Ljul AYu, Ivanova GM (1977) The signs of cosmochemical anomaly in the 1908 Tunguska explosion region: II. The research of chemical composition of silicate microspherules (in Russian). Astronomicheskij Vestnik 11:209–218

Kolesnikov EM, Boettger T, Kolesnikova NV (1995a) Isotopic composition of carbon and hydrogen in peat from the Tunguska Cosmic Body explosion area (in Russian). Doklady Akad Nauk 343:669–672

Kolesnikov EM, Kolesnikova NV, Boettger T, Junge FW, Hiller A (1995b) Elemental and isotopic anomalies in peat of the Tunguska meteorite (1908) explosion area. XIV INQUA Congress vol 34. Berlin

Kolesnikov EM, Boettger T, Kolesnikova NV, Junge FW (1996) The anomalies in isotopic composition of carbon and nitrogen in peat from the Tunguska Cosmic Body explosion area of 1908 (in Russian). Doklady Akad Nauk 347:378–382

Kolesnikov EM, Kolesnikova NV, Boettger T (1998a) Isotopic anomaly in peat nitrogen is a probable trace of acid rains caused by 1908 Tunguska bolide. Planet Space Sci 46:163–167

Kolesnikov EM, Stepanov AI, Gorid'ko EA, Kolesnikova NV (1998b) Element and isotopic anomalies in peat from the Tunguska explosion (1908) area are probably traces of cometary material. Meteoritics and Planet Sci 33: Suppl A85

Kolesnikov EM, Boettger T, Kolesnikova NV (1999) Finding of probable Tunguska Cosmic Body material: isotopic anomalies of carbon and hydrogen in peat. Planet Space Sci 47:905–916

Kolesnikov EM, Kolesnikova NV, Stepanov AI, Gorid'ko EA, Boettger T, Hou QL (2003) Isotopic and elemental anomalies in peat from the Tunguska explosion area are probable traces of cometary material (in Russian). In: Tungusskiy zapovednik.Trudy, vol 1. Izd-vo Tomskogo universiteta, Tomsk, pp 250–266

Kolesnikov EM, Longo G, Boettger T, Kolesnikova NV, Gioacchini P, Forlani L, Giampieri R, Serra R (2003) Isotopic-geochemical study of nitrogen and carbon in peat from the Tunguska Cosmic Body explosion site. Icarus 161:235–243

Krinov EL (1966) Giant Meteorites. Pergamon Press, Oxford, pp 125–265

Kulik LA (1939) Data on Tunguska meteorite up to 1939 year (in Russian). Doklady Akad Nauk SSSR 22: 520–524

Kundt W (2001) The 1908 Tunguska Catastrophe: an alternative explanation. Current Science 81: 399–407

Longo G, Serra R, Cecchini S, Galli M (1994) Search for microremnants of the Tunguska Cosmic Body. Planet Space Sci 42:163–177

L'vov YuA (1984) Carbon in Tunguska meteorite material (in Russian). In: Meteoritnye Issledovaniya v Sibiri. Nauka, Novosibirsk, pp 83–88

Mao XY, Chai CF, Ma SL, Yang ZZ, Xu DY, Sun YY, Zhang QW (1987) Determination of trace elements in Wuxi fallen ice by INAA. J Radioanal Nucl Chem, Articles 114:345–349

Nazarov MA, Korina MI, Barsukova LD, Kolesnikov EM, Suponeva IV, Kolesov GM (1990) Material traces of the Tunguska bolide (in Russian). Geokhimiya 5:627–638; or (in English) Geochemistry International 27:1–12

Petrov GI, Stulov VP (1975) Motion of large bolides in the atmosphere of planets. Cosmic Res 13:525–531

Plekhanov GF (1997) Results of investigations and paradoxes of the 1908 Tunguska catastrophe (in Russian). In: Tungusskiy vestnik KSE. Tomsk, pp 16–18

Rasmussen KL, Clausen HB, Risbo T (1984) Nitrate in the Greenland ice sheet in the years following the 1908 Tunguska event. Icarus 58:101–108

Rasmussen KL, Clausen HB, Kallemeyn GW (1995) No iridium anomaly after the 1908 Tunguska impact: evidence from a Greenland ice core. Meteoritics 30:634–638

Rasmussen KL, Olsen HJF, Gwozdz R, Kolesnikov EM (1999) Evidence for a very high carbon/iridium ratio in the Tunguska impactor. Meteorit Planet Sci 34:891–895

Rasmussen KL, Aaby B, Gwozdz, R (2000) The age of the Kaalijärv meteorite craters. Meteoritics and Planetary Science, 35:1067–1071

Rocchia R, Bonte P, Robin E, Angelis M, Boclet D (1990) Search for the Tunguska event relics in the Antarctic snow and new estimation of the cosmic iridium accretion rate. In: Global Catastrophes in Earth History. Boulder, Colorado, pp 189–193

Rocchia R, Robin E, De Angelis M, Kolesnikov E, Kolesnikova N (1996) Search for remains of the Tunguska event. International Workshop Tunguska 96. Abstracts. Bologna, Italy, pp 7–8

Serra R, Cecchini S, Galli M, Longo G (1994) Experimental hints on the fragmentation of the Tunguska Cosmic Body. Planet Space Sci 42:777–783

Soderblom LA, Becker TL, Bennet G, Boice DC, Britt DT, Brown RH, Buratti BJ, Isbell C, Giese B, Hare T, Hicks MD, Howington-Kraus E, Kirk RL, Lee M, Melson RM, Oberst J, Owen TC, Rayman MD, Sandel BR, Stern SA, Thomas N, Yelle RV (2002) Observations of comet 19P/Borelly by the miniature integrated camera and spectrometer aboard deep space 1. Science 296:1087–1091

Taylor SR and McLennan SM (1985) The continental crust: its composition and evolution. Blackwell Scientific Publications

Turko RP, Toon OB, Park C, Whitten RC, Pollack JB, Noerdlinger P (1982) An analysis of the physical, chemical, optical, and historical impacts of the 1908 Tunguska meteor fall. Icarus 50:1–51

Vasiljev NV (1998) The Tunguska meteorite problem today. Planet Space Sci 46:129–150

Whipple FJ (1930) The great Siberian meteor and the waves, seismic and aerial, which it produced. Q J R Meteorol Soc 56:287–304

Xie LW, Hou QL, Kolesnikov EM, Kolesnikova NV (2001) Geochemical evidence for the characteristics of the 1908 Tunguska explosion body in Siberia, Russia. Sci. China (Ser D) 44:1029–1037

Yeomans DK (2000) Small bodies of the Solar System. Nature 404:829–832

Chapter 18

The Tunguska Event

G. Longo

18.1
Introduction

In the early morning of 30th June 1908, a powerful explosion over the basin of the Podkamennaya Tunguska River (Central Siberia), devastated 2 150 ± 50 km² of Siberian taiga. Eighty millions trees were flattened, a great number of trees and bushes were burnt in a large part of the explosion area. Eyewitnesses described the flight of a "fire ball, bright as the sun". Seismic and pressure waves were recorded in many observatories throughout the world. Bright nights were observed over much of Eurasia. These different phenomena, initially considered non-correlated, were subsequently linked together as different aspects of the "Tunguska event" (TE).

Almost one century has elapsed and scientists are still searching for a commonly accepted explanation of this event. Several reviews and books summarize the results acquired by the intensive investigations of the last century, e.g. Kulik (1922, 1939, 1940), Landsberg (1924), Krinov (1949, 1966), Gallant (1995), Trayner (1997), Riccobono (2000), Bronshten (2000), Vasilyev (1998, 2004) and Verma (2005).

Despite great efforts, the TE remains a conundrum.

18.2
The Hypotheses

The most plausible explanation of the event considers the explosion in the atmosphere of a "Tunguska Cosmic Body" (TCB), probably a comet or an asteroid-like meteorite.

18.2.1
Comet or Asteroid?

From his first determination of the basin of the Podkamennaya Tunguska River as the explosion site, Kulik (1922 and 1923) used the term "Tunguska meteorite", for the TCB, and continued searching for an iron body, similar to one found in Arizona (Kulik 1939, 1940; Krinov 1949, 1966). Voznesenskij (1925) hypothesized an equal probability for a stony or an iron body composition. Shapley (1930) was the first to suggest that the Tunguska event was caused by the impact of a comet and Kresák (1978) indicated the comet Encke as the origin of the TCB. Fesenkov (1949), for many years, supported the stony object hypothesis. Later, Fesenkov (1961) worked out a definite model of an impact between a comet and the Earth's atmosphere. From that time onward, the majority of

Russian scientists followed the cometary hypothesis (see for example Grigorian 1998), whereas many western scientists preferred an asteroidal model (e.g. Sekanina 1983, 1998; Chyba et al. 1993).

For many reasons, these two "schools" practically ignored each other until the international workshop Tunguska96, held in Bologna (Italy) from 15th to 17th July 1996 (Di Martino et al. 1998). In the recent past the cometary hypothesis has been favored on the basis that a low-density object was needed to explain the Tunguska catastrophe (Petrov and Stulov 1975; Turco et al. 1982). Subsequently, to account for the concentration of energy release of the explosion, two sub-versions of this hypothesis have been developed, one introducing chemical reactions (Tsymbal and Shnitke 1986), the other nuclear-fusion reactions (D'Alessio and Harms 1989). On the other hand, it has been shown (Grigorian 1976; Grigorian 1979; Passey and Melosh 1980; Levin and Bronshten 1986) that the fragmentation of a normal density object can greatly increase the rate of energy deposition in a small region near the end of the trajectory, thus appearing as an atmospheric explosion. Detailed calculations which include the effect of aerodynamic forces that can fracture the object, and the heating of the bolide due to friction with the atmosphere, have recently been performed, showing that the TE is fully compatible with the catastrophic disruption of a 60-100 m diameter asteroid of the common stony class (Chyba et al. 1993; Hills and Goda 1993). However, due to the uncertainty of such input parameters as the energy and height of the explosion or the inclination angle and the encounter velocity of the impactor, the same calculations do not exclude the possibility that the TCB was a high velocity iron object, nor rule out a carbonaceous asteroid as an explanation of the event. Considering a "plume-forming" atmospheric explosion, Boslough and Crawford (1997) have suggested that the commonly accepted energy-yield is an overestimate and that a 3 megaton event could generate the observed devastation. Many of the phenomena associated with the TE can be related to the formation and collapse of an atmosphere plume, caused either by a comet or by an asteroid. For example, the predicted ejection at altitudes of some hundreds of kilometers of the impactor mass can explain the "bright nights" associated with the TE.

It is difficult to definitely support one or the other hypothesis. Therefore, one way to achieve certainty about the nature and composition of the TCB remains the search for some of its remnants. Numerous radiocarbon analyses of Tunguska wood samples (Nesvetajlo and Kovaliukh 1983), chemical analyses of soil and plants (Kovalevskij et al. 1963; Emeljanov et al. 1963; Kirichenko and Grechushkina 1963; Iljina et al. 1971), bed-by-bed chemical analyses of the peat formed by *Sphagnum fuscum* in 1850-1950 (Vasilyev et al. 1973; Golenetskij et al. 1977a; Golenetskij et al. 1977b; Kolesnikov et al. 1977), isotopic analyses of many different soil, peat and wood samples (Kolesnikov et al. 1979), as well as analyses of the spherules from Tunguska soil samples collected in a radius of several tens of kilometers from the epicenter (Florenskij et al. 1968; Jéhanno et al. 1989; Nazarov et al. 1990) have been completed. Nevertheless, many conclusions of this intensive work are still uncertain, so that further investigations are needed. Although almost every year there is an expedition to Tunguska, so far no typical material has permitted a certain discrimination to be made between an asteroidal or cometary nature of the TCB. Some papers report that hydrogen, carbon and nitrogen isotopic compositions with signatures similar to those of CI and CM carbonaceous chondrites were found in Tunguska peat layers dating from the TE

(Kolesnikov et al. 1999, 2003) and that iridium anomalies were also observed (Hou et al. 1998, 2004). Measurements performed in other laboratories have not confirmed these results (Rocchia et al. 1990; Tositti et al. 2006). Moreover, a concentration of microparticles of inferred cosmic origin was found in tree resins dating from the TE (Longo et al. 1994; Serra et al. 1994). Although these data are compatible with the hypothesis of the impact of a cosmic body, they are by no means conclusive and are not sufficient to prove the nature of the TCB. The same can be said about the lacustrine sediments of Cheko Lake (Sacchetti 2001) studied in the framework of the multidisciplinary investigation as carried out by the Italian scientific expedition Tunguska99 (see *http://www-th.bo.infn.it/tunguska/*) (Amaroli et al. 2000; Pipan et al. 2000; Gasperini et al. 2001; Longo et al. 2001; Longo and Di Martino 2002 and 2003; Longo et al. 2005). This field research has been strengthened by theoretical studies and modeling. In a recent paper (Farinella et al. 2001), a sample of possible TCB orbits has been constructed and a dynamic model was used to compute the most probable source of a TCB placed on each of these orbits. The results of calculations gave a greater probability for a TCB coming from an asteroidal source (83%), than from a cometary source (17%).

18.2.2
"Non-traditional" Hypotheses

Vasilyev (2004) states, "We should not exclude the possibility that the Tunguska phenomenon is a qualitatively new phenomenon for the science, that should be analyzed from non-traditional positions". These "non-traditional" approaches still consider an impact with the atmosphere of "something" coming from external space. Several of them, though published in scientific journals, were found to be technically groundless, e.g. the hypotheses involving near critical fissionable material (Zigel' 1983; Hunt et al. 1960), antimatter meteors (Cowan et al. 1965), and tiny black holes (Jackson and Ryan 1973). Others consider alien spacecrafts (Kazantsev 1946; Baxter and Atkins 1976). Kazantsev was the first who explained the lack of fragments or impact craters in Tunguska by an explosion in the atmosphere. Nevertheless, I think that here we can ignore such extremely "non-traditional" hypotheses.

18.2.3
Alternative Approaches

Recently, some "alternative" approaches were presented to explain the TE. Different from the above-mentioned traditional or non-traditional explanations, these alternative approaches deny an impact of an external body with Earth. They claim that the event was triggered by a terrestrial cause. I mention here two of the more discussed alternative interpretations.

The first is a tectonic interpretation (e.g. Ol'khovatov 2002), which considers the coupling between tectonic and atmospheric processes in a "very rare combination of favorable geophysical factors." Another recent work that should be mentioned is the "kimberlite interpretation" (Kundt 2001), which considers the TE as caused by the tectonic outburst of some 10 megaton of natural gas. For the volcanic (outflow) interpretation, Kundt presents the estimates of the involved mass and kinetic energy of the vented natural gas, of its outflow timescale, supersonic and subsonic ranges, and buoyant escape towards the exosphere.

The main idea of this latter work is contradicted by at least two facts. The first and more obvious point against the hypothesis of an explosion from the ground is that the eyewitness testimonies describe the trajectory of a *bolide crossing the sky* (see Sect. 18.3.2). Among these testimonies, the earliest, given a few days after the event by educated people, have a high trustworthiness. On that basis, the first Kulik expedition (1921–1922) gathered sufficient information to conclude, that "the meteorite fall in the neighborhood of the Ogniya river, a left tributary of the Vanavara river, which is a right tributary of the Podkamennaya Tunguska (Hatanga) river" (Kulik 1922, 1923, 1927; Landsberg 1924). The first expedition could not go farther than Kansk, about 600 km from the Tunguska explosion site. Five years later, Kulik discovered the site about 50 km from the mentioned tributary of the Vanavara River.

A second objection comes from the absence of debris clearly referred to the explosion in the epicenter area. If we assume that the anomalous optical phenomena observed after the TE, were due to particles released in the atmosphere by the explosion, we should find an increasing concentration of those particles (with grain-size progressively decreasing) toward the explosion epicenter. As also pointed out by Kundt, we do not observe a carpet of dust in the vicinity of the epicenter as we should observe for the explosion of a meteorite, but also (and more markedly) for the explosion of a diatreme or any volcanic emission. How could an explosion "from below" disperse dust in the atmosphere to an extent comparable to that of the Krakatau without leaving significant traces close to the epicenter? It seems more probable that an explosion "from above" could explain this occurrence.

Moreover, geological maps of the region (Sapronov 1986) and our own observations during the Tunguska99 expedition do not report the presence of mantle rocks, such as peridotites or eclogites, which are usually associated with kimberlites. Though the area is centerd on the roots of the lower Triassic Kulikovsky paleovolcanic complex (see Fig. 18.1), which extends over an area 25 × 20 km wide, displaying numerous, various sized craters, it is presently a tectonically stable cratonic region, as testified by the low intraplate seismicity. The map from the USGS catalog, which reports significant worldwide earthquakes during historical times, confirms this stability.

Finally, the "radonic storm" registered at our base camp (see Fig. 18.2) during the Tunguska99 expedition (Longo et al. 2000; Cecchini et al. 2003) has nothing to do with a "kimberlite" phenomenon, as suggested by Kundt. Indeed, we registered an intensity enhancement of gamma radiation during a thunderstorm (see Fig. 18.3) due to radon daughters, as observed in other parts of the world, where no "kimberlite interpretation" is possible. Though we cannot accept the main ideas of Kundt, the outflow theory can help us to understand some aspects of the TE. It is plausible, and even probable, that gas releases took place from the permafrost dissolution (caused by the impact of the TCB and not by a kimberlite outflow). For example, part of the multiple explosions heard for more than half an hour by many earlier trustworthy witnesses (Kulik 1922, 1927; Obruchev 1925; Voznesenskij 1925) might probably be due to a rapid release of gas (methane) from the permafrost layer as a consequence of the thermal burst related to

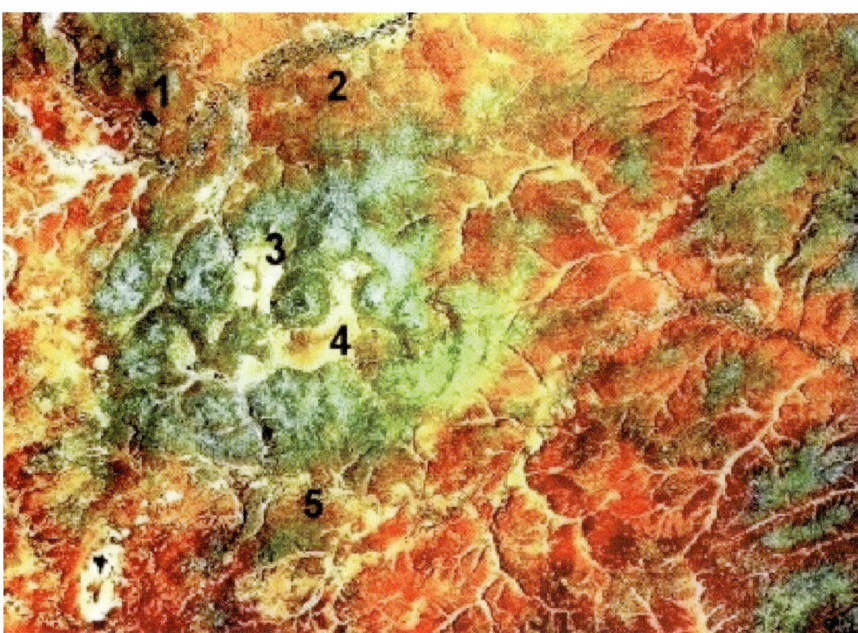

Fig. 18.1. Satellite view of the Kulikovsky paleovolcanic complex (*1* – lake Cheko, *2* – river Kimchu, *3* – Northern swamp, *4* – Southern swamp, *5* – river Khusma)

Fig. 18.2.
The base camp of the Tunguska99 expedition on the shore of the lake Cheko (drawing by Andrey Chernikov)

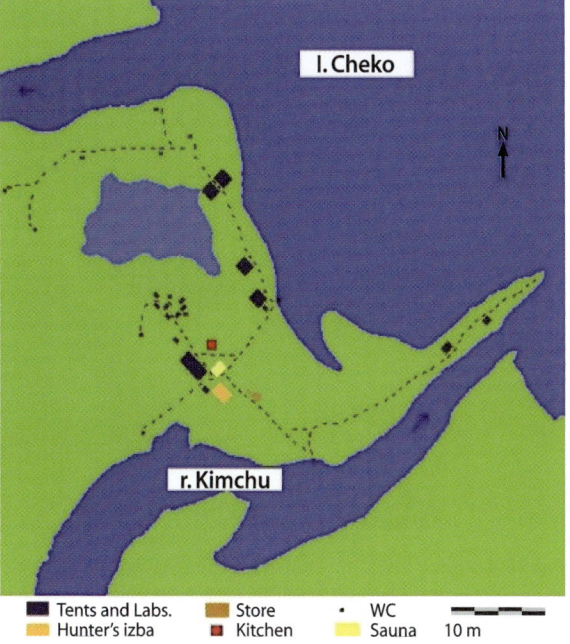

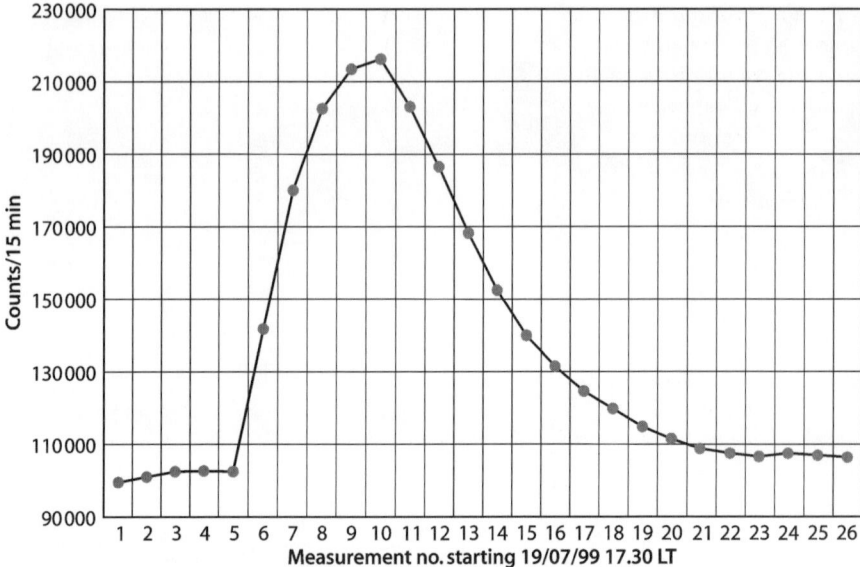

Fig. 18.3. Gamma-ray (25 keV – 3 MeV) intensity enhancement registered at the base camp of the lake Cheko during the thunderstorm of July 19, 1999. Note the steep rise of counting rate, while it is raining. It corresponds to gammas emitted by radon daughters

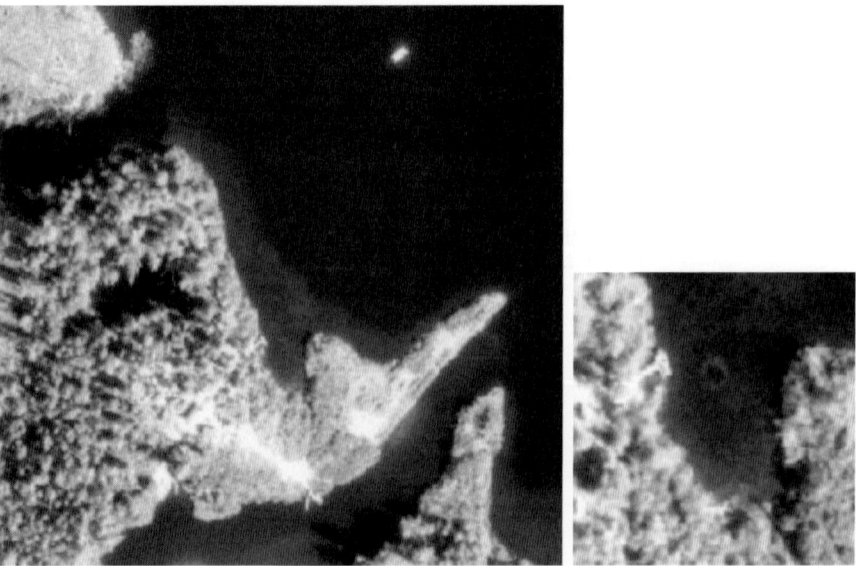

Fig. 18.4. Aerial view of the camp of the Tunguska99 expedition (23 July 1999). Near the shore, a hole with a few meters diameter resulting from a gas outflow can be seen on the lake bottom

the main event. Indeed, in July 1999, we observed a small "crater" originated "from below" on the Cheko Lake bottom (see Fig. 18.4). It could be due to methane emission from decaying organic matter in the *surface* layer of some tens of meters. Obviously, this does not contradict the known *tectonic stability* of the region.

18.3
Known Data

18.3.1
Objective Data

Three main kinds of objective data on the Tunguska explosion are available: seismic and barometric registrations, recorded immediately after the event, information on the bright nights, observed in Eurasia in July 1908, and data on forest devastation, systematically collected 50–70 years later and recently integrated with the data of the 1938 and 1999 aerial photographic surveys.

Seismic and Barometric Registrations

Seismic records from Irkutsk, Tashkent, and Tiflis were published together, two years after the event (Levitskij 1910), those from Jena, three years later. However, the first paper that connected to the TE the origin of these seismic waves was published only in 1925 (Voznesenskij 1925). Similarly, the barograms recorded in 1908 in a great number of observatories throughout the world, were associated with the TE some twenty years later (Whipple 1930; Astapovich 1933). From the analysis of the available seismograms and barograms, the time that the seismic and aerial waves started was calculated. The main results obtained are listed in Sect. 18.4.

Bright Nights Observed

In 1908, the attention of astronomers and geophysicists in Europe and Asia was drawn to some unusual phenomena, such as bright nights, noctilucent clouds, brilliant colorful sunsets and other observations. It is difficult to conclude that some of these phenomena are really "anomalous". For example, in June-July, the appearance of noctilucent clouds reaches its maximum and it is difficult to distinguish between "usual" and "unusual" noctilucent clouds. Therefore, I shall consider here only the bright nights phenomenon.

Bright nights ("at midnight, it was possible to read the newspaper without artificial lights"; see Figs. 18.5 and 18.6) were described in many papers (e.g., De Roy 1908; Shenrock 1908; Süring 1908; Svyatskij 1908). At that time, many explanations for the bright-nights phenomenon were proposed. Up to 1921, meager information about a great 1908 bolide was published only in some local Siberian newspapers. Nobody considered a link between these phenomena, although on 4 July 1908, the Danish astronomer Torwald Kohl wrote: "It would be advisable to learn whether in recent times some great meteorite has been seen in Denmark or elsewhere" (Kohl 1908). It was only in 1922, after his first recognition in Siberia that Kulik wrote about a probable link between the bright nights in Eurasia and the explosion in Central Siberia (Kulik 1922). From that time onward, such phenomena have been considered as two parts of the Tunguska event.

The phenomenon and its correlation to the TE, was thoroughly studied in the 1960s (Zotkin 1961; Vasilyev et al. 1965). The 4 March 1960 issue of *Science* published a letter from the Committee on Meteorites of the Academy of Science of the USSR addressed to foreign scientists and asking them to send all the information available on the optical phenomena of 1908 (Fesenkov and Krinov 1960).

Zotkin (1961) studied the bright nights, observed in 114 points of the globe. He distinguished observations following the 30 June from those preceding that date. He considers the latter poorly reliable and of "local character", whereas the events observed from the 30 June did not have a "local" character and were observed in more than a hundred points of Europe and Asia.

Vasilyev (1965) considered a more complete data set and referred to 86 communications and articles dated to 1908. He lists 14 cases of bright nights from 21 to 29 June 1908 and 159 cases from 30 June up to 3 July (in subsequent papers, he indicates about twenty other cases from 4 to 28 July). He considers *all* these cases related to the Tunguska event and this is not easy to explain.

It seems to me that Zotkin's approach is more acceptable. Only the bright nights following the 30 June should be related to the Tunguska event. This is confirmed by the *global* character of the phenomenon and by *polarization* measurements. The "global" character of the phenomenon, observed in the nights beginning on 30 June and 1 July 1908 are illustrated in Fig. 18.5 (Vasilyev and Fast 1976). As can be seen, the bright nights were observed on an area of about 12 million km^2, from the longitude 6.5° W (Armagh, Ireland; see Fig. 18.6) up to 92.9° E (Krasnoyarsk) and from the latitude 41° N (Tashkent) up to 60° N (Petersburg). If the bright nights are due to dust in the atmosphere, the light reflected should be polarized. Busch (1908a,b) measured the daylight polarization in Arnsberg (Germany). His results indicate an absence of the effect in the first half of 1908 up to 28 June, a strong effect the 1 July that gradually disappears up to 25 July. The conclusions of Zotkin were that it is difficult to accept that dust particles could reach Great Britain from Tunguska in 22 hours. Therefore, they were ice particles from the comet tail and the comet nucleus exploded in Tunguska. Bronshten (1991) hypothesized that the particles were transported from Tunguska by gravitational forces. In Boslough and Crawford model (1997), the mass of the impactor, as well as water from the humid lower atmosphere, are ejected above the top of the atmosphere and within 15 minutes can extend more than 2000 km from the impact site.

Data on Forest Devastation

The data on forest devastation are a second kind of objective information source about the event. The main part of these data refers to the tree fall and the direction of flattened trees. From these data we can obtain information on the coordinates of the wave propagation centers (often called "epicenter(s)") and on the final TCB trajectory.

Chapter 18 · **The Tunguska Event** 311

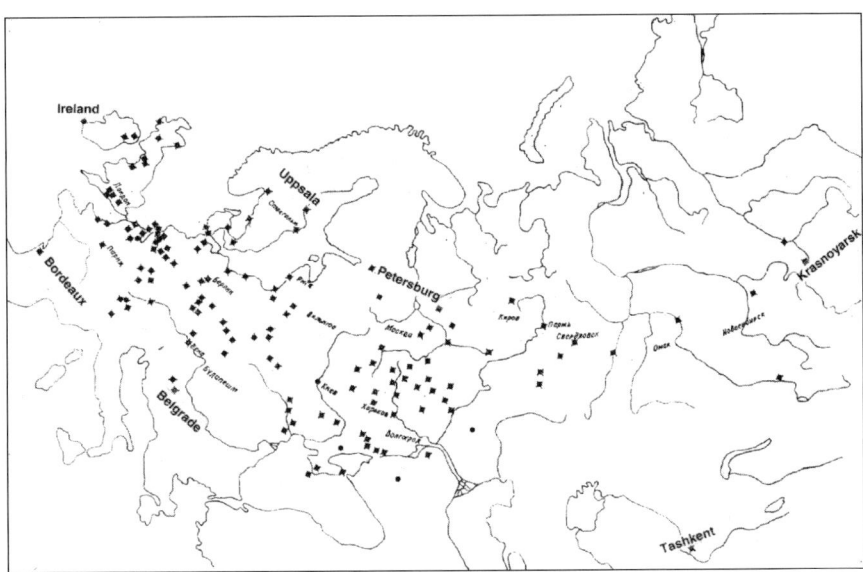

Fig. 18.5. Stations where anomalous bright nights were observed the 30 June/1 July 1908

Fig. 18.6. Photos taken during the bright night of 30 June 1908 in (**a**) Armagh, (**b**) Greenwich, and (**c**) Tambov. **d** The Irkutsk observatory at the beginning of the 20th century

Though Kulik discovered the radial orientation of fallen trees as early as 1927, systematic measurements of fallen tree azimuths were started only during the two great post-war expeditions organized by the Academy of Sciences in 1958 and 1961 (Florenskij et al. 1960; Florenskij 1963), and during the Tomsk 1959–1960 expeditions. Under the direction of Fast, with the help of Boyarkina, this work was continued for two decades during ten different expeditions from 1961 up to 1979. A total of 122 people, mainly from Tomsk University, participated in these on site measurements. The data collected have been published in a catalog in two parts: the first one contains the data obtained by six expeditions (1958–1965), which include the whole set of single-tree azimuths and the azimuths averaged on trial areas equal to 2500 m^2 or 5000 m^2, chosen throughout the whole devastated forest (Fast et al. 1967). In the second part, the data collected by the six subsequent expeditions (1968–1976) were given (Fast et al. 1983).

The data on forest devastation also give information on the energy emitted and on the height of the explosion. Indeed, these data include, not only fallen tree directions, but also the distances that different kinds of trees were thrown, the pressure necessary to do this, information on forest fires and charred trees, data on traumas observed in the wood of surviving trees and so on (e.g. Florenskij 1963; Vorobjev et al. 1967; Longo and Serra 1995; Longo 1996, 2005).

In order to correct, update and enlarge the fallen tree distribution data, we performed a new aero-photographic survey during the Tunguska99 expedition (Longo and Di Martino 2002, 2003) (see map on Fig. 18.7). This survey was needed to obtain a new unified catalog, which includes: (*1*) corrected Fast data (Fast et al. 1967, 1983), (*2*) data from Kulik's 1938 aerial photosurvey never previously analyzed, (*3*) never published data collected in 1967 by the Anfinogenov group in the central region of the site. These three datasets have been checked and completed with our on-site measurements carried out in July 1999 and 2002 to obtain the coordinates of different reference points in the same area. These data allowed us to recognize ground elements on the aerial pictures and to connect them to the regional topographic net.

Unfortunately, a map containing all the data from Fast's catalogs (Fast et al. 1967, Fast et al. 1983) has never been published. In the last 40 years, the map of fallen tree azimuths used for comparison with theoretical models (e.g. Korobeinikov et al. 1990; Boslough and Crawford 1997) was the one constructed by A. Boyarkina, V. Fast and co-workers (Florenskij 1963; Boyarkina et al. 1964). This map contains only the data on the azimuths measured in 1958–1961. The new unified catalog and the new map (Longo et al. 2005) have been constructed using a number of tree azimuths and trial areas several times larger than those considered in Fast's analyses. Moreover, we have introduced a reliability degree for each trial area averaged azimuth. The reliability degree has been assigned on the basis of the percentage of singletree azimuths that lay in a sector of 15° centered on the averaged azimuth. A good agreement between the new map and the horizontal aerodynamic pressure calculated on the basis of Korobeinikov et al. (1990) model has been obtained.

No Impact Craters or Meteorite Fragments

Data on forest devastation and records of the atmospheric and seismic waves have made it possible to deduce the main characteristics of the Tunguska explosion, i.e. its

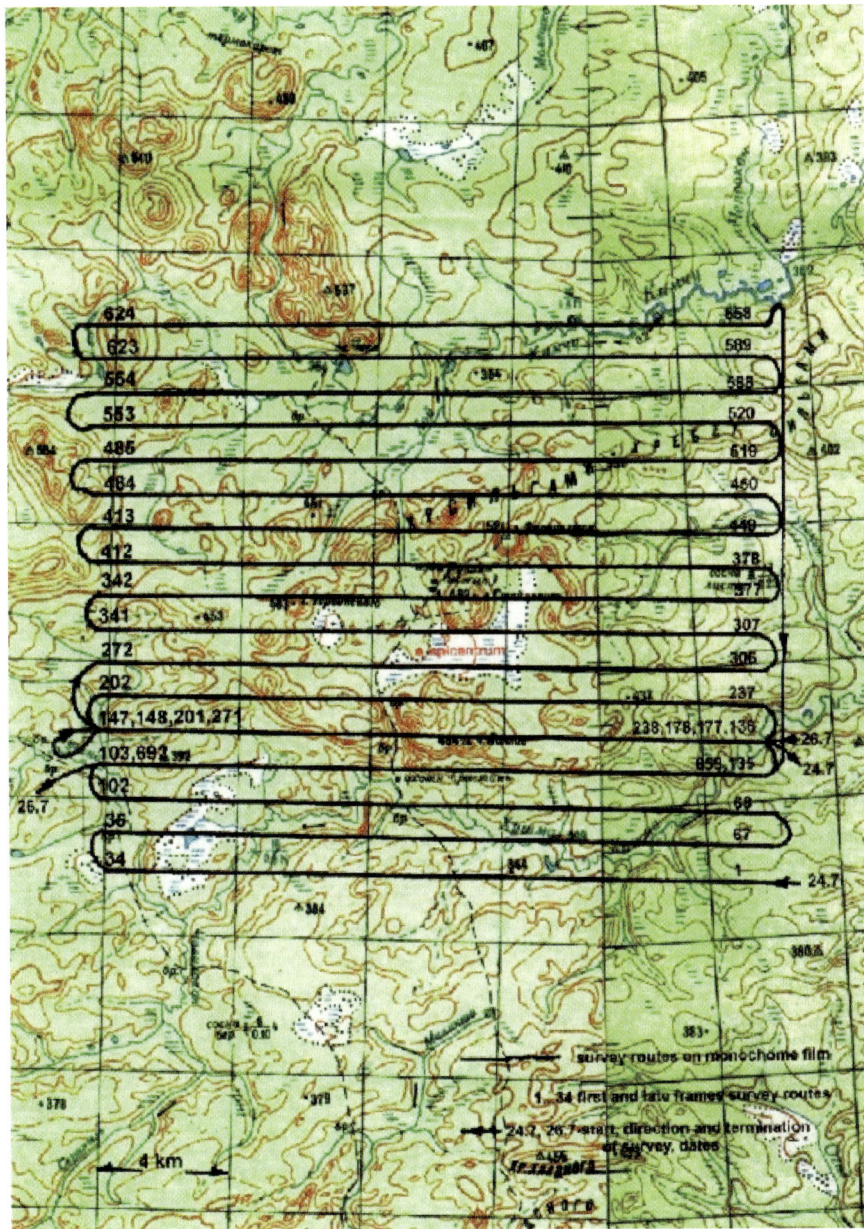

Fig. 18.7. Flight routes of the 1999 aero-photosurvey

exact time, $00^h 14^m 28^s$ UT (Ben-Menahem 1975), the coordinates of the point usually called epicenter, 60° 53' 09" N, 101° 53' 40" E (Fast 1967), the energy release, equivalent to 10–15 million tons of TNT (Megaton) that corresponds to about one thousand times the Hiroshima bomb energy, and height of the explosion (5–10 km), though the values for the last two parameters are estimated with great uncertainty. However, neither macroscopic fragments of the cosmic body, nor a typical signature of an impact, like a crater, have ever been found in an area of 15 000 km^2, so that the nature and composition of the TCB and the dynamic of the event have not yet been clarified.

18.3.2
Eyewitnesses Testimonies

There is a great number of eyewitness testimonies. The more complete collection of these testimonies is provided by Vasilyev et al. (1981). It contains direct observations of the Tunguska explosion from 386 different points and a list of the geographical coordinates of these points. To these observations, the authors have added news published in newspapers, reports and communications from many official employees for a total of 708 testimonies. It is easy to find contradictions in this material collected for more than 60 years by very different people. Sometime these contradictions are more apparent than real. As an example I can remember the contradiction recently removed by Fast VG[1] and Fast NP (2005). As is well known, two centuries before the TE, the czar Peter I introduced a reform in the Orthodox Church. Entire villages of people that did not recognize the reform were sent to Siberia. Therefore many Siberian regions and villages in 1908 were populated by people following the "old faith". For them, the daily timetable was regulated starting from the morning prayers at "*obied*", i.e. 8 o'clock in the morning. When asked about the explosion time, they answered that the explosion took place some time before the *obied*, which really corresponds to the seismic wave registrations after 7 o'clock local time. For the secular people that collected the testimonies, the word *obied* means lunch, i.e. about 12–14 o'clock. Therefore, they completed the forms by noting that the eyewitness stated that the explosion took place at noon, or even in the afternoon. These testimonies were considered not trustworthy due to the clear contradiction with instrumental registrations. A thorough statistical analysis performed by the Fasts (2005) has shown that the distribution of "midday eyewitnesses" correctly reproduces the distribution of the population following the "old faith".

To use them properly, it is important to take into account the different trustworthiness of the testimonies. I think that we can distinguish the following groups of testimonies in decreasing order of trustworthiness:

1. The testimonies collected *in the days immediately following the Tunguska explosion* by the director of the Irkutsk magnetic and meteorological observatory Voznesenskij (1925). Unfortunately, Voznesenskij published them only 17 years later due to an excess of scientific prudence. Immediately after the registration of the earthquake N° 1536, in the morning of 30 June 1908, Voznesenskij sent to all his correspondents a request to report what they or other people had observed on that morning. In his paper he gives a table with the results received from 61 correspondents and a map

---

[1] It was the last contribution to the Tunguska studies given by the great researcher Vilgem Genrikovich Fast (1936–2005).

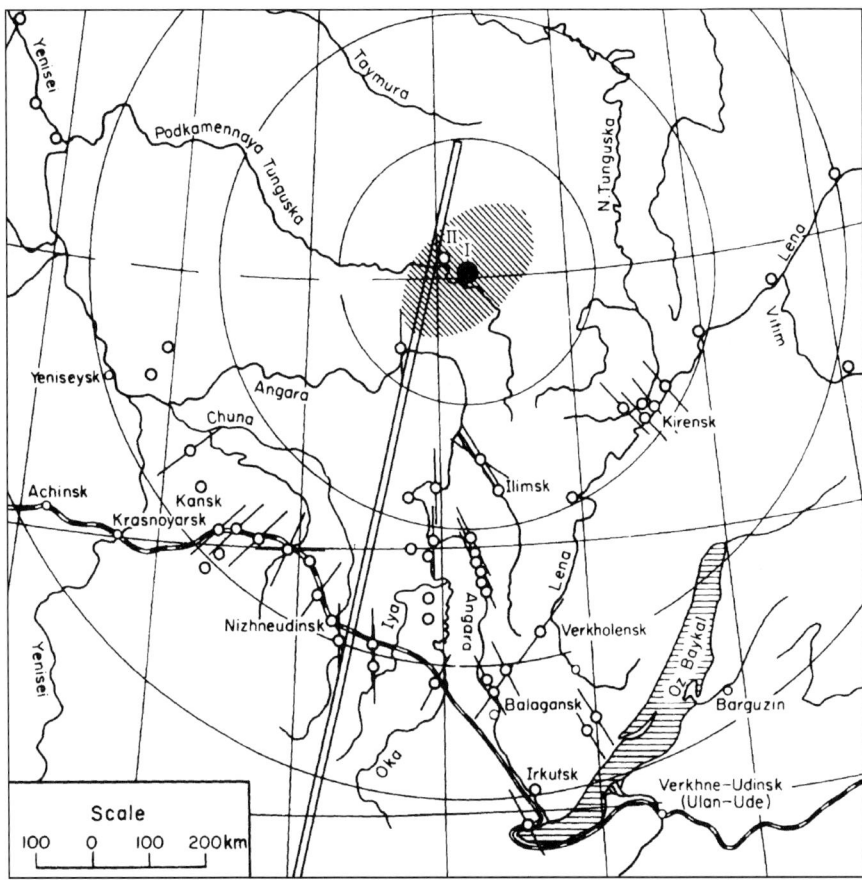

Fig. 18.8. A map with the dislocation of the correspondents that sent in July 1908 their reports to the Irkutsk Observatory. The map was published by Voznesenskij (1925) and reproduced by Krinov (1949, 1966)

showing their location on a very great territory (Fig. 18.8). Moreover, he refers to many individual testimonies from "cultured" people (chief of town post office, employees of meteorological observatories, agronomist and so on).

2. The testimonies collected before, during and immediately after (up to 1933) the expeditions of Kulik. They were collected mainly by Obruchev (1925), Suslov (1927) and Kulik (1922, 1923, 1927). I have mentioned in Sect. 18.2.3 that this primary information was sufficient to understand in 1922 that research had to be directed to the north of the Podkamennaya Tunguska River, in the neighborhood of the Vanavara River.

3. In the period from about 1933 up to 1958 practically no new eyewitness was questioned and, finally, in the 1960s, a massive material with hundreds of new testimonies from old people was collected in many regions. A thorough examination of these records can still be useful as Fasts's work shows.

No doubt that the more valuable testimonies are those written immediately after the fall by the correspondents of the Irkutsk observatory. They are not influenced by Voznesenskij who asks only about observations related to the *earthquake N° 1536*, without any reference to a flying body. Many of these reports are written before the publication in local newspapers of the first information on the event. These genuine reports, synthesized by Voznesenskij in 1925, are now stored in the Archives of the Meteorite Committee of the Russian Academy of Sciences hereafter referred as "Archive RAS". In the following paragraph, I give in brackets the page of the document N° 57 of Archive RAS in which these testimonies are gathered.

To describe what seen in the morning of 30 June 1908, no one of these reports testify something different from a flying object. Many reports are written after questioning a great number of persons. For example, the director of the meteorological station of Maritui states that his report is written after the interrogation of about 500 persons on a great territory around his station (19). I quote here some descriptions of the correspondents:

> "a large group of local inhabitants noticed a ball of fire in the north west coming down obliquely" (3); "the workmen saw a fiery block flying, it seemed, from south east to north west" (4); "in the north west a pillar of fire appeared about 8 meters in diameter... it was accurately established that a meteorite of very large dimensions had fallen" (5); "the local peasants told me that they saw some sort of fiery ball flying in the north" (6); "a loud noise was heard... probably from a passing meteor (aerolith)" (9); "some of local inhabitants had seen an elongated body narrowing towards one end, about one meter in length, torn as it were from the Sun...this body flew across the sky and fell in the north east" (16); "the fall of an aerolith was observed... a fiery streamer was seen" (26); "a ball of fire appeared in the sky and moved from south east to north west. As the ball approached the ground... it had the appearance of two pillars of fire" (36).

The testimony of page 16 was written the 30 June 1908 (the day of the event), that of page 6 – the 1 July 1908 (the day after the event), the others – from a few days up to six weeks after the event. Many correspondents could not understand what they have seen or heard. For example, in the letter referred to on page 6, the correspondent wrote to the Irkutsk observatory: "I have the honor to ask submissively the observatory to communicate and clarify what this means and could it be dangerous for human life".

These testimonies, and many others, contradict the "alternative" approaches (see Sect. 18.2.3) that deny the impact of an external body with Earth.

18.4
Parameters Deduced

18.4.1
Explosion Time

Studying the available seismic data, a first determination of the explosion time as $0^h 17^m 12^s$ UT was obtained by Voznesenskij (1925). This value was used up to the 1960s. The explosion time deduced from the barograms of 6 British meteorological stations, was equal to $0^h 15^m$ UT (Whipple 1930). The independent analysis of the barograms from 13 Siberian stations, gave an explosion time equal to $0^h 16^m 36^s$ UT (Astapovich 1933). These two sets of data were subsequently analyzed more carefully taking into account the exact distances and the properties of seismic and atmospheric waves. Pasechnik (1971) obtained a first result ($0^h\ 14^m\ 23^s$ UT), based solely on Jena and Irkutsk's seismic data. Two additional and more complete analyses were independently performed by Ben-Menahem (1975) and Pasechnik (1976). They found practically the same value for the time the seismic and aerial waves started (see Table 18.1, updated from Farinella et al. 2001).

Pasechnik (1976) calculated that the time of the explosion in the atmosphere was 7–30 seconds earlier depending on the height and energy of the explosion; this interval was subsequently reduced to 2–20 seconds (Pasechnik 1986). In the 1986 paper, however, Pasechnik revised his previous results obtaining a value equal to $0^h\ 13^m\ 35^s \pm 5^s$ UT. The commonly accepted explosion time is the time given by Ben-Menahem for the instant the seismic waves started, i.e. $0^h\ 14^m\ 28^s$ UT.

18.4.2
Coordinates of the Epicenter

The first contact point between the Earth surface and the shock wave from the airburst is commonly called "epicenter," though this term is not proper. From the data collected during the first three expeditions, Fast (1963) obtained the epicenter coordinates 60° 53' 42" N, and 101° 53' 30" E. These values are very close to the final ones 60° 53' 09" ± 06" N, 101° 53' 40" ± 13" E, calculated by Fast (1967) analyzing the whole set of data from the first part of the catalog (Fast et al. 1967). At about the same time, Zolotov (1969) performed an independent mathematical analysis of the same data and obtained the second values quoted in Table 18.1. The coordinates of Fast's epicenter with the uncertainties quoted, corresponding to about 200 m on the ground, were subsequently confirmed in all Fast's papers.

Examining the direction of fallen trees seen on the aerial photographic survey performed in 1938, Kulik suggested (1939, 1940) the presence of 2–4 secondary centers of wave propagation. This hypothesis was not confirmed, although neither was it definitely ruled out, by Fast's analyses and by seismic data investigation (Pasechnik 1971, 1976, 1986). Some hints of its likelihood were given by Serra et al. (1994) and Goldine (1998). This hypothesis is compatible with the recent reanalysis of the direction of fallen trees made on the basis of Fast's data integrated by those obtained from the 1938 and 1999 aerial photosurveys (Longo et al. 2005). The high trustworthiness of earlier eyewitnesses is also in favor of the multicenter hypothesis (Voznesenskij 1925, Archive RAS).

18.4.3
Trajectory Parameters, Height of the Explosion and Energy Emitted

The final TCB trajectory can be defined by its azimuth (α), here given from North to East starting from the meridian, the trajectory inclination (h) over the horizon and the height (H) of the explosion. These parameters can be estimated from the data on forest devastation, seismic records and eyewitness' testimonies.

The height of the explosion is closely related to the value of the energy emitted, usually estimated to be equal to about 10–15 MT (Hunt 1960; Ben-Menahem 1975), although some authors consider the energy value to be higher, up to 30–50 megaton

Table 18.1. Parameters deduced for the Tunguska explosion. In the last column, the sources used to find the given values are indicated: *SM:* seismic measurements; *BM:* barographic measurements; *FT:* fallen tree directions; *FD:* forest devastation data; *EW:* eyewitnesses

Source	Parameter	Remarks
	Time of the explosion (UT)	
Ben-Menahem (1975)	$0^h\ 14^m\ 28^s$	SM
Pasechnik (1976)	$0^h\ 14^m\ 30^s$	SM, BM
Pasechnik (1986)	$0^h\ 13^m\ 35^s$	SM
	Geographic coordinates of the epicentre	
Fast (1967)	60°53'09" N, 101°53'40" E	FT
Zolotov (1969)	60°53'11" N, 101°55'11" E	FT
	Height of the explosion, H (km)	
Fast (1963)	10.5	FD
Ben-Menahem (1975)	8.5	SM
Bronshten and Boyarkina (1975)	7.5	FD
Korotkov and Kozin (2000)	6 – 10	FD
	Trajectory azimuth, α (deg)	
Krinov (1949)	137	EW
Fast (1967)	115	FT
Zolotov (1969)	114	FT
Fast et al. (1976)	99	FT
Yavnel' (1988)	114 – 138	EW
Andreev (1990)	123	EW
Zotkin and Chigorin (1991)	126	EW
Koval' (2000)	127	FT, FD
Bronshten (2000)	122	EW
Bronshten (2000)	103	FT, FD
Longo et al. (2005) (single body)	110	FT
Longo et al. (2005) (multiple bodies)	135	FT
	Trajectory inclination, h (deg)	
Krinov (1949)	17	EW
Sekanina (1983)	< 5	EW
Zigel (1983)	5 – 14	EW
Yavnel' (1988)	8 – 32	EW
Andreev (1990)	17	EW
Zotkin and Chigorin (1991)	20	EW
Koval' (2000)	15	FT, FD
Bronshten (2000)	15	EW, FT
Longo et al. (2005) (single body)	30	FT
Longo et al. (2005) (multiple bodies)	30 – 50	FT

(Pasechnik 1971, 1976, 1986). In agreement with the first energy range, which seems to have more solid grounds, the height of the explosion was found equal to 6–14 km. A height of 10.5 ± 3.5 km was obtained by Fast (1963) from data on forest devastation. Using more complete data on forest devastation, Bronshten and Boyarkina (1975) subsequently obtained a height equal to 7.5 ± 2.5 km. From seismic data, Ben-Menahem deduced an explosion height of 8.5 km. Data on the forest devastation examined, taking into account the wind velocity gradient during the TCB flight (Korotkov and Kozin 2000), gave an explosion height in the range 6–10 km.

A close inspection of seismograms of Irkutsk station, made by Ben-Menahem (1975), showed that the ratio between East-West and North-South components is about 8:1, even though the response of the two seismometers is the same. Since the Irkutsk station is South of the epicenter, Ben-Menahem (1975) inferred that this was due to the ballistic wave and therefore the azimuth should be between 90° and 180°, mostly eastward. However, it is not possible to obtain more stringent constraints on the azimuth from seismic data.

It is not clear how Voznesenskij (1925) determined the direction of the bolide's flight given in Fig. 18.8. Using only the eyewitness data collected in 1908, Yavnel' (1988) obtained $\alpha = 114° - 138°$ and $h = 8° - 32°$. A critical analysis of the eyewitness reports written in 1908 together with those collected in the nineteen-twenties, made by Krinov (1949) gave an azimuth $\alpha = 137°$ with $h = 17°$.

Analysing the data on flattened tree directions from the first part of his catalog (Fast et al. 1967), Fast found a trajectory azimuth $\alpha = 115° ± 2°$ as the symmetry axis of the "butterfly" shaped region (Boyarkina et al. 1964; Fast 1967). The independent mathematical analysis of the same data gave $\alpha = 114° ± 1°$ (Zolotov 1969). Having made another set of measurements, Fast subsequently suggested a value of $\alpha = 99°$ (Fast et al. 1976). In this second work, the differences between the mean measured azimuths of fallen trees and a strictly radial orientation were taken into account. He gave no error for this new value, but a close examination of Fast's writings suggests that he considered an error of 2°. Koval' subsequently collected complementary data on forest devastation and critically re-examined Fast's work. He obtained a trajectory azimuth $\alpha = 127° ± 3°$ and an inclination angle $h = 15° ± 3°$ (Koval' 2000).

From a critical analysis of all the eyewitness testimonies collected in the catalog of Vasilyev et al. (1981), Andreev (1990) deduced $\alpha = 123° ± 4°$ and an inclination angle $h = 17° ± 4°$. Zotkin and Chigorin (1991) using the data in the same catalog obtained: $\alpha = 126° ± 12°$ and $h = 20° ± 12°$, whereas from partial data, Zigel' (1983) deduced $h = 5° ± 14°$. A different analysis of the eyewitness data (Bronshten 2000), gave $\alpha = 122° ± 3°$ and $h = 15°$. In the same book a mean value $\alpha = 103° ± 4°$ is given obtained from forest devastation data. Sekanina (1983, 1998) studied the TE on the basis of superbolide theories and the analysis of the data available and eyewitness testimonies. He suggested an inclination over the horizon $h < 5°$ and an azimuth $\alpha = 110°$.

From the data on fallen tree directions in our new unified catalog (Longo et al. 2005), we obtain a single-body trajectory azimuth $\alpha = 110° ± 5°$ and $h = 30°$. The same data are compatible with the hypothesis that the cosmic body was composed by at least two bodies, falling independently but very close one to the other, with a trajectory azimuth ~135° and an inclination of the total combined shock wave axis between 30° and 50°. The first body, with a greater mass, emitted the maximal energy at a height of about

6–8 km. The second, of minor mass, flew a little higher, on the right side and behind the first body, following the azimuth ~135° in the direction of the Lake Cheko. The last azimuth is in agreement with what found by Krinov (1949) and Yavnel' (1988) analyzing earlier eyewitness testimonies.

18.5
Tunguska-like Impacts

The Tunguska event is the only phenomenon of this kind that has occurred in historical time. The consequences of the event can be directly studied in situ. From such a study we can obtain a great amount of information useful to better understand and predict the characteristics of future Tunguska-like impacts, i.e. due to bodies with diameters equal to a few tens of meters. Many different models have been proposed to describe the impact with our planet by bodies having these dimensions. I mention here only some recent models, which imply a greater impact frequency and, therefore, a greater hazard.

18.5.1
Recent Models and Impact Frequency

The frequency of Tunguska-like impacts is highly dependent on the emitted energy, the explosion height and the entry angle.

Most of the published models for Tunguska have assumed that the explosion was essentially from a point source.

Recent models consider that such events are more analogous to explosive line charges, with the bolide's kinetic energy deposited along the entry column.

Plume-forming Impacts

Boslough and Crawford (1997) explain the TE as due to a "plume-forming" atmospheric explosion, i.e. as associated with the ejection and collapse of a high plume. I report here a brief description of the three overlapping phases of the plume formation, as summarized by Stokes et al. (2003):

1. *Entry phase*. When a bolide penetrates a planet atmosphere, it encounters gases at high speed that both slow it down and heat it up. A "bow shock" develops in front of the bolide where atmospheric gases are compressed and heated. Some of this energy is radiated to the bolide, causing ablation (i.e., melting and vaporization that remove material of the bolide's surface) and deformation. The rest of the energy is deposited along the long column created by the bolide's passage; much of the bolide's kinetic energy is lost in this manner. In some cases, aerodynamic stresses may overcome the bolide's tensile strengths and cause it to catastrophically disrupt within seconds of entering the atmosphere. Airblast shock waves produced by this sequence of events may reflect off the surface causing great devastation.
2. *Fireball phase*. The events taking place during the entry phase produce a hot mixture of bolide material and atmospheric gas called a fireball that is ballistically shot upward by the impact. Since it is incandescent, it radiates energy away in visible and

near infrared wavelengths. Buoyant forces cause the fireball to rise because it is less dense than the surrounding atmosphere.

The fireball's energy expands most easily along the low-density high sound speed entry column that was created by the bolide' passage.

3. *Plume phase.* The expanding fireball (and associated debris) rushes back out the entry column, ultimately reaching altitudes of many hundreds kilometers above the top of the atmosphere. After ~10 minutes of cooling and contracting at these heights, however, the plume splashes back onto the upper atmosphere, releasing additional energy as it collapses and impacts.

Boslough and Crawford (1997) re-examined the phenomena associated with the TE in the context of their model. They found that a 3 megaton plume-forming event could generate the seismic waves that were actually observed, whereas Ben-Menahem (1975) considering the waves generated by a point explosion has obtained the generally accepted value of about 12.5 megaton. Boslough and Crawford (1997) obtained a qualitative agreement between the calculated wind speed at different distances and the treefall shown on the map (Boyarkina et al. 1964) used for almost 40 years. This agreement would be improved using our new unified catalog and the corresponding map (Longo et al. 2005). The most convincing aspect of the plume-forming model is that it not only account for forest devastation and seismic and pressure waves but, for the first time, it gives a simple and reasonable explanation of the magnetic field disturbance and of the "bright nights" associated with the Tunguska event. The resulting plume, 100 seconds after the impact, is given in Fig. 18.9. As shown, a mixture of dust, water and tropospheric air is ejected above the top of the atmosphere. It is this material, transported westward rapidly enough, that caused the bright nights within 12 hours at distances up to 6 000 km.

Shuvalov (1999) developed a similar plume-forming model. Firstly, he considered a volumetric absorption in the projectile of the radiation emitted by shock compressed atmospheric gas. Subsequently, Shuvalov and Artem'eva (2002) improved the model considering a surface absorption of the radiation. They elaborated a 2D numerical model with radiation and ablation for the impact of Tunguska-like bodies and obtained results similar to those of Boslough and Crawford (1997) for the plume formation and the ejection in the upper atmosphere of hot vapor and air.

All the authors of plume-forming simulations consider their calculations as preliminary and underline the necessity of developing a totally self-consistent 3D numerical model using realistic topography and including simultaneously radiation and ablation, disruption of the bolide, formation and evolution of a fireball and of a plume.

Foschini Hypersonic Flow

Let me mention two other representations of impacts that consider the bolide energy deposited along an entry column. Foschini (1999, 2001) developed a model studying the hypersonic flow around a small asteroid entering the Earth's atmosphere. This model is compatible with fragmentation data from superbolides. Foschini considers a bow shock in the front of the cosmic body that envelops the body. As the air flows toward the rear of the body, it is re-attracted to the axis. Therefore, there is a rotation of the

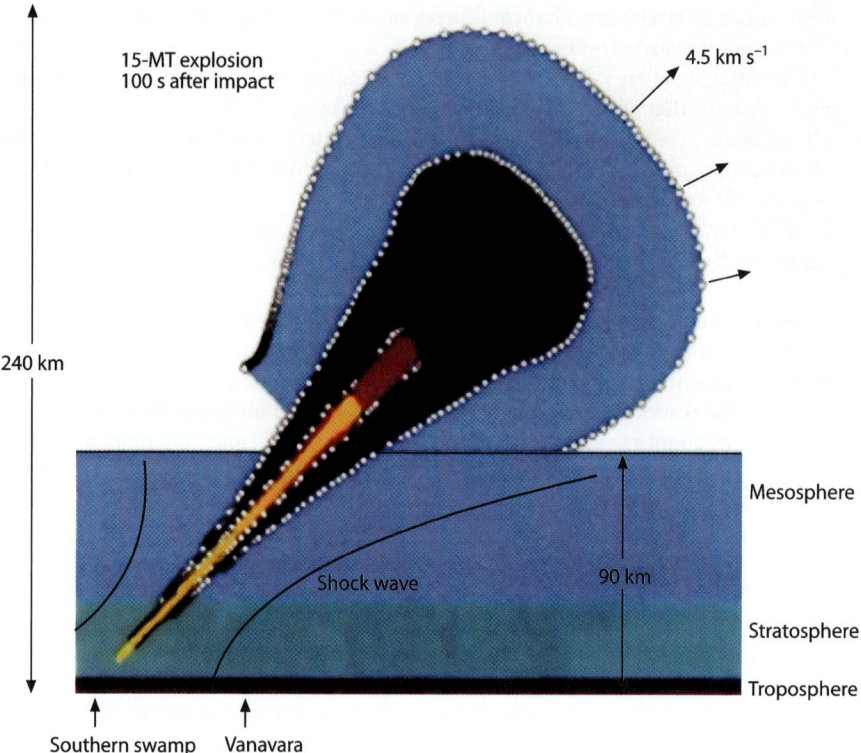

Fig. 18.9. A plume due to a total energy deposition of 15 megaton, 100 seconds after the impact of a stony asteroid (Boslough and Crawford 1977). Material within all but the outermost shell has been ejected from within the troposphere, and contains the mass of the impactor, as well as water from the humid lower atmosphere

stream in the sense opposite to that of the motion and this creates an oblique shock wave (wake shock). Since the pressure rise across the bow shock is huge when compared to the pressure behind the body, it can be assumed that there is a vacuum behind the cosmic body. According to the model, the condition for fragmentation depends on two regimes: steady state, when the Mach number does not change, and unsteady state, when the Mach number undergoes strong changes (Foschini et al. 2001). In the latter case, the distortion of shock waves causes the amplification of turbulent kinetic energy. So, a sudden outburst of pressure that can overcome the mechanical strength of the body, starting the fragmentation process is expected. On the other hand, in the first case – the steady state – the effect of compressibility suppresses the turbulence, and then the viscous heat transfer becomes negligible. The cosmic body is subjected to a combined thermal and mechanical stress.

The key point in fragmentation is how the ablation changes the hypersonic flow. The existence of asteroids with an extremely low density, such as Mathilde (~1300 kg m^{-3}), suggests that such a body could have an increased efficiency in deceleration. A possible process by means internal cavities could increase the deceleration and airburst effi-

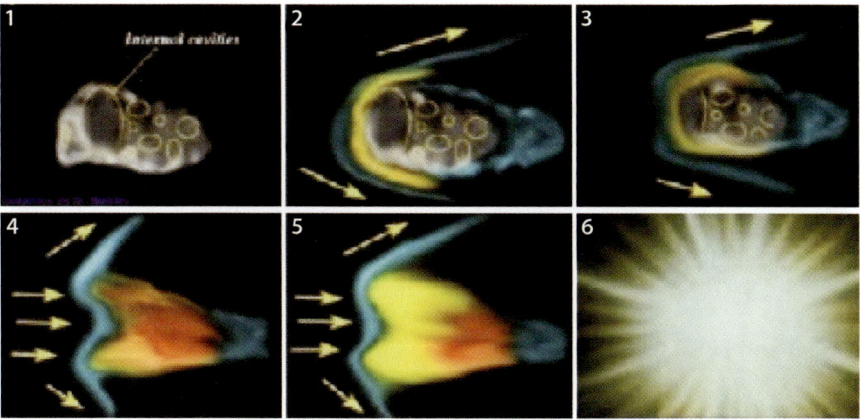

Fig. 18.10. In the Foschini model, as the cosmic body enters the Earth atmosphere, the ablation removes the surface, discovering the internal cavities, which act as something similar to a parachute, thereby increasing the deceleration

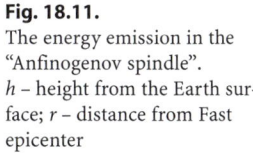

Fig. 18.11.
The energy emission in the "Anfinogenov spindle".
h – height from the Earth surface; r – distance from Fast epicenter

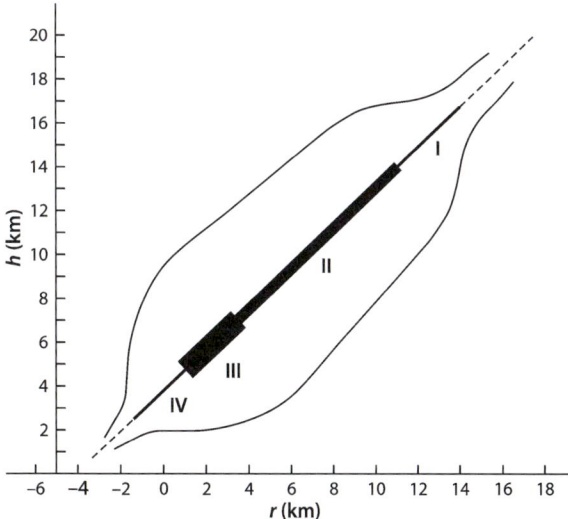

ciency is shown on Fig. 18.10. Following these lines Farinella et al. (2001) concluded that an object like asteroid Mathilde could explain the TE.

Anfinogenov Spindle

Anfinogenov (1966) and Anfinogenov and Budaeva (1998) proposed a qualitative model of the energy emitted by a "semi-infinite" linear source. The bolide begins disrupting and vaporizing when it enters the stratosphere and releases an increasing energy as it moves down. The energy emission is schematically described by the four cylinders shown in Fig. 18.11. In region I some 20% of mass and energy is lost, about 80% is emitted in regions II and III and less than 1% in region IV. The maximal energy emission is reached at a height of 6–8 km. The resulting shock wave has the form of the so-called "Anfinogenov spindle".

On the basis of the tree fall data and earlier eyewitness testimonies we consider that the TCB was a multiple bolide formed by at least two bodies of similar mass (Longo et al. 2005). They likely entered the atmosphere very close to each other following parallel trajectories with azimuths ~135°. The second body flew slightly higher, behind the first, and was decelerated by the shock wave. The resulting summary shock wave from the different spindles had an inclination angle of it symmetry axis ~45°.

18.5.2
Global and Local Damages

No doubt that a KT impact causes global damages, but the local character of damages from Tunguska-like events is questionable. It depends on the target. For the majority of the Earth's surface, which is water, there would be no damage (the lower limit for tsunami generation is about 10 times the Tunguska energy). Also, most of the land surface is still sparsely populated. The situation is quite different for a Megaton explosion in a large city or a populated region. Apart from the direct damages and casualties, we cannot exclude that some country could interpret that it had suffered a nuclear attack. Even at the time of the real Tunguska explosion, its consequences would have been very different, *if* the cosmic body would have reached the Earth about four hours later. Instead of hitting a non-populated forest at about 60° N, it could have impacted the Russian capital of St. Petersburg at the same latitude. Under these conditions, the Russian participation to World War I and the Russian Revolution would not have been possible. The whole history of humanity in the 20th century would be different. In short, the consequences, even of a "modest" impact are highly dependent on the target.

18.6
Concluding Remark

From the models mentioned in Sect. 18.5, it was deduced that the Tunguska explosive yield has been overestimated by a factor 3–4. This means that the interval between Tunguska-like events can be about three times less than usually expected. The expected frequency for such events, from the present value of about twice in a millennium can approach the century timescale. Therefore, the Tunguska-like impacts may present a more serious hazard than previously estimated. The real Tunguska event is the only phenomenon of this kind that happened during relatively recent time and that can be studied directly. The analyses of the data and samples collected during recent in situ expeditions have made it possible to check some characteristics of the Tunguska event. Many of its aspects are still unclear. Therefore, it is important to further both theoretical and experimental research on this phenomenon. For example, most of the scientists consider that the Tunguska event was due to the impact with the atmosphere of an asteroid or a comet. A clear choice between these two hypotheses has important practical consequences. The knowledge of the nature of the object, which explosion caused the devastation observed, will make it possible to verify and develop the models of the explosion mechanisms and fragmentation of cosmic bodies in the atmosphere. Broadening the study to the known impacts, will allow obtaining better estimates of the impact probability for cosmic bodies with different composition and dimensions.

Acknowledgments

Thanks are due to all the participants to the Tunguska91 and Tunguska99 expeditions (see full list in Serra et al. 1994 and Amaroli et al. 2000), whose work made it possible to write the present review. I have the pleasure to thank L. Gasperini and M. Pipan for their contribution in the analysis of the "kimberlite" hypothesis and M. Di Martino for discussions on the tree fall. I am grateful to I. Doroshin and his collaborators for generously sharing their copy of the Archives of the Meteorite Committee of the Russian Academy of Sciences containing the original documents from the eyewitnesses of the TE. I thank B. Bidiukov and his collaborators for providing an electronic copy of many Russian publications on the TE. I gratefully acknowledge the referees Eric W. Elst, Eugeny M. Kolesnikov and Wolfgang Kundt who provided useful comments and suggestions for improving the present paper. One of the referees (WK) disagrees with the infall interpretation; his objections stimulated integrations to better elucidate the interpretation favored by the author.

References

Amaroli L, Andreev G, Anfinogenov J, Baskanova T, Benati L, Biasini G, Bonatti E, Cancelli F, Casarini J, Chernikov A, Chernova T, Cocchi M, Deserti C, Di Martino M, Doroshin I, Foschini L, Gasperini L, Grechko G, Kolesnikov E, Kononov E, Longo G, Nesvetajlo V, Palazzo G, Pavlova L, Pipan M, Sacchi M, Serra R, Tsvetkova I, Vasiliev N, Vigliotti L, Zucchini P (2000) A multidisciplinary investigation in the site of the Tunguska explosion. In: Aiello S, Blanco A (eds) IX GIFCO: What are the prospects for cosmic physics in Italy? SIF Conference Proceedings 68, Bologna, pp 113–120
Andreev GA (1990) Was the Tunguska 1908 Event caused by an Apollo asteroid? In: Lagerkvist CI, Rickman H, Lindblad BA, Lindgren M (eds), Asteroids, comets, meteors III. Uppsala Astronomical Observatory, Uppsala, pp 489–492
Anfinogenov DF (1966) O Tungusskom meteoritnom dozhde. In: Uspekhi meteoritiki (Tezisy dokladov), SO AN SSSR, Novosibirsk, pp 20–22
Anfinogenov DF, Budaeva LI (1998) Tungusskie etiudy. Ed. Trots, Tomsk
Archives of the Meteorite Committee of the Russian Academy of Sciences, document N° 57, Moscow
Astapovich IS (1933) Novye materialy po poletu bolshogo meteorita 30 iunija 1908 g v Tsentralnoj Sibiri. Astron Zh 10(4):465–486
Baxter J, Atkins J (1976) The fire came by. Doubleday. Garden City, NY
Ben-Menahem A (1975) Source parameters of the Siberian explosion of June 30, 1908, from analysis and synthesis of seismic signals at four stations. Phys Earth Planet Inter 11:1–35
Boslough MBE and Crawford DA (1997), Shoemaker-Levy 9 and Plume-forming Collisions on Earth, in Near-Earth Objects. Annals of the New York Academy of Sciences 822:236–282
Boyarkina AP, Diomin DV, Zotkin IT, Fast VG (1964) Izucheniye udarnoj volny Tungusskogo meteorita po vyzvannymy yeyu razrusheniyam lesa. Meteoritika 24:12–140
Bronshten VA (1991) Priroda anomal'nogo svechenija neba, svjazannogo s Tungusskim javleniem. Astron Vestnik 25(4):490–504
Bronshten VA (2000) Tungusskiy Meteorit: Istoria Issledovaniya. Selyanov AD, Moskva
Bronshten VA, Boyarkina AP (1975) Raschety vozdushnyh voln tungusskogo meteorite. In: Problemy meteoritiki, Nauka, Novosibirsk, pp 47–63

Busch F (1908) Leuchtende Nachtwolken am Nordhorizont. Meteorol Z 314
Busch F (1908) Über die Lichterscheinung in den Nächten vom 30. Juni bis zum 2. Juli 1908, Mitt Verein, Freuden Astron u Kosmisch Phys 18:85
Cecchini S, Galli M, Giovannini G, Longo G, Pagliarin A (2003) Real-time monitoring of environmental radiation in Tunguska (Siberia). J of Geophys Res 108(D2):4057–4067
Chyba CF, Thomas PJ, Zahnle KJ (1993) The 1908 Tunguska explosion: atmospheric disruption of a stony asteroid. Nature 361:40–44
Cowan C, Atluri CR, Libby WF (1965) Possible anti-matter content of the Tunguska meteor of 1908. Nature 206:861–865
D'Alessio SJD, Harms AA (1989) The Nuclear and Aerial Dynamics of the Tunguska Event. Planet Space Sci 37:329–340
De Roy F (1908) Les illuminations crépusculaires des 30 juin et 1 juillet 1908, Gazette astronomique No 8
Di Martino M, Farinella P, Longo G (1998) Foreword of the Tunguska issue. In: Di Martino M, Farinella P, Longo G (eds) "Tunguska96." Planet Space Sci, special issue 46(2–3):125
Emeljanov M Jr (1963) Radiofotograficheskoje issledovanie srezov derevjev iz rajona padenija Tungusskogo meteorita. In: Problema Tungusskogo meteorita, Izdatelstvo Tomskogo Universiteta, Tomsk pp 153–158
Farinella P, Foschini L, Froeschlé Ch, Gonczi R, Jopek TJ, Longo G, Michel P (2001) Probable asteroidal origin of the Tunguska cosmic body. Astron. Astrophys 377:1081–1097
Fast VG (1963) K opredeleniyu epitsentra vzryva Tungusskogo meteorita. In: Problema Tungusskogo meteorita, Izdatelstvo Tomskogo Universiteta, Tomsk, pp 97–104
Fast VG (1967) Statisticheskij analiz parametrov Tungusskogo vyvala, in Problema Tungusskogo meteorita, Izdatelstvo Tomskogo Universiteta, Tomsk, part 2, pp 40–61
Fast VG, Fast NP (2005) Pokazanija ochevidtsev o momente Tungusskogo bolida. Tungusskij Vestnik, 16:4–12
Fast VG, Bojarkina AP, Baklanov MV (1967) Razrushenija, vyzvannyje udarnoj volnoj Tungusskogo meteorita. In: Problema Tungusskogo meteorita, Izdatelstvo Tomskogo Universiteta, Tomsk, part 2, pp 62–104
Fast VG, Barannik AP, Razin SA (1976) O pole napravlenij povala derevjev v rajone padenija Tungusskogo meteorita. In: Voprosy meteoritiki. Izdatelstvo Tomskogo Universiteta, Tomsk, pp 39–52
Fast VG, Fast NP, Golenberg NA (1983) Katalog povala lesa, vyzvannogo Tungusskim meteoritom. In: Meteoritnyje i meteornyje issledovanija. Nauka, Novosibirsk pp 24–74
Fesenkov VG (1949) Pomutneniye atmosfery, proizvedennoye padeniyem Tungusskogo meteorita 30 iyunya 1908 g. Meteoritika 6:8–12
Fesenkov VG (1961) O kometnoj prirode Tungusskogo meteorita. Astron Zhurn 38:577–592
Fesenkov VG, Krinov EL (1960) The meteorite of 30 June 1908. Science 131:652–653
Florenskij KP (1963) Predvaritelnyje rezultaty Tunguskoj meteoritnoj kompleksnoj ekspeditsii 1961 g. Meteoritika 23:3–29
Florenskij KP, Vronskij BI, Emeljanov M Jr, Zotkin IT, Kirova OA (1960) Predvaritelnyje rezultaty rabot Tunguskoj meteoritnoj ekspeditsii 1958 g. Meteoritika 19:103–134
Florenskij KP, Ivanov AV, Iljin NP, Petrikova MN, Loseva LE (1968) Khimicheskij sostav kosmicheskikh sharikov iz rajona Tunguskoj katastrofy i nekotorye voprosy differentsjatsii veshchestva kosmicheskikh tel, Geokhimija 10:1163–1173
Foschini L (1999) A solution for the Tunguska event. Astronomy and Astrophysics 342:L1–L4
Foschini L (2001) On the atmospheric fragmentation of small asteroids. Astronomy and Astrophysics 365:612–621
Foschini L, Longo G, Jopek TJ, Froeschlé Ch, Gonczi R, Michel P (2001) On the atmospheric dynamics of the Tunguska cosmic body. In: Proceedings of the Meteoroids 2001 Conference, Kiruna (Sweden). ESA SP-495, pp 371–376
Gallant RA (1995) The day the sky split apart. Atheneum books, New York
Gasperini L, Alvisi F, Biasini G, Bonatti E, Di Martino M, Morigi C, Longo G, Pipan M, Ravaioli M, Sacchetti F, Sacchi M, Vigliotti L (2001) Geophysical/sedimentological study of a lake close to the epicenter of the great 1908 Siberian (Tunguska) explosion. Abstracts and Proceedings of the Norwegian Geological Society 1:29–30

Goldine VD (1998) Search for the local centres of the Tunguska explosions, In: Di Martino M, Farinella P, Longo G (eds) Planetary and Space Science, special issue "Tunguska 96" 46(2–3):151–154

Golenetskij SP, Stepanok VV, Kolesnikov EM (1977a) Priznaki kosmokhimicheskoj anomalii v rajone Tungusskoj katastrofy 1908 g, Geokhimija 11:1635–1645

Golenetskij SP, Stepanok VV, Kolesnikov EM, Murashov DA (1977b) K voprosu o khimicheskom sostave i prirode Tungusskogo kosmicheskogo tela, Astronomicheskij Vestnik 11(3):126–136

Grigorian SS (1976) K voprosu o prirode Tungusskogo meteorita, Doklady Akad. Nauk SSSR 231(1):7–60

Grigorian SS (1979) Motion and destruction of meteorites in planetary atmospheres. Cosmic Res 17: 724–740

Grigorian SS (1998) The cometary nature of the Tunguska meteorite: on the predictive possibilities of mathematical models. In: Di Martino M, Farinella P, Longo G (eds) "Tunguska96." Planet Space Sci, special issue 46(2–3):213–217

Hills JG, Goda MP (1993) The fragmentation of small asteroids in the atmosphere. Astron J 105:1114–1144

Hou QL, Ma PX, Kolesnikov EM (1998) Discovery of iridium and other element anomalies near the 1908 Tunguska explosion site. In: Di Martino M, Farinella P, Longo G (eds) "Tunguska96." Planet Space Sci, special issue 46(2–3):179–188

Hou QL, Kolesnikov EM, Xie LW, Kolesnikova NV, Zhou MF, Sun M (2004) Platinum group element abundances in a peat layer associated with the Tunguska event, further evidence for a cosmic origin. Planet Space Sci 52: 331–340

Hunt JN, Palmer R, Penny W (1960) Atmospheric waves caused by large explosions, Philos. Trans Roy Soc Lond A252:275–315

Iljina LP, Slivina LM, Demin DV, Zhuravlev VK, Potekhina LP, Levchenko MN, Vronskij BI, Egorshin OA, Ivanova GM (1971) Rezultaty spektral'nogo analiza prob pochvy iz rajona padenija Tungusskogo meteorita. In: Sovremennoje sostojanie problemy Tungusskogo meteorita. Izdatelstvo Tomskogo Universiteta, Tomsk, pp 25–27

Jackson AA IV, Ryan MP (1973) Was the Tunguska event due to a black hole? Nature 245:88–89

Jéhanno C, Boclet D, Danon J, Robin E, Rocchia R (1989) Etudes analytique de sphérules provenant du site de l'explosion de la Toungouska. C R Acad Sci, Paris, 308(Serie II):1589–1595

Kazantsev AP (1946) Vzryv. Vokrug sveta 1:39–46

Kirichenko LV, Grechushkina MP (1963) O radioaktivnosti pochvy i rastenij v rajone padenija Tun-gusskogo meteorita, in Problema Tungusskogo meteorita. Izdatelstvo Tomskogo Universiteta, Tomsk, pp 139–152

Kolesnikov EM, Shestakov GI (1979) Izotopnyj sostav svintsa iz torfov rajona Tungusskogo vzryva 1908 g. Geokhimia 8:1202–1211

Kolesnikov EM, Ljul AL, Ivanova GM (1977) Priznaki kosmohimicheskoj anomalii v rajone Tungusskoj katastrofy 1908 g, II. Astronomicheskij Vestnik 11(4):209–218

Kolesnikov EM, Kolesnikova NV, Boettger T (1998) Isotopic anomaly in peat nitrogen is a probable trace of acid rains caused by 1908 Tunguska bolide. In: Di Martino M, Farinella P, Longo G (eds) "Tunguska96." Planet Space Sci, special issue 46(2–3):163–167

Kolesnikov EM, Boettger T, Kolesnikova NV (1999) Finfing of probable Tunguska cosmic body material: isotopic anomalies of carbon and hydrogen in peat. Planet Space Sci 47:905–916

Kolesnikov EM, Longo G, Boettger T, Kolesnikova NV, Gioacchini P, Forlani L, Giampieri R, Serra R (2003) Isotopic-geochemical study of nitrogen and carbon in peat from the Tunguska Cosmic Body explosion site. Icarus 161(2):235–243

Kohl T (1908) Über die Lichterscheinungen am Nachtimmel aus dem Anfang des Juli. Astron Nachr 178(4262):239

Korobeinikov VP, Chushkin PI, Shurshalov LV (1990) Tungusskij Fenomen: gazodinamicheskoje modelirovanije. In: Sledy Kosmicheskih Vozdejstij na Zemlju, Nauka, Novosibirsk, pp 59–79

Korotkov PF, Kozin VN (2000) Solar System Res 34:326

Koval'VI (2000) Meteorinyie issledovaniya molodezhnogo tvorcheskogo kollektiva Gea astrolaboratorii dvortsa tvorchestva na Miussah i usstanovlenie osnovnyh parametrov tungusskogo superbolida 1908 g. In: Tungusskij sbornik, MGDTDiJu, Moskva, pp 80–94

Kovalevskij AL, IV Reznikov NG Snopov, Osharov AB, Zhuravlev VK (1963) Nekotoryje dannyje o raspredelenii khimicheskikh elementov v pochvakh i rastenijakh v rajone padenija Tungusskogo meteorita. In: Problema Tungusskogo meteorita. Izdatelstvo Tomskogo Universiteta, Tomsk, pp 125–133

Kresák, L (1978) The Tunguska object: a fragment of comet Encke? Bull Astron Inst Czechosl 29:129–134
Krinov EL (1949) Tungusskij meteorit, Izdatelstvo Akademii Nauk SSSR, Moskva Leningrad
Krinov EL (1966) Giant meteorites. Pergamon Press, Oxford
Kulik LA (1922) Otchet Meteoritnoj ekspeditsii o rabotah, proizvedennyh s 19 maja 1921 g po 29 nojabrja 1922 g. Izvestija Ross Akad Nauk 6(16):391–410
Kulik LA (1923) Pervaja meteoritnaja ekspeditsija v Rossii i ocherednyie zadachi meteoritiki. Mirovedenie 1 (44):6–15
Kulik LA (1927) K istorii bolida 30/VI 1908 g., Doklady AN SSSR, seria A, 23, pp 393–398
Kulik LA (1939) Dannyje po Tungusskomu meteoritu k 1939 godu. Doklady Akad Nauk SSSR 22(8): 520–524
Kulik LA (1940) Meteoritnaja ekspeditsija na Podkamennuju Tungusku v 1939 g. Doklady Akad Nauk SSSR 28(7):597–601
Kundt W (2001) The 1908 Tunguska catastrophe: an alternative explanation. Current Science 81: 399–407
Landsberg D (1924) Prvni meteoritova vyprava ruske Akademie ved v roce 1921–1922. Rise hvezd, Praha 5:154–158, 6:179–181
Levin B Yu, Bronshten VA (1986) The Tunguska event and the meteors with terminal flares. Meteoritics 21(2):199–215
Levitskij G (ed) (1910) Bjulleten Postojannoj tsentralnoj seismicheskoj komissii za 1908 god, Sankt Petersburg
Longo G (1996) Zhivyie svideteli Tungusskoj katastrofy. Priroda 1:40–47
Longo G (2005) O tungusskoj zagadke: versii, gipotezi. In: Vyzovy XXI veka, Globalizaciya i problemy identichnosti v mnogoobraznom mire, Izd. Ogni, Moskva, pp 245–256
Longo G, Di Martino M (2002) Recalculation of the Tunguska Cosmic Body parameters on the basis of the 1938 and 1999 Aerophotosurveys. In: Warmbein B (ed) Asteroids comets meteors 2002. ESA-SP-500, Berlin, pp 843–846
Longo G, Di Martino M (2003) Remote sensing investigation of the Tunguska explosion area. In: Manfred O, D'Urso G (eds) Remote sensing 2002. Spie Press 4879, Washington, pp 326–333
Longo G, Serra R (1995) Some answers from Tunguska mute witnesses. Meteorite! 4:12–13
Longo G, Serra R, Cecchini S, Galli M (1994) Search for microremnants of the Tunguska Cosmic Body. Planet Space Sci 42:163–177
Longo G, Cecchini S, Cocchi M, Di Martino M, Galli M, Giovannini G, Pagliarin A, Pavlova L, Serra R (2000) Environmental radiation measured in Tunguska (Siberia) and during the flights Forli-Krasnoyarsk-Forli. In: IX GIFCO, SIF Conference Proceedings 68, pp 121–124
Longo G, Bonatti E, Di Martino M, Foschini L, Gasperini L (2001) Exploring the site of the Tunguska impact. Abstracts and Proceedings of the Norwegian Geological Society 1:48–50
Longo G, Di Martino M, Andreev G, Anfinogenov J, Budaeva L, Kovrigin E (2005) A new unified catalogue and a new map of the 1908 tree fall in the site of the Tunguska Cosmic Body explosion. In: Asteroid-comet Hazard-2005, Institute of Applied Astronomy of the Russian Academy of Sciences, St. Petersburg, Russia, pp 222–225
Nazarov MA, Korina MI, Barsukova LD, Kolesnikov EM, Suponev IV, Kolesov GM (1990) Veshchestvennye sledy Tungusskogo bolida. Geochimia 5:627–639
Nesvetajlo VD, Kovaliukh NN (1983) Dinamika kontsentratsii radiougleroda v godichnykh kol'tsakh derevjev iz tsentra Tungusskoj katastrofy. In: Meteoritnyje i meteornyje issledovanija. Nauka, Novosibirsk, pp 141–151
Obruchev SV (1925) O meste padeniya bolshogo Hatangskogo meteorita 1908 g., Mirovedenie 14:38–40
Ol'khovatov A Yu (2002) The Tunguska event – the facts are against an extraterrestrial impact and point to geophysical origin. Conference "Environmental Catastrophes and Recovery in the Holocene", Brunel Univ, Uxbridge, UK, Aug. 28–Sept. 2
Passey QR, Melosh HJ (1980) Effects of atmospheric breakup on crater field formation. Icarus 42: 211–233
Pasechnik IP (1971) Predvaritel'naya otsenka parametrov vzryva Tungusskogo meteorita 1908 goda po seismicheskim i barograficheskim dannym. In Sovremennoye sostoyanie problemy tungusskogo meteorita, Izdatelstvo Tomskogo Universiteta, Tomsk, pp 31–35

Pasechnik IP (1976) Otsenka parametrov vzryva Tunguskogo meteorita po seismicheskim i microbarograficheskim dannym. In: Kosmicheskoje veshchestvo na zemle. Nauka, Novosibirsk, pp 24–54

Pasechnik IP (1986) Utochneniye vremeni vzryva Tunguskogo meteorita 1908 goda po seismicheskim dannym. In: Kosmicheskoye Veshchestvo i Zemlya. Nauka, Novosibirsk, pp 62–69

Petrov GI, Stulov VP (1975) Motion of large bodies in the atmosphere of planets. Cosmic Res 13:525–531

Pipan M, Baradello L, Forte E, Gasperini L, Bonatti E, Longo G (2000) Ground penetrating radar study of the Cheko Lake area, Siberia. In: Noon DA, Stickley GF, Longstaff D (eds) Eighth International Conference on Ground Penetrating Radar. Spie Press 4084:329–334

Riccobono N (2000) Tunguska. Rizzoli, Milano

Rocchia R, Bonte P, Robin E, Angelis M, Boclet D (1990) Search for the Tunguska event relics in the Antarctic snow and new estimation of the cosmic iridium accretion rate. In: Sharpton VL, Ward PD (eds) Global catastrophes in Earth history. GSA Spec. Publ. 247, Boulder, Colorado, pp 189–193

Sacchetti F (2001) Studio della sedimentazione recente e della geochimica del lago Cheko (Siberia Centrale): eventuale correlazione con l'evento di Tunguska. Thesis, University of Bologna

Sapronov NL (1986) Drevnije vulkanicheskije struktury na juge tungusskoj sineklizy. Nauka, Novosibirsk

Sekanina Z (1983) The Tunguska event: no cometary signature in evidence. Astron J 88(9):1382–1414

Sekanina Z (1998) Evidence for asteroidal origin of the Tunguska object. In: Di Martino M, Farinella P, Longo G (eds) "Tunguska96." Planet Space Sci, special issue 46(2–3):191–204

Serra R, Cecchini S, Galli M, Longo G (1994) Experimental hints on the fragmentation of the Tunguska cosmic body. Planet Space Sci 42:777–783

Shapley H (1930) Flight from chaos. A survey of material systems from atoms to galaxies. Mc Graw Hill, New York

Shenrock AM (1908) Zarya 17 (30) iunya 1908 g, Ezhemesyacnyi biulleten' Nikolaevskoj glavnoj fizicheskoj observatorii (6):1

Shuvalov VV (1999) Atmospheric plumes created by meteoroids impacting the Earth. J Geophys Res 104:5877–5890

Shuvalov VV, Artem'eva NA (2002) Numerical modeling of Tunguska-like impacts. Planet Space Sci 50: 181–192

Stokes GH, Yeomans DK, Bottke WF, Chesley SR, Evans JB, Gold RE, Harris AW, Jewitt D, Kelso TS, McMillan RS, Spahr TB, Worden SP (2003) Appendix 4 of NASA report 030825

Süring R (1908) Die ungewöhnlichen Dämmerungserscheinungen im Juni und Juli 1908. Berichte der Presse, Meteorol. Inst, S 79

Suslov IM (1927) K rozysku bolshogo meteorita 1908 g. Mirovedenie 16:13–18

Svyatskij DO (1908) Illyuminatsiya sumerek. Priroda i Liudi No. 37

Tositti L, Mingozzi M, Sandrini S, Forlani L, Buoso C, De Poli M, Ceccato D, Zafiropoulos (2006) A multitracer of peat profiles from Tunguska, Siberia. Global and Planetary Change 53:278–289

Trayner C (1997) The Tunguska event. J Br Astron Assoc 107(3):117–130

Tsymbal MN, Shnitke VE (1986) Gazovozdushnaja modelj vzryva Tunguskoj komety, in Kosmicheskoje veshchestvo i zemlja. Nauka, Novosibirsk, pp 98–117

Turco RP, Toon OB, Park C, Whitten RC, Pollack JB, Noerdlinger P (1982) An analysis of the physical, chemical, optical and historical impacts of the 1908 Tunguska meteor fall. Icarus 50:1–52

Vasilyev NV (1998) The Tunguska Meteorite problem today. In: Di Martino M, Farinella P, Longo G (eds) "Tunguska96". Planet Space Sci, special issue 46(2–3):129–150

Vasilyev NV (2004) Tungusskij meteorit, kosmicheskij fenomen leta 1908 g., Moskva, Russkaya panorama

Vasilyev NV, Fast NP (1976) Granitsy zon opticheskikh anomalij leta 1908 goda. In: Voprosy meteoritiki. Izdatelstvo Tomskogo Universiteta, Tomsk, pp 112–131

Vasilyev NV, Zhuravlev VK, Zhuravleva RK, Kovalevskij AF, Plekhanov GF (1965) Nochnye svetyashiesya oblaka i opticheskie anomalii, svyazannye s padeniem Tunguskogo meteorita. Nauka, Mosca

Vasilyev NV, Lvov A Jr, Vronskij BI, Grishin A Jr, Ivanova GM, Menjavtseva TA, Grjaznova SN, Vaulin PP (1973) Poiski melkodispersionnogo veshchestva v torfah rajona padenija Tunguskogo meteorita. Mete-oritika 32:141–146

Vasilyev NV, Kovalevskij AF, Razin SA, Epitektova LA (1981) Pokazaniya ochevidcev Tunguskogo padeniya, registrirovano VINITI 24.11.81, N 10350-81.

Verma S (2005) The Tunguska fireball: solving one of the great mysteries of the 20th century. Icon books, Cambridge

Vorobjev VA, Iljin AG, Shkuta BL (1967) Izuchenie termicheskih porazhenij vetok listvennic, perezhivshih tungusskuyu katastrofu. In: Problema Tungusskogo meteorita, part 2, Izdatelstvo Tomskogo Universiteta, Tomsk, pp 110–117

Voznesenskij AV (1925) Padenie meteorita 30 iyunya 1908 g. v verkhoviyah reki Hatangi, Mirovedenie 14(1):25–38

Whipple FJW (1930) The great Siberian meteor and the waves, seismic and aerial, which it produced. Quart J Roy Meteorol Soc 56:287–304

Yavnel' AA (1988) O momente proleta i traektorii tungusskogo bolida 30 iyunya 1908 g. po nabliudeniyam ochevidcev. In: Aktualnyie voprosy meteoritiki v Sibiri, Nauka, Novosibirsk, pp 75–85

Zigel' F Ju (1983) K voprosu o prirode Tungusskogo tela. In: Meteoritnyje i meteornyje issledovanija. Nauka, Novosibirsk, pp 151–161

Zolotov AV (1969) Problema Tungusskoj katastrofy 1908 g., Nauka i tekhnika, Minsk

Zotkin IT (1961) Ob anomal'nyh optichekih yavleniah v atmosfere svyazannyh s padeniem tungusskogo meteorita. Meteoritika 20:40–53

Zotkin IT, Chigorin AN (1991) The radiant of the Tunguska meteorite according to visual observations. Solar System Res 25(4):490–504

Chapter 19

Tunguska (1908) and Its Relevance for Comet/Asteroid Impact Statistics

Wolfgang Kundt

19.1
What Happened North of the Stony Tunguska River in the Early Morning of 30 June 1908?

Depending on distance from the event – at (101° 53' 40" E, 60°53' 09" N) – the Siberian catastrophe of 30 June 1908 was reported as "*cannon shots*" (barisal guns, brontides: Gold and Soter 1979) and/or "*storms*" followed by "*columns of fire*", also described as "lightning" and "thunderclaps", after which an area of more than 2000 square kilometers, *diameter* some 50 km, had its trees debranched, felled, or their tops chopped off, varying with their distance from the center and/or height above the valleys, even with islands of tree survival near the center, and in the valleys. A few tents (tepees), barns (storage huts), and cattle (reindeer) were damaged, hurled aloft, and/or *incinerated*. The haunting took some *ten minutes*, variously reported between 2 min and an hour; one man even washed in the bath house to meet the death clean.

Clearly, the destruction took much longer than the impact time $\Delta l/v$ of a swarm of meteoritic fragments, which measures in seconds. And for an impact, the luminous infall trail would have to precede the sounds of the touchdowns. In his detailed 1966 book "*Giant Meteorites*", Krinov praises the reliability and uniformity of dozens of eyewitness reports on the Tunguska event, yet has to correct them repeatedly by pointing to the ease at which one's memory can get confused at later times; see also the reports by Gallant (1994), Zahnle (1996), Vasilyev (1998), and Ol'khovatov (1999).

Among the informative eyewitness reports on Tunguska (1908) are the *heat* felt in the faces of inhabitants of Vanavara, the nearest trading post, at a distance of 65 km from the epicenter. As is known from bonfires, you can only feel the heat of a chemical fire on your skin if a large fraction of the sphere of seeing around you is filled with hot matter (gas) of sufficient column density. The short-lived, narrow trail of a distant impactor cannot be sensed, whereas km-high gas flames – like occasionally at Baku – can.

Informative are also the "*bright nights*" in Europe and western Asia, starting late on 29 June, culminating on 30 June, and fading thereafter, witnessed last around midnight of 2 July: the sky did not fall dark, down to northern latitude of Tashkent, 42°. This phenomenon has only been reported one more time for the 1883 Krakatoa volcanic eruption. It requires transient scatterers of sunlight at great heights, in the thermosphere, above 500 km, at heights which only methane and hydrogen are light enough to reach in sufficient quantity: molecules whose weight does not exceed that of atomic oxygen. Bronshten (2000) tries to explain these bright nights by softly braked cometary dust, settling to mesospheric heights (of 50 to 70 km), but has to make a number of unrealistic assumptions – among them a twofold (in series!) sunlight reflection from dusty clouds – and still falls short of explaining the four successive bright nights which follow the explosion.

What else is known about the destruction? This part of the Siberian permafrost is not easily accessible; it is snow-covered throughout most of the year, and defended by clouds of mosquitos during the few summer months. The first expedition into the area, in 1910, was carried out by the wealthy Russian merchant and goldsmith Suzdalev who, on return, urged the local inhabitants to keep silent about it. Had he discovered *diamonds*? Thorough investigations of the site, headed by Leonid Kulik, had to wait for 20 years, and were aimed at finding iron-nickel meteoritic debris. According to native reports, a number of funnel-shaped *"holes"* had been blown on that morning, of diameters ≤ 50 m, as well as a *"huge dry ditch"*, probably ≥ 1 km long, with many small "stones" in it. That ditch has not been found by later expeditions. For the most conspicuous crater lake in the area, the *"Suslov hole"*, draining revealed a preserved *tree stump* at its bottom, ruling against an impact origin.

Kulik and his followers explored the morphology of the *treefall pattern*. He discerned a central *"cauldron"*, a few km across, characterized by multiple treefall directions with some *five centers*. In this cauldron were islands of *"telegraph poles"*: trees that had lost all their branches but survived, and sprouted again. Such telegraph poles have meanwhile recurred at Hiroshima, after the nuclear bomb in 1945; they require supersonic blasting, fast enough to break off the branches before the latter can transfer the impacting momentum to the stem. The cauldron had a specific, centered geometry described by Kulik as the *"Merrill circus"* inside an *"amphi-theatre"*; it can still be recognized today, even on near-infrared satellite photographs. Beyond the cauldron, the treefall pattern is coarsely *radial*, though following the ridges and valleys (see Fig. 2 in Serra et al. 1994) whereby the trees on the ridges tended to be felled, those on the slopes often only lost their tops, and those in the valleys often survived, see Krinov's sketch (1966). Obviously, the stormfield had blown horizontally, not from above.

Remarkably, none of the scientists involved in the reconstruction of the assumed impact seems to have considered momentum balance: an incoming atmospheric shock wave transfers its momentum to the trees. If it enters at a shallow angle, it creates a parallel treefall pattern, not a radial one. In Tunguska, we deal with a *zero-net-momentum* pattern, formed by an explosion at its center. But according to Krinov (1966), explosions after an impact are set up by massive meteorites only, with crater diameters larger than 100 m. This rules against an impact interpretation. Note that a simulation of the destruction by Zotkin and Tsikulin (1966) used an unrealistic input: They built a cable car with low (free-fall) kinetic energy, and ignited a chemical explosion close to the ground. Instead, the evaporation of an icy comet is an endergonic process which taps the huge infall energy.

Vasilyev (1998) discusses another inconsistency of the impact scenario: the various reconstructed *infall directions* do not agree; they range from 95° to 137°, or even 192°, North towards East, with similar inconsistencies in the inclination angle(!). A strict interpretation of the reports even implies a midcourse manoeuver! He also discusses the *"stony asteroid vs comet alternative"*, which has persisted for decades: A stony asteroid would have left craters and debris in the impact area, and certain specific elemental anomalies, while a cometary nucleus would have disintegrated too high for the tree destruction, both in intensity and in morphology, also would have been detected weeks before arrival. The suspicion has even shifted to a carbonaceous chondrite as the impactor.

Additional peculiarities of Tunguska's treefall pattern are dozens of detached *root stumps*, with no indication of their origins (pits), some of which still lying around today, as well as *"John's stone"*, weighing 10 tons, which landed on the slope of Mt. Stoikevich with at least sonic speed. Such heavy ejecta, hurled through hundreds of meters, argue against an impact interpretation; they require forces familiar from volcanic ejections.

The biggest problem for the impact interpretation has always been a complete *absence of debris*, by a factor of $10^{-8\pm2}$ in mass: The estimated kinetic energy in the storm field that felled the trees, $10^{24\pm0.3}$ erg, would correspond to an impacting mass of some 0.4 Mt (Svetsov 1996, Foschini 1999). This mass would have left a several-mm-thick layer in the epicenter area if distributed uniformly, easy to detect. (Debris have been found for impactors weighing much less than a ton. For the 1947 Sikhote-Aline meteorite, one third of its mass was recovered within less than four years). Because of this absence of debris, alternative interpretations have been proposed over the decades, such as impacting antimatter, a low-mass black hole, solar transients, extraterrestrials, or mirror matter (Foot 2002), none of them without problems. In their estimate of "the possible origin of the TCB", Farinella et al. (2001) leave these problems unsolved.

There has been an intensive search for chemical, and isotopic anomalies in the Tunguska area, in *magnetic microspherules*, *sphagnum peat* columns, and *resin* layers, for essentially all the chemical elements, from hydrogen (deuterium), carbon, and nitrogen all the way up to the platinum group elements (including iridium), with small enhancements around 1908 found for most of them, and a depression for deuterium, and with sometimes surprising inhomogeneities for different sites (Kolesnikov et al. 1999; Hou et al. 2004). None of them have been able to give an unequivocal answer for the complete evidence, in particular to discriminate against terrestrial outgassing (Longo et al. 1994). Vasilyev (1998), after a careful presentation, summarizes the evidence by "To this day, the matter which might be unambiguously assigned to the Tunguska Meteorite has not been found".

Among the further recorded evidences on the Tunguska event are an atmospheric *shock wave* racing around the globe, local *magnetic-field disturbances* lasting for more than 4 hours, many small local *earthquakes* thoughout the year, and optical anomalies measured by disturbances in the normal run of *Arago* and *Babinet neutral points* lasting for weeks. Their (long) durations favor a tectonic origin, but the evaluations are less straight-forward than those of the earlier facts.

19.2
The Tectonic Interpretation of the Tunguska Catastrophe

Once we have appreciated the many *difficulties* of the impact interpretation – viz the (*i*) sounds before the lights, (*ii*) their duration (several 100 times too long), (*iii*) columns (not streaks) of fire, (*iv*) discrepant infall directions, (*v*) heat sensible at large distances, (*vi*) four bright nights, (*vii*) radial tree-fall pattern (*viii*) following the surface topography, (*ix*) cauldron structure, (*x*) hurled root stumps and John's stone, (*xi*) absence of impact craters, but (*xii*) formation of funnel-shaped holes, and (*xiii*) absence of meteoritic debris – why not follow Ol'khovatov (1999, 2003) and Yepifanov (2002), and pursue the other alternative, a tectonic outburst?! Knowing from Shoemaker, and Alvarez (1997) that there is only one impact crater for 30 *volcanic craters* on Earth, this re-interpretation of Tunguska cannot even be considered unlikely, except for the missing lava.

But *volcanism* has many different faces, ranging from supersonic ejections, plate tectonics and the formation of mountains, "maars", and kimberlites through lava flows, mud volcanoes, burning torches, and solfataras to quasi-steady outgassings, depending on the *viscosity* of the magma, on the magma supply rate, and on the transmissivity of the surface layers. Driving – in all cases – is *natural gas*, dissolved in liquid magma, often from as deep as the molten core of Earth (Kundt und Jessner 1986; Kundt 1991, 2001; Gold 1999). Highly viscous (acid) magma leads to explosive eruptions (like Mt. St. Helens) whereas in rising low-viscosity (mafic) magma, the natural gas often separates from the melt before reaching the surface, forming a *mystery cloud*. In all likelihood, this is what happened at Tunguska.

More specifically, Tunguska may have been the present-day formation of a kimberlite. *Kimberlites* are called after the south-African town of Kimberley, where diamonds and gold have been found by digging. They are huge, narrow funnels, growing in diameter from a few meters, at a kilometer's depth, to a dome-shaped tuff ring at the top, some km across, and occasionally enclosing a shallow crater lake (Dawson 1980; Haggerty 1994). They occur in all continents, lie at the intersection of major fracture zones, in old, stable cratons, are intruded by ultra-alkaline rock types containing high amounts of volatiles, and show several spasmodic, often cold intrusions. An explosive injection from great depth is indicated, driven by volatiles. In Russia, the "*Zanitsa pipe*" was discovered in 1954, in the headwaters of the Markha river in Siberia. Gold (1999) mentions that there is no evidence of frozen lava in kimberlites.

In the case of Tunguska, I have estimated a natural-gas mass of 10 Mt, required both for blowing the funnel-shaped "holes" (like the Suslov hole; and the ditch?), ejecting the *root stumps*, and for setting up an overpressure dome – the cauldron – big enough to drive the storm field for felling the trees out to some 30 km distance (Kundt 2001). On venting (through some five of these holes), this expanding, initially liquidized gas – some 80% *methane* – escapes supersonically, thereby creating the "*telegraph islands*", until it has sufficiently expanded to be stalled by the ambient air mass. It then shoots up vertically, again supersonically, in the form of a giant *mushroom*, many times higher (200 km) than the mushrooms of nuclear explosions (30 km) because of its much lower molecular weight m and lower adiabatic index κ (both of which enter as $1/m[\kappa-1]$), while the surrounding air mass is pushed radially outward, in the form of a big storm field. This same gas will *burn* partially whenever it gets mixed with ambient oxygen and ignited (by self-generated lightning), and will continue burning at great height whenever it meets the surrounding atomic oxygen, thereby heating up and rising further. The newly formed water vapor will freeze out and remain frozen even when embedded in the hot thermosphere, because it gets radiatively cooled by seeing the cold night sky. In this way, snow clouds can reach the exosphere, and give rise to the bright nights, for a few days.

This alternative explanation of the Tunguska explosion – as the present-day formation of a kimberlite – is supported by the facts that (*j*) its epicenter coincides with the

250 Myr old *Kulikovskii volcanic crater*, forming a part of the *Khushminskii* tectono-volcanic complex, (*jj*) it lies near the crossing point of a number of *tectonic fault lines*, one of them running towards lake Baikal, (*jjj*) it sits at an Asian *geomagnetic* maximum, and also *heat flow* maximum, surrounded by ringlike *Moho isohypses* (Jerebchenko 2003; Kochemasov 2001), and (*jv*) it is located above a thick, sealing *basalt layer* (Yepifanov 2002), near the center of the Siberian *craton*. Moreover, (*v*) during their 1999 expedition to (the near) lake Cheko, Longo's group recorded a local *radon* outburst lasting four hours. And as mentioned above, (*vj*) tectonic outbursts are at least 20 times more *frequent* than meteoritic impacts at the same destruction energy (Kundt 2001, 2002).

Finally, (*vjj*) the *mystery clouds* mentioned above are likely to form the low-intensity, more frequent tail of the tectonic methane-outburst distribution, observed at a rate of many clouds per year – by airplane pilots and by satellite photography – and indirectly as "*pockmarks*" on 6% of the sea floor (Walker 1985; Kundt 2001; May and Monaghan 2003). The clouds rise rapidly from an unresolved spot on the surface – land or water – expand, hence cool, and bend downwind as they rise, looking whitish by condensing water vapor on their periphery. They tend to ignite near the ground when escaping from land, due to self-generated lightning, but rise unburnt when issuing from the sea, probably causing a threat to commercial sea and air traffic, at a rate of more than once per year.

19.3
(Other) Recorded Impact Events

What do we know about terrestrial impacts? Their *rates* have been estimated between monthly and 10^8-yearly events for impact masses between 10^{-1} t and 10^{12} t, in Kundt (2001, 2002), collected from Ol'khovatov (1999), Krinov, and other recordings. Well-studied cases are *Sikhote-Aline* (Siberia, 1947), *Gibeon* (Namibia, < 1838), Wabar (Saudi-Arabia, 1704), *Barrington* crater (Arizona, –50 ka), and *Chicxulub* (Yucatán, –65 Ma; Alvarez 1997, Melosh 1997), the latter's age being recently somewhat controversial. The list does not include *cometary* impacts, because they are estimated to be rare, and would likely not cause any lasting damage to the surface of Earth. Yet there is the 1994 crash of comet Shoemaker-Levy 9 onto Jupiter, whose probability must be multiplied by a factor of $10^{-5.0}$ when compared with terrestrial events – for equal embedding fluxes – because accretion rates at moderate approach velocities ($< v_{esc}$) scale as the square of the accretor's mass. As the best-studied case, I now turn to Sikhote-Aline.

On 12 February 1947, at 10:30 local time, an iron meteorite struck the easternmost edge of Siberia, in the western part of the *Sikhote-Aline* mountain range. Eyewitnesses reported a bolide crossing the atmosphere within ≤ 5 s, though noises were heard for 10 ± 5 minutes (Krinov 1966). The bolide left a gigantic trail, or smoke band which got increasingly wiggly but disappeared only towards the evening. According to eyewitnesses, the bolide split up successively at the four heights of 58, 34, 16, and 6 km, towards a final diameter of 0.6 km. From infall channels in the ground and tree destructions, its infall angle could be measured as 30 ± 8 deg w.r.t. the vertical.

Within the four succeeding years, over a hundred small *craters* were detected in that area, the largest of diameter 26.5 m. They formed three concentrations, spread over an ellipse of diameters 1 and 2 km. All craters were formed by meteoritic fragments whose impact channels penetrated between 1 and 8 m into the ground, depending on their shape and orientation. The summed weight of all the collected iron-rich fragments was 23 t, and estimates yielded about 70 t total for the impacted mass, corresponding to an iron bolide of diameter 6 m, some $10^{-3.5}$ in mass of the hypothetical Tunguska bolide. Even if a comparable amount of rocky material had been left behind in the atmosphere, in the shape of the dust trail, the Sikhote-Aline meteorite was still some 1000 times lighter than Tunguska's hypothesized one. No *impactites* were found at Sikhote-Aline: explosions after impact tend to occur (only) for crater diameters ≥ 100 m. Telegraph poles and snapped-off tree tops were plentiful. Trees were felled radially around craters, but only in directly adjacent ringlike domains, of width ≤ 30 m. Some of them took bizarre shapes.

19.4
(Likely) Tectonic Outbursts

The list of internal (tectonic) outbursts is less uniform than that of external (impact) catastrophes. It contains *volcanic eruptions*, like Mt. St. Helens (1980), Krakatoa (1883), Tambora (1815), and Santorin (Thera, 1400 BC), with their large outcrops of lava, in excess of km³ per event. It also contains the mountain-forming activities of the Eifel ($-\leq$ Myr), still ongoing, and of the Alps ($-\leq 10$ Myr). The disappearance of *Sodom and Gomorrah* at biblic times may also be due to tectonic events, making the two cities slide to the bottom of the Dead Sea. Finally, among the smaller, more recent events – compiled by Ol'khovatov (1999) – there are *Cando* (NW Spain, 1994), *Allende* (Mexico, 1969), the *Zanitsa pipe* (Siberia, $\leq$ 1954), and *Tunguska* (1908). These events are distributed between occurrence *rates* of yearly and millennial, and liberated energies corresponding to impact masses between 10^2 t and 10^6 t. (Impact velocities tend to be at least 10 times higher than outburst velocities, hence correspond to destruction energies at least 100 times larger [than outburst energies] for a given mass).

Let us now look at the *Cando* outburst, a more recent miniature Tunguska event. A destruction energy comparable to Sikhote-Aline was liberated by the bolide of 18 January 1994, seen and heard at 7:15 UT in the parish of Cando, NW of Spain, (Docobo et al. 1998). It took three months until a newly formed crater was reported, of size 29 × 13 m, 1.5 m deep, whose former (big) pine trees were hurled downhill through 50 to 100 m. An in-between road remained clear of soil from the ejection, eliminating the possibility of a landslide – which did, however, occur on the same day 300 m NW of the main crater, knocking down two pines. No meteoritic debris were recovered. The authors prefer a high-speed gas-eruption explanation.

19.5
How to Discriminate between Impacts and Outbursts?

Tunguska, Sikhote-Aline, and Cando are three catastrophical events of the last century – the first of them some 10^3 times more energetic than the two others – which have found quite different explanations in the literature. Whereas Krinov (1999) spends

129 pages of his 397-page book on giant meteorites on the "Tunguska meteorite", Ol'khovatov (1999) prefers a tectonic interpretation. Even Sodom and Gomorrah have been recently interpreted as former cities on the SE bank of the Dead Sea, blown up and/or slid to the bottom of the Sea by a volcanic eruption. How can we *discriminate* between the extraterrestrial and the terrestrial interpretation?

Whereas with the latter interpretation you can be rejected from peer-reviewed journals, even when based on sober and friendly arguments, the former interpretation may only apply to a 3% minority of all events. Eyewitnesses speak of bolides – or fireballs – in all cases, and of barisal guns lasting for many minutes. Trees are felled, or debranched, or their tops chopped off, craters are formed, and fires are ignited in all cases. What differs are the *details*, of which I have listed some 20 above, each of which can be used for a discriminaton. They read:

Volcanic *flames* in the sky can last for up to an hour whereas a meteoritic infall trail flashes only for a few seconds, and its *heat* cannot be sensed in the faces of eyewitnesses, because of too small an extent in space and time. But a meteoritic *trail* tends to stay visible for hours, unlike volcanic flames. *Barisal guns*, on the other hand, are heard for comparable times in both cases by distant eyewitnesses ($d \geq 70$ km) because sound echos from warm layers above the stratosphere take that long. For tree falls, their *pattern* matters: Absorbed *momentum*? How many *centers*? Telegraph poles require strong shock waves, hence trace the *supersonic* domain. *Craters*, if blown from below, can contain tree stumps, whereas those formed by infall show an impact channel plus debris. Volcanic outblows can *throw* trees, or root stumps, or rocks through several hundred meters, whereas non-explosive infalls (with small craters) redistribute the impacted soil in their immediate surroundings (≤ 30 m). Meteoritic *debris* tend to be recovered for impact masses in excess of fractions of a ton.

There are additional criteria. Volcanic blowouts require pressurized vertical exhaust pipes from a deep-lying fluid reservoir, which have their imprints on the local *geography*, like the Kulikovskii crater. Moreover, when megatons of natural gas – mainly methane – are suddenly released into the atmosphere, they will rise, burn, and form *clouds* in the thermosphere for several days, at heights above 500 km, where they scatter the sunlight. Such scattered sunlight at night is known as the bright nights of both Krakatoa (1883) and Tunguska (1908). We live on a tectonically active planet.

How to evaluate the impact risks? None of the published repetition rates I have seen have attempted to *discriminate* between external and internal hazards. Rather, a power law has been fit through Tunguska (1908) and Chicxulub (–65 Myr), starting with Shoemaker (1983), and continuing through Chapman and Morrison (1994), Jewitt (2000), and Atkinson (2001). Note that a determination of the density of near-Earth objects (*NEOs*), as by Rabinowitz et al. (2000), cannot reliably predict the collision rates with Earth because of the uncertain transfer function, which depends on their orbital parameters, in particular on their orbital inclinations. I therefore made an attempt in (Kundt 2001, 2002) to estimate the independent statistics of extraterrestrial and terrestrial events, and came up with a much less pessimistic prediction for the likelihood of harmful future impacts – in line with the fact that life on Earth has not been erased for much more than a Gyr. Even the great "*mass extinctions*" of the past were, in reality, only extinctions of species, not extinctions of life as a whole.

19.6
Conclusions

During the 1908 Tunguska catastrophe, trees were flattened over an area of more than 2 000 km^2 of Siberian taiga. For several decades, this destruction was thought to be caused by the impact of a sizable meteorite, some 60 m across, and served as a well-defined data point on the terrestrial impact spectrum (i.e. impact rate versus destruction energy). But doubts in the impact interpretation emerged, via a continued absence of detected impact grains, via controversies between the cometary and asteroidal proponents, via the net-zero-momentum treefall pattern with its multiple centers, the three European bright nights (illuminated by scattered sunlight), details of the eyewitness reports, and the geologically preferred site of the destruction, in the Kulikovskii volcanic crater with its intersecting fault lines. In this report, I shall analyze the Tunguska event, compare it with several similar catastrophes, and present more than a dozen criteria that can serve to discriminate between a meteoritic impact and a tectonic outburst (of the kimberlite type). Tectonic outbursts turn out to be (likewise) power-law distributed (as a function of destruction energy), with practically the same spectral slope as the asteroidal impacts, but are more frequent than the latter by a factor of at least 20.

Acknowledgments

My sincere thanks go to the members of my weekly seminar, in particular to Gernot Thuma and Hans Baumann, and to Günter Lay for help with the electronic data handling.

References

Alvarez W (1997) T. rex, and the Crater of Doom. Penguin Books
Atkinson H (2001) Risks to the Earth from impacts of asteroids and comets. Europhysics News 32(4):126–129
Bronshten VA (2000) Nature and destruction of the Tunguska cosmical body. Planetary and Space Science 8:855–870
Chapman CR, Morrison D (1994) Impacts on the Earth by asteroids and comets: assessing the hazard. Nature 367:33–39
Dawson JB (1980) Kimberlites and their xenoliths. Springer, Berlin
Docobo JA, Spalding RE, Ceplecha Z, Diaz-Fierros F, Tamazian V, Onda Y (1998) Investigation of a bright flying object over northwest Spain, 1994 January 18. Meteorites and Planetary Sciences 33:57–64
Farinella P, Foschini L, Froeschlé Ch, Gonczi R, Jopek TJ, Longo G, Michel P (2001) Probable asteroidal origin of the Tunguska cosmic body. A & A 377:1081–1097
Foschini L (1999) A solution for the Tunguska event. A & A 342:L1–L4
Foot R (2002) Shadowlands, quest for mirror matter in the universe, ISBN 1-58112-645-x; also: astro-ph/0407623
Gallant RA (1994) Journey to Tunguska. Sky and Telescope 87:38–43
Gold T (1999) The deep hot biosphere. Springer, New York
Gold T, Soter S (1979) Brontides: natural explosive noises. Science 204(4391):371–375
Haggerty SE (1994) Superkimberlites: a geodynamic diamond window to the Earth's core. Earth and Planetary Science 122:L57–L69

Hou QL, Kolesnikov EM, Xie LW, Kolesnikova NV, Zhou MF, Sun M (2004) Platinum group element abundances in a peat layer associated with the Tunguska event, further evidence for a cosmic origin. Planetary and Space Science 52:331–340, 773

Jerebchenko IP (2003) Geological and geophysical aspects of the Tunguska phenomenon. In: Theses of the jubilee scientific conference "95 years of the Tunguska problem 1908–2003", Moscow. GAIsh, June 24–25, 2003. The Moscow State University Publishing, pp 96–97

Jewitt D (2000) Eyes wide shut. Nature 403:145–147

Kochemasov GG (2001) On probable terrestrial origin of the 1908 Tunguska explosion. In: Reports of the jubilee international conference "90 years of the Tunguska problem", June 30–July 2, 1948, Krasnoyarsk, pp 208–212

Kolesnikov EM, Boettger T, Kolesnikova NV (1999) Finding of probable Tunguska cosmic body material: isotopic anomalies of carbon and hydrogen in peat. Planetary and Space Sci 47:905–916

Krinov EL (1966) Giant Meteorites. Pergamon, pp 125–265

Kundt W (1991) Earth as an Object of Physical Research. In: Latif M (ed) Strategies for future climate research. Klaus Hasselmann's 60th anniversary, Hamburg, pp 375–383

Kundt W (2001) The 1908 Tunguska catastrophe: an alternative explanation. Current Science 81:399–407

Kundt W (2002) Risks to the Earth from impacts of asteroids and comets. Europhysics News 33(2):65–66

Kundt W, Jessner A (1986) Volcanoes, fountains, earth quakes, and continental motion – what causes them? Journal of Geophysics 60:33–40

Longo G, Serra R, Cecchini S, Galli M (1994) Search for microremnants of the Tunguska cosmic body. Planetary and Space Science 42(2):163–177

May DA, Monaghan JJ (2003) Can a single bubble sink a ship? American J of Physics 71(9):842–849

Melosh HJ (1997) Multi-ringed revelation. Nature 390:439–440

Ol'khovatov AYu (1999) The tectonic interpretaion of the 1908 Tunguska event. Internet: *www.geocities.com/CapeCanaveral/Cockpit/3240*

Ol'khovatov AYu (2003) Geophysical circumstances of the 1908 Tunguska event in Siberia, Russia. Earth, Moon and Planets 93:163–173

Rabinowitz D, Helin E, Lawrence K, Pravdo S (2000) A reduced estimate of the number of kilometre-sized near-Earth asteroids. Nature 403:165–166

Serra R, Cecchini S, Galli M, Longo G (1994) Experimental hints on the fragmentation of the Tunguska cosmic body. Planetary and Space Sciences 42:777–783

Svetsov VV (1996) Total ablation of the debris from the 1908 Tunguska explosion. Nature 383:697–699

Shoemaker E (1983) Asteroid and comet bombardment of the Earth. Ann Rev Earth Planet Sci 11:461–494

Vasilyev NV (1998) The Tunguska Meteorite problem today. Planetary and Space Science 46:129–143

Walker DA (1985) Kaitoku Seamount and the mystery cloud of 9 April 1984. Science 227:607–611

Yepifanov V (2002) Degassing of the Earth. In: Conference of the Russian Academy of Sciences, Moscow

Zahnle K (1996) Leaving no stone unburnt. Nature 383:674

Zotkin IF, Tsikulin MA (1966) Modelling of the Tunguska meteorite explosion. Doklady AN SSSR 167:59–62

Chapter 20

Atmospheric Megacryometeor Events versus Small Meteorite Impacts: Scientific and Human Perspective of a Potential Natural Hazard

Jesús Martínez-Frías · José Antonio Rodríguez-Losada

20.1 Introduction

It is important to differentiate between a natural hazard and a natural disaster. A natural hazard is an unexpected or uncontrollable natural event of unusual magnitude that threatens the activities of people or people themselves (NHERC 2004). A natural disaster is a natural hazard event that actually results in widespread destruction of property or causes injury and/or death. Only a very small fraction of the actual meteorite events are observed as falls in any given year. It has been predicted that 5800 meteorite events (with ground masses greater than 0.1 kg) should occur per year on the total land mass of the Earth. In a recent work, Cockell (2003) emphasizes the scientific and social importance of giving a coordinated and multidisciplinary response to events related with the entrance of small asteroidal bodies that could potentially collide with the Earth. In fact, it can be said that the recovery of small meteorites between 1 kg to 200 kg is relatively common; in Spain alone there are four meteorites in the collection of the National Museum of Natural History, weighing more than 30 kg (e.g. Colomera iron meteorite). But what would happen if the impact bodies, despite weighing up to 200 kg, would melt?

From the 8th to the 17th January 2000, numerous big ice conglomerations (weighing from around 300 g to more that 3 kg) fell in different parts of the Iberian Peninsula under clear sky atmospheric conditions (Martinez-Frias et al. 2000, 2001; BAMS 2002; Brink et al. 2003; Martinez-Frias and Rodriguez-Losada 2004) resulting in damage to cars and an industrial storage facility (Figs. 20.1 and 20.2). Due to these unusually recurrent falls, a research program was initiated in Spain to accomplish the following: (*a*) confirm the atmospheric nature of the ice blocks, since some of them of were several hundred kilograms in weight, (*b*) to obtain an updated systematic database incorporating similar events around the world, (*c*) to have the samples well preserved in freezer rooms for future study, (*d*) to promote the creation of an international working group that can communicate through an electronic network, and (*e*) to inform the public about the importance of reporting the occurrence of new falls as a potentially underestimated natural hazard. The term *megacryometeor* was recently coined (Martinez-Frias and Travis 2002) to denote large atmospheric ice conglomerations which, despite sharing many textural, hydrochemical and isotopic features detected in large hailstones, are formed under unusual atmospheric conditions which clearly differ from those of the cumulonimbus clouds scenario (i.e. clear-sky conditions). The historical review and in-depth study of the atmospheric megacryometeor

Fig. 20.1. Megacryometeor fell in January 2000 in San Feliz, Lena (Asturias province, Spain). Note its textural variation that is also reflected in hydrochemical and isotopic heterogeneities

Fig. 20.2. Megacryometeor of approximately 18 kg that fell in the Soria province (Spain) on 27 January 2002

events show that many of their phenomenological aspects, as well as the human perception associated with the falls, do not differ much from the episodes of small meteorite falls, weighing between 0.5 kg to around 400 kg.

20.2
Megacryometeors

The fall of large ice blocks (from about 1 kg to hundreds of kilograms) from the clear sky is, in accordance with Meaden (1977) one of the most interesting and controversial issues in atmospheric sciences. Meaden used the term "ice meteors" to denote them and proposed that their origin had to be different from that of the large hailstones. Later, Corliss (1983) used the term "hydrometeors" also differentiating them from the classical hailstones and suggesting that they have an atmospheric origin but under different possible genetic scenarios. Historically, the falls were routinely assigned, without verification, to aircraft icing processes or simply to waste water from aircraft lavatories (blue ice). For instance, a recent example that has been well studied and recorded in the scientific literature is the blue ice mass of 2–3 kg, which fell in the Courel locality, Galicia (Spain) on 9 July 1996 (Docobo et al. 1997).

A detailed historical review of such events (Martinez-Frias and Lopez-Vera 2000, 2002) shows that there are many documented references of falls of large blocks of ice which go back to the first half of the 19th century ("pre-Wright" or previous to the invention of the airplane); for instance, in 1829, a block fell in Córdoba, Spain weighing 2 kg, and one in 1851 in New Hampshire weighing 1 kg. A block of 5 kg fell in October 1844 in Cette (France) and another one some 1 m × 1 m × 60 cm (surely much more massive) fell in 8 May 1802 in Hungary. Folkard (2003) describes the fall, in 13 August 1849, in Scotland of huge megacryometeors (of around 2 m in size) formed by hundreds of chunks of ice. The fall of a large ice conglomeration, which measured 26 × 14 × 12 cm and weighed 2.04 kg, is cited, in Germany, in 1936 (Talman 1936). More than 50 megacryometeors approximating 75 kg each (almost a ton in total weight) fell in Long Beach, California, on 4 July 1953 (McDonald 1960).

For many years the largest hailstone officially reported in the United States was one that fell at Potter, Nebraska, on 6 July 1928. It had a circumference of 43 cm and weighed 680 g. This record was surpassed on 3 September 1970 at Coffeyville, Kansas (USA). The giant hailstone measured 18 cm (7 inches) across about 44 cm (17.5 inches) in circumference, and weighed more than 750 grams (26 ounces) (NOAA 2004). Other cases of large hailstones include a large block of ice of almost 2 kg that fell in Kazakhstan, and another of almost one kg that fell at Strasbourg (Meaden 1977; Corliss 1983). Probably the best-documented fall of an ice chunk was April 2, 1973, in Manchester, England. The block weighed 2 kilograms and consisted of 51 layers of ice. Its origin was not determined (Griffiths 1975). Spectacular events in China and Brazil, fortunately all well studied by prestigious scientists, can be mentioned. In 1995, an ice block of about 1 m (Parker 1995) fell in Zhejiang (China) and in Campinas (Brazil) two huge megacryometeors of 50 and 200 kg (Pinto 1997) fell in 1997.

In the context of the study of ice fall events that have occurred in Spain, many of these scientists were personally contacted by one of the authors of the present paper (JMF) and are now co-ordinated in an international working group [see *http://tierra.rediris.es/megacryometeors/*]. The results of our studies indicate that megacryometeors are not the classical big hailstones, ice from aircrafts (waste water or tank leakage), or the simple result of icing processes at high altitudes. They are a different type of atmospheric meteors whose characteristics and atmospheric conditions of formation we are just starting to consider and understand. For this reason, although the analysis of the historical record of ice falls gathers many cases which are apparently similar, a simplistic analysis of these events, as a whole, could lead to confusion, as different types of ice falls correspond to different formation scenarios; hence the importance of defining differentiation criteria to distinguish between them.

20.2.1
Textural, Hydrochemical and Isotopic Characteristics

The systematic investigation of the ice fall events during these last five years has allowed us to define the following features and characteristics:

- The fall of blocks of ice weighing several kilograms has been reported in many regions around the world from the distant past to the present. There are verified records of similar ice fall events on all continents (except in Africa).
- Their textures and hydrochemical features are clearly of an atmospheric (tropospheric) nature. Textures of megacryometeors include zones of "massive ice", large isolated cavities, mm-sized oriented air bubbles and ice layering. The thickness of the layers ranges from less than 1 mm to more than 1 cm. Also, tiny solid particles can be found randomly distributed in the interior of the ice. Early chemical and isotopic analyses (Martinez-Frias et al. 2000) showed evidence of compositional heterogeneity with large densities of ions – up to five times larger than normal meteoric waters – and corresponding to solutions of halite, calcite, anhydrite and quartz or feldspar aerosols. New hydrochemical analyses, using the combination of capillary electrophoresis, molecular absorption spectrometry (UV-Vis) and ICP-AES (Santoyo et al. 2002), indicate that the blocks of ice are formed from waters of variable mineralization (between 106 and 858 µS cm^{-1}), with very low values of SiO_2 (< 0.7 ppm), and the presence of NH_4 (0.21 to 0.78 ppm) in some samples.
- $\delta^{18}O$ and δD (V-SMOW) of the megacryometeor samples fall into the Meteoric Water Line. The distribution of the samples on classical Craig's line (Martinez-Frias et al. 2001) suggests either a variation in condensation temperature and/or different residual fractions of water vapor (Rayleigh processes). The most positive values are typical of rainwater in Spain. Isotopic mapping of δD values in the hailstones display: (*a*) significant general variations from –24.4‰ to –126.4‰, and (*b*) specific variations of up to 25 δD within some individual blocks.
- Atmospheric soundings from NOAA were collected in the days prior to and during the occurrence of the megacryometeors in Spain (mainly 10–17 January) (Santoyo et al. 2002). Soundings from La Coruña, Santander, Zaragoza, Madrid, Palma (Balearic Islands), Murcia and Gibraltar were the closest available. The analysis of the soundings indicated that the tropopause sank from a level of 250 hPa ($\approx$ 10 500 m), on the days prior to the event, to a lower level of $\approx$ 400 hPa ($\approx$ 7 000 m) on the days of the events. This process was not observed simultaneously at all stations and seems to have propagated from northwest to east and then to south. Along with the amount of sinking, the other significant factor is the accompanying increase in humidity (near saturation) observed in all cases (except over Madrid). Ozone anomalies and wind shear were also found to exist occurring simultaneously with the tropopause

undulations, and data from the World Area Forecast Centre (London) (Martinez-Frias et al. 2002) also confirmed the low tropopause height values.

20.2.2
Theoretical Modeling

As previously defined the atmospheric (mainly tropospheric) nature of the ice blocks was clearly determined. In order to explain the causes that contribute to both the formation of first ice nuclei and their later growth, several hypotheses have been proposed alluding to both cosmic (Foot and Mitra 2002) and terrestrial (Martínez-Frías et al. 2000, 2001; Brink et al. 2003) origins. To date, a theoretical model was developed to try to explain the physical parameters, which could govern the formation of megacryometeors and suggests a plausible scenario for the events, based on the well-known theory of nucleation (Martinez-Frias et al. 2001). The free energy for nucleation in homogeneous media is:

$$G = -(S-1) \cdot \mu(T) \cdot n + \gamma(T) \cdot n^{2/3}$$

where μ and γ are the chemical potential of equilibrium and the surface free energy for a given temperature, respectively, and S and n are the supersaturation of the vapor and the number of molecules in the aggregate. The critical nuclei is given, taking

$$\frac{dG}{dn} = 0, \text{ by } n_c = \left(\frac{2\gamma}{3\mu(S-1)}\right)^{1/3}$$

Therefore, when $S \gg 1$ many nuclei are formed close to each other because the value G^c (critical free energy) is accessible by T fluctuations. The distance between the nuclei is small, and they cannot grow too much, because they are all competing for vapor molecules. When $S < 1$ there is no condensation. For $S \approx 1$ the value of n_c tends to infinity. However, the critical energy of nucleation is extremely high and condensation cannot take place due to temperature fluctuations alone.

Nevertheless, if an external perturbation (see below) is produced within the vapor volume (i.e. extra cooling, injection of ion concentration – heterogeneous nucleation –, irruption of a sound wave, etc.), then nuclei will form at large distances from each other. The particle condensed will be of ice if the temperature is well below zero, and they can grow large at the expense of surrounding molecules, as the nuclei are scattered. Theoretical estimations show that the radius R of the ice chunks is

$$R \approx \frac{\rho_v}{\rho_w}\left(1 + \frac{V_T}{V}\right)h$$

where ρ_v, ρ_w, V_T, V and h are the gas and water densities, the thermal and falling velocities and the height of the gas volume. This gives $R = 6$ cm, 4 cm and 2 cm for $S \sim 1$ at $T = -5, -10$ and -20 °C, and $h \sim 1$ km. Notice that the growing process is controlled by the thermal velocity of the water molecules. In the above calculations, the latent heat of the hailstone is removed by the carrier air molecules. Also, the nucleation at $S \geq 1$ is consistent with the large number of ions in the specimens detected by the chemical analyses. Another alternative possibility could be that an ice crystallite from the frozen stratosphere, where are known to exist in clouds, enters a region of larger humidity (from around 7 or 8 km down) and starts growing.

Since the numerous megacryometeor events, which occurred in Spain in 2000, other falls of large ice blocks have been recorded in Argentina, Australia, Austria, Canada, Colombia, Italy, Mexico, New Zealand, Portugal, Spain, Sweden, The Netherlands, UK and USA, clearly indicating the planet-wide nature of such events. The last spectacular case, whose circumstances of fall are very well documented, occurred at 21 July 2004 in the locality of Maqueda, Toledo (Spain). A huge mass of more than 400 kg fell very close to a 15-year-old girl, the niece of the Justice of Peace of Maqueda. Thanks to our recommendations (and hence the connection and importance of a social concern in relation with the fall of meteorites), a small piece of ice could be adequately preserved and recovered for investigation. First ICP-MS analysis indicates that its hydrochemical features resemble those determined in other previous similar cases. A much more detailed isotopic characterization of the megacryometeor fragment is in progress.

20.3
Megacryometeors versus Small Meteorite Impacts

During the last half century, more than 50 cases of megacryometeors of more than 1 kg in size have been recorded, of which about 11 of them, weigh between 20 and > 400 kg. It might appear, given their size, that these events are not truly significant but four main aspects must be considered: (*a*) the planet-wide character of the phenomenon, (*b*) the regional nature of the atmospheric anomalies, (*c*) its possible climatic implications, and (*d*) that these ice falls, together with the fall of meteorites and observation of fireballs, are the most frequent events noticed by people in the last century and, in a certain way, are marking social conceptions (and misconceptions) about meteorites and their impacts and hazards. As Chapman (2004) indicates "ways to eliminate instances of hype and misunderstanding involve public education about science, critical thinking and risk; familiarizing science teachers, journalists and other communicators with the impact hazard might be especially effective". Specifically focused on this human perspective, it is important to note how (primitive and inexperienced) were the responses and reactions of people, mass media and even some policy makers to the very recent and well-studied spectacular Iberian fireball of 4 January 2004 (Martinez-Frias and Madero 2004). This fireball and the accompanying and subsequent social and institutional reactions, provide a textbook case study in appropriate and inappropriate reactions to such an episode, of terminological confusion ('meteoroids,' 'bolides,' 'meteors,' 'meteorites,' etc.), of mistaken public conceptions and, in broad terms, of the need for a 'taskforce' capable of providing an appropriate and adequate response. As previously defined, Cockell (2003) emphasizes the scientific and social importance of a coordinated, multidisciplinary response to the atmospheric entry of small asteroidal or cometary bodies that have the potential to collide with the Earth. But scientific and human perception regarding meteorites has an advantage with respect to megacryometeors. Meteorites have been clearly recognized as real natural events for more than 200 years thanks to the efforts of Ernst Florens Chladni in 1794, who proposed his audacious thesis indicating that they actually represent genuine rocks from space. Of course his view received immediate resistance and mockery by the scientific community. It is science that decides what is real and what is not, what exists and what does not exist, and the meteorite falls passed through three stages: a stage of uncorrelated observations, a stage of intense controversy, and finally the stage of scientific acceptance (Westrum 2004).

In contrast, the scientific study of megacryometeors is extremely recent; the first scientific papers in peer-review journals were published in the 1970s (Griffith 1975), the first systematic study did not start until the Spanish ice falls of 2000, the term 'megacryometeor' was officially proposed in 2002 (Martinez-Frias and Travis 2002), and this term started appearing in the scientific dictionaries and encyclopaedias in 2003 (e.g. Dictionary of Weather, The FreeDictionary, WorldHistory, Nodeworks Encyclopedia, WordIQ). In addition, an extremely significant point to stress is that, whereas atmospheric origin is clearly demonstrated from hydrogeochemistry and isotopic signatures, the mechanism that is responsible for their formation is not yet understood.

Another aspect linked with the social misconception is that many people confuse both types of events (cosmic and atmospheric): In fact, the term "aerolito" (in Spanish) was erroneously assigned by the mass media to name the first cases. This fact has however the positive element that such confusion is helping society to gain interest in learning about the different types of "falls from sky", and the education about meteorites and their origin, risks and impact effects has benefited from the study of these unusual ice conglomerations.

20.3.1
Comparison of the Rate of Falls during Human Times (Historical Record)

The first problem to solve in determining the rate of meteorite falls is the practical lack of realistic astronomical observation and convincing records during the history of mankind. Also there is a significant human factor in gathering meteorite falls (Beech 2002). Perhaps China is an exception. As a tradition, the Chinese people hold the belief that the fortune of a king (emperor) or a general is revealed clearly in the sky and that one should follow universal principles. Therefore, the people there have a long history of astronomical observation. There were 359 historical records of meteorite falling events from July 2133 BC to late AD 1905. One entry described that five pieces of meteorites fell at Shanghiu, Henan on December 24, 645 BC. The meteorites were believed to be from space for the first time in human history, which was by far earlier than the famous Chladni (1794) publication on meteorite origin (Liu 2004).

It is well known that the fall of meteorites to the Earth's surface is part of the continuing process of accretion of the Earth from the dust and rock of space. The frequency with which meteorites fall decreases strikingly as the size of the meteorites increases. Dust sized particles fall regularly on every home in the world. Particles of a gram (about the size of a small pea) or more in weight, however, are estimated to fall at a rate of less than 8 per square mile per year (Nelson 2004). Similarly, objects over 10 grams (about the size of a quarter) fall at something less than the rate of 1 per 1000 square miles per year. As the size of the objects get larger, the rate of fall becomes exponentially smaller, so that we can expect that an object over 1 kg might fall in a given 1 square mile piece of land only once in every 50 000 to 100 000 years. Estimates of fall rates vary widely, and the above numbers may be overly generous. The best estimates of the total incoming meteoroid flux indicate that about 10 to 50 meteorite events occur over the Earth each day (Beauford 2004). However, some 2/3 of these events will occur over ocean, whereas another 1/4 or so will occur over very uninhabited land areas, leaving only about 2 to 12 events each day with the potential for discovery by people. Half of these again occur during the night, with even less chance of being noticed. Due to the combination of all of these factors, only a handful of witnessed meteorite falls occur each year. Roughly 500 meteorites larger than 0.5 kilograms are thought to fall on Earth every year, but only about 4 are actually observed because most fall in the ocean or sparsely populated areas. As an order of magnitude estimation, each square kilometer of the Earth's surface should receive on average one meteorite fall about once every 50 000 years. If this area is increased to 1 square mile, this time period becomes about 20 000 years between falls (see *http://planpro.jpl.nasa.gov/mrsrch3.html*).

In the recent review (Martin-Escorza 2004), studying the possible existence of historical periodicity of meteorite falls, 1700 documented events are recorded. Of these, 359 cases come from the compilation of Kumlehn (1987) relative to the historical ancient falls in China, including events that go back to several centuries before the birth of Christ. Branch (2004), in his revisit of all known cases of meteorites that have hit humans, animals and/or man-made objects, lists 102 documented events since 1800 to present (Table 20.1).

Regarding the fall of megacryometeors with similar sizes and weights and which also fell under similar circumstances (producing damages, social responses, etc.) the number of cases is much less (although it is important to note here the spectacular increase of megacryometeor events after the recognition of the Spanish cases of 2000). Something similar happened after the recognition of the meteorites as natural and scientifically significant specimens (Chladni's effect). The compilation of the megacryometeor events, also covering the period from 1800 up to the present day, is shown in Table 20.1. Most of references listed until 1982 were collected by Meaden (1977) and

Table 20.1. Chronological comparison of the impacts of small meteorites and megacryometeors from 1800 to the present (2004). As in the megacryometeor events, the meteorite falls listed here correspond to all known cases that have struck humans, animals and/or man-made objects and from a societal point of view the considerations and impressions about both phenomena should not differ very much. Based on the compilations of Branch (2004) and Martinez-Frias and Lopez-Vera (2000 and 2002)

Period	Frequency of hits (meteorites)	Frequency of hits (megcryometeors)
2000	5	42
1975–1999	28	11
1950–1974	26	25
1925–1949	14	1
1900–1924	10	0
1875–1899	7	4
1850–1874	6	0
1825–1849	3	1
1800–1824	4	3

Corliss (1983) and there exists information about their circumstances of fall, witnesses, effects of the impact, etc. Other cases of ice falls, although with inferior sizes and weights also occurred in France, Sweden, The Netherlands, Italy, Canada and Spain and further information about them can be found in Martinez-Frias and Lopez-Vera (2000, 2002). Our study indicates that, until recently (year 2004), 87 megacryometeor events witnessed and producing damages to houses, cars, etc., have been registered between 1800 and 2004 (see selected falls in Table 20.2).

20.4
Final Remarks

The historical review and in-depth study of the atmospheric megacryometeor events show that many of their phenomenological aspects, as well as human perceptions associated with the falls, do not differ very much from the episodes of small meteorite falls, weighing between 0.5 kg to around 400 kg. Given that ICSU recently recog-

Table 20.2. Representative chronological selection of 29 megacryometeor events that, either due to their size or to the effects of their impacts, are considered socially relevant

Country	Year	Description
Hungary	1802	Block of ice of around 1 × 1 × 0.6 m
France	1844	Block of ice of around 5 kg
United Kingdom	1849	Huge block of ice of more than 2 m
USA	1881	Block of ice of around 50 × 50 × 50 cm
USA	1882	Block of ice of around 40 kg
United Kingdom	1897	Block of ice of around 50 kg
Germany	1936	Block of ice of around 2 kg
USA	1949	Block of ice of around 20 kg
United Kingdom	1950	Various blocks of ice, the largest of around 7 kg
United Kingdom	1950	Block of ice of around 60 kg
USA	1953	More than 50 blocks of ice, some around 75 kg each
USA	1955	Various blocks of ice from 3 to 13 kg
USA	1957	Block of ice of around 70 cm
USA	1959	Block of ice of around 15 kg
United Kingdom	1959	Block of ice like of around 45 cm
Australia	1960	Various blocks of ice of around 40 to 60 cm
USA	1965	Block of ice of around 25 kg
USA	1972	Block of ice of around 20 kg
United Kingdom	1975	Block of ice weighing more than 20 kg
United Kingdom	1977	Block of ice of around 50 kg
USA	1982	Block of ice of around 15 kg
China	1995	Block of ice of around 1 m
Brazil	1998	Two ice blocks of 50 and 200 kg
Spain	2000	Numerous cases weighing several kilograms
Mexico	2002	Block of ice of around 10 kg
Portugal	2002	Block of ice of around 20 kg
Canada	2002	Block of ice of around 5 kg
New Zealand	2004	Block of ice weighing around 5 kg
Spain	2004	Block of ice of more than 400 kg

nized that the societal implication of a comet/asteroid impact on Earth warrants an immediate consideration by all countries in the world (see ICSU *http://www-th.bo.infn.it/tunguska/tenerife.doc*), we recommend to monitor and evaluate the risks and effects linked to such cosmic events considering: (*a*) the different scales of the problem and (*b*) carrying out comparative analyses (scientific, social, cultural, etc.) with respect to other similar phenomena (e.g. megacryometeors). This will provide a better knowledge and management of the sociological consequences and communication of the impact risks to the public, as well as other issues connected with an adequate response of institutions and policy makers. As was recently stated (Morrison et al. 2003) "once it is accepted that the impact hazard is a social and not just a scientific problem, it is a short step to allow that considerations of maximum social benefit may well constrain the scope and form of scientific investigation."

The study of the megacryometeor events and their comparison with the rate of small meteorite falls indicates that, mainly after 1950, the number of hits has spectacularly increased affecting practically the whole planet. It is still soon to ascertain whether there is a real multiplication effect of the number of megacryometeor events due to natural causes or simply now the information circulates very fast and we can know rapidly what is happening in different parts of the world. In either case, we suggest one monitor these phenomena because they cannot only be a potential natural hazard for people, aviation, etc., but perhaps they are also signals of more serious environmental problems.

Acknowledgments

Thanks to ICSU, CAB and CSIC for their institutional support and David Hochberg for the revision of the English version. Also thanks to an anonymous referee and Dr. Richard E. Spalding for their revision, corrections and scientific remarks which have greatly improved the original manuscript. Thanks to Dr. Nicolás García (CSIC) for proposing the original idea of the theoretical model.

References

Beech M (2002) The human factor in gathering meteorite falls. Meteorite Magazine 8:1–4
Branch W (2004) *http://www.branchmeteorites.com/metstruck.html*
Brink K, Travis D, Martinez-Frías J (2003) Upper tropospheric conditions associated with recent clear-sky ice falls. 57th Annual Meeting, The Wisconsin Geographical Society, September 19–20, UW-Eau Claire
Beauford R (2004) *http://www.enchanted-treasures.com/Meteor/General/general.html*
Bosch X (2002) Great balls of ice. Science, News Focus 297:765
Chapman CR (2004) The hazard of near-Earth asteroid impacts on Earth. Earth and Planetary Science Letters 222:1–15
Cockell CS (2003) A scientific response to small asteroid and comet impacts Interdisciplinary Science Reviews, 2003, 28:74–75
Corliss WR (1983) Ice falls or hydrometeors. In: Tornados, dark days, anomalous precipitation and related weather phenomena. A catalog of geophysical anomalies. The Sourcebook project. PO Box 107, Glen Arm, MD 21057:40–44
Docobo JA, Tamazian V, Guitian F (1997) Investigación sobre un hielo azulado caído en Galicia en julio de 1996. Ibérica 330:320–321
Folkard C (2003) Guinness World Records. Bantam

Foot R, Mitra S (2002) Ordinary atom-mirror atom bound states: A new window on the mirror world. Phys Rev D 66:061301
Griffiths RF (1975) Observation and analysis of an ice hydrometeor of extraordinary size. Met Mag 104: 253–260
Kumlehn M (1987) Transcripcion of the names of China's ancient meteorites. Meteoritics 22(2):137–149
Liu P (2004) *http://www.greatwallct.com/meteorit.htm*
Martín-Escorza C (2004) Caidas de meteoritos ¿periodicidad en su variación histórica? In: Martinez-Frias J, Madero J (eds) Meteoritos y geología planetaria. Junta de Comunidades de Castilla-La Mancha (in press)
Martínez-Frías J, López-Vera F (2000) Los bloques de hielo que caen del cielo: antecedentes y fenomenología reciente. Rev Ens Cien Tierra 8(2):130–136
Martínez-Frías J, López-Vera J (2002) Grandes bloques o meteoros de hielo In: Ayala-Carcedo FJ, Olcina Santos J (eds) Riesgos Naturales Ariel Ciencia, pp 1141–1148
Martínez-Frías J, Madero J (2004) The Iberia fireball event of 4 January 2004. Interdisciplinary Science Reviews 29(2):1–6
Martínez-Frías J, Travis D (2002) Megacryometeors: fall of atmospheric ice blocks from ancient to modern times. In: Leroy S, Stewart IS (eds) Environmental catastrophes and recovery in the Holocene, abstracts volume. Brunel University, West London (UK), pp 54–55
Martínez-Frías J, López-Vera F, García N, Delgado A, García R, Montero P (2000) Hailstones fall from clear Spanish skies. Geotimes, News Notes. American Geological Institute June/2000:11–12
Martínez-Frías J, Millán M, García N, López-Vera F, Delgado A, García R, Rodríguez-Losada JA, Reyes E, Martín Rubí JA, Gómez-Coedo A (2001) Compositional heterogeneity of hailstones: Atmospheric conditions and possible environmental implications. Ambio 30(7):452–455
McDonald JE (1960) The ice-fall problem. In: Meaden GT (1977) The giant ice meteor mystery. J Met 2(17):137–141
Morrison D, Harris A, Sommer G, Chapman C, Carusi A (2003) Dealing with the impact hazard. In: Bottke W, Cellino A, Paolicchi P, Binzel RP (eds) Asteroids III, University of Arizona Press, Tucson pp 1–46. *http://128.102.38.40/impact/downloads/NEO_Chapter_1.pdf?ID=113*
Meaden GT (1977) The giant ice meteor mystery. J Met 2(17):137–141
Nelson SA (2004) *http://www.tulane.edu/~sanelson/geol204/impacts.htm*
NHERC (2004) *http://www.naturalhazards.org/discover/*
NOAA (2004) *http://www.crh.noaa.gov/ict/swaw/recordhail.pdf*
Parker J (1995) *http://www2.jpl.nasa.gov/sl9/news56.html*
Pinto HS (1997) *http://www.cpa.unicamp.br/gelo/gelo.html*
Reynertson G (2004) *http://www.nctc.net/~hazard/origin/*
Talman CF (1936) Ice from thunderclouds. Nat Hist 3(8):109–19
Westrum R (2004) *http://www.n2.net/prey/bigfoot/biology/scientists.htm*

Part III Socio-Economic and Policy Implications

Chapter 21 Social Science and Near-Earth Objects: an Inventory of Issues

Chapter 22 Perception of Risk From Asteroid Impact

Chapter 23 Hazard Risk Assessment of a Near Earth Object

Chapter 24 Social Perspectives on Comet/Asteroid Impact (CAI) Hazards: Technocratic Authority and the Geography of Social Vulnerability

Chapter 25 May Land Impacts Induce a Catastrophic Collapse of Civil Societies?

Chapter 26 The Societal Implications of a Comet/Asteroid Impact on Earth: a Perspective from International Development Studies

Chapter 27 Disaster Planning for Cosmic Impacts: Progress and Weaknesses

Chapter 28 Insurance Coverage of Meteorite, Asteroid and Comet Impacts – Issues and Options

Chapter 29 The Economic Consequences of Disasters Due to Asteroid and Comet Impacts, Small and Large

Chapter 30 Communicating Impact Risk to the Public

Chapter 31 Impact Risk Communication Management (1998–2004): Has It Improved?

Chapter 32 Towards Rational International Policies on the NEO Hazard

Chapter 33 A Road Map for Creating a NEO Research Program in Developing Countries

Chapter 21

Social Science and Near-Earth Objects: an Inventory of Issues

Lee Clarke

21.1 Introduction

It would have been ridiculous, not too long ago, to admit openly that you were thinking about asteroids and comets slamming into the Earth. Such events could mean the end of the world as we know it – TEOTWAWKI as millenialists call it – and that kind of talk is often ridiculed. Then again, it would have been ridiculous, not too long ago, to think that two hijacked 767s would slam into the World Trade Center and make both towers fall. Thinking about NEOs is becoming more commonplace, although not entirely normal.

Respectable people are pondering the issues. For example, S. Pete Worden, who is a Brigadier General in the US Air Force and Deputy Director for Command and Control Headquarters at the Pentagon, has said that he believes "we should pay more attention to the 'Tunguska-class' objects – 100 meter or so objects which can strike up to several times per century with the destructiveness of a nuclear weapon" (*http://abob.libs.uga.edu/ bobk/ccc/ceo20700.html*). The General is referring to a meteorite impact near the Tunguska River, in Siberia, in 1908. It was discovered in 1927 by Russian scientist Leonid Kulik. The object exploded with the force of 15 megatons of TNT and flattened trees for tens of kilometers in the vicinity. Of course, when space debris that size comes around again it will most likely have no effect. That's because most of Earth's surface is uninhabited and so any random 15 megaton explosion would most likely devastate little if anything at all. But if it exploded over Manhattan, a large part of the city would disappear.

21.2 Globally Relevant Disasters

The scientific discovery that near Earth objects pose catastrophic potential was not enough to establish inquiry into NEOs as a legitimate activity. Science can only produce knowledge. It can not produce the larger social and political conditions that give scientific findings urgency, and currency. I cannot know for sure, but it seems doubtful that NEO research would have been possible in the 1950s. But AIDS, the 2001 terrorist attacks on the United States, globalization of capitalism, the Internet, and a truly worldwide, non-stop media barrage broaden our horizons and stretch our imaginations beyond the local milieu. The time is right to gain a larger audience for ideas, evidence, and theories about what I call *Globally Relevant Disasters*. Such disasters have effects far beyond their immediate environs.

Globally relevant disasters aren't new. The 1883 eruption of Krakatau in Indonesia killed perhaps 35 000 people, most from resultant tsunamis, one of which reached the Arabian Peninsula, some 7 000 kilometers away (*http://www.meteo.mcgill.ca/195-250/ tsunami/index.htm*). The Tambora eruption in 1815 was much worse, with more global consequences (*http://volcano.und.nodak.edu/vwdocs/volc_images/southeast_asia/ indonesia/tambora.html*). It was 150 times larger than Mt. Saint Helens, and ejected a volcanic column 40 kilometers in the air. It darkened the skies entirely, over a distance of 500 kilometers (Stothers 1984). Estimates are that 92 000 people died from Tambora, 82 000 of these from starvation caused by cooler temperatures – 1816 is known as the "year without a summer" – across the northern hemisphere. The combination of plagues that we call the Black Death emerged in Europe in 1347 and within four years had wiped out two thirds of the populations of many cities. The catastrophe continued for 300 years, claiming nearly *one-third* of the European population and touching every part of society. And of course, Earth's collision with an asteroid 65 million years ago was globally catastrophic, for about ½ of the world's species, especially the dinosaurs.

But if globally relevant disasters aren't new, we *do* have new ways to bring catastrophe to more people in places far removed from the point of threat. Time and globalization processes have brought with them new "disaster vectors," connecting people to damage in unprecedented ways. The obvious disaster-vector is interdependence, which means that people's social networks provide mechanisms for the transmission of harm. Faster and cheaper modes of transportation, for example, can potentially spread diseases exponentially. AIDS wouldn't have taken nearly as large a toll 100 years ago. Somewhat less obviously, modern social organization and technologies bring with them new ways to harm people who are far away in both time and space. Nuclear explosions, nuclear accidents, and global warming are examples. We are increasingly "at risk" of global disasters, most and probably all of which would qualify as worst cases. This situation presents us with unprecedented challenges both in terms of anticipating worst cases and responding to them.

GRDs, and the possibility of GRDs, pose new challenges for international relations. What happens if our skills become honed to such a degree that we can predict precisely where the Earth would be struck by an NEO? Should an asteroid be headed for the middle of sub-Saharan Africa, I wonder what resources the rich countries would be willing to devote to the rescue effort. The UN would undoubtedly try to mobilize support for rescue but possible recipient countries would wrangle over who should shoulder the greater responsibility. If the impending threat were serious enough, the force of the blast may destroy the better part of an entire country. Should that happen, the surviving countries would be faced with either helping the country that was destroyed rebuild or with providing the refugees a permanent home. It's hard to imagine scenarios with satisfactory outcomes.

There are other daunting issues regarding GRDs. We are ignorant about many things we need to know. Scholars have not been thinking about disaster at this scale for very long. Is it like predicting a hurricane? A nuclear meltdown? Super-volcanic eruptions? To what class of events should we look for intellectual and practical guidance? We have the 2004 tsunami, which killed perhaps 250 000 people in a handful of countries, but not much else. We have a number of analytic tools, ranging from case studies to counterfactuals. In using those tools, it may very well be that moving farther away from known realities moves closer to what it would actually be like to suffer a GRD. Events that earn the designation of "worst case" are ones that are beyond our imagination: a nuclear explosion in downtown London, a liquefied natural gas explosion near Tokyo. GRDs will probably be outside our imagined purview.

Very few, if any, policy makers are thinking about such issues; nor are they likely to give GRDs the attention they deserve, I argue below. This embryonic state of affairs means that we often don't have clear parameters for what counts as truth and knowledge. Consequently, we don't enjoy set standards for what constitutes expertise in an area. An epidemiologist may know a lot about how a disease spreads through a population, but that is no guarantee of expertise in modeling how the disease might spread *across* populations, or continents. A social scientist might know how people and organizations respond to earthquakes, floods, and the like but the truth-value of extrapolations to the entire world is hard to estimate.

Our ignorance concerning knowledge about globally relevant disasters has the blessing that there are no disciplinary boundaries. There are, as of yet at least, no claimants to intellectual property, no chest-beaters crowing about how their own outlook is so much better than all the others. This is a good thing, because understanding a disaster of global consequence would seem to require many intellectual talents. Hopefully, thinking about and study of GRDs will keep its interdisciplinary character.

A caveat to the above: scholarship on NEOs seems not to have involved social science very much, as yet. Clark Chapman, and coauthors, said in a recent paper that:

> ... essentially no analysis has been done of how to mitigate other repercussions from predictions of impacts (civil panic), how to plan for other kinds of mitigation besides deflection (e.g. evacuation of ground zero, storing up food in the case of a worldwide breakdown of agriculture, etc.), or how to coordinate responses to impact predictions among agencies within a single nation or among nations (*http://www.internationalspace.com/pdf/NEOwp_Chapman-Durda-Gold.pdf*).

Indeed, to date the field seems dominated by those with non-social scientific backgrounds. That's natural because it is from disciplines such as biology, astronomy and the like that many threats are discovered in the first place. But ultimately any disaster is interesting because it involves humans, and we'll have to turn to the social sciences for that.

In this paper I can provide no definitive answers, partly because we don't have enough direct data points – incidents of near Earth object impacts – on which to base conclusions. I intend this paper to lay out an agenda of things we need to know about, think about, and research, if we are to begin to understand the limits and possibilities of social science research relevant to the NEO threat.

21.3
Preparation and Response: General Issues

We should think carefully about both preparation and response issues. It may be tempting to try to think of these two moments of societal response as separate, but for some purposes we should avoid that temptation. The chief reason for this is that how leaders, organizations, and experts prepare for disasters has important implications for how the objects of their efforts – the general public – will respond afterward.

In the interest of analytic clarification, imagine two models of response: one that's driven from the top, and one that's driven from the bottom. The first model – let us call it the Official Response model – presumes that those on the top of the response effort will determine the course of events. Leaders and planners will have made forecasts and plans for how their organizations will respond. They will have created lists of resources and contacts. They will specify executive succession, which would be necessary should a top official be lost to the effort. The plans that the experts have created will specify what technologies will be necessary to maintain communications and coordination.

The Official Response model concentrates on officials and formal organizations. In this view, communication, cooperation, and coordination are the primary problems that need to be solved for effective preparation and response. The Official Response model is well represented in accident reports from the US National Transportation Safety Board and other such agencies, in journalistic accounts of accidents and disasters, in investigatory commissions, and in the courts.

Essential here is the idea that tight *coordination* among organizations, clear *communication* among officials, and *cooperation* among leaders *and* organizations make disaster response more productive, saving lives and property. These precepts, incidentally, would seem to apply to all sorts of planning, not just planning for accidents. If planning and response are indeed always and everywhere so tightly, causally coupled then coordination, communication, and cooperation should be related to each other as follows. Good communication leads to cooperation between organizations which then enables coordination of efforts and thus successful response. Successful emergency response is usually attributed to the planning and foresight of responsible organizations. In failures it is typical to find people who are touched by the disaster and subsequent evacuation complaining about chaos and uncertainty, and placing responsibility for those unhappy conditions at the door of poor planning.

Two experienced disaster researchers, for instance, draw strong conclusions on how officials, organizations, and networks of organizations ought to behave in response to impending doom or catastrophe. They say that when organizations are flexible, have good information, clearly defined positions for responding to disaster, and good communication then responses to disaster warning will be more effective. Further, "organizations must be able to see emergency response as their job, and have clearly defined roles to play. Emergency experience or planning can help fulfill this need." Coordination "is enhanced through preparedness ..." especially if authority relations within and between organizations are clear and uncontested (Mileti and Sorensen 1987).

An example of how this work is the Livingston train wreck. On 28 September 1982, before the sun came up in Livingston, Louisiana, 43 cars of a 101 car Illinois Central Gulf Railroad train derailed. All of Livingston had to be evacuated because of the explosions and fires; "a virtual inferno" is how an authoritative report on the accident described it (White et al. 1984; I will refer to this as the Livingston Report). Several homes were destroyed and 17 others were severely damaged. Over four million pounds of chemicals breached containment (NTSB 1983). There were huge fireballs, a lot of black smoke, and considerable angst as officials and experts tried to figure out how to handle several very toxic chemicals.

It could have been worse, because the train derailed about ¼ of a mile from the town's main intersection. Massive devastation, toxic chemicals, and conflagration. We might have expected chaos, but chaos did not happen. While the evacuation was going on state and federal agencies worked together, coordinating their efforts in a cooperative manner. The Louisiana State Police followed its legal mandate to coordinate official efforts after a hazardous materials spill until other, more appropriate organizations, could assume responsibility. The state's Office of Environmental Quality was responsible for managing hazardous wastes and arrived quickly, as did the Office of Health Services and Environmental Quality. Rather than proclaiming no risk to the general population, which is what we often see from officials, officials immediately assumed the potential for major harm. On that assumption, the agencies monitored people's health, adopted an aggressive stance for protecting the environment, and assumed responsibility for informing residents when they could return to their homes. State agencies were thus surprisingly willing to take on the difficult task of defining acceptable risk.

The official model is often the correct one, because ours is an organizational society. We look to experts, officials, and organizations for preparation and response to disasters. Organizations are the best tools we've yet developed for solving complex problems. But it would be a mistake to rely too heavily on officials and organizations. The Official Response Model is often wrong. For it is often the case that the action in a disaster is *unofficial* and *unorganized*. For example, one of the reasons that more lives weren't lost in the World Trade Center collapse is that people took it upon themselves to get out of the buildings. In the south tower, which was struck second, people ignored the official who was yelling at them through a bullhorn that they would be safer sitting at their desks. The 150 000 people who evacuated from around Three Mile Island ignored the reassurances of officials and got themselves out of what might well have become harm's way. In any earthquake or tornado, it is the response of the person-in-the-street that will most likely result in the saving of self, and others.

So the second model – the Unofficial Response model – is as important as the first. Russell Dynes, a long-time contributor to research on disasters, has pointed out repeatedly that the centralized, military style organization of emergency services (Official Response) is contradicted by research on disaster response (1993, 1994). For instance, community emergency planning often recommends highly centralized control of resources after disasters, but ethnographies and surveys of people in the throes of disaster, and in the immediate aftermath, show that the most consequential actions happen "in the field." If that's so, then centralizing resources in formal organizations can be counter-productive, in spite of everyone's good intentions.

Another example, to use a globally relevant disaster, is AIDS. There has been much that drug companies, governments, churches, and NGOs can and should do to help prevent the spread of AIDS. But ultimately it is, and will be, the response of people closest to the threat – the needle user, those having unprotected sex – who will stem the disaster's tide.

21.4
Preparation and Recovery: Planning

Planning will obviously be crucial for any program of preparation and recovery. Research, not to mention good sense, tells us that planning can go a long way in mitigating the effects of disasters and in restoring life to normalcy. Critical infrastructure needs to be conceptualized and protected. "Life line" organizations have to be identified and protected. Coordination between organizations must be facilitated. Effective disaster response is often facilitated by well organized communities. And that is key. Disaster response, in the United States, has historically been a local responsibility. Disasters happen in particular places, after all.

But GRDs, and NEOs in particular, pose new problems for preparation and recovery. For any significant NEO event will likely have effects that go far beyond the local environment. Federal involvement will be crucial. GRD planning will require a lot of resources. It will require a lot of money if we're going to protect large numbers of people. Officials and planners will have to work with state and local governments, non-governmental organizations, just to mention a few, many of whom will have nothing in their budgets for programs that may never have a payoff. The political commitment and person-power required just to think through and research what *might* constitute an adequate plan seem large. If planning for an NEO strike happens at all, it will be one of the most extensive exercises in foresight ever conducted.

To get started, let us ask the simple question, where do we look for planning models? Instead of starting with planning generically, let us start with types of calamities and then look at the sorts of plans that go along with them. Consider four categories: disasters with precedent, unprecedented disasters, globally relevant disasters, and counterfactuals. Examples of the first are hurricanes and earthquakes; an example of the second is nuclear holocaust; an example of the third is super-volcanism; an example of the last is the CIA using Hollywood directors think up terrorist attacks.

Disasters with precedents are those with which we have the most experience. Because of that experience it is here that we can most clearly see the connection between planning and effective response. In *Mission Improbable* I called such plans functional plans (Clarke 1999). They are functional plans because extensive experience means that planners' abilities to control the untoward have a reasonable chance of success. For example, after officials order people on the North Carolina coastline to evacuate in front of hurricane that's bearing down on them, they put into action contingency plans that have been used before. Emergency personnel are extremely effective at staving off hurricane threats to lives and property largely because they've done it many times before. Superficially, at least, it appears that disasters with precedent, and the planning that goes along with them, are categorically different than globally relevant disasters. To the extent that is true then our greatest storehouse of knowledge may be only broadly useful.

Disasters without precedent, but for which planning has occurred, are another possible source of models for thinking about an NEO kind of event. These kinds of events were the main cases I dealt with in *Mission Improbable*. As noted, I distinguished between functional plans and symbolic plans. Symbolic plans, which I call "fantasy documents" are those where the ratio of actual utility to symbolic utility is low. In other words, fantasy documents are more useful as instruments of rhetoric, or hopeful statements of how things *might* happen, rather than actual blueprints for action. For example, Alaskan regulators and oil industry officials once claimed that they could respond effectively to a 8 400 000 gallons barrel spill (the *Exxon Valdez* spill was 11 million gallons). That was an impossible task, and had never been done before. The Alaskan oil spill contingency plan was a fantasy document.

Functional plans are grounded in extensive experience. Fantasy documents, by contrast, are not. They are constructed to respond to anticipated events, and in that way they are like any other plan. But unlike other, functional, plans, fantasy plans can't be built with other such events in mind. What experience can you appeal to if you're trying to plan to respond to the total obliteration of a society's energy sources? Because direct experience isn't available, fantasy documents are characterized by metaphors, similes, and extrapolations as substitutes. Consider the plans of civil defense and nuclear war fighting. Civil defenders claimed they could save 80 percent of the US population and they had plans for recovery after general nuclear war. Those plans were fantasy documents because there wasn't enough knowledge or experience available to create a realistic plan. My argument is not that the plans were worthless but that there was so little experience, and so much uncertainty, revolving around the key issues that the plans could only be symbolic. The planners had no way to know how a nuclear war would actually be conducted (assuming something less than total obliteration), although such knowledge would be central to actually saving so many people. Short of a full-fledged general war in which everything of value is targeted twice, the uncertainties were just too great: what would the targets be? how about the blast yields? the number of warheads? airbursts or ground bursts? time of year? These uncertainties, along with others, meant that plans for societal recovery were shear guesses and not based on expert knowledge, although they were dressed up to appear as if they were. Fantasy documents symbolize the promise of control more than actual control. If unprecedented disasters are more similar to NEO-level disasters than disasters that we have experience with, the kinds of plans that generally go along with them should give us little solace. Symbolic plans may be better than nothing when the disaster comes, but we have no way to know how much better.

Third, we can consider actual globally relevant disasters such as the major volcanic eruptions of 1258, 1815 (Tambora), and 1883 (Krakatau). We might add a Tunguska-like event, should it happen over Washington, D.C. These are disasters with precedent but for which planning has never been attempted. What can we learn from such cases? We learn that we are vulnerable to worst-case natural killers, because they could happen again. We can point to the famines and pestilence that such eruptions caused. For example, the 1258 eruption resulted in "severe crop damage and famine throughout much of Europe," says NASA scientist and veteran detective of super-volcanism Richard B. Stothers (2000, p 361). It might have spewed as much as 600 megatons of sulfuric acid into the sky. It also resulted in a large and constant "dry fog" across Europe and the Middle East. So much aerosol was in the atmosphere that the moon completely disappeared in a lunar eclipse. "Both hemispheres," says Stothers, "were obscured by aerosols," which means the event was truly global (Stothers 2000, p 264).

While such investigations are helpful, society is probably sufficiently different now compared to the 13th or 19th centuries that trying to draw direct lessons is dangerous. At the end of the day we must admit that our "expertise" in such cases is highly circumscribed. There's just not enough meaningful experience on which to base our judgments.

The matter needs more scholarly attention. Consider the single issue of interdependence. Modern societies are more interdependent than they were 700 years ago. We are dependent on world trade in food and energy, to mention the two most important commodities. Imagine that there's a super-volcanic eruption or a large NEO-induced disaster. What would we expect to happen? We know that the volcanoes of antiquity led to famines and flagellations. Can we confidently predict the same for modern times? We can postulate that people living 1000 years ago lived closer to the edges of starvation and disease than at least rich countries do today. Thus it may not have taken very much to push them over the edge into societal catastrophe. On the other hand, in subsistence societies that planted a variety of crops, starvation rarely occurred even with crop failure (Granovetter 1979) The tropospheric cooling caused by some large volcanic eruptions usually led to crop failures and disease fairly quickly (Stothers 1998). Perhaps we moderns have more resources, more buffers, and more fat to sustain us through the hard times.

Then again, perhaps we would suffer more. After all, if there were widespread failure of crops in America's heartland large parts of the world would suffer – most likely the rich peoples of the world who can afford to buy American products; poor peoples are less dependent on us. People in the cities would have no chance to rely on husbandry. It's not as if the 20 million people in the New York City metropolitan area could feed themselves by fishing in the Hudson or the East River (which wouldn't be very safe anyway). Most of those people probably don't even having fishing rods. And the guns that city people have are not helpful in killing food.

There is one final source of models to which we might turn: counterfactuals. This is the disciplined speculation about alternative futures. What happens if Britain is destroyed? What happens if the nuclear winter affects only the Northern Hemisphere? What happens if a giant tsunami in the Atlantic wipes out the coastlines of the Americas and Europe?

Let us briefly consider the possibility of whether a bolide explosion could spark a nuclear war. In June 2002 a bolide exploded over the Mediterranean with the force of a ten kiloton bomb. To consider the matter deeply, we would have to anticipate the extent of damage but also people's responses. Explosions are more predictable than people. Imagine the following scenario. India and Pakistan are involved in a standoff, each with nuclear weapons at the ready. Tensions are high and both sides have a strong incentive to fire first. Only a few countries have technology sophisticated enough to distinguish a naturally occurring explosion from a non-natural one. A bolide explodes and US military intelligence detects it, but can't decide whom to warn. Millions of people are incinerated in the ensuing five minute war. Clearly there is much to learn from using counterfactuals. The problem with using them is that of not knowing where to stop. At what point does scenario-building become so unrealistic that it becomes unproductive? I can not develop an answer to this question here, but note that there is a technical literature in political science, history, and philosophy on how to distinguish useful from useless counterfactuals (Hawthorn 1991; Ferguson 1999; Cowley 2000).

21.5
Preparation and Response: the Problem of Trust

Preparation and response are part and parcel of effective post-traumatic response. What people are told by their leaders, and how, matters for how actions and statements are interpreted after the disaster comes. For analytic purposes, the same general issues and processes are involved regarding leaders' behavior. The key issue regarding public behavior is the question of panic, on which more below. Straightforwardly, what can be expected from leaders in disseminating information, both before and after a crash? And what can be expected from members of the public? Because we've had a lot of disasters, and because social scientists have been studying disasters for a long time, we know a lot about how to answer these questions. But I repeat the caveat that what we know from a local or even national disaster may not be directly applicable to globally relevant disasters. They are probably in the same class of events, but they may not be.

People carry around with them theories about how people and society work. This is no less true of leaders than of others. So an important issue to understand is how emergency response leaders theorize how the world works and how it breaks down. We have reams of literature on how regular people perceive risk, and on how they respond in disaster. But we have little on how elites and planners think about the same issues.

One reason that the issue of leader response is important is that if they are to be able to lead in a crisis then people must trust what they are saying. Sociologist William Freudenburg has labeled this the problem of recreancy (Freudenburg 1993). Initially interested in the problem of risk perception, Freudenburg looked at a lot of different kinds of evidence and discovered that age, sex, political party affiliation, and self-assessed political ideology do not predict risk perception very well. This contradicted much of the mainstream literature on risk perception, as well as common sense. He found that the best predictors of what people fear are the degree to which they trust science and business and governmental ability to manage danger. When officials are recreant they fail to fulfill their duties. This may involve outright lying but more commonly it involves over-promising safety or condescending approaches to risk communication. For example, after accidents at nuclear plants we can expect officials to say that "there was never any danger to the public." After airplane accidents usually come the platitude that "it's safer to fly than to drive." After mishaps at chemical plants we hear that common refrain that "without risk there is no progress." And so on. These sorts of bumper-sticker slogans are unlikely to engender trust from the public. They are unlikely to reassure people because they are easily recognized as clichés. Too, they are transparently premised on the idea that people are fearful in the same that children are.

Leaders seem to say such things because they think that misperception of risk will lead to panic, will lead people to reach way beyond the bounds of reason and perspective, to do things that are destructive for themselves and their communities. As an idea, panic is widespread in society. And "panic" has long figured in policy battles over technology – such as those over environmental regulations, toxic chemicals, or nuclear power.

21.6
Preparation and Response: the Problem of Panic

Any consideration of post-traumatic response must consider the question of public panic. Should an NEO with real possibilities of striking be discovered, officials and the media would certainly predict widespread panic. There are really two questions here, because we're interested in the post-prediction pre-impact response as well as the post-impact response. In large measure the post-prediction pre-impact issue is a matter of risk communication. What can we expect from the public?

One model, specifically related to NEOs, is given to us by movies. In the films, Armageddon and Deep Impact, panic was rampant. As pieces of space debris crashed through the poor Chrysler Building in Deep Impact *and* Armageddon, people ran hysterically through the streets, pushing others aside to save themselves. In such visions of panic, well socialized Jekyls transmogrify into raving Hydes. Disaster movies suggest a tipping point beyond which people are so overcome with fear that they will put self-interest over regard for others. We think it's wrong to yell "fire" in a crowded theatre – even if the theatre is on fire – because of worry that panic would cause more death than the fire itself. Would the same be true if we knew an NEO was coming our way? Perhaps.

One reason we need to think through the panic issue is not just that we don't want to alarm people unnecessarily. Officials often believe that people are highly prone to panic and so another reason is that we want to predict *official* behavior. Before the Y2K rollover, for example, politicians and business managers urged people not to panic, if there were computer failures. Alan Greenspan worried that there might be runs on the banks. John Koskinen, chair of the President's Commission on Year 2000 Conversion, became less concerned about failing machines and more about panic: "As it becomes clear our national infrastructure will hold, overreaction becomes one of the biggest remaining problems." Decision-makers also sometimes withhold information because they claim to be convinced that panic will ensue. For example, in the Three Mile Island crisis utility officials failed to tell people, and even government officials, how serious the situation was because they were trying to "ease the level of panic and concern."

What does the empirical record show about panic? We have two places to look for evidence. First, we can look to cases where impending doom is predicted by experts and leaders. Second, we can look to cases of actual disasters, to see how people behave when their worlds fall apart.

Starting with the second set of cases first, because they are the easiest to detail. Here, the record is clear. Panic is quite rare in actual disaster situations. Let us define panic, with the Oxford English Dictionary, an "excessive feeling of alarm or fear … leading to extravagant or injudicious efforts to secure safety." While disaster victims often report feelings of great fear or alarm, the behavioral consequences of those feelings are usually just the opposite of injudicious effort.

Panic was rare even among residents of German and Japanese cities that were bombed during World War II. The U.S. Strategic Bombing Survey, established in 1944 to study the effects of aerial attacks, chronicled the unspeakable horrors, terror, and anguish of people in fire stormed cities and even in the nuclear attacks. Researchers found that, excepting some uncontrolled flight from the Tokyo firestorm, little chaos occurred.

Researchers at the Disaster Research Center, now at the University of Delaware, have been investigating people's responses to extreme events for nearly 50 years. They have looked at literally hundreds of disasters and one of the strongest findings is that people rarely lose control of themselves. When the ground shakes, sometimes dwellings crumble, fires rage, and people are crushed. Yet people do not run screaming through the streets in a wild attempt to escape the terror, even when they are terrorized. Tornadoes come to wreak havoc on neighborhoods or even entire communities. Yet people do not usually turn against their neighbors or suddenly forget personal ties and moral commitments. Rather, the more consistent pattern is that people bind together in the aftermath of disasters, working together to restore their physical environments and their cultures to recognizable shapes.

The non-finding of panic is robust, though not exclusive. Rather than panic, it is generally true that even when people confront what they consider the worst case, they organize themselves to provide succor and salvation to their friends and even to complete strangers. This is called a "therapeutic community" or "altruistic community." The therapeutic community is characterized by, as disaster scholar Russell Dynes conceives it, "the development of an emergency consensus, the development of altruistic norms and behavior, the expansion of the citizenship role, the minimization of community conflict, and the generation of hostility toward outsiders" (1970, p 101; cf. Erikson 1994). I do not mean to say, of course, that people never panic in the traditional sense of that word. There *are* soccer riots and trampling, the existence of which shows that the phenomenon is real enough. Yet the bulk of available *evidence* is that even under very threatening circumstances people generally behave with consideration and good sense.

The evidence on panic (more precisely, the lack of panic) suggests that policies regarding risk should be constructed to trust people. People are more likely to distrust high level decision makers when they think they're not being told the truth, or when they think they're being condescended to, than when they hear bad news. It is true that a policy based on full disclosure rather than soothing slogans will sometimes generate opposition to official positions and even complaints about leaders. But *that*, indeed, should be seen as an acceptable risk of democracy.

Thus on the basis of available evidence, we would have to predict that people will not panic after the disaster. But all predictions from the social sciences must be highly provisional. My no-panic prediction assumes that an NEO-induced disaster is the same kind of disaster that we have a lot of research on. That assumption may be wrong. What would panic look like in the event of the known threat of a city-killer? Planet killer? We have no reliable way to answer those questions.

The second class of relevant events to which we can look for guidance regarding the panic question is where some disaster or hazard has been predicted. How do people respond then? I am still researching this question.

21.7
Conclusions

Niels Bohr, presaging Yogi Bera, is said to have said that "prediction is very difficult, especially about the future." We predict nevertheless. I predict that policy makers will systematically neglect and even ignore the risk from NEOs. The main reason for this is the long lead-time that is characteristic of NEO risks. Policy makers spend the most time worrying about problems that will affect their political careers, especially whether they will be re-elected. This is the trouble we're seeing with the issue of "long term stewardship" of Department of Energy sites that have been contaminated with uranium, plutonium, thorium, and volatile organic compounds. A real commitment to long term stewardship would mean payoffs for people far into the future. National leaders don't often allow themselves that luxury.

I am a pessimist regarding planning for a disaster induced by near Earth objects. But not because of the uncertainties. The uncertainties are interesting intellectually but they are not obstacles to actually preparing and, perhaps as important, simply thinking through the issues required before preparation could even begin. I am pessimistic for the practical reason that it's hard to identify who could profit from it. Those who want to build missile defense shields, and especially those who want to have space-based nuclear weapons would be clear beneficiaries of a national commitment to planning for NEO-level events; and I predict they will be strong supporters. But beyond that it's hard to identify potential champions with sufficient economic and political power to get the issue/s on the public agenda.

Beyond such crass reasoning, I am also pessimistic because it's simply hard to engage in political debate about near Earth object dangers. In considering the NEO issue there will inevitably be a lot of doomsday talk. Politicians would likely face considerable ridicule should they propose spending large amounts of resources on preparing for doomsday. Certainly they would make themselves targets for derision in an electoral campaign. Too, it takes rare and extraordinary courage for leaders to propose actions that may generate considerable benefit long after they're out of office, let alone long after they're dead. True, a lot of attention, planning, and remediation were brought to bear on the Y2K problem. But the consequences of those failures – some of them anyway – were readily identifiable: businesses could have lost a lot of money. I see no such powerful incentive at work with the NEO issue. Saving the world isn't good enough.

Acknowledgments

For help thanks to Ivan Bekey, Clark Chapman, Fadi Essmaeel, Al Harris, Al Harrison, Kathleen Hollingsworth, David Morrison, Dana Rohrabacher, Geoffrey Sommers, Richard Stothers, and Harvey Wichman. Some material comes from my books, *Mission Improbable* and *Worst Cases*.

References

Chapman C., Durda, D.D (2001) The comet/asteroid impact hazard: a systems approach. http://www.internationalspace.com/pdf/NEOwp_Chapman-Durda-Gold.pdf
Clarke L (1999) Mission improbable: using fantasy documents to tame disaster. Chicago, University of Chicago Press
Clarke L (2005) Worst cases: inquiries into terror, calamity, and imagination. Chicago, University of Chicago Press
Cowley R (Editor) (2000) What if? The world's foremost military historians imagine what might have been. New York, Penguin Putnam
Dynes RR (1970) Organized behavior in disaster. Lexington, D.C. Health
Dynes RR (1993) Disaster reduction: the importance of adequate assumptions about social organization. Sociological Spectrum 13:175–192
Dynes RR (1994) Community emergency planning: false assumptions and inappropriate analogies. International Journal of Mass Emergencies and Disasters 12(2): 141–158

Erikson K (1994) A new species of trouble: explorations in disaster, trauma, and community. W.W. Norton, New York
Ferguson N (1999) Virtual history. Basic Books, New York
Freudenburg WR (1993) Risk and recreancy: Weber, the division of labor, and the rationality of risk perceptions. Social Forces 71:909–932
Granovetter M (1979) The idea of "advancement" in theories of social evolution and development. American Journal of Sociology 85(3):489–515
Hawthorn G (1991) Plausible worlds: possibility and understanding in history and the social science. Cambridge University Press, Cambridge
Mileti DS, Sorensen JH (1987) Determinants of organizational effectiveness in responding to low probability catastrophic events. Columbia Journal of World Business 22:13–22
National Transportation Safety Board (1983) Final report: ICA derailment, September 28, 1982, Washington, DC, August 10.
Stothers RB (1984) The great Tambora eruption in 1815 and its aftermath. Science 224(4654):1191–1198
Stothers RB (1998) Far reach of the tenth century Eldgja eruption, Iceland. Climatic Change 39:715–726
Stothers RB (2000) Climatic and demographic consequences of the massive volcanic eruption of 1258. Climatic Change 45:361–374
White LE, Bock SF, Englande AJ (1984) The Livingston derailment. A report for the Office of Health Services and Environmental Quality, Department of Health and Human Resources and the Department of Environmental Quality, State of Louisiana

Chapter 22

Perception of Risk from Asteroid Impact

Paul Slovic

22.1
Early Work: Decision Processes, Rationality, and Adjustment to Natural Hazards

Perhaps the earliest studies of risk perception with regard to natural hazards were conducted by geographer Gilbert White (1945, 1964) and his students (e.g. Burton and Kates 1964). Later, in 1974, this author joined with White and economist Howard Kunreuther to review this early work in the context of new research in cognitive psychology (Kahneman and Tversky 1972; Tversky and Kahneman 1971, 1973) describing the idiosyncratic ways human minds think about probability, uncertainty and risk (Slovic et al. 1974). This research illustrated the workings of Herbert Simon's theory of "bounded rationality" (1959), which asserts that human cognitive limitations force decision makers to construct a simplified model of the world in order to deal with it.

One way in which bounded rationality was evident was in the limited range of alternatives perceived by resource managers trying to cope with natural hazards. Another indication of limited perception was systematic misperception of risks and denial of uncertainty. For example, residents on floodplains viewed floods as repetitive and even cyclical phenomena. In this way, the randomness that characterizes the occurrence of the hazard is replaced by a determinant order in which history is seen as repeating itself at regular intervals (Burton and Kates 1964). Another common view was the "law of averages" approach, in which the occurrence of a severe flood in one year made it unlikely to recur the following year. Other floodplain occupants reduced uncertainty by means of various forms of denial. Some thought that new protective devices made them 100 percent safe. Others attributed previous floods to a freak combination of circumstances unlikely to recur. Still others denied that past events were floods, viewing them instead as "high water". Another mechanism was to deny the determinability of natural phenomena. For these people, all was in the hands of a higher power (God or the government). Thus, they did not need to trouble themselves with the problem of dealing with the uncertainty.

Another important tendency, as evident today as it was when White observed it 60 years ago, is crisis orientation: "National catastrophes have led to insistent demands for national action, and the timing of the legislative process has been set by the tempo of destructive floods" (White 1945, p 24). Burton and Kates (1964) commented that, despite the self-image of the conservation movement as a conscious and rational attempt at long-range planning, most of the major policy changes have arisen out of crises generated by catastrophic natural hazards. After interviewing floodplain residents,

Kates (1962) concluded that it is only in areas where elaborate adjustments have *evolved* by repeated experiences that experience has been a teacher rather than a prison. He added: "Floods need to be experienced, not only in magnitude, but in frequency as well. Without repeated experiences, the process whereby managers evolve emergency measures of coping with floods does not take place" (Kates 1962, p 140).

Psychological studies of probabilistic information processing by Amos Tversky and Nobel laureate Daniel Kahneman provided further evidence for bounded rationality. They found that people do not follow the principles of probability theory in judging the likelihood of uncertain events. Instead, people replace the laws of chance by intuitive heuristics, which sometimes produces good estimates, but all too often yield large and systematic biases. For example, Tversky and Kahneman (1973) proposed that people estimate probability and frequency by a number of heuristics, or mental strategies, which allow them to reduce these difficult tasks to simpler judgments. One such heuristic is that of availability, according to which one judges the probability of an event (e.g. snow in November) by the ease with which relevant instances are imagined or by the number of such instances that are readily retrieved from memory. Our everyday experience has taught us that instances of frequent events are easier to recall than instances of less frequent events, and that likely occurrences are easier to imagine than unlikely ones; thus mental availability will often be a valid cue for the assessment of frequency and probability. However, availability is also affected by recency, emotional saliency, and other subtle factors, which may be unrelated to actual frequency. If the availability heuristic is applied, then factors that increase the availability of instances should correspondingly increase the perceived frequency and subjective probability of the events under consideration. Thus, use of the availability heuristic results in predictable systematic biases in judgment.

The notion of availability is potentially one of the most important ideas for helping us understand the distortions likely to occur in our perceptions of natural hazards. For example, Kates (1962, p 140) writes:

> A major limitation to human ability to use improved flood hazard information is basic reliance on experience. Men on flood plains appear to be very much prisoners of their experience... Recently experienced floods appear to set an upward bound to the size of the loss with which managers believe they ought to be concerned.

Kates further attributes much of the difficulty in achieving better flood control to the "inability of individuals to conceptualize floods that have never occurred" (p 92). He observes that, in making forecasts of future flood potential, individuals "are strongly conditioned by their immediate past and limit their extrapolation to simplified constructs, seeing the future as a mirror of that past" (p 88). In this regard, it is interesting to observe how the purchase of earthquake insurance increases sharply after a quake, but decreases steadily thereafter, as the memories become less vivid (Steinbrugge et al. 1969). Similarly, Kunreuther et al. (1985) found that people will purchase flood insurance in the aftermath of a disaster but then cancel it several years later if no floods have occurred.

The availability hypothesis implies that any factor that makes a hazard highly memorable or imaginable – such as a recent disaster or a vivid film or lecture – could considerably increase the perceived risk of that hazard.

22.2
Stage 2: Psychometric Studies of Risk Perception

Late in the 1970s, researchers began to study perceived risk by developing taxonomies for hazards that could be used to understand and predict responses to their risks. A taxonomic scheme might explain, for example, people's extreme aversion to some hazards, their indifference to others, and the discrepancies between these reactions and experts' opinions.

The most common approach to this goal has employed the psychometric paradigm (Fischhoff et al. 1978; Slovic et al. 1984), which uses psychophysical scaling and multivariate analysis techniques to produce quantitative representations of risk attitudes and perceptions. People's quantitative judgments about the perceived and desired riskiness of diverse hazards and the desired level of regulation of each are related to their judgments about other properties, such as (i) the hazard's status on characteristics that have been hypothesized to account for risk perceptions and attitudes (e.g. voluntariness, dread, knowledge, controllability), (ii) the benefits that each hazard provides to society, (iii) the number of deaths caused by the hazard in an average year, (iv) the number of deaths caused by the hazard in a disastrous year, and (v) the seriousness of each death relative to a death due to other causes.

Numerous studies carried out within the psychometric paradigm have shown that perceived risk is quantifiable and predictable. Psychometric techniques seem well suited for identifying similarities and differences among groups with regard to risk perceptions and attitudes. When experts judge risks, their responses correlate highly with technical estimates of annual fatalities. Lay people can assess annual fatalities if they are asked to (and produce estimates somewhat like the technical estimates). However, their judgments of risk are related more to other hazard characteristics (e.g. catastrophic potential, fatal outcomes, lack of control) and, as a result, tend to differ from their own (and experts') estimates of annual fatalities.

Psychometric studies show that each hazard has a unique pattern of qualities that appears to be related to its perceived risk. Figure 22.1, for example, shows the profile across nine characteristic qualities of risk for the public's perception of the risk posed by nuclear power and medical X-rays (Fischhoff et al. 1978). Nuclear power was judged to have much higher risk than medical X-rays and to be in need of much greater re-

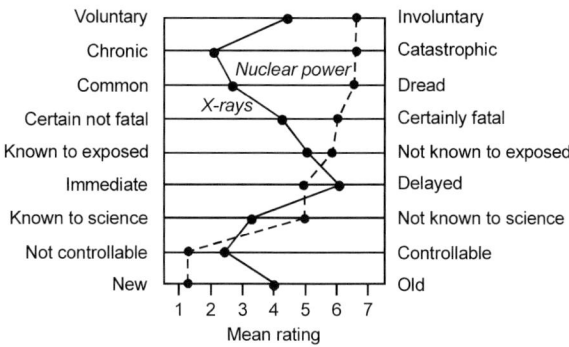

Fig. 22.1.
Qualitative characteristics of perceived risk for nuclear power and X-rays across nine risk characteristics. Source: Fischhoff et al. (1978)

duction in risk before becoming "safe enough." As the figure illustrates, nuclear power also had a much more negative profile across the nine risk characteristics.

Many of the qualitative risk characteristics that make up a hazard's profile tend to be highly correlated with each other, across a wide range of hazards (e.g. hazards rated as "voluntary" tend also to be rated as "controllable" and "well-known"; hazards that appear to threaten future generations tend also to be seen as having catastrophic potential). Factor analysis can be used to reduce the identified set of risk characteristics to a smaller set of higher-order factors. Each factor is made up of a set of correlated characteristics.

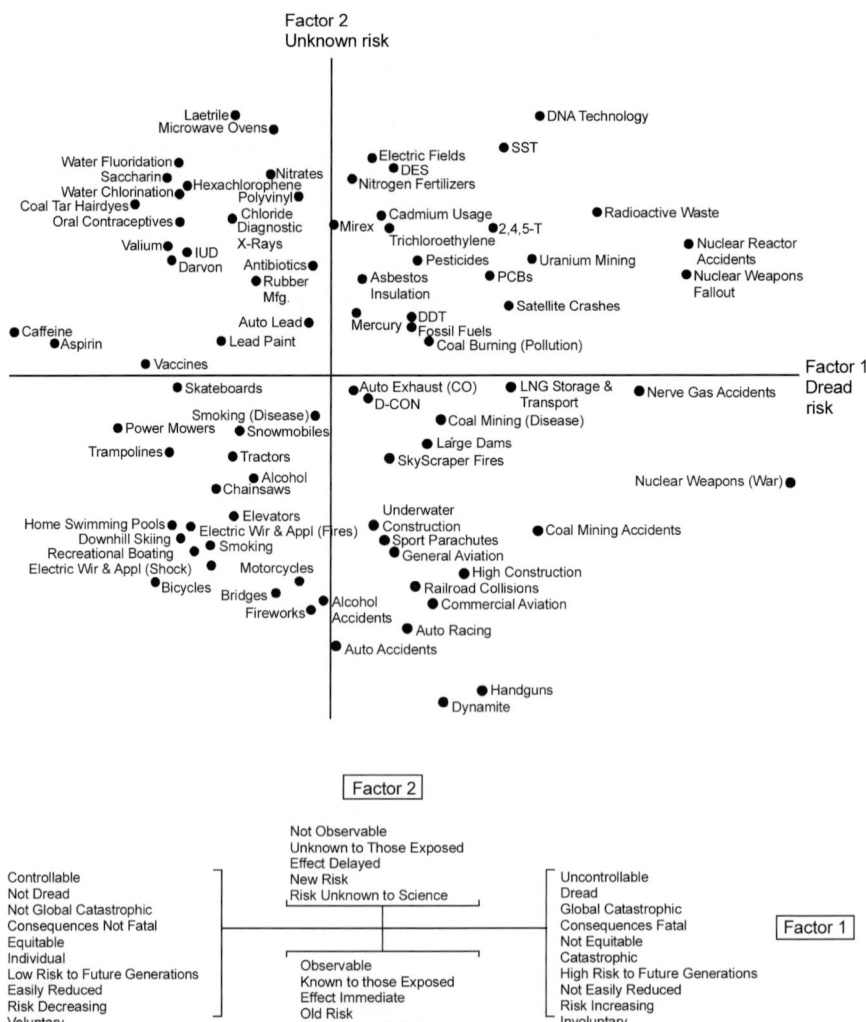

Fig. 22.2. Location of 81 hazards on Factors 1 and 2 derived from the interrelationships among 15 risk characteristics. Each factor is made up of a combination of characteristics, as indicated by the lower diagram. Source: Slovic (1987)

The factor space presented in Fig. 22.2 has been replicated across groups of lay people and experts judging large and diverse sets of hazards. Factor 1, labeled "dread risk," is defined at its high (right hand) end as perceived lack of control, dread, catastrophic potential, fatal consequences, and the inequitable distribution of risks and benefits. Nuclear weapons and nuclear power score highest on the characteristics that make up this factor. Factor 2, labeled "unknown risk," is defined at its high end by hazards judged to be unobservable, unknown, new, and delayed in their manifestation of harm. Chemical and DNA technologies score particularly high on this factor. Given the factor space in Fig. 22.2, the perceived risk of the terrorist attacks of September 11 and the subsequent anthrax attacks in the U.S. would almost certainly place them into the extreme upper-right quadrant. Later in this paper, I discuss where the asteroid impact hazard might fall in the risk perception factor space.

Laypeople's risk perceptions and attitudes are closely related to the position of a hazard within the factor space. Most important is the "Dread" Factor. The higher a hazard's score on this factor (i.e. the further to the right it appears in the space), the higher is its perceived risk, the more people want to see its current risks reduced, and the more they want to see strict regulation employed to achieve the desired reduction in risk. In contrast, *experts*' perceptions of risk are not much related to these factors, but, instead, are closely related to expected annual mortality (Slovic et al. 1979). Many conflicts between experts and laypeople are the result of these differences in the characteristics that are seen as important in defining risk.

22.3
Perceptions have Impacts: the Social Amplification of Risk

Perceptions of risk and the location of hazard events within the factor space of Fig. 22.2 play a key role in a process labeled the *social amplification of risk* (Kasperson et al. 1988). Social amplification is triggered by the occurrence of an adverse event (e.g. an accident, the outbreak of a disease, or an incident of sabotage) that falls into the risk-unknown or risk-dreaded category and has potential consequences for a wide range of people. Risk amplification is analogous to dropping a stone in a pond as shown in Fig. 22.3. The ripples spread outward, encompassing first the direct victims, but then

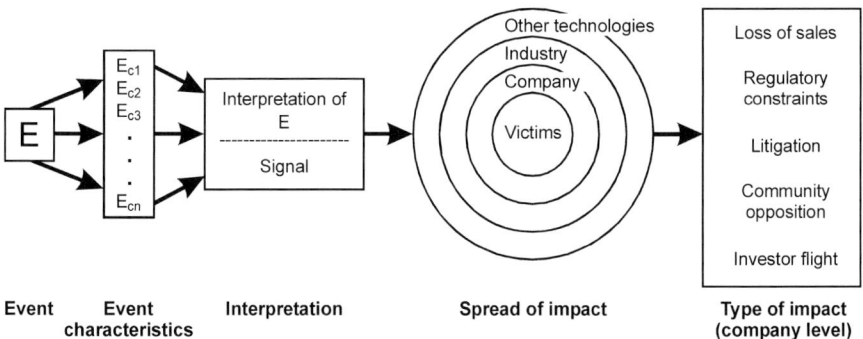

Fig. 22.3. A model of the social amplification of impact for unfortunate events. Source: Slovic (1987)

reach the responsible company or agency, and, in the extreme, other companies, agencies, or industries.

Multiple mechanisms contribute to the social amplification of risk. One such mechanism arises out of the interpretation of adverse events as clues or signals regarding the magnitude of the risk and the adequacy of the risk-management process and is thus related to the "Unknown Risk" factor (Burns et al. 1990; Slovic 1987). The signal potential of a mishap, and thus its potential social impact, appears to be systematically related to the risk profile of the hazard. An accident that takes many lives may produce relatively little social disturbance (beyond that caused to the victims' families and friends) if it occurs as part of a familiar and well-understood system (e.g. a train wreck). However, a small incident in an unfamiliar system (or one perceived as poorly understood), such as a nuclear waste repository or a recombinant DNA laboratory, may have immense social consequences if it is perceived as a harbinger of future and possibly catastrophic mishaps.

The concept of accidents or incidents as signals helps explain the strong response to terrorism. Because the risks associated with terrorism are seen as poorly understood and catastrophic, terrorist incidents anywhere in the world may be seen as omens of future disaster everywhere, thus producing responses that have immense psychological, socioeconomic, and political impacts.

One implication of the signal concept is that effort and expense beyond that indicated by a cost-benefit analysis might be warranted to reduce the possibility of "high-signal events." Adverse events involving hazards in the upper right quadrant of Fig. 22.2 appear particularly likely to have the potential to produce large ripples. As a result, risk analyses involving these hazards need to be made sensitive to the possibility of higher order impacts. Doing so would likely bring greater protection to potential direct victims as well as indirect victims such as companies and industries.

22.4
Stage 3: Risk as Feelings

Most recently, risk researchers have begun to examine the role that emotions and feelings play in risk perception and response to threats. This work is closely linked to modern theories in psychology which indicate that there are two fundamentally different ways in which human beings process information about the world when they make judgments or arrive at decisions (Chaiken and Trope 1999; Epstein 1994; Sloman 1996; Slovic et al. 2002). One processing system is evolutionarily older, fast, mostly automatic, and hence not very accessible to conscious awareness and control. It works by way of similarity and associations, including emotions, often serving as an "early-warning" system. The other processing system works by algorithms and rules, including those specified by normative models of judgment and decision making (e.g. the probability calculus, Bayesian updating, formal logic), but is slower, effortful, and requires awareness and conscious control. For the rule-based system to operate, we need to have learned the rule explicitly. The association/similarity-based processing system requires real world knowledge (i.e. experienced decision makers make better decisions using it than novices), but its basic mechanisms seem to be innate. These two processing systems often work in parallel and, when they do, more often than not result in identical judgments and decisions. We become aware of their simultaneous presence and operation in those situations where they produce different output. Thus, the question of whether a whale is a fish produces an affirmative answer from the similarity-based processing system ("a whale sure looks like a big fish"), but a negative response from the rule-based system ("it can't be a fish because it is warm blooded").

Experience or association-based processing in the context of risk, because of its automaticity and speed, has enabled us to survive during the long period of human evolution and remains the most natural and most common way to respond to threat, even in the modern world (Slovic et al. 2002). Recall also Kates' (1962) observation that people on floodplains were "prisoners of their experience." Experiential thinking is intuitive, automatic and fast. It relies on images and associations, linked by experience to emotions and affect (feelings that something is good or bad). This system transforms uncertain and threatening aspects of the environment into affective responses (e.g. fear, dread, anxiety) and thus represents *risk* as a *feeling*, which tells us whether it's safe to walk down a dark street or drink strange-smelling water (Loewenstein et al. 2001). The psychological risk factors such as dread, catastrophic potential, uncontrollability, risk to future generations, etc. (see Fig. 22.2) clearly are affective in nature and likely have their impact on perceived risk as the result of association-based processing. Loewenstein et al. (2001) document that risk perceptions are influenced by association- and affect-driven processes as much or more than by rule- and reason-based processes. They show that in those cases where the outputs from the two processing systems disagree, the affective, association-based system usually prevails.

Proponents of formal risk analysis tend to view affective responses to risk as irrational. Current wisdom suggests that nothing could be further from the truth. The rational and the experiential system not only operate in parallel, but the former seems to depend on the latter for crucial input and guidance. Sophisticated studies by neuroscientists have demonstrated that logical argument and analytic reasoning cannot be effective unless it is guided by emotion and affect (see Damasio 1994). Rational decision making requires proper integration of both modes of thought. Both systems have their own set of advantages, as well as biases and limitations. The challenge before us is to design risk assessment methods and procedures that capitalize on the advantages and minimize the limitations while integrating the outputs of the two systems.

The relationship and interplay between the two processing modes is further complicated by the fact that it seems to be contingent on the way people receive information about the magnitude and likelihood of possible events (Hertwig et al. 2004; Weber et al. 2004). Experimental studies of human reaction to extreme and usually rare events reveal two robust but apparently inconsistent behavioral tendencies. When decision makers are asked to choose between risky options based on a description of possible outcomes and their probabilities (provided either numerically [a 0.01 chance of losing $1000, otherwise nothing] or in the form of a graph or pie chart), rare events tend to be overweighted as predicted by prospect theory (Kahneman and Tversky 1979). This happens at least in part because the affective, association-based processing of described extreme and aversive events (such as losing $1000) dominates the analytic processing that would and should discount the affective reaction in proportion to the (low) likelihood of the extreme event's occurrence (Loewenstein et al. 2001; Rottenstreich and Hsee 2001). When people, on the other hand, learn about outcomes and their likelihood in a purely experiential way (by making repeated choices, starting out under complete ignorance and basing subsequent decisions on previously obtained outcomes), they tend to underweight rare events (Erev 1998; Hertwig et al. 2004; Weber et al. 2004). This happens in part because, with small samples, rare events often are not experienced in proportion to their theoretical likelihood. (In those instances where a rare and extreme event *is* experienced in a small sample, one expects decision makers to overweight it.) It also happens because experiential learning places greater weight on recent rather than more distant events, and rare events have a small chance of occurring in the recent past (Hertwig et al. 2004). There are obvious implications here for the difficulty in getting people to take seriously hazards such as asteroid impact, for which they have no experiential referent.

An important consequence of reliance on affect and experience to create "risk as feelings" is that the way that risk information is communicated may have a large effect on how people respond to that information. For example, Slovic and colleagues (2000) demonstrated that probabilistic risk framed as a relative frequency (the bad event will occur with a chance of 1 in 10) seems much more risky than that same event described as having a 10% or 0.10 chance of occurring. This appears to result from the frequency format triggering affect-laden images of the event, which are not evoked by the probability numbers.

Increasing the time horizon of an event is another way to get people to pay attention to probabilities. Slovic and others (1978) showed this in their study of attitudes towards wearing a seatbelt while driving. Individuals were much more likely to consider wearing a belt when the probability of a serious accident was cumulated over a lifetime of driving than when the probability was reported for a single trip.

Another consequence of "risk as feelings" is extreme insensitivity to probability when the adverse event carries sharp and strong affective meaning, as is the case with a lottery jackpot or a cancer. In such situations, variation in probability often carries too little weight. As Loewenstein et al. (2001) observe, one's images and feelings toward winning the lottery are likely to be similar whether the probability is one in 10 million or one in 10 000 (see also Kunreuther et al. 2001). Loewenstein et al. (2001) further note that responses to uncertain situations appear to have an all or none characteristic that is sensitive to the *possibility* rather than the *probability* of strong positive or negative consequences, causing very small probabilities to carry great weight. This, they argue, helps explain many paradoxical findings such as the simultaneous prevalence of gambling and the purchasing of insurance. It also explains why societal concerns about hazards such as nuclear power and exposure to extremely small amounts of toxic chemicals fail to recede in response to information about the very small probabilities of the feared consequences from such hazards. Support for these arguments comes from Rottenstreich and Hsee (2001), who show that, if the potential outcome of a gamble is emotionally powerful, its attractiveness or unattractiveness is relatively insensitive to changes in probability as great as from 0.99 to 0.01.

Sunstein (2003) argues that this insensitivity, which he labels "probability neglect," explains overreaction to certain rare but emotionally powerful events such as terrorist acts. As a result of probability neglect, people are often much more concerned about risks from terrorism than about risks from statistically greater risks that they confront in ordinary life. Chapman and Harris (2002) make a similar point.

Probability neglect has an important implication for risk perception and communication regarding asteroids. Suppose that credible observers identify an object large enough to cause catastrophic consequences *if* it impacts Earth, and suppose those consequences are described in an affectively powerful way. Assume further that the probability of impact is uncertain but thought to be very small (e.g. 10^{-4} or lower). The sense of danger (and possibly despair and social disruption) might well be as great as if the assessed probability was much greater (e.g. 10^{-2} or higher).

Another idiosyncrasy of affective thinking that is relevant to rational thinking about asteroid impact is that it does not do well with statistical descriptions of catastrophic outcomes. This could result in desensitization rather than the hypersensitivity associated with probability neglect. For example, the affective system seems designed to sensitize us to small changes in our environment (e.g. the difference between 0 and 1 deaths) at the cost of making us less able to appreciate and respond appropriately to large changes further away from zero (e.g. the difference between 500 deaths and 600 deaths). Fetherstonhaugh and colleagues (1997) referred to this insensitivity as "psychophysical numbing." Albert Szent-Gyorgi put it another way: "I am deeply moved if I see one man suffering and would risk my life for him. Then I talk impersonally about the possible pulverization of our big cities, with a hundred million dead. I am unable to multiply one man's suffering by a hundred million."

These two examples suggest that reaction to an identified probabilistic threat may depend greatly upon whether the potential consequences are described vividly (e.g. visual images, narratives), thus sparking affect, or numerically, thus dampening affect.

22.5
Public Perceptions of the Impact Hazard

Whereas a great deal of study has been devoted to defining and characterizing the impact hazard, little effort has thus far been spent to understand how laypeople perceive this threat. Public perceptions of impact risks are important for several reasons. First, our general understanding of human response to risk will undoubtedly benefit from studies of people's response to this hazard, which is unique in its combination of very low probability and very great consequence. Second, public attitudes and perceptions influence government policies toward risk management. Third, understanding how the impact hazard is perceived is essential for effective education and communication efforts.

22.5.1
Will the Public be Concerned about the Impact Hazard?

If a credible prediction of an imminent catastrophic impact were made, the public would undoubtedly be quite frightened, much as we have seen from recent, though not always specific, predictions of terrorist attacks. Absent such a specific prediction, how might we expect people to respond to the statistical threat of impact, as it is reported in the news media? Will their level of concern be great enough to induce them to support expenditure of public funds to detect threatening asteroids or comets? As noted above, we can find reasons from previous experience and research to predict both lack of concern and a high degree of concern. Reasons for expecting lack of concern and possible opposition to large expenditures are the following:

1. Natural hazards such as impacts tend to be less frightening than technological hazards (Erikson 1990). People perceive nature as benign and react rather apathetically to the threat from natural hazards (Burton et al. 1978). Personal experience of a natural disaster is usually necessary to motivate action to reduce future risks.
2. Probabilities are typically more important than consequences in triggering protective actions (Kunreuther et al. 1978; Slovic et al. 1977); hence the impact probabilities may be too low and the risk apparently too remote in time to trigger concern, in spite of their high consequences.
3. As discussed earlier, people are often insensitive to very large losses of life. We will expend great effort to save an individual life, but in a context of impersonal numbers or statistics, the lives of individuals lose meaning.
4. People tend to prefer 100% insurance against a threat (Kahneman and Tversky 1979; Slovic et al. 1977). If impact defense systems cannot provide 100% protection, they may be undervalued.

On the other hand, there are also reasons to expect the public to be concerned enough about impact hazards to support action:

1. The risk is demonstrable (sizable asteroids have hit the Earth) and is endorsed by credible scientists.
2. The potential consequences of large impacts are uniquely catastrophic and are qualitatively different from other natural hazards.
3. The probabilities of catastrophic impacts, while small, are not trivial. Considerable public funds are already being spent to deal with risks of even lower probability, such as death or injury from tornadoes or terrorist attacks. Careful analysis of costs and benefits can be shown to justify many actions that reduce asteroid risks (Posner 2004).
4. Unless action is taken to identify potential Earth-crossing asteroids, the risk is unknown and uncontrollable. Lack of control, dread, and catastrophic potential are all qualities associated with high risk perception and strong desire for action to reduce risk (Slovic 1987), even when the probabilities are miniscule (Rottenstreich and Hsee 2001; Sunstein 2003).

In addition, we can expect increased public awareness of the hazard as the media report discoveries of new asteroids and comets and more frequent "near misses." The collision of comet Shoemaker-Levy 9 with Jupiter in July of 1994 drew great public attention to the general impact issue (Chapman 1994). This awareness may lead to diverse reactions, ranging from incredulity that anyone could be concerned about such unlikely events to calls for public action to avert or reduce the risks.

22.5.2
Exploratory Research on Public Attitudes and Perceptions

Just as astronomers need observational data to determine the probability of asteroidal impact, social scientists need data to improve their understanding of risk perception and to forecast public attitudes toward impact detection and defense policies. Fortunately, data on public perceptions are relatively easy to acquire, by means of survey techniques. We now summarize the results from a two-part survey of attitudes and perceptions of the impact hazard (Slovic and Peterson 1993), carried out with a sample of 200 college students shortly after a *Newsweek* cover story on the impact hazard (November 1992). The participants were students at the University of Oregon with a median age of 20. This sample of students has been shown in previous studies to respond rather similarly to broader demographic samples of American adults. Before answering, each respondent was asked to read a seven-page briefing consisting of media articles on the impact threat. Despite the extensive recent media coverage of this topic, only about 25% of the respondents said they had heard about this hazard prior to participating in this study.

Part A of the survey asked people to rate 24 hazards on each of the 11 scales. The hazards included cigarette smoking, motor vehicle accidents, AIDS, floods, earthquakes, nuclear power plant accidents and an asteroid hitting the Earth. The scales included perception of risk to the American public, immediacy of risk, severity of consequences, ability of scientists to control the risk, threat to future generations and potential for global catastrophe. Part B asked respondents to agree or disagree with a wide range of statements about the impact hazard dealing with such items as perceived risk, immediacy of the threat, support for establishing a tracking network and attitudes toward development of a defense system.

The results of this survey showed that the impact risk ranked 14th out of 24 with regard to mean rating of risk to the American public. The impact risk was judged higher than the risks from prescription drugs, medical X-rays, bacteria in food, floods and air travel, but lower than risks from earthquakes and hurricanes. Impact risks were rated as extreme with regard to being unknown to scientists and the public, distant in time (non-immediate), uncontrollable, and catastrophic. There was modest support for detection efforts but considerable opposition to the use of weapons in space, even to deflect a threatening asteroid. The survey respondents indicated a strong preference for collecting more data on the risk before developing a defense system. Support for asteroid tracking and defense systems was greatest among those who tended to trust both the scientific community and the government, and was lowest among those concerned about militarization of space and those who felt that the next major impact is likely to occur very far in the future.

It is of interest to look at the attitudes of the respondents toward the immediacy of the impact threat. Only 6% believed that a catastrophic impact would occur in the next 50 years; 35% believed that one would occur more than 1000 years from now, and 34% denied that a threatening asteroid "could appear within the next 20 years." When asked to interpret the statement that "scientists say that a civilization-threatening asteroid impact can be expected every 300 000 to 1 000 000 years," 56% felt that "we don't really have to worry about this threat in our own lifetimes," 38% agreed with the assertion that "no such asteroid will appear for thousands of years," and 33% agreed with the assertion that "this statement is not believable because no one can predict the future for hundreds of thousands of years."

While these results must be interpreted with caution because of the small and non-representative sample used, they appear to demonstrate at least two important results. First is the high degree of credibility afforded the scientists who were expressing concern about this threat. Apparently the media had treated their activities in a positive light and had not interpreted their public statements as particularly ill-founded or self-serving. Second is the general problem of appreciating the possible imminence of low-probability events. Considering the number of scientists who have interpreted long average spacing between impacts to mean that this is a problem for future (rather than present) generations, it is not surprising that a sample of laypersons, most of whom were being exposed to this discussion for the first time, should be similarly disposed.

22.6
Where Next?

Throughout the past century, in a world beset by all manner of hazards and catastrophes, destructive natural hazards have elicited far less concern than risks created by humans, such as nuclear power, chemicals, biotechnology, war and terrorism. Even the serious threat posed by global climate changes has received scant attention. After a major natural disaster, concern rises but eventually returns to its prior apathetic state. The psychological processes described in this paper indicate why it will be hard to generate concern about asteroids unless there is an identifiable, certain, imminent, dreadful threat.

Understanding perception of risk from asteroid impact can serve two objectives. One is to provide guidance to risk-communication and crisis-response efforts prior to or after a serious impact. Realistically, I doubt that any meaningful sustained progress can be made towards this objective in the absence of a credible, imminent threat. Even the threat of terrorism, which has led, in many ways, to exaggerated protective actions in the United States, has failed to stimulate any meaningful program on risk communication – witness the inane color-coding scheme created to represent threat levels.

A more achievable, yet still challenging, objective for research and education programs is to create a realistic appreciation for the risk of asteroid impact so that decisions are made by evaluating the risks by analysis and not just by feelings. Posner (2004), for example, uses cost/benefit analysis to cut through psychological barriers to action against rare catastrophes. He concludes that the probable costs of catastrophic risks such as those posed by asteroids, when compared with the probable costs of efforts to minimize such risks, indicate that greater investment in asteroid detection and impact prevention would be justified.

Bringing the right mix of analytic and experiential thinking to bear upon decisions about asteroid risks will require a collaborative research effort between astronomers and social scientists. I believe this volume has taken an important first step toward creating the foundation for this effort.

Acknowledgment

This material is based upon work supported by the National Science Foundation under Grant No. SES-0241313. Any opinions, findings, and conclusions or recommendations expressed in this material are those of the author and do not necessarily reflect the views of the National Science Foundation.

References

Burns W, Slovic P, Kasperson R, Kasperson J, Renn O, Emani S (1990) Social amplification of risk: an empirical study. Carson City, NV: Nevada Agency for Nuclear Projects Nuclear Waste Project Office

Burton I, Kates RW. (1964) The perception of natural hazards in resource management. Natural Resources Journal 3:412–414

Burton I, Kates RW, White GF (1978) The environment as hazard. Oxford University Press, New York

Chaiken S, Trope Y (1999) Dual-process theories in social psychology. Guilford, New York

Chapman CR (1994) The scars of Comet Shoemaker-Levy 9. Geotimes 39(12):18–19

Chapman CR, Harris AW (2002). A skeptical look at September 11th: how we can defeat terrorism by reacting to it more rationally. Skeptical Inquirer. Retrieved September 17, 2004 from http://www.findarticles.com/p/articles/mi_m2843/is_5_26/ai_91236225

Damasio AR (1994) Descartes' error: emotion, reason, and the human brain. Avon, New York

Epstein S (1994) Integration of the cognitive and the psychodynamic unconscious. American Psychologist 49:709–724

Erev I (1998) Signal detection by human observers: a cutoff reinforcement learning model of categorization decisions under uncertainty. Psychological Review 105:280–298

Erikson K (1990) Toxic reckoning: business faces a new kind of fear. Harvard Business Review 68(1):118–126

Fetherstonhaugh D, Slovic P, Johnson SM, Friedrich J (1997) Insensitivity to the value of human life: a study of psychophysical numbing. Journal of Risk and Uncertainty 14:283–300

Fischhoff B, Slovic P, Lichtenstein S, Read S, Combs B (1978) How safe is safe enough? A psychometric study of attitudes toward technological risks and benefits. Policy Sciences 9:127–152

Hertwig R, Barron G, Weber EU, Erev I (2004) Decisions from experience and the effect of rare events in risky choices. Psychological Science 15:534–539

Kahneman D, Tversky A (1972) Subjective probability: a judgment of representativeness. Cognitive Psychology 3:430–454

Kahneman D, Tversky A (1979) Prospect theory: an analysis of decision under risk. Econometrica 47:263–291

Kasperson RE, Renn O, Slovic P, Brown HS, Emel J, Goble R, Kasperson JX, Ratick S (1988) The social amplification of risk: a conceptual framework. Risk Analysis 8:177–187

Kates RW (1962) Hazard and choice perception in flood plain management. University of Chicago Department of Geography, Chicago

Kunreuther HC, Ginsberg R, Miller L, Sagi P, Slovic P, Borkin B, Katz N (1978) Disaster insurance protection: public policy lessons. Wiley, New York

Kunreuther HC, Sanderson W, Vetschera R (1985) A behavioral model of the adoption of protective activities. Journal of Economic Behavior and Organization 6(1):1–15

Kunreuther H, Novemsky N, Kahneman D (2001) Making low probabilities useful. Journal of Risk and Uncertainty 23(2):103–120

Loewenstein GF, Weber EU, Hsee CK, Welch ES (2001) Risk as feelings. Psychological Bulletin 127(2):267–286

Posner RA (2004) Catastrophe: risk and response. Oxford University Press, Oxford

Rottenstreich Y, Hsee CK (2001) Money, kisses, and electric shocks: on the affective psychology of risk. Psychological Science 12(3):185–190

Simon HA (1959) Theories of decision making in economics and behavioral science. American Economic Review 49:253–283

Sloman SA (1996) The empirical case for two systems of reasoning. Psychological Bulletin 119(1):3–22
Slovic P (1987) Perception of risk. Science, 236, 280–285
Slovic P, Peterson K (1993) Perceived risk of asteroid impact. Unpublished manuscript
Slovic P, Kunreuther H, White GF (1974) Decision processes, rationality and adjustment to natural hazards. In: White GF (ed) Oxford University Press, New York, pp 187–205
Slovic P, Fischhoff B, Lichtenstein S, Corrigan B, Combs B (1977) Preference for insuring against probable small losses: insurance implications. Journal of Risk and Insurance 44(2):237–258
Slovic P, Fischhoff B, Lichtenstein S (1979) Rating the risks. Environment 21(3):14–20, 36–39
Slovic P, Fischhoff B, Lichtenstein S (1984) Behavioral decision theory perspectives on risk and safety. Acta Psychologica 56:183–203
Slovic P, Monahan J, MacGregor DG (2000) Violence risk assessment and risk communication: the effects of using actual cases, providing instructions, and employing probability vs. frequency formats. Law and Human Behavior 24(3):271–296
Slovic P, Finucane ML, Peters E, MacGregor DG (2002) The affect heuristic. In: Gilovich T, Griffin D, Kahneman D (eds) Heuristics and biases: the psychology of intuitive judgment. Cambridge University Press, New York, pp 397–420
Steinbrugge KV, McClure FE, Snow AJ (1969) Studies in seismicity and earthquake damage statistics. U.S. Department of Commerce, Washington, DC
Sunstein CR (2003) Terrorism and probability neglect. The Journal of Risk and Uncertainty 26:121–136
Tversky A, Kahneman D (1971) Belief in the law of small numbers. Psychological Bulletin, 76, 105–110
Tversky A, Kahneman D (1973) Availability: a heuristic for judging frequency and probability. Cognitive Psychology, 5:207–232
Weber EU, Shafir S, Blais A-R (2004) Predicting risk sensitivity in humans and lower animals: risk as variance or coefficient of variation. Psychological Review 111:430–445
White GF (1945) Human adjustment to floods: a geographical approach to the flood problem in the United States. University of Chicago Department of Geography, Chicago
White GF (1964) Choice of adjustment to floods. University of Chicago Press, Chicago

Chapter 23
Hazard Risk Assessment of a Near Earth Object

Roy C. Sidle

23.1
Background

Estimation of the risk of any natural hazard is problematic when occurrences are very rare and predictions are based on sparse data. While some natural hazards are perceived as totally random phenomenon, in some cases improved monitoring techniques and models have heightened awareness and allowed for better disaster mitigation strategies (e.g. alerts, evacuations, long-term best management practices) to be implemented (e.g. Thouret et al. 1995; Wu and Sidle 1995; La Delfa et al. 2001). Volcanic eruptions are examples of hazards where improved techniques for monitoring dome growth, seismic conditions, air chemistry and even groundwater can help forecast the onset of a major eruption (e.g. Miller and Chouet 1994; Miyabuchi 1999; La Delfa et al. 2001). Now it is often the very infrequent hazards related to volcanic eruptions (e.g. pyroclastic flows, lahars, dome collapses) that inflict the most damage due to their lower predictability (Major et al. 2001; Reid et al. 2001; Sheridan et al. 2001). For most natural disasters, such 'secondary' hazards must be considered in hazard risk assessments and mitigation measures. Although it is known that the Earth has been impacted by asteroids in the past large enough to annihilate most life on the contemporary planet (Sleep et al. 1989; Pope et al. 1994; Tate 2000; Paine, 2001; Chapman 2004), many of these isolated occurrences remain undiscovered.

Recent heightened awareness of a possible collision of a Near Earth Object (NEO; i.e. asteroid/comet) with Earth was generated by the impacts of Comet Shoemaker-Levy 9 on Jupiter in 1994 and progressive acceptance of the mass extinctions of about half the living animals and plants at the Cretaceous-Tertiary (K-T) boundary of geological and paleontological records associated with the impact of a 10–15 km NEO in Mexico about 65 million years ago (Alvarez et al. 1980; Hildebrand and Boynton 1990; Pope et al. 1994; Chapman 2004). The most recent significant impact of an NEO with Earth occurred in 1908 when an asteroid of about 60 m in diameter disintegrated in the atmosphere about 8 km over Tunguska, Siberia, flattening forest vegetation in a radius of more than 20 km (Morrison 1992). Although the probability of such a small NEO striking an urban or industrial setting is very remote, the consequences would be high.

Determining the risk of damage or near total destruction of the Earth by a Near Earth Object (NEO) is complex because of the lack of past evidence and the drastically different potential outcomes, some of which could be catastrophic. Much of the recent research has rightly focused on the identification (Tedeschi and Teller 1994; Milani et al. 2000; Tate 2000; Giorgini et al. 2002) and potential impact physics (Hills and Goda 1999; Poveda et al. 1999a; Ward and Asphaug 2000; Chelsey et al. 2002) of asteroids, with some recent emphasis on possible mitigation strategies for larger (predictable) impacts (Garshnek et al. 2000; Carusi et al. 2002; Chapman et al. 2001; Spitale 2002). Nevertheless, it is difficult to find papers that address the continuum from the potential hazard to the disaster(s) at all possible scales. In fact, most papers focus on either land impacts or ocean impacts alone and these involve drastically different mitigation and response strategies. Even recent textbooks on natural hazards and hazard risk assessment often do not mention asteroid hazards, attesting to the problems in addressing risks related to such rare events in an age when related information is accumulating at a rapid pace. At present, large uncertainties are associated with the probability and location of NEO impacts as well as the nature of the consequences, leading to speculations that range from denial to hysteria associated with asteroid hazard risk (Chapman 2004). Nevertheless, new advances in prediction of future NEO impacts hold substantial promise for long-term advanced warnings and development and implementation of disaster planning and mitigation measures (Garshnek et al. 2000; Chapman et al. 2001).

To lend more balance to the overall process of NEO hazard risk assessment, the problem needs to be considered from a wide perspective that includes the ontology of the disaster, the nature of conditional uncertainties of various causal factors and consequences, and the evolving prediction and mitigation methods. Full recognition of NEO hazard risk requires the assessment of related and indirect hazards. In this paper a general framework is proposed for such a hazard risk assessment.

23.2
Defining Risk

A natural hazard can be defined as a natural phenomenon that originates in the lithosphere, atmosphere, hydrosphere or biosphere that has the potential to exert either an extreme impact on humans and/or the natural environment or a more progressive, cumulative impact. Human activities can exacerbate natural hazards or lower the threshold for hazard occurrence. Specific risk (R_s) is the expected degree of loss due to a particular natural hazard, and is the product of the hazard probability (P_h) and vulnerability (V) for a particular 'element'. Thus, total risk (R_t) is defined as

$$R_t = \sum E \cdot R_s = \sum E(P_h \cdot V) \tag{1}$$

where E represents the elements at risk (e.g. humans, property, land productivity), P_h is the probability of the impact of the hazard in a defined area over a given time period, and V is the vulnerability expressed in terms of magnitude of losses. However, risk is also subject to stochastic factors, including changes in the exposure to the hazard over time and as incoming information becomes available, changes in the vulnerability of elements at risk, and changes in the probability that a hazard will strike and the way (or area) in which it may strike. All of these factors introduce uncertainties into any natural risk assessment. Additionally, the 'supply' of particularly smaller (< 100 m) NEOs is believed to change with time related to the disintegration of larger NEOs (Asher et al. 1994). Thus, the probability of a hazard (P_h) may need to be assessed in various ways. Also, vulnerability will not likely be constant with time because, at any point in time, total vulnerability is governed by actions that increase risk levels minus actions that mitigate them, but tempered by factors of perception. In the case of NEOs we must consider scenarios of centuries or millennia; obviously vulnerability will change in that period in response to changes in patterns or break up of NEOs, evolving detection and mitigation technologies, changing global demographics and socio-economic conditions, and education. If the timing of a potential NEO hazard can be predicted far in advance, then a partly deterministic hazard assessment is possible; for predicting the hazard related to a random collision, a purely stochastic approach is necessary.

In the context of NEO collisions with Earth, assessment of hazard risk involves at least three distinct steps. Firstly, identification of the suite of natural hazards that will likely result in a disaster – i.e. what hazardous events may occur? Secondly, it is necessary to estimate the probability of occurrence for each event and determine whether they are independent or inter-dependent. Finally, the social consequences of the derived risk need to be evaluated – i.e. what is the expected loss created by each event? These three components provide the basis for assessing what types of precautionary actions (if any) are most appropriate at any particular time and what are the best contingent plans for response in the event of an NEO collision.

Changes and updates of beliefs (based on accumulating information) related to an impending NEO collision can be evaluated using Bayesian decision analysis. Bayes' theorem provides a general method for updating or adjusting uncertainty in light of new evidence and is expressed as:

$$P(H_0|E) = \frac{P(E|H_0)P(H_0)}{P(E)} \tag{2}$$

where, $P(H_0)$ is the prior probability of the hypothesis (H_0) that was developed based on earlier observations, $P(E|H_0)$ is the conditional probability of seeing observation E given that H_0 is true, $P(E)$ is the 'marginal probability' (a normalizing function that achieves a probability density function), and $P(H_0|E)$ is the posterior probability which takes new evidence into consideration. Such an approach may be useful in dealing with the dynamic information environment related to NEO discoveries as well as new developments in mitigation strategies. Similar applications of decision analysis have recently been used for other hazard related issues, such as landslides, debris flows and earthquakes where quantitative information is sparse (Allison et al. 2004; Antonucci et al. 2004; Jiménez et al. 2004; Scott and Rogova 2004). Such approaches can rely on expert opinion in the absence of 'hard data'; these 'opinions' can then be updated as information or evidence accumulates. Although many scientists still do not accept the qualitative aspects of such analysis, it holds some promise for NEO collision scenarios where direct evidence is extremely sparse, related information is accumulating and systems may be dynamic.

23.3
Ontology of NEO Hazards

To simplify hazard risk assessment for NEOs, four impact levels based on size of the asteroid are proposed: Level 1 – extremely localized damage caused by very small asteroids or fragments (0.1–5 m); Level 2 – site-specific damage on Earth by a small (5–100 m diameter) asteroid (including the break up of a larger NEO); Level 3 – impact by a relatively large NEO (0.1–1 km) that may cause changes in climate/air quality, large-scale destruction, and/or other significant life-threatening environmental changes without obliterating life on Earth; and Level 4 – impact by a very large (> several kilometers in diameter) asteroid that causes major obliteration of human life on Earth and total destruction of civilization as we know it (Table 23.1). These four levels strongly differ not only in terms of probability of occurrence, impact energy, and damages, but also in the methods by which risk assessment can be best conducted and mitigation strategies can be planned and implemented. This classification focuses on impacts; thus, another approach would be to characterize impact *per se* irrespective of size (this may be better in assessing the remote chance of a comet collision).

Of these four scenarios, the smaller size classes (i.e. < 50 m) are currently not detectable until days before their arrival near Earth or not at all (Chapman et al. 2001; Chapman 2004). The numbers of asteroids that have some potential to impact Earth follow an exponential distribution related to size (cf. Poveda et al. 1999a). As a result, it is much more difficult to completely inventory the more numerous smaller asteroids. Thus, the probability that one of these would strike the Earth with little or no advanced warning is relatively high. Longer-term data on the larger sizes of asteroids are now becoming available at a relatively fast pace due to the initiation of NASA's Spaceguard Survey (Morrison 1992; Tate 2000; Chapman 2004). At the onset of this program in 1998, only about 7% of the Earth crossing asteroids > 1 km in diameter were documented (Hills and Goda 1999); as of February 2004 about 55% of these NEOs have been cataloged and all asteroids > 3 km in diameter appear to have been found (Chapman 2004). Nevertheless, records of such larger impacts on Earth are extremely sparse making probabilistic predictions difficult.

For smaller asteroids (< 50 m), there is a much higher incidence of fracture and break up in the atmosphere compared to larger asteroids; however, this break up process is poorly understood (Hills and Goda 1999). Thus, energy transmitted to Earth by larger asteroids is much less subject to dissipation due to break up prior to impact, even though it is generally felt that all NEOs will fracture in the atmosphere. In contrast, there appears to be an increasing probability of break up (and thus impact energy) in the spectrum from 50 m to fragments < 1 m in diameter. The degree to which asteroids (or fragments) in this smaller size range are attenuated is unclear and it is not evident how this affects risk analysis.

23.3.1
Level 1 NEOs

The smallest size class of NEO (10 cm to 5 m) listed in Table 23.1 is within a range that is usually ignored by scientists. Poveda et al. (1999b) calculated the annual probability

Table 23.1. Probable associated damages for various size classes of near earth object collisions with Earth

NEO impact level	Immediate expected natural disasters (global average, events yr^{-1})	Indirect effects and hazards	Overall impact area	Severity[a]
Level 1 Very small (fragments) (0.1–5 m)	Average globally estimated collisions with cars (0.06) and planes (0.0001) – 10 cm particle	Minor inconveniences	≤25 m^2	1. minor 2. very minor 3. negligible
Level 2 Small to moderate size (5–100 m)[b]	5 m NEO striking earth (4 yr^{-1}); ≈8 kT (TNT-equivalent energy): small-scale destruction (e.g., may destroy a small village); wildfire	Local dust and chemical emissions persisting for several days; local short-term pollution of water supplies; temporary displacement of people and loss of work	Several km radius	1. moderately severe 2. minor-moderate 3. minor
	100 m NEO (0.00223 yr^{-1}); ≈70 MT (TNT-equivalent energy): city-scale destruction, large-scale fires; 2–5 m tsunami possible; rock falls, ice avalanches, landslides, seismic shock	Regional release of dust particles/gasses that may persist for weeks; large-scale loss of forests/crops; long-term displacement of many people/job loss; infrastructure disruption	≈50 km radius; or moderate shorelines	1. very severe 2. moderately severe 3. moderate
Level 3 Large (100 m–1 km)	1 km NEO (0.0002 yr^{-1}); ≈250 000 MT: regional explosive impact; millions of people killed; hemisphere-wide gas/dust ejection; widespread wildfires; >50 m tsunami could occur; regional seismic shock; widespread landslides, ice falls, rock falls; threshold for global scale effects	Enhanced greenhouse effect after initial atmospheric clearing; widespread and sustained crop loss and polluted water supplies; disease outbreaks; conflict and societal breakdown; mass displacement of people; global economic chaos; river rerouting; topsoil loss; sea level rise	Continental to global scale	1. catastrophic 2. nearly catastrophic 3. very severe (hydrogen bomb scale of destruction without radioactivity)
Level 4 Very large (>several km)	(0.0000033); catastrophic global gas/dust ejection; large percentage of world's people killed or severely injured; huge (≥100 m) tsunami possible; total destruction of many coastlines; worldwide crop failures; global wildfires; sudden darkness and temperature drop	Global warming after several months; strongly polluted water bodies; sea level rise; global societal/institutional collapse or the end of civilization	Global	Destruction of much of civilization as we know it

[a] 1: urban; 2: sparse habitation; 3: rural. [b] Note that for these smaller size NEOs, many will lose considerable mass prior to impact, thus energy estimates are absolute maxima.

of very small (10 cm) meteorites to impact automobiles and aircraft on a global scale as about 0.06 and 0.0001, respectively, for an assumed equal distribution of type S (siliceous) and type C (carbonaceous) meteorites. Estimated numbers of meteorites in this study were based on an exponential distribution of the cumulative luminosity function versus absolute magnitude; this function was then transformed to a frequency distribution of diameters (Poveda et al. 1999a). Disintegration of these small NEOs was not accounted for in this study and this could be significant (Morrison 1992; Hills and Goda 1999). Obviously, such small meteorite impacts are trivial compared to most other types of natural hazards and represent only occasional inconveniences. The damages would be much less than those incurred during hail-storms. At the larger end of this category ($\approx$ 5 m), local damage and even isolated loss of life could occur via a direct hit on a village or urban area. A 5-m NEO would release about 8 kT of TNT-equivalent energy – about half the energy released during by the Hiroshima atomic bomb without the associated radioactivity. Such an impact could severely damage a small village or block within a city, however given the small chance of a direct strike on such developed areas, the risk is extremely low. The unique characteristics of Level 1 NEOs include their almost total unpredictability, high frequency, and the relatively small and localized damage they inflict.

23.3.2
Level 2 NEOs

Small to moderate-sized NEOs (5 to 100 m) release about 8 kT of energy at the lower end of this range to about 70 MT at the upper end (Morrison 1992; Chapman 2004). In addition to the direct impact shock wave, the primary hazard associated with the smaller end of this range would be fire (Garshnek et al. 2000). Additionally, localized, short-term air and water pollution would occur due to particles and chemicals released during the explosion. While it has been proposed that almost no direct impact damage would occur at ground level for stony asteroids smaller than 10–25 m (Hills and Goda 1999; Tate 2000), iron or stony-iron particles may reach the ground and form small craters (Morrison 1992; Tate 2000). Nevertheless, this size range of NEOs would produce much panic in the event of a land impact near populated areas and would set off local fires if fuel supplies exist on the ground (Tate 2000). As the size of the NEO increases, fires would become more severe and widespread, and other associated disasters would occur. If an asteroid in the larger end of this size category impacts the sea, tsunami waves of 2 to 5 m (depending on impact location and shoreline conditions) could occur (Ward and Asphaug 2000). However, the extent of tsunami is not fully understood and the hazard may be quite lower depending on near coastal conditions or much higher if the impact is within a confined water body. For example, a 525 m tsunami run-up occurred following a large earthquake-triggered rockfall in Lituya Bay, Alaska, in 1958; if an asteroid struck such a confined water body in a populated area, much local damage would be incurred. In mountainous impact areas, landslides, rockfalls, and ice avalanches could be triggered. These secondary hazards in mountainous terrain, such as breakage of ice or landslide-dammed lakes, could be much more catastrophic than the NEO impact itself. Other effects associated with a larger size (50–100 m), Level 2 NEO impact include local to regional short-term air and water pollution by dust particles and vaporized chemicals and large-scale loss of forests and crops. Localized acidification of soils would also occur. If this size of asteroid struck a major urban area, widespread property damage would occur and many lives would be lost (Tate 2000; Chapman 2004); however, such an impact would be very improbable. A recent analog for such an impact on uninhabited land is the 1908 Tunguska event in Siberia where a forested area of $\approx 1\,300$ km^2 was completely destroyed by the blast and collision of a ≈ 60 m asteroid with damage extending outwards an additional 3 800 km^2 (Morrison 1992). The estimated probability of a 100 m asteroid striking a city is 1.4×10^{-8}; the probability of striking some inhabited area is more than 4 orders of magnitude higher, but such an impact may kill as many as 3 million people (Garshnek et al. 2000). For this size of asteroid, a 1–2 km crater would be produced if the impact occurred on land and ejected material would be spread up to a 10 km radius. Given present capabilities it is not possible to detect most of these relatively smaller NEOs (Chapman 2004). The defining characteristics of Level 2 NEOs are their modest frequency of occurrence, relatively local but potentially acute damages, and unpredictability.

23.3.3
Level 3 NEOs

Relatively large (100 m to 1 km) asteroids release energies in the range of 70–250 000 MT – from about 5 000 times the energy released during the Hiroshima bomb to approximately the energy released by a hydrogen bomb explosion, without the radioactivity (Morrison 1992; Tate 2000). The areas devastated by such blasts and land impacts in the lower half of this size class would range from 7 200 to 70 000 km^2; asteroids near one km in size would devastate areas at regional scales far beyond the direct impact zone (Garshnek et al. 2000; Chapman et al. 2001). It has been speculated that stony and metallic asteroids in this size range may strike the Earth about once in 5 000 yr (Morrison 1992), although with limited records it is very difficult to state this in probabilistic terms. Furthermore since some of the asteroids in the upper end of this size class are being discovered (Chapman 2004), there may be advanced warning for some of the future impacts. Thus, some of the larger potential collisions can be predicted deterministically, whereas others will remain in the domain of 'random' events unless better detection methods are developed.

Both metallic and stony Level 3 asteroids that are on a collision course with Earth will generally penetrate the atmosphere and produce craters varying in size from 1.5 to 6 km (Morrison 1992; Tate 2000). For these relatively large impacts in rural land areas, the primary damage would be associated with the regional to hemisphere-scale wildfires and seismic shock, as well as widespread ejection of dust, water vapor, and chemicals into the atmosphere. For a one km asteroid, even a rural land impact would have devastating regional and possibly hemispheric consequences related to secondary effects such as global ozone loss resulting from the emission of greenhouse gases following atmospheric clearing, widespread crop loss and associated food shortages, sustained water pollution, and disease outbreaks in a stressed economic environment (Garshnek et al. 2000). In the higher probability scenario of an NEO impact at sea, tsunami ranging from tens of meters to 100 m could be produced depending on impact location and shoreline conditions; in the upper end of this NEO size category, tsunami could extend to hemispheric scales prompting unprecedented evacuations of coastal and low elevation inland areas, as well as mass hysteria (Hills and Goda 1999; Tate 2000; Ward and Asphaug 2000; Chapman et al. 2001). It should again be noted that the extent of tsunami related to an NEO impact is not fully understood and the hazard may actually be lower. Countries and regions most at risk from a tsunami catastrophe are those with long coastlines and low relief for long distances inland (e.g. Sumatra, Netherlands, Atlantic coastal plains of North and South America, most island nations); particularly susceptible developed cities include Hilo (Hawaii), San Francisco and Tokyo (Garshnek et al. 2000; Ward and Asphaug 2000). In the highly improbable but extremely unfortunate case of a one km collision in an urban area, tens of millions of people could be killed and the entire metropolitan area would be immediately destroyed with extensive damage incurred in the surrounding region (Garshnek et al. 2000). Additionally, depending on the economic stability and preparedness of the impacted area, similar or even greater levels of consequences may be experienced at regional or hemispheric scales via long-term food and potable water shortages due to crop loss, climate change, and water pollution. Level 3 NEOs are characterized by some evolving degree of predictability (related to increased detection of NEOs), major regional to hemisphere scale damages and secondary disasters (but not destruction of entire civilizations), catastrophic impacts in the greater area of the impact, probable occurrence in a time frame that can be imagined by most people, and numerous short to long-term environmental and socio-economic consequences. Due to the many uncertainties, high potential destruction, and the believable, albeit remote chance of such an NEO impact, Level 3 poses the greatest challenges for risk analysis. Detection of more and smaller NEOs in this size range will reduce these uncertainties (and thus the risk), as will better understanding of tsunami propagation from NEO impacts at sea.

23.3.4
Level 4 NEOs

Very large asteroids (> several km) have impacted Earth in the past, but never in the short history of human habitation. Such catastrophic impacts on Earth are believed to occur on average once in about 300 000 yr (Morrison 1992), although it is difficult to express such infrequent occurrence in terms of probability. The energy released by a 3 km asteroid striking land (1 million MT) would probably be capable of destroying civilization (Morrison 1992; Chapman 2004). This global catastrophic threshold would be reached primarily by the massive ejection of dust into the atmosphere that would depress temperatures for a least a growing season, leading to global scale crop failures and widespread starvation. Ballistic ejecta re-entering the atmosphere would ignite firestorms throughout areas $> 10^7$ km^2, which would further reduce incoming solar radiation (Garshnek et al. 2000; Chapman 2004). Nitrous oxide produced by the burning of atmospheric nitrogen would destroy much of the ozone layer and the resulting nitric acid produced would pollute soils, lakes, oceans and streams. Following the clearing of the atmosphere (months after impact), the release of large quantities of water vapor and carbon dioxide would strongly enhance global warming (Morrison 1992; Garshnek et al. 2000). Agriculture and forests would largely be destroyed worldwide, leaving few materials for the survivors, and mass extinctions of plant and animal species would occur. Geomorphic hazards would increase both as the direct result of the impact (e.g. earthquake shock, landslides, rockfalls, ice falls, jökulhlaups, coastal flooding), as well as long after the impact due to widespread devastation of vegetation cover, climate change and other indirect effects (e.g. massive soil erosion, landslides, glacial hazards, permafrost melting, localized flooding). Although any estimates of loss of life in such a global catastrophe are totally speculative, it is conceivable several billon people could die from the initial impact of the disaster together with the resulting secondary impacts and global socio-economic collapse (Chapman 2004).

In the case of an ocean impact, huge tsunami would occur globally; heights of several hundreds of meters are likely within impacted ocean basin shorelines (Hills and Goda 1999; Garshnek et al. 2000; Ward and Asphaug 2000; Tate 2000). Many inland areas would be inundated and destroyed, and massive erosion, coastline changes, river rerouting and island destruction would occur. The only survivors would be people living far inland or who have been safely evacuated to such higher elevation areas. Such an impact on ice caps could cause sea level rise and regional coastal flooding.

One aspect of a Level 4 NEO that strongly differs from the other NEO levels is that most of these large asteroids have already been discovered or are likely to be discovered in the next decade (Chapman 2004). Thus, deflection of the trajectory of such an Earth-bound asteroid would be possible if the probability of impact was sufficiently high. The social vulnerability of such an impact may differ from smaller impacts. For small NEO impacts it is generally the poor regions and nations that are most vulnerable; however, if the impact was in a rural region the inherent survival skills of farmers and villagers may benefit them in such a cataclysm. Unique attributes of Level 4 NEOs are their catastrophic potential to destroy civilization; dramatic impacts on land cover, geomorphic processes and coastlines (in the case of tsunami); and almost certain forewarning in a timeframe that may range from centuries to decades for asteroids and possibly only months for comets.

23.4
Dynamic Hazard Risk Assessment and Possible Mitigation and Preparedness Strategies

The four-level structure for NEO impacts was incorporated to provide a useful analog for risk assessment and identification of potential precautionary, disaster planning, and mitigation strategies. An existing metric for communicating the severity of potential NEO impacts is the 10-category Torino Scale, which examines the increasing probability and severity of Earth collisions based on impact energy and probability (Binzel 2000). The Torino Scale, somewhat analogous to the Richter Scale for earthquakes, was primarily designed as a tool for public communication and assessment for NEO impact predictions in the next century. The four-level scale presented herein is simpler (in some aspects) and designed as a general template for risk and hazard analysis professionals as well as government officials; it addresses both land and water impacts specifically, as well as other issues (e.g. related hazards) inherent in risk analysis. Updating hazard probability is fundamental in both systems, not only the probability of an NEO impact, but also the consequences of the impact. For example, as better knowledge is generated concerning tsunami propagation following an NEO impact at sea, this can be used to refine impact scenarios and warnings.

Dynamic risk assessment of NEO hazards can be conducted within each of the four impact levels. As mentioned earlier, risk related to NEOs is complicated due to the very large but improbable damages that can be incurred by objects > 100 m (Levels 3 and 4) and the evolving nature of the predictability of NEO hazards, especially Level 3 and the upper range of Level 2. A general conceptual model for risk analysis consists of determination of the NEO impact magnitude, the probability of occurrence (or a predictive estimate of occurrence), the estimated location of impact, the advanced warning time (for updated response), assessment of the probability of direct hazards associated with the NEO impact, and vulnerability analysis of both direct and indirect effects (Fig. 23.1). Very important issues for future risk assessments are whether it is possible to predict the approximate impact area of an NEO and within what timeframe prior to impact and with what degree of certainty.

Hazard risk assessment of NEO impacts should follow several sequential considerations or steps. Some of these considerations must be updated as new information arrives about the impending hazard. These considerations are important not only for outlining possible precautionary and mitigation measures, but also to identify the most critical data needs related to risk assessment.

- Consideration 1 – Establish the hazard level and determine the predictability of the hazard. This procedure should be instituted for each level of NEO impact – at present, the difficulties in detecting both Level 1 and 2 asteroids telescopically (Chapman 2004) would relegate these hazards to more or less random occurrences for the purpose of risk assessment. In the lower range of Level 3, uncertainty certainly exists

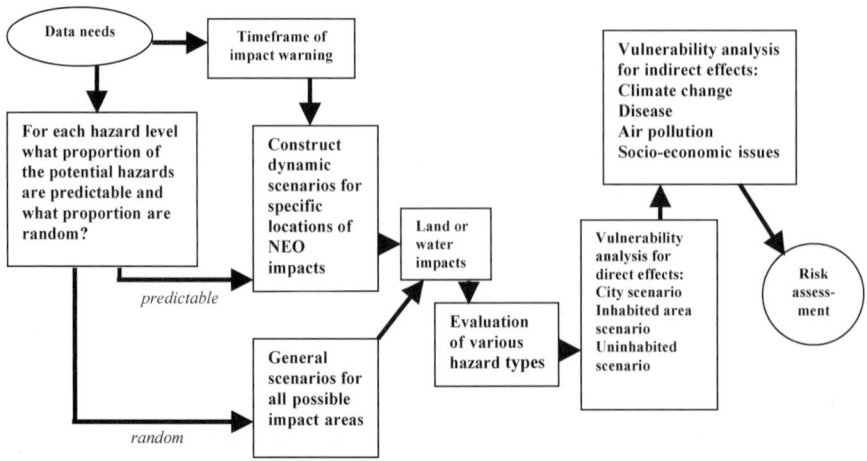

Fig. 23.1. Conceptual model of risk analysis of a NEO impact

and would have to be incorporated into risk assessment for that category – e.g. there may be a 0.2 probability that the approach of a 200 m asteroid will be undetected (at least within a few hours of impact – see Consideration 3). Thus, for 80% of the possible impacts of this size, long-range plans for evacuation and emergency response may suffice, but the 20% probability of a "random" strike must be considered in the risk assessment, and decisions must be made whether any type of contingent strategy is appropriate.

- Consideration 2 – Estimate the location of the impact. This cannot be done far in advance and it is currently uncertain what if any relationship exists between accuracy of predicted location and time to impact. Nevertheless, this knowledge is *critically important* for dynamic risk analysis. For long-term advance warnings of NEO impacts, currently we must assume a ≈ 0.3 probability of a land strike and a ≈ 0.7 probability of a water strike. For a nearer-term impact, one possible futuristic scenario might consist of: (*1*) Earth impact of a 0.4 km NEO is predicted at a probability of 0.1, 10-years prior to impact; (*2*) 1 year prior to impact the probability increases to 0.8 and emergency measures are now fully implemented (still no approximate impact location can be determined); (*3*) 6 months before impact the probability increases to 0.9 and it is estimated that there is a 0.8 chance that the impact will occur north of the equator in the Western Hemisphere; at this point estimates can be made of the probability of the asteroid striking land or water for this region; appropriate tsunami warnings can be made and evacuation *plans* for all susceptible coastal sites are completed; (*4*) 1 month prior to impact (now virtually certain) it is established with a probability of 0.8 that the collision will occur in the Atlantic Ocean between 300–500 km east of Nova Scotia; potentially impacted zones are specifically identified and evacuated in an orderly manner according to previous plans with adjustments made for the potential size of the tsunami.

- Consideration 3 – Not all "predictable" NEO impacts are equal – the advance time of forewarning is an important consideration. Forewarnings less than a day are probably not very useful in highly populated areas, due to the time to disseminate warnings and the obvious chaos inherent in short-notice evacuations. Consideration 2 outlines an optimistic scenario for a 'planned' evacuation based on newly updated information. Forewarnings of decades or even centuries need to be periodically updated for dynamic risk assessment. Long-term lead times are essential in the unlikely case of a catastrophic Level 4 asteroid so that deflection strategies can be considered. Likewise, dynamic risk assessment can rule out earlier impact predictions that later prove to be unwarranted.

- Consideration 4 – Consider the suite of direct hazards that will possibly occur and their respective probability for each level of collision. In the 'early' stages of risk assessment the advent of a water or land impact will be unknown, thus "water impact" hazards could be weighted by 0.7 and land hazards by 0.3, except for cases where one would induce the other. Hazards directly related to NEO impacts include: (*1*) the explosive impact; (*2*) wildfire; (*3*) seismic shock waves; (*4*) landslides and rockfall; (*5*) massive chemical and dust ejection into the atmosphere; (*6*) ice falls and jökulhlaups; and (*7*) tsunami and related coastal flooding. This information also must be updated as the impact zone becomes better defined. Special consideration must be made for particularly high risk sites (e.g. cities, low lying coastlines, nuclear power plants, dams, radioactive waste depositories), once these areas are targeted as potential impact sites.

- Consideration 5 – Conduct a vulnerability analysis for various impact levels and location scenarios; this involves consideration of *all of the direct hazards* that occur in selected core impact regions (probabilities of each hazard should be weighted in the assessment). One early stage approach would be to list scenarios of habitation levels in the designated core impact zones and multiply these by a factor that reflects the intensity of the impact to account for deaths (see hypothetical examples below):
  - City: affected population in core zone x (1.0 for level 4 NEO; 0.95 for level 3 NEO; 0.4 for level 2; 0.0001 for level 1).
  - Inhabited region: affected population in core region x (1.0 for level 4 NEO; 0.95 for level 3 NEO; 0.4 for level 2; 0.0001 for level 1).
  - Uninhabited region: no deaths via direct impact.

  Each of these values would be multiplied by a correction factor for 'safe evacuation': e.g. in a city of 200 000 people, maybe 20% of the people could be safely evacuated from an unexpected Level 3 impact; then the estimated deaths would be $200\,000 \cdot 0.95 \cdot 0.8 = 152\,000$ deaths. This very simple analysis lumps post-hazard response into the 'impact factor' – e.g. possibly 50 000 deaths occurred instantly, but the remaining 102 000 fatalities occurred because severely injured people could not be efficiently transported to treatment facilities and treatment facilities were damaged or destroyed.

  Vulnerability analysis also needs to consider the property and environmental values at risk. Property damage in direct impact area can hypothetically be categorized as before:
  - City – list all property replacement values in the core region x (1.0 for level 4; 0.95 for level 3; 0.25 for level 2; 0.0001 for level 1).
  - Inhabited region – affected property in the core region x (1.0 for level 4 NEO; 0.9 for level 3 NEO; 0.25 for level 2; 0.00001 for level 1).
  - Uninhabited region – affected property in the core region x (1.0 for level 4 NEO; 0.9 for level 3 NEO; 0.25 for level 2; 0 for level 1).

  Total property values in each affected region are then multiplied by the appropriate damage factor: e.g. for an inhabited area, property value in the core region of an unexpected level 2 impact may be 3 homes and related personal property (= US$ 600 000) × 0.25 (partial damage) = US$ 150 000 total property damage. Additional impacts will also occur from associated hazards outside the core impact zone and the following tangibles should be assessed separately for all impact scenarios of direct strikes in cities, inhabited areas and uninhabited areas: (*1*) possible loss of life; (*2*) property damage; and (*3*) environmental damage. In this case, environmental damage consists of loss of crops and forests, polluted water and soils, coastline destruction, river rerouting, enhanced soil erosion, and landslides.

  Finally, vulnerability analysis must address the indirect effects of NEO impacts. For a land impact, air pollution, including related climate change (especially for large Level 3 and Level 4 impacts) is a major consideration. At these high-impact levels, short-term cooling effects due to massive dust and gas ejections as well as ash from widespread fires would destroy crops in wide regions or globally, and subsequent global warming would result from huge releases of carbon dioxide and water vapor into the atmosphere. In the upper half of Level 3 impacts and certainly for Level 4 disasters, these short and long-term atmospheric perturbations and the resulting crop loss, water and air pollution, and famine would be the most significant effects, albeit indirect. Other indirect effects that are important include disease outbreaks, prolonged pollution of water supplies, loss of land productivity, prolonged high levels of erosion, social collapse, lost jobs, mass hysteria, infrastructure breakdown, fear of a nuclear attack by governments, economic upheaval and shifting demographics. In the more probable case of a Level 4 water impact, the indirect effects on climate change due to the huge discharge of water vapor and gases into the atmosphere have apparently not been studied in detail. For such a large impact on ice caps, sea level rise will occur, resulting in widespread coastal flooding.

23.5
Potential Mitigation, Data Needs, Response, and Prognosis

The present risk for Level 1 NEOs (0.1–5 m) is very small and would not warrant any direct mitigation measures or even disaster planning. Even if the probability of the hazard increased by an order of magnitude, it would not equal the risk posed by hail (smaller, but similarly impacting particles) worldwide (Smith 1996). Given the possibility of higher densities of small NEOs in future millennia (Asher et al. 1994), it is advisable to continue inventories and attempt to improve detection of smaller asteroids. Additionally, the nature of the disintegration process of asteroids related to material composition needs to be investigated in more detail. These studies would provide long-range forecasts of impending susceptibility to the currently random, smaller NEO hazards. Due to the small size of these NEOs, consequences would be restricted mainly to minor property damage. Damage related to water impacts would be negligible. Even in a millennium of significantly higher asteroid density; the likelihood of direct impacts with aircraft, boats, cars, and humans would still be minimal.

Level 2 NEO impacts are of considerable concern because they are currently poorly documented even in the upper range of this size category (Chapman et al. 2001). Thus, current risk assessment for this impact group would largely assume 'random' collisions based on past occurrence estimates (ranging from 4 impacts per year at the lower end of this level to 0.00223 impacts per year for 100 m asteroids). Such a random component in the risk analysis (as opposed to a predictable impact) greatly compromises preparedness, as governments are reluctant to invest financial resources into such 'highly improbable' ventures. A realistic long-range (century scale) goal would be to document a large proportion of the potential asteroid hazard in the upper half of this impact level (50–100 m); such asteroids are capable of massive local to regional destruction in the event of a land strike and possibly significant tsunami in the case of a water impact. Level 2 impacts currently represent the *greatest uncertainty* for potentially significant, but not catastrophic, NEO disasters. Hazard risk scenarios should be developed for present scenarios of information and updated as significant research findings accumulate.

The present risk status for Level 3 NEO impacts is the most complex because some of these NEOs have been or will be discovered in the next decade (Tate 2000; Chapman

2004), but large numbers of NEOs in this size class have gone undetected (particularly in the lower end of this size range), and there is uncertainty regarding the proportion of detected versus undetected potential impacting objects. Furthermore, while the incidence of Level 3 NEOs is rare (0.0002 per year for 1 km asteroids), the hazard probability is 'believable' within the perception of most humans. Thus, current hazard risk assessment for Level 3 impacts should include 'educated' assumptions about the proportion of 'documented' and 'random' potential collisions, possibly in the framework of Bayesian decision theory. These estimates would need to be updated frequently as information accumulates. For 'documented' occurrences of potential impacts, the largest benefits to human welfare will be derived by accurately establishing both the time (this now seems to be available) and probability (this needs to be updated on a regular basis, see example in "Consideration 2") of impact. In addition to improving probability of impact estimates, the weakest link in risk assessment is the unknown location of the impact. Even a crude approximation of impact area (e.g. Atlantic Ocean, Indian Ocean, China) can greatly benefit pre-disaster response, especially if the impacted nations are properly prepared. Response to a large tsunami triggered by an asteroid impact in the ocean is drastically different from the response required for an impact on land. Additionally, the extent of tsunami propagation following NEO impacts need to be more clearly articulated. With greater accuracies of spatial predictions and longer pre-disaster response times, optimum strategies for disaster mitigation can be implemented in the least chaotic atmosphere. Such decisions may include no action (in outlying areas where the risk is not considered significant), short-term retreat to underground shelter with adequate food and water supplies, temporary evacuation, and regional relocation of the population to higher elevation terrain.

Because of the catastrophic, long-term global consequences of a Level 4 impact, it is unimaginable that any disaster mitigation efforts could successfully deal with this magnitude of collision. Response efforts would obviously break down in most regions, especially in the continents directly impacted by a land or water impact. Given these catastrophic consequences of either land or water impacts, it would be imperative that all available international resources be pooled to mount a series of mitigation efforts (not all relying on the same strategy if possible) to deflect the asteroid from its approach to Earth. Because most all Level 4 asteroids have already been discovered, there should a century or more to develop plans for such an NEO deflection, thus resources could be mobilized over a long time frame in an atmosphere of improving technology. Some contingent evacuation strategies for remote inland (possibly underground) evacuation would be possible in the event that all deflection attempts failed (or in the very unlikely case of miscalculated impact), but these would undoubtedly lead to total chaos and social breakdown prior to the impact, as well as poor chances for survival. At present there appears to be only one possible Earth-impacting Level 4 NEO in the coming millennium: asteroid 1950 DA that has a 0.00327 probability of impacting Earth on 16 March 2880 (Giorgini et al. 2002). If in the coming decades this unlikely 2880 impact can be discounted and if there appears to be no potentially hazardous NEO in the next 10 000–15 000 yr, it would be advisable to devote resources strictly to other levels of NEO impacts for hazard risk analysis and not to 'deflection technology'. In time frames on the order of 10 000 to 20 000 years, other significant climate and environmental changes would pose much greater risks for civilization compared to very large NEO impacts.

Of the four categories of NEO hazards, Levels 2 and 3 pose the greatest challenges for risk assessment. For disaster planning purposes, it may be possible to combine these two hazard levels. Given the high potential damage related to Level 3 collisions, a more complete inventory of these hazards will substantially reduce their uncertainty and allow for temporally targeted disaster planning and mitigation efforts. More significant benefits for disaster response would be derived if approximate spatial impact zones could be identified, even with only a few days of advanced warning. The less destructive but approximately 100 times more frequent Level 2 NEO impacts must be analyzed as totally random occurrences; thus, preparedness and mitigation strategies currently need to be developed for general impact scenarios as outlined herein. Additionally, the break up potential of Level 2 (and smaller Level 3) asteroids of different compositions needs further investigation, given the exponentially decreasing impact energy with decreasing projectile size. The development of improved technologies for detecting 50–100 m asteroids in coming centuries would greatly reduce hazard risk by allowing more targeted response efforts for this most destructive size Level 2 NEOs. Pre-disaster planning and post-disaster response are critical for Level 2 and 3 hazards, including establishment of a clear line of communication for managing response protocol, determining evacuation plans (timing, routes, transport methods and destination), preparing temporary safe shelter areas (possibly underground), establishing alternative potable water supplies for impacted areas as well as for areas where people will be evacuated, ensuring food supplies for the crisis period, determining 'helpful' roles of the media, triage and emergency medical care for the injured, psychological issues, and restriction of access related to unsafe areas. For NEOs > 50 m, multiple disaster planning and mitigation strategies need to be considered and implemented. In the very unlikely chance of a Level 4 impact, several deflection strategies could be considered and implemented if necessary because of the cataclysmic consequences of either a land or water impact.

References

Allison C, Sidle RC, Tait D (2004) Application of decision analysis to forest road deactivation in unstable terrain. Environ Manage 33: 173–185

Alvarez LW, Alvarez W, Asaro F, Michel HV (1980) Extraterrestrial cause for the Cretaceous–Tertiary extinction. Science 208: 1095–1108

Antonucci A, Salvetti A, Zaffalon M (2004) Hazard assessment of debris flows by creedal networks. In: Pahl-Wostl C, Schmidt S, Jakeman T (eds) iEMSs 2004 International Congress: Complexity and Integrated Resources Management. International Environmental Modelling and Software Society, Osnabrueck, Germany

Asher DJ, Clube SVM, Napier WM, Steel DI (1994) Coherent catastrophism. Vistas in Astronomy 38(1): 1–27

Binzel RP (2000) The Torino Impact Hazard Scale. Planet Space Sci 48:297–303

Chapman CR (2004) The hazard of near-Earth asteroid impacts on Earth. Earth Planet Sci Letters 222: 1–15

Chapman CR, Durda DD, Gold RE (2001) The comet/asteroid impact hazard: a systems approach. Southwest Research Institute White Paper, Boulder, Colorado

Carusi A, Valsecchi GB, D'Abramo G, Boattini A (2002) Deflecting NEOs in route of collision with the Earth. Icarus 159:417–422

Chesley SR, Chodas PW, Milani A, Valsecchi GB, Yeomans DK (2002) Quantifying the risk posed by potential Earth impacts. Icarus 159:423–432

Garshnek V, Morrison D, Burkle FM Jr. (2000) The mitigation, management, and survivability of asteroid/comet impact with Earth. Space Policy 16:213–222

Giorgini JD, Ostro SJ, Benner AM, Chodas PW, Chelsey SR, Hudson RS, Nolan MC, Klemola AR, Standish EM, Jurgens RF, Rose R, Chamberlin AB, Yeomans DK, Margot J-L (2002) Asteroid 1950 DA's encounter with Earth in 2880: physical limits of collision probability prediction. Science 296:132–136

Hildebrand AR, Boynton WV (1990) Proximal Cretaceous-Tertiary boundary impact deposits in the Caribbean. Science 248:843–847

Hills JG, Goda MP (1999) Damage from comet-asteriod impacts with Earth. Physica D 133:189–198

Jiménez A, Posadas AM, Hirata T, Garcia JM (2004) Probabilistic seismic hazard maps from seismicity patterns analysis: the Iberian Peninsula case. Natural Hazards Earth Sys Sci 4:407–416

La Delfa S, Patané G, Clocchiatti R, Joron J-L, Tanguy J-C (2001) Activity of Mount Etna preceding the February 1999 fissure eruption: inferred mechanism from seismological and geochemical data. J Volcan Geochem Res 105:121–139

Major JJ, Schilling SP, Pullinger CR, Escobar CD, Howell MM (2001) Volcano-hazard zonation for San Vicente volcano, El Salvador. U.S. Geol. Surv. Open-file rep. 01-367, Washington, DC

Milani A, Chesley SR, Valsecchi GB (2000) Asteroid close encounters with Earth: risk assessment. Planet Space Sci 48:945–954

Miller TP, Chouet BA (1994) The 1989-1990 eruptions of Redoubt Volcano: an introduction. J Volcan Geotherm Res 62:1–10

Miyabuchi Y (1999) Deposits associated with the 1990-1995 eruption of Unzen volcano, Japan. J Volcan Geotherm Res 89:139–158

Morrison D (ed) (1992) The Spaceguard Survey: report of the NASA international near-Earth object detection workshop. Jet Propulsion Laboratory, California Inst. Tech., Pasedena, CA, January 25, 1992

Paine M (2001) Source of the Australasian tektites? Meteorite 7:34–37

Pope KO, Baines KH, Ocampo AC, Ivanov BA (1994) Impact winter and the Cretaceous/ Tertiary extinctions: results of a Chicxulub asteroid impact model. Earth Planet Sci Letters 128:719–725

Poveda, A, Herrera MA, Garcia JL, Curioca K (1999a) The diameter distribution of Earth-crossing asteroids. Planet Space Sci 47:679–685

Poveda, A, Herrera MA, Garcia JL, Hernández-Alcántara A, Curioca K (1999b) The expected frequency of collisions of small meteorites with cars and aircraft. Planet Space Sci 47:715–719

Reid ME, Sisson TW, Brien DL (2001) Volcano collapse promoted by hydrothermal alteration and edifice shape, Mount Rainier, Washington. Geology 29:779–782

Scott PD, Rogova GL (2004) Crisis management in a data fusion synthetic task environment. In: Proc. 7th International Conf. on Information Fusion, 28 June – 1 July 2004, Stockholm, Sweden. Swedish Defense Research Agency, Stockholm, pp 330–337

Sheridan MF, Hubbard B, Bursik MI, Abrams M, Siebe C, Macías JL, Delgado H (2001) Gauging short-term volcanic hazards at Popocatépetl. Eos Trans Am Geophys Union 82:185, 188–189

Sleep NH, Zahnle KJ, Kasting JF, Morowitz HJ (1989) Annihilation of ecosystems by large asteroid impacts on the early Earth. Nature 342:139–142

Smith K (1996) Environmental hazards. 2nd edn. Routledge, London

Spitale JN (2002) Asteroid hazard mitigation using the Yarkovsky effect. Science 296:77

Tate J (2000) Avoiding collisions: the Spaceguard Foundation. Space Policy 16:261–265

Tedeschi W, Teller E (1994) A plan for worldwide protection against asteroid impacts. Space Policy 10:183–184

Thouret J-C, Gourgaud A, Uribe M, Rodriguez A, Guillande R, Salas G (1995) Geomorphological and geological survey, and spot remote sensing of the current activity of Nevado Sabancaya stratovolcano (south Peru): assessment for hazard zone mapping. Z Geomorph. NF 39:515–535

Ward SN, Asphaug E (2000) Asteroid impact tsunami: a probabilistic hazard assessment. Icarus 145:64–78

Wu W, Sidle RC (1995) A distributed slope stability model for steep, forested basins. Water Resour Res 31(8):2097–2110

Chapter 24

Social Perspectives on Comet/Asteroid Impact (CAI) Hazards: Technocratic Authority and the Geography of Social Vulnerability

Kenneth Hewitt

24.1 Introduction

Until quite recently, research into comet and asteroid hazards was focused on establishing the scale and scope of past impacts, credible estimates of their recurrence, and models for physical impact scenarios. If there is still much to be done, the threat does seem convincingly demonstrated. CAI hazards have moved well beyond the realm of ungrounded speculation and apocalyptic visions. The results represent more than just new findings. They revolutionize, or are about to revolutionize, some basic understandings about the Earth, its history, biological evolution and future. Although human life has had a tiny place in the story so far, our longer term fate seems to be challenged by these forces and may be decided by them.

However, the task of addressing and situating the dangers among modern risk profiles and priorities has hardly begun. Their place in international disaster measures remains to be decided. In particular, there has been very limited input of social understanding, or even recognition that such threats could have important social dimensions. There is often a sense, in the literature on CAI hazards, that the most destructive impact processes and events with globally catastrophic potential are all-important. It has been claimed that the destructive potential of CAI far exceeds that of any other natural forces or disasters in recent, if not all, human history (Morrison and Teller 1994). In such visions experience with known risks and disasters may appear to be of little help. Human diversity, social concerns and changes can seem irrelevant – except for their likely annihilation or a precarious biological survival. And it does seem there is a potential for catastrophic, global impacts that would overwhelm any existing social response capacities. Apparently, however, these are the least likely events. They are thought to recur at intervals not less than 100 000 years, possibly millions or tens of millions. Obviously this is longer than the history of civilizations, if not *Homo sapiens*.[1]

Yet, the hazards attending impacts from intermediate and even relatively small, atmosphere-penetrating bodies are far from trivial. They threaten local and regional disasters, perhaps multi-nation or widely dispersed patterns of damage. However, in such cases, impacts, their severity and reach, are constrained by the specifics and di-

---

[1] Later, I will argue against risk estimations that calculate and project backwards the assumed damages in such multi-million year events as though they happened in today's world and the same risk space – current populations etc – as everyday damages and recent disasters.

versity of Earth environments. This applies especially to secondary hazards such as tsunami and fires, generated by asteroid impacts. These dangers, and possible human responses, are not utterly different from more familiar natural disasters. It is likely that the geography of human settlement and vulnerability, related habitats, will be as decisive in losses and responses, as in other modern calamities. For international disaster preparedness, these more likely calamities seem the appropriate place to begin.

We need to ask whether and how CAI concerns would enter into the agendas of existing organizations charged with disaster preparedness. Are there specific tasks to be added to the mandates of national emergency measures agencies and international preparedness? What can be said to gain the interest of other organizations with relevant responsibilities? 'Moderate' impacts threaten key areas involving the FAO, UNEP and 'Catastrophic Loss' insurance, as well as those with front line response duties such as UNHCR. How does one articulate the relevance of CAI for leaders and advocacy groups currently preoccupied with the HIV/AIDS pandemic, Darfur, Iraq or Chechnya; with climate change, disappearing rain forests, or terminally depleted fisheries?

To address social perspectives I ask if there are issues relating to CAI comparable to existing material in, say, the annual "World Disasters Reports" of the Red Cross and Red Crescent Societies, or publications of Doctors Without Borders, Oxfam, the Mennonite Central Committee and other humanitarian relief organizations? Do we have anything to say to groups working on the roles, in disasters, of economic development strategies, urbanization, gender, armed violence, child malnutrition or poverty? If not, perhaps CAI do lie in a futuristic realm beyond the bounds or grasp of civil care and crisis management. Alternatively, as will be argued here, we have failed to recognize how social vulnerability and public policy apply.

Since most of us are 'knowledge workers' ideas about disasters need to be examined, the assumptions driving international preparedness and response. There are on-going debates over the interpretation of risk and disaster. Various contributing fields and constituencies tend to have very different proposals for what should be done to mitigate and prepare for destructive events. Nor is this simply a matter of the quality of investigation or modeling, but also of contested social perspectives on risk and disaster. In the case of CAI, a prevailing focus on the physical hazard, typical of techno-scientific fields, can be prejudicial to social awareness and understanding.

24.2
The Perspective of Social Vulnerability

The position taken here is that calling something a risk, indeed a hazard, is a social construct and necessarily engages social issues. Risks are 'social' firstly because they involve conditions that vary greatly within and between societies, and according to a range of material, institutional, political and cultural conditions. Secondly, social risk is not just a function of exposure to a dangerous agent. Major consequences in all known modern disasters derive from or involve the social order and human activities. The latter affect endangerment in ways that go well beyond, and differ in character and genesis from, whatever can be learned about, say, hurricanes, or weapons of mass destruction. Thirdly, the preoccupations and preferences that enter into assessments of a danger are themselves socially constrained. They differ widely in different constituencies. Economic, technological, cultural, and governmental or ethical concerns, not only affect how a society will respond but what we need to know about a hazard for risk assessment and disaster preparedness.

A social perspective begins on the ground in human settlements, rather than in the lithosphere, atmosphere or space. It gives priority to the places and predicaments where destructive impacts may occur, rather than where hazardous phenomena originate and their overall extent. In the Earth environment and experience to date, a majority of potentially destructive natural events do little or no harm. Most of the geophysical cycle and space over which potentially destructive forces operate lie outside where the disaster itself occurs. This applies in earthquake and volcanic eruption, hurricanes, blizzards and tornadoes, river floods and tsunami, catastrophic landslides and avalanches, forest fires and insect infestations. There is a quite weak relationship between any measures of the size or intensity of these hazards, and the scales, let alone forms, of disaster (Hewitt 1997).

The empirical basis for these assertions comes mainly from distributions of loss in recent disasters. In them, the forms and intensities of damage reflect *and reveal* patterns of pre-existing social vulnerability more than any other risk variables. This applies to the worst of examples of 'man-made' disasters such as Chernobyl or Bhopal, and the most costly natural disasters such as the earthquakes at Mexico City in 1985, Kobe, 1995 or Izmit, Turkey, 1999; hurricanes "Andrew" in Florida, 1992, or "Mitch" in Central America, 1998. In each case, *who* lived and died, how people died or were injured, their material losses turned overwhelmingly on their social status and influence.[2] In turn, these were critical for available or absent protections (Hewitt 1997; Enarson and Morrow 1998; Steinberg 2000; Dauphine 2001; Shaw and Goda 2004).

The primary *social* elements of public safety involve people's exposure to given threats, the vulnerability of their bodies, homes and livelihoods, and their response capacities. In modern societies, the provision of organized protections has assumed overwhelming importance in relation to all of these, as well as hazard mitigation and disaster response. While each element is important and needs careful study, they ultimately contribute to and define *degrees* of social vulnerability. 'Degrees' are emphasized because vulnerability is significant less in terms of abstract or absolute safety, than the huge variations within and between societies. The variations are context, and society-specific. They can change dramatically over time. This is also a way to express civil security from the perspective of its weakest or more insecure areas (Bohle 1993: Blaikie et al. 1994). It reflects each person's and group's place in their own society, each community and region's influence in their own country, and in relation or comparison to the wider world (Burton et al. 1978; Sen 1981; Kreps 1989; Alabala-Bertrand 1993).

It is a common place that large differences in personal vulnerability apply to traffic accidents, cancers, heart or occupational disease, as a function of age, gender, occupation, ethnicity, wealth or life-style. Yet, such social 'discrimination' in risk or harm is not

---

[2] Two events, shortly after our meeting but before the revised version of the paper was submitted, gave horrendous reinforcement of these points: the Indian Ocean tsunami of Christmas 2004, and Hurricane "Katrina" impacting the Gulf Coast of the USA, September, 2005. Since the most likely result of a CAI is an ocean impact generating great tsunami it is especially relevant how both events revealed the ecological and social vulnerability coastal zones. And they did so in poor and rich nations, but singling out particular sectors of their populations, and coastal zone developments of recent decades.

confined to chronic biological, 'consumer' and life-style hazards. The widely accepted separation of natural disasters as a class of 'indiscriminate' events or accidental 'bolts-from-the blue' is a social fiction. Their damages tend to be highly discriminating usually, though not always, favoring the wealthier and more influential groups, and harming the already disadvantaged.

However, vulnerability is not a passive or inevitable condition awaiting an impact. Such ideas are just another carry over from the view of disasters as 'Acts of God' or accidents in which death or survival are matters of luck. Rather, society intervenes in fundamental ways to influence or construct exposure to hazards and peoples' capacities to respond, as well as in the provision or absence of social protections. Moreover, while our focus may be on hazards and disaster, for the most part security or endangerment arise *indirectly* within political and economic systems. The social and economic conditions governing land use, housing, employment, education and health services, access to information and so forth, largely allocate degrees of vulnerability to each type of hazard.

For such reasons, social scientists and humanitarian agencies see the space of risk as rooted in social values, a social responsibility maintained, or changed for better or worse, by the social order. Social conditions are an integral part of endangerment rather than, as in so many geophysical analyses, entirely secondary and essentially dependent upon understanding, predicting and responding to floods, volcanic eruptions or industrial explosions. Thus we have to ask not just how large or how frequent dangerous forces are, but how they enter into social systems, their relations to the habitat and capacities to respond.

Effective public security measures certainly depend upon understanding the nature and incidence of hazardous processes. Communities where such knowledge is poor, neglected or forgotten, are uniquely vulnerable. In each case, however, such knowledge is seen through the lens of on-going material and cultural life, its deployment depending upon the values, expectations and experience of responsible persons and institutions. Meanwhile, a particular hazard or risk is unlikely to be dealt with in isolation. It will be treated in relation to the range of dangers, concerns and priorities of a society or, at least, of the institutions and groups who dominate its policies.

With such concerns in mind, we can now return to the question of evaluating CAI hazards. I will focus on CAIs below a global catastrophe threshold, and give as much emphasis to secondary hazards such as wildfires or tsunami, as to primary impacts. Comparative hazard analysis, an essential part of risk assessment generally, is relevant here, including disaster experience with other hazards that generate comparable forces or situations. In the longer frame of Earth history there are other dangers of comparable severity. Comparisons with these hazards and known disasters suggest what features of CAI may be unique, which ones are more likely to affect different regions, countries and habitats, and which may be dealt with by integrated disaster preparedness.

24.3
Regional and Comparative Aspects of CAI Hazards

Anticipated impacts of CAI below a global catastrophe threshold apparently involve bodies between about 50 m and 2 km in diameter – perhaps as small as 30 m for the rarer metallic objects.[3] Impact energies are rated between 10 and 10^5 megaton yield, and recurrence intervals from a few centuries up to 100 000, perhaps a million years (Morrison et al. 1994; Verschuur 1996, p 165; Abbott 2004). In terms of the recently developed 'Torino Scale' for CAI, events of interest seem to lie in categories 8 and 9 for "certain collisions". However, categories 4–7 – close encounters having a low but definite probability of collision – imply a need for vigilance and preparations for crisis management (Morrison 2004).

24.3.1
Regional CAI Risks and the Role of Secondary Hazards

Astrophysical factors decide whether CAIs occur on or over land, ocean, or in shallow waters. However, the nature of these environments constrains impact processes and the triggering of secondary hazards. Terrestrial or marine geography, and the regional habitats involved, act as intervening variables of risk. Even the global catastrophe known as the "K/T event", commonly identified with the demise of the dinosaurs, seemingly depended in various critical ways on a terrestrial and shallow water impact over the Yucatan Peninsula, and the geology of that area (Robertson et al. 2004). Again, this is a well-known feature of other hazards. Earthquake risk at any given place is strongly constrained by topography, geology, alluvial deposits, water tables and vegetation cover – the so-called micro seismic environment (Hewitt 1984; Degg 1992). Coastal zone conditions can be decisive in the scope of storm surge or tsunami hazards.

As the size of Earth-penetrating bodies decreases, the importance of where impact occurs becomes increasingly important. The potential for generating different secondary hazards can be decisive in the scope of the risk (Table 24.1). Over and above the properties of the asteroid body, very different dangers arise from an ocean versus terrestrial impact. A CAI in the ocean can result in tsunami, massive water injection into the atmosphere, marine pollution and ecocide. Pacific, Indian or Atlantic Ocean impacts in the lower range of penetrating asteroids are thought to be capable of generating mega-tsunami. Such waves could kill thousands, if not millions, along vulnerable, heavily populated coastlines (Hills et al. 1994). Impacts on land threaten extensive wild fires, induced seismic activity, landslides, and injections of dust and combustion gases into the atmosphere. Of special concern is atmospheric pollution sufficient to cause short-term climate catastrophes – untimely 'cosmic winters' of global, hemispheric or, at least, regional scales. An impact in a coastal zone, shelf sea or island arc might result in a combination of marine and terrestrial secondary hazards.

Some think such processes can be triggered by adverse fall of bodies less than 1 km in diameter, perhaps as small as 100 m. These are the most numerous of NEOs capable of penetrating to the surface. Explosion and break-up high in the atmosphere involves another range and balance of hazards (Avdushkin and Nemchinov 1994; Toon and Zahnle 1994). Destruction or impairment of the protective ozone layer is also a concern, through the secondary hazard from ultraviolet radiation.

---

[3] The strictly CAI aspects of my analysis are, of course, second hand, based on a survey of the more accessible literature, but trying to determine from it which matters seem most likely to have societal implications.

Table 24.1. A summary of *(i)* the main secondary hazards that may be triggered by an asteroid impact, *(ii)* the mechanisms or condition by which they can cause harm, *(iii)* the environmental variables which could influence their occurrence, intensity and reach and, *(iv)* the kinds of habitat, land use, settlement and social conditions influencing vulnerability to these hazards (hazards after Avdushkin and Nemchinov 1994; Toon and Zahnle 1994)

Secondary hazard	Dangerous mechanisms	Earth environment variables	Social vulnerability concerns
Dust injections and dispersal	Atmosphere effects: – Cooling – Blocking sunlight – Reduced visibility – Toxicity	$A/L/O^a$ – Stratosphere/troposphere loadings – Inversion layers	– Biotic resources – Crop loss – Food security – Heating buildings – Respiratory health
Water injections	Atmosphere effects: – Troposphere warming – Stratosphere cooling (ice clouds) – Transparency – Chemistry	O/A – Regional air mass – Circulation – Latitude – Hemisphere	– Agriculture – Energy supplies – Transportation
NO_2, SO_2 injections	Atmosphere effects: – Cooling – Chemistry – Acid rain – Toxicity	$A/O/L$ – Bedrock, soils – Vegetation – Land use – Industries	– Agriculture – Water quality – Aquatic life – Lake/river amenities
Fires	– Combustion – Soot (cooling) – Pyrotoxins – Acid rain	L/A Land cover: – Natural vegetation – Crops – Built-up – Toxic industries and materials	Natural resources: – Forest/Grasslands – Crop lands – Food security – Water contamination – Respiratory health
Water waves and tsunami	– Impact waves – Long-distance waves – Inundation – Destructive impact – Sedimentation – Contaminant transport	O – Water depth – Basin configuration – Coastal configuration – Islands – Estuarine environments	– Coastal occupancy – Fisheries – Maritime trade – Tourism and recreation – Island peoples – Drownings
Geomorphic hazards	– Landslides – Flooding – Sedimentation	L – Topography – Drainage systems – Regolith	– Land use – Settlements – Water supply
Induced seismicity	– Earthquakes	L – Tectonics	– Macro- and micro-seismic vulnerability
Untimely ('Nuclear') winter'	– Zero or limited photosynthesis – Surface cold – Impaired hydrological cycle	A/L – Atmospheric circulation – Hemisphere	– 'Dyings' extinctions? – Food security – Energy supplies – Natural resources

[a] O = Oceanic environment constraint/effect; L = terrestrial environment constraint/effect; A = atmospheric environment constraint/effect.

The spread and intensity of secondary hazards will depend upon the shape of ocean basins, coastal zones, vegetation belts, planetary wind systems and air mass climatology. As was seen with the Chernobyl disaster, a specific synoptic situation may shape the geography of dangers from material injected into the atmosphere. Given the greater extent of the hydrosphere, ocean impacts are more likely. Several analyses find the most lethal threat from low to intermediate impacts is from tsunami (Morrison et al. 1994; Verschuur 1996, p 165).

Earth environment and location-dependent impacts not only constrain the impact and introduce other kinds of destructive forces. They also transform the space of dangers. Assuming a random distribution of space object impacts, hazard profiles should mainly reflect the distribution of habitats on Earth. Different impact scenarios will apply in given countries and regions. They depend upon regional patterns of land cover, land use, settlement and human activity. This is where the analysis engages with social vulnerability and human geographies of risk. The full importance of secondary hazards and habitats emerges in relation to patterns of human exposure and intervening social variables such as degree of urbanization or coastal occupancy. Moreover, likely consequences here introduce the threat of 'tertiary hazards'. They include dangers related to release of toxic materials, refugee crises, social unrest or political instability, famine and epidemic disease. Experience in known disasters shows that such hazards can cause more grief than the original impact. Meanwhile, attention to risks from secondary and tertiary hazards prepares the ground for evaluating CAI relative to other natural hazards.

24.3.2
Comparative Threat Evaluations

Other environmental processes that can cause extreme devastation on comparable time frames to those envisaged for CAIs are of particular interest. A summary of the scope of impacts, and recurrence interval estimates found in the literature, gives some sense of the broader hazard prospect (Table 24.2). There are many uncertainties concerning the incidence of rarer, more catastrophic events, but it seems CAI hazards are not, after all, unparalleled at *any* scale or time frame. Impacts with estimated recurrence intervals between 10^3 and 10^5 years share the risk space with calamitous volcanic eruptions known from Santorini (4500 BP), Mazama (7600 BP), and Toba (75000 BP) and volcanic island flank collapses. The dynamic waves and tsunami they would trigger offer equally sobering prospects. This is also the time frame of major glaciations, and rise or fall of sea level by 10s of meters, huge if not such sudden threats (McGuire et al. 1996; McCoy and Heiken 2000).

Moving to longer time frames, there are caldera collapses as identified at Yellowstone, USA (±750 000 yr) and the immense climatic and ecological changes over the course of an Ice Age. Some think reversals of the Earth's geomagnetic polarity, with recurrence intervals of 10^5 to 10^7 years, could be singular threats for living things. Over time frames of 10^6 to 10^8 years, identified with mass extinctions caused by the largest CAI, the geological record indicates a number of other great global crises associated with 'Flood basalts', Kimberlite or diatreme explosive events, perhaps Super novas in our part of the Galaxy (Kusky 2000; Abbott 2004). Epicontinental sea episodes may occur quite slowly, but do reduce the land area of Earth by half or more. They imply drastic

Table 24.2. Comparison of return periods estimated for CAI and selected other hazards. Since many of the rare cases are based on just one or two events, or assumptions about distributions with no event dates, their true temporal distributions are not known and they hardly meet minimal criteria as probability estimates

Recurrence estimates (yr)	CAI hazards (diameter)	Volcanic hazards	Climate change and weather hazards	Other
10^7+	>7 km: Global catastrophe certain, "global extinctions"	Flood basalts, Kimberlite events	Ice ages	'Close' supernova, epicontinental inundation
10^6+	>3 km: Global catastrophe, Tunguska-type, USA or European etc, cities impact	Caldera collapse, global climate catastrophes	Ice age	New and fading tectonic belts, rearrangement of continents, human ecological 'footprint'
50^5+	>1.5 km: Major continental catastrophe	Yellowstone caldera collapse	Glaciations, monsoons shift/stop	Major changes in shelf seas
10^5+	>500 m: Regional catastrophe, Tunguska-type urban center	Toba-type eruptions	Glacial-interglacial	Major coastal zone changes, shifts in biogeographical regions
10^4+	>300 m: Impact + mega-tsunami catastrophe	Mazama-type eruptions	Next glaciation	Glacial lakes and megafloods
10^3+	>100 m: Major local disaster a/o mega-tsunami	Mt. Rainier-, Monserrat-, Santorini-type eruptions; Flank collapse mega-tsunami	Global warming or cooling episodes of 5–10 °C	Rise and fall of civilizations
500+	>70 m: Local a/o tsunami disaster			'Black death' type pandemic
100+	>50 m: 'Tunguska' type event	Tambora-type eruption, 1–2 year climate effect, tsunami (100 000+ d)		Shanxi, type earthquake (500 000 to 1 m d) HIV/AIDS pandemic 'death cycle' (100 m + d)
<100	none	Several eruptions of St. Helen's or Pinatubo, scales	El Nino oscillation	Local, regional and multination disasters from a full range of natural, disease, technological and social hazards

climatic and hydrological change and drowning of most of today's productive land and cities.

At the other end of the scale various hazards threaten great catastrophes over shorter time frames than any yet established for CAI. They include examples of explosive volcanism from Krakatau to Pinatubo, or earthquakes from Tangshan, China or Sumatra in December 2004. Pandemic diseases such as the 1918 influenza pandemic and ongoing HIV/AIDS calamity, or the world wars and quite real threat of a nuclear one, involve millions to tens of millions of deaths, untold misery and destroyed livelihoods. Each of the hazards involved has had several events over the past century capable of causing catastrophes but in uninhabited areas – placing them in the risk envelop of the Tunguska 1908 event or so-called 'near miss' CIAs.

In general, a comparative approach provides a less 'exceptionalist' view of CAI hazards. It is also valuable because of the lack of experience with CAI disasters in the modern world. Apparent parallels among known disasters seem to be associated with volcanism. Where they occur, the primary impacts of explosive eruptions, edifice collapses, pyroclastic flows, lahars and lava flows are totally devastating. A variety of secondary hazards may be triggered that can cause the bulk of the economic loss and death tolls, threatening people far beyond the eruption zone (Blong 1984; McCall et al. 1992). Tsunami, crater lake outburst floods, catastrophic melting of ice caps or jokulhlaups, hazardous dust and pollutants, and short-term but intense climate change, involve risks similar to those CAI may generate (McGuire et al. 1996). Disasters due to large earthquakes, storm surges, catastrophic rock slides and wild fires, offer other parallels. Meanwhile, as noted, the geological record reveals long term prospects of vastly more destructive volcanism.

A crucial point to make is that each type of hazard is not merely an added class of danger, but compounds the possible scope of disasters. Having a variety of rarer hazards, in effect, *compresses* the time and space frame of catastrophic risk in general. When the focus shifts to questions of public policy, it is both unnecessary and dangerous to treat these different hazards as competing dangers. For social vulnerability and international disaster preparedness they comprise a spectrum of dangers. It makes a lot more sense to look at integrating or sharing assessment, monitoring, and preparedness as far as possible recognizing what is common to each danger and, to be sure, what is unique.

At the very least, risk management needs to be informed by comparative analysis of hazards. An 'all-hazards' approach is increasingly favored by emergency measures agencies. Most social responses to disaster are common across many if not all dangers. There are common requirements for getting warnings to communities in an understandable and responsible fashion. Key measures have much in common regardless of the source of disaster, for example, evacuation procedures and emergency shelter, providing basic necessities from clean water to cooked food, acute medical care, moving to restore services and livelihoods as efficiently as possible. Of course, many disasters also call for highly technical and hazard-specific knowledge. But this too is successful or fails, to the extent it is, or is not, sensitive to and well-coordinated with the fundamental requirements of civil security.

The situation and the everyday needs of the elderly, infants, the sick and disabled, of women in themselves and as the care-givers in most contexts, are of special concern. They do not go away just because there is a disaster, but often involve the very requirements relief measures are least effective in meeting. Especially when combined with low income and lack of a voice in public affairs this is often the focus of the worst misery in disasters. Moreover, it tends to reveal historical patterns of neglect or worse, such that, in a great many cases, whether there is a disaster and its scale mainly reflects patterns of social disadvantage more than the relative impact of geophysical forces. To a quite remarkable extent, the public face of disasters and the development of emergency measures are dominated by male voices, usually specialists analyzing situations to which they were not, themselves, exposed. At the very least this requires them to listen to others and pay much closer attention to the condition and social contexts of vulnerable groups. It can be anticipated that such social conditions will also intervene to affect responses and consequences of CAI hazards.

24.3.3
Uncertain Uncertainties

Having chosen to concentrate on lesser impacts, more likely in a time frame of a few centuries, one must recognize that far larger events can still occur. Betting on it may be like staking one's life on winning the lottery but, if real, even very rare events must happen sometime. They have a marginal probability of doing so tomorrow, next year or in the lifetime of many now living. Meanwhile, estimates such as those assembled here are fraught with more or less great uncertainties. For CAIs, parameters employed often depend upon what has happened on the Moon and other planets, ancient and incomplete samples for Earth impacts. Impact forces are estimated using data from non-CAI events like nuclear weapons tests, or laboratory experiments and computer simulations. It is not unusual for such simulations to differ from, or miss critical factors in, actual events. The uncertainties seem likely to be even greater in estimating the scale and scope of secondary hazards (Avdushkin and Nemchinov 1994).

A social perspective also raises doubts about an approach adopted in the CAI literature that grew up around technological hazards in the mid-twentieth century. It mainly addressed problems in the nuclear industry, the tobacco and toxic chemical debates, or relating to insurance of catastrophic risks (Starr 1969; Kates et al. 1986; Cutter 1994). Unfortunately, while 'socially constructed' within technocratic institutions, the approach involved very dubious notions of social content. What does it mean to compare the statistics of death from smoking or driving, with a major nuclear 'accident', or any of these with earthquake losses, or deaths from heart disease, in wars or criminal activity – let alone speculative events that have no precedents? The statistics can provide useful profiles or background for social monitoring and general policy. To compare them as though they have a common social ground ignores serious conceptual, technical and ethical questions.

24.4
Conceptual Issues

To date, work on CAI hazards has focused almost wholly on the astrophysical and geophysical processes that could cause or influence destructive events. This makes sense when there has been inadequate evidence and understanding of the phenomena. It is an essential ingredient for risk assessment and disaster preparedness. Nevertheless, for the latter, other and different ingredients enter the picture. It will be useful to review what they are (Table 24.3).

Table 24.3. The main sets of elements whose interactions determine damage in disasters and identify risk variables (after Hewitt 1997). Each enters into the profile of loss and survival in any given disaster. Actions under *(e)*, especially, may serve to modify one in relation to the others. However, each set can and does vary independently of the others between events, places and over time. This requires attention to the fact that they are governed by, and change in response to, different systems of control. For technological and social hazards, and *(b)–(e)* generally, the main controls are social

a	Dangerous agent(s) or condition(s), commonly referred to as hazard(s)	These threaten harm, initiate and cause much of the damage in disasters. They include earthquake, storms and fire, CAI, communicable diseases, toxic chemicals, equipment failure, impaired operatives, criminal acts and armed violence. Understanding of their specific, dangerous properties plays a key role in identifying possible social responses
b	Human exposure	All sorts of conditions have the potential to cause harm. They only become actual dangers if persons or property are exposed to them. Exposure may be of life and limb, or of property, livelihoods, investments, cultural heritage or all of these. The extent and forms of exposure arise mainly from social activity, settlement and land use conditions. They vary greatly in given cases, and between events and places
c	Vulnerabilities	These comprise weaknesses and susceptibilities, lack of protections or options, in persons and communities exposed to dangerous agents. (*Differential* social vulnerability is a particular focus of the paper)
d	Social contexts and intervening conditions	There are always societal and environmental circumstances that indirectly, but often decisively, influence risk, yet exist and are carried forward with little or no regard to some or all hazards. Specifically, context involves conditions that intervene between people and dangerous forces to buffer or magnify impacts. The severity of impact of forces such as earthquake depends upon a range of variables where impacts occur – land versus sea, topography, vegetation cover, building density, design, and materials. These habitat or social conditions are critical for the development and reach of secondary and tertiary hazards (see text discussion). They may or may not reflect experience with given hazards or conscious efforts to influence risk, hence are distinct from *(e)*
e	(Organized) risk-reducing and disaster-mitigating measures	This refers to deliberate actions aimed at forecasting and reducing dangers, protecting those at risk, responding to crises, or off-setting damages. They may be directed either at controlling hazards or reducing vulnerabilities, social uplift or emergency preparedness, or a mix of these factors

In many ways the central problems of risk and disaster response are how to address and balance these five elements, or aspects of them that enter into particular cases. On the one hand, they are recognized as distinctive in character and origin. Each can, and often does, vary from place to place and event to event, independent of the others. On the other hand, all have a part in determining whether, where, and for whom disaster occurs. However, many disciplines, professions and institutions specialize on just one of the elements. They mostly ignore the other elements or treat them in a casual way. The CAI community is currently identified with the most pervasive basis of these specializations, a focus on the hazard or 'agent-specific' approach (Gilbert 1998; Hewitt 1998).

24.4.1
Limitations of the Agent-specific Approach

From a social perspective it is generally a mistake to base risk assessment primarily on scientific, let alone popular, treatment of the hazardous agent. This problem has emerged as a major flaw in most work on disasters involving natural, technological, disease and social violence hazards (Hewitt 1983, 1997). Much of it ignores societal and context-specific considerations, or subordinates them to properties of the dangerous agent.

This is not about bad versus good science, nor to suggest agent-specific knowledge is not important. To neglect it or fail to take it seriously, as also happens, can be fatal too. Nevertheless, neglect or subordination of the other, especially social factors and contexts, is more widespread and leads to most of the common empirical and interpretive illusions about disasters (Green 1997). Work on CAI hazards is typical of the narrowest agent-specific approaches. It leads to what I would characterize as flawed preoccupations that in summary are:

1. Risk assessment dominated by the space hazard and primary impact parameters.
2. A focus on global and totalizing or 'worst-case' scenarios in which the full, most destructive, scope of the CAI hazards is expected, but neglecting (much more likely) regional disasters and less-than-calamitous global ones, and neglect of Earth environment and settlement conditions that will influence them, as outlined above.
3. Ignoring, or treating the geography of social exposure, vulnerability and capacities, as subordinate issues, at best summarized in global and impersonal probabilities of population exposure.
4. The assumption that only defenses directed against the Near Earth Objects (NEOs) themselves, using more or less advanced, mostly space technologies, offer ways to reduce risk and prevent disasters.
5. Proposals for responses, including disaster preparedness and allocation of resources, that are entirely about the hazard, the conditions and technologies associated with astrophysical and geophysical sciences.

The kinds of responses assumed and advocated in the CAI literature, indicated in points 3) through 5) above, are in accord with the agent-specific view. They indicate a state of affairs such as Gilbert F. White (1945) critiqued so long ago in relation to the flood hazard. He showed the dangers as well as the limitations of response strategies based only on flood monitoring and forecasting, and engineering control works. Agencies and communities relying on that strategy continually made the flood problem worse. In many places the situation has not changed very much (Steinberg 2000). However, White's influence relates especially to his seminal idea of 'alternative adjustments' to hazards. In the original case it included such things as watershed management and land use planning for flood plains, flood proofing, insurance and the like. He did not saying hydro science is unimportant, or hydraulic engineering never a valid or even, sometimes, offering the best solution. He made it clear that these need to be balanced against understanding of the full range of options. Moreover, when that is done, the other elements of risk loom large and we see how the sources of social vulnerability become keys to disaster mitigation (Burton et al. 1978; Blaikie et al. 1994).

I would advocate a similar approach to assessing CAI, and what to do about them. No one can yet say there will not be destructive impacts. Space technologies may never eliminate all of them, and basing these solely on agent-specific targets is not without its dangers. At a minimum we still lack adequate, socially aware analysis in the matter of CAI forecasts, warnings and emergency responses. However, it is the secondary and tertiary hazards that highlight the importance of other adjustment options. The public perception of Hollywood leading the way here should give pause for reflection.

The two right hand columns in Table 24.2, relating to 'Earth environment variables', and sources of social vulnerability, speak directly to questions of social and ecological risk profiles and methods of improvement. Moreover, each one is already the subject of concern in relation to many other hazards. There is a rich literature on the sorts of responses required to reduce exposure and vulnerability, albeit not so widely adopted.

However, the agent-specific focus in the CAI community raises another, essentially social concern, so often taken for granted. It is the role of organizations and institutional agendas in modern responses to disaster.

24.4.2
Organizational Risk

There has been a tendency, especially in the West, to see the world as governed, on the one hand, by individual or 'private' choice, life-styles, consumer demand, and, on the other, by impersonal forces or controls – 'the market', genetics, climates, population, scarce resources, ideology or religion, stage of development, race or 'manifest destiny'. Organizations, even governments, are regarded essentially as instruments to achieve particular ends in relation to people's desires and needs (March and Olsen 1989). Technical and managerial arrangements are treated as somehow 'transparent' functions responding to established duties and needs. In such a view, if things go wrong, it must be because of imperfect information, lack of the latest knowledge, 'human error' or failure to specify duties carefully enough. At least, one finds such assumptions throughout the literature of hazards and disasters. It is, however, quite misleading from a social perspective; a curious approach to a world in which organizations, governmental and non-governmental, international, corporate and professional, have acquired unprecedented powers and influence.

In fact, one of the most useful learning activities for risk managers is to study investigations of the role of responsible institutions in past disasters (Turner 1978; Toft and Reynolds 199; McClean and Johnes 2000). Since we have no CAI disaster to examine, let us look at a series of recent ones that occurred in other contexts (Table 24.4).[4]

What do the disasters listed have in common? No doubt they seem to be quite different kinds of threats and events, very different in scale and scope, occurring in diverse, widely separated contexts. Firstly, however, each was the subject of major public inquiries and/or of extensive multi-disciplinary and independent investigations. Secondly, and the reason for raising them here, each of the inquiries attributed the disasters, primarily, to failures within the relevant systems of public care and security. Nor

---

[4] This is based on a study prepared for the Canadian Risk and Hazards Network (CRHNet), presented at their Annual Symposium in Winnipeg, November, 2004 and to be included in a published volume.

Table 24.4. Organizational risk: some events with reports and inquiries that highlight the role of institutions with safety responsibilities, and institutional failures, in whether a disaster happens, or its form and severity

Event	Country/Region	Date
'Blackout'	USA and Canada	14 August, 2003
SARS outbreak	Ontario	2002–2003
Earthquake	Gujarat	2001
Contaminated drinking water	Walkerton, Ontario	2000
Earthquake	Izmit, Turkey	2001
'9/11' terrorist attacks	USA	2001
Kobe ('Great Hanshin') earthquake	Japan	1995
Sampoong Mega-department store collapse	Seoul, South Korea	1995
'Estonia' ferry sinking	Baltic Sea	1994
Westray mine explosion	Nova Scotia, Canada	1992
Floods, Indus valley	Pakistan	September 1992
'Tainted blood' scandal	France	1990s
'Tainted blood' scandal	Canada	1980s–1990s
BSE outbreak	UK	1990s
'Exxon Valdez' oil spill	Alaska	1989
Hillsborough soccer stadium crush	UK	1989
'Marchioness' pleasure boat sinking	London, UK	1989
Dryden, Air Ontario crash	Canada	1989
'Piper Alpha' oil rig explosion and fire	North Sea	1988
'Herald of Free Enterprise' ferry sinking	Zeebrugge	1987
Challenger Space Shuttle accident	USA	1986
Chernobyl nuclear plant explosion and fire	Ukraine	1986
Earthquake	Mexico City	1985
Bhopal toxic chemical leak	India	1984
'Ocean ranger' drilling rig loss	Grand Banks	1982
Political terror in the 'dirty wars'	Argentina, Chile and Guatemala	1970s 1980s
Teton Dam failure and flood	Idaho, USA	1976
Flixborough chemical plant explosion	UK	1974
'Bloody Sunday' massacre	Northern Ireland	1972
Buffalo Creek flood	USA	1972
Aberfan coal tip collapse	S. Wales, UK	21 October, 1966
'Civil disorders'	USA	1967
Vaiont Dam landslide and flood	Italy	1963
'Titanic' sinking	Grand Banks	1912

[a] To keep the number of references reasonable I have not cited most of the relevant reports but can provide them to individuals who are interested.

were the reasons found in some *potentially* available safety measures, but specific forms of action and inaction violating accepted practices or a reasonable performance of their mandates, usually in organizations distracted by other priorities. Most surprising, perhaps, is how many conclude that the deaths and damages, or most of them, could have been prevented: not just 'mitigated' or made less painful, but *prevented*.

In general, the inquiries suggest a quite different interpretation of these events from that of the 'agent-specific' approach. They do not give much or any importance to the 'usual suspects' of hazards discourse – 'unscheduled' events, Mother Nature's powers or surprises, 'Acts of God' or a disembodied human nature, not even merely aberrant, criminal or incompetent persons. And it is not because they pay no attention to damaging agents. On the contrary, almost all expend more ink on analyses and expert testimony relating to objective hazards than anything else. Yet, their conclusions do not give primacy to an agent-specific explanation. Rather, they emphasize one or more failures of civil care and emergency measures. They come close to Ulrich Beck's "risk society" notions and, if on different ground, his view that "… the production of risks is [now] the consequence of scientific and political efforts to control or minimize them …" (1992, p 12).

Especially dismaying is the repeated citing of previous inquiries whose recommendations would have mitigated or prevented the later disasters – had they been adopted and enforced! Indeed, a whole series of other disasters are identified for which previous inquiries made recommendations that should have prevented the later disasters, but did not. They include multiple events involving passenger ships and ferries, trains, soccer stadiums, coal mines, toxic storage facilities, weapons systems, dams, 'race riots', school shootings, earthquakes, floods and storms, diseases transmitted in and by medical facilities.

Of course, inquiries have their problems too. Toft and Reynolds (1994) provide a useful account of their limitations. Nevertheless, many of them provide quite the best post-mortems of disasters that we have; comprehensive, wide-ranging and, on balance, independent of interests that might favor or promote a self-serving view. At the least, any contribution to public policy and disaster management needs to take seriously the conclusions they support.

Some recent developments in disaster theory and management that do take account of organizational risk may seem more directly relevant to proposed responses to CAI; namely, studies of the ways in which catastrophic risks arise from failures of otherwise sophisticated and advanced techno-scientific organizations. The 'normal accidents' model of Perrow (1997) is an example, applied mainly to nuclear facilities, space technology and weapons systems (Sagan 1993). Much of the science, the evidence and experiments of which CAI event reconstructions are based, have depended upon national space agencies and, hence, are never far removed from military-strategic concerns. The arrangements for watching and reporting CAIs do reflect a remarkable and open cooperation between amateur and professional observers. Yet, most options proposed for forecasting and other responses to threatening NEOs involve technologies and institutions where national security, secrecy, and strategic interests loom large. Debates within the CAI community raise warnings about the dangers of technocratic and, especially, military-type space defenses (Harris et al. 1994). Morrison and Teller (1994) urged a greater 'democratizing' of the debate and management options in this context.

The CAI field has been shaped largely by men like me or, at least, most of you – in suits or uniforms, lab coats or hard hats, spending a disproportionate part of our day in front of computers, massaging data, designing and managing machines, testing equipment, or in professional meetings. As argued above, disasters are primarily about the plight of families, neighborhoods, communities, homelands, livelihoods, cultures, habitats – and conditions on the ground where safety measures fail. The victims tend to be 75% or more women, children and the elderly.

Support from international agencies may hinge on defining a more sensitive social and humanitarian role within, perhaps helping to improve, disaster mitigation more generally. Wider public support may depend upon how far perceived CAI threats and proposed responses remain identified with a few powerful states, 'big science' and strategic defense systems. It is difficult to see how this can be achieved without reaching out to a wider public, recognizing common cause with other, if not closely parallel, geophysical dangers. The CAI community should be talking to the tsunami, storm surge and sea level change communities about coastal zone hazards. Among other things they will quickly learn of common dangers relating more to what is happening to coastal communities, cities, resource development and ecosystem degradation, than which geophysical process might trigger a tsunami. Again, the essence of the problem is what to do about social vulnerability, and that starts with the communities at risk.

24.5
Concluding Remarks

The International Red Cross and Red Crescent Societies recently made a review of disaster priorities and concluded that:

- "Disaster mitigation and preparedness must form part of the wider context of risk reduction – relevant to all those working in hazardous regions, whether in relief, development, business, civil society or government
- Long-term partnerships based on good governance across many sectors and disciplines will provide the best basis for tackling threats posed by disasters; and
- Setting targets for risk reduction could provide a way to focus political will and adequate resources on the problem." (IFRCRCS 2002, p 9).

These conclusions arose from experience in responding to crises world wide, and testimony to the overwhelming roles of differential social vulnerability in human casualties and damage. Like many others with such experience, they also support a much greater emphasis on reducing vulnerability. Humanitarian assistance that merely reacts to dangers, without considering how society, its history and preoccupations, sets the scene for the security of civil populations, can find itself on a treadmill of eroding effectiveness and credibility. Risk assessment which focuses only on the hazardous agents and what to do about them in possible emergencies, suffers from similar limitations.

Can one see any relevance of these points for CAI hazards, or contributions we might make in the course of trying to mitigate risks? The overview above positions CAI within a broad range of physical hazards, if emphasizing those that threaten catastrophic loss.

Before deciding the best way forward, it seems essential to develop common assessments of the risks, drawing upon and integrating knowledge from each hazard type. Even more important, most or all these hazards relate to similar or overlapping problems of social vulnerability and its geography. The main conditions that affect the wellbeing of people generally in each region and country seem much more important in differentiating risk 'scenarios' than individual agents.

In the modern world CAI and many other risks do, and seemingly must, rely heavily on scientific and technical expertise, and oversight by technocratic institutions. That does not mean we have to ignore social and ethical concerns. When we go to the international community it will ask about guarantees or forms of preparedness that will not simply draw resources away from other, seemingly more urgent, needs and existing victims. Not only are they stretched to the limit in actual epidemic diseases, wars and deadly legacies such as landmines, actual or emerging famines. In some of the worst of these responses have and are clearly failing tens of million of people. This creates a different sense of rare and probably very distant dangers, however catastrophic.

Wider public support will also depend upon how far proposed responses remain identified with a few powerful states and strategic defense systems. In spite of the militarizing of civil security systems in the wake of 9/11, we need to make an effort to reach out to other kinds of institutions involved in environmental and civil care. They could range from the Red Cross to the Gender and Disaster Network. Support from them may hinge on our defining a more sensitive social and humanitarian role within, perhaps helping to improve, disaster mitigation more generally. In this case national and international agencies will or should ask how proposed responses can help the most vulnerable and needy rather than promoting high profile projects for wealthy states or enclaves.

We may well feel these are matters best, and properly, left to elected officials and public debate. Yet, it is not merely that few areas of modern science are independent of political agendas and professional ethics. By the time we pose questions such as getting wider support to address the CAI threat, our institutions and research activities have already been shaped by the 'culture' and interest of those who have promoted and funded the work. In the area of humanitarian assistance be prepared for 'culture shock'! Yet, here we are back with the original and compelling arguments for rational inquiry to be as independent of 'interests' as possible. Disaster is one of those topics that needs an environment that promotes open inquiry, reaching out to different constituencies, listening to those most at risk, and an ethos encouraging constant review of our findings and reflection on their significance (Beck 1992).

References

Alabala-Bertrand JM (1993) Political economy of large natural disasters: with special reference to developing countries Clarendon Press, Oxford
Abbott PL (2004) Natural disasters, 4th edn. McGraw Hill, New York
Adushkin VV, Nemchinov IV (1994) The consequences of impacts of cosmic bodies on the surface of the Earth. In: Gehrels T (ed), pp 721–775
Beck U (1987) Risk society: towards a new modernity (translated: M. Ritter). Sage, London
Beck U (1992) Politics of risk society. In: Franklin J (ed) The politics of risk society polity press, London, pp 9–22

Blaikie P, Cannon T, Davis J, Wisner B (1994) At risk: natural hazards, people's vulnerability and disasters. Routledge, London
Bohle HG (ed) (1993) World's of pain and hunger: geographical perspectives on disaster vulnerability and food security. Freiburg Studies in Development Geography. Verlag Breitenbach, Saarbrücken
Burton I, Kates, RW, White, GF (1978) The environment as hazard. Oxford University Press, Oxford
Cutter S (ed) (1994) Environmental risks and hazards. Prentice-Hall, Englewood Cliffs, NJ
Dauphine A (2001) Risques et catastrophes: observer – spatialiser – comprendre – gerer. Armand Colin, Paris
Degg MR (1992) Some implications of the 1985 Mexican earthquake for hazard assessment. In: McCall et al. (ed), pp 105–114
Dressler BO, Sharpton VL (eds) (1999) Large meteorite impacts and planetary evolution II. Geological Society of America, Special Paper #339, Boulder, Colorado
Enarson E, Morrow BH (1998) The gendered terrain of disaster: through women's eyes. Praeger, Westport, Conn
Gehrels T (ed) (1994) Hazards due to comets and asteroids. University of Arizona Press, Tucson
Gilbert C (1998) Studying disasters: changes in the main conceptual tools. In: Quarantelli E (ed), pp 11–18
Green J (1997) Risk and misfortune: the social construction of accidents. University College of London Press, London
Grieve RAF, Shoemaker EM (1994) The record of past impacts on Earth. In: Gehrels T (ed), pp 417–462
Harris AW, Canavan GH, Sagan C, Ostro SJ (1994) The deflection dilemma: use versus misuse of technologies for avoiding interplanetary collision hazards. In: Gehrels T (ed), pp 1145–1155
Hewitt K (ed) (1983) Interpretations of calamity from the viewpoint of human ecology. George Allen and Unwin, London
Hewitt K (1997) Regions of risk: hazards, vulnerability and disasters. London
Hewitt K (1998) Excluded perspectives in the social construction of disaster. In: Quarantelli E (ed), pp 75–91
Hewitt K, Burton I (1971) The Hazardousness of a place: a regional ecology of damaging events. Department of Geography Research Publication #6, University of Toronto Press, Toronto
Hills JG, Nemchinov IV, Popov SP, Teterev AV (1994) Tsunami generated by small asteroid impacts in Gehrels T (ed), pp 779–789
IFRCRCS (2002) World disasters report International Red Cross and Red Crescent Societies. Oxford University Press, Oxford
Kates RW, Hohenemser C, Kasperson JR (1985) Perilous progress: managing the hazards of technology. Westview Press, Boulder, COL
Kreps G (1989) Social structure and disaster. University of Delaware Press, Newark DE
Kusky TM (2000) Geological hazards: a sourcebook. Greenwood Press, Westport, Conn
March JG, Olsen JP (1989) Rediscovering institutions: the organizational basis of politic. The Free Press, New York
Maskrey A (1989) Disaster mitigation: a community-based approach. Oxfam, Oxford
McCall GJH, Laming DJC, Scott SC (1992) Geohazards: natural and man-made. Chapman and Hall, London
McClean I, Johnes M (2000) Aberfan: disasters and government. Welsh Academic Press, Cardiff
Morrison D (2004) Torino Impact Scale: asteroid and comet impact hazards. NASA Ames Research Center
Morrison D, Teller E (1994) The impact hazard: issues for the future. In: Gehrels T (ed), pp 1135–1143
Morrison D, Chapman CR, Slovic P (1994) The impact hazard. In: Gehrels T (ed), pp 59–91
Oppenheimer C (2003) Climatic, environmental and human consequences of the largest known historic eruption: Tambora volcano (Indonesia) 1815. Progress in Physical Geography 27/2:230–59
Peiser B (2002) Preparing the public for an impending impact. Cambridge Conference Correspondence, http://abob.libs.uga.edu/bobk/ccc/ceo72802.html
Pelanda C (1982) Disaster and sociosystemic vulnerability. In: Jones B, Tomazevic M (eds) Social and economic aspects of earthquakes. Cornell University, Ithaca, New York, pp 67–91
Perrow C (1984) Normal accidents: living with high risk technologies. Basic Books, New York
Quarantelli EL (ed) (1998) What is a disaster? Perspectives on the question. Routledge, New York

Rampino MR (1999) Impact crises, mass extinctions, and galactic dynamics: the case for a unified theory. In: Dressler BO, Sharpton VL (eds), pp 241–248

Robertson DS, McKenna MC, Toon OB, Hope S, Lillegraven JA (2004) Survival in the first hours of the Cenozoic. Geological Society of America Bulletin 116(5/6):760–768

Sen A (1981) Poverty and famines: an essay on entitlement and deprivation. Clarendon Press, Oxford

Shaw R, Goda K (2004) From disaster to sustainable society: the Kobe experience. Disasters 28/1:16–40

Steinberg T (2000) Acts of God: the unnatural history of natural disaster in America. Oxford University Press, New York

Toft B, Reynolds S (1994) Learning from disasters. Butterworth/Heinemann, Oxford

Toon OB, Zahnle K (1994) Environmental perturbations caused by asteroid impacts. In: Gehrels T (ed), pp 791–826

Turner BA (1978) Man-made disasters. Wykeham Publications, London

Verschuur GL (1996) Impact! The threat of comets and asteroids. Oxford University Press, New York

Weissman PR (1994) The comet and asteroid impact hazard in perspective. In: Gehrels T (ed), pp 1191–1212

Whelan F, Kelletat D (2003) Submarine slides on volcanic islands – a source for mega-tsunamis in the Quaternary. Progress in Physical Geography 22/2:198–216

White GF (1945) Human adjustment to floods: a geographical approach to the flood problem in the United States. University of Chicago

Wisner B (1993) Disaster vulnerability: geographical scale and existential reality. In: Bohle H (ed) World's of pain and hunger: geographical perspectives on disaster vulnerability and food security. Freiburg Studies in Development Geography. Verlag Breitenbach, Saarbrücken, pp 13–54

Chapter 25

May Land Impacts Induce a Catastrophic Collapse of Civil Societies?

Andrea Carusi · Alessandro Carusi · Luca Pozio

25.1 Introduction

The possible influence of impacts of celestial bodies on the evolution of life on Earth has been brought to the attention of the scientific community in 1980, with the publication of a famous paper by Alvarez et al. (1980) on the event that caused the mass extinction at the boundary between the Cretaceous and the Tertiary, 65 million years ago.

The problem of quantifying the impact hazard posed by Near-Earth Objects (NEOs) to the Earth and its inhabitants has been extensively studied in the past decade (see, e.g. Morrison et al. 2002, and references therein). The risk associated with major impacts is now quite well established and the related probability of occurrence rather well understood. Moreover, it is now clear to scientists that this problem is only partially of a scientific nature. There is a growing awareness among students of NEOs, and in the larger international community, that impacts represent a natural threat at least comparable to more familiar threats, like volcanic eruptions and earthquakes, which are given special attention and monitoring efforts by the civil defense organizations in most countries. It has also been stated many times that the human societies, because of their complexity, may be particularly vulnerable to cosmic impacts.

A number of international organizations have already paid attention to the impact problem (Council of Europe 1996; United Nations 1999; OECD 2003). What is lacking at this time, however, is an investigation of the real effects on a civil society of the impact of an object in the small-medium size range (50 to 500 meters). Such events have already been indicated as potentially very damaging should they occur in the oceans, because of the ensuing tsunamis but, to the best of our knowledge, no detailed studies have been performed on the consequences, for the structures of a society, of impacts on land.

We present here an attempt to examine the consequences of an impact on a global scale, taking into account as much as possible the mutual interactions of the various sectors of the civil life. It is a typical problem of the growing science of "complexity", and the human society can well be considered a complex system.

In Sect. 25.2 we will present our case study – impacts over a European country – and in Sect. 25.3, after a brief summary of the most recent understanding of complex systems and complexity in general, we provide a classification of the structures of a civil complex system. In Sect. 25.4 we then present a qualitative analysis of the possible consequences in this case study. In the final section we finally discuss the results.

25.2
Medium–Small Scale Impacts on a European Country: a Case Study

As a test case for our study we have chosen a European country: Italy. The reasons for this choice were basically two: (*i*) it was rather easy for us to collect the relevant information on the structures of the Italian society, and (*ii*) the historical, cultural and geophysical properties of Italy make it a special case among the western countries, providing us with the opportunity to study also aspects that may be not much relevant in other cases.

The purpose of the study has been to *qualify* the effects of impacts on the political, economic and social structures of a western society. The study was conducted taking into consideration two types of events: (*i*) the impact of a Tunguska-class object and (*ii*) the impact of a 200–300 meter object. We have considered a terminal energy of 13 MT (megaton) in the first case and of 1000 MT in the second one. It is assumed that no forecast of these events is made, i.e., the society is completely unaware and unprepared.

We have chosen three locations at random within the national boundaries and have drawn contours of the areas subject to relevant damages for the two types of impact for each location, resulting in 6 events (an example of this procedure is given in Fig. 25.1 for location 2).

Fig. 25.1. Distance contours around the impact location 2 (Lazio) The radius of the inner circle is 50 km, that of the outer circle 100 km. The small *inset* in the lower-right shows the three impact locations

For each of the 6 events we have examined:

- the probability of impact
- the area subject to relevant damages and the level of damage within the area, depending on the distance from the sub-impact point
- the possible consequences (only in a qualitative form)

25.2.1
Probability of Impact and Objects Properties

Italy is a rather small country, with an area of 3.01×10^5 km². The ratio of its surface to the total area of the Earth (5.1×10^8 km²) is therefore 6×10^{-4}. This number coincides basically with the probability that an impact (of any object) may happen on Italy and must be combined with the annual probability of impacts at the specified levels. This is possible using the most recent evaluations of the cumulative population of NEOs, as reported for example by Morrison et al. (2002). The results are contained in Table 25.1.

Our fictitious objects are both stony asteroids. The smaller one has a diameter, before the entry in the atmosphere, of about 60 m and the larger one of about 280 m. They both fall on Earth at a speed of 20 km s^{-1} and with an impact angle of 45°. These numbers represent average cases; different circumstances may have some relevance, especially the relative velocity at impact.

Unless otherwise stated, computations of the events associated to these impacts have been performed using the *Earth Impact Effects Program* by Marcus et al. (see Collins et al. 2004), an interactive program accessible on the web. Multiple interrogation of the program allows one to compute the major effects at increasing distances from the sub-impact point. Doing that, it is possible to map the most important effects on the whole area affected by the impact. Table 25.2 lists the most important parameters associated to the impacts and the physical characteristics of the two projectiles.

25.2.2
Level of Damage

There have been a number of studies on the effects of a NEO impact. In a good fraction of cases these studies rely on the published reports concerning nuclear tests, whereas in other cases they are based mainly on modeling of the related phenomena. A good review of these issues is contained in Toon et al. (1997), and an extensive review of the effects of nuclear explosions is contained in Glasstone and Dolan (1977).

Table 25.1.
Probability (P) and recurrence (R, yr) of impacts on Earth and Italy

Impact (MT)	P_{Earth}	R_{Earth}	P_{Italy}	R_{Italy}
13	1.3×10^{-3}	800	8×10^{-7}	1.3×10^6
1 000	3.3×10^{-5}	30 000	2×10^{-8}	5.0×10^7

Table 25.2. Parameters of the impacts and physical properties of the projectiles

Parameter	13 MT	1 000 MT
Diameter (m)	60	280
Density (g cm^{-3})	2.7	2.7
Impact velocity (km s^{-1})	20	20
Impact angle (degrees)	45	45
Target density (g cm^{-3})	2.5 (sedimentary)	2.5 (sedimentary)
Energy before atmospheric entry (MT)	14.6	1 480
Altitude of initial break-up (km)	59.9	59.9
Terminal energy (MT)	13.1	988
Altitude of airburst (m)	10 900	–
Diameter of the final crater (km)	–	4.44

Table 25.3. Parameters characterising the effects of a 13-MT impact according to distance

Parameter	Distance from ground zero (km)				
	10	20	30	50	70
Thermal exposure (kJ m^{-2})	11.92	5.02	2.56	1.00	0.52
Radiant flux (solar units)	2.87	1.21	0.62	0.24	0.13
Timing of the air blast (s)	44.8	69.0	96.7	155.1	214.7
Overpressure[a] (psi)	6.0	3.0	1.8	0.8	0.6
Wind velocity (m s^{-1})	85.4	45.8	28.4	13.0	9.8

[a] The overpressure has been computed directly using the formulae and graphs by Glasstone and Dolan (1977).

In our study we have primarily considered four major effects, namely:

1. Thermal emission (T);
2. Seismic wave (S) (only for the 1 000 MT impact);
3. Fallout of ejecta (E) (only for the 1 000 MT impact);
4. Air blast (A).

The weight of these effects is strongly dependent upon the energy of the event and the distance from the sub-impact point. Tables 25.3 and 25.4 report the parameters associated with the 13 MT and 1 000 MT events according to the distance.

Other effects are probably important, although their relevance cannot be easily determined: (*i*) electromagnetic pulse (EMP), (*ii*) dust loading into the atmosphere,

Table 25.4. Parameters characterising the effects of a 1 000-MT impact according to distance

Parameter	Distance from ground zero (km)				
	10	20	30	50	70
Thermal exposure (MJ m^{-2})	1.970	0.488	0.213	0.073	0.034
Radiant flux (solar units)	472.00	117.00	51.10	17.50	8.19
Timing of seismic wave (s)	2	4	6	10	14
Magnitude (Richter scale) (–)	6.6	6.6	6.6	6.6	6.6
Magnitude (Mercalli scale) (–)	7–8	7–8	7–8	7–8	6–7
Ejecta arrival time (s)	45.2	64.0	78.5	101.0	120.0
Ejecta mean thickness (m)	1.990	0.249	0.074	0.016	0.004
Average fragment size (m)	12.200	1.940	0.664	0.171	0.070
Timing of the air blast (s)	30.3	60.6	90.9	152.0	212.0
Overpressure (psi)	99.30	22.40	9.88	3.80	2.14
Wind velocity (m s^{-1})	623.0	243.0	130.0	56.9	33.4

(*iii*) nitric and acid rains, (*iv*) water injection in the atmosphere, (*v*) sulfate aerosol formation, and (*vi*) any other geophysical effect (influence on local tectonics and volcanism).

25.2.2.1
Effects of the 13 MT Impact

It is convenient, as reported in Table 25.3, to analyze the effects in annular rings of increasing distance from the sub-impact point. The 13 MT explosion takes place at an altitude of 10 900 m and there is no crater formation. However, the effects are relevant up to a distance of 30 km from ground zero (about 32 km from the air burst), i.e. in the area where the overpressure is larger than 2 psi. The major effects are reported in Table 25.5.

25.2.2.2
Effects of the 1 000 MT Impact

It is again convenient to analyze the effects in annular rings of increasing distance from the impact point. The supposed object of 280 m size certainly reaches the ground, although fragmented, at a speed of 16.3 km s^{-1}, and therefore produces a crater of diameter 4.44 km (rim diameter, after the relaxation of the crater walls, see Table 25.2). The explosion gives rise to a fireball whose angular diameter, at a distance of 50 km, is still about 14 times that of the sun. The major effects are reported in Table 25.6.

Concerning the other effects listed above, the most serious one in this case study seems to be the Electro-Magnetic Pulse (EMP), originated by the extremely hot fire-

Table 25.5. Effects of the 13-MT impact according to the distance from ground zero

Distance (km)	Type	Effect
10	T	Most combustible material will ignite, flash blindness
	A	Houses collapse, many commercial buildings collapse, heavier constructions severely damaged, masonry collapse, up to 60% trees blown down, wind velocity 85 m s^{-1}
20	T	Easily ignitable and dry materials will ignite
	A	Walls of a typical steel frame building blown away, masonry collapse, up to 30% trees blown down, wind velocity 46 m s^{-1}
30	T	Occasional fires
	A	Severe damage to structures, masonry collapse, wind velocity 28 m s^{-1}
50	T	Occasional fires
	A	Light damages to poorly built structures, wind velocity 13 m s^{-1}

ball. It seems to have the potential to cause a complete black-out of all electronic devices in an area much larger than that considered here. However, this estimate is based on data reported by Glasstone and Dolan (1977) and referring to nuclear explosions that will produce a lot of X- and γ-rays. To our knowledge there is no good estimate of the importance of this phenomenon in a context like ours. The EMP could put out of order many computers and other processor-based devices including, among others, the control devices of most of the communication and energy production/transportation lines. This effect may be relevant also for the less powerful impact.

25.3
The Civil Society as a Complex System

25.3.1
Recent Developments in the Science of Complexity

No universally accepted definition of a complex system exists nowadays; theories and researches of diverse character have been built under the heading *complexity*. Although valuable, they have produced around this term different definitions and different methodological approaches. Complexity studies the common aspects of organization phenomena; in particular, the research is based upon the non-equilibrium states from which self-organising systems arise. Prigogine (Prigogine and Stengers 1979; Prigogine and Nicolis 1989) calls this process "thermodynamics of irreversible processes": the most peculiar of such processes can be identified in the origin of life and, consequently, of all those biological systems and organisms, and of all the social and sociocultural systems characterized, as every irreversible process, by self-organization. There is, in other words, a *constructive* role of non-equilibrium. Far from equilibrium, states and complex structures are created which could not exist in a reversible world.

There is a coordinated effort to investigate the natural events that give rise to this "emergent order." Every self-organized system is a complex system: biology, sociol-

Table 25.6. Effects of the 1 000-MT impact according to the distance from the impact point

Distance (km)	Type	Effects
10	T	3° degree burning, trees and grass ignite, wooden materials ignite (gas, gasoline, and other explosive materials will contribute), flash blindness, coretinal damages
	S	Collapse of all non anti-seismic buildings
	E	2 m deposit, fall of 12 m fragments at $v = 313$ m s^{-1}
	A	Collapse of multi-storey wall-bearing buildings, collapse of highway truss and girder bridges, up to 90% trees blown down, wind velocity 623 m s^{-1}
20	T	1° degree burning, widespread fires (gas, gasoline, and other explosive materials will contribute), flash blindness
	S	Serious damages to most buildings, collapse of poorly built edifices and structures, collapse of walls, monuments, chimneys
	E	25 cm deposit, fall of 2 m fragments at $v = 443$ m s^{-1}
	A	Collapse of multi-storey wall-bearing buildings, collapse of highway truss bridges, up to 90% trees blown down, wind velocity 243 m s^{-1}
30	T	Widespread fires (ejecta will contribute, as well as explosion of gas and gasoline stations and pipelines), flash blindness
	S	Serious damages to many buildings, collapse wooden structures
	E	7 cm deposit, fall of 66 cm fragments at $v = 542$ m s^{-1}
	A	Collapse of multi-storey wall-bearing buildings, up to 90% trees blown down, wind velocity 130 m s^{-1}
50	T	Occasional fires (mainly induced by secondary sources), flash blindness at night
	S	Partial collapse of poorly built structures
	E	2 cm deposit, fall of 17 cm fragments at $v = 700$ m s^{-1}
	A	Collapse of roofs and poorly standing monuments, wind velocity 57 m s^{-1}, glasses will shatter
70	T	Occasional fires (mainly induced by pipelines break-up), flash blindness at night
	S	Partial collapse of poorly built structures
	E	Fall of 7 cm fragments at 826 m s^{-1}
	A	Glasses will shatter, minor damages, wind velocity 33 m s^{-1}

ogy, economics and physics are all disciplines working on the same subject, contributing to an ensemble of inevitably multi-disciplinary researches.

Numerous centers for the study of complex systems were already active in the 1980s (Waldrop 1992), and several tens of them were active at the end of the century in almost all the western countries. Although mainly studied by scholars from the scientific domain, complexity is nowadays effectively considered the "third culture" (Brockman 1995), at the crossover of the scientific and humanistic disciplines. It tries to keep a unitary character and to tend towards a common language, but it still lacks the conceptual and terminological systematization necessary to obtain a universally accepted academic status.

For an overview of different approaches to complexity the reader is also referred to Morin (1977), Varela (1979), Maturana and Varela (1980), Maturana and Varela (1987), Casti (1994), Gell-Mann (1994), Luhmann (1995), Maynard-Smith and Szathmàry (1999).

25.3.2
What is a Complex System?

The organization of every dynamical system is based on the ability of its components to interact and communicate by exchanging energy and information. Complex systems are organized by an internal network of *communication channels*, also exchanging information with the environment (here called the *context*), which provides the system with the fundamental capacity to adapt to perturbations. Though continually adapting, the system usually does not suffer drastic modifications in this mutual exchange. At the same time, it increases its information content, giving rise to new structures and properties, or to gains or losses of its flexibility, which ultimately determines its survival or collapse. Through the analysis of flexibility and organization, complexity is devoted to the investigation of the evolutionary behaviors of the systems and, on the other hand, it tries to understand whether (and how) would it be possible to handle them. Because of their chaotic and structured nature, complex systems are:

- *Irreducible.* They exhibit a behavior that cannot be deduced from the analysis of their components alone. Therefore, a holistic approach is necessary. In complex systems the whole is greater than the sum of its parts.
- *Sensitive to the initial conditions.* Complex systems are deterministic, like all macroscopic systems, but they are also largely chaotic inasmuch as small local variations, in time and space, may lead to large effects, that can be global and/or on long time scales, not foreseeable and often abrupt (Ruelle 1991).
- *Structured.* Every complex system is structured and every structure plays a precise role and is potentially connected with every other structure of the system. This network may be subject to modifications, although keeping its state of dynamical equilibrium.
- *Interconnected through communication channels.* The communication channels allow the exchange of information among the structures and between the system and the context.
- *Self-organization, adaptivity* and *flexibility* are among the most peculiar properties of a complex system: when it is subject to a strong perturbation (like a sudden climatic change or the impact of a large NEO in the case of ecological systems, or an economic crisis in case of a human system) it tries to reorganize itself and to reach a new dynamical equilibrium. These changes, however, may deeply transform the system and, in extreme cases, may not be sufficient to avoid a catastrophic collapse.

25.3.3
The Phase of Catastrophe

The term *collapse* refers to a radical transformation of a system, where the majority of the structures lose or change drastically their functionality; when this happens in a short time through a cascade event, we have a catastrophic collapse. The collapse of a social system is mainly due to the break-up of communication channels, more than to the physical destruction of structures. It is not frequent that in a social system macrostructures exist handling a large portion of the global information; in most cases this is distributed among many different structures that are able, in case of loss of part of them, to allow the system to survive.

The communication channels, on the other hand, are difficult to rebuild in a short time, and this *segregation* may cause serious damages to some structures and to the system itself, with the possibility of a catastrophic collapse. The break-up of communication channels prevents the exchange of information among structures and blocks the self-organization of the system, opening the way to the formation of evolutionary niches. Indeed, as the system is no more an interconnected whole, the isolated structures gain a larger flexibility: this self-sufficiency may protect them from a collapse, but may also avert the system from recovering the initial organization. In case of strong perturbations, it is fundamental for the system to maintain control on the isolated structures.

Management and control of social systems are then possible when, in case of a possible partial collapse, a program exists for learning contextual information for reorganization. In other words, the system must be induced to resume spontaneously its equilibrium in order to regain the flexibility necessary for its survival. It is clear the advantage of those systems in which the control centers and the vital communication channels are not put all together in one or a few structures: on one hand it is improbable that they become all lost simultaneously, on the other hand the assortment of centers produces a variety of control nodes, and this may allow to restore the damaged channels without losing too many structures.

25.3.4
Main Structures of the Country Social System

Based on the principles of complexity just mentioned, we have examined three main aspects of our sample:

1. the type and relevance of each structure in the system,
2. the existence of communication channels, both from a physical point of view (roads and the like) and from a civilian point of view (telephone lines, etc.),
3. the consequences of the disruption of structures and communication channels and the possible set-up of "cascade" events.

We have identified 15 structures that include all the major activities within the country; they are listed in Table 25.7 and for each of them we have indicated the most important elements and components likely to be affected by an impact.

Table 25.7. The major structure of the civil complex system

Structure	Related components and elements
Political, decisional	National and local governments and parliaments and associated activities. Buildings, libraries, networks, databases.
Administration and law	National and local administrations. Buildings, networks, databases. Tribunals, prisons. Lawyer offices and facilities. Criminal investigation departments. Notary offices and databases.
Finance, credit	Banks, insurances, stock exchange, financial companies. Buildings, networks, databases.
Agriculture	Farming, breeding, fishing. Industrial and food cultivations, cattle. Structures, farms, ships, equipment, irrigation systems.
Handicraft and trade	Workshops, laboratories, equipment, shops, stores, means of delivery. Trade-related companies, offices, buildings.
Industry	Key-industry. Consumer goods and food industries. Plants, equipment, facilities.
Energy production and distribution	Power plants and lines. Oil refineries, oil and gas pipelines, gas stations. Hydroelectric infrastructures and dams. Aqueducts. Renewable energy plants. Nuclear power plants (currently none active in Italy) and radioactive waste. Mining resources.
Infrastructures, transportation	Roads, railways, bridges, tunnels. Local transport lines, subways. Railway stations, harbours and airports. Trains, mercantile fleet, aircrafts. Cars and trucks. Underground distribution systems, sewer systems. Buildings and houses, public and private. Bus systems, traffic control systems.
Communications	Radio, TV, newspapers, telephone/fax, internet, mail. Telecommunication systems, computer systems, mobile phone systems. Cabling, fibre optics, radio-links, antennas, space communications.
Homeland, environment	Forests, woods, national and regional parks, biodiversity. Lakes, rivers, shores, mountains, underground waters. Waste: dangerous and non-dangerous. Recycling plants, water purifying plants.
Society and culture	Art and architecture, libraries, museums, monuments, archaeological sites, churches. Landscapes. Cinemas, theatres, sport structures. Local cultural resources. Social services, tourism.
Health care	Hospitals and clinics, public and private. National health care system institutes. Laboratories, medical centres and offices, pharmacies and chemist shops. Medical facilities, networks, databases. Ambulances and first aid equipment.
Education and research	Schools, university campuses, public and private. Education facilities, networks, databases. Research centres, laboratories, offices, public and private. Large research facilities: nuclear laboratories, astronomical observatories. Seismic and volcanologic monitoring systems.
Civil and military defence	Civil protection structures, national and local. Equipment and stores. Police, national and local, networks and databases. Intelligence Services. Army, Air Force, Navy, Carabinieri Corps. Military bases, airports, harbours, military equipment, aircrafts and ships, weapons, radio-links. Administration and chain of command.
International Relationships	Embassies and consulates (of Italy and the Vatican State). Industrial, commercial and cultural foreign offices. International organisations centres. Customs and coastal patrol services.

25.4
Results

The three impact points chosen for our analysis are located in regions of the country with very different characteristics. The two types of impacts have also very different consequences; it is therefore useful to examine them separately. The first important result is that the 13 MT event, although impressive from a "human" point of view, is not so much different from a more usual natural disaster like an earthquake or a flood, events rather common in Italy. We will describe this event for all the three locations in the next paragraph, whereas the more energetic impact will be described separately for the three points in the subsequent paragraphs.

25.4.1
Consequences of the 13 MT Impact on the three Points

The most important consequences of this impact are related to the thermal exposure and to the air blast (and possibly to the occurrence of an EMP). The object explodes at about 11 km of altitude and the fireball is several times more brilliant than the Sun in an area that extends to about 10 km from ground zero (the point on the surface directly below the air burst location). According to Glasstone and Dolan (1977), the overpressure originated by the shock wave is above two psi (pounds per square inch) up to a distance of about 30 km from ground zero. These data are consistent with the Tunguska explosion, possibly a little less energetic. The major effects would be the collapse of many structures, especially ancient buildings, and the extended fires. Probably the air blast would contribute to extinguish fires, at least partially, but it is particularly difficult to judge the relevance of other events, like the explosion of gas and oil pipelines, that could greatly contribute to make the situation very difficult. The power of the air blast is sufficient to flatten trees up to a distance of about 20–30 km from ground zero. These fallen trees would block most of the roads leading to the center of the area. When coupled to probable car accidents caused by the flash blindness, to the damages to the electric network, to the probable collapse of old bridges on the rivers, and to the ash and dust put in the air by the fires, this estimate seems to show that the point of the explosion, up to a distance of about 30 km, would be very difficult to reach both with surface vehicles and by air. The first emergency operations would face tremendous problems.

However, none of the three events would represent a serious danger for any of the social structures on a national scale. Medium and long-term effects are likely to occur for the national economy, but not much different from those experienced in the past for other natural events. The most difficult situation is probably represented by the event on Point 3 (Sicily), where two local administrative centers (Caltanissetta and Enna) would be seriously damaged.

25.4.2
Point 1 Consequences of the 1000 MT Impact

Point 1 is in the Regione Veneto, not far from the city of Belluno and at about 80 km from Venice and Vicenza. The distribution of damages for the 1000 MT impact according to distance is reported in Table 25.8.

Table 25.8. The most important events connected to a 1 000-MT impact on point 1 (Veneto)

Distance (km)	Major effects	Notes
10	Total destruction of all buildings and other structures; global fires; casualties of about 100%; landslides; EMP	Complete destruction of the city of Belluno and its surroundings. All roads interrupted, collapse of bridges on the river Piave and the last portion of the A27 highway. Complete obliteration of flora and fauna. Strong perturbations to the hydrologic regime. This area will be inaccessible even from the air (because of fires). The area is mostly covered by the crater itself and its ejecta blanket
20	Collapse or very serious damages to most buildings; collapse of bridges, railways and road system; global fires; landslides and outflows; EMP	Many small towns involved. Interruption of all roads and railways. Collapse of most private and public buildings. Complete obliteration of flora and fauna. Problems to the lakes S. Croce and Vaiont. If in winter, avalanches probable. This area will be inaccessible even from the air
30	Collapse or serious damages to most of the non anti-seismic buildings; effects typical of a strong tornado (wind velocity 470 km h^{-1}); landslides; up to 90% of trees and electric lines flattened; "bombs" of 70 cm at more than 500 m s^{-1}; EMP	Very serious problems to the towns of Vittorio Veneto, Feltre and Pieve di Cadore. Collapse of serious damages to a large fraction of civil buildings. All roads interrupted (also because of fallen trees). Damages to the bridges on river Piave and of its tributaries. Problems to the lakes of Pontesei, Barcis and Alleghe. Landslides and avalanches will seriously damage all tourist structures. This area will be inaccessible by surface vehicles
50	Serious and partial damages to most building, collapse of monuments and roofs; wind velocity 200 km h^{-1}, "bombs" of 20 cm at more than 700 m s^{-1}; large fraction of trees and electric lines flattened; EMP	Major towns involved: Cortina d'Ampezzo, S. Stefano di Cadore, Maniago, Pordenone, Conegliano, Porcia, Sacile, Cordenons, Valdobiadene, Montebelluna. Problems to the A27 and A28 highways. All minor roads interrupted because of fallen trees. Serious problems to the many tourist resorts in Valle d'Ampezzo, Cadore and Val Settimana
70	Partial fires (mainly due to explosions); collapse of poorly built buildings; scattered interruptions of roads, railways and electric lines; EMP	Problems to the highways A22, A23, A4 and to the river systems of Piave, Tagliamento and Brenta. Involved valleys: Val Sugana, Altopiano dei Sette Comuni, Val Pusteria. Major cities involved: Treviso, Bassano del Grappa, Bressanone, Tolmezzo, Gemona del Friuli, Portogruaro, Sandona di Pieve and Castelfranco Veneto. Effects will reach Austria

This is a highly industrialized region, full of very small but very specialized companies. The impact point is on the Northern side of the Padana Valley, which is full of large and small rivers, often subject to outflows. These rivers, and the many natural and artificial lakes, are of great importance for the production of electricity (about 32% of the total energy in the region). Major highways and railways cross the region, connecting the North-East to the North-West and to the South of the country, and to Austria. The level of destruction in this area would imply a very serious and long-lasting stop to the production activity. Given the relevance that North-East of Italy has on the national budget, this would certainly cause a serious depression of the national economy, to be added to the loss of lives and properties (difficult to quantify). Moreover, the Northern part of this region is occupied by the Alps, where many summer and winter resorts are located, and the whole region is rich in art masterpieces; so, the second most important resource of the region (tourism) would also suffer a tremendous loss. The geographical structure with many mountains and valleys, together with the destruction of bridges and the obstruction by landslides and fallen trees, would render almost impossible to reach a large portion of the area for a rather long time.

25.4.3
Point 2 Consequences of the 1000 MT Impact

Point 2 is in the Regione Lazio, at about 50 km from Rome. The place, called Sabina, is mainly rural, with well-developed and important production of olive oil and wine. The valleys towards Rome are highly industrialized (especially high-tech companies) and represent important interchange locations where the major roads, railways and highways connecting the North to the South are located. Important rivers are Tiber and Aniene, plus a number of minor rivers often closed by dams to produce electricity. The entire region is obviously full of archaeological and other art sites of extreme importance. The effects of the 1000 MT impact on this area are shown in Table 25.9.

The major problem here would be the involvement of the city of Rome. Although not completely destroyed, the city would suffer very serious damages. Most of the structures concerning the political and administrative management of the country are in Rome, as well as the embassies and other foreign institutions connected with Italy and the Vatican State. It is probable that serious problems in the administration of the country would arise, although possibly not catastrophic. The economy of the entire area, both from industrial and rural sources, would greatly suffer for the event. Another important consequence of an impact in this region is that the ground transportations between the North and the South would be practically interrupted, although the roads and railways along the Adriatic coastline would continue to be efficient. Due to the historical relevance of Rome and its archaeological sites, the loss for art and culture (and therefore for tourism) would be enormous. Apart from Rome, other important locations would be severely affected. The city of Terni, for example, produces a good fraction of the Italian steel, and the Fucino and Agro Pontino plains are very productive regions for food. Fucino also hosts important space communication systems.

25.4.4
Point 3 Consequences of the 1000 MT Impact

Point 3 is in the center of Sicily, not far from the city of Caltanissetta. It is a basically rural region: Sicily has 65% of the Italian citrus plantations, and produces a good fraction of wheat, wine and olive oil. At about 35 km from the impact point there is the Blufi water system, which supplies water to the provinces of Caltanissetta, Enna, Agrigento. Important archaeological sites are scattered in the region, especially towards the plain of Agrigento. Table 25.10 reports the most important effects of the 1000 MT impact in this area.

Table 25.9. The most important events connected to a 1 000-MT impact on point 2 (Lazio)

Distance (km)	Major effects	Notes
10	Total destruction of all buildings and other structures; global fires; casualties of about 100%; landslides; EMP	Salaria road interrupted. Serious problems to the rivers and lakes Turano and Salto. About 15 small towns and many minor villages destroyed. Complete obliteration of flora and fauna. This area is inaccessible even from the air (because of fires). The area is mostly covered by the crater itself and its ejecta blanket
20	Collapse or very serious damages to most buildings; collapse of bridges, railroads and road system; global fires; landslides and outflows; EMP	Larger towns are involved (Palombara Sabina, Poggio Mirteto). Complete obliteration of flora and fauna. Destruction of crops. Problems to the A24 highway and the local railways. This area is inaccessible even from the air
30	Collapse or serious damages to most of the non anti-seismic buildings; effects typical of a strong tornado (wind velocity 470 km h^{-1}); landslides; up to 90% of trees and electric lines flattened; "bombs" of 70 cm at more than 500 m s^{-1}; EMP	Very serious problems to large towns (Rieti, Tivoli, Guidonia, Monterotondo). Collapse or serious damages to many civil and military assets. Interruption of the A1 and A24 highways and of the Salaria and Tiburtina Roads. Most minor roads interrupted for the fall of trees. High-speed train connection to Florence and the North interrupted. This area is inaccessible by surface vehicles
50	Serious and partial damages to most building, collapse of monuments and roofs; wind velocity 200 km h^{-1}; "bombs" of 20 cm at more than 700 m s^{-1}; large fraction of trees and electric lines flattened; EMP	This area includes major cities (Rome, Terni, L'Aquila, Frascati, Avezzano, Narni). All roads are interrupted (fall of bridges and trees) or seriously damaged. The rail node of Rome is blocked. Very serious damages to archaeological sites. Problems to the Tiber and Aniene Rivers, including possible outflows. Problems for the gas, oil and water pipelines. Problems to the Vigna di Valle Air Force base
70	Partial fires (mainly due to explosions); collapse of poorly built buildings; scattered interruptions of roads, railways and electric lines; EMP	The area includes most of the regions Lazio and Abruzzo. Problems to the cultivations in Fucino and Agro Pontino plains. Problems to the lake, dam and hydro-electric power station of Campotosto. Problems to the Marcia and Peschiera aqueducts to Rome. Problems to the space antennas of Telespazio, to the National Parks, to the Physics Laboratories under the G. Sasso tunnel. Problems to the Air Force base of Pratica di Mare

Again, the interruption of most of the communication lines (roads, highways, railways) would severely affect the region, because the two most important cities, Palermo and Catania, would be disconnected for a rather long time. The food production would be drastically reduced, with serious consequences for the economy of the region and with a probable negative influence on the national economy. Also in this case the important resource of tourism would be very much reduced.

Table 25.10. The most important events connected with a 1 000-MT impact on point 3 (Sicily)

Distance (km)	Major effects	Notes
10	Total destruction of all buildings and other structures; global fires; casualties of about 100%; landslides; EMP	Complete destruction of Caltanissetta. Interruption of the A19 highway (Palermo-Catania). River Salso deviated. Complete obliteration of flora and fauna. This area is inaccessible even from the air (because of fires). The area is mostly covered by the crater itself and its ejecta blanket
20	Collapse or very serious damages to most buildings; collapse of bridges, railroads and road system; global fires; landslides and outflows; EMP	Seven larger towns involved. Complete destruction of flora and fauna. Crop destruction. This area is inaccessible even from the air
30	Collapse or serious damages to most of the non anti-seismic buildings; effects typical of a strong tornado (wind velocity 470 km h^{-1}); landslides; up to 90% of trees and electric lines flattened; "bombs" of 70 cm at more than 500 m s^{-1}; EMP	Very serious problems to the city of Enna and other large towns (Canicattì, Leonforte). Probable destruction of the Piazza Armerina archaeological site. This area is inaccessible by surface vehicles
50	Serious and partial damages to most building, collapse of monuments and roofs; wind velocity 200 km h^{-1}; "bombs" of 20 cm at more than 700 m s^{-1}; large fraction of trees and electric lines flattened; EMP	This area practically cuts Siciliy in two. Larger towns involved: Caltagirone, Miscemi, Licata, Favara. Serious problems to the Madonie Natural Park. Serious problems to the Blufi water system and dam. Very serious losses for art and archaeology
70	Partial fires (mainly due to explosions); collapse of poorly built buildings; scattered interruptions of roads, railways and electric lines; EMP	Effects down to Agrigento and its archaeological area. Problems to Porto Empedocle, Catania plain, Cefalù. Problems to the industrial cities of Termini Imerese and Gela (oil industry). Problems to the Nebrodi Natural Park

25.5 Discussion

The first problem that would face any country in an impact event is the identification of the immediate emergency actions to be taken. In our view, however, this is a "minor" problem, because we are more interested in the analysis of the medium-long term consequences, those that could cause a collapse of the social system. Although the death toll may be tremendous, this is not the major point of concern: it is more relevant to understand whether the system would maintain its capabilities, in terms of self-organization and adaptation, to recover from the crisis. It seems probable, as already stated in Sect. 25.4.1, that a 13 MT event would not seriously endanger the global organization: the political and decisional centers would be still active and the country has certainly the capability to tolerate a disaster of this size.

The larger impact could be much more serious. The level of destruction is in itself so high that an exceptional effort would be necessary just to assure assistance to the survivors. Many local administrative structures would be destroyed and the central government would be pressed by the necessity to confront with many urgent needs, with the risk of being overwhelmed by the task. From the point of view of complexity, the major danger comes from the break-up of virtually all the communication channels in a portion of the national territory. This applies not only to roads and power lines, but also to the exchange of information and to the channels devoted to the transmission of directives. Thus, there is a potential risk that some of the structures would not be able to communicate and to reorganize.

However, the destruction would still be local and would not affect the entire territory; distant regions would supply support and assistance. Most of the Regional Administrations would still maintain their capabilities and could provide an "emergency network" capable to face the first difficult moments. It is important, in this respect, to note that the national civil defense organization (Protezione Civile) is scattered over the whole country and would be able to react even if the headquarters in Rome would be damaged. Moreover, if we consider that Italy is part of the European Union, and that certainly many other countries would provide support and help, including remote sensing and communication from space, we do not think that the country as a system would collapse. Certainly, the consequences of the impact would affect the national economy and welfare for a long time, but none of the main structures of the system would be completely destroyed. Thus, provided that the central authority might maintain control over the global situation, the system would slowly but continuously recover.

We can compare the events depicted in this paper with the devastating tsunami in the Indian Ocean in December 2004. The energy released in that event, i.e. the energy released by the earthquake that caused the tsunami, was probably equivalent to some hundreds of MT (this information is derived from the USGS site *http://earthquake.usgs.gov/faq/meas.html*). Both the level of devastation and the amount of help and assistance from outside could be in roughly the same range.

It should also be noted that the analysis done here does not take into account an important factor: that the impact could be anticipated, even if not avoided. This information is in itself extremely relevant, potentially allowing the establishment of emergency plans before the event. A detailed analysis of the consequences of a specific, well known event would also provide (time permitting) the opportunity to plan measures to decrease the break-ups of communication channels and to reduce the concentration of vital functions in individual structures. It would also make a difference if contingency plans had already been established in anticipation of a similar event, because they could become operative in a very short time (hours). In the case of Italy, for example, an evacuation plan has been studied and tested, and is ready to be used, in case of a big eruption of the Vesuvius, that would affect directly a region with more than half a million inhabitants. To the knowledge of the authors no country in the world has ever studied a similar plan for asteroid impacts.

This study has shown, although still only qualitatively, what would be the weak points for a social system in an impact event. The most important, in our opinion, is the vulnerability of the communication channels, many of which find their ways through "hubs". In our case, an example of this kind are the railroads and highways networks, with two basic nodes in Rome and Bologna: if either of the two cities is damaged, ground transportations would become a serious problem. On the contrary, telephone and electric networks are already well distributed. The nation capital is another "hub", at least from a political and administrative point of view. The indication that comes out naturally from an analysis in terms of complexity is that these structures should not be concentrated in a single place or arranged in a rigid hierarchical way. In many biological systems (circulatory and neural systems, for example) when a major channel is blocked the rest of the network is able to develop spontaneously alternative ways in order to assure the survival of the whole system. Social systems should be structured in the same way.

The purpose of this paper, as stated at the beginning, was to ascertain whether an impact of objects in the small-medium size range might induce the catastrophic collapse of a social system. The data gathered in our analysis seem to suggest that this is not the case, at least for a western country. Different situations in different countries, however, may lead to different results, and we do not pretend to have answered this question in a general way. Moreover, our analysis has been only qualitative, and we plan to examine this test case also from a quantitative point of view.

Acknowledgments

The authors are indebted to Giovanni B. Valsecchi for a critical review of the manuscript. They also wish to thank the two referees for very useful comments and suggestions, especially on the difficult task of interfacing the sciences of physics and complexity.

References

Alvarez LW, Alvarez W, Asaro F, Michel HV (1980) Extraterrestrial cause for the Cretaceous-Tertiary extinction. Science 208:1095–1108
Brockman J (1995) The third culture. Simon and Schuster, New York
Casti JL (1994) Complexification: explaining a paradoxical world through the science of surprise. Harper Collins, New York
Collins GS, Melosh HJ, Marcus R (2004) Earth impact effects program: a web-based computer program for calculating the regional environmental consequences of a meteoroid impact on Earth. Preprint: *http://www.lpl.arizona.edu/impacteffects*
Council of Europe (1996) Report on the detection of asteroids and comets potentially dangerous to humankind. Doc 7480, 9 February 1996, Council of Europe, reference motion 1080
Damasio AR (1994) Descartes' error, emotion, reason, and the human brain. Putnam, New York
Gell-Mann M (1994) The quark and the jaguar: adventures in the simple and the complex. WH Freeman & Co, New York
Glasstone S, Dolan PJ (1977) The effects of nuclear weapons. United States Department of Defense and Energy Research and Development Administration, Washington
Kauffman S (1995) At home in the universe: the search for the laws of self-organization and complexity. Oxford Univ Press, New York
Luhmann N (1995) Social systems. Stanford Univ Press, Stanford
Maturana H, Varela F (1980) Autopoiesis and cognition: the realization of the living. D Reidel, Dordrecht
Maturana H, Varela F (1987) The tree of knowledge: the biological roots of human understanding. Shambhala, New York

Maynard-Smith J, Szathmàry E (1999) The origin of life: from the birth of life to the origins of language. Oxford Univ Press, Oxford

Morin E (1977) La méthode. Le Seuil, Paris

Morrison D, Harris AW, Sommer G, Chapman CR, Carusi A (2002) Dealing with the impact hazard. In: Bottke W, Cellino A, Paolicchi P, Binzel RP (eds) Asteroids III. Univ Arizona Press, Tucson

Organisation for Economic Co-operation and Development (OECD) Global Science Forum (2003) Final report workshop on near-Earth objects: risks, policy and actions. *http://www.oecd.org/dataoecd/39/40/2503992.pdf*

Prigogine I, Stengers I (1979) La nouvelle alliance: métamorphose de la science. Gallimard, Paris

Prigogine I, Nicolis G (1989) Exploring complexity, an introduction. Freeman, New York

Ruelle D (1991) Hasard et chaos. Editions Odile Jacob, Paris

Toon OB, Zahnle K, Morrison D, Turco RP, Covey C (1997) Environmental perturbations caused by the impacts of asteroids and comets. Rev Geophys 35:41–78

United Nations (1999) Report of the third United Nations conference on the exploration and peaceful uses of outer space. United Nations Publication A/CONF 184/6, Vienna

Varela F (1979) Principles of biological autonomy. Elsevier, New York

Waldrop MM (1992) Complexity, the emerging science at the edge of order and chaos. Simon and Schuster, New York

Watzlawick P, Beavin JH, Jackson DD (1967) Pragmatics of human communication: a study of interactional patterns, pathologies, and paradoxes. WW Norton & Co, New York

Chapter 26

The Societal Implications of a Comet/Asteroid Impact on Earth: a Perspective from International Development Studies

Ben Wisner

26.1
A Mighty Heuristic: Scale, Space and Time

It is important not to let the potential magnitude of the impact from a comet or asteroid impact (CAI) skew discussion. Without doubt the energy released, hence consequences, from an ocean or terrestrial impact would be very large (McGuire 1999, pp 231–235; McGuire et al. 2002, pp 133–158). An impact in the world ocean (approximately 71% of our planet's surface), could affect much of humanity living in large coastal cities and other coastal settlements. Recent trends in urbanization and migration to coastal areas have placed many hundreds of millions of people in harm's way (Wisner et al. 2004, Chap. 2 and 7). <1> A terrestrial impact on a heavily populated area is highly unlikely since humanity's cities cover such a very small percentage of the Earth's surface (only about 2–3%). Yet their "ecological footprint" is many times greater – 15 times as great in the case of greater Vancouver (Canada), 13 times in the case of the whole of the densely populated Netherlands (Wackernagel and Rees 1996). So in both the case of destruction of coastal cities by large tsunami and an impact on a major urban area, there arises the question of providing for survivors and displaced evacuees (if current or future tsunami warning systems can provide sufficient warning). The challenge of immediate relief (provision of water, food, shelter, sanitation, and medical assistance to survivors) following a tsunami produced by a CAI can be imagined by multiplying the logistical efforts required by the Asian tsunami (December 2005) or the impact of hurricane Katrina on New Orleans and the Gulf Coast (August 2005) by an order of magnitude.

Any terrestrial impact will introduce huge quantities of dust into the atmosphere that could have the effect predicted in the 1980s when "nuclear winter" was modeled (Ambio 1982). The eruption of the Tambora volcano in 1815 did, in fact, cause very cold summers and crop failures, provoking what Post (1977) described as the last great subsistence crisis of Europe.

How does one think beyond questions of search and rescue, immediate shelter and care for the survivors (Wisner and Adams 2003; Wisner et al. 2005), and even beyond the difficult years of food insecurity and disrupted economic activity? What are the civilizational implications of such an event?

Thinking about a large comet or asteroid impact invites us, in fact *forces* us to think back in time, searching for comparable challenges and the lessons to be learned from human adaptability and resilience. We are also forced to think forward in time to a world as it likely to be when such CAI takes place. That world will probably have a relatively stable population of about 8 billion human beings. It will be one, at best, with significantly reduced *in situ* biodiversity and diminished fossil fuel reserves. Humanity will likely be concentrated even more than it is today in large urban regions, and these regions will be connected by evermore sophisticated and complex networks of trade, information exchange, and financial transaction.

What I have just described in the *best* case scenario. A worse one is to imagine humankind still engaged in the kind of warfare that afflicts us at this time. War and other kinds of large-scale violence considerably complicate any disaster management picture (Wisner 2002; Wisner 2003a).

CAI provides an awesome heuristic, a challenge to our imagination. In scale, have we ever seen anything like this? Humanity did, in fact, survive the last Ice Age, the Black Death, the influenza pandemic that swept the globe during World War I. The food shortages produced in Europe by the cold growing seasons that followed the 1815 eruption of the Tambora stressed the ability of nation state to provide a dietary safety net, but did not break these systems (Post 1977). What lessons can we learn about adaptation and resilience? In the face of CAI, what kind of resources should the world invest in preparing?

26.1.1
Assumptions

Throughout these reflections I have made certain assumptions.

- I have assumed that CAI will be without long forewarning. At present only limited resources are directed toward monitoring potentially threatening comets and asteroids. I am assuming considerably increased resources are *not* forthcoming, and that there is not early warning (months or years) of impending impact. Another paper would be required to map out the likely plans, actions, and institutional arrangements necessary to evacuate (or permanently re-locate) coastal populations given sufficient warning time.
- I have thought most about a Pacific Ocean impact, given the relative size of the world's various oceans. Tsunami would not affect *all* exposed coasts in all scenarios, of course.
- Another assumption is that CAI occurs within the next 100–150 years, beyond which the outlines of the world's energy system and related urban/industrial geography are very hard to predict.
- Finally, I have taken for granted that within this time scale, international order has not decayed as a result of nuclear war or the continued proliferation of many smaller and larger wars. If CAI occurs when humanity is living, as Thomas Hobbes once put it, in warfare of "all against all", then life will already be "nasty, brutish, and short," and CAI would amount only to an additional threat.

26.2
Do CAI-Scale Events have any Precedents?

Such region-wide impacts and challenges of various kinds are not unprecedented in the history of our species. First, we should not forget that for most of our history since the invention of agriculture and a settled way of life, in Ponting's words (1991, p 88), "[a]bout 95 per cent of the people in the world were peasants; directly dependent on the land and living a life characterized by high infant mortality, low life expectancy, chronic under-nourishment and with the ever present threat of famine and the outbreak of virulent epidemics." The human population reached its first billion in 1825. It had taken about two million years to get to this point. The second billion was reached in 1925, after only one hundred more years. The three billion mark came in 1960; four billion in 1975, five billion in the late 1980s, and around 2001 humanity welcomed its six billionth member (Ponting 1991, p 240).

Despite this seemingly triumphant colonization of the planet, our species experienced many challenges and setbacks. Whole civilizations have, in fact, disappeared such as the Maya and the irrigators of Sumer and other early cities in the Near East. Famine stalked humanity until well into the 19th century (Davis 2001), and, for many in Africa still today, regional food security can easily be undermined by desert locust infestation and drought, especially when populations are displaced or weakened by war.

The Mogol invasions of China killed 35 million, and the epidemics in China during 1586-89 and 1639-44 caused a fifth of the population perish on each occasion (Ponting 1999, p 95). The Black Death in the 14th century killed a quarter of Europe's population. Millions also died in Europe during the "Little Ice Age" (1430-1850). The history of famine shows that until the 20th century, food insecurity was, in fact, the rule for most of humanity (Ponting 1991, pp 103-110). <2>

Even in the 20th century, a massive famine in China in the years of intensive industrialization (1958-61) may have killed as many as 30 million people (Wisner et al. 2004, Chap. 4; Yang 1996).

Epidemic disease has also taken very large numbers of lives, and perhaps is a better model for the kind of stress that CAI would produce (Wisner et al. 2004, Chap. 5). This is because of the large regional and even world-wide scope of pandemics of plague, cholera, and influenza. At the end of World War I the great influenza pandemic that swept the globe may have taken as many as 50 million lives (Kolata 1999).

26.2.1
Adaptation and Resilience

How has humanity responded to these large mortality events and regional stresses? On the whole, the historical record shows that population numbers rebound quickly (Clarke et al. 1989). There has often been violent conflict at these times, as people move into new territories and struggle for control of resources. Social and economic changes often occur. Institutions (secular and faith-based) are challenged, but generally adapt.

So, perhaps, we needn't worry that much about CAI? The problem, however, is that the ecological context of all previous challenges and responses was different. Recall the likely state of humanity and planet Earth when CAI will occur. *In situ* biodiversity will be eroded. Fresh water resources will have been diverted from irrigated agriculture to meet growing urban industrial needs. Fossil fuels used to synthesize artificial fertilizers and other agricultural chemicals will be much more expensive and scarce – in competition with end uses for generation of energy and mobility. And there will be 8 billion of us. In short, humanity will not have the luxury of "starting over" with the domestication of plants and animals and creation of agriculture. British economist Malcolm Caldwell wrote a book, entitled *The Wealth of Some Nations* in which he demonstrated that the petroleum and other fossil fuel resources of this planet are not sufficient for a second great agricultural and industrial revolution as we saw in Europe in the past few centuries. At that time, Europe had benefited since the 1500s from import of wealth (gold, silver) and later massive amounts of organic matter (guano) and energy. Such "primitive accumulation" cannot be repeated, Caldwell argued.

A more optimistic view might suggest that the CAI might occur once the post-petroleum transition has been successfully accomplished, and there is less conflict over scarce mineral and energy resources, or – to take optimism to near Panglosian limits – even struggles over water or arable land. However, a post-petroleum spatial organization of humanity might require coastal cities to be even larger and more densely population to minimize the sprawl now supported by petroleum intensive transport. In addition, whatever the major energy sources at that point in our future, complex coordination and communication will be required, and those links will doubtless be severed by the CAI.

26.3
The perspective of International Development Studies

Current thinking recognizes a large overlap between sustainable human development policy and disaster risk reduction policy (UNDP 2004; Wisner et al. 2004). The "big" idea that has emerged by this cross-fertilization is that sustainable human development itself is the single best way to prepare for disaster. This idea underlies the draft Program of Action of the World Conference on Disaster Reduction, held in Kobe, Japan in January 2005 (UNISDR 2004a) and its final outcome – the Hyogo Framework of Action (UNISDR 2005).

Increased polarization between rich and poor, marginalization and displacement of the poor and rapid urbanization combine to place vulnerable people in the way of hazard events. Over the past few decades the numbers killed by such events have continued to rise. Between 1993–2002, there were more than 600 000 people killed and more than 1.5 billion affected by natural events that cost $ 700 million (UNISDR 2004b, p 3).

The model used by most natural hazard researchers is that risk is a function not only of the hazard event (its intensity, duration, location, frequency) but also of the potential for loss, that is "vulnerability" (Hewitt 1997, pp 21–39; Alexander 2000, pp 7–22; Wisner et al. 2004, Chap. 1 and 2; Wisner 2004). In short hand form this is:

$R = H \times V$

The conclusion of a number of studies have converged on the need to do a series of ambitious but necessary actions to bring disaster vulnerability (V) under control, since efforts to control H (hazard) along (through hydro engineering works, etc.) have not been sufficient. These measures include (UNDP 2004; Wisner 2003b; Wisner et al. 2004, Chap. 9):

- Reduction of violent conflicts that displace persons (making them more vulnerable) and get in the way of other effort to build the base for sustainable economic activity and land use (cf. Wisner 2002, 2003a).
- Encouragement of accountable and competent governance at all levels from nation to locality.
- Expansion and strengthening of the public health network and infrastructure (Wisner et al. 2005).
- Controlling unplanned urban growth and sprawl and expansion of efforts to upgrade squatter settlements.
- Expansion of efforts to protect wild biodiversity, hedging our bets on the future utility of this DNA.
- Control of global warming – a major likely factor increasing H in the future.
- Expansion of livelihood opportunities for the rural and urban poor, a key to reducing V (Wisner et al. 2004, Chap. 3).

These measures are generally agreed to be necessary pre-requisites for the achievement of the Millennium Development Goals, that is, sustainable human development *and also* disaster risk reduction.<3>

26.3.1
Would "Sustainable Development" be Enough?

Achieving significant progress in the seven measures mentioned above would have stunning impact on the ability of humanity to respond to a CAI and to adapt to conditions afterwards. For example, resolution of current violent conflicts and development of a truly international and efficient mechanism for preventing future conflicts would provide the next step in the gradual development of humanitarianism and international cooperation. Already regional peace keeping forces have been developed in Africa and the NATO countries, for example. A famine in 1991–92 that could have threatened the lives of 17–20 million people in southern Africa was avoided through many-sided cooperation (DeRose et al. 1998). Food security was restored through the actions of the countries involved in the region, many international organizations such as the World Food Program, bi-lateral donors, and non-governmental organizations (NGOs). This gives a hint of the potential effectiveness of international humanitarian action. However, such successes are still too few, and the international system is still at a primitive stage of its development. It is not well coordinated. Different U.N. bodies respond to different kinds of emergencies – OCHA to natural disasters and internal displacement of persons, the UNHCR to the needs of refugees that cross international borders, UNEP and WHO to technological disasters such as Chernobyl and Bhopal. UNDP and ISDR focus on building capacity to plan for disasters, prepare for them, and to prevent or to mitigate their impacts. Still other international agencies (e.g. IPCC; IHDP) deal at the moment with the human dimensions of climate change with less than adequate coordination with those working on "other" potential disasters. The concerns of small island independent states (SIDS) are often treated separately a "special case." While SIDS do, indeed, face special risks (Pelling and Uitto 2001; Kelman 2004) – and could be catastrophically affected by CAI – such "special treatment" can also close off important cross-linkages to other programs and institutions.

However, even if we imagine a relatively peaceful world with a well-connected and well-financed international disaster response mechanism, would that be enough to cope with a CAI?

The answer is probably no. The scale of urban destruction would be great. It would include the obliteration of port facilities that are the still the heart of international trade. International financial transactions would be disrupted for a period. The cost and logistical requirements to meet the needs of displaced persons would be great, but that is not the main problem. The Marshall Plan dispensed $ 13 billion between 1947–1953 to feed and clothe a large part of the European population following World War II and to begin to rebuild livelihoods (US Department of State 2004). This is approximately $ 238 billion in the value of 2004 dollars, a considerable investment by the Marshall Plan in a kind of disaster response.

The problem is that now, and certainly by the time we suffer a CAI, the urban industrial system will be (*a*) larger and more mutually interdependent and (*b*) already stressed during the final decades of petroleum availability (Heinberg 2003; Shah 2004). One estimate of the impact of a recurrence of the 1923 earthquake in Tokyo produced by the consulting firm Risk Management Solutions considers the cost of disrupted markets plus the cost of the physical damage to a much larger metro area. The number they got was $ 2.1–3.3 *trillion* (Stanford 1996). The knock on effect of such an event – a simple earthquake of known size and location – would be world wide. This loss estimate dwarfs even the considerable economic destruction caused by the 26 December 2004 tsunami that affected 11 countries in Southeast Asia, South Asia, the Indian Ocean and coastal East Africa. Even a very rich country such as the U.S. has a hard time absorbing the economic shock of a single large hurricane when it hits the heart of one of its main petroleum production regions and a major city. Hurricane Katrina, which flooded New Orleans and did catastrophic damage to the Gulf Coast of Mississippi, is at the time of writing likely to be the most costly disaster triggered by a natural event in U.S. history, surpassing the $ 48.4 billion cost of hurricane Andrew that devastated Miami in 1992 (Fields and Rogers 2005).

This leads me to the conclusion that the precautionary principle (Harremoes et al. 2002) demands more than the seven measures already on the agenda of international development and disaster management experts and policy makers (as ambitious as they already seem). Lateral thinking is required, as Foster puts it, "to survive change" (1997). Thus three additional measures are required:

- Roll out more rapidly alternative energy sources, so that we save petroleum as a future feedstock for pharmaceuticals and other useful things. This would also reduce the potential impact of disrupted international oil shipments due to CAI/tsunami damage of oil terminals and refineries near the coasts. It would also reduce the economic stress of the next century and a half that otherwise will have to cope with ever increasing oil prices.
- Legislate at national level incentives to decentralize megacity populations to regional growth centers. This would have the benefit of producing much needed employ-

ment and releasing productive forces while it also reduced the numbers of people in the big coastal megacities. Decentralized, renewable energy, and decentralized, smaller towns and settlements are a key to resilience in the face of a CAI. On a modest scale, planners in the U.S. have already put forward the notion that control of sprawl and more ecologically sound cities are essential for disaster risk mitigation (Burby 1998).
- Accelerate the acceptance of low input sustainable agriculture (LISA) through research, incentives, and subsidies. It is possible to wean agriculture off of high energy inputs in the form of agro-chemicals and unnecessary mechanization. Zero-tillage techniques, biological control of pests, and many other aspects of LISA are well studied, but they have not spread rapidly. Cuba, for example, is a natural laboratory for LISA since it was forced by the end of the cheap oil it received from the USSR and the U.S. embargo to grow much of the food for its population using low input techniques (Rosset and Benjamin 1994). At present food in the U.S. travels very long distances before it is consumed – for example, 1494 miles from production to consumer in the case of Iowa (a farm state) (Pirog and Benjamin 2003). Food systems that are, at least for non-luxury items, more reliant on local or closer regional sources and that use LISA techniques would be far less disrupted by the effect of a CAI on trade and petroleum availability.

26.3.2
A Remaining Big Worry

Even in a world characterized by considerable progress on the ten measures I have proposed, there would be an additional concern – climate disruption caused by the millions of tons of dust. The renewable sources of energy one might like to see as a transition from petroleum dependence are solar dependent. But there will be far less sunlight for several years following a CAI. Also, whether one is growing food with LISA or conventional techniques, plants need sunlight and warmth to grow. Is there a case to be made for development of varieties of crops that contain lots of energy (tubers of some kind) that grow under harsh conditions and reduced light? Will we be reduced to growing barley adapted to Tibet and eating *momos*, or, perhaps taking a page from Roald Dahl's *Big Friendly Giant* (1998), and cultivating snozzcumbers? <5>

Might CAI therefore be a factor that would lean society toward retaining nuclear energy as an option? The trade off between the current enormous environmental cost of the whole uranium fuel cycle from mining through disposal of high level waste is so great, it is hard to imagine the calculation that would balance these costs against the benefit of having nuclear power to fall back upon in the case of CAI (Cutter et al. 1985, pp 378–391).

26.4
Some Tentative Conclusions

1. Disasters of regional scope are known and, in fact, not uncommon (e.g. those triggered by drought, population displacement, fires, tsunami, oil spills, and epidemic outbreaks). This means that CAI would not present a unique and unprecedented

situation for humanity. Lessons can be learned from the more remote past (e.g. the 14th century plague in Europe). They can also be inferred from the more recent past. Examples include the 19th century food shortages in Europe following the 1815 Tambora volcano eruption (Post 1997), as well as 20th century African famines and floods in Bangladesh and China, all of which affected tens of millions of people on each occasion (Wisner et al. 2004, Chap. 4 and 6).

2. Current thinking about the management of the "normal" range of natural hazard risks considers good governance and investments resilient infrastructure (public health, water, power, communications) to be central to risk reduction (Twigg 2004; UNISDR 2004a; UNDP 2004). Thus implementation of the U.N.'s Millennium Development Goals (by 2015) and the program of action of the World Conference on Disaster Reduction (2005–2015) would simultaneously help to provide humanity with more resilience in the face of CAI (UNISDR 2004b, 2005).

3. Also the international community is presently re-appraising the international humanitarian assistance system (Sphere Project 2004). Incremental improvements of that system which facilitate rapid deployment, cutting through red tape, resolving civilian/military competition, etc. are not only valuable in dealing with conflict and post-war situations. In the case of a large disaster, including CAI, improved international response capability would be crucial.

4. However, it should be noted that nearly everything I have proposed (the ten items in the two lists above) *have many benefits besides* preparation of humanity for CAI. This point is crucial. I am not convinced that sums of money spent exclusively for CAI preparation or mitigation (besides further scientific study and astronomical observation, discussed below) can be justified. My belief is based on the critical state of the human population at this moment. There is a great need for investments in maintenance of the childhood vaccination system world wide, reduction of maternal mortality rates, provision of safe drinking water, etc. – in short, implementation of the Millennium Development Goals. Since one no longer hears of a "peace dividend" (Brown and Wolf 1988), such money will be scarce and must compete with investment in preparing uniquely for CAI, that is, preparing in ways that do not have collateral benefits.

5. Finally, I believe it may difficult to build an international consensus around policies for mitigating harm from CAI in a bi-polar world, where not only wealth and income are highly unequal, but also where access to scientific resources are uneven. The continuing gap between more and less developed nations has two implications. Firstly, as just noted, whatever is done to prepare for CAI must have additional, complementary benefits for "normal" disaster risk reduction and sustainable human development. Secondly, scientific capacity in astronomy/planetary science and Earth sciences in less developed countries (LDCs) should be reinforced. This should be part of a general effort to spread educational benefits more evenly throughout the world. <4> The more we have good, solid scientific capacity in a large number of nation states, the easier it will be to reach a consensus on what to do about CAI. There has already been some complaint from scientists and policy makers in LDCs that they are dependent on wealthier countries for data on global climate change. This sensitivity is bound to show up again in the case of the scientific basis for CAI policy.

Notes

<1> An overview of the growth of coastal exposure is provided by Wisner et al. (2004, Chap. 2) and Wisner and Ahlinvi (2001). Evacuation would depend on the lead-time provided by a tsunami warning system such as the pan-Pacific system that now exists, as well as on the availability of evacuation routes and modes of transport sufficient to cope with the population of coastal megacities. A means of rapid communication with the people would be necessary and a great deal of discipline and self control. Evacuation of a major city such as Los Angeles, Mexico City, or Tokyo *even if several days advance warning* of an earthquake would be impossible (Mitchell 1999). On the other hand, with 2–4 days advance warning, 1–2 million people have been successfully evacuated from coastal Florida, Alabama, and Louisiana when warned of hurricanes, but many do not manage (Alexander 2002, pp 149–155). Cuba is extremely effective in organizing evacuations before hurricanes hit, and generally does not lose human life as a result (Thompson 2004). Approximately 50 000 people or 10% of New Orleans' population did not leave the city prior to the impact of hurricane Katrina and subsequent breeching of the city's protective levees. When a total evacuation of the city was officially ordered on 31 August 2005, the best estimate was that only 10 000–15 000 of these remaining people could be evacuated per day.

If astronomical observations give a more or less precise window of time for CAI and thus provide several months' or even years' warning, it might be possible to organize an orderly temporary retreat from coastal settlements, securing and "mothballing" immobile assets and infrastructure. Of course, it would be better to de-densify and decentralize such coastal settlement for a variety of reasons discussed below.

<2> See also Aykroyd (1974), Rotberg and Rabb (1983), Arnold (1988), and Newman (1990).

<3> The Millennium Development Goals (MDGs) are a set of eight objectives adopted by the General Assembly of the United Nations in 2000. Since then they have been re-affirmed and operationalized in a series of targets and benchmarks to be achieved by 2015–2020. See: *http://www.un.org/millenniumgoals/* and *http://www.developmentgoals.org/*

<4> One of the MDGs is to get approximately 100 million children of school age – many of them girls – in school. Opportunities for secondary school education even more skewed toward citizens of industrial countries, and the chances of going to university or developing a career in science for someone from one of the HIPC countries is very slim (Highly Indebted Poor Countries are a U.N. category.)

<5> Mushrooms, of course, grow quite well in reduced light. They are quite nutritious when considered in units per dried weight unit of mushroom. One source states: "Mushrooms are relatively high in protein, averaging about 20% of their dried mass. They contribute a wide range of essential amino acids, are low in fat (0.3–2.0%), high in fiber and provide several groups of vitamins, particularly thiamine, riboflavin, niacin, biotin, and ascorbic acid. While nutrients vary from one kind of mushroom to the next, many contain protein, vitamins A and C, B-vitamins and minerals including iron, selenium, potassium and phosphorus." (*http://whatscookingamerica.net/Q-A/PortabellaMushrooms.htm*). At least one commercial product (Quorn) is made from the mold, *Fusarium venenatum*, from which is extracted a mycoprotein. However, it is hard to imagine large populations subsisting on such products during the "CAI win-

ter" hiatus of farming. Fungus-based cuisine may suit "bogeymen" (Briggs 1977), but not most human beings. One might, however, be tempted to dull the suffering of the long "CAI" twilight with the consolation of "magic mushrooms."

The sea might be considered another source of food as agricultural production falls. However, ocean fishing would also be crippled by reduced numbers of useable boats and harbors (destroyed by the tsunami) and limited petroleum. In addition, CAI is likely to have severe impacts on ocean ecosystems and possibly even currents, so that food from the sea would not necessarily be plentiful.

References

Alexander D (2000) Confronting catastrophe: new perspectives on natural disasters. Oxford University Press, Oxford

Alexander D (2002) Principles of emergency planning and management. Harpenden, Hertfordshire, Terra Publishing, UK

Ambio (1982) Nuclear war: the aftermath. Theme issue of Ambio: A Journal of the Human Environment 11(2–3):76–176

Arnold D (1988) Famine: social crisis and historical change. Basil Blackwell, Oxford

Aykroyd WR (1974) The conquest of famine. Chatto and Windus, London

Briggs R (1977) Fungus the Bogeyman. Penguin, London

Brown L, Wolf E (1988) Reclaiming the Future. In: Brown L et al. (eds) State of the World 1988. WW Norton, New York, pp 170–188

Burby R (ed) (1998) Cooperating with nature: confronting natural hazards with land-use planning and sustainable communities. Joseph Henry Press, Washington DC

Caldwell M (1977) The wealth of some nations. Zed Press, London

Clarke J, Curson P, Kayastha SL, Nag P (eds) (1989) Population and disaster. Basil Blackwell and IGU Commission on Population Geography, Oxford

Cutter S, Renwick Hl, Renwick W (1985) Exploitation, conservation, preservation: a geographic appraisal of natural resource use. Rowman and Allanheld, Totowa NJ

Dahl R, Blake Q (illustrator) (1998) The BFG. Puffin, London

Davis M (2001) Late Victorian holocausts: El Nino famines and the making of the Third World. Verso, London

DeRose L, Messer E, Millman S (1998) Who's hungry? And how do we know? Food shortage, poverty, and deprivation. United Nations University Press, Tokyo. http://www.unu.edu/unupress/unupbooks/uu22we/uu22weob.htm

Fields G, Rogers D (2005) Already under scrutiny, FEMA is now in the spotlight. The Wall Street Journal, 31 August, p B1

Foster HD (1997) The Ozymandias principles: thirty-one strategies for surviving change. Southdowne Press, Victoria BC

Harremoes P et al. (2002) The precautionary principle in the 20th century. Earthscan, London

Heinberg R (2003) The party's over: oil, war and the fate of industrial societies. New Society Publishers, Gabriola Island BC

Hewitt K (1997) Regions of risk: a geographical introduction to disasters. Longman, Harrow UK

Kelman I (2004) Small island vulnerability project. Cambridge UK. http://www.arct.cam.ac.uk/islandvulnerability/projects.html

Kolata G (1999) Flu: the story of the great influenza pandemic of 1918 and the search for the virus that caused it. Touchstone, New York

McGuire J (1999) Apocalypse: a natural history of global disasters. Cassell, London

McGuire J, Mason I, Kilburn C (2002) Natural hazards and environmental change. Arnold, London

Mitchell JK (ed) (1999) Crucible of hazard: megacities and disaster in transition. United Nations University Press, Tokyo

Newman LF (ed) (1990) Hunger in history: food shortage, poverty, and deprivation. Basil Blackwell, Oxford

Pelling M, Uitto J (2001) Small island developing states: natural disaster vulnerability and global change. Environmental Hazards 3(2 September):49–62
Pirog R, Benjamin B (2003) Checking the food odometer: comparing food miles for local versus conventional produce sales to Iowa institutions. Leopold Center for Sustainable Agriculture Iowa State University, Ames, Iowa: *http://www.leopold.iastate.edu/pubs/staff/files/food_travel072103.pdf*
Ponting C (1991) A green history of the world. Penguin, New York
Post JD (1977) The last great subsistence crisis in the western world. Johns Hopkins University Press, Baltimore MD
Rosset P, Benjamin M (eds) (1994) The greening of the revolution: Cuba's experiment with organic agriculture. Food First Books, Oakland CA
Rotberg R, Rabb TK (eds) (1983) Hunger and history: the impact of changing food production and consumption patterns on society. Cambridge University Press, Cambridge
Shah S (2004) Crude: the story of oil. Seven Stories Press, New York
Sphere Project (2004) Humanitarian charter and minimum standards in disaster response, 2nd edn. Geneva: The Sphere Project, *http://www.sphereproject.org/*
Stanford News (1996) Casualty, damage estimates of great quakes revised upward. Stanford News (1 October 1996), Stanford University, Stanford CA. *http://www.stanford.edu/dept/news/pr/96/960110greatquake.html*
Thompson M, Gaviria I (2004) Cuba, weathering the storm: lessons in risk reduction from Cuba. Oxfam America, Boston. *http://www.oxfamamerica.org/workspaces/newsandpublications/public/publications/research_reports/art7111.html*
Twigg J (2004) Disaster risk reduction: mitigation and preparedness in development and emergency programming. Good Practice Review No 9. ODI/ Humanitarian Practice Network, New York, March 2004
United Nations Development Programme (UNDP) (2004) Reducing disaster risk: a challenge for development. UNDP, New York. *http://www.eldis.org/static/DOC14322.htm*
UNISDR (United Nations Inter-Agency Secretariat for the International Strategy for Disaster Reduction) (2004a) World conference on disaster reduction. *http://www.unisdr.org/eng/wcdr/wcdr-index.htm*
UNISDR (2004b) Living with risk: a global review of disaster reduction initiatives. Version vol 1. United Nations, New York
UNISDR (2005) Hyogo Framework of Action 2005–2015: building the resilience of nations and communities to disasters (HFA). *http://www.unisdr.org/eng/hfa/hfa.htm*
US Department of State (2004) The Marshall Plan. *http://usinfo.state.gov/usa/infousa/facts/democrac/57.htm*
Wackernagel M, Rees W (1996) Our ecological footprint. New Society Publishers, Philadelphia
Wisner B (2002) RADIX – violent conflict and disasters. RADIX: *http://online.northumbria.ac.uk/geography_research/radix/violent-conflict.html*
Wisner B (2003a) Swords, plowshares, earthquakes, floods, and storms in an unstable, globalizing world. Invited Keynote Address DPRI – IIASA 3rd International Symposium on Integrated Disaster Risk Management (IDRM-2003) Kyoto international conference hall. Kyoto, Japan, 3–5 July, 2003
Wisner B (2003b) Sustainable suffering? Reflections on development and disaster vulnerability in the post-Johannesburg world. Regional Development Dialogue 24(1; Spring):135–148
Wisner B (2004) Assessment of capability and vulnerability. In: Bankoff G, Frerks G, Hilhorst T (eds) (2004) Vulnerability: disasters, development and people. Earthscan, London, pp 183–193
Wisner B, Adams J (eds) (2003) Environment health in emergencies and disasters. Geneva: WHO (for WHO/ IFRC/ UNHCR), *http://www.who.int/water_sanitation_health/hygiene/emergencies/en/*
Wisner B, Ahlinvi M (2001) Natural disasters and their impact upon the poorest urban populations. Report prepared for UNESCO, Social Sciences Division. International Social Science Council, Paris. *http://www.unesco.org/most/isscreport.htm*
Wisner B, Blaikie P, Cannon T, Davis I (2004) At risk: natural hazards, people's vulnerability and disasters, 2nd edn. Routledge, London
Wisner B, Adams J, Alexander D (2005) Environmental health in disasters: water, sanitation, and shelter. In: Noji E (ed) The public health consequences of disasters, 2nd edn, Chap. 7. Oxford University Press, forthcoming, New York
Yang D (1996) Calamity and reform in China: state, rural society and institutional change since the great leap famine. Stanford University Press, Stanford CA

Chapter 27

Disaster Planning for Cosmic Impacts: Progress and Weaknesses

Harold D. Foster

> What plagues and what portents, what mutiny
> What raging of the sea, shaking of the earth,
> Commotion in the winds, frights, changes, horrors,
> Divert and crack, rend and deracinate
> The unity and married calm of states.
>
> Ulysses in *Troilus and Cressida*
> Act 1, Scene iii
> William Shakespeare (1564–1616)

27.1 Introduction

On the evening of June 18, 1178, several witnesses near Canterbury, England saw a spectacular night sky event (Ingram 1999). These observers reported directly to a monk who was keeping detailed records of events occurring in or around Christ Church Cathedral. Fortunately, this diary, the *Chronicles of Gervase* has survived and provides a detailed description of the strange events of 1178:

> This year, on the Sunday before the Birth of Saint John the Baptist, after sunset when the moon had first become visible, a marvellous phenomenon appeared to five or more men while sitting facing it. Now there was a bright new moon, and as usual the horns protruded to the east; and lo, suddenly, the upper horn split in two. From the middle of this division a firebrand burst forth, throwing over a considerable distance fire, hot coals and sparks. Meanwhile the body of the moon which was lower [than this] writhed as if troubled, and in the words of those who told this to me and who saw it with their own eyes, the moon throbbed as a beaten snake. It then returned to its former state. This phenomenon was repeated twelve times and more, the flame assuming various twisting shapes at random then returning to normal. And after these vibrations it became semi-dark from horn to horn, that is, throughout its length. Those men who saw this with their own eyes reported these things to me who writes them; [they are] prepared to give their word or oath that they have added nothing false to the above.

Hartung (1976) has argued that this was the first and only sighting, in recorded history of a large asteroid striking the moon and contended that this collision created the twenty-two kilometer-wide crater, known as Giordano Bruno. In contrast, Nininger and Huss (1977) postulated that the twelfth century English eyewitnesses had seen a meteor in the Earth's atmosphere that happened to be in the line of sight of the moon. Calame and Mulholland (1978), however, strongly support Hartung's position, arguing that the moon was still reverberating from the collision and ringing like a bell. If these authors are correct, the Canterbury eyewitnesses saw an event releasing some

100 000 megatons of energy, that is an event that was ten million times more powerful than the atomic bombs that destroyed Hiroshima and Nagasaki (Ingram 1999).

In July, 1994 Comet Shoemaker-Levy 9 (S-L9) fragmented as it entered the dense atmosphere of Jupiter, creating impact scars the size of the Earth. There is no doubt about the subsequent comet-planet collisions. These events were the most widely witnessed in astronomical history (Morrison 1996).

On December 8, 1994, less than a day before it was expected to strike the Earth, astronomers discovered a new asteroid, 1994 XM1 that had the mass of a large house and was moving at 108 000 kilometers per hour. Fortunately, it missed, but only by some 105 000 kilometers (Wood 2000). More recently, another asteroid of similar size, 2003 SQ222, came even closer, avoiding our planet by only 88 000 kilometers (Knocke 2003).

Clearly, not all encounters with near-Earth objects have ended so fortuitously. Unlike the Moon, the Earth has retained only a small sample of its population of impact structures as the result of geomorphological processes. Beyond that, since the oceans occupy about 70 percent of the planet's surface, many other near-Earth objects must have struck these areas. Nevertheless, over 160 impact craters have so far been identified on Earth. A complete listing of their size and location is available at the Earth Impact Database (2004). A further 15 or so major impacts can be recognized in the stratigraphic record (Grieve 1997; Kaiho et al. 2001). Impact scars range in size from the Vredefort (South Africa), Sudbury (Canada) and Chicxulub (Mexico) craters that are respectively 300, 250 and 170 kilometers in diameter to the 1.5 m Haviland crater in Kansas (Earth Impact Database 2004, Grieve and Kring n.d.).

Given such enormous range in scale, the consequences of impact must also have differed dramatically. The Chicxulub crater, located under Mexico's Yucatán Peninsula, is thought to have been created by an asteroid that was roughly 10 kilometers in diameter. It is estimated that it hit the Earth with the energy equivalent to more than 5 billion Hiroshima atom bombs, that is 100 million megatons (Morrison 1996). Aside from the initial concussion and heat, two major post-impact events caused massive secondary planetary damage. Large quantities of rock and dust blown out of the crater subsequently rained down as meteors, heating the atmosphere and creating worldwide forest and grassland fires. Not all the dust returned to Earth quickly, however, a finer layer remained suspended in the atmosphere for months, blocking photosynthesis and causing plummeting surface temperatures. It is likely also that the ozone layer was seriously damaged (Birks et al. 2006). These events triggered massive global terrestrial and marine extinctions, bringing to a close the domination of the dinosaurs and, with it, the end of the Cretaceous Period and Mesozoic Era. It is possible that a similar collision, creating what is now a buried impact crater offshore of Northwestern Australia, may have marked the end of the Permian (Becker et al. 2004). The Bedout impact may have triggered the Permian-Triassic extinction in the same way that the Chicxulub impact terminated the Cretaceous era (Kerr 2004). Indeed, based on variations in sulfur isotopes and the presence of a nickel-rich layer in end-Permian limestone, marl and shale in southern China, Kaiho and colleagues (2001) previously had postulated such an extinction event, caused by a meteorite of up to 60 kilometers in diameter (Ball 2001). However, controversy continues over whether, or not, the end-Permian extinction event had an extraterrestrial cause (Koeberl and Farley 2004).

While it is apparent that on rare occasions in the geological past, huge devastating asteroids have collided with the Earth, it is probably more relevant to ask the question "What is the minimum sized near-Earth object that has the capability of causing serious damage?" This question has been addressed by Hills and Mader (1997) who wrote:

> The fragmentation of a small asteroid in the atmosphere greatly increases its cross section for aerodynamic braking, so ground impact damage (craters, earthquakes, and tsunami) from a stone asteroid is nearly negligible if it is less than 200 meters in diameter. A larger one impacts the ground at nearly its velocity at the top of the atmosphere producing considerable impact damage. The protection offered by Earth's atmosphere is insidious in that smaller, more frequent impactors such as Tunguska only produce air blast damage and leave no long-term scars on the Earth's surface, while objects 2.5 times larger than it, which hit every few thousand years, cause coherent destruction over many thousands of kilometers of coast. Smaller impactors give no qualitative warning of the enormous destruction wrought when an asteroid larger than the threshold diameter of 200 meters hits an ocean. A water wave generated by an impactor has a long range because it is two-dimensional, so its height falls off inversely with distance from the impact. When the wave strikes a continental shelf, its speed decreases and its height increases to produce tsunamis. The average run-up in height between a deep-water wave and its tsunami is more than an order of magnitude. Tsunamis produce most of the damage from asteroids with diameters between 200 meters and 1 km. An impact anywhere in the Atlantic by an asteroid 400 meters in diameter would devastate the coasts on both sides of the ocean by tsunamis over 100 meters high. An asteroid 5 km in diameter hitting in mid Atlantic would produce tsunami that would inundate the entire upper East Coast of the United States to the Appalachian Mountains.

Even though smaller, more frequent impactors do not create large tsunamis or long-preserved impact craters, they are far from harmless. On June 30, 1908 a near-Earth object, some 50 to 70 meters in diameter, exploded 8 km above the Stony Tunguska River, in Siberia. Whether it was an asteroid or comet is still in dispute, but the resulting air blast devastated an area of some 2 150 square kilometers. In the hot central epicenter the forest flashed into a huge ascending column of flame that was visible for several hundred kilometers. Fires burned for weeks destroying 1 000 square kilometers of forest. Ash and powdered fragments of tundra were drawn skywards by the fiery vortex and carried around the world by the global air circulation (Gallant n.d.). The blast felled trees outwards in a radial pattern over an area half the size of Rhode Island. The mass of the object involved was probably about 100 000 tons and the explosion's force some 40 megatons of TNT, that is 2 000 times the energy of the Hiroshima atomic bomb. St. Petersburg seismograph station, 4 000 kilometers to the west recorded tremors associated with the blast.

Fortunately, the Tunguska region was a very sparsely inhabited. Nevertheless, the event instantly incinerated a local herdsman, Vasily Dzhenkoul, together with his hunting dogs, and 600 to 700 reindeer (Gallant n.d.). Despite the extraterrestrial object's relatively small size, as Chapman (1998) has pointed out, its associated destruction covered an area larger than either New York City or Washington, D.C. Had such a cosmic body exploded over a densely populated area of Europe instead of the desolate region of Siberia, the number of human victims would have been 500 000 or more, not to mention the ensuing ecological catastrophe and geopolitical ramifications (Galland 2004).

27.2
Probabilities

Every significant hazard has its own lobby groups consisting of those who have the most to gain from various levels of mitigation. Such organizations compete to increase, or decrease, government attention to particular threats. Clearly, before logical mitigation strategies can be implemented, a hazard hierarchy must be established. Cosmic impacts can be realistically compared with thousands of other natural and man-made hazards only after their frequency of occurrence and associated damage consequences have been established. Chapman (2003) has attempted to do this and Table 27.1 draws heavily on his assessment.

Earth is constantly being bombarded with cosmic debris. While estimates of scale and frequency should not be treated as exact, it is known that some ten pea-sized meteoroids and one walnut-sized impactor enter our atmosphere every hour. These are followed by one grapefruit-sized meteoroid every 10 hours. A basketball-sized impactor enters the Earth's atmosphere roughly once a month, whereas a rock with a diameter of 50 meters can be expected once a century (Gallant 2004). During the next century there is also a 0.2 percent chance of a cosmic impact with a near-Earth object having a diameter greater than 300 meters. In contrast, the probability of a collision with an object over 1 kilometer in diameter, during the next one hundred years is roughly 0.02 percent (Chapman 2003).

Although they can damage satellites and spacecraft, small meteoroids burn up in the atmosphere and so cause no problems on the Earth's surface. From a disaster planning point of view, the most worrisome meteoroids are those that range in size from greater than ten to hundreds of meters in diameter. As pointed out by Chapman (2003), although impact rates and their consequences vary enormously, they have several important characteristics in common. Whether explosion occurs in the atmosphere, ground surface or ocean they can have devastating consequences. Despite this threat, they are too small to be easily detected or tracked by existing telescope programs, and their impacts are too infrequent and too unpredictable to be studied in detail. As a consequence, their nature and effects are not well understood. This means that "scientific uncertainties are greatest for just those objects whose sizes and impact frequencies should be of greatest practical concern to public officials" (Chapman 2003).

In 1994, Chapman and Morrison compared the chance of being killed directly or indirectly by the impact of an asteroid or comet, in the United States, to those of other potential causes of death. This is a useful concept, although it must be admitted that it lacks precision. The average American has a 1 in 100 chance of dying in a motor vehicle accident. Other hazards with high probability include homicides, fires and firearm accidents and are likely to be the cause of death of 1 in 300, 800 and 2500 Americans respectively. Americans have a 1 chance in 20 000 of being killed directly, or indirectly, by the impact of an asteroid or comet. A similar probability is given for the likelihood of death in a passenger aircraft crash. In contrast, floods and tornadoes can be expected to kill roughly 1 in 30 000 and 1 in 60 000 Americans respectively (Chapman and Morrison 1994). If these figures are even of the right order of magnitude, it can be argued that mitigating the adverse impacts of cosmic impacts should be paid at least as much attention as reducing flood and tornado losses.

Table 27.1. Frequency of cosmic impacts of various magnitudes

Asteroid/comet diameter	Energy and where deposited	Chance this century (world)	Potential damage and required response
>10 km	100 million MT; global	<1 in a million[a]	Mass extinction, potential eradication of human species; little can be done about this extraordinarily unlikely eventuality, except the establishment of bases on other planets
>3 km	1.5 million MT; global	<1 in 50 000[a]	Worldwide, multi-year climate/ecological disaster; civilization destroyed (a new Dark Age), most people killed in aftermath; chances of having to deal with such a comet impact are extremely remote
		↑ Of no practical concern ↑	
>1 km	80 000 MT; major regional destruction; some global atmospheric effects	0.02%	Destruction of region or ocean rim; potential worldwide climate shock – approaches global civilization-destruction level; consider mitigation measures (deflection or planning for unprecedented world catastrophe); probable collapse of global economy
>300 m	2 000 MT; will form local crater, and cause regional destruction	0.2%	Crater ~5 km across and devastation of region the size of a small nation or unprecedented tsunami; advance warning or no notice equally likely; internationally coordinated disaster management required; probably exceed current capacity to effectively respond
>100 m	80 MT; lower atmosphere or surface explosion affecting small region	1%	Low-altitude or ground burst larger than biggest-ever thermonuclear weapon, regionally devastating, shallow crater ~1 km across; after-the-fact national crisis management
>30 m	2 MT; stratosphere	40%	Huge stratospheric explosion; shock wave topples trees, wooden structures and ignites fires within 10 km; numerous deaths likely if in populated region, especially an urban area (Tunguska, in 1908, was several times more energetic); advance warning unlikely, advance planning for after-event local crisis management desirable
>10 m	100 kT; upper atmosphere	6 per century	Extraordinary explosion in sky; broken windows, but little damage on ground
>3 m	2 kT; upper atmosphere	2 per year	Blinding explosion in sky; could be mistaken for atomic bomb triggering retaliation
		↓ Of no practical concern ↓	
>1 m	100 t TNT; upper atmosphere	40 per year	Bolide explosion approaching brilliance of the Sun for a second or so; harmless
>0.3 m	2 t TNT; upper atmosphere	1 000 per year	Dazzling, memorable bolide or "fireball" seen; harmless

[a] Frequency from Morrison et al. (2002); but no asteroid of this size is in an Earth-intersecting orbit; only comets (a fraction of the cited frequency) contribute to the hazard, hence "<." This table is based on that of Chapman (2003).

27.3
Goal Setting

The Earth is an intricate risk mosaic. On a daily basis, television and radio broadcasters and newspapers provide a deluge of information about recent disasters. From epidemics to invasions, each headline is accompanied by graphic descriptions of death, suffering and destruction. Since it is impossible to avoid all risk, societies have evolved to permit operation within specific levels of tolerance for natural and anthropogenic events. Typically limits to what can be successfully accommodated are defined either by law or by common practice. Usually regulations, such as building or public health codes, identify the maximum event that must be guarded against. As a result, the level of socially accepted safety reflects such factors as needs, wants, wealth and past experience (Foster 1980). This process works quite well for repetitive hazards, like earthquakes, heavy rainfall, tornadoes or fires. It does not necessarily provide an adequate level of safety for those hazards, such as moderate or large asteroids, that may rarely but catastrophically impact with the Earth.

Mitigation costs money and this is generally allotted by politicians and bureaucrats who have to select which hazards will be given the most attention and where related mitigation effects will take place. Unfortunately, all too often, decision-makers respond to more exotic threats only after a disaster has occurred. Even major ongoing catastrophes, such as the global spread of HIV-1, Hepatitis B and C viruses and the Coxsackievirus B that are currently killing some 7 million people annually and have infected over 2 billion in total have been very inadequately addressed (Foster 2002, 2004).

What is needed in the near-Earth-object debate is a comprehensive plan for risk reduction. At the very least a safety program should include six major elements: risk mapping, greater safety by improved design, disaster simulation and prediction, adequate warning systems, disaster planning and planning for reconstruction (Foster 1980). Naturally, few if any of these strategies will be adequately implemented until those in power can be convinced of the reality of the dangers of cosmic impacts.

27.4
Risk Mapping

Most natural hazards are spatially selective so there is nothing random about the deaths, injuries and damage they cause. While the chaos brought about by river floods, seiches, avalanches, storm surges, earthquakes and tsunamis traditionally has been a stimulus for belief in the supernatural, such decimation reflects differences in the distribution of factors controlling risk rather than any plan of divine retribution. Mapping risk factors that are often geological, geomorphological or hydrological in nature, allows spatial predictions of future destruction and so plays a key role in disaster planning.

Cosmic hazards are unusual in that they are not spatially selective. They will either miss the Earth, or they will not. In the latter case, the location of the impact will be random. This makes traditional risk mapping of the land surface irrelevant since any point on the planet appears to have a similar chance of being struck by a near-Earth object. Naturally, the larger the country, the greater its chance of being impacted. This means, of course, that the next asteroid striking the planet is more likely to crater Canada, the United States, Brazil, Australia and Russia than it is Luxemburg or Switzerland.

If one takes the fraction of the Earth that was badly damaged by the Tunguska impact, about one-millionth of the surface area of the planet, and multiply it by the global population, it can be argued that such a relatively small impact would, on average, kill about 10 000 people (Harris n.d.). This figure, however, is meaningless because if such an air blast occurred above New York, London or Tokyo, millions would probably die. In contrast, if it took place above the Sahara Desert, there might be no casualties. However, given that the oceans cover the majority of the Earth's surface, and that they are interconnected, it is quite possible that the next hit by a near-Earth object could generate a tsunami.

Numerous tsunami risk maps already have been produced. Typically they portray the areas that have been, or will probably be, inundated by earthquake-generated waves. They can be used, for example, as a tool to reduce construction in low-lying zones at high risk, plan evacuation routes and model expected damage for tsunamis of differing magnitudes. As part of the activities of the U.S. National Tsunami Hazard Mitigation Program (2004), for example, such maps are being produced for communities in Alaska, Washington, Oregon, California and Hawaii. THAMS, (Tsunami Hazard Assessment and Mitigation Studies) is a collaborative effort, begun in 1992, among three European institutes and Tohoku University, Japan. Much of THAMS effort has been directed towards identifying European tsunami risk and the improvement of tsunami mapping methodology (THAMS n.d.).

Certainly, tsunamis are not rare events. The Global Tsunami DataBase Project covers the period from 1628 BC until the present (Gusiakov 2003, 2006). It contains evidence of almost 2 250 tsunami or tsunami-like events, 1 206 of which occurred in the Pacific Ocean. A further 263 and 126 have been experienced in the Atlantic and Indian Oceans respectively, whereas 545 have occurred in the Mediterranean Sea. Beyond this, Bryant (2004) has provided depositional and erosional geographic evidence from the South Coast of New South Wales, North-eastern Queensland and Northwest Australia that is suggestive of cosmogenic mega-tsunamis.

There is roughly a 1-in-1 000 chance of an asteroid, with a diameter greater than 200 meters, striking the Earth during the 21st century. If it does, the most likely point of impact would be the Pacific Ocean. While there is still disagreement about the size of the resulting tsunami, there can be no doubt that it would cause immense damage around the ocean's rim and beyond (Hills and Mader 1977; Ward and Asphaug 2000). If the impact point of the asteroid were in the center of the Pacific Ocean, within twenty-four hours or so, hundreds of port cities, ranging from Melbourne and Sydney through Hong Kong, Shanghai and Tokyo to Vancouver, Seattle, Portland, San Francisco and Valparaiso would have been seriously damaged, if not completely destroyed. Financial losses would inevitably be in the tens if not hundreds of trillions of dollars, causing a collapse of the world's economy. If the Tsunami Warning System in the Pacific and its 26 member states functioned exceptionally well, life loss might be kept in the millions but, if not, or if the impact site was close to one shore or the other, relative mortality rate in coastal areas could exceed that of the Black Death. Obviously, computer generated tsunami risk maps, showing potential inundation from an asteroid strike should be prepared and used to plan evacuation routes and reduce construction on high risk sites. They would not completely prevent either large-scale destruction or loss of life, but they would help in their reduction. They also may be useful tools in encouraging politicians to take cosmic threat seriously. A lack of such tools and associated mitigation planning were responsible for much of the life loss around the Indian Ocean, caused by the Great Sumatra-Andaman earthquake on the 26th December 2004 (Lay et al. 2005).

27.5
Safety by Improved Design

Given the enormous kinetic energy of an impacting asteroid or comet, none of the standard architectural and engineering techniques for increasing integrity, improving operational compatibility, or for creating forgiving environments appear relevant to the debate (Foster 1980). Improved building codes to strengthen roofs, for example, may reduce hurricane damage, but are hardly relevant to discussions of a flying mountain, bigger "than the world's largest domed stadium ... crashing to Earth at a speed of a hundred times faster than that of a jet airliner" (Chapman 2003).

Nevertheless, for many reasons including, but not limited to cosmic impacts, society should pay far more attention to the ways in which our increasingly integrated, technological-dominated world is becoming more susceptible to catastrophic failures. *The Ozymandias Principles* (Foster 1997), outlines thirty-one dimensions of resilience (Table 27.2) and describes how their application can produce systems that are far less subject to dramatic collapse. Those dimensions that seem most pertinent here are the need for functional redundancy, the requirement for rapid response to stimuli, autonomous operation, mobility and early fault detection.

There seems to be a 1-in-1000 chance that, during this century, many of the major coastal cities of the planet will be badly damaged, if not destroyed, by what would be, by astronomical standards, a relatively small near-Earth-object. If this is the case, then electrical power systems, oil and gas pipelines, telecommunications grids and other social networks should be designed so that, given such a cosmic impact, they can still function. That is, those parts of these grids that are unlikely to suffer tsunami damage should be capable of autonomous operation. Such design would make them far less susceptible to other hazards, including earthquakes, hurricanes and terrorist attack. Greater functional redundancy would also help to protect against total collapse given serious damage to coastal areas. Beyond this, as little as possible that is irreplaceable should be immobile, especially if it is normally located in a high risk zone. Early detection of near Earth objects speaks for itself. The greater the length of forewarning of an impending impact, the more time society has either to prevent it, or at least to prepare to reduce its associated damage. In summary, it is not the strongest or the most intelligent species that ultimately survives, but rather the one that is most adaptable (Foster 1997). For this reason, it is suggested that the first Moon, or other extraterrestrial base include an egg and sperm bank for humans and other animals, and a seed depository.

27.6
Disaster Simulation and Prediction

Attempting to predict and respond to potential disasters is essentially a branch of futurology. There are at least 27 methodologies that have been used to predict the fu-

Table 27.2. Dimensions of Resilience (after Foster 1997)

Social dimensions	1. Compatibility with diverse value systems 2. Capacity to satisfy several goals 3. Equitable distribution of benefits and costs 4. Generous compensation for major losers 5. Accessibility
Systems characteristics	1. Significance of internal variables 2. Impact of external variables 3. Diversity of components 4. Functional redundancy
Economic dimensions	1. Incremental funding 2. Wide range of potential financial support 3. High benefit-cost ratio 4. Early return on investments 5. Equitable division of benefits and costs
Environmental characteristics	1. Minimal adverse impacts 2. Replenishable or extensive resource base
Time and timing	1. Short lead time and rapid response to stimuli 2. Open-end life span
Operational characteristics	1. Efficient 2. Reversible 3. Incremental operation 4. Autonomous operation
Physical dimensions	1. Not site specific 2. Fine grained and modular 3. Standardization 4. Mobile 5. No esoteric components 6. Unique skills unnecessary 7. Stable 8. Fail-safe design 9. Early fault detection

ture (Foster 1980). Many of them, for example, scenario building, the Delphi technique, scale modeling and computer simulation could be applied in efforts to understand the implications of cosmic impacts more fully. To illustrate, simulation models are important methods of investigating the development of potential disasters through time. These are normally of three types: scale, analog and mathematical (Chorley and Kennedy 1971). In 1970, for example, Whalin and coworkers described a scale representation of the harbor at San Diego, California. This model was built to investigate the impact of deep-water wave heights from about 4 to 15 meters. Such waves could be generated by localized seismic disturbances, an explosion, a massive landslide, or the impact of a meteorite. They concluded, as the result of experiments conducted with their scale model, that waves of this magnitude would cause extensive inundation of the Silver Strand, the city of Coronado, and parts of the North Island. It was thought unlikely that any vessel would survive them in the surf zone.

Computer simulations that permit relatively accurate predictions of potential disaster losses are extremely valuable managements tools. Regardless of the hazard involved, the construction of such models require four common steps. The first is an analysis of the physical characteristics of the hazard. This allows the subsequent development of a mathematical model capable of forecasting the severity and frequency of its impact. The approach taken is to develop a model that can produce a spatial representation of intensities with properly spaced contours, which are consistent with the size, shape and configuration of observed patterns. This distribution will be controlled by the magnitude of the event, modified by the impact of certain local variables. In the case of a tsunami generated by an asteroid, the scale of inundation and associated damage would reflect size and speed of the impactor, its location in the ocean, and the presence or absence of local features such as bays, reefs, submarine ridges, canyons and the width of the continental shelf.

To predict the damage and casualties caused by such an event, it is also necessary to know the geographical location and characteristics of the population, and the type and value of the infrastructure at risk. Such information is used to produce a geographical representation of the society threatened by the hazard. In the United States, for example, the Travelers Insurance Company collected such information for some 85 000 grid areas that completely covered the 48 contiguous states of the United States. These data were used in computer simulations that permitted the setting of realistic premiums for policies covering a variety of natural hazards (Friedman 1973).

Once these first two steps have been taken, the models of the disaster agent and of the infrastructure and its inhabitants must be linked by a matrix representing the loss relationship between property type and intensity of impact. This is usually designed by historical research, based on known disasters and the damage caused by hazard impacts of differing magnitude. Foster and Carey (1976), for example, produced such a matrix for the simulations of earthquake damage in Victoria, British Columbia. Given the completion of these three steps, it is possible to apply the mathematical representation of the hazards to the geographical distribution of inhabitants and infrastructure. This produces a synthetic, computer simulation of the disaster experience that can be represented in terms of economic loss, degree of damage to particular buildings, and fatalities and injuries sustained.

Risk Management Solutions Inc. (1995 a, b and c) for example, has produced computer simulations for major earthquakes striking Los Angeles, San Francisco Bay and the Tokyo region that seem particularly relevant in this discussion. They concluded that if an earthquake having the same characteristics as that occurring in 1923 were again to strike Tokyo, it would cause between 30 000 to 60 000 deaths and 80 000 to 100 000 serious injuries. Total expected economic losses would range from US$ 2 100 000 to US$ 3 300 000 million and undermine the entire global economy. Dore (2006) has begun this simulation process for cosmic hazards by examining the economic impact on the global economy of potential strikes by asteroids and comets of differing sizes.

Computer simulations of tsunami damage associated with cosmic impacts could be used to argue for greater investment in mitigation strategies, better design of evacuation routes and near-Earth-object and tsunami warning systems, and more realistic disaster exercises and gaming. They would be relatively easy to produce, especially for cities such as Los Angeles, San Franciso and Tokyo for which earthquake models already exist (Risk Management Solutions Inc. 1995 a,b and c). The only major obstacle seems to be the great difference in opinion expressed by researchers about the size of the deep ocean waves likely to be generated (Ward and Asphaug 2000; Hills and Mader 1997). Clearly, this issue should be resolved before computer simulations of damage can be realistically attempted.

27.7
Warning Systems

During the Cold War, in the event of a nuclear attack, Canadians were advised to "duck, hide, hope and pray" (Stirton 1971). They were expected to take these actions only after sirens have sounded and every radio and television station in the country had broadcast the Attack Warning. How effective this would have been following this advice is highly debatable, but it does illustrate that, like chains, warning systems are only as strong as their weakest links. For this reason, such networks have to be designed with great care. Attention should be paid to both technical and social components, to their interactions and to the networks' roles in the social system of which they are merely a small part.

Arthur C. Clarke, noted for his excellence as a science fiction writer, introduced the concept of a "Spaceguard Survey" in his 1973 novel *Rendezvous with Rama*. This system searched the heavens for asteroids that threatened the Earth (Chapman 1998). Since then, progress by small under funded search groups, like that at the University of Victoria, British Columbia, has been slow. In 1990, Congress requested that NASA speed up discovery of potentially threatening near-Earth asteroids, beginning with those larger than 1 kilometer in diameter that were considered the most dangerous (Morrison 1996). A team of international astronomers suggested setting up a program to obtain a complete census of these larger asteroids called the Spaceguard Survey. In 1992 these results were reported to Congress and NASA provided $1 million in additional funds so that existing search programs could be updated. Simultaneously, the International Astronomical Union appointed a working group to promote more cooperation in the search for cosmic threats.

After the dramatic impacts of fragments of Comet Shoemaker-Levy 9 into Jupiter in 1994, public awareness and support for a cosmic impact warning system increased and, as a result, the Spaceguard Survey was formally endorsed by NASA in 1998. The goal was set of discovering, within a decade, 90 percent of near-Earth asteroids larger than one kilometer in diameter.

Currently, Spaceguard consists of a network of professional laboratories, dominated by two 1-meter aperture telescopes near Socorro, New Mexico (operated by MIT Lincoln Laboratory) and numerous amateur and professional observers who follow up discoveries and attempt to refine knowledge of their orbits. Members of the Spaceguard search programs include the Lowell Observatory's LONEOS in Flagstaff, Arizona, Jet Propulsion Laboratory's near-Earth Asteroid Tracking [NEAT] facility, located in Maui and on Mt. Palomar, California and Spacewatch on Arizona's Kitt Peak (Chapman 2004). In addition, the International Spaceguard Foundation is centerd in Italy. This consists of a team of astronomers who collaborate by e-mail whenever one discovers a particularly threatening Near-Earth-asteroid. This global network of professional and amateur observers continues to discover a new Near-Earth-asteroid every few days. As of February 2004, almost 2 670 have been found, some 600 of which are potentially hazardous. As Chapman (2004) points out, this compares with only 18 that were known in 1981. It is believed that the census is complete for near-Earth-asteroids greater than 3 kilometers in diameter. The estimated number of near-Earth-asteroids greater than one kilometer in diameter is some (1 100 ± 200) (Bottke 2006). About 55 percent of this total had been identified by early 2004. NASA also supports a Near-Earth Object Program that was established in 1998 to help coordinate and provide a focal point for research into asteroids and comets that approach the Earth's orbit. It operates from the Jet Propulsion Laboratory and provides data on the recent approaches to the Earth, including the name of the object, its closest approach date, miss distance, estimated diameter and relative velocity. On June 27, 2004, when the author visited this website (NASA 2004) 40 such objects were listed, varying in size from an estimated 900 m – 2.0 km to 15 m – 34 m in diameter, with miss distances reaching a minimum of 1.5 LD (1 LD [lunar distance] = ~384 000 kilometers).

While, clearly, a great deal of warning system progress has been made in the past decade, there are still some very obvious weaknesses. When assessing any natural hazard warning system several key questions must be asked. These, for example, include "Are all threats from this type of hazard being adequately monitored?" Others include, "Is it clear who will issue warnings and will they be believed?" It also is extremely important that, where a threat is perceived, the public is sufficiently aware of its consequences to react in a manner that reduces risk in a cost-effective way.

Clearly, Spaceguard and the Near-Earth Object Program do not yet seek to identify and monitor all cosmic threats. NASA, however, has had a Science Definition Team studying the benefits and costs of extending the program to search for, and monitor, smaller asteroids (Morrison 2004a).

> Even using a conservative approach to estimating the losses that would be expected from impacts by sub-km asteroids, the annualized losses are much greater than the costs of mitigating the hazard by a more capable survey. The sub-km hazard has two peaks, one for land impacts (near 200 m) and one for tsunamis from ocean impacts (near 350 m). The total cost to carry out surveys that are 90% complete for NEA [Near-Earth asteroids] larger than 140 m is less than $ 400 million, with both ground-based and space-based options possible.

As things stand, a highly dangerous near-Earth-object could remain undetected until all chance of altering its course has passed. To rectify this deficiency, Safeguard needs to be expanded. Morrison (2004a), for example, has suggested the necessary addition of an LSST-type telescope with an 8 m aperture and wide field of view.

Technology is important but there is much more to a well designed warning system than merely hardware. A warning is a recommendation based on a prediction, to take precautionary, protective, or defensive action. The decision to warn, therefore, carries with it a great deal of responsibility. Once any organization has issued such a public pronouncement, especially if it is based upon the prediction of an event of great destructive potential, that agency, and the public's response to it will never be the same again. This is true, whether or not the warning proves correct. For this reason, the decision to warn cannot be taken lightly. Spaceguard's record to date has not been good.

In early 1998, the global media announced that a huge asteroid might strike the Earth in 2028. The next day, astronomers claimed that new data proved that there was no such danger of cosmic impact. This chain of events was not true, as described by Chapman (1998):

> That's what was reported in the press, but it is not exactly what happened. We now realize that data were already collected two-and-a-half months before March 11th, and published on the Internet, which were sufficient to demonstrate that the asteroid called 1997 XF11 was certifiably safe: it simply could not, realistically, impact the Earth. But months went by and the few astronomers who are funded, part-time if at all, to study all the new asteroid discoveries never had a chance to examine the data in detail. When one under funded astronomer suddenly noticed quirky data about 1997 XF11 in early March, his hasty response was to announce a possible impact. Within hours, his colleagues finally looked at the data and concluded – as they just as well could have done months earlier – that the object could not possibly strike Earth in 2028.

This was only one of several impact scares between 1998 and 2004 (Marsden 2006). Clearly, the Spaceguard Survey requires a firm chain of command and a well-established procedure for issuing warnings. After all, imagine what would be required if an official warning of an imminent collision with even a 400 meter diameter asteroid were issued. If such an impactor were to strike the ocean and generate an enormous tsunami, every port and low lying region of the planet would require evacuation. Safe havens would be required for shipping; but where? All works of art and other articles and equipment of value would have to be moved inland to higher altitudes. Possible toxic and dangerous substances would require removal from threatened areas. These tasks, and many others, would stretch mankind's capacity to adequately react up to and probably beyond its limits. The social and financial costs would be enormous. Now consider the political implications of an error in issuing such a warning. On the other hand, imagine refusing to issue such a warning and having such an impactor strike the planet, destroying every major coastal city around the Pacific.

As shown in Table 27.3, a well-designed natural hazard warning system has sixteen main components, most of which are social not technical (Foster 1980). Beyond issues already discussed, these include provisions for the education of user groups, procedures for testing and revision of the warning process, and the creation of feedback loops that ensure that reactions to warnings will be those intended. Unfortunately, many of these dimensions are, as yet, missing from the Spaceguard Survey. This seems to be largely because it is under-funded and understaffed. It seems enigmatic that while three space agencies can cooperate and spend over $ 3 billion on the Cassini-Huygens mission to Saturn and Titan (Jet Propulsion Laboratory 2004) they are unwilling to provide a fraction of this to greatly increase the safety of the Earth.

27.8
Disaster Planning

Disasters are characterized by an urgent need for rapid decisions, accompanied by acute shortages of the necessary trained personnel, materials and time. To help mitigate these difficulties, disaster plans should be drawn up and tested long before they are needed. Such plans can be prepared at every level from the international to the local long be-

Table 27.3. Sixteen steps in the design of the "ideal" warning system (after Foster 1980)

1. Recognition by decision makers that there is the possibility of danger from a particular source.
2. Design of a system to monitor changes in the hazard and issue warnings if danger increases beyond certain thresholds.
3. Installation and operation of the system.
4. Education of the user group, often the general public so that should a warning be issued, responses will be appropriate. The infrastructure may also have to be modified to permit effective operation.
5. Testing the system, when there is little danger, to ensure that it is technically sound and that those involved in issuing and receiving its warnings act as required.
6. Modifying the system if test results indicate that changes are necessary.
7. Detection and measurement of changes in the hazard that could result in increases in death, injury and/or property damage.
8. Collation and evaluation of incoming information.
9. Decisions as to who should be warned, about what damage, and in what way.
10. Transmission of a warning message, or messages, to those whom it has been decided to warn.
11. Interpretation of the warning messages and action by the recipients.
12. Feedback of information about the actions of message recipients to issuers of the warning messages.
13. Transmission of further warning messages, corrected in terms of the user groups responses to the first and subsequent messages and noting any secondary threats.
14. Transmission of an all-clear when danger has passed.
15. Hindsight review of the operation of the warning system during potential disaster situation and the implementation of any necessary improvements.
16. Testing and operation of the revised system.

fore any cosmic impact disaster (Foster 1980). All should seek to identify the problems that are likely to occur and the decisions that probably will have to be made as a result. Good disaster plans are essential if decision making is to be anywhere near optimum under crisis conditions. These plans typically consider 25 significant aspects of disaster.

One key aspect of planning for disaster is identifying a chain of command. Whom for example will be in charge of global response if the Spaceguard Survey issues a warning of an imminent cosmic impact? What will be the responsibilities of major international and national agencies? How will these responses be funded? There are numerous large and small scale issues that should be addressed and settled now. To wait until a significant threat has been identified is to wait too long.

Two of the most important planning issues are briefly examined here. Firstly, the possibility of deflecting or destroying smaller comets or asteroids, so that an Earth impact is prevented, needs detailed consideration. A wide range of approaches to impact prevention has been put forward in the literature. Mitigation subsystems might involve rocket propulsion, rocket-delivered nuclear warheads, kinetic energy systems using projectiles, directed energy from lasers, mass drivers, solar sails and biological, chemical or mechanical asteroid and/or comet "eaters". Suggestions have been made also of super magnetic field generators and futuristic force fields, tractor beams and gravity manipulation (Morrison 1996, 2004a; Simon 2002). Considerable progress has been made very recently in this area. The NASA Institute for Advanced Concepts has just announced five Phase II awards for the further development of revolutionary advanced concepts to help protect the Earth from cosmic collision (NIAC 2004). Beyond this, the European Space Agency has given priority to "Don Quijoté", selected from six potential asteroid protection missions. This will involve an asteroid 500 meters in diameter and two spacecraft, Sancho and Hidalgo. Sancho will arrive first and orbit the asteroid for several months, deploying penetrating probes to form a seismic network. When this is ready, and adequate data has been collected, Hidalgo will arrive, crashing into the asteroid at about 10 kilometers per second. Sancho would then study the changes in the asteroid's orbit, rotation and structure caused by Hidalgo's impact. This information will give insights into what is needed to modify the orbit of any similar asteroid that may threaten Earth (Morrison 2004b). The United States is currently installing a missile defense system (Missile Defense Agency 2005). With greater international cooperation, this might be expanded to provide the capacity to protect the planet against errant near-Earth-objects, including medium-sized asteroids.

Disaster plans should be tested long before they are needed in earnest. After an expert panel has evaluated such potential technologies for impact mitigation, two or three of the most promising should be tested on small, non-threatening asteroids. The sooner the planet has a functional defense system, the better. The technology required to provide one already exists. What is lacking is the political will and the financing required.

A second key disaster planning issue that should be addressed now is "How should we respond to the threat of very large tsunamis generated if there is an oceanic cosmic impact?" Obviously, given the great differences of opinion concerning the magnitude of such potential tsunamis (Hills and Mader 1997; Ward and Asphaug 2000), modeling has to be improved. Once an expert consensus has been reached, the major issue of adequate tsunami warnings and associated evacuations must be addressed. While there is an effective tsunami warning network for the Pacific, nothing comparable exists elsewhere. How then could warnings and evacuation be effectively organized for the populations of low lying coastal areas around the Atlantic, Arctic and Indian oceans? How could the evacuation of the total populations of low altitude countries like the Netherlands and Bangladesh be organized? What about that around the Mediterranean or the Great Lakes? What about islands without central mountain cores, such as the Marshall and Tokelau Islands? The issues are enormous and the logistics far beyond anything humanity has ever attempted. It seems much more likely that, rather than face up to these problems, many decision-makers would prepare to issue warnings of impending impact together with the advice to 'duck, hide, hope and pray' (cf. Stirton 1971). Table 27.4, of course, provides an insight into a little of what is really needed.

27.9
Reconstruction

Given their roles within economic regions, speed of population re-growth and the psychological impact of abandonment, few cities fail to recover from major disasters (Kates and Pijawka 1977). In the twentieth century, for example, only two were perma-

Table 27.4. Typical contents of a disaster plan (after Foster 1980)

1. Policy statement on value of disaster planning by chief executive officer
2. Legislative authority for the design of the disaster plan and for the steps it contains
3. Aims of the plan and conditions under which it comes into force
4. Assessment of community disaster probabilities
5. Disaster scenarios
6. Relationships with other levels of government, particularly emergency-related agencies
7. Authority organization chart
8. List of names, addresses, and telephone numbers of all relevant agencies, their heads and deputies
9. Operation of warning systems:
    - types of warnings
    - distribution
    - obligations on receiving warnings
10. Pre-impact preparations:
    - relationships between type of disaster agent and necessary preparations
    - responsibilities of different agencies
    - location of greatest risk sites
11. Emergency evacuation procedures:
    - conditions under which evacuation is authorized
    - routes to be followed and destinations
    - accommodating the special needs of the elderly, ill, or institutionalized
12. Shelters:
    - locations
    - facilities
13. Disaster control center and subcenters:
    - location(s)
    - equipment
    - operation
    - staffing
14. Communications
15. Public information
16. Search and rescue:
    - responsibilities
    - equipment
    - areas most likely to require servicing
17. Community order
18. Medical facilities and morgues:
    - location
    - transportation
    - capacity
    - facilities
19. Restoration of community services:
    - order of priorities
    - responsibilities
20. Protection against continuing threat:
    - the search for secondary threats
    - actions to be taken if discovered
21. Continuing assessment of total situation:
    - responsibilities
    - distribution
22. Reciprocal agreements and links with other municipalities
23. Testing the plan:
    - disaster simulations
    - simulation evaluations
24. Revision and updating of the plan
25. Plan distribution

nently destroyed by natural hazards, St. Pierre, Martinique and Yungay, Peru. The former was completely demolished by a nuée ardente, a glowing avalanche of gas and debris ejected from Mount Pelé and the latter buried beneath sediments deposited by an earthquake-triggered avalanche (Griggs and Gilchrist 1983; Office of Emergency Preparedness 1972). Nevertheless, civilizations have been destroyed by natural hazards. To illustrate, the tsunami that swept the lowlands of the Mediterranean Sea (circa 1450 to 1480 BC) generated by the eruption of the volcano of Santorini, likely decimated the Minoans (Foster 1980).

Whether port cities, destroyed by tsunamis generated by a near-Earth-object, would be quickly rebuilt is uncertain and dependent upon the size and location of the impactor and scale of its associated destruction. Nevertheless, it is well known that the degree of uncertainty occurring after any major disaster plays a significant role in influencing the speed of rebuilding (Kates and Pijawka 1977). To avoid unnecessary delays, it is important to pre-plan for recovery. Unfortunately, the widespread unwillingness to face up to the possibility of major disasters increases the suffering associated with such adverse events when they occur. This was extremely obvious when the tsunami, generated by the Great Sumatra-Andaman earthquake of December 26, 2004 swept the Indian Ocean (Lay et al. 2005). It also reduces society's chances of benefiting from the opportunities for creative reconstruction that they offer.

"Think tanks" should be set up to review what should be done to speed human recovery from a significant cosmic impact. At the very least, the horrifying economic and social scenarios that they would generate might encourage politicians to take the need for a global extraterrestrial defense system more seriously.

27.10
Summary and Conclusions

Archaeological records show that civilizations may fail to recover from major catastrophes. Societies operate within specific levels of tolerance for repetitive natural hazards and anthropogenic modification. Catastrophic events, like cosmic impacts, lie outside the realm of human experience and so are difficult to plan for and respond to.

All near-Earth objects with the ability to cause serious damage and potential hazards that might result from a large impact event (e.g., tsunamis, earthquakes, volcanism, secondary impacts, wildfires, climate change, orbital and axial changes, economic collapse, disease, famine and war) must be identified. The most worrisome objects, ranging from 10s to 100s of meters in diameter, are not easily detected by earthbound platforms. There is a 1 in 1000 chance of an object greater than 200 m impacting this century, with the most likely target being an ocean. The resulting tsunamis would devastate coastal zones and hundreds of port cities, leading to an unprecedented mortality rate and global economic collapse. Current architectural and engineering designs will be unable to cope with the kinetic energy released by a cosmic impact.

During most disasters, an urgent need for rapid decisions is confounded by the lack of trained personnel, materials and time. Mitigation for on-going catastrophes (e.g., famines, droughts) receive the most attention from decision makers, the response to exotic threats occurs only after the disaster has occurred. However, mitigating for cosmic impacts should be seriously considered given that the chances of being killed by a meteorite impact are similar to the chances of being killed in an airplane disaster (i.e., 1 in 20 000). Unfortunately, current Space Guard programs have limited success because of under funding and technological limitations, but also socio-economic factors and human error.

Well-designed hazard warning systems should include provisions for education, testing and revising the warning process, and feedback loops to ensure that responses to warnings are valid. It is essential to establish a chain of command for issuing warnings of impact, delineating responsibilities and funding of various agencies. Pre-planning is essential to reduce societal losses and increase the chances of benefiting from reconstruction. Disaster mitigation options include deflecting or destroying smaller objects to prevent significant impact, enhancing existing missile defense systems, and responding to the threat of very large tsunamis following an oceanic impact. Computer-generated risk maps need to accurately predict the potential inundation from tsunamis to calculate disaster losses, plan evacuation routes and mitigate loss of life and destruction.

Reconstruction is dependent on the size and location of the impactor and scale of destruction. Infrastructure and other social networks should be designed to function autonomously. Functional redundancy and adaptability is also recommended to protect against societal collapse. Extraterrestrial genetic ark repositories should also be established to ensure species continuity following a global catastrophic event.

Although the last decade has seen some progress in preparing for the possibility of cosmic collision, we continue to be very ill-prepared for such an event. What is needed is the political will to cooperate and dedicate adequate financial and human resources to mitigate the threat posed by cosmic impacts, and studies of how the human race can recover from such hazards.

References

Ball P (2001) Brimstone pickled Permian. Nature, Science update. *http://www.nature.com/nsu/010920/010920-6.html*

Becker L, Poreda RJ, Basu AR, Pope KO, Harrison TM, Nicholson C, Iasky R (2004) Bedout: a possible End-Permian impact crater offshore of Northwestern Australia. Science Express. *www.sciencexpress.org*

Birks JW, Crutzen PJ, Roble RG (2007) Frequent ozone depletion resulting from impacts of asteroids and comets. Chapter 13 of this volume

Bottke WF (2007) Understanding the near-Earth object population: the 2004 perspective. Chapter 9 of this volume

Bryant E (2004) Geological and cultural evidence for cosmogenic tsunami. Conference Comet/Asteroid Impacts and Human Society. The International Council for Science Workshop, Santa Cruz de Tenerife, Canary Islands, November 30, 2004

Calame O, Mulholland JD (1978) Lunar crater Giordano Bruno AD 1178: impact observations consistent with lunar ranging results. Science 199:875–877

Chapman CR (1998) The risk to civilization from extraterrestrial objects. Chance Lectures, December 11 1998, Dartmouth College, Hanover, New Hampshire. *http://www.boulder.swri.edu/~cchapman/chance.html*

Chapman CR (2003). How a near-Earth object impact might affect society. Commissioned by the Global Science Forum, OECD, for the Workshop on near Earth Objects: Risks, Policies, and Actions. January 2003, Frascati, Italy. *www.tpg.com.au/users/horsts/chapman4oecd.pdf*

Chapman CR (2004) The hazard of near-Earth asteroid impacts on Earth. Earth and Planetary Science Letters 222:1–15

Chapman CR, Morrison D (1992) Asteroid threat! 1993 Science Yearbook. Franklin Watts, New York, pp 41–45
Chapman CR, Morrison D (1994) Impacts on the Earth by asteroids and comets: assessing the hazard. Nature 367:33–40
Chorley RJ, Kennedy BA (1971) Physical geography: a systems approach. Prentice-Hall International, London
Dore MHI (2007) The economic consequences of disasters due to asteroid and comet impacts, small and large. Chapter 29 of this volume
Earth Impact Database (2004) *http://www.unb.ca/passc/ImpactDatabase/*
Foster HD (1980) Disaster planning: the preservation of life and property. Springer, New York
Foster HD (1997) The Ozymandias principles: thirty-one strategies for surviving change. Southdowne Press, Victoria
Foster HD (2002) What really causes AIDS. Trafford Publishing, Victoria
Foster HD (2004) How HIV-1 causes AIDS: implications for prevention and treatment. Medical Hypotheses 62(4):549–553
Foster HD, Carey RF (1976) The simulation of earthquake damage. In: Foster HD (ed) Victoria: physical environment and development. Western Geographical Series 12:221–240
Friedman DG (1973). Computer simulation of natural hazard effects. The Travelers Insurance Company, Hartford, Connecticut
Gallant RA (2004) Tunguska: the cosmic mystery of the century. Soultworth Planetarium, University of Southern Maine. *http://www.usm.maine.edu/~planet/tung.htm*
Grieve RAF (1997) Target Earth: evidence for long-scale impact events. In: Remo, JL (ed) Near-Earth objects: The United Nations International Conference. Annals of the New York Academy of Sciences 822:319–352
Grieve RAF, Kring DA (2007) The geologic record of destructive impact events on Earth. Chapter 1 of this volume
Griggs G, Gilchrist JA (1983) Geologic hazards, resources, and environmental planning. Wadsworth, Belmont, California
Gusiakov VK (2003) NGDC/HTDB meeting on the historical tsunami database proposal. Tsunami Newsletter 35(4):9–10
Gusiakov VK (2007). Tsunamis as a destructive aftermath of oceanic impacts. Chapter 14 of this volume
Harris A (n.d.) The large and the small impact debate: Part III. Astrobiology Magazine. *www.astrobio.net*
Hartung JB (1976) Was the formation of a 20-km-diameter impact crater on the Moon observed on June 18, 1178? Meteoritics 2:187–194
Hills JA, Mader CL (1997) Tsunami produced by the impacts of small asteroids. In: Remo JL (ed) Near-Earth Objects: The United Nations International Conference. Annals of the New York Academy of Sciences 822:381–394
Ingram J (1999) The barmaid's brain and other strange tales From science. Penguin Books Canada, Toronto
Jet Propulsion Laboratory (2004) Cassini-Huygens mission to Saturn and Titan. *http://www.jpl.nasa.gov/stars_galaxies/*
Kaiho K, Kajiwara Y, Nakano T, Miura Y, Kawahata H, Tazaki K, Veshima M, Chen ZQ, Shi GR (2001) End-Permian catastrophe by a bolide impact: evidence of a gigantic release of sulphur from the mantle. Geology 29:815–818
Kates RW, Pijawka D (1977) From rubble to monument: the pace of reconstruction. In: Haas JE, Kates RW, Bowden MJ (eds) Reconstruction following disaster. MIT Press, Cambridge, Massachusetts, pp 1–23
Kerr RA (2004) Evidence of huge, deadly impact found off Australia Coast? Science 304(5673):941
Knocke MM (2003) Asteroid's flyby of Earth closest yet. The Planetary Society. *http://www.planetary.org/html/news/articlearchive/headlines/2003/asteroid2003-sq222.html*
Koeberl C, Farley A, Psyckas-Ehrenbrink B, Sephton MA (2004) Geochemistry of the end-Permian extinction event in Austria and Italy: No evidence for an extraterrestrial component. Geology 32(12):1053–1056
Lay T, Kanamori H, Ammon CJ, Nettles M et al. (2005) The Great Sumatra-Andaman earthquake of 26 December 2004. Science 308(5725):1127–1133

Marsden BG (2007) Impact risk communication management (1988–2004): has it improved? Chapter 31 of this volume

Missile Defense Agency (2005) Making ballistic missile defense a realilty. *http://www.mda.mil/mdalini/html/mdalink.html*

Morrison D (1996) Target: Earth. In: Castagno JM (ed) 1997 Science Yearbook. Grolier, New York, pp 94–99

Morrison D (2004a). Planetary defense conference: protecting the Earth from asteroids. Notes from AIAA Planetary Defense Conference, 23–26 February 2004. *http://nai.arc.nasa.gov/impact/news_detail.cfm?ID=136*

Morrison D (2004b) Don Quijoté, Toutatis and Sagan. Asteroid and Comet Impact Hazards, NASA Ames Research Center. *http://128.102.38.40/impact/news_detail.cfm?ID=144*

Morrison et al. (2002), cited by Chapman (2003)

NASA Institute for Advaced Concepts (NIAC) Phase II Awards (2004). *http://www.niac.usra.edu/.//home/*

National Aeronautics and Space Administration (2004) Near Earth objects program. *http://neo.jpl.nasa.gov/welcome.html*

Nininger HH, Huss GI (1977). Was the formation of Giordano Bruno witnessed in 1178? Look again. Meteoritics 12:21–25

Office of Emergency Preparedness (1972) Report to the Congress: disaster preparedness. US Government Printing Office, Washington, DC

Risk Management Solutions Inc (1995a) What if a major earthquake strikes the Los Angeles area? Risk Management Solutions Inc, Menlo Park, California

Risk Management Solutions Inc (1995b). What if the 1906 earthquake strikes again? A San Francisco Bay Area scenario. Risk Management Solutions Inc, Menlo Park, California

Risk Management Solutions Inc. (1995c). What if the 1923 earthquake strikes again? A five-prefecture Tokyo region scenario. Risk Management Solutions Inc, Menlo Park, California

Shakespeare W (1564–1616) Troilus and Cressida. Act 1, Scene III

Simon (2002) Earth impact. *http://www.s-d-g.freeserve.co.uk/intro.html*

Stirton AM (1971). Emergency public information planning in Canada. EMO National Digest 11(5):1–6

THAMS (n.d.) List of joint publications arising from the research of THAMS. *http://yalciner.ce.metu.edu.tr/thams/thams-publications.htm*

The National Tsunami Mitigation Program (2004) *http://www.pmel.noaa.gov/tsunami-hazard/*

University of Bologna (Italy), Department of Physics. Tunguska Home Page. *http://www-th.bo.infn.it/tunguska/index.html*

Ward SN, Asphaug E (2000) Asteroid impact tsunami: a probabilistic hazard assessment. Icarus 145: 64–75

Whalin RW, Bucci DR, Strange JN (1969) A model study of wave run-up at San Diego, California. Tsunamis in the Pacific, Proceedings of the International Symposium on Tsunamis and Tsunamis Research, University of Hawaii, Honolulu, October 7–10, 1969

Wood D (2000) Disasteroids! Reader's Digest Canada. *http://astrowww.phys.uvic.ca/media/press/1.htm*

Chapter 28

Insurance Coverage of Meteorite, Asteroid and Comet Impacts – Issues and Options

Paul Kovacs · Andrew Hallak

28.1 Introduction

An asteroid or comet will threaten a major urban center sometime in the future. It is very unlikely to happen this year, but some day it will happen. The potential damage will be catastrophic. A typical property insurance policy promises coverage for damage caused by such an impact, but there are limits to the capacity of insurance to pay. Moreover, damage from an asteroid or comet strike in a major urban center does not fit the principles of insurance coverage, so insurers may use the months or years between detection and impact to exclude this peril in insurance policy renewals that take place before the strike occurs. National and international policy makers should develop preparedness plans assuming that they will manage society's recovery from an asteroid or comet strike in a major urban center, including responsibility for financial matters.

Most property insurance policies also cover damage caused by meteorites or a small asteroid. These risks are consistent with the principles of insurability, and can continue to be affordably managed by the insurance industry over the longer term.

28.2 A Brief History of Insurance

The basic concept of insurance involves many policyholders pooling their modest premiums to cover the random and often significant losses that affect a few. This concept has been in practice for a long time, and was in place prior to the founding of the modern insurance industry. For example, to reduce the risk of loss due to theft in ancient times, the Babylonians devised a system of contracts in which the supplier of capital for a venture agreed to cancel the loan if the trader was robbed of his goods. The trader borrowing capital paid an extra amount for this protection (a premium). As for the lender, collecting these premiums from many traders made it possible for him to absorb the losses of the few (Kathy Bayes Insurance Agency n.d.).

The modern insurance industry developed after the Great Fire of London in 1666. Fire swept through nearly 80 percent of the largely wooden city, destroying more than 13 000 homes and 100 churches including St. Paul's Cathedral (Insurance Bureau of Canada 2003). Following the fire, demand arose for fire suppression and insurance protection. Insurance grew over the next three hundred years to cover a remarkably broad range of perils. By 1706, the Sun Fire Office in London was offering coverage on contents and dwellings. Insurance companies soon opened in Scotland (by 1720), Germany (1750), the United States (1752), and Canada (1804) (Insurance Bureau of Canada 2003). Insurance is now available around the world. The United Nations has described the industry as an essential foundation for a nation's economic success.

During the early 20th century, there was a major reform in typical coverage. Policies covering named perils, such as fire and theft, were largely replaced by comprehensive, multi-peril or all-risk policies. This included homeowners and commercial insurance coverage. These policies cover all risks that are not specifically excluded. In addition to property insurance, insurance has become a remarkably flexible mechanism to protect oneself from a wide variety of threats.

The property and casualty insurance industry is largely independent from the life insurance industry. Although the impact of an asteroid or comet would undoubtedly have dramatic effects for both industries, it is the property and casualty insurance industry that is more actively engaged in the assessment and management of natural hazard risk. This paper focuses on the impact of an asteroid strike on the property and casualty insurance industry.

28.3
Insurance and Natural Hazards

Insurance protection is available for damage caused by most natural hazards. Coverages differ somewhat around the world, but a typical all-perils insurance policy in North America and Europe provides coverage against damage caused by hazards that include severe wind, tornado, hurricane, hail, freezing rain, lightning, heavy snowfall, freezing pipes and falling objects including meteorites, asteroids and comets. Additional coverage can often be purchased for sewer back-up and earthquake damage if requested (Insurance Bureau of Canada 1994).

Some hazards, such as flood and landslide damage, are often not covered by most standard insurance policies, or endorsements, because they do not satisfy the underwriting requirements set out below. Government agencies may provide insurance-like coverage for these risks but they typically are not covered by private insurers.

Risks should meet three broad criteria before they are accepted as insurable:

- There is a random occurrence of loss;
- A relatively large population is exposed to a risk and is willing to pay for coverage; and
- A relatively small share of the exposed population is likely to incur a loss at any particular time (Insurance Bureau of Canada 1994).

Flood and landslide losses are not random. Properties located in areas of high risk are more likely to experience damage. Private insurance is largely not available in such instances.

28.4
Do Asteroid Impacts Fit within the Principles of Insurance?

An impact by a meteorite, asteroid or comet may or may not fit the principles of insurance. Consider some scenarios to assess this question further.

28.4.1
Scenario 1: Asteroid Impact

NASA established the Spaceguard program to discover by 2008 at least 90 percent of all Near Earth Objects (NEOs) whose diameters are larger than one kilometer (Morrison 2005). Currently, the best estimate of the total population of NEOs larger than one kilometer is about 1100 (Bottke Jr. 2005; Morrison 2005).

The impact of an asteroid of roughly one kilometer in diameter could create energy equivalent to one million megatons and would lead to a global catastrophe that would kill a substantial fraction of the Earth's human population (NASA n.d.). For the insurance industry, an impact of such a magnitude would lead to catastrophic losses. So does the impact of a large asteroid fit within the principles of insurance?

Principle: There is a random occurrence of loss
NASA states that tracking the orbit of a NEO can be quite complex. This is supported by the fact that there have been instances in the past where predictions of an Earth impact were re-evaluated and changed. Such was the case for the 1997 FX$_{11}$ asteroid (Marsden 2005). In discussing the issue of how much warning there would be in the event of an asteroid strike, NASA states that "with so many of even the larger NEOs remaining undiscovered, the most likely warning today would be zero – the first indication of a collision would be the flash of light and the shaking of the ground as it hit."(NASA n.d.)

Another consideration with respect to the randomness of an asteroid strike would be how closely it is tracked. Larger size NEOs including asteroids could be detected from their motion using modest-sized ground-based telescopes based on a single night's sighting (Morrison 2005). If an asteroid or comet is sighted several times during its orbit, the track that it is following becomes more and more clear, thus reducing the true randomness of the orbit's track.

In summary, the majority (perhaps 90 percent by 2008) of the risk of a large asteroid or comet impact is not random as the objects have been detected and their orbit is known with increasing accuracy.

Principle: A large population is exposed to risk
Every part of the Earth is vulnerable to meteorite, asteroid and comet impacts. It is certainly the case that should a large asteroid (> 1 kilometer) make a land impact with the Earth, a very large population would be affected. In fact, an impact of this size would have global consequences (Melosh 2005; MacCracken 2005) and put an entire nation or nations at risk.

Principle: Small share of the population is likely to incur the loss
As Melosh and MacCracken point out in their respective studies, an asteroid with a diameter of one kilometer striking the Earth would affect large segments of the world's population. The affected regions could transcend international borders, not only creating logistical nightmares with respect to saving human lives but also with respect to clean up and recovery processes including insurance coverage. It is thus clear that this principle of insurance would fail in such a circumstance.

Summarizing, in the case of an impact from an asteroid of at least one kilometer in diameter two of the three principles of insurability are not met, making a large asteroid or comet impact ultimately an uninsurable event. Coverage is presently in force but this is likely unsustainable.

28.4.2
Scenario 2: Meteoroid Impact (Meteorite)

Principle: There is a random occurrence of loss
A more frequent concern for those tracking and recording the orbits of NEOs are those objects under one kilometer in diameter. It has been suggested that there could be as many as one billion NEOs greater than four meters but less than one kilometer in diameter (Chapman 2005). A meteoroid under several tens of meters in diameter is likely to break up before making contact with the Earth's surface due to the stresses of the atmosphere. Objects smaller than one kilometer in diameter are difficult to detect in space and track using existing programs. Their impacts are infrequent and unpredictable (Foster 2005). Furthermore, the location of telescopes tracking NEOs is critical in the success of locating these objects (Bottke Jr. 2005). This inability to track coupled with the unpredictability of timing can easily result in no warning before a strike with the Earth's surface. We simply would not see it until impact has been made.

Principle: A large population is exposed to risk
A meteorite or small asteroid can strike anywhere on Earth. Everyone is exposed to this risk. The likelihood of impact is small but the exposed population is extremely large.

Principle: Small share of the population is likely to incur the loss
An object of several meters, although devastating to those within the immediate vicinity, would not have the global impacts of an object one kilometer in diameter or greater. As a result, damage would be confined to a regional level. It follows that with a limited scope of damage, those affected would appear to be classified appropriately as "the few", and could be covered by the larger pool of policyholders. A meteorite that resulted in the destruction of a house would be an example of a strike that would be covered by the industry.

In summary, the risk of a meteorite or small asteroid impact satisfies the three principles of insurance and could be covered by a typical insurance policy. In light of the above discussion on the insurability of an asteroid and meteorite strike, this paper next explores the limits of the insurance industry's financial capacity to cover this peril.

28.5
Insurance Coverage of Asteroid and Meteorite Damage

Damage due to the direct impact of a falling object is currently covered under property insurance policies. This may be the result of the shift from named-peril policies to all-risk coverage. The terms and conditions of an all-risk policy do not specify any exclusion of falling objects per se, therefore insurance coverage will exist. Named-perils coverage is no longer common, and it would be important to determine if such policies state coverage for falling objects. Only with time have insurers begun to consider and sometimes exclude coverage of certain hazards. For example, recently many insurers have begun to exclude damage caused by terrorism from standard coverage.

The analysis above however shows that without the exclusion of asteroid or comet impact coverage, insurance companies may be in a vulnerable position that could result in having to compensate policyholders in the event of a large asteroid or comet impact. As was discussed above, such an impact could cause a global catastrophe and could cause the collapse of the insurance industry.

In addition to the direct impact, additional destruction may take place due to the characteristics and nature of the impact. With respect to secondary effects of an asteroid or meteorite impact, Munich Reinsurance Company analyzed a typical insurance policy and described the coverage as follows (Munich Re 2001):

- *Fire.* As a meteoroid or asteroid enters the Earth's atmosphere, the object heats up. In the event of an impact on land or explosion, there is a likelihood that fires to nearby buildings or forests may occur (Chapman 2005). If a fire results from a meteorite or asteroid impact, this is usually covered under all-perils policies and at present is not excluded (Insurance Bureau of Canada 1994).
- *Explosion.* Depending on the size and density of a meteorite or asteroid, it is possible that the object may explode prior to actually impacting the surface. Such was the case for the object that exploded over Siberia in 1908. If a meteorite or asteroid does reach the Earth's surface or explodes in the atmosphere, this is viewed as an explosion under a typical property insurance policy. Again, unless specifically noted, this peril is covered in most general property policies across Europe and North America.
- *Tsunami.* Most of the world's surface is covered by ocean, approximately 71 percent (Wisner 2005), so a meteorite or asteroid impact may generate a tsunami. Garshnek et al. (2000) point out that a tsunami resulting from an ocean impact could cause fatalities and damage around the continental margins. Populated areas most immediately at risk include low-lying areas like the Netherlands, Bangladesh, and the Atlantic coastal communities in North and South America. Major cities at risk due to their elevations include Halifax, Honolulu, Tampa Bay, New Orleans, Calcutta and Amsterdam (Garshnek et al. 2000). The catastrophic asteroid impact that struck a shallow sea near Chixulub, Mexico left tsunami deposits in Haiti, Texas and Florida (Ward and Asphaug 1999).
- *Flood.* Flood damage is not covered under a typical property insurance plan in many parts of the world. Most insurance exclusions refer to the rise of a river or overflow of a body of water. There is a general consensus, however, that flooding caused by the impact of an object striking a body of water would be covered by a typical property insurance policy. Referring to a 2003 study conducted by Chesley and Ward, Bottke (2005) points out that the highest risk from flooding comes from small but more frequent impact events, where waves of just a few meters could cause considerable property damage.

- *Earthquake.* In an all-perils policy, earthquake (or shake) coverage exists or an endorsement is available. With the exception of pure impact and pressure wave losses, the destructive results of an asteroid impact are by and large included in the scope of coverage of the terms and conditions of insurance generally used throughout the world. It only takes an impact of an object of several meters in diameter to create a severe shake of the ground (Chapman et al. 2001).

In summary, various direct and indirect impact damage is currently covered by a typical insurance policy (Munich Re 2001).

28.6
Assessing the Potential for Damage

More than 100 meteorites are known to have impacted the Earth during the past century and more than 160 impact craters have been identified, but most of these dating back 100 of millions of years (Foster 2005). The largest event occurred in 1908 when an estimated thirty to fifty meter meteoroid exploded over Siberia on June 30th. That event devastated an area of 2 200 square kilometers, felling or seriously damaging all the trees and leaving the area scarred.

Space observation has revealed that of the objects in orbit around the Earth, more than 200 are between 10 and 30 meters in size. The impact of an asteroid can range from minor to catastrophic, depending on the size of the object, density, potential and capacity for detection and deflection, effectiveness of warning systems and the location of impact. A large strike would likely prompt other hazards including floods, fires, earthquakes and tsunami. These secondary effects will compound the initial destructive force of the original strike and could have devastating impacts on the infrastructure of one or possibly more countries (Chapman 2003).

NASA has undertaken to find, by 2008, 90 percent of the objects near Earth that are larger than one kilometer in diameter. The probability of a substantial impact this century is generally regarded as being low, but it is widely accepted that "a future collision of an asteroid or cometary nucleus with the Earth with catastrophic effects is inevitable unless technology is developed to modify the orbit of such bodies" (NASA 2002).

If an NEO larger than a few dozen meters in diameter strikes a major urban center, the insurance industry would sustain losses unlike anything it has experienced in its history. Fires, earthquakes, tsunamis and direct impact damage could overwhelm the capacity of the insurance industry to cope with the number and value of claims. In fact, the industry likely would not be able to cope if a meteorite strikes a major urban center without the aid and intervention of government and international agencies. Dore (2005) points out that there is the possibility that insurance companies could declare bankruptcy and default on their payment obligations.

While scientists assess how to defend the Earth from the threat of an impact, there is no consensus on how this could or should be achieved. If we are not able to deflect the object and have it bypass the Earth, then we may seek to break it up. This will increase the likelihood of an impact with a densely populated region.

28.7
Insurers need to Prepare

The insurance industry needs to assess its preparedness for an asteroid or comet strike. Many insurers were not fully prepared when Hurricane Andrew struck Southern Florida in 1992, and again when terrorists attacked the World Trade Center on September 11th 2001. These events have led the industry to re-evaluate its preparedness and capacity to deal with major events. The industry has begun to establish partnerships with national governments and international agencies to ensure appropriate preparedness and capacity to respond to major events. An asteroid or comet impact is a further example of a low probability/high cost event that must be addressed, and ideally this should take place well before the strike occurs.

Of immediate concern for the industry is the question related to whether or not the damage sustained by an NEO should continue to be covered or not. In particular, if an asteroid or comet has been detected and the orbit is largely determined, then if it becomes highly likely that the object will impact with the Earth. In this case, the industry should no longer treat this as a random event. Will the insurance industry choose to use the size of the object as a condition? As discussed earlier, asteroids and comets do not fit the model used to insure against damage whereas smaller objects such as meteoroids would. Does the industry use probability of impact as a measure of whether or not to insure the damage as a result of an impact? Perhaps insurability should depend on whether or not the impact was one that was known to occur for a period of time versus the threat of impacts from undetected objects. These are questions that decision-makers in the global insurance and reinsurance industries should tackle.

28.8
The Cost of an Impact

Trying to assess the costs of a strike on the Earth requires numerous assumptions. There have been few strikes of significant impact and size during periods of our recorded history, and meteorites have differing physical make-ups. This adds to the challenge of assessing the possible insured damage. In addition, those impacts that have occurred have been in several international jurisdictions creating virtually no effective insurance model for pricing or valuation. While insurance typically values premiums using historical events as a benchmark, this is not possible for an asteroid strike. We can compare the possible devastation in terms of costs of a strike with that of other disasters to create a rough assessment of the possible insurable costs of a significant asteroid strike in an urban area.

For this analysis we use the September 11th 2001 terrorist attacks in New York and compare the impact with the Meteor Crater in Arizona and the Tunguska Incident in Siberia, to assess potential insurance claims with respect to an asteroid strike.

The terrorist attack on the World Trade Center represents the costliest disaster ever faced by the insurance industry. Total claims paid for property damage and business interruption was approximately US$ 21 billion. The area of devastation as a result of the twin tower attacks was roughly 0.25 square kilometers (FEMA 2002).

The Meteor Crater in Arizona was caused by a meteorite approximately 30–50 meters in diameter. The crater that was created as a result of the impact has a diameter of 1200 meters and an area of 1.13 square kilometers.[1] This is almost five times the area devastated by the World Trade Center attacks. Severe debris-pressure wave damage occurred over a much larger area. In their study, Garshnek et al. (2000), note that an asteroid or comet with a diameter of 50 meters could potentially devastate up to 1900 square kilometers, an area 7600 times larger than that damaged in New York. The Tunguska Incident, for example, resulted in severe damage over an area 8800 times larger than that in New York and was also caused by a meteorite 30 to 50 meters in diameter. According to Demographia, that level of devastation is larger in size than the total urban land area for cities such as Toronto, London, Paris, New York and the Tokyo metropolitan area. Thus overall damage that will result with the strike of an asteroid of similar size to the Meteor Crater in Arizona would lead to insurance losses far beyond anything with which the industry could cope.

If an asteroid with a diameter of 30–50 meters had struck the World Trade Center in 2001, then using the level of devastation discussed by Garshnek et al. (2000), we might estimate that the direct damage and insurance claims may have approached US$ 2–4 trillion. Such losses are well beyond anything the industry has ever faced, and it is unclear how the industry could continue to function. Furthermore, an impact on New York would have a devastating impact on domestic and international capital markets and could lead to a stock market crash in the United States similar to that of the 1930s (Dore 2005).

It is important to acknowledge that this scenario combines the very low probability (ranging from one-tenth to one percent on an annual basis) that the Earth is struck by an asteroid larger than 30 meters in diameter, coupled with a low probability that the impact occurs on land (only 29 percent of the world's surface is land mass) and that the impact strikes an urban center (perhaps less than two percent based on the urban area of the United States) (see www.demographia.com). The random nature of these events means that the impact will likely be with the ocean, or in a remote region, where the fatalities and insured losses would be greatly reduced. Nevertheless, the example illustrates that a 30 meter object could have a catastrophic impact on the global insurance industry. An asteroid larger than the example cited may result in exponentially larger damage.

28.9
Insurers' Capacity to Pay

The total capital in the world's non-life insurance industry in 2003 was US$ 1.3 trillion (Swiss Re 2004). As set out above, an object of a few dozen meters in diameter could lead to damage claims ranging from US$ 2–4 trillion[2] if the impact occurs directly in the heart of a major urban center like New York, Tokyo or London. This is clearly beyond the capacity of the industry to manage.

---

[1] Relates only to the size of the crater; excludes damage due to ejected material as well as pressure shock and fires.
[2] Based on total damage values of NYC terrorist attacks × 1500 (the factor by which the area of devastation in NYC must be multiplied by equal 1900 km^2 of devastation as proposed by Garshnek et al. 2000).

Garshnek et al. (2000) point out that there has been some analysis of methods for diverting these objects away from the Earth, but very little effort has been devoted to the idea of implementing a disaster management plan with respect to an asteroid impact. Since insurers presently offer to cover damage caused by asteroid impacts, they will be motivated to participate in this planning.

28.10
Conclusions

Meteorites, asteroids and comets have struck the Earth in the past and will do so again. Insurance policies promise to pay for damage caused by an impact. The impact however, of an asteroid or comet larger than 2000 meters would likely cause so much damage that civilization as we know it would come to an end, and few would think about insurance issues. It has been demonstrated that an object larger than a few dozen meters that strikes a major urban center would likely overwhelm the insurance industry. There is a very low probability that this will occur, but the high consequences imply that the insurance industry should pay more attention to this hazard. Some specific actions the industry should consider:

- Should insurers and re-insurers continue to cover the damage from asteroid or comet impact?
- How can insurers encourage loss prevention and preparedness initiatives?
- How can insurers work with governments and international agencies to manage threats like asteroids that are beyond the financial capacity of the insurers to address alone?

References

Available: *http://www.ci.nyc.ny.us/html/dcp/home.html*
Available: *http://www.demographia.com*
Available: *http://www.investorwords.com*
Available: *http://www.worldwidewebfind.com/encyclopedia/en/wikipedia/n/ne/new_york_new_york.html*
Banham R (2004) Catastrophe insurance – masters of disaster. U.S. Insurer, Summer Edition, pp 14–17
Bottke Jr W (2005) Understanding the near-Earth object population: the 2004 perspective. ICSU Workshop, Comet/Asteroid Impacts and Human Society, Santa Cruz de Tenerife, November 27 – December 2 2004
Chapman CR (2003) How a near-Earth object might affect society. Commissioned by the Global Science Forum, OECD, for the Workshop on Near Earth Objects: Risks, Policies and Actions, January 2003, Frascati, Italy
Chapman CR (2005) The asteroid impact hazard and interdisciplinary issues. ICSU Workshop, Comet/Asteroid Impacts and Human Society, Santa Cruz de Tenerife, November 27 – December 2, 2004
Chapman CR, Durda D, Gold RE (2001) The comet/asteroid impact hazard: a systems approach SwRI White, February 2001. *http://www.boulder.swri.edu/clark/neowp.html* accessed September 5, 2004
Cushman & Wakefield (ed) (2001) Marketbeat series – Manhattan, mid-year 2001. Cushman & Wakefield Inc
Dore M (2005) The economic consequences of disasters due to asteroid and comet impacts, small and large. ICSU Workshop, Comet/Asteroid Impacts and Human Society, Santa Cruz de Tenerife, November 27 – December 2, 2004
Federal Emergency Management Agency (FEMA) (2002) World Trade Center – building performance study: data collection, observations and recommendations, 2nd edn. Greenhorne & O'Mara Inc, New York

Felsted A (2002) Survey – world insurance – looking after number 1. Financial Times, May 24, 2002
Foster HD (2005) Disaster planning for cosmic impacts: progress and weaknesses. ICSU Workshop, Comet/Asteroid Impacts and Human Society, Santa Cruz de Tenerife, November 27 – December 2, 2004
Garshnek V, Morrison D, Burkle Jr FM (2000) The mitigation, management, and survivability of asteroid/comet impact with Earth. Space Policy (16):213–222
Glanderton PT, Brookshire DS, McKee M, Steward S, Thurstand H (2000) Buying insurance for disaster-type risks: experimental evidence. Journal of Risk and Uncertainty 20(3):271–289
Institute for Catastrophic Loss Reduction (2004) Ottawa Ontario Presentation, February 26, 2004
Insurance Bureau of Canada (1994) A statement of principles regarding insurance and natural hazards, February 1994
Insurance Bureau of Canada (2003) Facts of the general insurance industry. Toronto, Canada
Insurance Bureau of Canada (2003) Fire-following – options for ensuring insurance availability and affordability for homeowners and businesses in Ontario, April 2003
Kathy Bayes Insurance Agency (n.d.) http://www.insurance4texas.com/history0.htm
London is the world's most expensive city for office space, April 17, 2001, http://www.cushwake.com/cw/news/media.cfm?artcl _id=2078
MacCracken M (2005) The climatic effects of asteroid and comet impacts: consequences for an increasingly interconnected society. ICSU Workshop, Comet/Asteroid Impacts and Human Society, Santa Cruz de Tenerife, November 27 – December 2, 2004
Marsden B (2005) Impact scare management, 1998–2004: has it improved?. ICSU Workshop, Comet/Asteroid Impacts and Human Society, Santa Cruz de Tenerife, November 27 – December 2, 2004
Melosh H (2005) Indirect physical effects of comet and asteroid impacts. ICSU Workshop, Comet/Asteroid Impacts and Human Society, Santa Cruz de Tenerife, November 27 – December 2, 2004
Morrison D (2005) The impact hazard: advanced NEO surveys and societal responses. ICSU Workshop, Comet/Asteroid Impacts and Human Society, Santa Cruz de Tenerife, November 27 – December 2, 2004
Munich Reinsurance Company (ed) (2001) Topics annual review: natural catastrophes. Munich, Germany
National Aeronautics and Space Administration (NASA) (n.d.) http://128.102.32.13/impact/intro_faq.cfm
National Aeronautics and Space Administration (NASA) (n.d.) http://128.102.32.13/impact/intro_faq.cfm
National Aeronautics and Space Administration (NASA), (2002) Final report – NASA workshop on scientific requirements for mitigation of hazardous comets and asteroids. Arlington, Virginia, September 30, 2002
Near Earth Object Program, August 24, 2004. http://neo.jpl.nasa.gov/images/meteorcrater.html
Noble JW (1998) What if huge asteroid hits atlantic? You don't want to know. New York Times, January 8, 1998
Simmons KM, Kurse J, Smith DA (2002) Valuing mitigation: real estate market response to hurricane loss reduction measures. Southern Economic Journal 3(68):660–571
Swiss Reinsurance Company (ed) (2003) Sigma – natural catastrophes and man-made disasters in 2003: many fatalities, comparatively moderate insured losses. Economic Research and Consulting, Zurich Switzerland
Swiss Reinsurance Company (ed) (2003) Sigma – world insurance in 2003: insurance industry on the road to recovery. Economic Research and Consulting, Zurich Switzerland
Ward SN, Asphaug E (1999) Asteroid impact tsunami: a probabilistic hazard assessment. Institute of Tectonics, University of California, Santa Cruz, USA
Why is Manhattan So Expensive? http://www.manhattan-institute.org/html/cr_39.htm
Wisner B (2005) The social implications of a comet/asteroid impact on Earth: a perspective from international development studies. ICSU Workshop, Comet/Asteroid Impacts and Human Society, Santa Cruz de Tenerife, November 27 – December 2, 2004

Chapter 29

The Economic Consequences of Disasters due to Asteroid and Comet Impacts, Small and Large

Mohammed H. I. Dore

29.1 Introduction

The objective of this paper is to investigate the economic consequences of asteroid or comet impacts, referred to here as *near Earth objects* (NEO). As of September 8, 2005, according to the Near Earth Objects Program of NASA (NASA 2005), there are 3535 NEOs, of which asteroids (NEAs) are 3438. NEAs greater than 1 km in diameter are represented by 797. The number of potentially hazardous asteroids (PHAs) is 720, of which PHAs greater than 1 km are 146. Of these, 3 appear (on September 8, 2005) on the NASA "impact risk" page. An NEO that is less than 50 meters would have a 5 megaton energy impact, although NEOs of even less than 30 meters could be damaging, depending on their composition and density. From about 50 meters to about 1 km diameter, an impacting NEO can do tremendous damage on a local scale. With an energy level above a million megatons (diameter about 2 km), an impact will produce severe environmental damage on a global scale. Still larger impacts can cause mass extinctions, such as the one that ended the age of the dinosaurs 65 million years ago (15 km diameter and about 100 million megatons). Table 29.1, reproduced from Chapman (Chap. 7 of this volume) is instructive; it summarizes impact energies and possible physical damages.

On June 30, 1908, an object 55 meters wide is believed to have exploded some 10 km above the Tunguska region of Siberia with the force equivalent to a 3 megaton bomb, although some have thought that it was in the 10 to 20 megaton category. One thousand square kilometers of forest were flattened, untold numbers of reindeer were roasted and a man standing 100 kilometers away was knocked unconscious. On the other hand, the Shoemaker-Levy 9 comet collision with Jupiter in July 1994 had an estimated impact of 100 million megatons of TNT. This latest collision has convinced the world that such collisions are indeed real and possible.

But as Table 29.1 shows, the probabilities of impacts are small. A more complete picture of the probabilities is given in Fig. 29.1, prepared by William F. Bottke Jr., who estimates the frequency of 1 km-sized bodies striking the Earth is about once every half a million years, whereas an NEO with a diameter > 50 meters might strike once every thousand years or so.

The timescales involved are indeed long and the probabilities are small. But the Indian Ocean earthquake and tsunami of December 26, 2004 was also considered to be a very low risk, which is why there was no tsunami early warning system comparable

Table 29.1. NEO sizes and possible physical damages

Object diameter	Impact energy	Chance per 100 yr	Character of damage
>3 km	1.5 mil. MT	<1 in 50 000	Global climate disaster, most killed, civilization destroyed
>1 km	80 000 MT	0.02%	Devastation of large region or an entire ocean rim
>300 m	2 000 MT	0.2%	5 km crater; huge tsunami or destruction of small nation
>100 m	80 MT	1%	Exceeds greatest H-bomb; 1 km crater; locally devastating
>30 m	2 MT	40%	Stratospheric explosion; possible damage within ten km
>10 m	100 kT	6 per century	Aerial burst, little damage below (e.g. broken windows)
>3 m	2 kT	2 per year	Blinding flash, could be mistaken for atomic bomb

Source: Chapman, this volume (reproduced with permission).

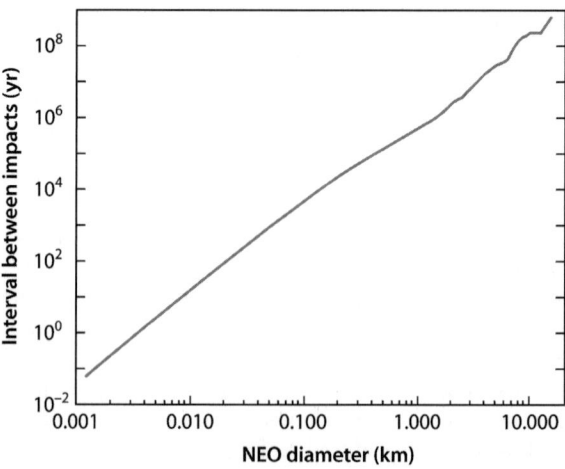

Fig. 29.1.
Diameter of NEOs and the interval between impacts in years

to the one in the Pacific. However, perhaps the reason why there was no warning system may simply be the cost involved for the poorer nations. The 2004 Tsunami is estimated to have caused over 200 000 deaths and over half a million people were injured in twelve countries. Perhaps this is the most devastating international tsunami disaster experienced so far.

This paper is confined largely to the *economic and financial* effects of NEO impacts, large and small. For the indirect physical effects of NEO impacts see the excellent paper by H. J. Melosh (Chap. 12 of this volume). The indirect physical effects can also have economic, social, and psychological effects. But space does not permit their adequate treatment here. See also the paper by Andrea Carusi et al. (Chap. 25 of this volume), in which the authors simulate the social and economic effects of two scenarios: (*a*) the consequences of a Tunguska-class impact of 60 meter diameter, with a terminal energy impact 13 megatons, and (*b*) a 200 to 300 meter object, with an energy impact of 1000 megatons, affecting an industrialized country like Italy. This is a sophisticated analysis of a society as a complex system but does not cover the detailed economic and financial consequences as simulated in this paper.

As there are no historical data on the economic consequences of such impacts, this paper uses exogenous disturbances as analogs to study the possible consequences. The historical record thus serves to discriminate between the consequences of large and small exogenous disturbances. Using these data, several scenarios are constructed, ranging from the historical record of manageable natural disasters to larger impacts with deadlier consequences. But in each scenario the inquiry is guided by economic history. Otherwise the exercise would be complete speculation. Section 29.2 begins by discussing the necessary and sufficient conditions for a stable, reasonably well functioning capitalist economy, periodically subject to recessions and expansions. Section 29.3 is devoted entirely to scenario construction.

29.2
Necessary Conditions

It may be helpful to begin by considering the *necessary conditions* for the orderly functioning of a modern market economy. These are: (*a*) production of goods and services under stable conditions, (*b*) transmission of demand information to production units, (*c*) orderly flow of payments through a well-functioning banking and credit system, (*d*) the pooling of financial resources for investment through financial intermediaries, and (*e*) the efficient and orderly valuation of claims on assets (stock markets, commodity exchanges). All these are accompanied by *information flows,* which in the developed countries are largely electronic. The above five conditions can also have what economists call *real* counterparts, in the form of real flows such as goods and services. Real flows require a transportation network to move goods and people. The financial flows require a network of markets, and informational flows require electronic communications, personnel, the necessary hardware and energy (electricity).

In addition, it should be noted that the real and financial flows function in a context that requires an array of economic institutions, such as (*a*) courts, and the rule of law and its enforcement mechanisms (*b*) Central Banks to issue a fiat currency, and the Securities and Exchange Commission (SEC) to regulate financial flows that constitute a claim on real assets (buildings, factories, etc.). Whereas the courts enforce economic contracts, the Central Bank holds international reserves and issues currency. But fiat currencies work because of confidence, which at times can be fragile.

Although the five conditions and two institutional features are necessary, they are by no means sufficient. While economists have produced sufficient conditions for an abstract model for a "competitive" economy, their relevance to actual economies is far from clear, as in real economies the sufficiency conditions are just not possible. The historical record of economies is characterized by business cycle fluctuations, accompanied by panics and manias, currency collapses and financial crises. The fiduciary role of economic institutions such as Central Banks and the SEC demonstrates that modern macro economies function on the basis of expectations and trust. When expectations are not fulfilled, or when trust has been breached there have been crises, as the historical record shows. Economists such as Hyman Minsky and others concluded that there is no set of sufficient conditions to guarantee the stability of a modern market economy; indeed it may be this very instability that gives the economy certain dynamism. If there are no sufficient conditions, it follows *a fortiori* that there can be no necessary *and* sufficient conditions for stability. Consequently a modern market economy reacts to exogenous *news*, from news of a frost in Brazil, to news of wars and civil wars. News of death and destruction due to some natural or unnatural disaster can buffet a modern economy. News of what has happened in the past is often "digested" well. Hence, hurricanes, floods and other small magnitude disasters can be handled by the capital markets. But major earthquakes and major wars can shock the markets severely.

In the next section, we examine with the help of historical analogs how the magnitude of a shock may affect a market-based economy. Six scenarios are constructed in order to assess the possible repercussions of impacts of asteroids and comets, both large and small.

29.3
Scenario Construction

Any exogenous shock can disrupt the functioning of a market based on production and exchange, which is facilitated by some means of payment. A shock can be a typical natural disaster, such as a flood, hurricane, earthquake, famine or drought. The economic cost of these disasters depends on their location and magnitude and the vulnerability of the affected population. The US economy copes regularly with a number of disasters, both large and small. For example, in the period 1980–2003, there were 45 disasters producing under $5 billion in damage costs, nine in the $5–20 billion range, two in the $20–40 billion range, and two over $40 billion. On the other hand, Hurricane Mitch of 1998 affected mainly the developing countries of Central America. Damages were only $8.5 billion, but virtually the entire infrastructure of roads, buildings, hospitals and schools was destroyed, with 9000 confirmed deaths and three million left homeless. It was the deadliest hurricane to hit the western hemisphere in two centuries (results of Hurricane Katrina are not yet known). The classification system of natural disasters mentioned here is that used by NOAA; it will be used loosely below to guide the construction of some scenarios.

29.3.1
Scenario 1

With the NOAA classification of natural disasters, we now construct six scenarios, in order of increasing severity.

A Near Earth Object (NEO) of say 30–35 meters diameter hits a populated area. It knocks out power for a few days and disrupts transportation of goods. This would be comparable to an average disaster causing less than $ 5 billion in damage costs. Most such disasters are below the FEMA threshold, and are handled by public assistance at the State level and by private insurance. The typical death toll may be between ten and one hundred. At the high end in this scenario may be something such as the 1989 San Francisco earthquake that killed 63 people, injured 3757 and led to property damage of $ 5.9 billion.

At this scale, recovery and return to some semblance of normality could take two weeks to a month to resettle people, and perhaps six months to rebuild damaged infrastructure. If this NEO struck a developing country, the damage costs might be lower, but the death toll higher; and there might be outbreaks of disease due to a fragile water and sanitation infrastructure. For example, the Bangladesh floods of 1998 cost $ 3 billion, but affected thirty million people. A NEO impact that damaged food production in a rural area of a developing country through collateral damage (e.g. impact on dams resulting in flooding), could cause severe hardship. These same Bangladesh floods of 1998 led to a food deficit of 2.2 million tonnes of rice, or 7% of the country's output. In contrast, flooding in the Midwestern US in 1993 caused forty-eight deaths and damage of $ 27 billion.

29.3.2
Scenario 2

This is a NEO impact falling within the $ 5–20 billion range of damage costs. For a developed industrialized country, it may cause deaths running into the thousands, but would still be manageable, with no need for external assistance. Disruption would be localized, and would be comparable to the worst natural disasters in NOAA's databank, with costs given (as above) in 2003 constant dollars. The 1980 drought in the U.S. cost $ 49 billion and caused 10 000 deaths. A more costly drought occurred in 1988, producing costs of $ 62 billion, but a slightly lower death toll of 7500. By way of comparison, Hurricane Andrew in 1992 cost $ 36 billion, but only 61 deaths. While Andrew affected infrastructure, the 1980 drought did not. So another example in this category is worth considering.

The 1995 Kobe (Japan) earthquake ($M = 7.2$) killed some 6300 people, and caused damage of the order of $ 200 billion, of which $ 150 billion was for state-owned buildings alone. Some 300 000 people were made homeless. A developed, industrialized country could cope with this disaster. Now consider a developing country example: that of the 1976 Guatemala City earthquake ($M = 7.5$). It killed 80 000 and made one million people homeless. Damage cost estimates are not available, but external assistance amounted to just under $ 20 million.

Another good historical analog is war damage and costs associated with wars. Table 29.2 gives some cost figures for wars for the USA. As the table shows, WW II was the most costly war in which the US has been involved. It cost the US more that $ 20 000 per person (2002 dollars) and caused a huge government deficit that took nearly 20 years to eliminate. All other wars had a cost that was a mere tenth of the cost of this war. In terms of lives lost on both sides, WW II has a total death toll of 55 million (22 million is the USSR alone), compared to the death toll of WW I of 15 million (Wallechinsky 1999). Of course other countries bore greater costs.

Table 29.2. Financial cost to the USA of major wars

Conflict	Cost in billions: current dollars	Cost in billions: 2002 dollars	Per capita cost: 2002 dollars
Civil War (1861–1865)	5.20	578	1 683
World War I (1917–1918)	26.00	255	2 485
World War II (1941–1945)	288.00	2 719	20 352
Korea (1950–1953)	54.00	343	2 262
Vietnam (1964–1972)	111.00	450	2 200
Gulf War (1990–1991)	61.00	79	305

Sources: United States Civil War Center; computed by the author using data from: http://www.cwc.lsu.edu/cwc/other/stats/warcost.htm.

Table 29.3. Most populous cities of the world: 2004

Rank	City[a]	Population
1	Shanghai, China	13 278 500
2	Mumbai (Bombay), India	12 622 500
3	Buenos Aires, Argentina	11 928 400
4	Moscow, Russia	11 273 400
5	Karachi, Pakistan	10 889 100
6	Delhi, India	10 400 900
7	Manila, Philippines	10 330 100
8	Sao Paulo, Brazil	10 260 100
9	Seoul, South Korea	10 165 400
10	Istanbul, Turkey	9 631 700
11	Jakarta, Indonesia	8 987 800
12	Mexico City, Mexico	8 705 100
13	Lagos, Nigeria	8 682 200
14	Lima, Peru	8 380 600
15	Tokyo, Japan	8 294 200
16	New York City, US	8 091 700
17	Cairo, Egypt	7 609 600
18	London, United Kingdom	7 593 300
19	Teheran, Iran	7 317 200
20	Beijing, China	7 209 900

[a] Refers to the city proper, as opposed to an urban agglomeration, which would also count the surrounding urban areas in the total.
Source: Stefan Helders, World Gazetteer, 2004.

29.3.3
Scenario 3

In this scenario, we consider a NEO impact that is significantly larger than the previous scenarios. Again, its economic consequences depend on the economic importance of its location. As in Table 29.1, all dollar figures are given in 2002 constant dollars.

Consider an average-sized city such as San Diego, California, with a civilian labor force of 1.5 million, and per capita personal income of $ 34 872. Assume that about 5% of the civilian labor force is unemployed, and that the city's annual contribution of income is about $ 49.7 billion. The loss of this city would be a blow, but of the same order of magnitude as the 1980 drought in the US, assuming this one time loss was restored after a short period of adjustment. Of course the total loss would depend on how long it takes for the city to return to its pre-disaster income levels. But in the year of the impact, no doubt a major relief operation would have to be mounted that could cost possibly another $ 100 billion. Rebuilding the city, or re-settling the population would have the same scale of problems as the Kobe earthquake reported above. Even with (say) a death toll of under 10 percent of the city (say around 100 000 people), we can expect injuries to be 20 or 50 times that number, because the injuries typically also affect areas adjacent to the city. Some important dead or injured personnel would have been in charge of *key public utility infrastructure*: medical care and hospitals, water supply and sewage treatment, electricity generation and distribution, heating plants and care of the aged. This entire critical infrastructure would be seriously jeopardized even in a city like San Diego. In a developing country there would be the additional problem of waterborne diseases, hunger and a severe threat to the lives of children and the aged. (We have seen these emerge as major factors as a result of the Indian Ocean seaquake and tsunami of December 2004 and Hurricane Katrina).

29.3.4
Scenario 4

Consider next the loss of New York City. As Table 29.3 shows, it is the 16th largest city in the world, but is ranked first in the US. With its population of 8.1 million, it has a civilian labor force of 4.8 million, and the city's GDP in 2004 was $ 414.1 billion. Personal per capita income in 2002 dollars was $ 40 680. The September 11 attack led to a 20% decline in the city's GDP in that quarter, with a severe shock to the financial sector, a 12% decline. Total one-time capital and human loss was estimated at $ 45 billion, and for the fiscal year 2002–03, the total economic loss was estimated to be between $ 45 and $ 60 billion (The City of New York, 2001). However, the financial sector is now recovering.

Table 29.3 also gives the population of other large cities in the world. Whereas the loss of life may be of the same order of magnitude, none of the other cities would have as a large a global economic impact as New York City. With the exception of London, Tokyo and New York, all the other cities are in developing countries.

It is the economic importance of the city that governs the loss. New York is by far the largest capital market in the world. The market capitalization of the two New York Exchanges (NYSE and Nasdaq) is around $ 17 trillion. The loss of New York City would be a blow to the world economy as well as to the US. This loss could disrupt the global capital market and lead to a stock market crash of 1930s proportions, as stockholders try to move out of equity and into cash, gold or non-US dollar assets. The stock market crash of 1939–42 led to a 40% decline in the Dow Jones Industrial Average (DJIA). But the 1929 stock market crash was even worse: see Figs. 29.2 and 29.3, using DJIA and the S & P 500 respectively. The S & P 500 declined from a peak of 300 in June 1929 to a trough of just 50 in June 1932. Such stock market crashes not only affect the perceived real wealth

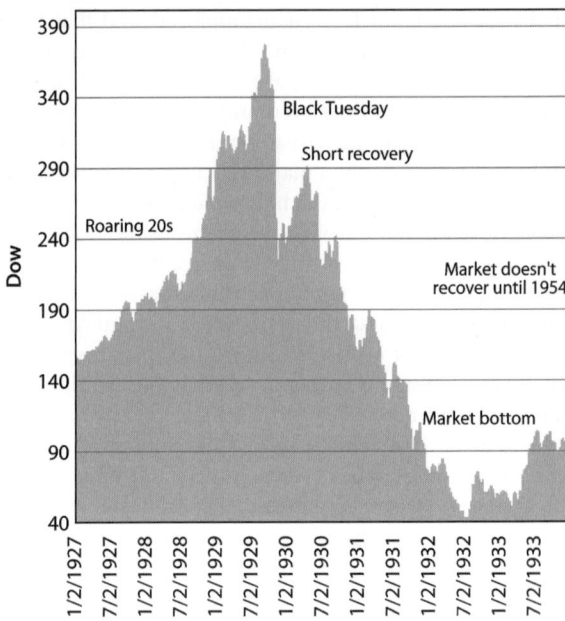

Fig. 29.2.
The 1929 stock market crash: The DJIA Index. Source: *http://mutualfunds.about.com/cs/history/l/bl1929graph.htm*

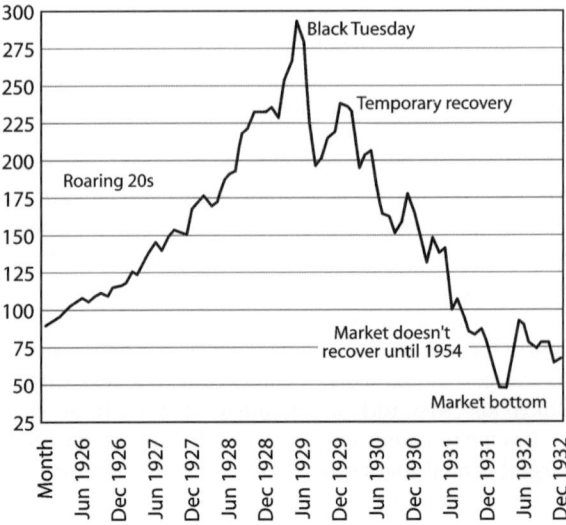

Fig. 29.3.
The 1929 stock market crash: The S&P 500 Index. Source: *http://mutualfunds.about.com/cs/1929marketcrash/l/bl1929graph.htm*

of stockholders but are also accompanied by large reductions in output and employment. The real hardships of populations in the US after the 1929 crash are well known.

The 1939–42 decline of 40% was not unprecedented; a similar decline had occurred between November 21st 1916 and December 19th 1917. History shows that stock markets recovered – eventually. But serious recovery did not take place until the 1950s. Now there is a view, promulgated largely by the US Federal Reserve, that such crashes cannot happen again: look at the 1987 crash (36% decline), from which the Fed was able to engineer a recovery by pumping unlimited liquidity, i.e. credit. But the causes of the 1987 crash were rather technical and not rooted in the goods markets.

Although this scenario concentrates on the economic and financial effects, needless to say what was said about the loss of critical public infrastructure (see Scenario 3 above) also applies here, perhaps with added force. The adverse consequences of the loss of New York would also affect cities in the adjacent states of New Jersey, Maryland, etc. A cascading effect can be expected, compounding the problems.

A NEO impact will be accompanied not only by a financial crisis, but also by a loss of real assets. It would therefore be instructive to consider the loss of London, on which more data is available. If the city of London were lost, the impact on financial markets would not be as large as that of New York. London has become more specialized; while just under 20 percent of international bank lending is done through London, more than a third (actually 36 percent) of all derivative trades is carried out there. But it has the largest foreign exchange market in the world, with an average daily turnover of $ 504 billion, which is more than all foreign exchange traded in New York and Japan combined. It is also the world's largest insurance market, covering international trade and ocean tanker traffic.

The main global impact of the loss of London would be disruption of the foreign exchange market, international trade and insurance. With global integration, more and more countries are dependent on the smooth flow of international trade and its payment through orderly foreign exchange markets. Its disruption would affect hospitals, water and sanitation and could trigger disease outbreaks that could then spread.

There could be secondary or other long term consequences of the large NEO impact, such as the much talked about nuclear winter associated with sunlight becoming blocked for months on end due to particles in the atmosphere (see Chapman, Chap. 25 of this volume). In that case we can also expect large-scale population movements – a flow of refugees further complicating the difficulties.

29.3.5
Scenario 5

Suppose the capital market disruption was accompanied by a more serious financial crisis, perhaps because of the destruction of a government along with its Central Bank. Following such an impact a government might be perceived to be financing recovery by issuing more currency. A loss of some minor currency would not be a problem. Even the loss of an internationally traded currency such as the Canadian dollar would not be a problem, as those affected would merely shift to one of the major currencies such as the US dollar, which is now the reserve currency of the world. But if the magnitude of the NEO impact were large enough to destroy confidence in a major world currency, such as the Japanese Yen, the euro or the US dollar, then its consequences could be serious on a global level.

Loss of confidence in a major currency can be rapid. The worst decline in a major currency occurred in Germany between January 1921 and December 1923, as shown in Table 29.4. It shows a rapid decline in the external value of the currency, as domestic prices rose in response to the German government printing more and more currency to meet its reparation obligations. This occurred after it lost a large part of its productive capacity after WW I along with the destruction of the state apparatus and its ability to levy taxes.

With regard to the social cost of this hyperinflation, it is worth noting that the *real* quantity of money (after adjusting for inflation) declined by 92 percent, indicating a

Table 29.4.
German exchange rates, 1921 to 1923

Year	Month	US cents per Mark
1921	January	1.60
	February	1.64
	March	1.60
	April	1.57
	May	1.63
	June	1.44
	July	1.30
	August	1.19
	September	0.96
	October	0.68
	November	0.39
	December	0.53
1922	January	0.52
	February	0.48
	March	0.36
	April	0.35
	May	0.34
	June	0.32
	July	0.20
	August	0.10
	September	0.07
	October	0.03
	November	0.01
	December	0.01
1923	January	0.007
	February	0.004
	March	0.005
	April	0.004
	May	0.002
	June	0.001
	July	0.000 3
	August	0.000 033 9
	September	0.000 001 88
	October	0.000 000 068
	November	0.000 000 000 043
	December	0.000 000 000 022 7

virtual collapse of monetary exchange and a return to barter (Keynes 1923). After this loss of confidence in the German Mark, the government introduced a new currency, called the Rentenmark, and declared that one unit of the new currency would be equal to 1 trillion (i.e. 10^{12}) old marks! Hungary, Poland and Austria had similar experiences, although the German case was the most spectacular. Thus history shows that a fiat currency can collapse, and collapse most dismally, under the right conditions.

A dramatic decline in the external value of the US dollar could occur even if there were not a catastrophic NEO, but some "large" event that shakes the confidence of the international community, especially as the US has now become a debtor country and does not have enough reserves to cover its external debt, as the Table 29.5 shows.

The comparison of Reserves of Foreign Exchange and Gold with external debt should be carried out with care. The external debt can be held in government bonds, or it can be held in the form of equity in real investments in corporations in the form of plant, machinery and equipment. The external debt results in an *annual flow* of dividends and interest payments out of the indebted countries. It is this flow that can be jeopardized when the level of reserves and holdings of foreign currencies is inadequate to meet the flow. A depreciation of the currency (in terms of its external value) can follow. If a NEO impact that interferes with this annual flow could have a serious effect on that particular currency, it could be "dumped" in the market, leading to its precipitous decline.

Another set of data that is of relevance is what is called the Net Investment Position: this is in fact the total external debt of the country. That information is given in Table 29.6. The international comparison in Table 29.6 is an indicator of indebtedness and hence of vulnerability. The English speaking countries and Finland show heavy foreign investments into their countries and net indebtedness. The level of indebtedness approaches the debt of some developing countries. Of particular concern is the US debt ratio (−22.2%). Fortunately for the US, its currency is a world reserve currency, but it is all a matter of confidence. As soon as some major player, such as OPEC, prices its oil in euros, the US dollar as a reserve currency can lose ground in a matter of years. The debt service would then become burdensome and a rapid decline would be a real possibility.

Table 29.5. Reserves and external debt of some countries

Country	Reserves of foreign exchange and gold	External debt (US$)
USA	$85.94 billion (2003)	$1.4 trillion (2001 est.)
Japan	$827.95 billion (2004)	−$1.36 (US trillion, 2001)
Canada	$36.27 billion (2003)	$1.9 billion (2000)
European Union	366.1 (euro, billions, 2002)	289.6 (euro, billions, 2002)
China	$412.7 billion (2003)	$197.8 billion (2003 est.)
India	$102.3 billion (2003)	$101.7 billion (2003 est.)
Brazil	$49.3 billion (2003	$214.9 billion (2003)

Note on Japan: Its negative debt indicates a *surplus* with the rest of the world.

Table 29.6. Net investment position of major countries in billions of Japanese Yen

	Year	Assets	Liabilities	Net assets	Ratio to nominal GDP (%)
Japan	2001	379 781	200 524	179 257	35.6
Switzerland	2000	155 396	121 292	34 103	120.1
Germany	2000	290 736	282 813	7 924	3.7
France	2000	276 587	269 670	6 917	4.6
Belgium	1999	67 040	60 781	6 260	25.8
Italy	2000	128 052	123 048	5 004	4.0
Canada	2001	74 855	91 690	−16 835	−18.8
United Kingdom	2001	600 710	618 526	−17 816	−9.4
Finland	2000	17 023	37 892	−20 869	−148.2
Australia	2000	23 997	47 898	−23 901	−57.6
United States	2000	826 108	1 077 446	−251 338	−22.2

Source: Bank of Japan, Net Investment Position, 2002.

A NEO impact in the indebted countries would have major implications for the main creditor nations that are Japan and Switzerland, followed by countries in Western Europe. A major catastrophic impact of a NEO in the indebted countries, with destruction of factories or real estate could be viewed as a *force majeure*, and an *act of God*. When real capital is wiped out, it is the insurance companies who will have to meet the costs. Of course the insurance companies could declare bankruptcy and default on their payment obligations. Such a financial disaster would have cascading multiplier effects. Its global spread would be rapid.

It is important to note that whereas financial disasters could have major consequences, the damage or destruction of real capital assets such as power stations could have enormous consequences, as the world has become more reliant on large power sources such as nuclear power stations and hydroelectric dams. Table 29.7 gives information on the world's 11 largest hydroelectric dams. The destruction of any one of them could cause massive floods as well as disruption of economic life.

29.3.6
Scenario 6

The worst global human catastrophe in recorded history was perhaps the Black Death of 1347–53. In that disaster Europe lost one-third of its population. The Black Death was a very slow onset disaster, as compared to the scenario here. This gave people much more time to respond and adapt. Furthermore, it affected human health but not infrastructure, of which there was not much anyway. We consider one final scenario, worse than the most catastrophic global disaster yet encountered. Suppose over 3 billion people are killed and more than half the world's capital stock is destroyed: electricity-produc-

Table 29.7. World's largest hydroelectric plants (over 4 000 MW capacity)

Name of dam	Location	Rated capacity (MW)		Year of initial operation
		Present	Ultimate	
Itaipu	Brazil/Paraguay	12 600	14 000	1983
Guri	Venezuela	10 000	10 000	1986
Grand Coulee	Washington	6 494	6 494	1942
Sayano-Shushensk	Russia	6 400	6 400	1989
Krasnoyarsk	Russia	6 000	6 000	1968
Churchill Falls	Canada	5 428	5 428	1971
La Grande 2	Canada	5 328	5 328	1979
Bratsk	Russia	4 500	4 500	1961
Moxoto	Brazil	4 328	4 328	n.a.
Ust-Ilim	Russia	4 320	4 320	1977
Tucurui	Brazil	4 245	8 370	1984

Notes: MW: megawatts. *n.a.:* not available. *Source:* International Hydropower Association, Sutton, England.

ing dams, grain elevators, factories, etc., are all destroyed and land is poisoned or becomes barren. Production becomes impossible; there is no accepted fiat currency. There is no banking system and no mechanism to enforce commercial contracts. In this case, production and exchange (the core of the economy) collapse. The international economy disintegrates, perhaps to be replaced by some local markets that depend on barter, or some medium of exchange – perhaps gold or grain. Social and economic anthropologists might argue that there are two possibilities here: (*a*) a stable socialized economy to manage a long transition back to normal market economy, or (*b*) an anarchic economy.

A stable socialized economy would be comparable to what has happened in time of war. A remnant state is organized or re-emerges with some degree of authority and power of coercion, backed by military force. There may be voluntary association, or some form of dictatorship (with abuses) is imposed. Food rationing is organized; a rudimentary health care system is re-organized and care is given on the basis of needs. Work brigades might be formed to carry out whatever is required to enable the remaining population to survive on a day-to-day basis. Eventually law and order might re-emerge.

In the case of an anarchic economy, bands of people selfishly loot and pillage what food or other useful things remain, such as tools and also implements of coercion, such as guns, knives, and ropes. Wars or other conflicts between groups become the norm as a way of making a living. Even in this situation, neighboring bands could barter some surplus goods, or even agree on a medium of exchange. But the struggle over the control of limited resources could lead to the re-emergence of forms of slavery or serfdom.

29.4
Summary and Conclusions

There is no data on how the global economy might react to impacts of NEOs of various magnitudes. In the absence of such data it was necessary to carry out some thought experiments. The paper began by considering the necessary conditions for the well functioning market economy. These are: (*a*) production of goods and services under stable conditions, (*b*) transmission of demand information to production units, (*c*) orderly flow of payments, (*d*) the pooling of financial resources for investment, and (*e*) the efficient and orderly valuation of claims on assets (stock markets, commodity exchanges). In addition, it was argued that the necessary conditions should also include economic institutions, such as courts, and central banks to issue fiat currencies. However, the necessary conditions do not assure *stability*. Indeed the historical record of market economies shows considerable instability. Modern markets function on expectations and confidence. When the expectations are not realized, or when trust is breached, panic ensues. Hence the possible consequences of NEO impacts could vary considerably. In order to discern the possible consequences, six thought experiments were carried out, supported by data from historical analogs. These are guided in part by the classification by NOAA of natural disasters.

The first thought experiment (Scenario 1) concerns disasters that cost less than $ 5 billion, which covers the typical manageable hurricane or earthquake. The second posits a moderately larger loss of the order of $ 40 billion. The third considers the loss of a city in a developed country of about 1.5 million people. A good analog here is the Kobe earthquake; this is still manageable although it obviously is accompanied by considerable hardship. The fourth considers the loss of New York or London, as they both play a special economic role in the global economy. As financial crisis is of special importance, the fifth scenario is devoted entirely to the detailed examination of a financial crisis triggered by a large NEO. The sixth scenario is about a major global catastrophe, with the death of 50 percent of global population. This could lead to the disintegration of the economy founded on production and exchange. The theory of self-organization suggests the possibility of a socialized economy with state control, or the emergence of lawless marauding bands appropriating what is left by force. Which of the two is more likely cannot be answered by economic theory. The main lesson is that economies are fragile and global integration (also called "globalization") increases the fragility. Unlike the Black Death in the 14th century, major shocks will have rapid international transmission effects.

Is there a policy conclusion emerging from these simulations? When the return periods for asteroid impacts are so long, as shown by Fig. 29.1, is there a rational policy that might prevent catastrophic losses from very rare events? I believe that the December 2004 earthquake and tsunami and Hurricane Katrina in 2005 drive home an important public policy lesson: the international community must take adequate measures to discover fully the level of threats from possible asteroid impacts. The census for the search for NEOs larger than 3 km is essentially complete. Now the goal of the U.S. Spaceguard Survey is to have a 90% census of NEOs greater than 1 km completed by 2008. Much more expensive surveys, contemplated to begin within the next decade but as yet not funded, could do a 90% complete census of bodies down to maybe 150 meters in size during the ensuing decade. It is essential to implement a global "no regrets" policy in the identification of all the main possible threats from NEOs. As this paper shows, the consequences of a major asteroid impact are so horrendous that they must be taken seriously. *The very existence and survival of human civilization may be at stake and no price can be put on that.* When an event is considered rare, a probabilistic calculation might suggest that it might be "rational" to do nothing, but that would be tantamount to gambling with *all* existence.

Acknowledgments

This revised version owes much to help from William F. Bottke Jr., Clark R. Chapman, John Davis, Robert Dimand, David Etkin, and David Morrison. However, all remaining deficiencies are my own.

References

Bank of Japan (2001) Japan's net investment position at year-end. Bank of Japan Quarterly Bulletin August 2002. Also available online at: *http://www.boj.or.jp/en/ronbun/02/data/ron0208c.pdf*
Central Intelligence Agency (USA) The word factbook. Available online at: *http://www.cia.gov/cia/publications/factbook/geos/xx.html*
International Monetary Fund: International financial statistics, various issues
Keynes JM (1923) A tract on monetary reform. Macmillan, London
The City of New York, Office of the Comptroller, Alan G. Hevesi, Comptroller (2001) The impact of the September 11 WTC attack on NYC's economy and city revenues, October 4, 2001. Available at: *http://www.comptroller.nyc.gov/bureaus/bud/reports/WTC_Attack_Oct_4-final.pdf*
United States Civil War Center, Louisiana State University. Data available from: *http://www.cwc.lsu.edu/cwc/other/stats/warcost.htm*
NASA (2005) See: *http://neo.jpl.nasa.gov/*
Wallechinsky D (1999) The twentieth century: history with the boring bits let out. The Overlook Press, New York
Woodard D (n.d.) Website. Data available at: *http://mutualfunds.about.com/cs/history/l/bl1929graph.htm*

Chapter 30

Communicating Impact Risk to the Public

Michel Hermelin

30.1
Introduction

The first conscious recollection I have from my childhood was an aerial bombing. It was a beautiful summer afternoon in June 1940, in a small French village east of Paris. Fortunately no one in my family was hurt. During the following four years, with other children of my age, I was often pulled out from home and school by siren whistles announcing airplanes approaching. In none of these cases was there panic: the adults and children had been trained to react instantaneously and to seek refuge in vaulted cellars or in trenches.

A few months after the war was over, in about 1946, I witnessed another interesting situation. The French national radio transmitted a version of Orson Welles' 1938 CBS broadcast "Mercury Theater on the Air", based on H. G. Wells book "The War of the Worlds". At that time I really felt what panic was: terrified people were running from one house to another, looking for information and advice. Fortunately the situation did not last for a long time, as the radio operators revealed the phoney nature of the broadcast. I learned later (Rubin 2002) that after the 1949 replay of this broadcast in Ecuador, an angry mob reaction caused six deaths.

The conclusion is that even a populace exposed to the terrible issues of four years of war and trained to face that reality may react very poorly to another kind of menace or alert: one that they are not familiar with.

I live in a country where natural disasters are very common: we are used to earthquakes, tsunamis and volcanic eruptions. Many people have lost their properties and relatives in less extended – but also deadly – events as landslides and flash floods. After becoming a little more familiar with the topic of NEO, I realized that my previous experience with helping people to face "classical" natural risks was to be of little value. I will however attempt to present some ideas on the subject, trying not to lose my own Ariadne's thread: the fact that the real world of people living in underdeveloped conditions may be quite different from what it is generally thought to be in middle latitudes.

30.2
Our Present World: Brief Considerations

Since the publications of the conclusions of the Club of Rome and of books such as "Our Common Future" (CMMA&D, 1988), concern about the future of the planet has reached the scientific community. Oddly enough, the main problem –overpopulation –

goes now almost unmentioned (Sartori and Mazzoleni 2003). Of a total of 6 400 million of inhabitants, about 80% live in underdeveloped nations, and the participation of most of these countries to international wealth is progressively lower. Colombia, which has an annual income per capita of about US$ 2 000, is far from the bottom; however it has telling statistics:

- The population is now about 44 million, more than four times that of 1950.
- live under the poverty level and half of this percentage under the misery level, which means that people belonging to this category do not eat properly every day. This situation is not very different from that of the other Andean nations.
- Another appalling fact is the global rate of urbanization. Only 25% of Colombian people lived in urban areas 50 years ago. Now this proportion is about 75%.

This last point case illustrates very well the tendency in most "developing" countries. There are today an estimated 405 cities with populations of more than one million and some 28 cities with more than 8 millions inhabitants (Heiken et al. 2003). Table 30.1 (McGuire et al. 2002) gives a clear description of the tendency.

Another problem arises from the fact that about 40% of the world population lives in coastal areas (Shi and Singh 2003; WRI 2001), a fact that could amplify exposure to tsunamis generated by NEO impacts.

A final consideration is that underdevelopment and poverty mean not only a lower level of education, but also that in many cases the presence and managerial capacity of government representatives may be very low. This may occur parallel to a low level of community organization, particularly in expanding cities. This is obviously not the most favorable scenario for special information, particularly if it is unpleasant and difficult to assimilate.

Table 30.1. The ten most populous cities in the World in 1950 and 2015 (adapted from McGuire et al. 2002)

	1950		2015	
	City	Population	City	Population
1	New York	12 300 000	Tokyo	28 900 000
2	London	8 700 000	Mumbai	26 900 000
3	Tokyo	6 900 000	Lagos	24 600 000
4	Moscow	5 400 000	Sao Paulo	20 300 000
5	Paris	5 400 000	Dhaka	19 500 000
6	Rhine-Ruhr (Essen)	5 300 000	Karachi	19 400 000
7	Shanghai	5 300 000	Mexico City	19 200 000
8	Buenos Aires	5 000 000	Shanghai	18 000 000
9	Chicago	4 900 000	New York	17 600 000
10	Calcutta	4 400 000	Calcutta	17 300 000

30.3
Principal Characteristics of NEO Impact Risks

Excluding the Tunguska event, (Rubin 2002) which was witnessed by a very limited number of people, we must content ourselves with proxy evidence and calculations based on remote sensing observations of NEOs to predict the consequences of their impacts. This fact does not cast any doubt on their validity, at least for the scientific community. Tables 30.2 and 30.3 (adapted from McGuire et al. 2002, adapted from Chapman and Morrison 1994, respectively), briefly present what the effects of an impact would be for the planet. These effects become planetary for asteroids with a diameter from 0.6 to 1.5 kilometers and for comets from 0.4 to 1 kilometers, with typical recurrence intervals of 7×10^4 years for the smaller and 5×10^5 years for the larger. In these cases the impact is accompanied by large-scale tsunamis and destruction of the ozone layer.

Larger sized impacts would progressively trigger impact winters, photosynthesis disruption, widespread fires, reduced light levels and mass extinction. Larger impacts are unique – so unique that they can reach a level where life may totally disappear from the planet. On the other hand, even a smaller impact has characteristics that differ from what we have learned until now from more common natural risks:

- Asteroids and comets may be detected and their trajectories calculated precisely enough so that their impact place and time can be predicted months in advance. In this sense the term "Early Warning" used for other natural risks acquires a very different meaning (Twigg 2003). This assessment does not exclude the possibility of impacts occurring without detection and this situation will diminish with the improvement of exploration systems as Spaceguard Survey, NASA Near Earth Asteroid Tracking (NEAT) and the European Near Object Search Project (Gritzner 2003). Such detection nets are operated through a limited number of observatories and their equipment is still extremely expensive for most countries of the world. In a way, this is a situation similar to what happens at present with hurricanes.

Table 30.2. Impact scales and energies: different estimates of the global threshold (adapted from McGuire et al. 2002, after Chapman and Morrison 1994).

Type of event	Diameter of impactor	Energy (MT)	Frequency (yr)
Tunguska-scale event	50–300 m	9–2 000	250
Large sub-global events (3 estimates)	300–600 m 300 m to 1.5 km 300 m to 5 km	2 000–1.5 × 10^4 2 000–2.5 × 10^5 2 000–1 × 10^7	35 × 10^3 25 × 10^3 25 × 10^3
Low global threshold	>600 m	1.5 × 10^4	7 × 10^4
Nominal global threshold	>1.5 km	2 × 10^5	5 × 10^5
High global threshold	>5 km	1 × 10^7	6 × 10^6
Rare K/T-scale events	>10 km	1 × 10^8	1 × 10^8

Table 30.3. Environmental effects of impacts (adapted from McGuire et al. 2002, after Morrison et al. 1994, and Toon et al. 1994)

Energy (MT)	Impactor size	Crater (km)	Effects
10^1–10^2	75 m	1.5	Iron objects make craters, stone objects produces airbursts (Tunguska). Land impacts destroy area the size of a city (e.g. London, Moscow)
10^2–10^3	160 m	3	Iron and stone objects produce groundbursts; comets produce airbursts. Land impacts destroy area the size of a large urban area (e.g., New York)
10^3–10^4	350 m	6	Impacts on land produce craters; ocean tsunamis become significant. Land impacts destroy area the size of a small state (e.g., Wales, Estonia)
10^4–10^5	700 m	12	Tsunamis from marine impacts reach global scales and exceed damage from land impacts. Land impacts destroy an area the size of a moderate state (e.g., Virginia, Taiwan)
10^5–10^6	1.7 km	30	Land impacts raise enough dust to affect climate and trigger impact winter. Ocean impacts generate hemispheric-scale tsunamis. Global destruction of ozone shield. Land impacts destroy area the size of a large state (e.g., U.K.)
10^6–10^7	3 km	60	Both land and ocean impacts raise enough dust to trigger impact winter. Photosynthesis ceases. Impact ejects are global, triggering widespread fires. Land impacts destroy areas the size of a large state (e.g., Mexico)
10^7–10^8	7 km	125	Vision impossible due to reduced light levels. Global conflagration. Probable mass extinction. Direct destruction approaches continental scale (e.g., Brazil)
10^8–10^9	16 km	250	Large mass extinction (e.g., K/T)
>10^9			Threatens survival of all life

- NEOs may eventually be detected or destroyed, but at present this decision and operation can only be taken on and carried out by one country: the USA, through two of its agencies, NASA and the Department of Defense.
- For direct and most secondary effects, vulnerability of human beings and of human constructions and activities is almost 100%. No mitigation is possible from this standpoint.
- No place on the planet is free of a possible future impact. Coastal areas would be exposed to collision generated tsunamis; areas with local poor ground characteristics which make them more susceptible to earthquakes would also be more affected. These considerations and the low recurrence of impacts make the inclusion of this risk in planning schemes impossible, thus suppressing the best possibility of reducing their consequences.
- The direct effects of impact are so drastic that very little likelihood exists for considering later use of the affected area.

30.4
Previous Experiences in Disaster Prevention

To be effective, communication about natural disasters should be only a part of a much greater process destined to foster the awareness of entire countries in this respect. In Colombia, real interest in prevention began in 1985, after a volcanic eruption that produced 23 000 victims. Numerous previous events, with high death tolls and heavy losses were apparently not enough to inspire decision-makers to create the National System for Prevention and Relief (República de Colombia 1989), which now covers the entire country: at the national level, with the participation of institutes dedicated to geology, hydrology and meteorology; on the regional level, with the support of corporations created for environmental management; on the municipal level, with groups comprising people trained to watch the development of hazard phenomena and to participate in relief duties. This organization is backed by national technical committees, educational programs from elementary to university level and a national research program oriented toward risk mitigation. Furthermore, Colombian law requires that natural hazards considerations be included in municipal development plans (República de Colombia 1997). Of course this ideal organization suffers from many problems: the level of effectiveness and reliability of the system varies from one town and region to another, depending very much on the human quality and sense of belonging in the individuals in charge of its management; investments earmarked for disaster prevention are not so popular among decision-makers as are those for distribution of relief to disaster victims; law enforcement in the municipal planning process is far from perfect. Finally, the inclusion of natural disaster topics in environmental education programs is not entirely satisfactory.

One positive aspect is the involvement of local communities in the processes of detection and prevention (Wilches-Chaux 1998), which include hydrometeorological measurements taken by children under teachers' supervision in urban and rural primary schools (Mejía et al. 2003). Similar experiences are known from Costa Rica and in other countries from Latin America and Asia. Cultural aspects are very important in the perception of risk and they must be taken into account in the warning and managing processes (Saenz-Segreda 2003).

30.5
To Communicate or to Educate?

The international scientific community, as represented by ICSU, has reached the conclusion that NEO impact is a non-negligible risk for the inhabitants of Earth. It has at the same time acquired the obligation to inform people about this hazard, as old as the planet but which only recently has earned our concern.

In fact, moviemakers have already discovered the bounty which sensationalism can draw from NEO impacts and have produced at least two memorable films, Armageddon and Deep Impact, which grossly falsify many of the real aspects of the problem. Documents such as the one prepared by the Discovery Channel are of course much more recommendable, but are of relatively limited reach.

The challenge for ICSU is the immediate preparation of a worldwide program designed to give people an objective vision about what NEO impacts really mean for humanity. It does not seem convenient to wait until a hazardous one appears to start acting. If this idea is acceptable, ICSU should become the "official voice" and assume responsibilities implied by this role: this would include the fact that information to be broadcast must be "understandable, credible, solicit the proper response and not confuse the public". Precautions must be taken to ensure that this information can reach marginal and remote populations; this is of great importance in underdeveloped countries (Landis 2003). Finally, once the program has started, communication should be continuous in order to warrant its effectiveness and also the confidence of those who receive the information (Gross 2003).

Considering that the challenge is more an education goal than a merely informative one, the following scheme is proposed; it must be clearly stated that what is being considered in the actual context is an educational approach, very different from the diffusion of a "before the event" notice implying well-defined measures to be taken by authorities and populace. The expected results would be the motivation on the part of the natural authorities to act in order to implement better comprehension of the NEO problem by their countries' populations.

Of course it cannot be expected that the results will be the same everywhere. Countries with very low standards of living or under the effects of national or foreign conflicts will tend to ignore or minimize the information due to the necessity of solving other urgent problems.

On the other hand it must be considered that some nations may simply ignore the message or use it to their own advantage. The Kyoto agreement on greenhouse gas emissions is an example of what governments are able to do with international treaties established on scientific and political consensus in order to keep their own internal supporters happy. At any rate, recommendations by Steel (2002) provide a good philosophical background to face the problem.

30.6
A Scheme for Transmission of Information

In order to be successful, a program designed to communicate NEO impact risks to the public should involve participation of international organizations, governments, scientific communities, educators and the media (Fig. 30.1). Even after the crisis produced by the invasion of Iraq, the United Nations still retained, for most countries, the prestige required to back the initiative of diffusing knowledge about world issues as the hazards related to NEO impact.

The natural channel for ICSU is UNESCO, which could convey the idea to the Secretary-General and eventually to the General Assembly and to the Security Council, as the use of nuclear explosions might be required to attempt NEO deflection or destruction. A decision taken by the UN General Assembly would be crucial to convince national governments that the hazard, although remote, is real. It would also mean that UN representatives in each country would cooperate with diffusing and participating in the relevant committees. UNESCO would act not only as a bridge between ICSU and

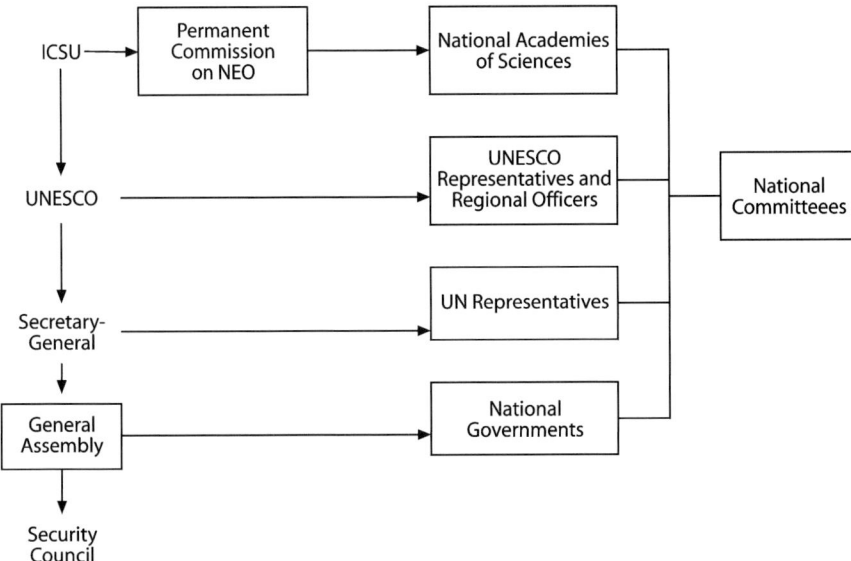

Fig. 30.1. An organizational chart for a hypothetical international response to NEO hazards

other UN organizations: it has its own world network of regional offices and national delegations that might also participate in national committees.

The other way to reach national governments is through national Academies of Sciences in countries where they exist. In advanced countries, the prestige of these academies is so well established that they probably could, by themselves, convince their governments to participate in the educational effort on NEO impacts. In less developed countries this could by proposed to governments through a national coordinating committee made up of the scientific community (Academy of Science, AAAS national equivalents, scientific societies, etc.), and by UN and UNESCO representatives. This integration is perhaps too numerous, but it would probably motivate a more receptive attitude. In countries with no organized scientific community, this task should rely mainly on the influence of international entities.

On the other hand, ICSU should consider the convenience of establishing a permanent commission on education for NEO impact risks. It would be responsible for fostering contacts with UNESCO and national academies and encouraging and providing scientific advice for the preparation of audiovisual documents designed for the public and for teaching purposes. The ICSU book on Global Change (ICSU 1994; CIAC 1994) is an excellent example of what can be done in this respect for effective diffusion at high school and college levels. The construction of a website would also be very useful. National Committees could be integrated by:

- national scientific community representatives (Academy of Sciences, AAAS equivalents, etc.);
- UN, UNESCO and UNDP delegates;

- government representatives (at least a delegate from the Ministry of Education and from the National Office for Disaster Prevention and Relief);
- a representative of the national association of science-related journalism or an equivalent.

Basic duties of National Committees would be as follows:

a contact high level government officials (President of the country, Prime Minister or equivalent);
b contact specific government officials:
   - Ministries of Education,
   - National Office for Disaster Prevention and Relief,
   - Geological Survey, etc.;
c propose an educational program consisting of:
   - inclusion of NEO basic knowledge in educational programs at all levels,
   - publication or distribution of related material produced by ICSU Commission and by local experts;
d foster participation of national observatories in NEO detection programs;
e encourage research on evidence of past NEO impacts in the entire territory;
f establish a national website in connection with ICSU site;
g deliver lectures for scientific societies, universities and the educated public;
h organize interviews with local media and to supply educational materials.

Concerning the last point it is appropriate to remember some necessary rules that have been stressed by experts in the field of public communication (Gross 2003; Niedek 2003; Montgomery 2003):
- Modern media, particularly television, have an enormous influence on society. Thus it is very important that the information released be objective and totally reliable;
- ICSU and National Committees must totally coincide in the information they release to the media.

30.7
Conclusions

The scheme presented here tries to take into account the difficulties that a truly international project on education about NEO impact risks may encounter. Although it may appear cumbersome, such a project would have the advantage of encouraging national scientific communities in the diffusion of an interesting -if mortally dangerous- topic. It would also strength the contacts between ICSU and national academies of science.

The special context of NEO detection, in the hands of a relatively small number of observatories and organizations, makes the endorsement of an international organization such as ICSU recommendable to the end of becoming truly accepted by all nations. Agreement on the deviation or destruction of an NEO would possibly pose a much more complicated problem, as these operations are in the hands of two agencies of a single country. It is difficult to anticipate how specific countries would react to the education/diffusion program launched by ICSU. Its guarantee by the United Nations would certainly enhance its credibility and favor its success. National committees would, for this specific topic, accomplish what permanent organizations of this type should carry on in many countries, with or without the support of international organizations.

Results obtained would probably depend on the level of education and organization of the particular nations. Countries affected by hunger, war or other problems would be too busy solving their immediate problems to feel particularly motivated by the NEO impact risk. However, this risk exists and it is our duty as members of the scientific community to communicate such information to all people. Nothing would be more difficult to handle than a NEO alert without previous basic knowledge on the effects of such an impact.

References

Chapman LR, Morrison D (1994) Impact on the Earth by asteroid and comets assessing the hazard. Nature vol. 367: 33–39

CMMA&D (Comisión Mundial del Medio Ambiente y del Desarrollo) (1988) Nuestro futuro común. Alianza Editorial Colombiana/Colegio Verde Villa de Leyva, Bogotá

CIAC (Consejo Internacional de Asociaciones Científicas) (1994) Cambio global. Guadalupe Ltda, Bogotá

Gritzner C (2003) Early warning system for asteroid impacts. In: Kueppers JZ, Andreas N (eds) Early warning systems for natural disaster reduction. Springer, Berlin, pp 625–632

Gross EM (2003) Public communications of warnings. In: Kueppers JZ, Andreas N (eds) Early warning systems for natural disaster reduction. Springer, Berlin, pp 67–69

Heiken G, Fakundiny R, Sutter J (2003) Earth science in the city. American Geophysical Union

ICSU (1994) Global change. ICSU, Paris

Landis RC (2003) Public communications of early warnings: role of government. In: Kueppers JZ, Andreas N (eds) Early warning systems for natural disaster reduction. Springer, Berlin, pp 71–72

McGuire B, Mason I, Kilburn C (2002) Asteroid and comet impacts as initiators of environmental change. In: McGuire B, Mason I, Kilburn C (eds) Natural hazards and environmental changes. Arnold, London, pp 133–158

Mejía LJ, Lopez J, Hermelin M (2003) Importancia de la instalación de redes meteorológicas densas en las ciudades localizadas en montañas del trópico húmedo. In: Meteorología Colombiana 5:115–122

Montgomery SL (2003) The Chicago guide to communicating science. Chicago University Press, Chicago

Morrison DE, Chapman CR, Slovic, P (1994) The impact hazard. In: Gehrels T (ed) Hazards due to comets and asteroids. University of Arizona Press, Tucson, pp 59–91

Niedek I (2003) The role of media in public communication of early warning. In: Kueppers JZ, Andreas N (eds) Early warning systems for natural disaster reduction. Springer, Berlin, pp 73–75

República de Colombia (1997) Planes de ordenamiento territorial. Ley 388 de 1997

República de Colombia (1989) Sistema nacional para la prevención y atención de desastres. Decreto 919 de 1989

Rubin AE (2002) Disturbing the solar system. Princeton University Press, Princeton

Saenz-Segreda L (2003) Psychological Interventions in Disasters Situation. In: Kueppers JZ, Andreas N (eds) Early warning systems for natural disaster reduction. Springer, Berlin, pp 119–126

Sartori G, Mazzoleni G (2003) La tierra explota: superpoblación y desarrollo. Taurus, Santillana, Buenos Aires

Shi H, Singh A (2003) Status and interconnections of selected environmental issues in the global coastal zones. Ambio 32:145–152

Steel D (2002) NEO impact hazard: the cancer metaphor. Workshop on Scientific Requirements for Mitigation of Hazardous Comets and Asteroids. NASA, Arlington

Twigg J (2003) The Human factor in early warnings: risk perception and appropriate communications. In: Kueppers J Z, Andreas N (eds) Early warning systems for natural disaster reduction. Springer, Berlin, pp 19–26

Wilches-Chaux G (1998) Auge, caída y levantada de Felipe Pinillo, mecánico y soldador o yo voy a correr el riesgo. Guía de la Red para la Gestión Local del Riesgo. LA RED ITDG, Lima

World Resources Institute (WRI) (2001) Pilot analysis of global ecosystems. In: World Resources Institute. World Resources Institute, Washington, DC

Chapter 31

Impact Risk Communication Management (1998–2004): Has It Improved?

Brian G. Marsden

31.1
Introduction

Although scares associated with potential Earth impacts by specific comets and asteroids date back to quite early in the eighteenth and twentieth centuries, respectively (e.g. Marsden 2004), the era of "modern" impact scares is frequently considered to have begun with the 1997 XF_{11} incident in early 1998 (cf. Morrison et al. 2004). Both cited papers discuss that particular incident, as well as some subsequent scares, but in my opinion the second account contains errors. In fact, I published a very detailed "discourse" on 1997 XF_{11} several years ago (Marsden 1999a). That paper fully acknowledges that mistakes were made, by several people, in the manner the 1997 XF_{11} situation was handled at that time.

31.2
1997 XF_{11}

The Morrison et al. (2004) statement that "Marsden ... suggested that the probability of an impact [by (35396) 1997 XF_{11} in 2028] could be as high as one-in-a-thousand" is incorrect. Whereas "other astronomers" may have come up with this figure, I certainly did not. I made *no quantitative statement whatsoever* regarding the impact probability in 2028. This was because I made no computation at the time that would have the object come any closer to the Earth's center than the nominal miss distance of 0.00031 AU specified on the *IAU Circular* that was the prime announcement of the putative encounter (Marsden 1998a). Obviously, this was unlikely to be the minimum *possible* miss distance, but I had no time that day to investigate the matter further. I did make some attempts to estimate how much *larger* than 0.00031 AU the miss distance might be, and all these gave distances smaller than that of the moon, although a few of the results were in fact quite close to the distance of the moon. Indeed, when other colleagues repeated the computation using the same observational data, *all* obtained nominal miss distances that were substantially smaller than that of the moon. But when, the next day, precovery observations of 1997 XF_{11} were identified from eight years earlier, it immediately became apparent that the 2028 miss distance would be more than *twice* that of the moon. The initial underestimate was the one and only shortcoming to my calculations. It meant that my suggestion (Marsden 1998a) "that passage within 0.002 AU is virtually certain" was incorrect, a point I subsequently acknowledged (Marsden 1998b).

However, given that the maximum possible miss distance was heavily dependent on how accurate one judged the available observations to be, one cannot say that it was a particularly damning error. If time were available, I would clearly have explored the matter further. In his later statistical analysis, Muinonen (1999) confirmed that the data available at the time indicated a strong bias towards very small miss distances in 2028. He found a 72-percent probability that 1997 XF_{11} would pass closer than the moon – and a 99-percent probability that it would pass closer than it actually does.

Given more time, however, I should have more appropriately spent it examining the *minimum* possible miss distance. Some might condemn me for not doing this, but given that passage at the moon's distance was clearly a possibility, I never really felt the need to do this calculation. It should be understood that my whole emphasis was on the need to obtain additional observations, for only they would ultimately allow us to say what would actually happen. Recent observations had been few and far between, and the object was fading and becoming more difficult to observe. After all, the principal purpose of the *IAU Circulars* is to draw new astronomical results to the attention of observers able to obtain further data. It was also obvious that the recognition of observations of 1997 XF_{11} from earlier years would resolve the matter of the 2028 miss distance once and for all. To include the necessary past predictions on the *IAU Circular* was impractical, however, so I simultaneously prepared these (notably for the 1990 opportunity that was so quickly and successfully taken up) for a "Press Information Sheet" that was posted on the internet.

Since the *IAU Circular* itself attracted the attention of the press, one could argue that the sense of urgency would thereby be conveyed to astronomers in a position to search for old observations, and that it would therefore have been sufficient to provide the past ephemerides in a more matter-of-fact manner on a *Minor Planet Electronic Circular*. I did not know that at the time, however, so, for better or for worse, I took the Press-Information-Sheet route, and this implied some need for a more popular description of the situation. So the relatively sober *IAU-Circular* remark about a possible 0.002-AU miss distance became first-paragraph newspeak that it was "virtually certain that it will pass within the moon's distance of the Earth a little more than 30 years from now". This was followed up with the unfortunate sentence about the unlikelihood of a collision, but that "one [not "it", as Morrison et al. write] is not entirely out of the question". I say that it was "unfortunate", because the significance of such a sentence tends to be regarded differently in the U.S. and the U.K. For example, U.S. commentators are quite likely to take this sentence literally (i.e., to jump immediately to the conclusion that "You'd better watch out!"). Those familiar with the U.K.'s more mandarin style of writing would be more likely to consider the sentence as simply meaning "Let's just take this under advisement". This may seem a strange point, but since part of the thrust of the Morrison et-al. article is concerned with differences in U.S. and U.K. journalistic practices – and after all, I spent the first 22 years of my life in the U.K. - it is in fact a point worth making.

Be that as it may, I also point out that, relegated to the sixth paragraph of the Press Information Sheet was the perfectly reasonable amplification that "There is still some uncertainty to the computation. On the one hand, it is possible that 1997 XF_{11} will come scarcely closer than the moon. On the other hand, the object could come significantly closer than 30 thousand miles." Of course, one of the problems with press stories in both the U.S. and the U.K. is that readers' attention spans rarely seem to extend beyond the first paragraph. In any case, to quote Blair (2004), "I accept full personal responsibility for the way in which the issue was presented and therefore for any errors that were made" in the material on the Press Information Sheet, to which I appended only my name and the date. These remarks were certainly not intended as an official statement of the International Astronomical Union, for example. Nevertheless, in response to a request from the American Astronomical Society for material to be released formally to the press, I provided a copy of the Press Information Sheet. In adapting this to their purpose, the Society gratuitously added, between my name and the date, the words "International Astronomical Union".

Morrison et al. (2004) next assert that, "unlike Marsden", the group at the Jet Propulsion Laboratory "had the software to estimate the actual odds of hitting". This is untrue with regard to the first point and misleading with regard to the second. As noted above, I made no initial attempt to compute the minimum possible miss distance of 1997 XF_{11} in 2028. A couple of days later, when I actually did this computation, I readily derived – like everyone else – the value of 0.0002 AU. Since the Earth's radius is 0.00004 AU, this is hardly an exercise in "estimating the odds", at least if one goes by the "normal rules" and assumes that there are no unknown perturbations, such as those due to cometary-type non-gravitational forces or to the gravitational effect of a very close approach to another asteroid that had not been considered. To estimate the odds, or as Morrison et al. (2004) go on, to undertake the "rapid *calculation* [my emphasis] of impact odds" is a very different enterprise. This exercise is meaningful only if one cannot obviously exclude the possibility of impact – and here one could. With this understanding, to say that JPL and two other teams "quickly demonstrated this capability [i.e. to calculate the impact probability]" is quite wrong.

The point is that, *without the 1990 precovery data*, it could be shown that 1997 XF_{11} was in fact a potential danger to the Earth on a number of occasions a decade or so *after* 2028. I began to suspect this already a few days after the precovery observations were secured, but, given the pressure of other duties, it took me almost three months to find such an instance in 2037 (Marsden 1998c) and another month to establish a couple more, the best example – which I first presented orally at an IAU Colloquium in Namur, Belgium – being on 2040 October 26 (Marsden 1999a). My reasoning was that (*a*) a general systematic decrease in the minimum distance between the orbits of 1997 XF_{11} and the Earth meant that this distance could be less than the radius of the Earth during several years around that time, and (*b*) the uncertainty in the minimum distance between the bodies themselves in 2028 greatly augmented the uncertainty in the asteroid's subsequent motion, notably in terms of orbital period and location in its orbit. On arranging things so that the asteroid would emerge from the 2028 encounter near an appropriate low-order resonance with the Earth, the circumstances of any possible impact could then be studied in detail. In retrospect, this seems a very straightforward idea, but it did not occur to the research teams to whom Morrison et al. attribute sophisticated impact-calculation software, although Muinonen (1999) did (already in April 1998) extrapolate variant orbits from the 1997–1998 observations alone through 2034. On presenting my 1997 XF_{11} results at the Namur colloquium I in fact made a point of recommending to the Pisa team that they look into performing similar computations routinely for other near-Earth asteroids. The JPL team also later looked specifically into the 1997 XF_{11} case, confirming and amplifying my results from the initial 98-day arc, to the point that they assigned a 1-in-50 000 probability to the potential 2040 impact event (Chodas and Yeomans 1999).

But Morrison et al. (2004) ignored these results, choosing instead to make their arguments that "the problem with [1997] XF_{11} was premature announcement without calculating a formal impact probability or consulting with colleagues" and that "The solution seemed to be better software and more consultation before making announcements". These comments quite miss the point that the whole reason for the "announcement" was to secure further observations that might otherwise not have been forthcoming for a long time. Furthermore, although I did in fact specifically present my "post-2028" impact calculations for peer review (Marsden 1998c), no confirmation of them – or, indeed, any acknowledgment that this kind of computation was of any significance – was forthcoming for some nine months.

31.3
1999 AN_{10}

So along came 1999 AN_{10}, and the Pisa group was by then ready with software to carry out precisely the kind of calculation I had advocated and, indeed, performed in a much more laborious fashion three-quarters of a year earlier for 1997 XF_{11}. Given the magnified uncertainty following a near miss in 2027, the possibility that the subsequent orbital motion of 1999 AN_{10} would be resonant with that of the Earth allowed a 1-in-a-billion impact probability in 2039. The production of this software (with similar advances by then developing at other institutions) was indeed one of the "great strides" mentioned by Morrison et al. Convinced that their computations were essentially correct – quite understandably, for the possibility of significant computational errors was no more of an issue here than it was for 1997 XF_{11} – Andrea Milani and his Pisa colleagues submitted their work for publication in a refereed journal and specifically requested comments from a dozen experts. Although he informed these experts that he had every intention of presenting the work at upcoming scientific conferences anyway (and presumably whether or not the journal he had selected accepted the manuscript), about half of them responded, all of these favorably, at which point he "posted his paper with no fanfare on his website", as Morrison and colleagues remark. Indeed, the manuscript was later accepted and published in the journal selected (Milani et al. 1999).

The question is whether the course of events was acceptable, given that there was, for the very first time, a credible prediction of a possible Earth impact only decades hence that *had not yet been eliminated* by the acquisition of further observations. On the one hand, if the press were to hear of the impact prediction, it would likely be sensationalized. On the other, as long as the prediction remained viable, the more time that elapsed before it ultimately became public, the more the scientists would be accused of cover-up. Although Morrison et al. (2004) acknowledge this dilemma, very properly brought to the fore by attention being drawn to Milani's website (Peiser 1999), their view that "The solution seemed to be to formalize the review and let the IAU, through its technical review, provide an international, professional context for any released information" is not the answer. Technical disagreements about the calculations themselves are minor and are not the issue. The problem is entirely about what is written and said, and the review process does not address that.

Morrison et al. (2004) do not mention the all-important point about the acquisition of further observations. Unlike the situation regarding 1997 XF_{11}, it was impossible to make further observations when Milani quietly posted the manuscript. Milani was fully aware of this fact, and it contributed to his presentation. Of course, further observations would be possible by the time the manuscript was actually published, so some thought would surely have had to be given to ensuring that any potential observer, professional or amateur, would be aware of the webpage posting by then. But Peiser's publicizing of the situation took care of this point. Indeed, when the first observations were obtained following the object's conjunction with the sun, it was immediately reported (Chodas 1999) that not only was the 2039 impact opportunity still "on", but its calculated probability had increased one-hundredfold. Milani then found an even greater impact possibility in 2044. The latter is the "less than one-in-a-million chance ... that it would impact the Earth" mentioned by Morrison et al., but their abbreviated version of this circumstance distorts how it came to be known.

Finally, of course, as happened with 1997 XF_{11}, the recognition of precovery observations, this time as far back as 1955 (Gnädig et al. 1999), reduced both the 2039 and 2044 impact probabilities to zero, and we are indeed safe from 1999 AN_{10} for at least the next century.

31.4
2000 SG_{344}

By the time 2000 SG_{344} was found, not only had the Torino Scale (Binzel 2000) become an official "tool" of the IAU, but the "IAU 72-hour technical review" was then a fact of life. The review was to be a voluntary process, carried out in secrecy as a free service by the IAU for impact scenarios where the TS gives a value greater than zero. While a computed "impact probability as high as 1 in 500" would indeed trigger a review, with a TS value greater than zero, even for objects as small as 2000 SG_{344}, the TS is defined in such a way that there would be a review in the case of impact probabilities as small as 1 in a million, if the potential impactor were as large as 1999 AN_{10} or 1997 XF_{11}. The reason the impact probability – on 2030 September 21 – was as high as 1 in 500 for 2000 SG_{344} (actually suspected to be an old rocket upper stage) was that it had been linked to an object observed on a single night 17 months earlier.

As Morrison et al. (2004) remark, "The review was completed on a Friday afternoon". They add that "Within a few hours, new observations were available that showed no impact was possible". Actually, this is not quite true, because there developed instead a probability of 1-in-1400 for impact in 2071 (and, indeed, dozens of other lower-probability impact scenarios during the last third of this century), but, given the object's small size, the TS value would still be zero. More troublesome, however, was the fact that at least some members of the panel had been warned that these "new observations" (actually, measurements of images recognized on frames that had been deliberately exposed for the object two nights after the earlier discovery 17 months before) would shortly become available. It was not that the review panel "did not anticipate the availability of new data", but rather that a deliberate and ill-advised decision was made not to wait for them.

So, indeed, this case where Morrison et al. (2004) concede that "Once again the astronomers looked foolish" seemed to set the stage for some relaxing of the IAU rules, perhaps even the ill-considered one with regard to secrecy. After all, following their success with 1999 AN_{10}, the Pisa group had established an organized webpage updating on a daily basis the situation with regard to impact risks for the next 80 years. Morrison et al. mention a suggested remedy to the effect "that no information should be released until all possible data were collected". How one achieves this is not too clear, even if this is intended just to apply to the report that is being put together by the IAU panel. And, as noted above, 2000 SG_{344} may be tiny and artificial, but it still poses some small threat later in the century, a situation that will remain unchanged until observations next become possible during a close approach to the Earth in 2028.

31.5
2002 MN

This fourth case considered by Morrison et al. (2004) indeed "made the evening news", but since nothing untoward was either computed or stated by the astronomical community, no further comment on it is necessary.

I can, however, take the opportunity to note that there had been two new developments since the 2000 SG_{344} episode – though not in fact inspired by that "unfortunately, on the weekend no-one was correcting the statements of alarm" fiasco. One development was the invention of the Palermo Scale (Chesley et al. 2002), designed to address some of the shortcomings of the TS, which does not differentiate between immediate impact possibilities and those far in the future. The TS consists merely of somewhat subjective integral values from 0 to 10. In that it leads to larger numbers for more imminent events, the PS is a continuous function, positive or negative, again basically arranged so that a value of zero corresponds to an impact probability equal to that of impact by a comparably sized unknown object between the present and the time of the putative event. At the end of 2001 a positive value of the PS, rather than the TS, was adopted as necessary for triggering an IAU technical review. The second new development was that by early 2002 JPL was ready with its *Sentry* system that provided with its own webpage an assessment of impact dangers (for the next 100 years), essentially independent of the Pisa "riskpage".

31.6
2002 NT_7

Here the point is that, according to the Pisa group, 2002 NT_7 was the first case for which the PS value was positive. The reason involved a combination of relatively large size and impact probability but was mainly leveraged by the fact that the projected impact date was less than 17 years into the future. And, contrary to what Morrison et al. write, it is my recollection that this potential event *did* generate a value of 1 on the TS. Indeed, an important reason for not having an IAU technical review was for the first time stated to be that "new data were coming in". The actual motivation for this, however, was that a 72-hour embargo would not allow the Pisa and JPL "risk pages" to be updated during this time, a circumstance guaranteed to raise suspicions in the press and among the public. There was also the point that since, after all, there *were* now the two independent "risk pages" that essentially confirmed each other (even though slight differences meant that the PS value computed at JPL was actually very slightly negative, but still a record to date), a formal review process was no longer needed. Of course, this had become obvious more generally when JPL initiated its "risk page" four months earlier, yet, curiously, it remains the official IAU policy to this day.

Purely and simply, the fact is that 2002 NT_7 was a news story *because* it was the first case where astronomers had indicated a positive value on the PS. As with my "one is not out of the question" from four years earlier, there is really nothing wrong with the use of phrases like "the most threatening object yet detected in space" and "on a collision course with Earth", provided the news stories also objectively describe the actual situation. Lighten up! The BBC story in question (Whitehouse 2002) includes these phrases in its first two sentences, the second of which reads in its entirety "A preliminary orbit suggests that 2002 NT_7 is on an impact course ['a collision course' was in the story's title] with Earth and could strike the planet on 1 February, 2019 – although the uncertainties are large". Then the third sentence explains the whole rationale for the story: "Astronomers have given the object a rating on the so-called Palermo technical scale of threat of 0.06, making [2002] NT_7 the first object to be given a positive value". What is "especially provocative" about this? Personally, I'm much more impressed (and pleased!) that the story won the prize for "Best news story broken on the net" at the annual "NetMedia European Online Journalism Awards" in 2003, in a competition adjudicated by 118 judges from European media organizations and journalism schools in 20 different countries. And, as for the "stylistic ocean that separates American and British media", can one honestly claim that the editorial two days later in *The New York Times* that starts out "Thank goodness! Another killer asteroid is on the way, just in time to take our minds off the stock market and foreign affairs" contributes less to the "media fuss"?

31.7
2004 AS_1

Although Morrison et al. (2004) discuss only five representative impact-scare stories during 1998–2002, there were quite a few others, and there have been more since. I therefore now want to discuss the case of 2004 AS_1, which in some respects is the most bizarre of all, almost making one wonder whether *anything* had been learned since 1997 XF_{11}.

Late on the afternoon of 2004 January 13, following its usual practice, the LINEAR (Lincoln Laboratory Near-Earth-Asteroid Research) team sent to the Minor Planet Center its observations from the previous night. As was the case on several nights that month, the observing conditions had been quite poor. Also following his usual practice, Tim Spahr at the MPC picked out as candidate NEOs several of the objects noted by LINEAR as having apparent motions greater than those of main-belt minor planets, and he placed nominal ephemerides for these candidates for the following couple of days, together with estimates of the ephemeris uncertainties, on "The NEO Confirmation Page" the MPC maintains in the internet. The purpose of the NEOCP is to alert observers who may be able, not only to confirm or deny the actual existence of the objects mentioned on the NEOCP, but also to provide follow-up data allowing the MPC to conclude whether they are or are not NEOs. If an object is indeed confirmed as an NEO, a *Minor Planet Electronic Circular* is then formally issued with all the relevant information. The point is that users should not draw unwarranted conclusions from the necessarily limited NEOCP information.

On this occasion, Spahr had already worked a ten-hour day and was able to spare only some 35 minutes more before going out to dinner with a visiting colleague. Nevertheless, he picked out five candidate NEOs from the LINEAR submission and provided the usual predictions on the NEOCP. Unfortunately, in his haste, he had not noticed that the nominal ephemeris for one of the objects indicated that it would collide with the Earth some 27 hours later! Of course, the accompanying uncertainty plot covered a rather considerable amount of sky, but that was ignored by the amateur astronomer who drew the apparently distressing state of affairs to the attention of participants in the Internet "Minor Planet Mailing List". This quickly precipitated one of those semi-informed discussions for which the Internet is renowned. Curiously, nobody thought to draw the attention of anyone at the MPC to the situation for more than two-and-a-half hours! This included one relatively inexperienced observer in the U.K. who merely remarked, without explanation, that he had failed to find this object near the nominal prediction.

All this was therefore news to me when I – also having been in the office for the preceding ten hours – received a telephone call from Steve Chesley of JPL. It would also be news to Spahr, who, having returned from dinner, had only minutes earlier started on his one-hour drive home. Indeed, before leaving, he had checked his e-mail and, as a result, removed one of the other new objects from the NEOCP on advice from LINEAR that it was not real.

I found the observations of the object now causing the consternation and checked to see what Spahr had done. Although I did not immediately know the precise basis for the 450 variant orbits he had computed to define the ephemeris uncertainty, I could see that the Earth-impacting orbit he had selected as nominal had the object only 0.03 AU away from the Earth during the 70 minutes it had been under observation. Since I could also verify that an assumption of a greater distance did not satisfy the observations as closely, the best "quick fix" was obviously to adopt the 0.03-AU distance but to change the velocity so that the object would instead be *receding* from the Earth. It seemed to me that the simplest way to defuse a situation that ought never to have arisen was by substituting such a computation for Spahr's nominal orbit, while retaining his assessment of the uncertainty. Although I quickly made this change, it was clear that the damage had been done, because many people had got the idea that we might seriously be in danger from the object.

But the principal need was to get some observations! By this time I realized that the fact that we had received no follow-up observations of any of the NEOCP objects from anywhere in continental Europe meant that there was widespread bad weather. One of the most reliable amateur astronomers in the U.K. had sent some observations shortly earlier, but not of the worrisome object. A quick exchange of e-mail brought the unfortunate news that at the object's rather extreme northerly declination his telescope's fork mounting obstructed the view of the camera. It was still rather early for observations from North America, but I also alerted a few of the more reliable observers in North America to the problem, only to find that bad weather was also very prevalent; this would particularly affect the observers in the western U.S. on whom we principally depend. Although actual, positive observations were to be preferred, negative searches at sky positions consistent with impact would at this stage perhaps provide some reassurance, even if, for various reasons, they were necessarily inconclusive, as appeared to be the case with the aforementioned MPML report.

Although I knew that the JPL team had little or no experience in computing orbits from scratch (without being supplied with starting values that can be differentially adjusted when further observations become available), I was not aware that Chesley spent several hours that evening preparing a computer program to fill this gap. Partly in consultation with Spahr, he used the LINEAR observations to compute many different orbits for the new object. From the fraction that yielded collisions with the Earth, Chesley concluded that, within the next day or two, it had an impact probability that was as high as 25 percent – perhaps even 40 percent! I was also unaware that, prompted by the MPML speculations, a small group that included the first two authors of Morrison et al. was discussing what should be done to inform authorities in the U.S. of the possible impending impact. This matter came to a head when they heard of the high probability calculated by Chesley. A potentially embarrassing outcome was avoided, however, when they learned, a couple of hours later, of a negative search by an amateur astronomer in Colorado at the position given by the "impact" ephemeris.

At last, the next morning, the LINEAR team reported to the MPC some positive second-night observations. These showed conclusively that the object was as far away as 0.2 AU and certainly no immediate danger to the Earth. The NEOCP prediction was updated, clear weather in the Czech Republic allowed further observations a few hours later, and the object was formally announced with the designation 2004 AS_1. The ephemeris supplied with this formal publication (Spahr 2004) indicated a minimum distance from the Earth of 0.08 AU in mid-February, and the JPL and Pisa extrapolations of the orbit for the next century revealed absolutely no impact possibilities.

So what went wrong? The principal problem was much the same as the one that had affected the calculations for 1997 XF_{11}, namely, underestimation of the uncertainty, coupled with an unfortunate bias in the available data that led to the conclusion that the object would be much closer to the Earth than it really was. But, whereas there was little consequence to the factor of two or three in the misestimate of the maximum possible distance of 1997 XF_{11} on its 2028 passage, the range indicated for the distance of 2004 AS_1 at discovery (i.e., one that was too small by a factor of at least ten and maybe one hundred) meant everything.

The four LINEAR observations spanned just 110 *arcseconds* of sky, pretty much along a great circle. There is no way one can actually *determine* an orbit from such information! One essentially has the positions of the beginning and end of the arc, i.e., just four pieces of information, whereas six elements are required to define a unique orbit. Two elements (or related quantities) therefore need to be *assumed*, and it then becomes possible to compute values for the other four that are consistent with the data. While it is usual to define one orbit to be the "nominal" one, it is reasonable to make several choices of values for the two quantities over some appropriate range, so as to construct a series of viable orbits, the ephemerides computed from these then being used to delineate the uncertainty. One particularly useful pair of choices for the assumed quantities is the object's topocentric distance and heliocentric velocity at the beginning or end time of observation. Specifically, Väisälä (1939) recommended that it was sufficient to take the heliocentric radial velocity to be zero in all cases. While the "perihelic" (or "aphelic") orbits so derived have considerable merit, particularly for asteroids that are clearly in the main belt, my more generalized procedure (Marsden 1999b), which involves rotation into a particular coordinate system, not only allows the Väisälä case to be generated using the appropriate value of a single component of the heliocentric velocity, but it has the added advantage that the choice of zero for this component gives the orbit with the smallest possible semimajor axis for the selected topocentric distance. There is also a specific maximum choice (in absolute value) that yields limiting parabolic solutions.

After fitting the first and last observations of 2004 AS_1 exactly one could inspect for each orbit the residuals of the other two observations. Indeed, differential corrections could be made from the four observations with some of the orbital elements held fixed.

What Chesley and Spahr did was reject all the orbits with residuals beyond some specified limit, i.e., they did not count the cases of significantly larger topocentric distance in their estimate of the impact probability. But how can one reliably specify such a limit? In a case like this one can't, particularly with the appreciation that poor observing conditions had led to the rejection of one of what are normally five LINEAR observations. The only orbits that should have been completely dismissed are those that would be physically very improbable – i.e. hyperbolic solutions. Given that there were possible retrograde parabolic orbital solutions for 2004 AS_1 (and, after all, given the short exposures and poor conditions, one did not know that it was not a comet) out to a topocentric distance of 2.6 AU, the actual impact probability must have been extraordinarily small.

Of course, given a sky uncertainty as large as that of 2004 AS_1, there is the chance that even an experienced follow-up observer would not be able to cover the whole area. To try and accommodate this problem, the MPC made a cosmetic change in the NEOCP set-up, during the weeks after the 2004 AS_1 fiasco. The uncertainty plots are now color-coded, essentially so that the regions corresponding to passages of the objects close to the Earth are in red, while those where the objects would be far away are in green. Observers who are interested only in helping demonstrate that there is no imminent danger need therefore consider only the former regions, whereas those who wish to make a more concerted effort actually to find the objects can concentrate more on the latter. And already on the evening of January 13, another quick change was made so that the MPC staff member attending to the NEOCP receives a warning if he were inadvertently being led to post a nominal orbit that yields an imminent collision with the Earth.

Finally, it should be noted that the 2004 AS$_1$ situation came and went in mid-January 2004 without any mention whatsoever in the world's press. It became a media issue some six weeks later, just because it was specifically included for discussion at a professional conference on NEOs in California.

31.8
2004 MN$_4$

The most recent case I want to discuss, that of 2004 MN$_4$, was in fact handled superbly by the press, at least in part because it was a story for just two days on each side of Christmas. Its significance is that it sounds the death knell to the Torino Scale.

During the two days in December 2004 that this object was on the NEOCP, it became clear that it was the same object that had been designated on the basis of observations on two consecutive nights six months earlier but then lost – at least partly on account of initial errors in both the measurements and the timings. Because of the long observed arc and the resulting absence of any possibility of intervening large perturbations, the first potential Earth impact, for 2029 April 13, immediately showed up with probability 0.0005; with a further week's worth of data the probability marched up to almost 0.03, corresponding to TS = 4. Before the potential significance of this could sink in (after all, those elements of the science media not on holiday were by then occupied with the tragedy of the Indian Ocean tsunami), the recognition of observations from a single night *nine* months before its discovery dropped the probability to zero, giving instead a clear, but extraordinarily near miss. Because it was such a near miss, the expected magnification of the uncertainty allowed the subsequent orbital period to range between 1.12 and 1.26 years, thereby allowing possible first-order-resonance impacts each April from 2034 through 2038 (some of them having TS = 1), as well as possible impacts from higher-order resonances with the Earth during the following two decades. It was also obvious that at least the more entrenched impact possibilities would not go away quickly. Given the absence of prediscovery observations prior to 2004, coupled with the fact that 2004 MN$_4$ will be located essentially behind the sun from 2007 through 2011, no further clarification of the impact situation (one way or the other) will be possible until observations are made during the next modest approach to the Earth during 2012–2013.

This is a wonderful example of a possible impactor – everything that 1997 XF$_{11}$, 1999 AN$_{10}$ and others might have been, but weren't. Sure, with just a slight shift in the geometry, 1997 XF$_{11}$ is large enough that it could have registered at TS = 6 or even 7 – for 1-percent impact probability – for 2028, but the precovery observations were in any case so decisive that there can be no danger from that object for millennia to come. In this sense, one can say that, despite their larger potential TS values, the intrinsically bright, kilometer-sized NEOs are actually *less* threatening than those in the 300–400-meter size range of objects like 2004 MN$_4$, because clarifying precovery photographs are more likely to be found of the former. Given the range of temporary enhancements the TS can show, it is not difficult to conclude that the only values of significance (apart from 0) are 8, 9 and 10. But these require an impact probability in excess of 99 percent! Although the threshold probability desired for a possible response to a potential impact threat has never actually been specified, conventional wisdom seems to be that this is in the range 10–40 percent. Given the jump from 1 to 99 percent, the TS therefore does not help on this important point.

31.9
2003 QQ$_{47}$

For my final example I go back a year or more in time to 2003 QQ$_{47}$, which was actually the last occasion on which the press could reasonably be criticized (Morrison 2003). This time it was the U.K. Near-Earth Object Information Centre that bore the brunt of the fuss about a possible impact in little more than ten years. Although this potential event – again quickly dismissed with the availability of more data – remained negative on the PS, it did register as a 1 on the TS, and the Information Centre merely pointed out that, as the TS description sheet specified, it was therefore "an event meriting careful monitoring".

If the IAU review committee is now using the PS, however, why – one may ask – should one pay attention anymore to nonzero values on the TS? The answer is that the IAU continues to recommend the TS as *the* tool for communication between astronomers and the press. Some astronomers feel that the PS, despite its built-in allowance for the time remaining until the event and its general improved definition, is too "complicated" for the press and public to understand!

The significance of the 2003 QQ$_{47}$ case is that it inspired some modification to the Torino Scale. Unfortunately, this modification did not address the scale's basic rationale and was merely a matter of rewording some of the descriptive remarks associated with the TS numbers. Most notably, rather than "meriting careful monitoring", TS 1 is now classed as "normal" (Morrison et al. 2004).

In reality, however, the problem with the TS (as well as with the PS) is that it combines two quite unrelated quantities – impact probability and impact energy. By equating the zero-point of the scale to the background of objects of similar size, the TS is effectively saying that "what we don't know *can* hurt us". While that is an understandable message, I think it is the wrong message, at least when it is given in the context of a news story about a specific potential impactor. The rationale is also sometimes given that the TS is analogous to the Richter Scale for earthquakes. But the RS is principally used for assessing events that have already occurred, whereas the TS is to be applied to future predictions. In making such predictions, the point is that the first TS ingredient changes all the time, while the other ingredient is quite irrelevant – because, if we can help it, we're not going to allow an impact to occur! If an impact is in the cards, it surely makes much more sense to consider, not the impact energy, but the energy that will have to be expended to deflect the object away from the Earth (Remo 2004).

For a particular object, the component of the impact calculation that tends to be the most stable is the *date* of a potential event. The "risk pages" currently contain about 70 objects with one or more nonzero (in practice, that means greater than something like 1 in a billion) impact probabilities on particular dates. New discoveries – including the ones that have caused scares – regularly go on to the "risk pages", but they are also removed, at a rate of about 100 per year, when it is recognized that there is no longer any impact possibility for these objects during the next century. Clearly, the monitoring process is proving effective. Actually, since some 95 percent of the removals occur within a month of their arrivals on the "risk pages", one might almost say that the process is working *too* well. It is difficult to avoid the impression that, if one waited a couple of weeks after discovery before placing objects on the "risk pages" (rather than racing to place them there after a couple of days), humanity would be just as "safe" for a lot less effort.

31.10
Purgatorio Ratio

When the 2003 QQ_{47} TS = 1 situation arose, I suggested (Marsden 2003) that the communication of impact threats to the public could be managed very simply by the use of what I shall call here – rather in line with the practice of adopting the names of Italian cities – the "Purgatorio Ratio". By this I mean the ratio of an NEO's observed arc, or time between the first and last observation, to the time between the present and the next possible impact date. This is a simple enough concept, even for editors of tabloids, my point being that there is good PR value in being able to say that, if the PR value is less than 0.01, say, any possible threat should be considered as of *absolutely no consequence* to the public.

Applying this principle to the objects on the JPL "risk page", we can see that the object that is currently the most "dangerous" is in fact the aforementioned 2004 MN_4, which has PR = 0.039. Next comes 2000 SG_{344} with PR = 0.022 (given that its next impact possibility is in 2068). Although it is not actually listed on the "risk page" (because the suggested impact date is not until the year 2880), the largest PR value, 0.063, applies to (29075) 1950 DA, the observations of which now span more than half a century.

With a PR of 0.011, there is currently just one other object on the "risk page", 2003 DW_{10}, that might conceivably be of public interest, solely because the observations made on a single night five months after the original five-day span yield the calculation of a small impact possibility in 2046. Even if 1997 XF_{11} really *had* been a threat for 2028 given the data at our disposal on 1998 March 11, its PR would have been only 0.0088 (and for the first real threat it was only 0.0068), so, I therefore now say, it was inconsequential anyway. Indeed, the two other persistent cases of nonzero TS, 2004 VD_{17} and 1997 XR_2, currently weigh in only at PR = 0.0037 and 0.0008, respectively, and the median value for *all* the objects on the JPL "risk page" is just PR = 0.0005!

By considering only the four high-PR objects – 1950 DA, 2004 MN_4, 2000 SG_{344} and 2003 DW_{10} – what we are saying, in effect, is that "what we don't know *can't* hurt us". But, bearing in mind my remark in the previous section about the persistence of impact-possibility dates, the astute reader might say: "It's all very well to concentrate only on these cases, but what happens as we get closer in time to the calculated possible impact dates for the others?" In due course, the PR will not only exceed 0.01, but it will eventually become infinite. So should we worry about this? For example, as I write in early June 2005, 2004 FU_{162} appears at a completely negligible PR = 0.0001. But its first possible impact date is less than ten months away, on 2006 April 1. The PR is currently very small because the observed arc for this object is only 0.03 day. Since this is shorter than the interval covered by the initial observations of 2004 AS_1, the answer is obvious.

The median observed time span for the entries on the JPL "risk page" is little more than 8 days. Indeed, most of the objects are there *precisely because* their observed arcs are short. When further observations are obtained, it (almost) invariably happens that the impact possibilities will disappear. Of course, in most of the short-arc cases further observations were not forthcoming, and the objects are now lost. Again, it is still very likely that *if* further observations had been made, the impact possibilities would have disappeared – and one should therefore not worry ... too much. (If one is worried, he/she can always consider the possibility of searching for such *virtual impactors*, as Milani has called them – and as was considered in the rather silly case of 2004 AS_1 – and hoping not to find them.) Only a dozen or so of the objects on the "risk page" have been observed for longer than a month, which is the kind of arc one needs reasonably to guarantee the success of planned observations of these objects in the future. For these objects, if ongoing observations do not eliminate the possibility of impact as the date approaches, there may indeed be cause for worry – a point reflected in increases in both the PR and the impact probability. After all, if the 2036 (for example) impact possibility for 2004 MN_4 still exists after the completion of the 2012–2013 observations, the PR will be up to 0.4 and the impact probability substantially larger than the present 0.00006.

Of course, we also need to question the reliability of the orbit determination when there are isolated observations, like those of 2003 DW_{10}. But if we ignore the data on this single night, so long after the confirmed observations, the PR drops to only 0.0003, rendering the calculated 2046 event inconsequential. Of course, in this case, there would likely be additional potential events before 2046, not shown on the "risk pages". While it might be useful to know if any of these are imminent (e.g., such that the PR would have been above 0.01 already in 2003), this 5-day-arc case would likely, in time, become inconsequential anyway. Again, "what we don't know *can't* hurt us".

As for the 2000 SG_{344} case, we know that the PR will have increased from the present 0.022 to 0.034 shortly before the next observations are made in 2028. What actually happens then remains to be seen, although most would hazard that any impact possibilities for the following century would vanish. But since such computations are routinely computed only a century ahead, one could argue that, with its then 0.29-century observed arc, the PR *might* be a whopping 0.29, if the termination of the calculations caused us just to miss an impact event! As with 1950 DA (and, indeed, with other NEOs observed for many decades), in such a case there may be value in extending the impact calculations for more than a century into the future.

Where the use of the PR becomes particularly illuminating is for an object that is discovered only shortly before it hits the Earth – the kind of situation 2004 AS_1 might have offered. Particularly as the interest evolves now toward smaller NEOs, it seems almost inevitable that at some point an object will be detected in space only a short time before it enters the atmosphere. Although the object is likely to be small enough that it is harmlessly destroyed in the upper atmosphere, the novelty of this experience guarantees that it will attract attention. If the second night of observations of 2004 AS_1 had shown an impact still to be viable (and, obviously, with high probability) 48 hours, say, into the future, we should have had PR = 0.5. After 24 hours (with further observations presumably being obtained feverishly in the mean time), the PR would be 2.0, a change that stresses the imminence of the hazard at least as well as its static appearance at TS = 8.

The "Spaceguard Survey" was established on the premise of finding 90 percent of the kilometer-sized NEOs by the end of 2008. Costing at most some 3-4 million per year, and whether or not it actually completes its task before the specified time limit, this has been an extraordinarily inexpensive enterprise, given also the extent to which many amateur astronomers participate with follow-up observations. A serious extension to smaller NEOs will cost considerably more, as activities in preparation of new programs such as Pan-STARRS are starting to demonstrate, and it is unlikely that amateurs will be able to participate in a significant way – other than perhaps by making last-minute searches, hopefully negative, using impact trajectories shortly before the impacts might be predicted to occur.

Much of the 2004 AS_1 press emphasis was on the complete absence of any coherent plan to inform authorities of an impending possible NEO danger. As a result of this, some minimal steps towards such a plan have now been made, at least in the U.S., where the view is that the astronomers should inform authorities at NASA. Whether this really represents a step toward improved impact risk communication management, particularly at an international level, largely depends on what those authorities would do, and that is far from clear.

References

Binzel RP (2000) Planet. Space Sci 48:297–303
Blair T (2004) U.K. House of Commons Hansard, July 14, column 1435
Chesley SR, Chodas PW, Milani A, Valsecchi GB, Yeomans DK (2002) Icarus 159:423–432
Chodas PW (1999) http://neo.jpl.nasa.gov/news017.html
Chodas PW, Yeomans DK (1999) BAAS 31:1227
Gnädig A, Doppler A, Williams GV, Marsden BG (1999) MPEC N21
Marsden BG (1998a) IAUC 6837
Marsden BG (1998b) IAUC 6879
Marsden BG (1998c) CCNet Digest, June 8, item 1
Marsden BG (1999a) J Br Interplanetary Soc 52:195–202
Marsden BG (1999b) In: Fiala AD, Dick SJ (eds) Proc. Nautical Almanac Office Sesquicentennial Symp., U.S. Naval Observatory, pp 333–351
Marsden BG (2003) CCNet 73, Sept. 12, item 4
Marsden BG (2004) Adv. Space Res. 33:1514–1523
Milani A, Chesley SR, Valsecchi GB (1999) AAp 346:L65–L68
Morrison D (2003) NEO News, Sept. 3
Morrison D, Chapman CR, Steel D, Binzel RP (2004) In: Belton MJS, Morgan TH, Samarasinha N, Yeomans DK (eds) Mitigation of hazardous comets and asteroids. Cambridge University Press, pp 353–390
Muinonen K (1999) In: Steves BA, Roy AE (eds) The dynamics of small bodies in the solar system: a major key to solar system studies. Kluwer
Peiser B (1999) CCNet Special, Apr. 13
Remo J (2004) Acta Astronautica 54:755–762
Spahr TB (2004) MPEC A56
Väisälä Y (1939) Mitt. Sternw. Univ. Turku No. 1
Whitehouse D (2002) http://news.bbc.co.uk/2/hi/science/nature/2147879.stm

Chapter 32

Towards Rational International Policies on the NEO Hazard

Johannes Andersen

32.1 Introduction

Astronomy may be the purest of sciences, but even astronomy must interface with the rest of society. First, society influences astronomy: Astronomical research is largely supported by public funds, and political priorities decide which of our favorite projects may become reality. And waste from human activity increasingly limits our ability to distinguish the faint signals from the Universe from such human-generated interference as light pollution, space junk, and radio noise from the ground and from space.

But certain astronomical phenomena may also influence human life. Solar activity and impacts of planetary fragments on Earth are the most important examples. Solar activity has immediate effects on humans and instruments in space and on power lines on Earth, and the study of Solar-terrestrial relations is a booming field of research. Asteroid impacts have vastly greater potential for damage, but occur at such long intervals that they are difficult to place in context with other, more familiar natural hazards.

Because no serious harm to humans from an asteroid impact has ever been recorded, the danger is easily brushed off as negligible. However, impacts do continue to occur in the Solar System, as the entire world witnessed in 1994 when Comet Shoemaker-Levy 9 slammed into Jupiter. Finding out if, when, and where something like this might happen to Earth in the near future is a task for astronomers. And if astronomers want the support of society, they will be wise to also listen to society's concerns regarding hazards of astronomical origin and convey their findings in a form society can understand and use. Doing so effectively requires a different set of skills than does astronomical research.

The IAU effort to develop rational policies on the impact hazard offers some useful illustrations of the assets that professional scientific unions may bring to such issues, and the limitations they meet when venturing outside the familiar territory of pure science. The following summary is offered as the personal view of someone who was General Secretary of the IAU in 1997–2000 when this process was initiated.

32.2 "The 1997 XF11 Affair"

The NEO issue hit the headlines with a vengeance on March 11, 1998: The IAU *Minor Planet Center* noted that an impact on Earth by an object called 1997 XF11 in 2028 was not totally excluded, and the news leaked to the press. Such alarms had been heard before, but having the IAU back such an observation was new, and criticism was widespread when the risk of impact was quickly shown to be negligible. Journalists were indignant, many colleagues also, and NASA, which shouldered most of the actual effort to detect potentially hazardous asteroids (PHAs in community jargon), was 'not amused.'

The author's background in addressing the situation consisted of: *(i)* A solid lack of knowledge of Solar-System objects in general and NEOs in particular; *(ii)* a strong conviction that a science-based NEO IAU policy was an absolute necessity; and *(iii)* experience from representing the IAU in ICSU, where one is introduced to the other scientific unions and to international science policy making in general. Invaluable advice was given by the Assistant General Secretary (and designated successor), Prof. Hans Rickman, who knew both the science and the personalities involved.

Within the IAU, scientific work on the NEO hazard was given a focus with the creation of a *Working Group on NEOs* in 1991, but defining a formal IAU policy on the issue had not been considered. An initiative to do so was in fact taken already at an *Officers' Meeting* a month before the '1997 XF11 affair': Regardless of the circumstances of that event, NEOs were clearly of considerable interest to both laypeople and at least some politicians, and it appeared fundamentally unacceptable to the Officers that the IAU should be unable to comment on an astronomical matter of such perceived importance to society. Hence, first steps were taken to approach the relevant IAU bodies for advice, i.e. Division III, "*Planetary Systems Sciences*", Commission 20 "*Positions and Motions of Minor Planets, Comets & Satellites*", and the "*Working Group on NEOs*". But the '1997 XF11 affair' precipitated immediate and more focused action.

32.3
Putting the Astronomers' House in Order

The traditional IAU structure is well adapted to handle developments on time scales of years or decades, but not hours, days, or weeks. Moreover, the NEO community hosts a remarkable diversity of views, both scientific and personal, and achieving consensus on any contentious issue is difficult. So it appeared to be easier for an outsider to plan action on the main issues of principle, which in my view were the following: *(i)* define rules of operation for the Minor Planet Center and the degree of control over it by the IAU; *(ii)* define a satisfactory procedure for dealing with predictions of an imminent impact; *(iii)* define lines of communication in case a prediction of an imminent impact were to be substantiated; and *(iv)* enlist the relevant scientific unions in a comprehensive, interdisciplinary, and impartial study of the likely consequences of asteroid impacts in a plausible range of magnitudes.

It was realized from the outset that a scientific union, such as the IAU, has no means of forcing its decisions or opinions on the scientific community or the world at large. To the argument that, "The IAU should stick to its guns!", the only answer is that, "The IAU has no guns to stick to!" The only power a union such as the IAU has is that of being right on the science, as proven by past performance. Attempts to enforce a view through dictate or arm-twisting will not only be ineffective in the short term, but also damage the IAU's credibility and ability to act in the long term. IAU arguments must be absolutely bullet-proof, or the effort will be counterproductive in the end.

32.3.1
The Minor Planet Center

The IAU Minor Planet Center (MPC) has been hosted by the Smithsonian Astrophysical Observatory (SAO) and directed by Dr. Brian Marsden for over 25 years. During that time the numbers of observations, identifications, and orbit computations processed annually by its small staff have grown enormously, and the MPC has in fact become about equal to the rest of the IAU in terms of both budget and public visibility. Yet, in 1998 not a single page existed to define the relations between the MPC, the SAO, and the IAU itself.

The IAU Officers and Executive Committee agreed that an IAU MPC must have a set of *Terms of Reference* to define its top-level policies on data checking and acceptance, orbit computations and data products, and data access and intellectual property rights. After lengthy negotiations, Terms of Reference were agreed upon in 2000 and a contract for the MPC to implement these policies was eventually signed with SAO. Through these, apparently formal steps, the IAU had defined its own procedures on a clear scientific basis.

32.3.1.1
Reviewing Asteroid Impact Predictions

From time to time, the initial observations of a minor planet lead to a provisional orbit that passes near the Earth at some time in the future. Typically, the object is then followed and the orbit determination refined until any risk of an impact is either substantiated or certified to be negligible.

This is simple in theory, but involves several complex scientific and non-scientific questions: How reliable is the computed orbit? How does the uncertainty in the data translate into positions of the object relative to the Earth during a passage that may be decades into the future? How is an impact probability derived from these data? Are there positions of the object during a close passage which – while safe in themselves – might cause the orbit to change and the object to impact at a later passage? What threshold should be adopted for alerting astronomers to observe the object immediately? What (higher) threshold should be adopted for alerting the outside world to the existence of a non-negligible risk? Who should then be alerted – governments, the media, or both – and in which order? Should this be done immediately, or only when new observations have confirmed a substantial risk? How is the news presented in understandable, yet non-sensational terms?

The precise computation of asteroid orbits, including orbital resonances with the Earth and the realistic mapping of observational errors onto position space of the object years, decades, or even centuries ahead is an exciting field of research in its own right, but a detailed description is not needed here. The immediate concerns for the IAU were: How do we react if/when an impact prediction is made? How does one describe the hazard associated with the putative impact? How do we contact the relevant authorities and/or the media? And will they know how to react and be authorized to take action if needed? Answers had to be developed on the background that, as NEO searches ramp up, potentially hazardous objects will be discovered frequently in the future, and crying Wolf! every time would be counterproductive.

It was also soon realized that, while an authoritative IAU statement on any impact claim was expected, a mandatory pre-publication review procedure would not only be ineffective – sensations hit the headlines regardless of IAU rules – but would also lead to accusations of secrecy and cover-up. Instead, a procedure was established by which an impact prediction could be submitted voluntarily to the IAU. The IAU *NEO Technical Review Team* would deliver its assessment rapidly (within 72 hours) and make it public if the claim was substantiated. The procedure was voluntary, but the IAU would decline any comment on impact predictions that did not pass it. And this worked.

32.3.1.2
Describing the Impact Hazard

Three issues of communication remained to be addressed: First, how do we describe the danger associated with a potential impact in terms that are both neutral and readily understandable to the public? Discussions in the community led to the definition of two numerical scales expressing the range of energy release and resulting damage, from zero to global ecological disaster. The *Torino Scale* (Binzel 2000) describes only the impact probability and potential damage, whereas the *Palermo Scale* (Chesley et al. 2002) takes the warning time before the impact into account. A grade of zero or below means, *"Don't worry, anyone!"*, but a positive value on the Palermo Scale signals that astronomers should observe the object as a matter of priority.

Timing any information to the press is a second thorny issue. Spectacular impact scares that prove unfounded a few days later lead to accusations of sensationalism by colleagues and the press alike. On the other hand, withholding the news until substantiated leads to accusations of secrecy, cover-up, and conspiracy – often by the same individuals. The solution is to place the news on an open web page where anyone can access it and refer to it freely, but without 'juicy' comments to attract non-professional attention while verification is ongoing.

The final question is how governments and/or authorities are notified in case a real danger would be substantiated. The early discussion focused on the lines of command to governments (Direct? Through the UN? Via space or defense agencies? etc.). It ended without conclusion, because the real problem is far more complex than just having a telephone number to call: Unless the official receiving the call is familiar with the issue and knows his/her own lines of command and what preventive action may be taken, the message may not be useful at all. Thus, finding out how to turn an astronomical alert into useful action is the final, and as yet unfinished, part of our story.

32.4
From Pure Science into the Real World

Assuming a credible impact prediction was made, how would one assess its consequences on the environment, humans, and society? This is territory where astronomers are no longer experts, so the task must be passed to those who are. To model the effects of an impact on land we need geophysicists and geologists; for impacts in the sea we need oceanographers and hydrologists; and computing consequences for the atmosphere and the climate is the task of meteorologists and geophysicists. A major impact will have consequences for plant and animal life in smaller or larger regions, and biodiversity will be affected. This, in turn, will impact the food supply for humans – perhaps for humanity. Finally, the reaction of people and human society to disaster scenarios of various magnitudes, such as mass emigration or panic, and maintaining basic structures and services of society, etc, call for contributions from the social sciences. A model for this is ICSU's *International Human Dimensions Program on Global Environmental Change*.

The NEO impact hazard is thus a textbook example of ICSU's key role in providing the international, interdisciplinary, and impartial scientific advice needed for rational policy setting by governments – 'Science for Policy' indeed, identified as a unique role in the 1996 review of ICSU's mission and priorities for the future. The present volume is a step towards this goal, and it *must* be a success – there is no alternative to ICSU for just this task!

But contacts are also needed to layers of government other than those reachable through learned academies or national research organizations. The OECD *Global Science Forum* gathers high-level science policy makers from most developed nations, and NEO impacts are truly a global science issue. Through this channel, contacts have been made to civil protection and emergency response managers from several countries, who will be in charge of reacting to NEO impacts, whether large or small, and many of them heard of this scenario for the first time. These are key people, not only in the unlikely event of an impact in the near future, but above all in specifying what information they need, and in which format, in order to assess the NEO impact hazard in advance on a similar basis as more familiar events such as earthquakes, tsunamis, volcanic eruptions, hurricanes, forest fires, etc.

Another forum for consulting governments is the United Nations, which actually held a conference on the NEO impact hazard in 1995 (Remo 1997). The IAU entry point to the UN system is the *Committee on the Peaceful Uses of Outer Space* (UN-COPUOS), where the IAU has permanent observer status. The main IAU interest in the committee is the possibility to alert space-faring nations to the need to reduce the production of space debris and radio noise emissions from satellites. But in keeping with normal rules for developing good working relations, the IAU has also paid close attention to the Committee's concerns on other astronomy-related subjects, notably international education in basic space science, including astronomy, where the IAU has a proud record. But also NEOs turned out to be of interest to delegates, and several presentations in lay terms have been given and much appreciated.

However, the UN also works in ways that are unfamiliar to astronomers. In particular, one-off shows may generate passing interest, but results are only obtained by working patiently with the committee for a long period and develop relations of trust with committee members. As an example, careful preparation resulted in sensible recommendations on NEO research being included in the *Vienna Declaration* (United Nations 1999), which set priorities for international space activities over the next 15–20 years. And contacts at UN-COPUOS enabled the IAU to assist the UK Task Force reporting on the NEO impact hazard in 2000 in ways that would not have happened otherwise. But persistence is needed: As for radio frequency interference, timescales for progress are decades rather than years. And unless the IAU and other interested organizations continue to push for action, nothing will happen. However, given the likely timescales of NEO impacts, this is reasonable as long as progress is actually made. And the present volume is a sign of such progress.

32.5
Epilog: the True Mess

With the benefit of hindsight, this story has been structured so the rationale behind the initiative and the relations between the players would hopefully become clear. Real life, of course, is different: Surprises spring at any moment from any corner, the different strands of the story unfold in parallel if not in antiphase, and only gradually are arguments developed and refined, potential allies identified, and plans and policies prepared to the point when decisions can be taken. A strictly chronological, and possibly more amusing, version of the same story could also have been written, but the underlying message would have been harder to extract.

The common thread in the real and idealized stories is, however, the basic recognition of the 'market forces' in the situation: The key – and only – asset of a scientific union is its scientific credibility, based on the fact that it represents a large fraction of the world's best experts in its field. Outside its field of science allies must be sought, and they must be treated as equals – even politely if a favor is sought. And in the real world, good working relations are built on an understanding of mutual benefits. Nothing more than common sense, in other words, but common sense appears to also have a place when discussing the NEO hazard – possibly the reason why minor planet (9300) was unexpectedly named 'Johannes' in the year 2000.

References

Andersen J (ed) (2000) IAU transactions. Astron Soc Pacific XXIVA:85–140
Binzel RP (2000) Planetary and Space Science 48:297, 303
Chesley SR, Chodas PW, Milani A, Valsecchi GB, Yeomans DK (2002) Icarus 159:423–432
Remo J (ed) (1997) Near-Earth objects – the United Nations International Conference. Ann NY Acad Sci, vol 822
Rickman H (ed) (2001) IAU transactions. Astron Soc Pacific XXIVB:115, 127, 139
Rickman H (ed) (2003) IAU transactions. Astron Soc Pacific XXVA:127, 139, 187
United Nations (1999) Report of the third United Nations conference on the exploration and peaceful uses of outer space. UN Publ A/Conf 184/6, Resolution 1.1 c i, iii, and iv

Chapter 33

A Road Map for Creating a NEO Research Program in Developing Countries

Wing-Huen Ip

33.1 Introduction

In the last six years, COSPAR has organized consecutively three NEOs-related meetings in its General Assemblies. The main purpose was to focus the attention of the scientific community to the potential impact hazards of NEOs to the global society. If we look back at the presentation materials in these meetings, they could be collected into several categories mirroring the responses on this critical issue. They are: (1) awareness of the threat; (2) analysis of the threat; (3) mitigation of the threat; and (4) utilization of the threat. The last item came about at the end of the NEOs session in COSPAR's General Assembly in Paris on July 21, 2004. As a final round up of the meeting, several young scientists and PhD students were invited to a panel discussion on the study of NEOs in year 2030. Not surprisingly, the younger the researchers the more optimistic were their opinions. Instead of the roaming catastrophe brought about by an asteroidal or cometary impact, these young researchers were considering topics such as mining of the asteroids and how to build large-scale structures in space to accommodate such an enterprise. Perhaps there is a lesson to be learned here. As we have heard from a Chinese saying that crisis could also mean opportunity, it might be of interest to assess what are the benefits (instead of gloomy images) to be derived from the present discussions in this volume.

This point is particularly important for the developing countries since they have only been marginally involved in the assessments of this world-wide threat. On the other hand, their poverty and social fragility would imply that these societies are least able to stand-off the damages to be incurred by a deep impact event. Even now the effects of global warming has already brought low-lying countries like Bangladesh and some island states in South Pacific to high alert of environmental disasters. The atmospheric dust clouds – which caused one million deaths each year according to some statistics – are another emerging risk factor in human and social terms. It is on this basis that space technology developed to manage global disasters caused by natural catastrophes under the charter of UNISPACE III could be considered to be a dress rehearsal for the Extinction Level Event (ELE) that might eventually happen. It is on this basis that the emerging natural hazards probably related to climate change are briefly described in Sect. 33.1. In this paper I discuss in Sect. 33.2 how the current study activity on a space technology and disaster management system might be expanded to cover the NEO impact hazards. In Sect. 33.3, I give an example on how the NEO threat could be further utilized to promote space research and astronomy education in developing countries so that they could become a important force in the analysis and defense against NEO impact hazards. A summary is given in Sect. 33.4.

33.2
The Crisis

On October 24, 2004 Niigata City in Japan was rocked by a strong earthquake of magnitude 7 on the Richter scale. In the next few days, a number of recurrent aftershocks followed. Even though the loss of life and material damages were moderate in comparison to the Kobe earthquake in 1995 and Taiwan's Chi Chi earthquake in 1999 (see Fig. 33.1). Some new phenomena of note should be mentioned here. First, Japan was hit by the worst typhoon in many decades just two days earlier. There was thus the general fear that a new typhoon with huge rainfall would probably cause severe landslides thus multiplying the damages by a huge factor. Indeed, Taiwan has been subjected to an increasing frequency of major landslides as one of the aftermaths of the Chi Chi earthquake (see Fig. 33.2). Things will probably get worse if the high rates of typhoons and hurricanes experienced in 2004 continue. The damages of flooding inflicted to Haiti due to hurricane Jeanne in September 2004, were another case in point. Second, the situation is worsening for perennial ecological nightmares like the annual flooding in Bangladesh and China that have plagued the economic development for centuries not to mention the loss in human lives in these countries.

Fig. 33.1. Buildings damaged in the great Chi Chi earthquake on September 21, 1999 in Taiwan. More than two thousand lives were lost

The source of these chain reactions could probably be traced to the dramatic decays in the global environment as a result of climate change, deforestation and desertification. Perhaps there is no need for an asteroid impact to bring havoc to these regions since they are already on the eve of environmental destruction. On a short-term scale (~tens of years) such occurrence could be even more devious and damaging to mankind than a sub-kilometer sized asteroid collision. What it means is that in the near future some natural hazards could take on a completely new dimension with the possibility of mega-deaths as depicted in the scenarios of NEO collision. As a warning shot, the Sumatra Earthquake/Indian Ocean Tsunami on December 26, 2004 demonstrated most vividly this point. A death toll of over 250 000 within a few hours is close to the doomsday scenario depicted for the NEO impact. How should we prepare for this?

Part of the answer might be found in the Space Technology and Disaster Management System to be facilitated by UN's Office for Outer Space Affairs (OOSA). The introduction of global early warning systems on natural disasters and coordination of massive rescue operations by space-born platforms would be absolutely necessary. In turn, the OSSA plan could serve as the best training ground for mitigation of the NEO impact risks – if it were to occur. But the developing countries could do much more than just serve as a passive participant in this endeavor. They could first use the crisis to facilitate the buildup of their environmental science and technology crucial to their future survival. They could next turn themselves into an active partner of the international NEOs study campaign. For this, COSPAR and IAU would have to help.

Fig. 33.2. A site of landslides in middle Taiwan showing the frightening power of the force of nature with a combination of reconfiguration of landforms by seismic activities and flash floods brought on by typhoons. From *http://home.kimo.com.tw/homework1026/web/05.htm*

Fig. 33.3. Two examples of space projects to study the physical properties of comets and asteroids. **a** The Deep Impact mission of NASA to Comet Tempel 1; **b** the Hayabusa mission of JAXA to Asteroid Itokawa. Both missions will reach their targets in 2005

33.3
The Opportunity

Discussions in the COSPAR sessions on NEOs and on space projects such as NASA's Deep Impact mission and JAXA's Hayabusa asteroid sample return mission reflected the urgent need for extensive ground-based observations of comets, near-Earth asteroids (NEAs) and the main-belt asteroids – since they are the parent bodies of NEAs (see Fig. 33.3). Spacecraft in-situ measurements of just a few asteroids and comets alone are not sufficient to provide comprehensive knowledge of this diverse group of small stray bodies in the solar system. To understand in detail the surface properties, interior structures, rotation states and shapes of asteroids and comets of various sizes and different taxonomic types would require long-term photometric and spectroscopic monitoring measurements. This task that can be performed by using one-meter class telescopes turns out to be not as easy as it first appears. The recent work by Yoshida et al. (2004) on the Karin family asteroids demonstrated this point quite clearly (see Fig. 33.4). It also shows that a global program should be initiated taking advantage of the availability of small telescopes in many developing countries. One problem here is that some of these observatories are in a state of dilapidation. Why shouldn't the United Nations provide support and guidance with the assistance of IAU in establishing a network of small telescopes for asteroidal study – if the NEOs have been considered to be such a major threat to global security.

For space research, the developing countries are definitely in a less privileged position even though some of them do have emerging space capabilities. But even without a full-fledged space science program, scientists in developing nations can nevertheless participate in research projects of scientific significance. The knowledge and expertise achieved from the ground-based observation program as mentioned earlier will necessarily path the way to the interest in joining the data analysis effort in space projects of technologically more advanced countries. This is one area in which COSPAR could

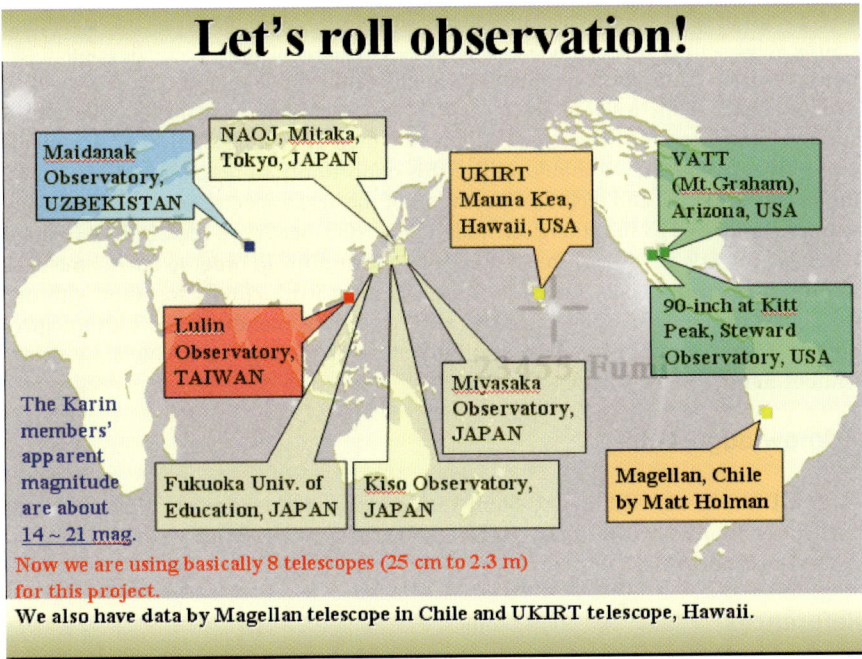

Fig. 33.4. A schematic view of the asteroid observation program led by Dr. Fumi Yoshida (of NAOJ) which covers three continents. Figure courtesy of Dr. Fumi Yoshida

play an important role, namely, the capacity building in planetary research starting with the investigations of NEOs. In the last few years, COSPAR has been successive in promoting scientific interests in X-ray astronomy and magnetospheric physics by holding workshops in these fields in developing countries. It is time to organize similar capacity-building workshops on asteroidal and cometary science.

33.4
Conclusion

The above analysis thus indicates that a significant research program ranging from hazard mitigation to advanced NEO study could be pursued in a concerted manner by UN's OSSA, IAU and COSPAR. Briefly speaking, we can see the following main lines of approach:

- UN/OSSA: Space technology in global and environmental disaster mitigation and management systems in developing countries;
- IAU: World-wide network in asteroid observations with special emphasis on partnership with developing nations;
- COSPAR: Capacity-building programs on planetary science using NEOs as the centerpiece.

A program of Earth defense against the threat of NEO impacts would have long-lasting value only if this is not conceived as a unilateral agenda of the developed nations. The leaderships and scientific community of the developing countries should be involved in a systematically manner. After all, they represent the majority of the global society. In addition to the spiritual support by their rich philosophical and religious views on such matter, we should utilize this opportunity to become a champion on mining the scientific and technical potentials of the broad spectrum of the developing nations. As advised by the young scientists at the NEO meeting at the COSPAR General Assembly in Paris in 2004, we must move on to the future. In order to have something to protect in 2030, COSPAR, IAU and the UN must work closely together with our colleagues in the developing countries with emerging economic and technological powers. The potentials of their young generation in solving this problem for us are without limits.

Acknowledgments

I thank Drs. Alberto Cellino and Fumi Yoshida for useful comments. This work was partially supported by NSC 93-2112-M-008-006, NSC 93-2112-M-008-001 and NSC 93-2752-M-008-001-PAE.

References

Yoshida F (2004)

Index

A

abyssal impact structure 61
accelerator mass spectrometry radiocarbon method 269
acid rain 16, 220, 265
acidification 220
acquisition 203
activity
 –, of ^{14}C 297
 –, of ^{39}Ar 291
adaptation 439
adaptivity 426
aerial photographic survey 309, 313, 317
aerolito 347
aerosol 127, 220
affective thinking 377
age distribution 4
agent-specific approach 410
agriculture 18
Agro Pontino 431
air
 –, blast 15–17, 212, 213, 216, 422, 429, 451, 455
 –, burst 32, 46, 145, 150, 423
 –, temperature 282
Akkadian Empire 34
Alaska 248, 251–253
Aleutians 251
Alice Springs 31
Allende 336
all-risk coverage 472
Alps 431
alternative energy source 442
altruistic community 365
American National Academy of Sciences 160, 501
Amors asteroid group 175
Amsterdam 473
Anfinogenov spindle 323
Aniene 431
annual growth indices 106
Antarctic ozone hole 234, 235

Antarctica 12
anthropology 25
apocalyptic 108, 109
Apollos asteroid group 175
archaeological 26
 –, record 72, 465
 –, research 73
archaeology 25, 71
architecture 72
Argentina 30–33
Argentine Pampas 33
art 76
artifact 73
asteroid 18, 145, 175, 225, 233, 254, 303, 479
 –, 1997 XF11 461, 505–508, 517
 –, 1997 XR2 517
 –, 1999 AN10 508
 –, 2000 SG344 509, 510, 517
 –, 2002 MN 510
 –, 2002 NT7 510
 –, 2003 DW10 517
 –, 2003 QQ47 516
 –, 2004 AS1 511, 513, 515, 517
 –, 2004 FU162 517
 –, 2004 MN4 515–517
 –, 2004 VD17 517
 –, belt 176
 –, deflection 170
 –, diameter 235
 –, impact 155, 233–235, 369, 471, 523
 –, impact hazard 145
 –, Mathilde 253 299
 –, observation program 531
astrology 74, 75
Astronomical Unit 204
astronomy 76, 160, 521
astrophysical
 –, observation 190
 –, process 408
Atens asteroid group 175
Atlantic 247, 473

atmosphere 13, 54, 219, 225, 235, 277
atmospheric
    –, anomaly 346
    –, impact 280
    –, pollution 403
    –, radiation 281
    –, temperature 238
AU (see *Astronomical Unit*)
Australia 10, 12, 31
automated survey 206

B

backscattering 196
Baker nuclear test 217
Baku 331
Baltic Sea 267–269
Bangladesh 473, 527
Barringer
    –, Crater 14, 18, 211
    –, impact event 14
Barrington 335
basaltic achondrite 9
Bavarian Crater Field 36
Bayes' theorem 385
Bayesian decision analysis 385
Bellingshausen Sea 38
Belluno 429
Bethlehem Star 76
Bhopal 401
Bible 78
biblical quotes 109
biodiversity 439
biological extinction, indirect effects of 221
biosphere 13, 17
Black Death 490
bolide 150, 158, 228, 291, 323
Bolivia 36
Bologna 435
bomb 8
boundary 265, 383
bounded rationality 369
breccia 6, 8, 11
bright night phenomenon 309, 311, 331
bromide 231
bromine 227, 233
bronze age 95, 272
bulk density 190, 191
Burckle Crater 61, 62
burial 17

C

CAI (see *comet-asteroid impact*)

Calcutta 473
caldera 126
California 253
Cambrian 145
Campo del Cielo 30
Canada 253
Cando 336
Canterbury 449
carbon isotope study 17
Carthage 72
Cassini-Huygens mission 461
cataclysmic
    –, eruption 126
    –, event 46
Catalina 206
    –, sky survey 147
catalyst 234
Catania 432
catastrophe 427, 454, 465
catastrophic event 225
cauldron 332–334
celestial object 71
CFC (see *chloro-fluoro carbon*)
Chaco Canyon 73
Cheko Lake 305, 307, 309
chemical perturbation 227
Chernobyl 401, 405
Chesapeake 12
    –, structure 12
Chi Chi earthquake 528
Chicxulub 3, 4, 7, 9, 15, 19, 214, 216, 218, 259, 335, 473
    –, crater 175, 450
    –, impact 450
    –, impactor 212
Chiemgau 36
Chile 248, 251
China 105, 528
chloride 231, 232
chlorine 227, 231–233
chloro-fluoro carbon 232
chondrite 292, 294, 295, 298
chondritic material 298
cinema 79, 80
city 496
civil
    –, complex system 428
    –, society 419, 424
civilization-killing impact 156
climate 265, 403
    –, change 124, 125
    –, system 280
climatic effect 277, 280
Club of Rome 495

coastal city 456
Coconino sandstone 10
coesite 9, 10
cognitive psychology 369
collapse 427
collision 203
 –, possibility computation 203
Colombia 499
comet 145, 175, 225, 479, 303
 –, -asteroid impact 123, 136, 138, 399, 400, 408, 411, 413, 437–440, 444
 –, -asteroid impact hazard 399, 408
 –, -asteroid impact hazard threat 136
 –, -asteroid impact hazard scale event 438
 –, debris 105
 –, disruption 193
 –, fragmentation 193
 –, impact 54
 –, -planet collision 450
 –, Shoemaker-Levy 9 150, 459
Comet Nucleus Sounding Experiment by Radiowave Transmission 198
cometary
 –, impact 335
 –, nuclei 195, 198
 –, showers 5
comet-planet collision 450
Committee on Space Research 527, 529–531
Committee on the Peaceful Uses of Outer Space 525
communication 358, 499
 –, channel 426, 427
comparative threat evaluation 405
complex system 424, 426
complexity 424, 434
compression wave 251
computer simulation 458
condensation nuclei 230
CONSERT (see *Comet Nucleus Sounding Experiment by Radiowave Transmission*)
cooperation 358
coordination 358
Coquimbo 260
corrupted air 112
cosmic
 –, hazard 454
 –, impact 25, 453
cosmogenic tsunami 260
COSPAR (see *Committee on Space Research*)
counterfactual 362
country social system 427
crater 6–8, 95, 150, 213, 268, 269, 281
 –, field 32
cratering 265

–, mechanics 7
–, rate 5
Cretaceous 145, 225, 419, 450
Cretaceous-Tertiary
 –, boundary 3, 9, 11, 12, 15, 59, 151, 259, 265, 296, 383
 –, extinction 212, 219
 –, impact 19, 151, 225, 282, 324
 –, event 218
critical nucleus 345
crust 4
crystalline rock 7
C-type asteroid 299
cultic activity 266
Cumbre Vieja volcano 133, 134

D

damage 421
 –, potential assessment 474
debris 248, 282
decision making 375
 –, process 369
Deep Impact mission 149, 160, 198, 530
deep-sea core 12
deep-water wave height 457
deflection 154, 155, 208
dendrochronology 111
density 190
Department of Energy 159
destruction 150
destructive impact event 3
devastation 212
 –, local 212
 –, regional 212
developing country 527
diamond 9
diaplectic glass 9
diffusive resonance 177, 181
disaster
 –, management 154, 161
 –, mitigation 410, 531
 –, plan 464
 –, planning 449, 454, 461
 –, prevention 499
 –, priority 414
 –, simulation 454–456
 –, vector 356
Disaster Management Research Center 364
disruption 194
disturbance event 265
Doctors Without Borders 400
Don Quijote 209
Dow Jones Industrial Average 486

dust 112, 127, 151, 220, 283, 403
   –, loading 219
dynamic hazard risk assessment 391
dynamical
   –, origin of near-Earth object 176
   –, system 426
dynasty 105

E

early
   –, detection 203
   –, historic culture 72
   –, life 13
   –, warning 497
Earth 226, 227, 419
   –, -colliding orbit 209
   –, -crosser 175
   –, -crossing orbit 175
   –, environment variable 411
   –, radius 205
   –, science 18
Earth Impact Database 27, 450
earthquake 135, 151, 214, 252, 265, 401, 406, 458, 474, 483
   –, magnitude 260
ecliptic comet 179
ecological footprint 437
economic 420
   –, consequence 479
economy 278, 481, 491
education 499
Egypt 72
ejecta 8, 16, 151, 213, 333, 422
   –, blanket 215
   –, deposition 215
   –, plume 216
   –, rain back 218
   –, time of arrival 216
ejection 7, 304
   –, angle 230
El Chichon 18
El Hierro 134
El Niño Southern Oscillation 125
ELE (see *extinction-level event*)
electro-magnetic pulse 423, 429
Eltanin 38
emergent order 424
emotion 374
EMP (see *electro-magnetic pulse*)
engineering
   –, design 465
   –, technique 456
ENSO (see *El Niño Southern Oscillation*)

entry phase 320
environmental
   –, damage 16
   –, downturn 105, 111
   –, effect of impact 498
   –, perturbation 17
   –, process 405
   –, stress 277
   –, threat 203
epicenter 260, 291
Eros 149
erosion 248
eruption 107, 125, 129
ESA (see *European Space Agency*)
escape velocity 226
Estonia 29, 265–267
ethnographic 73
Euphrates River 35, 91, 97
Eurasian plate 90
European Near Object Search Project 497
European Space Agency 148, 198, 208, 463
evolution 179, 225
explosion 303, 332, 473
   –, -generated tsunami 253
   –, time 316
explosive volcanism 406
extinction 3, 18, 265
   –, event 3
   –, -level event 527
extra-terrestrial impact 254
eyewitness 314

F

FAO (see *Food and Agriculture Organization*)
fauna 14, 17
Federal Emergency Management Agency 155
feeling 374
feldspar 9, 10
Fennoscandia 27
financial disaster 490
fire 473
fireball 151, 213, 216, 226, 227, 346, 423, 429
   –, phase 320
firestorm 151
flash flood 529
flexibility 426
flood 473
   –, comet impact 58, 60
   –, mythology 47
   –, myth 48, 60
flora 17
Food and Agriculture Organization 400
food security 441

forest
    –, devastation 310–312
    –, fire 286
forewarning 393
Foschini
    –, hypersonic flow 321
    –, model 323
fossil
    –, fuel 439
    –, record 225
free energy 345
frequency 347
fresh water resource 439
Fucino 431
Fuerteventura 132
fused glass 9

G

gabbroic anorthosite 229
GDP (see *gross domestic product*)
Gee-Gee (see *global geophysical event*)
Gelasian stage 25
geochemical data 292
geographical location 458
geologic
    –, process 7, 18
    –, record 3
geology 7
geophysical
    –, anomaly 4, 11
    –, cycle 401
    –, process 408
geophysics 11
GGE (see *global geophysical event*)
Gibeon 335
Gilgamesh 50
Giordano Bruno Crater 449
glass
    –, melt 33
    –, particle 8
global
    –, consequence 218
    –, cooling 150
    –, ecological disaster 265
    –, environmental crisis 145
    –, environmental downturn 105, 112–114
    –, geophysical event 123, 124
    –, seismic network 257
    –, warming 527
Global Tsunami DataBase 249, 455
globally
    –, averaged ozone concentration 242
    –, relevant disaster 355–357, 360, 361

good governance 444
Gosses Bluff 10
Gran Canaria 132
Gran Chaco 43–46
granitic rock 10
graphite 9
Graphite Peak 12
gravitational force 231
gravity survey 100
GRD (see *globally relevant disaster*)
great flood 46
    –, myth 46–48
Great Kanto Earthquake 135
Great Sumatra-Andaman Earthquake 456, 465
Greece 72
greenhouse effect 265
Greenland GISP2 ice core 127, 135
GRIP ice core 116
gross domestic product 135, 485
growth reduction 106
Guatemala City 483

H

hailstone 343
halide ion 231
Halifax 473
    –, harbour 253
Halley's Comet 298
Halley-type comet 179, 184
halogen 227, 230, 231
Han 105
Hawaii 40, 130
Hayabusa mission 198, 530
hazard 399, 452, 525
    –, level 392
    –, risk assessment 383, 386, 392
    –, warning system 466
    –, research 161
heat 282
    –, flow 335
Heaven's Gate cult 82
Henbury 31
    –, crater field 31
Hidalgo mission 209
high-frequency gee-gee 134
high-pressure polymorphism 9
Hiiumaa Island 269
historical record 108, 347, 439
history 115
Hollywood 83
Holocene 25, 26, 34, 38, 57, 89, 265, 266
Honolulu 473
horoscope 76

Index

HTC (see *Halley-type comet*)
human
   –, civilization 18
   –, culture 265
   –, -meteorite interaction 265
   –, population 471
   –, society 83, 265
humanitarian assistance 414
humanity 18, 439
hurricane 401
   –, Katrina 482, 485, 492
   –, Mitch 482
hydrocarbon deposit 19
hydrochemical 344
hydrodynamics 254
hydroelectric plant 491
hydrogen oxide 232
hydrometeor 343
hydrosphere 13
hypervelocity impact 3

I

IAU (see *International Astronomical Union*)
ice
   –, block 343
   –, conglomeration 341
   –, core 107, 116, 243, 291, 292
   –, core Greenland GISP2 127, 135
   –, core GRIP 116
   –, fall 344–346
   –, meteor 343
ICG-ITSU (see *International Co-ordination Group for the Tsunami Warning System in the Pacific*)
ICSU (see *International Council for Science*)
IEO (see *inside Earth's orbit*)
ignimbrite 127
Ilumetsa crater 29
image 76
immediacy 379
impact 153, 155, 156, 175, 203, 225, 227, 233, 280, 336, 420–423, 437
   –, cost of 475
   –, crater 89, 312, 266
   –, damage 472
   –, ejecta 17
   –, event 18, 19
   –, frequency 124, 233, 320
   –, glass melt 32
   –, hazard 145, 146, 156, 163, 164, 211, 419, 524
   –, -induced ozone depletion hypothesis 243
   –, kinetic energy 256
   –, -melt rock 8
   –, -melt sheet 8
   –, melting 8
   –, physical effect of 211
   –, plume 213
   –, possibility 206, 207, 509
   –, probability 206
   –, process 12, 254
   –, record 3, 27
   –, risk 379, 479
   –, risk communication 495
   –, risk communicaton management 505
   –, scale 497
   –, statistics 331
   –, structure 4–8, 18
   –, velocity 147, 226, 229
Indian Ocean 247
Indonesia 125
infall angle 335
information 481, 500
   –, flow 481
infrastructure 458
injection 236, 280
   –, from fire 282
   –, of asteroidal and cometary material 281
   –, of climatically active gas 220
   –, of dust 281
   –, of nitrogen oxide 284
   –, of sulfur dioxide 283
   –, of water 283
inside Earth's orbit 175
   –, object 175
   –, population 175
in-situ measurement 530
insurance 469
   –, coverage 469, 472
   –, industry 475
Intergovernmental Oceanographic Commission 257
international
   –, cooperation 170
   –, development study 437, 440
   –, disaster measure 399
   –, humanitarian assistance system 444
   –, response 501
International Astronomical Union 168, 169, 522–525, 529–531
International Coordination Group for the Tsunami Warning System in the Pacific 257
International Council for Science 83, 500, 525
International Spaceguard Foundation 459
International Tsunami Information Center 257
Internet 82
inundation of salt water 248
IOC (see *Intergovernmental Oceanographic Commission*)

iodide 231
IP (see *impact possibility*)
Iran 90
Iraq 35, 89–92
iron meteorite 30, 335
isotopic
 –, analysis 304
 –, characteristic 344
 –, composition 292, 295, 296, 304
 –, dating system 4
 –, investigation 295
 –, method 295
 –, shift 295
Italy 36, 420, 421
ITIC (see *International Tsunami Information Center*)
Iturralde 36

J

Jalisco 257
Japan 490, 528
 –, Sea 252
Japan Aerospace Exploration Agency 530
JAXA (see *Japan Aerospace Exploration Agency*)
 –, Hayabusa mission 198
Jet Propulsion Laboratory 459
Jupiter 149, 450, 459

K

Kaali 29, 265–268, 270
 –, crater field 265
 –, event 273
 –, meteorite 29, 266
 –, meteorite crater field 267
 –, meteorite impact 265
 –, meteorite impact, age 268
 –, meteorite impact, effects of 271
Kamchatka 248, 251, 253, 259
Kärdla 265
Karikioselkä impact 27
Kilauea 130
kimberlite 305, 306, 334, 405
kinetic
 –, deflection 208
 –, energy 211, 213, 332, 333
Kobe 483
 –, earthquake 528
Kohala volcano 134
Kõivasoo bog 266, 269
Krakatau (Krakatoa) 18, 253, 258, 331, 356
Kulikovsky paleovolcanic complex 307
K-T (see *Cretacious-Tertiary*)

L

lahar 253
Lake Cheko 307
Lake Chiemsee 37
Lake Kaali 266
land impact 419
landslide 401, 529
Large Synoptic Survey Telescope 148, 167
Lazio 432
 –, region 431
L-chondrite 13
level of damage 421
life 225
Lincoln Laboratory Near Earth Asteroid Research 147, 154, 159, 206, 511, 513
 –, survey 165
LINEAR (see *Lincoln Laboratory Near Earth Asteroid Research*)
LISA (see *low input sustainable agriculture*)
literature 76, 77
Lituya Bay 252
Livermore 159
local 17
lofting 220
 –, of water 229
LOI (see *loss on ignition*)
LONEOS (see *Lowell Observatory Near-Earth Object Search*)
long-period comet 179
long-term 530
Los Alamos 159
loss on ignition 270
low input sustainable agriculture 443
Lowell Observatory Near Earth Object Search 147, 206, 459
LSST (see *Large Synoptic Survey Telescope*)
lunar record 13

M

Madagascar 61
Mahuika 39
Main Belt 204
 –, asteroid 206
Manicouagan 8, 9, 16
man-made disaster 401
marker horizons 105
Mars 149
 –, -crossing orbit 178
marshland 91, 92
Martinique 253
mass 190
 –, extinction 17, 265

–, extinction event 157
–, movement 129
Massachussetts Institue of Technology Lincoln Laboratory 459
Mathilde 253 asteroid 299
Matuyama-Brunhes boundary 34
Mauna Loa 130
Maya 72
Medicine 75
Mediterranean Sea 247
megacryometeor 341–343, 346, 348, 349
megalithic structure 72
mega-tsunami 455
–, formation 129
–, risk 133
melt lithology 8
melting 9
Mennonite Central Committee 400
Mesolithic period 272
Mesopotamia 72, 91, 93, 95
Mesopotamian basin 90
mesosphere 230
metamorphism 7–9
Meteor Crater 211–214, 475, 476
Meteoric Water Line 344
meteorite 145, 150, 472
–, fall 347
–, impact 265
–, impact crater 89
–, utilization 266
meteoritic
–, debris 332
–, fragment 331
–, material 9
meteoroid 226, 452
–, impact 472
meteorological tsunami 253
methane 334
–, outburst 296
microspherule 29, 35, 269, 270
microtektite 33
Middle East 34
middle ocean ridge 251
migration 437
military 157, 159
Mindoro 257
minimum orbital intersection distance 204, 205
Minor Planet Center 511–513, 522, 523
miss distance 505, 506
Mistastin 10
mitigation 154, 391, 454
–, strategy 189
MMB (see *Matuyama-Brunhes boundary*)
Modified Mercalli Intensity Scale 214

Moho isohypse 335
MOID (see *minimum orbital intersection distance*)
monitoring 530
Mono Lake 253
monolith 191
moon 6, 449, 450
Morphology 6
mortality event 439
motion resonance 177
Mount Pele 253
Mount St. Helens 126, 129, 356
Mount Wilson Observatory 291
MPC (see *Minor Planet Center*)
multi-ring form 7
myth 39–42
mythology 26, 39, 40, 73, 110, 115

N

narrowest ring event 105
NASA (see *National Aeronautics and Space Administration*)
Nasdaq 485
National Aeronautics and Space Administration 148, 154, 155, 459, 460, 474
–, Institute for Advanced Concepts 463
–, Near Earth Asteroid Tracking 147, 206, 497
–, Science Definition Team 166
National Research Council 160
natural
–, disaster 341
–, hazard 161, 247, 341, 369, 378, 383, 384, 454
–, hazard warning system 461
–, oscillation 279
NEA (see *near Earth asteroid*)
near Earth asteroid 147–151, 163, 165, 167, 168, 176, 184, 204, 207, 208
–, impact 150, 153, 154, 158, 160
–, research 160
–, size frequency distribution 184
near Earth comet 178
near Earth object 84, 124, 164, 175, 178, 182, 189, 204, 355, 357, 360, 383, 385, 419, 471, 479, 521
–, collision 385
–, diameter 480
–, dynamics 180
–, hazard 175, 385, 386, 397, 521
–, impact 209, 525
–, impact hazard 525, 527
–, impact risk 497
–, level 1 386, 395
–, level 2 388
–, level 2 impact 395

Index   541

-, level 3 389
-, level 3 impact 395
-, level 4 396
-, level 4 impact 390
-, orbital distribution 175, 181
-, physical property of 189
-, population 175, 181
-, size 480
-, space 176, 179
-, survey 163, 204, 205
-, Technical Review Team 524
Near Earth Object Confirmation Page 512, 513, 515
Near Earth Object Information Center 155
Near Earth Object Program 460
Near Earth Object Research Program 527
Near Earth Object Science Definition Team 185
near infrared 196
NEAR Shoemaker mission 149, 160
nearly isotropic comet 179, 183
-, population 183
nearly Earth-crossing orbit 175
near-miss 158
NEAT (see *National Aeronautics and Space Administration Near Earth Asteroid Tracking*)
NEO (see *near Earth object*)
Neolithic 272
Neotethys Ocean 90
Net Investment Position 489
Netherlands 473
Neugrund 265
New Mexico 73
New Orleans 473
New York 475
New York City 485
New York Stock Exchange 485
New Zealand 39
news media 157, 158
NIC (see *nearly isotropic comet*)
Nicaragua 257
Niigata City 528
nitric acid 151
nitric oxide 227, 228, 231, 235, 236, 238, 240, 241, 284, 286
-, production 228
nitrogen oxide 231, 232, 281
noctilucent cloud 309
non-equilibrium 424
North American Craton 5
northern hemisphere polar region 236
Norway 253
Nova Scotia 253

nuclear
-, weapon 155, 286
-, winter 287
nucleation 345
numerical
-, algorithm 254
-, simulation 198

O

ocean floor 4
-, impact 391
oceanic impact 38, 217, 247, 254
ocean-island volcano 129
-, collapse 130, 131
OECD Global Science Forum 525
Office for Outer Space Affairs 529
Office of Space Science and Applications 529, 531
official response model 358
Okushiri 252
ontology 386
OOSA (see *Office for Outer Space Affairs*)
OPEC (see *Organization of the Petroleum Exporting Countries*)
operational tsunami warning 257
oral
-, history 73
-, tradition 26, 39–41, 73
orbit
-, determination 203, 205, 206
-, Earth-colliding 209
-, Earth-crossing 175
-, Mars-crossing 178
orbital
-, distance 147
-, distribution 182
-, eccentricity 176
organization 359
Organization of the Petroleum Exporting Countries 489
organizational risk 411
oscillation 248
OSSA (see *Office of Space Science and Applications*)
outburst 336
Ovid 61
Oxfam 400
ozone 241
-, column 243
-, concentration 240
-, depletion 225, 230, 232–235, 240
-, depletion catalyst 234
-, depletion frequency 233
-, hole 232
-, layer 287, 403

P

Pacific Ocean 38, 247
Padana Valley 430
painting 76
paleoclimatic record 287
paleoenvironmental record 26
paleotsunami 259, 260
Palermo 432
–, Scale 511, 514, 524
Pampas 33
panic 363, 364
Pan-STARRS (see *Panoramic Survey Telescope and Rapid Response System*)
Panoramic Survey Telescope and Rapid Response System 147, 167
Papua New Guinea 252, 257
paragenesis 9
partial fragmentation 194
peat 291, 292, 304
–, bog 29, 269, 293
–, column 292, 294, 296
–, sample 292, 296
Pelisoo 267
–, mire 269
perception 156, 379
perihelia 147
perihelion distance 204
periodic impact 5
periodicity 5
permafrost 332
Permanent Scatterer Interferometric Synthetic Aperture Radar 137
Permian-Triassic boundary 12, 265
perturbation 280, 426
PGE (see *platinum group element*)
Philippines 257
PHO (see *potentially hazardous object*)
Piila 267
–, bog 271
Pitkasoo 267
–, mire 269
plague 112
planar microstructure 11
planning 359
–, model 360
platinum group element 292, 293, 295, 297, 298
Pleistocene 25, 27
Pliocene 25
plume-forming impact 320, 321
Podkamennaya Tunguska 291
–, River 303
polarization 196
policy 469

pollen 270, 271
–, analysis 268
–, DNA 243
–, influx 271
Polynesia 40
Popigai 9, 12
–, structure 12
popular culture 71, 74
population 175, 442, 471, 472
–, decentralization of 442
porosity 191, 192
post-impact sediment 4
post-petroleum transition 440
post-traumatic response 362
potential mitigation 395
potentially hazardous object 147
power law 5, 233
PR (see *Purgatorio Ratio*)
precession frequency 176
predictability 392
prediction 454, 456, 523
prehistoric society 73
preparation 359, 362, 363
–, and response issue 357
preparedness strategy 391
principle of insurance 470, 471
priority 521
probability 378
–, distribution 182
–, neglect 377
projectile 422
property insurance 469, 473
Protezione Civile 434
PS (see *Palermo Scale*)
PS InSAR (see *Permanent Scatterer Interferometric Synthetic Aperture Radar*)
psychological process 380
psychometric paradigm 371
public 377
–, attitude 379
–, awareness 83
–, education 84
–, perception 377
–, perception of impact hazard 377
–, policy issue 169
–, safety 401
–, support 414
Purgatorio Ratio 517, 518
pyroclastic flow 253

Q

quantitative modeling of near Earth object population 180

quartz 9–11
Quaternary 25–27, 33, 34
  –, record 26

R

radiocarbon 269, 291
  –, dating 30
radonic storm 306
rare earth element 293–295
reconstruction 454
recovery 359
recreancy 363
recurrence interval 260
Red Crescent 400, 414
Red Cross 400, 414
REE (see *rare earth element*)
regional stress 439
regolitic layer 197
religion 75, 157, 159
remote sensing 497
Renaissance 76
resilience 439, 457
resilient infrastructure 444
resonance 176, 177
  –, diffusive 177, 181
  –, v_6 181
  –, 3:1 181
response 362, 363
  –, strategy 410
Richter Scale 125, 214
Ries Crater 15, 25
rim diameter 4
Rio Cuarto 32
risk
  –, analysis model 392
  –, as feeling 374, 376
  –, assessment 161, 410
  –, communication 161
  –, definition 384
  –, estimate 208
  –, evaluation 203
  –, information 376
  –, management 407, 444
  –, mapping 454
  –, model 440
  –, of impact 189
  –, perception 369, 374
  –, perception, psychometric study of 371
  –, reduction 203
  –, society 413
Ritter Island 134
rock art 72
Rome 72, 105, 431, 435

Rosetta probe 198
rubble pile 191, 192
Russian Academy of Sciences 316

S

Saaremaa 266–268, 272
  –, Island 269
salt
  –, condensation nucleus 230
  –, particle 229–231
San Diego 485
San Francisco 483
Sancho 209
Santorini 253
satellite imagery 89, 95
Saudi Arabia 30
scale 437
science 83, 157, 160
  –, fiction 78
Science Definition Team 166, 460
SDT (see *Science Definition Team*)
sea
  –, level change 130
  –, -salt 229, 234
  –, -salt particle 231
  –, -surface temperature 131
  –, -water 231
secondary hazard 403, 405
security measure 402
sedimentary rock 7
segregation 427
seismic
  –, activity 17
  –, record 309
  –, shaking 213, 214
  –, wave 312, 422
seismogram 319
seismotectonic tsunami 251
self-organization 426
September 11th 2001 475
shape 192
shatter cone 11
shock
  –, fused glass 9
  –, compression 8
  –, -metamorphic effect 7, 8
  –, metamorphism 7, 8
  –, pressure 4, 7
  –, wave 7, 17, 212, 216, 226, 227, 229, 285, 322, 332
Shoemaker-Levy 9 comet 136, 450, 479
shooting star 145
Siberia 32, 148, 291, 303, 335, 479

Index

Sicily 429, 431, 433
siderophile 9
Sikhote Alin 27, 335, 336
siliceous microspherule 269
simulation model 457
Sirente 36
size
 –, class 386
 –, distribution 4, 175, 233
 –, -frequency distribution 5
 –, -frequency relationship 148
slide-generated tsunami 252
slope failure 253
small meteorite impact 341, 346
social
 –, amplification of risk 373
 –, dimension 399
 –, effect 265
 –, perspective 399
 –, science 355
 –, structure 420
 –, system 433, 434
 –, vulnerability 399, 400, 410
societal
 –, development 277
 –, impact 157
 –, implication 437
 –, response 163
 –, vulnerability 279
society 157, 265, 285
Socorro 459
soil sample 304
solar radiation 283
Soloviev-Imamura Scale 249, 250
song 78
soot deposition 295
sound wave 216
space 437
 –, mission 198
Space Technology and Disaster Management System 529
Spaceguard 147, 154, 471
 –, Survey 153, 159, 163, 164, 167, 459, 461, 462, 497, 519
 –, Survey Report 163
Spacewatch 147, 206, 459
Spain 346
spatial distribution 4
spherule 291, 304
Spitzer Space Telescope 131
SST (see *Spitzer Space Telescope*)
statistical frequency 271
stishovite 9
Stonehenge 72

Stony Tunguska River 331
strain rate 7
stratigraphic
 –, date 4
 –, record 12
stratosphere 233, 283
structural
 –, failure 129
 –, trough 90
structure 192
subduction zone 251
sublimation 231
submarine
 –, earthquake 251
 –, landslide 252, 260
Sudbury Igneous Complex 9
sulfuric acid 232
sulfate 151
sulfur
 –, dioxide 283
 –, oxide 16
Sumatra 125
Sunda Strait 253
super-eruption 127
 –, threat 128
supersonic blasting 332
surface
 –, free energy 345
 –, impact 286
 –, property 195, 196
Surusoo 267
sustainable development 441
Switzerland 490
Sydney Basin 12

T

taiga 303
Taiwan 528, 529
Tambora
 –, blast 134
 –, eruption 356
Tampa Bay 473
TCB (see *Tunguska cosmic body*)
TE (see *Tunguska event*)
technocratic authority 399
technological hazard 161, 378
tectonic
 –, fault line 335
 –, outburst 305, 336
 –, stability 309
tectosilicate 11
tektite 32, 33
telegraph pole 332

television 79, 80
Tempel 1 mission 149
temperature anomaly 107
Tenochtitlan 72
tephra 127
Terni 431
terrestrial
 -, cratering rate 5
 -, dust 295
 -, impact 437
 -, impact structure 6, 7, 11, 18
 -, planet 3
 -, record 5
Tertiary 419
textural 344
therapeutic community 365
thermal
 -, emission 422
 -, infra-red 98
 -, pulse 218
 -, radiation 213
thermohaline circulation 288
Thermosphere-Ionosphere-Mesosphere-Electrodynamics General Circulation Model 235, 240
Third United Nations Conference on the
 Exploration and Peaceful Uses of Outer
 Space 527
three-body resonance 177
Tiber 431
tide gauge
 -, network 258
 -, record 259
Tigris 35, 91, 97
time 437
TIME-GCM (see *Thermosphere-Ionosphere-
 Mesosphere-Electrodynamics General
 Circulation Model*)
Toba 18, 126, 127
Tofino 259
Torino Scale 168, 403, 509, 515, 516, 524
total risk 384
transform fault 251
transportation network 481
tree
 -, -fall pattern 332, 333
 -, -growth 112
 -, -ring 105, 107, 291
 -, -ring chronology 105
Triassic-Jurassic boundary 265
tropopause 229, 236
troposphere 281, 283
TS (see *Torino Scale*)
tsunami 57, 131, 137, 150, 151, 217, 247-250, 265, 391,
 401, 437, 455, 456, 458, 463, 465, 473, 480, 485, 492

 -, damage 458
 -, generation 131
 -, geological trace of 259
 -, geographical distribution of 249
 -, height 261
 -, intensity 250
 -, occurrence 250
 -, risk map 455
 -, source 251
 -, temporal distribution of 249
 -, warning system 257, 261, 455
tsunamigenic potential 254
Tunguska 14, 18, 19, 32, 148, 151, 211, 213, 259, 261,
 265, 271, 291, 294, 296, 304, 309, 310, 331-334,
 336, 451, 475, 479
 -, -class impact 481
 -, cosmic body 291, 292, 295, 297, 298,
  303-306, 310, 314
 -, cosmic body material 292
 -, cosmic body remnant 291
 -, cosmic body trajectory 317
 -, epicentre 317
 -, event 285, 291, 303-306, 310, 320
 -, explosion 312, 318
 -, explosion time 316
 -, impact 455
 -, impactor 291
 -, meteorite 303
turbidity 291
TWS (see *tsunami warning system*)

U

Umm al Binni 35, 89, 95
 -, lake 89
 -, structure 94, 96-100
UN (see *United Nations*)
UN-COPUOS (see *United Nations Committee
 On the Peaceful Use of Outer Space*)
underwater earthquake 247
UNEP (see *United Nations Environment Program*)
UNESCO (see *United Nations Educational,
 Scientific and Cultural Organization*)
UNHCR (see *United Nations High Commis-
 sioner for Refugees*)
UNISPACE III (see *Third United Nations
 Conference on the Exploration and Peaceful
 Uses of Outer Space*)
United Nations 525, 529-531
 -, Committee On the Peaceful Use of Outer
  Space 525
 -, Educational, Scientific and Cultural
  Organization 257
 -, Environment Program 400

–, High Commissioner for Refugees 400
–, Office for Outer Space Affairs 160
United States of America 253
  –, Air Force 159
unofficial response model 359
urbanization 437

V

VA (see *virtual asteroid*)
Vanavara 331
vaporization 227
vegetation 14, 16, 283
VEI (see *Volcanic Explosivity Index*)
velocity 247
Veneto 430
  –, region 429
Venice 429
VI (see *virtual impactor*)
Vicenza 429
video 79
virtual
  –, asteroid 206, 207
  –, impactor 206, 207
volcanic
  –, -acid signal 107
  –, crater 334
  –, eruption 253, 336, 401
  –, landslide 129
  –, super-eruption 123, 125
  –, tsunami 253
  –, winter 127
Volcanic Explosivity Index 125, 126, 128, 134
volcanism 334
volcano 107
  –, instability 129
vulnerability 401, 440
  –, analysis 394

W

Wabar 30, 335
  –, impact 30
warning system 454, 459, 462
water 227, 229
  –, current 248
  –, diversion 91
  –, injection 55
  –, vapor 231, 233, 235, 236, 241
wave
  –, height 256
  –, train 131
wildfire 15, 16, 219, 265
World Disasters Report 400
world trade 361
World Trade Center 475, 476
World Wide Web 82

X

X-ray 371
  –, astronomy 531
  –, diffractometry 37, 271

Y

Yellowstone 126
  –, caldera 126
Yokohama 135
Younger Dryas 60
Yucatán 175, 212
  –, Peninsula 450

Z

Zagros Mountains 90
Zanitsa Pipe 334, 336